B. L. van der Waerden

Zur algebraischen Geometrie

Selected Papers

Mit einem Geleitwort von F. Hirzebruch

Springer-Verlag
Berlin Heidelberg New York Tokyo
1983

Prof. Dr. B. L. van der Waerden
Mathematisches Institut der Universität Zürich

CIP-Kurztitelaufnahme der Deutschen Bibliothek
Waerden, Bartel Leenert van der:
Zur algebraischen Geometrie: selected papers / B. L. van der Waerden.
Berlin; Heidelberg; New York; Tokyo : Springer, 1983.
ISBN-13: 978-3-642-61783-6   e-ISBN-13: 978-3-642-61782-9
DOI: 10.1007/978-3-642-61782-9

Offsetdruck: J. Beltz, Hemsbach. Bindearbeiten: J. Schäffer OHG, Grünstadt
2141/3140-543210

B. L. van der Waerden, April 1982

Geleitwort

Als ich 1945 in Münster zu studieren begann, war van der Waerdens „Moderne Algebra" eines der wenigen Bücher, die ich mir in diesen schwierigen Zeiten leihen konnte. Wie vielen Studenten so war also auch mir „der van der Waerden" vertraut von Anfang des Studiums an. Ich lernte van der Waerden einige Jahre später kennen, und er sagte mir, wie merkwürdig es sei, daß alle Mathematiker ihn wegen dieses Buches kennen, das Vorlesungen von Emil Artin und Emmy Noether benutzt, während seine wirklichen mathematischen Leistungen ganz woanders lägen. In dem Gespräch zeigte sich dann, daß van der Waerden seine Arbeiten zur algebraischen Geometrie und insbesondere die in den Mathematischen Annalen erschienene Reihe „Zur algebraischen Geometrie", die es in weiteren Jahren bis zur Nr. 20 bringen sollte, für das Wichtigste hielt. (Etwa 30 Jahre später war ich zu einem Essen zu Ehren der Träger des Ordens Pour le Mérite für Wissenschaft und Künste eingeladen. Die beiden Ordensträger van der Waerden und Elias Canetti unterhielten sich. Canetti bedauerte, daß man ihn hauptsächlich wegen seines Buches „Die gerettete Zunge" kenne, während andere Schriften doch viel wichtiger seien. Van der Waerden rief aus „Aber mir geht es doch ganz genau so mit meinem Algebra-Buch".)

Van der Waerden ist ein so ungewöhnlich vielseitiger Mathematiker mit bedeutenden Büchern und Arbeiten aus zahlreichen weit von einander entfernten Gebieten, daß die Entscheidung des Verlages, diese Selecta der algebraischen Geometrie zu widmen, sicherlich nicht selbstverständlich war. Aber die Entscheidung war gut und wird viele algebraische Geometer begeistern, die wichtige Entwicklungslinien der algebraischen Geometrie seit den zwanziger Jahren genauer kennenlernen wollen.

Jeder algebraische Geometer begegnet in seiner täglichen Arbeit Fragen der Spezialisierung und der Entartung algebraischer Varietäten, der Schnittbildung von Untervarietäten, der Multiplizität von Schnittkomponenten, des Multiplizitätsbegriffes in anderen Zusammenhängen, der verschiedenen Äquivalenzbegriffe für Untervarietäten und algebraische Zyklen beliebiger Kodimension, der Parametrisierung (Moduln) algebraisch-geometrischer Objekte. Diese Theorien standen vor fünfzig Jahren auf schwankenden Füßen. Der vorliegende Band zeigt wieviel van der Waerden für die Fundamente des bereits damals stolzen Gebäudes der algebraischen Geometrie getan hat. Neben den Beiträgen zu den Fundamenten enthält der Band viele schöne Einzelresultate, die man mit Vergnügen liest.

Dem Band ist ein historischer Artikel van der Waerdens vorangestellt, der die Entwicklung des Multiplizitätsbegriffs und der Schnitt-Theorie darstellt: "Just one line of development, that from Severi to A. Weil". Die Einleitung zu A. Weils "Foundations of Algebraic Geometry" (geschrieben 1944) zeigt in fast dramatischer Form, wie dringend die exakte Grundlegung damals war: "to consolidate shaky foundations, to provide roofs where they are missing, to finish, in harmony with the portions already existing, what has been left undone". Die Einleitung zeigt, wie stark van der Waerden an dieser Grundlegung beteiligt war.

Der Band bietet spannende Lektüre und kann dem angehenden algebraischen Geometer als Einführung in manche Arbeitsrichtung empfohlen werden. Nur wenige Höhepunkte sollen hier kurz angesprochen werden. Fundamental ist die Arbeit „Der Multiplizitätsbegriff der algebraischen Geometrie" (1927), die Severis Ansatz von 1912 unter geeigneten Voraussetzungen streng begründete, und die dann zu vielen anderen Arbeiten van der Waerdens zur Schnitt-Theorie führte. Die Arbeit „Zur algebraischen Geometrie XIV" (1938) soll besonders hervorgehoben werden. Der Schnittring der *algebraischen* Äquivalenzklassen algebraischer Zyklen wird eingeführt. Dabei ist natürlich die gemeinsame Arbeit ZAG IX mit W.-L. Chow „Über zugeordnete Formen und algebraische Systeme von algebraischen Mannigfaltigkeiten" (1937), in der die Chowschen Koordinaten algebraischer Zyklen eingeführt wurden, von grundsätzlicher Bedeutung. Auf den Abschnitt "Cycles and their coordinates" in van der Waerdens historischem Artikel sei besonders hingewiesen. Er schreibt dort "In 1932, Severi opened a new field of research by considering rational and algebraic systems of cycles C^d on a variety C^r". Wie aufregend dieses "new field of research" in den folgenden Jahrzehnten geblieben ist, ersieht man zum Beispiel aus den Proceedings der Internationalen Kongresse. In Amsterdam (1954) gab es eine intensive Diskussion im Anschluß an Severis Vortrag, an der sich u.a. Samuel, Severi, van der Waerden und Weil beteiligten. Es ging um den Begriff der *rationalen* Äquivalenz algebraischer Zyklen und den Schnittring der rationalen Äquivalenzklassen. (Vgl. van der Waerdens Artikel "On the definition of rational equivalence of cycles on a variety" in Band III der Proceedings. Van der Waerden hat sich mit der rationalen Äquivalenz in der Arbeit 32 dieser Selecta nochmals auseinandergesetzt.) In Edinburgh (1958) sprach P. Samuel über "Relations d'equivalence en Géométrie Algébrique". (Die präzise Einführung des Schnittringes der rationalen Äquivalenzklassen algebraischer Zyklen, heute als Chowscher Ring bekannt, verdankt man Chow, Chevalley und Samuel.) In Nizza (1970) sprach S. L. Kleiman über "Finiteness theorems for algebraic cycles". In den Proceedings von Vancouver (1974) findet man J. Tates Bericht über Mumfords Arbeiten (rationale Äquivalenz von 0-Zyklen auf einer algebraischen Fläche).

Besonders schön ist van der Waerdens Arbeit ZAG XI aus dem Jahre 1937, wo mit algebraischen Methoden die Existenz der $(3p-3)$-dimensionalen Modulvarietät für die algebraischen Kurven von Geschlecht $p > 1$ nach-

IV

gewiesen wird. Heute ist die Untersuchung von Modulvarietäten verschiedenster Art ein unerschöpfliches Thema der algebraischen Geometrie. Man hat große Fortschritte erzielt, aber als Einführung in die Problemstellung ist diese Arbeit immer noch zu empfehlen.

Die algebraische Geometrie ist heute ein überaus aktives Gebiet mit großem theoretischen Apparat, aber auch vielen konkreten Einzelresultaten, mit Verzweigungen in viele Gebiete von der Zahlentheorie bis zur Theorie der partiellen Differentialgleichungen. Für den angehenden Mathematiker ist es schwer, an wenigstens einer Stelle zur Grenze der Forschung vorzustoßen, für den erwachsenen Mathematiker fast unmöglich, nur ungefähr den Überblick zu behalten. Gerade in dieser Situation kann es anregend und beruhigend sein, die Arbeiten eines Pioniers zu lesen.

F. Hirzebruch

Inhaltsverzeichnis

1.

The Foundation of Algebraic Geometry from Severi to André Weil

Archive for History of Exact Sciences 7, 3 (1971) 171–180

From Max Noether to Severi

Algebraic geometry was created by MAX NOETHER. The Italian school, headed by CORRADO SEGRE, CASTELNUOVO, ENRIQUES and SEVERI, erected an admirable structure, but its logical foundation was shaky. The notions were not well-defined, the proofs were insufficient. And yet, as BERNARD SHAW puts it: "There is an Olympian ring in it. It must be true, for it is fine art."

Severi's Paper on the Conservation of Number

A first step towards a rigid foundation was taken by FRANCESCO SEVERI in 1912 in a paper "Il principio della conservazione del numero" [1]. SEVERI considers a correspondence between varieties U and V, *i.e.* a set of pairs of points (x, y), x in U, y in V, defined by homogeneous equations $H(x, y) = 0$. He supposes that U is irreducible and that to a generic point ξ of U corresponds a finite number of points $\eta^{(1)}, \ldots, \eta^{(h)}$ of V. If ξ tends to x, the points $\eta^{(\nu)}$ tend to limit points $y^{(1)}, \ldots, y^{(h)}$. If μ of the points $\eta^{(\nu)}$ tend to one and the same limit point y, this limit point is said to have *multiplicity* μ. Assuming the number of solutions of the equations $H(x, y) = 0$ for a given x to be finite, SEVERI states first, that the multiplicities are uniquely defined. While he states this, he does not prove it. Second, if the given correspondence is irreducible, every solution y of the equations $H(x, y) = 0$ occurs at least once among the $y^{(\nu)}$. Finally: the sum of the multiplicities of all solutions y for a given x is equal to the number of solutions of the set $H(\xi, \eta) = 0$ for a generic ξ. This is the *Principle of Conservation of Number*, first enunciated in a crude form by H. SCHUBERT [2]. Counterexamples to SCHUBERT's original statement were given by G. KOHN [3] and K. ROHN [4]. KOHN's counterexamples incited SEVERI to introduce the condition that the correspondence be irreducible.

In a later paper [29], SEVERI admitted that his first assertion, saying that the multiplicities are unique, required the additional assumption: x is a simple point of U. If this assumption is made, SEVERI's three assertions are indeed correct, but the algebraic tools needed to prove them were not yet available in 1912, when SEVERI wrote his earlier paper. Anyhow, SEVERI was the first to give an exact definition of "multiplicity" and to formulate exact assertions.

1

Algebraic Tools: From Dedekind to Emmy Noether

The algebraic tools needed for a rigorous foundation of algebraic geometry were: Theory of Ideals, Elimination Theory, and Theory of Fields. They were developed in the schools of DEDEKIND, of KRONECKER, and of HILBERT. A most important paper of DEDEKIND & WEBER [5] in which DEDEKIND's Theory of Ideals was applied to fields of algebraic functions, appeared as early as 1882. Elimination theory was developed by KRONECKER and his school [6]. Systems of resultants were studied by MERTENS [14] and HURWITZ [7]. Most important work on the theory of Polynomial Ideals was done by LASKER [8], the famous chess champion, who had got his problem from HILBERT, and by MACAULY [9], a schoolmaster who lived near Cambridge, England, but who was nearly unknown to the Cambridge mathematicians when I visited Cambridge in 1933. I guess the importance of MACAULAY's work was known only in Göttingen. Finally, we must mention the fundamental paper of STEINITZ [10]: "Algebraische Theorie der Körper", which appeared in 1910.

When I came to Göttingen in 1924 and showed EMMY NOETHER some work I had done on generalizations of MAX NOETHER's "fundamental theorem", she said "What you have done is all right, as far as it goes, but LASKER and MACAULAY have obtained much more general results." She told me to study the paper of STEINITZ and MACAULAY's tract on Polynomial Ideals, and she gave me some of her papers on Ideal Theory [11] and on HENTZELT's Elimination Theory [12].

Generic Points

Thus, armed with the powerful tools of Modern Algebra, I returned to my main problem: to give Algebraic Geometry a solid foundation. I asked myself: What do the Italian geometers mean by a generic point on a variety (punto generico), a generic plane in 3-space, *etc.*? Obviously, a generic point is supposed not to have certain undesirable properties. For instance, a generic plane is not tangent to a given curve, it does not pass through a given point, *etc.* I asked: is it possible to find, on a given variety U, a point ξ not having any special properties except those which hold for all points of U? Of course, I restricted myself to algebraic properties, which can be expressed by algebraic equations.

If U is the whole space, it is easy to construct such a point ξ. One just takes as coordinates of ξ indeterminates. If an equation $f(X) = 0$ with coefficients from the ground field holds for indeterminates $X_1, \ldots, X_m$, it holds for all special values x' as well. —Now let U be any irreducible variety. Can one find a point ξ on U which is "generic" in the sense that if an equation $F = 0$ with constant coefficients holds for ξ it holds for all points of U? EMMY NOETHER's paper [11] on HENTZELT's elimination theory showed me the way. In order to understand the fundamental ideas of this paper, we have to go back to KRONECKER's elimination theory.

KRONECKER had developed a method to find all solutions of a set of algebraic equations by successive eliminations [6]. Using this method, he had proved: First, every variety is a union of irreducible varieties. Second, after a suitable linear transformation of coordinates, all points of an irreducible variety U of dimension d can be obtained as follows: the first d coordinates $x_1, \ldots, x_d$ are arbitrary, and the others are algebraic functions of them.

EMMY NOETHER [12] had obtained the same results anew by a more elegant elimination method due to her pupil HENTZELT, who had died in the first World War. She worked out HENTZELT's results and added new ideas of her own. One of these ideas was to replace the coordinates $x_1, \ldots, x_d$ by indeterminates $\xi_1, \ldots, \xi_d$ and to determine the other ξ_i as algebraic functions of these indeterminates in an algebraic extension field $k(\xi)$. This field she called the *Nullstellenkörper* of the prime ideal $\mathfrak{p}$ belonging to the variety.

When I studied EMMY NOETHER's paper, I saw that the point ξ with coordinates $\xi_1, \ldots, \xi_m$ was just the generic point I was looking for. I also saw that EMMY NOETHER's Nullstellenkörper was isomorphic to the quotient field of the residue class ring $\mathfrak{o}/\mathfrak{p}$, where $\mathfrak{o}$ is the polynomial ring $k[X]$. Hence it was not necessary to go through HENTZELT's elimination procedure; one could start with any prime ideal $\mathfrak{p} \neq \mathfrak{o}$, construct the residue class ring $\mathfrak{o}/\mathfrak{p}$ and its quotient field $k(\xi)$, and thus find a generic point ξ of the variety of $\mathfrak{p}$.

I wrote a paper [13] based upon this simple idea and showed it EMMY NOETHER. She at once accepted it for the *Mathematische Annalen*, without telling me that she had presented the same idea in a course of lectures just before I came to Göttingen. I heard it later from GRELL, who had attended her course.

Specializations

The next problem I studied was the definition of multiplicity. Let us recall that SEVERI had considered a correspondence between U and V, defined by a set of homogeneous equations

$$(1) \qquad\qquad H(x, y) = 0.$$

SEVERI's assumptions are

1) U is irreducible.

2) If ξ is a generic point of U, the set of equations

$$(2) \qquad\qquad H(\xi, \eta) = 0$$

has a finite number of solutions $\eta^{(1)}, \ldots, \eta^{(h)}$.

3) For the given point x, the set of equations (1) has a finite number of solutions.

SEVERI now constructs specialized solutions $y^{(1)}, \ldots, y^{(h)}$ by a passage to the limit:

$$(3) \qquad\qquad (\xi, \eta^{(1)}, \ldots, \eta^{(h)}) \to (x, y^{(1)}, \ldots, y^{(h)}).$$

If the ground field k is a field without topology, such a passage to the limit makes no sense. Therefore, I introduced the notion of *specialization* or, as I first called it, *relationstreue Spezialisierung*. The set $(x, y^{(1)}, \ldots)$ is called a specialization of $(\xi, \eta^{(1)}, \ldots)$, if all homogeneous equations $F(\xi, \eta^{(1)}, \ldots) = 0$ remain valid for $(x, y^{(1)}, \ldots)$.

I now had to prove the existence and uniqueness of the specialization (3) under appropriate conditions. This I did by means of elimination theory.

3

The Existence of Extensions of Specializations

Mertens had already constructed, for any set of homogeneous equations $F_j(y) = 0$, a set of resultants $R_1, \ldots, R_s$ depending only on the coefficients of the forms F_i, such that the equations have a non-zero solution if and only if all resultants R_j are zero. In a paper presented to the Amsterdam Academy in 1926 [15], I gave a simpler derivation, based upon Hilbert's Nullstellensatz.

No matter whether one uses Mertens' resultants or mine, it is easy to see that every specialization $\xi \to x$ can be extended to a specialization

$$(\xi, \eta^{(1)}, \ldots, \eta^{(h)}) \to (x, y^{(1)}, \ldots, y^{(h)}).$$

Later on Chevalley proved the possibility of extending specializations without making use of elimination theory. André Weil included Chevalley's proof in his book [16], expressing the hope that this device "finally eliminates from algebraic geometry the last traces of elimination theory". Obviously Weil and Chevalley do not like Elimination Theory.

Uniqueness of Specialization Multiplicities

More difficult was the proof of the uniqueness of the specialization (3). In 1927, in a paper "Der Multiplizitätsbegriff der algebraischen Geometrie" [17], I made the following assumptions:

First, the coordinates $\xi_1, \ldots, \xi_m$ are independent variables.

Secondly, the equations (1) are assumed to have, for a given x, a finite number of solutions only.

From these hypotheses I proved that the specialized solutions $y^{(1)}, \ldots, y^{(h)}$ are uniquely determined but for their order. Hence the multiplicity μ of any solution y can be defined as the number of times it appears among the $y^{(\nu)}$. Obviously, the sum of the multiplicities of all solutions y is equal to h. This is the Principle of Conservation of Number. Moreover, if the correspondence (2) is irreducible, every solution of the specialized problem (1) occurs at least once among the $y^{(\nu)}$. Thus all assertions made in Severi's paper of 1912 [1] had been proved rigorously, under appropriate assumptions, by 1927.

In a later paper of mine [18] the assumption that the ξ_i be independent variables (which implies that the variety U is the whole affine or projective space) was replaced by the weaker assumption that x be a simple point of U. As we shall see later, André Weil replaced my second assumption, which says that the specialized problem (1) has a finite number of solutions, by the weaker assumption that one solution y be isolated. The assumptions were weakened still more by Northcott [19] and Leung [20].

Intersection Multiplicities

My next problem was, to define *intersection multiplicities* and to generalize Bézout's theorem on the number of points of intersection of two plane curves.

Intersection of Hypersurfaces

The case of the intersection of n forms or hypersurfaces in P^n is very easy. The multiplicities can be defined as exponents in the factorization of the u-resultant of the forms, *i.e.* of the resultant of the n forms plus one generic linear form

$\sum u_k y_k$. All this is explained on one or two pages in § 7 of my paper "Multiplizitätsbegriff" [17].

Intersection of a Curve and a Hypersurface

Next consider the intersection of an irreducible curve C with a form F. (A hypersurface, *i.e.*, a cycle of dimension $n-1$, will be called a form F.) How can we define the multiplicities of the points of intersection?

The problem can be solved by several methods. First one can use the classical theory of algebraic functions. The ratios ξ_i/ξ_0 of the homogeneous coordinates of a variable point ξ of C are algebraic functions of one complex variable z, univalent on a Riemann surface. To every point on the closed Riemann surface corresponds a point on the curve C. Vice versa, to every point on the curve corresponds at least one point on the Riemann Surface. Consider the rational function

$$F(\xi_0, \ldots, \xi_n)/L(\xi)^g,$$

L being a generic linear form and g the degree of F. The poles of this meromorphic function correspond to the points of intersection of L and g, each pole having order g. The zeros correspond to the points of intersection of F and C. Now define the *multiplicity* of any point of intersection to be the sum of the orders of the corresponding zeros on the Riemann surface. The total sum of the orders of all zeros is equal to the sum of the orders of all poles, hence the sum of the multiplicities of the meeting points of C and F is equal to the product of the degrees of C and F. This generalizes BÉZOUT's theorem. The idea of this definition and proof can be found in papers of the French geometer HALPHEN.

This method of proof works just as well for an arbitrary ground field, for one can use, instead of classical function theory, the arithmetical theory of DEDEKIND & WEBER [5].

Even simpler is the following, second definition, which was systematically used in my series of papers "Zur algebraischen Geometrie", beginning in 1933 [21]. Consider the form F as a specialization of a generic form F^* with indeterminate coefficients, which meets the curve C in the points $\eta^{(1)}, \ldots, \eta^{(h)}$. The specialization $F^* \to F$ can be extended to a specialization

$$(F^*, \eta^{(1)}, \ldots, \eta^{(h)}) \to (F, y^{(1)}, \ldots, y^{(h)}).$$

The $y^{(\nu)}$ are the points of intersection of F and C, and their intersection multiplicities can be defined to be equal to their specialization multiplicities.

A third possibility is offered by the theory of polynomial ideals. Introduce non-homogeneous coordinates. Let $\mathfrak{p}$ be the prime ideal belonging to the curve. The ideal $(F, \mathfrak{p})$ is an intersection of zero-dimensional primary ideals $\mathfrak{q}$ corresponding to the single points of the meet of F and C. For every $\mathfrak{q}$, the residue class ring $\mathfrak{o}/\mathfrak{q}$ has a finite rank. This rank can be defined to be the multiplicity of the point of intersection.

All three definitions are equivalent. For a proof see, *e.g.*, LEUNG's paper [20].

Intersection of Two Varieties in P^n

Much more difficult is the case of an intersection of two varieties A and B of arbitrary dimensions r and s in a projective space P^n. The general case can

5

easily be reduced to the case $r + s = n$, hence we shall suppose A and B to be varieties of dimensions d and $n - d$ in P^n, for instance two surfaces in 4-space, which meet in a finite number of points.

Lefschetz' Topological Definition

Lefschetz [22] gave in 1924 a topological definition of the intersection multiplicity, valid for the complex ground field. He considers the surfaces A and B as four-dimensional cycles in the complex projective space P^4. This space is an 8-dimensional orientable manifold. The space itself and the cycles A and B can be oriented in a fixed, canonical way. If y is a point of intersection and if the varieties have simple points and no common tangent at y, their topological intersection multiplicity, defined by simplicial approximation, is always $+1$. For the general case, in which there may be common tangents, I proved that the topological intersection multiplicity, or *index* of an isolated point of intersection is always positive. If this index is defined to be the intersection multiplicity, a perfectly satisfactory theory is obtained [23], and Schubert's "Calculus of Enumerative Geometry" [2] can be justified.

Applying a Projective Transformation

For an arbitrary ground field, this method does not work. I therefore proposed in 1928 [24] to bring A and B into a generic relative position by applying to one of them a projective transformation T with indeterminate coefficients. The transformed variety TA intersects B in a finite number of points. If T is specialized to the identity, the points of intersection of TA and B specialize to points of intersection of A and B. If A and B meet in a finite number of points, each of these appears with a certain specialization multiplicity, which may be defined to be the intersection multiplicity. In the complex field case, this definition is equivalent to that of Lefschetz. By a method due to the Dutch geometer Schaake, I proved the generalized theorem of Bézout.

In my paper of 1928, the proofs were very complicated, but in a later paper [25] I gave a simple proof based on the same idea.

A and B on a Variety U: Severi's Definitions

At the Zürich International Congress in 1932 I met Severi, and I asked him whether he could give me a good algebraic definition of the multiplicity of a point of intersection of two varieties A and B, of dimensions d and $n - d$, on a variety U of dimension n, on which the point in question is simple. The next day he gave me the answer, and he published it in the *Hamburger Abhandlungen* in 1933 [26]. He gave several equivalent definitions; I shall explain one of them for the case of two curves A and B lying on a surface U in 3-space and meeting in a simple point y of U.

Let S be a generic point in 3-space. Connecting S with A, one obtains a cone K. The point y has, as a point of intersection of the curve B with the cone K, a certain multiplicity μ. This is defined to be the multiplicity of y as a point of intersection of A and B on U.

This definition can easily be extended to intersections of varieties A and B of arbitrary dimensions r and s on a variety U of dimension n. If $r+s$ exceeds n, the normal dimension of the intersection is $r+s-n$. If a component C of the intersection has just this dimension, it is called a *proper component*. If all components are proper, each of them has, in SEVERI's theory, a well-defined multiplicity, and one can define an *intersection cycle* $A \cdot B$ as the formal sum of the components C, counted with their multiplicities μ:

$$A \cdot B = \sum \mu C .$$

Cycles and Their Coordinates

In explaining SEVERI's results, I have used the expressions *cycle* and *intersection cycle*, defining a cycle to be a formal sum of irreducible components C multiplied by integers μ.

To be sure, these expressions do not occur in SEVERI's paper. In SEVERI's papers (and in mine too) the same word "variety" (Varietà, in German, Mannigfaltigkeit) was used for two different notions, *viz.*

1) a point set in affine or projective space, defined by algebraic equations,

2) a formal sum of irreducible varieties C of the same dimension d with integer coefficients μ.

In the present paper, the word "variety" will only be used in the sense 1), whereas the formal sums 2) will be called *cycles*, following A. WEIL [16].

If the coefficients μ are arbitrary integers, the cycle will be called a *virtual cycle* (SEVERI: Varietà virtuale). The virtual cycles of any dimension d form an additive group, the free generators of the group being the irreducible varieties. If the integers μ are non-negative, we have a *positive cycle* (Varietà effettiva). In the sequel only positive cycles will be considered.

Varieties of cycles, e.g. congruences and complexes of lines, varieties of conics, *etc.* have been objects of investigation already in the 19th century. A variety of lines in 3-space is defined by a set of homogeneous equations in the PLÜCKER coordinates, and a linear system of conics by linear equations in the coefficients of a conic.

Linear systems of plane curves and their birational transformations were studied by CASTELNUOVO in a classical paper [27]. The theory was generalized to linear systems of curves C^1 on a surface V^2, and to linear systems of cycles C^{r-1} on a variety V^r, by CASTELNUOVO, ENRIQUES, and SEVERI. An excellent exposition of these classical theories was given by ZARISKI [28].

The generalization to non-linear systems of curves on a surface was inaugurated by SEVERI [29] in 1903; for an account of this theory see Chapter V of ZARISKI's report [28].

In 1932, SEVERI opened a new field of research by considering rational and algebraic systems of cycles C^d on a variety C^r [29]. A complete exposition of this theory was given by SEVERI in three volumes [30].

In all these theories, a correct definition of the notion "variety of cycles" (sistema di varietà) was missing. It was first given by CHOW & VAN DER WAERDEN in 1936 [31]. Our idea was, to assign to every cycle C^d a form $F(u)$ in $u^{(0)}, u^{(1)},$

..., $u^{(d)}$, called the *associated form* (zugeordnete Form, or Cayley-form) of the cycle, and to use the coefficients of this form as coordinates (Chow-coordinates) of the cycle. Varieties of cycles are now defined by homogeneous equations in these coordinates. The main theorem, proved in our joint paper [31], was: The totality of all cycles C^d in P^n of given dimension d an degree g is a variety of cycles, *i.e.* it can be defined by homogeneous equations in the coordinates of the cycles. The proof of this theorem is due to Chow.

The general theory of Varieties of Cycles was further developed in my paper, "Zur algebraischen Geometrie 14" [25]. In this paper, I showed that Severi's definition of the intersection cycle $A \cdot B$ has all desirable properties, including the commutative and associative laws for intersection cycles. Also: If A varies in a variety of cycles and if B is fixed, the intersection cycle $A \cdot B$ varies in a variety of cycles. Just so if A and B both vary in varieties of cycles.

The intersection theory developed in "Zur algebraischen Geometrie 14" was a global theory. The first to develop a *local* algebraic theory of intersections was André Weil, in his book "Foundations" [16]. To understand his line of thought, we must first return to the notion of *specialization multiplicity.*

Weil's Specialization Multiplicity

Weil considers a specialization of a pair of points in affine spaces:

$$(4) \qquad (\xi, \eta) \to (x, y')$$

and makes the following assumptions:

1) $\xi_1, \dots, \xi_m$ are independent variables,

2) the η are algebraic functions of the ξ,

3) y' is an isolated specialization over the specialization of ξ to x. This means that if we consider all equations $H(\xi, \eta) = 0$ which hold for ξ and η, then the specialized set of equations

$$(5) \qquad H(x, y) = 0$$

is supposed to define, for given x, a variety in which y' is an isolated point. Weil does not suppose, as I had done, that the set (5) has only a finite number of solutions.

Weil now considers the conjugates $\eta^{(\nu)}$ of η and proves (Chapter III, § 3, Proposition 7) that there exists a unique integer μ, called the *specialization multiplicity* of y', such that y' appears exactly μ times in any specialization

$$(\xi, \eta^{(1)}, \dots, \eta^{(h)}) \to (x, y^{(1)}, \dots, y^{(h)}).$$

The proof is given by means of the analytic theory of specializations, which is explained in Chapter III of Weil's book. It seems that Weil developed this analytic theory mainly or exclusively for the purpose of proving the uniqueness of the multiplicity μ, for on p. 61 (first edition) he writes:

> The reader may now forget the whole of the contents of §§ 1–3 of this chapter, except prop. 7 of § 3, which ... will form the corner-stone for our theory of multiplicities in algebraic geometry.

Weil's Intersection Multiplicities

WEIL's theory of intersection multiplicities proceeds in two steps. First, WEIL considers two varieties A and B in affine space, one of which, say B, is a *linear* subspace. Let C be a *proper* component of the intersection, *i.e.* a component of dimension $d = r + s - n$. It is easy to reduce d to zero by intersecting all varieties with a generic set of d linear forms. Hence we may suppose d to be zero, and C to be an isolated point of the intersection of A and B. Now the linear variety B can be obtained from a generic linear variety by specialization. It follows that the intersection multiplicity can be defined as a specialization multiplicity.

The second, decisive step is taken in Chapter 6 of WEIL's book. WEIL's idea is, to reduce the general case of an intersection $A \cdot B$ to the special case in which B is linear. Let A and B be two subvarieties of an n-dimensional variety U in the affine space S^N, and let C be a proper component of their intersection. Suppose a generic point of C is simple on U. WEIL now intersects the product variety $A \times B$ with the diagonal Δ of the product space $S^N \times S^N$. This diagonal is a linear subspace, defined by N linear equations $x_i - y_i = 0$. The number of equations may be too large; therefore WEIL forms n generic linear combinations $F_j(x - y) = 0$. They define a linear subspace L in $S^N \times S^N$. The diagonal Δ_C of $C \times C$ is contained in $A \times B$ and in L, hence in their intersection. WEIL now proves that Δ_C is a proper component of the intersection of $A \times B$ with L. As such, it has a definite multiplicity μ, because L is linear. This number μ is defined to be the multiplicity of C as a proper component of the intersection $A \cdot B$.

My time is up. Further important investigations on intersection multiplicities are due to SAMUEL, CHEVALLEY and SERRE. Completely new points of view were introduced by WEIL, by ZARISKI, by CHOW and in the school of GROTHENDIECK. However, in this lecture I have had to restrict myself to just one line of development, that from SEVERI to ANDRÉ WEIL.

The foregoing is an elaboration of a lecture delivered at Nice on September 5, 1970 at the International Congress of Mathematicians to the Section on the History of Science.

References

1. SEVERI, F., Il Principio della Conservazione del Numero. Rendiconti Circolo Mat. Palermo **33**, 313 (1912).
2. SCHUBERT, H., Kalcül der abzählenden Geometrie. Leipzig 1879.
3. KOHN, G., Über das Prinzip der Erhaltung der Anzahl. Archiv der Math. und Physik (3) **4**, 312 (1903).
4. ROHN, K., Zusatz zu E. STUDY, Das Prinzip der Erhaltung der Anzahl. Berichte sächs. Akad. Leipzig **68**, 92 (1916).
5. DEDEKIND, R., & H. WEBER, Theorie der algebraischen Funktionen einer Veränderlichen. Journal f. reine u. angew. Math. **92**, 181 (1882).
6. KÖNIG, J., Einleitung in die allgemeine Theorie der algebraischen Größen. Leipzig: Teubner 1903.
7. HURWITZ, A., Über die Trägheitsformen eines algebraischen Moduls. Annali di Mat. pura ed applic. (3) **20**, 113 (1913).
8. LASKER, E., Zur Theorie der Moduln und Ideale. Math. Annalen **60**, 20 (1905).
9. MACAULAY, F. S., Algebraic theory of modular systems. Cambridge Tracts **19** (1916).
10. STEINITZ, E., Algebraische Theorie der Körper. Journal f. reine u. angew. Math. **137**, 167 (1910).

11. NOETHER, E., Idealtheorie in Ringbereichen. Math. Annalen **83**, 24 (1921).
12. NOETHER, E., Eliminationstheorie und allgemeine Idealtheorie. Math. Annalen **90**, 229 (1923).
13. VAN DER WAERDEN, B. L., Zur Nullstellentheorie der Polynomideale. Math. Annalen **96**, 183 (1926).
14. MERTENS, F., Zur Theorie der Elimination. Sitzungsber. Akad. Wien **108**, 1173 und 1344 (1899).
15. VAN DER WAERDEN, B. L., Ein algebraisches Kriterium für die Lösbarkeit homogener Gleichungen. Proc. Acad. Amsterdam **29**, 142 (1926).
16. WEIL, A., Foundations of algebraic geometry (1st. ed., 1946). Amer. Math. Soc. Colloquium Publications **29**.
17. VAN DER WAERDEN, B. L., Der Multiplizitätsbegriff der algebraischen Geometrie. Math. Annalen **97**, 756 (1927).
18. VAN DER WAERDEN, B. L., ZAG 6. Math. Annalen **110**, 134 (1934).
19. NORTHCOTT, D., Specializations over a local domain. Proc. London Math. Soc. (3) **1** (1951).
20. LEUNG, K.-T., Die Multiplizitäten in der algebraischen Geometrie. Math. Ann. **135**, 170 (1958).
21. VAN DER WAERDEN, B. L., ZAG 1, Math. Annalen **108** (1933) up to ZAG 13, Math. Annalen **115** (1938). With ZAG 14 [25] a new development begins.
22. LEFSCHETZ, S., L'Analysis situs et la géométrie algébrique. Collection Borel 1924.
23. VAN DER WAERDEN, B. L., Topologische Begründung des Kalküls der abzählenden Geometrie. Math. Annalen **102**, 337 (1930).
24. VAN DER WAERDEN, B. L., Eine Verallgemeinerung des Bézoutschen Theorems. Math. Ann. **99**, 497 (1928).
25. VAN DER WAERDEN, B. L., ZAG 14. Math. Annalen **115**, 621 (1938).
26. SEVERI, F., Über die Grundlagen der algebraischen Geometrie. Abh. Math. Sem. Hamburg **9**, 335 (1933).
27. CASTELNUOVO, G., Ricerche generali sopra i sistemi lineari di curve piane. Mem. Accad. Sci. Torino, 2nd series, **42** (1892).
28. ZARISKI, O., Algebraic Surfaces. Ergebnisse der Math. III, 5 (1935).
29. SEVERI, F., Un nouvo campo di ricerche nella geometria sopra una superficiee sopra una varietà algebrica. Mem. Accad. Ital. Mat. **3**, No. 5 (1932).
30. SEVERI, F., Vol. I (1942): Serie, sistemi d'equivalenza e correspondenze algebriche sulle varietà algebriche. Vol. II (1958) and III (1959): Geometria dei sistemi algebrici sopra una superficie e sopra una varietà algebrica.
31. CHOW, W.-L. & VAN DER WAERDEN, ZAG 9. Math. Annalen **113**, 692 (1937).

Mathematisches Institut
Universität Zürich

(Received October 26, 1970)

2.

Zur Nullstellentheorie der Polynomideale

Mathematische Annalen 96, 2 (1926) 183–208

Die exakte Begründung der Theorie der algebraischen Mannigfaltig-keiten in n-dimensionalen Räumen kann nur mit den Hilfsmitteln der Idealtheorie geschehen, weil schon die Definition einer algebraischen Mannig-faltigkeit unmittelbar auf Polynomideale führt. Eine Mannigfaltigkeit heißt ja algebraisch, wenn sie durch algebraische Gleichungen in den n Koordi-naten bestimmt wird, und die linken Seiten aller Gleichungen, die aus diesen Gleichungen folgen, bilden ein Polynomideal (Def. § 1, 4).

Diese Begründung kann nun einfacher gestaltet werden als es bisher geschehen ist[1]), nämlich ohne Hilfe der Eliminationstheorie, ausschließlich auf dem Boden der Körpertheorie[2]) und der allgemeinen Idealtheorie in Ringbereichen[3]). Das zu zeigen ist das erste Ziel dieser Arbeit[4]). Die Eliminationstheorie hat in diesem Schema nur die Aufgabe, zu unter-suchen, wie man (bei gegebener Idealbasis) in endlichvielen Schritten

---

[1]) Kronecker, Grundzüge einer arithmetischen Theorie der algebraischen Größen. Journ. f. Math. **92** (1882), S. 1–122; J. König, Einleitung in die Theorie der algebraischen Größen (Leipzig 1903); F. S. Macaulay, Modular Systems (Cambridge Tracts **19** (1916)) und die dort angegebene Literatur; K. Hentzelt, Zur Theorie der Polynomideale und Resultanten (bearbeitet von E. Noether), Math. Ann. **88** (1922), S. 53–79; E. Noether, Eliminationstheorie und allgemeine Idealtheorie, Math. Ann. **90** (1923), S. 229–261.

[2]) E. Steinitz, Algebraische Theorie der Körper, Journ. f. Math. **137** (1910), S. 167–309.

[3]) E. Noether, Idealtheorie in Ringbereichen, Math. Ann. **83** (1921), S. 24–66.

[4]) Wie Frl. Noether mir mitteilte, hat sie in ihren Vorlesungen Winter 1923/24 und Sommer 1924 schon im wesentlichen mit den gleichen Methoden die Theorie aufgebaut. Einem Vorlesungsheft aus dieser Zeit verdanke ich eine Anzahl Ver-besserungen der Darstellung. Auch schulde ich Fräulein Noether Dank für viele nütz-liche Ratschläge bei der Abfassung der vorliegenden Arbeit.

die Nullstellenmannigfaltigkeiten eines Ideals und die Basis der zugehörigen Primideale und Primärideale finden kann[5]).

Das arithmetische Problem, dessen Diskussion die Lösung unserer geometrischen Probleme ergeben wird, lautet folgendermaßen. Sei P ein Körper, Ω ein Erweiterungskörper, der durch die Adjunktion endlichvieler Elemente $\xi_1, \ldots, \xi_n$ (unter denen sich überflüssige befinden dürfen, ja sogar Größen aus P) erzeugt wird. Läßt sich der Körper Ω durch algebraische Relationen zwischen den ξ_i charakterisieren? Wir werden zu dem Zweck die Gesamtheit der Polynome $f(x_1, \ldots, x_n)$ mit Koeffizienten aus P, für welche $f(\xi_1, \ldots, \xi_n) = 0$ ist, betrachten. Diese Gesamtheit ist, wie sich leicht zeigt, ein Primideal (Def. § 1, **12**) im Polynombereich, und die Struktur des Körpers Ω ist durch das Primideal nach einer formalen Regel (§ 3, **2**) eindeutig bestimmt. Umgekehrt läßt sich zu jedem vom Einheitsideal verschiedenen Primideal mittels dieser formalen Regel ein Körper Ω mit den oben angegebenen Eigenschaften bilden. Diese Reziprozität liefert die Eigenschaften der Primideale als einfache Folgen bekannter Körpereigenschaften (§ 3).

Diese arithmetischen Betrachtungen ergeben „geometrische" Tatsachen, wenn unter „Punkt" verstanden wird ein System von n Elementen aus dem Körper P, der dann als algebraisch-abgeschlossener Körper vorausgesetzt wird (etwa der Körper der komplexen Zahlen). Diejenigen Punkte, welche Nullstellen für alle Polynome des Primideals sind, bilden eine irreduzible algebraische Mannigfaltigkeit. Die Körperelemente $\xi_1, \ldots, \xi_n$, als algebraische Funktionen einer Anzahl unabhängiger Elemente aufgefaßt, ergeben eine „Parameterdarstellung" der Mannigfaltigkeit. Durch diese doppelte Erzeugungsweise ergeben sich ohne Schwierigkeit die (bekannten) Eigenschaften der irreduziblen Mannigfaltigkeiten (§ 4).

In § 5 wird der Zerlegungssatz der allgemeinen Idealtheorie herangezogen, mit dessen Hilfe auch die Nullstellen der Nichtprimideale der Untersuchung zugänglich werden. Es ergibt sich ohne weiteres die Zerlegung einer beliebigen algebraischen Mannigfaltigkeit in irreduzible; weiter der Dimensionsbegriff für beliebige Ideale und seine Eigenschaften, endlich der Hilbertsche Nullstellensatz und ein neuer Satz über Ideale von $n\text{-}1$-Dimensionen.

Sodann (§ 6) gestatten die entwickelten Methoden einen neuen Beweis einer von K. Hentzelt[6]) gefundenen n-dimensionalen Verallgemeinerung des M. Noetherschen „Fundamentalsatzes der Theorie der algebraischen

---

[5]) Grete Hermann, Die Frage der endlichvielen Schritte in der Theorie der Polynomideale, Math. Ann. **95** (1926), S. 736.

[6]) K. Hentzelts Dissertation ist nicht gedruckt. Sein Beweis findet sich aber bei Grete Hermann, a. a. O.

Funktionen". Der Beweis ist bedeutend einfacher als der Hentzeltsche, dafür aber nicht konstruktiv im Sinne der endlichvielen Schritte.

Die größere Einfachheit und zugleich größere Allgemeinheit der vorliegenden, stark an E. Noether (Math. Ann. **90**, s. o.) anlehnenden Theorie gegenüber den älteren Theorien konnte nur erreicht werden durch äußerste Arithmetisierung aller Begriffe und Operationen.

Ich konnte dabei anschließen an die Körpertheorie und an die allgemeine Idealtheorie, die in völlig arithmetisierter Gestalt schon vorliegen in den oben erwähnten Arbeiten von E. Steinitz und E. Noether (Math. Ann. **83**), und deren Grundbegriffe und Sätze, soweit sie hier nötig sind, in den §§ 1, 2 zusammengestellt sind. Nur ein Hilfssatz aus der Körpertheorie (§ 2, **7**), der von E. Noether in ihrer erwähnten Wintervorlesung gebracht wurde, ist noch nicht publiziert; mit wesentlichen Vereinfachungen wird hier der Noethersche Beweis wiedergegeben werden.

§ 1.
Ringe und Ideale.

1. Ein *Ring* ist eine Menge von Elementen $a, b, \ldots$, für welche eine reflexive, symmetrische und transitive Gleichheitsrelation definiert ist, ferner eine Addition, die eindeutig (im Sinne der Gleichheitsrelation), kommutativ, assoziativ und eindeutig umkehrbar ist, und eine eindeutige, assoziative und gegenüber der Addition distributive Multiplikation. Es folgt, daß ein Element 0 existiert, so daß $a + 0 = a$; weiter folgt $a \cdot 0 = 0$; $0 \cdot a = 0$. Der Ring heißt kommutativ, wenn die Multiplikation es ist. Der Ring hat keine Nullteiler, wenn aus $a\,b = 0$ und $a \neq 0$ folgt $b = 0$.

Sind diese beiden Eigenschaften erfüllt für einen Ring R, der ein von Null verschiedenes Element enthält, und hat außerdem die Gleichung $a\,x = b$ für $a \neq 0$ immer eine Lösung (und, da keine Nullteiler vorhanden sind, auch höchstens eine Lösung), so heißt der Ring ein *Körper*. Ein Körper enthält immer ein Einheitselement, das mit ε bezeichnet werden soll, mit der Eigenschaft $\varepsilon \cdot a = a$.

2. Aus einem kommutativen Ring R bildet man einen *Polynombereich* oder *Polynomring* $R\,[x_1, \ldots, x_n]$, indem man die Polynome $f(x_1, \ldots, x_n) = \Sigma\, a_{p_1, \ldots, p_n} x_1^{p_1} \ldots x_n^{p_n}$ als Elemente einführt (wo über endlichviele verschiedene Indizeskombinationen summiert wird, und wo die $a_{p_1, \ldots, p_n}$ Ringelemente sind), und für sie Addition und Multiplikation in der üblichen Weise definiert. Zwei Polynome heißen gleich, wenn die gleichnamigen Koeffizienten gleich sind. Ist $f(x_1, \ldots, x_n) \neq 0$, und hat der Koeffizientenring R unendlich viele Elemente und keine Nullteiler, so ist es immer möglich, den „*Unbestimmten*" $x_1, \ldots, x_n$ spezielle Werte $a_1, \ldots, a_n$ aus R zu geben, so daß $f(a_1, \ldots, a_n) \neq 0$.

3. Aus einem kommutativen Ring R ohne Nullteiler bildet man den *Quotientenkörper*, indem man die Quotienten $\frac{a}{b}$ (wo $b \neq 0$) rein formal als Elementenpaare definiert und dem Ring adjungiert. Gleichheit, Summe und Produkt werden definiert durch:

$$\frac{a}{b} = \frac{c}{d}, \quad \text{wenn} \quad a\,d = b\,c \qquad\qquad \frac{a}{b} = c, \quad \text{wenn} \quad a = b\,c$$

$$\frac{a}{b} + \frac{c}{d} = \frac{a\,d + b\,c}{b\,d} \qquad\qquad \frac{a}{b} + c = \frac{a + b\,c}{b}$$

$$\frac{a}{b}\,\frac{c}{d} = \frac{a\,c}{b\,d} \qquad\qquad \frac{a}{b}\,c = \frac{a\,c}{b}.$$

Man zeigt leicht die Körpereigenschaften.

4. Ein *Ideal* in einem *kommutativen* Ring R ist eine Untermenge von R derart, daß mit a und b immer $a - b$ (und folglich auch 0, $- b$ und $a + b$), und mit a immer $r\,a$ in der Untermenge enthalten sind, wo r ein beliebiges Ringelement ist. Aus einer beliebigen Untermenge von R wird ein Ideal *erzeugt*, nämlich die Gesamtheit der Elemente die entstehen, indem man die Elemente der Untermenge zunächst mit beliebigen Ringelementen multipliziert, sodann die Elemente selbst und ihre Vielfachen in allen möglichen Weisen addiert und subtrahiert. Hat ein Ideal $\mathfrak{a}$ endlichviele Erzeugende $a_1, \ldots, a_r$, so bilden diese eine *Basis*, und man schreibt $\mathfrak{a} = (a_1, \ldots, a_r)$.

5. Ist R insbesondere ein Polynombereich mit Koeffizienten aus einem Körper, so gilt der *Hilbertsche Basissatz*: *Jedes Ideal hat eine Basis* [7]).

6. Man nennt zwei Ringelemente a, b *kongruent* nach einem Ideal $\mathfrak{c}$, und schreibt:

$$a \equiv b \ \operatorname{mod} \mathfrak{c}$$

oder kurz

$$a \equiv b \, (\mathfrak{c}),$$

wenn die Differenz $a - b$ dem Ideal angehört.

7. Sind die Glieder zweier Summen kongruent nach einem festen Ideal $\mathfrak{a}$, so sind die Summen es auch, und das gleiche gilt für Produkte. Weiter ist die Kongruenzrelation reflexiv, symmetrisch und transitiv. Faßt man also die Kongruenz als eine neue Gleichheitsdefinition in R auf, so bilden die Ringelemente gegenüber der neuen Gleichheitsdefinition wieder einen Ring, der der *Restklassenring* nach dem Ideal $\mathfrak{a}$ (in Zeichen $R/\mathfrak{a}$) genannt wird.

---

⁷) D. Hilbert, Über die Theorie der algebraischen Formen. Math. Ann. **36** (1890), S. 473—534. Die beiden Beweise, die Hilbert gibt, der erste für den Fall, daß der Koeffizientenbereich ein Zahlkörper ist, der zweite für den Fall von ganzzahligen Koeffizienten, gelten allgemeiner, nämlich der erste für jeden Körper mit unendlich vielen Elementen, der zweite u. a. für jeden Körper als Koeffizientenbereich.

8. Ist $b \, \varepsilon \, \mathfrak{a}$ [8]), so ist $b \equiv 0 \, (\mathfrak{a})$, und umgekehrt. Ist ein Ideal $\mathfrak{b}$ Untermenge von $\mathfrak{a}$, so schreibt man $\mathfrak{b} \equiv 0 \, (\mathfrak{a})$, und nennt $\mathfrak{a}$ einen *Teiler* von $\mathfrak{b}$, $\mathfrak{b}$ ein *Vielfaches* von $\mathfrak{a}$.

9. Die *Summe* oder der *größte gemeinsame Teiler* $(\mathfrak{a}, \mathfrak{b})$ zweier Ideale $\mathfrak{a}, \mathfrak{b}$ ist das Ideal aller Summen $a + b$, wo $a \, \varepsilon \, \mathfrak{a}$, $b \, \varepsilon \, \mathfrak{b}$. Ist $\mathfrak{a} = (a_1, \ldots, a_s)$, $\mathfrak{b} = (b_1, \ldots, b_s)$, so ist $(\mathfrak{a}, \mathfrak{b}) = (a_1, \ldots, a_r, b_1, \ldots, b_s)$.

10. Das *Produkt* zweier Ideale ist das von allen Produkten $a \, b$, wo $a \, \varepsilon \, \mathfrak{a}$, $b \, \varepsilon \, \mathfrak{b}$, erzeugte Ideal. Die Multiplikation von Idealen ist kommutativ und assoziativ. Was eine Potenz eines Ideals ist, ist demnach klar. Ist $\mathfrak{a} = (a_1, \ldots, a_r)$, $\mathfrak{b} = (b_1, \ldots, b_s)$, so ist $\mathfrak{a} \, \mathfrak{b} = (a_1 b_1, a_1 b_2, \ldots, a_r b_s)$.

11. Der mengentheoretische Durchschnitt $[\mathfrak{a}, \mathfrak{b}]$ zweier Ideale wird auch ihr *kleinstes gemeinsames Vielfaches* (K. G. V.) genannt.

12. Ein Ideal $\mathfrak{p}$ heißt *prim*, wenn es aus $a \, b \equiv 0 \, (\mathfrak{p})$ und $a \not\equiv 0 \, (\mathfrak{p})$ folgt $b \equiv 0 \, (\mathfrak{p})$. Ist $\mathfrak{p}$ prim, so hat der Restklassenring $R/\mathfrak{p}$ keine Nullteiler. Sein Quotientenkörper heißt der *Restklassenkörper* von $\mathfrak{p}$.

13. Ein Ideal $\mathfrak{q}$ heißt *primär*, wenn aus $a \, b \equiv 0 \, (\mathfrak{q})$ und: keine Potenz von a ist $\equiv 0 \, (\mathfrak{q})$, folgt $b \equiv 0 \, (\mathfrak{q})$. Zu jedem Primärideal $\mathfrak{q}$ gehört ein Primideal $\mathfrak{p}$, nämlich die Gesamtheit aller Ringelemente, von denen eine Potenz in $\mathfrak{q}$ vorkommt. Offenbar ist $\mathfrak{q} \equiv 0 \, (\mathfrak{p})$. *Ist $a \, b \equiv 0 \, (\mathfrak{q})$ und $a \not\equiv 0 \, (\mathfrak{p})$, so folgt $b \equiv 0 \, (\mathfrak{q})$.* Hat insbesondere $\mathfrak{p}$ eine Basis, so gibt es eine kleinste Zahl ϱ, der *Exponent* von $\mathfrak{q}$, so daß $\mathfrak{p}^\varrho \equiv 0 \, (\mathfrak{q})$.

14. Ein Körper heißt *von der Charakteristik p*, wenn p die kleinste natürliche Zahl ist, für die $p \varepsilon = 0$ [8a]), wo ε die Einheit ist, und *von der Charakteristik Null*, wenn eine solche Zahl nicht existiert. Im ersten Fall ist p eine Primzahl.

§ 2.

Einige Sätze aus der Körpertheorie.

1. Zwei Körper Ω, Ω' heißen *isomorph*, wenn eine eineindeutige Zuordnung ihrer Elemente existiert, so daß Summe oder Produkt in Ω wieder in Summe oder Produkt in Ω' übergehen. Die Zuordnung selbst heißt *Isomorphismus*. Ist $\Omega' = \Omega$, so hat man einen *Automorphismus* von Ω vor sich.

2. Hat ein Körper Ω einen Unterkörper P, so heißt Ω eine Erweiterung von P. Die Erweiterung heißt *algebraisch*, wenn jedes Element von Ω einer algebraischen Gleichung mit Koeffizienten aus P genügt; sonst *transzendent*. Die algebraische Gleichung läßt sich immer durch eine in P irreduzible ersetzen. Sind alle Elemente von Ω rational durch

---

[8]) ε heißt: ist Element von.

[8a]) Das Symbol $p \varepsilon$ ist dabei wie üblich definiert durch die Rekursionsformeln $1 \cdot \varepsilon = \varepsilon$; $(n + 1) \varepsilon = n \varepsilon + \varepsilon$.

15

$\alpha_1, \ldots, \alpha_n$ mit Koeffizienten aus P ausdrückbar, so schreibt man $\Omega = \mathsf{P}(\alpha_1, \ldots, \alpha_n)$.

3. Es ist möglich, eine algebraische Erweiterung $\mathsf{P}(\alpha)$ eines Körpers P rein formal zu konstruieren, wenn die irreduzible Gleichung $f(z) = 0$ gegeben ist, der das zu adjungierende Element α genügen soll. Der Körper $\mathsf{P}(\alpha)$ muß nämlich immer isomorph dem Restklassenring des Polynombereichs $\mathsf{P}[z]$ nach dem Ideal $(f(z))$ sein, und das genügt um ihn formal zu bestimmen. Bezeichnet man mit $\mathsf{P}[\alpha]$ den Ring aller Elemente von $\mathsf{P}(\alpha)$, die sich als Polynome in α mit Koeffizienten aus P schreiben lassen, so folgt aus der genannten Isomorphie $\mathsf{P}[\alpha] = \mathsf{P}(\alpha)$. Durch Induktion folgt, wenn $\mathsf{P}(\alpha, \beta, \ldots, \delta)$ eine algebraische Erweiterung von P ist,

$$\mathsf{P}[a, \beta, \ldots, \delta] = \mathsf{P}(\alpha, \beta, \ldots, \delta).$$

4. Man kann durch fortgesetzte algebraische Erweiterung eines beliebigen Körpers P immer zu einem algebraisch-abgeschlossenen Körper Ω übergehen, d. h. zu einem solchen, in dem jedes Polynom $f(z)$ einer Unbestimmten z in Linearfaktoren zerfällt[9]). Ω ist bis auf Isomorphie eindeutig bestimmt.

5. Zwei Elemente α, β einer algebraischen Erweiterung eines Körpers P heißen *konjugiert in bezug auf* P, wenn es einen Automorphismus des algebraisch-abgeschlossenen Umfassungskörpers gibt, der P elementweise invariant läßt, und α in β überführt. Dazu ist hinreichend, daß es einen Isomorphismus der beiden Unterkörper $\mathsf{P}(\alpha)$ und $\mathsf{P}(\beta)$ gibt, der P elementweise invariant läßt, und α in β überführt. Alle Wurzeln einer in P irreduziblen Gleichung $f(z) = 0$ sind konjugiert. Ist $f(z)$ ein Polynom aus $\mathsf{P}[z]$, und sind α und β konjugiert, so folgt aus $f(\alpha) = 0$ auch $f(\beta) = 0$.

6. Eine *rein transzendente* Erweiterung $\mathsf{P}(\xi_1, \ldots, \xi_n)$ eines Körpers P kommt zustande, indem man den Quotientenkörper des Polynomrings $\mathsf{P}[\xi_1, \ldots, \xi_n]$ bildet. Seine Elemente heißen *rationale Funktionen* von $\xi_1, \ldots, \xi_n$.[10]) *Algebraische Funktionen* sind die Elemente einer algebraischen Erweiterung Ω von $\mathsf{P}(\xi_1, \ldots, \xi_n)$.

Daß man hier wirklich von (mehrdeutigen) Funktionen reden kann, deren Existenzgebiet eine Untermenge der Menge aller Systeme von n Elementen des algebraisch-abgeschlossenen Erweiterungskörpers Γ von P ist, ergibt die folgende Betrachtung, die wir sogleich für ein System von Funktionen $\eta_1, \ldots, \eta_m$ anstellen.

---

[9]) Dieser Steinitzsche Satz leistet für die Algebra das gleiche wie der „Fundamentalsatz der Algebra", der nicht der Algebra, sondern der Analysis angehört. Der Beweis setzt den Zermeloschen Wohlordnungssatz voraus.

[10]) Allgemeiner kann man eine beliebige (endliche oder unendliche) Menge von Unbestimmten adjungieren, indem man aus dem Ring der Polynome in beliebig vielen dieser Unbestimmten den Quotientenkörper bildet.

Wir adjungieren sukzessive die Größen $\eta_1, \ldots, \eta_m$ aus Ω dem Körper $\mathsf{P}(\xi_1, \ldots, \xi_n)$. Die jeweils irreduzible Gleichung für η_k mit Koeffizienten aus $\mathsf{P}(\xi_1, \ldots, \xi_n, \eta_1, \ldots, \eta_{k-1})$ laute:

$$(1) \qquad h_k(\eta_k) = \varepsilon\, \eta_k^{\varrho_k} + \alpha_{k,1}\, \eta_k^{\varrho_k-1} + \ldots + \alpha_{k,\varrho_k} = 0.$$

Die Koeffizienten $\alpha_{k,i}$ kann man schreiben als Quotienten von Polynomen in $\xi_1, \ldots, \xi_n, \eta_1, \ldots, \eta_{k-1}$; sogar kann man voraussetzen, daß die Nenner nur $\xi_1, \ldots, \xi_n$ enthalten (**3**). Diejenigen Elementsysteme $\xi_1', \ldots, \xi_n'$ aus Γ, für welche das Produkt sämtlicher Nennerpolynome von 0 verschieden ist, heißen *reguläre Argumentwerte* für das Funktionensystem $\eta_1, \ldots, \eta_m$. Da Γ, als algebraisch-abgeschlossener Körper, unendlich viele Elemente enthält, so gibt es immer reguläre Argumentwerte (§ 1, **2**). Jedes Elementsystem $\eta_1', \ldots, \eta_m'$, das sich aus den Gleichungen (1) für reguläre Argumentwerte ergibt, heißt ein zugehöriges *Funktionswertsystem*.

7. Mit den Bezeichnungen der vorigen Nummer gilt der Satz:

Ist f ein Polynom aus $\mathsf{P}[x_1, \ldots, x_n, y_1, \ldots, y_m]$, und ist

$$f(\xi_1, \ldots, \xi_n, \eta_1, \ldots, \eta_m) = 0, \ \text{ so ist } \ f(\xi_1', \ldots, \xi_n', \eta_1', \ldots, \eta_m') = 0$$

für alle regulären Argumentwerte $\xi_1', \ldots, \xi_n'$ und zugehörigen Funktionswerten $\eta_1', \ldots, \eta_m'$, und umgekehrt.

Beweis. Der erste Teil des Satzes ergibt sich durch vollständige Induktion, indem wir ihn als bewiesen annehmen für Polynome aus $\mathsf{P}[x_1, \ldots, x_n, y_1, \ldots, y_{k-1}]$, und beweisen für Polynome aus

$$\mathsf{P}[x_1, \ldots, x_n, y_1, \ldots, y_k].$$

Für Polynome aus $\mathsf{P}[x_1, \ldots, x_n]$ ist der Satz trivial.

Es habe $h_k(\eta_k)$ die Bedeutung (1). Dann ist die Voraussetzung

$$f(\xi_1, \ldots, \xi_n, \eta_1; \ldots, \eta_k) = 0$$

nach **3.** äquivalent mit

$$f(\xi_1, \ldots, \xi_n, \eta_1, \ldots, \eta_{k-1}, z) \equiv 0 \ \ (h_k(z)).$$

Diese Gleichung besagt, daß die Division der linken Seite durch $h_k(z)$ ohne Rest aufgeht. Nun sind die Koeffizienten von $h_k(z)$ Quotienten von Polynomen in $\xi_1, \ldots, \xi_n, \eta_1, \ldots, \eta_{k-1}$, wo die Nenner nur $\xi_1, \ldots, \xi_n$ enthalten, und für alle regulären Argumentwerte $\xi_1', \ldots, \xi_n'$ von Null verschieden sind. Außerdem ist der erste Koeffizient die Einheit ε. Führt man die Division wirklich aus, so kommen im Nenner des Quotienten niemals andere Polynome vor als die, welche in $h_k(z)$ schon im Nenner standen. Also wird:

$$f(\xi_1, \ldots, \xi_n, \eta_1, \ldots, \eta_{k-1}, z) - a_k(z)\, h_k(z) = 0,$$

wo sowohl $a_k(z)$ wie $h_k(z)$ im Nenner nur solche Polynome $q(\xi_2, \ldots, \xi_n)$ haben, für die $q(\xi_1', \ldots, \xi_n'') \neq 0$ ist. Multipliziert man die Gleichung mit dem Hauptnenner $n(\xi_1, \ldots, \xi_n)$ des zweiten Gliedes, so sind die Koeffizienten der Potenzen von z Polynome in $\xi_1, \ldots, \xi_n, \eta_1, \ldots, \eta_{k-1}$. Da für diese nach Voraussetzung der Satz gilt, so darf man spezialisieren $\xi_i = \xi_i'$ $[i = 1, \ldots, n]$, $\eta_j = \eta_j'$ $[j = 1, \ldots, k-1]$. Dividiert man schließlich durch die von Null verschiedene Größe $n(\xi_1', \ldots, \xi_n')$, so kommt:

$$f(\xi_1', \ldots, \xi_n', \eta_1', \ldots, \eta_{k-1}', z) - a_k'(z) h_k'(z) = 0,$$

wo a_k' und h_k' aus a_k und h_k durch die obige Spezialisierung entstanden sind.

Nun waren die Funktionswerte η_k' definiert durch die Gleichung $h_k'(\eta_k') = 0$, folglich ist

$$f(\xi_1', \ldots, \xi_n', \eta_1', \ldots, \eta_{k-1}', \eta_k') = 0, \qquad \text{q. e. d.}$$

Um den zweiten Teil zu beweisen, nehmen wir an, es wäre $f(\xi_1', \ldots, \xi_n', \eta_1', \ldots, \eta_m') = 0$ für alle regulären Argumentwertsysteme, und dennoch $f(\xi_1, \ldots, \xi_n, \eta_1, \ldots, \eta_m) \neq 0$. Setzen wir dann

$$\frac{\varepsilon}{f(\xi_1, \ldots, \xi_n, \eta_1, \ldots, \eta_m)} = \eta,$$

so können wir $\xi_1', \ldots, \xi_n'$ als reguläres Argumentwertsystem der Funktionen $\eta_1, \ldots, \eta_m, \eta$ bestimmen. Diese Argumentwerte sind dann sicher für $\eta_1, \ldots, \eta_m$ regulär. Aus

$$f(\xi_1, \ldots, \xi_n, \eta_1, \ldots, \eta_m)\, \eta - \varepsilon = 0$$

folgt nach dem ersten Teil des Satzes

$$f(\xi_1', \ldots, \xi_n', \eta_1', \ldots, \eta_m')\, \eta' - \varepsilon = 0$$

entgegen der Voraussetzung $f(\xi_1', \ldots, \xi_n', \eta_1', \ldots, \eta_m') = 0$.

8. Sei Ω ein Erweiterungskörper von P. Eine Teilmenge U von Ω heißt *irreduzibel* in bezug auf P, wenn eine Gleichung $f(\xi_1, \ldots, \xi_n) = 0$ zwischen endlichvielen Elementen von U mit Koeffizienten aus P nur dann bestehen kann, wenn das Polynom f identisch verschwindet. Die Elemente eines irreduziblen Systems sind also unabhängige transzendente (oder unabhängige Unbestimmte) in bezug auf P.

Zwei Teilmengen U, V von Ω heißen *äquivalent*, wenn jedes Element von U algebraisch in bezug auf P(V), und jedes Element von V algebraisch in bezug auf P(U) ist.

In jedem wohlgeordneten Erweiterungskörper Ω eines Körpers P läßt sich ein dem Körper Ω äquivalentes irreduzibles System konstruieren. Die Mächtigkeit dieses Systems ist von der gewählten Wohlordnung unabhängig, und heißt der *Transzendenzgrad* von Ω in bezug auf P. Auch in jedem Teilsystem V gibt es ein zu V äquivalentes irreduzibles System, dessen Mächtigkeit der *Transzendenzgrad* von V in bezug auf P heißt.

Ist der Transzendenzgrad von Ω in bezug auf P endlich und gleich n, so sind alle Elemente von Ω algebraische Funktionen von endlich vielen Unbestimmten $\xi_1, \ldots, \xi_n$. Ist er Null, so ist Ω algebraisch über P.

§ 3.

Der Nullstellenkörper eines Primideals [10a]).

1. *Ist $\Omega = \mathsf{P}(\xi_1, \ldots, \xi_n)$ ein Erweiterungskörper eines Körpers* P, *so bilden die Polynome f aus $R = \mathsf{P}[x_1, \ldots, x_n]$, für die $f(\xi_1, \ldots, \xi_n) = 0$, ein Primideal in* R.

Beweis. Aus $f(\xi_1, \ldots, \xi_n) = 0$ und $g(\xi_1, \ldots, \xi_n) = 0$ folgt $f(\xi_1, \ldots, \xi_n) - g(\xi_1, \ldots, \xi_n) = 0$.

Aus $f(\xi_1, \ldots, \xi_n) = 0$ folgt $f(\xi_1, \ldots, \xi_n)\, g(\xi_1, \ldots, \xi_n) = 0$.

Also bilden die betrachteten Polynome ein Ideal.

Aus $f(\xi_1, \ldots, \xi_n)\, g(\xi_1, \ldots, \xi_n) = 0$ und $g(\xi_1, \ldots, \xi_n) \neq 0$ folgt $f(\xi_1, \ldots, \xi_n) = 0$, da ein Körper keine Nullteiler hat.

Also ist das Ideal prim.

Beispiel. Seien $\xi_1, \ldots, \xi_n$ lineare Funktionen einer Unbestimmten λ mit Koeffizienten aus dem Körper P der komplexen Zahlen:

$$(1) \qquad \xi_i = \alpha_i + \beta_i \lambda.$$

Dann besteht das gemeinte Primideal aus allen Polynomen $f(x_1, \ldots, x_n)$, so daß $f(\alpha_1 + \beta_1 \lambda, \ldots, \alpha_n + \beta_n \lambda)$ identisch in λ verschwindet, oder (geometrisch ausgedrückt) aus allen Polynomen, die verschwinden in allen Punkten der Geraden, welche durch die Parameterdarstellung (1) im n-dimensionalen Raum bestimmt wird. Dieses Beispiel möge zur Veranschaulichung aller Sätze dieses und des folgenden Paragraphen dienen.

2. *Bedeutet $\mathfrak{p}$ das unter* 1. *konstruierte Primideal, so ist Ω dem Restklassenkörper Π von $\mathfrak{p}$ ($\S\,1$, 12) isomorph, und zwar so, daß den Elementen $\xi_1, \ldots, \xi_n$ die Elemente $x_1, \ldots, x_n$ entsprechen.*

Beweis. Sei Ω' der Ring derjenigen Elemente von Ω, die als Polynome in $\xi_1, \ldots, \xi_n$ geschrieben werden können. Ω ist, wie man leicht sieht, Quotientenkörper von Ω'. Wir ordnen jedem Element $f(\xi_1, \ldots, \xi_n)$ von Ω' das Element $f(x_1, \ldots, x_n)$ des Restklassenrings $R/\mathfrak{p}$ zu. Da aus $f(\xi_1, \ldots, \xi_n) - g(\xi_1, \ldots, \xi_n) = 0$ folgt $f(x_1, \ldots, x_n) - g(x_1, \ldots, x_n) \equiv 0\,(\mathfrak{p})$ oder $f(x_1, \ldots, x_n) \equiv g(x_1, \ldots, x_n)\,(\mathfrak{p})$ und umgekehrt, so ist die Zuordnung eineindeutig. Daß Summe und Produkt in Summe und Produkt übergehen, ist klar. Also sind die Ringe Ω', $R/\mathfrak{p}$ isomorph, Dann müssen auch die Quotientenkörper Ω und Π isomorph sein.

[10a]) Vgl. zu diesem Paragraphen E. Noether, a. a. O. (Math. Ann. 90).

3. *Zu jedem von R verschiedenen Primideal $\mathfrak{p}$ in R gibt es einen Körper $\Omega = \mathsf{P}(\xi_1, \ldots, \xi_n)$, so daß $\mathfrak{p}$ besteht aus allen Polynomen f aus R, für die $f(\xi_1, \ldots, \xi_n) = 0$.*

Beweis. Den Polynomen aus R ordnen wir Elemente einer neuen Menge R' zu, die den Koeffizientenkörper P umfaßt, wobei zweien nach $\mathfrak{p}$ kongruenten Polynomen das gleiche Element entsprechen soll, zweien inkongruenten aber verschiedene Elemente, und wobei die Elemente von P sich selbst entsprechen. Das ist immer möglich, denn zwei Elemente von P sind wegen $\mathfrak{p} \neq R$ dann und nur dann kongruent nach $\mathfrak{p}$, wenn sie gleich sind. Die den Elementen $x_1, \ldots, x_n$ entsprechenden Elemente nennen wir $\xi_1, \ldots, \xi_n$.

Die Menge R' ist auf den Restklassenring von R nach $\mathfrak{p}$ eindeutig abgebildet. Definieren wir in R' also eine Addition und eine Multiplikation, die der Addition bzw. Multiplikation im Restklassenring entsprechen, so ist R' dem Restklassenring isomorph, hat also keine Nullteiler, und gestattet die Bildung eines Quotientenkörpers Ω. In Ω ist $f(\xi_1, \ldots, \xi_n) = 0$ dann und nur dann, wenn $f(x_1, \ldots, x_n) \equiv 0 \ (\mathfrak{p})$, q. e. d.

4. Der nach **3** für jedes von R verschiedene Primideal $\mathfrak{p}$ konstruierbare, nach **1** auch *nur* für Primideale existierende, nach **2** bis auf Isomorphie eindeutig bestimmte Körper $\Omega = \mathsf{P}(\xi_1, \ldots, \xi_n)$, dessen Erzeugende ξ_i die Eigenschaft haben, daß $f(\xi_1, \ldots, \xi_n) = 0$ dann und nur dann, wenn $f \equiv 0 \ (\mathfrak{p})$, heißt *Nullstellenkörper* von $\mathfrak{p}$; das Elementsystem $\{\xi_1, \ldots, \xi_n\}$ heißt *allgemeine Nullstelle* von $\mathfrak{p}$. Unter *Nullstelle* schlechthin eines Ideals m verstehen wir jedes Elementsystem $\{\eta_1, \ldots, \eta_n\}$ eines Erweiterungskörpers von P, so daß $f(\eta_1, \ldots, \eta_n) = 0$, wenn $f \equiv 0 \ (\mathfrak{p})$. Jede nicht-allgemeine Nullstelle eines Primideals heißt *speziell*[11]).

5. Der Transzendenzgrad (§ 2, **8**) des Nullstellenkörpers Ω in bezug auf P heißt die *Dimensionszahl* des Primideals $\mathfrak{p}$.

6. *Sind $\mathfrak{p}, \mathfrak{p}'$ Primideale der Dimensionszahlen μ, μ', und ist $\mathfrak{p}' \equiv 0 \ (\mathfrak{p})$, so ist $\mu' \geqq \mu$, und das Gleichheitszeichen gilt nur dann, wenn $\mathfrak{p}' = \mathfrak{p}$.*

Beweis. Seien $\Omega' = \mathsf{P}(\xi_1, \ldots, \xi_n)$ und $\Omega' = \mathsf{P}(\xi_1', \ldots, \xi_n')$ Nullstellenkörper von $\mathfrak{p}$ bzw. $\mathfrak{p}'$. Ist f ein Polynom aus R, so folgt aus $f \equiv 0 \ (\mathfrak{p}')$ auch $f \equiv 0 \ (\mathfrak{p})$, m. a. W. aus $f(\xi_1', \ldots, \xi_n') = 0$ folgt $f(\xi_1, \ldots, \xi_n) = 0$.

Sei nun, evtl. nach Umnennung der Indizes, $\xi_1, \ldots, \xi_\mu$ ein mit Ω äquivalentes irreduzibles System (§ 2, **8**). Dann muß auch $\xi_1', \ldots, \xi_\mu'$ ein irreduzibles System in Ω' sein, denn jede algebraische Relation zwischen

---

[11]) Dieser Sprachgebrauch deckt sich, wie wir im § 4 sehen werden, mit der in der Geometrie üblichen Redeweise von allgemeinen und speziellen Punkten einer algebraischen Mannigfaltigkeit.

$\xi_1', \ldots, \xi_\mu'$ würde die gleiche Relation zwischen $\xi_1, \ldots, \xi_\mu$ nach sich ziehen. Daraus folgt die erste Behauptung: $\mu' \geqq \mu$. Ist aber $\mu' = \mu$, so ist Ω' algebraisch über $P(\xi_1', \ldots, \xi_\mu')$. Wir behaupten: aus $f(\xi_1, \ldots, \xi_n) = 0$ folgt $f(\xi_1', \ldots, \xi_n') = 0$. Wäre nämlich $f(\xi_1', \ldots, \xi_n') \neq 0$, so könnten wir nach § 2, 3 das Element $\dfrac{\varepsilon}{f(\xi_1', \ldots, \xi_n')}$ in der folgenden speziellen Form schreiben:

$$\frac{\varepsilon}{f(\xi_1', \ldots, \xi_n')} = \frac{g(\xi_1', \ldots, \xi_n')}{h(\xi_1', \ldots, \xi_\mu')}.$$

Daraus folgt:

und weiter:
$$h(\xi_1', \ldots, \xi_\mu') = f(\xi_1', \ldots, \xi_n') \, g(\xi_1', \ldots, \xi_n'),$$

$$h(\xi_1, \ldots, \xi_\mu) = f(\xi_1, \ldots, \xi_n) \, g(\xi_1, \ldots, \xi_n).$$

Da das Polynom h nicht identisch verschwinden kann (es stand vorhin im Nenner!) und da $\xi_1, \ldots, \xi_\mu$ ein irreduzibles System bilden, so ist die linke Seite dieser Gleichung $\neq 0$, also muß $f(\xi_1, \ldots, \xi_n) \neq 0$, entgegen der Voraussetzung. Also in der Tat: aus $f(\xi_1, \ldots, \xi_n)$ folgt $f(\xi_1', \ldots, \xi_n') = 0$. Oder: aus $f \equiv 0 \,(\mathfrak{p})$ folgt $f \equiv 0 \,(\mathfrak{p}')$. Das heißt $\mathfrak{p} \equiv 0 \,(\mathfrak{p}')$, mithin $\mathfrak{p} = \mathfrak{p}'$.

7. *Ist ein Ideal $\mathfrak{p}'$ der Dimensionszahl μ' gegeben, so hat jede Nullstelle einen Transzendenzgrad $\leqq \mu'$, und wenn der Transzendenzgrad genau μ' ist, so ist die Nullstelle allgemein.*

Beweis. Jede Nullstelle von $\mathfrak{p}'$ bestimmt nach **1** ein Ideal $\mathfrak{p}$; wenden wir auf $\mathfrak{p}$ und $\mathfrak{p}'$ den obigen Satz (**6**) an, so folgt die Behauptung unmittelbar.

Folge. *Ist $\mathfrak{p}'$ ein nulldimensionales Primideal, so ist jede Nullstelle algebraisch und allgemein.*

8. *Jedes μ-dimensionale Primideal hat einen $(\mu - 1)$-dimensionalen Primteiler.*

Beweis. Sei $\mathfrak{p}$ das gegebene Ideal, $\Omega = P(\xi_1, \ldots, \xi_n)$ sein Nullstellenkörper, $\xi_1, \ldots, \xi_\mu$ ein zu Ω äquivalentes irreduzibles System, also Ω algebraisch über $P(\xi_1, \ldots, \xi_\mu)$.

$\xi_{\mu+1}, \ldots, \xi_n$ sind als algebraische Funktionen von ξ_μ aufzufassen, wenn der Körper $P(\xi_1, \ldots, \xi_{\mu-1})$ als Grundkörper angenommen wird. Also gibt es in einem algebraischen Erweiterungskörper Γ von $P(\xi_1, \ldots, \xi_{\mu-1})$ (mindestens) einen regulären Argumentwert ξ_μ'; ein System zugehöriger Funktionswerte sei $\xi_{\mu+1}', \ldots, \xi_n'$. Das Elementsystem $\{\xi_1, \ldots, \xi_{\mu-1}, \xi_\mu', \ldots, \xi_n'\}$ von Γ hat in bezug auf P den Transzendenzgrad $\mu - 1$; das aus ihm nach **1** konstruierbare Primideal $\mathfrak{p}_1$ hat also die Dimensionszahl $\mu - 1$. Aus $f \equiv 0 \,(\mathfrak{p})$ folgt $f(\xi_1, \ldots, \xi_n) = 0$, also (§ 1, **7**): $f(\xi_1, \ldots, \xi_{\mu-1}, \xi_\mu', \ldots, \xi_n') = 0$, also $f \equiv 0 \,(\mathfrak{p}_1)$. Mithin ist $\mathfrak{p} \equiv 0 \,(\mathfrak{p}_1)$. Damit ist der Satz bewiesen.

In Verbindung mit **6** folgt: *Die Dimensionszahl eines Primideals* $\mathfrak{p}$ *ist zwei weniger als die maximale Gliederzahl einer von* $\mathfrak{p}$ *ausgehenden Primteilerkette*

$$\mathfrak{p}, \mathfrak{p}_1, \ldots, \mathfrak{p}_\mu, R. {}^{12})$$

9. Sei $\mathfrak{p}$ ein Primideal der Dimensionszahl μ, und seien die Unbestimmten $x_1, \ldots, x_n$ so numeriert, daß im Nullstellenkörper $\mathsf{P}(\xi_1, \ldots, \xi)$ die Größen $\xi_1, \ldots, \xi_\mu$ ein irreduzibles System bilden. Sei Ω ein algebraisch-abgeschlossener Erweiterungskörper von $\mathsf{P}(\xi_1, \ldots, \xi_n)$.

Dann bilden die mit $\{\xi_1, \ldots, \xi_n\}$ *in bezug auf* $\mathsf{P}(\xi_1, \ldots, \xi_\mu)$ *konjugierten Elementsysteme* $\{\xi_1, \ldots, \xi_\mu, \xi'_{\mu+1}, \ldots, \xi'_n\}$ *genau diejenigen Nullstellen von* $\mathfrak{p}$ *in* Ω, *deren erste* μ *Bestimmungszahlen die Werte* $\xi_1, \ldots, \xi_\mu$ *haben.*

Beweis. Daß die Elementsysteme Nullstellen bilden, ist klar, denn aus $f(\xi_1, \ldots, \xi_n) = 0$ folgt $f(\xi_1, \ldots, \xi_\mu, \xi'_{\mu+1}, \ldots, \xi'_n) = 0$, wenn f ein Polynom mit Koeffizienten aus P ist (§ 2, **5**). Andererseits aber hat jede Nullstelle $\{\xi_1 \ldots \xi_\mu, \eta_{\mu+1} \ldots \eta_n\}$ in Ω einen Transzendenzgrad $\geqq \mu$, ist also allgemeine Nullstelle (**7**); also gibt es immer einen Isomorphismus von $\mathsf{P}(\xi_1 \ldots \xi_n)$ und $\mathsf{P}(\xi_1 \ldots \xi_\mu, \eta_{\mu+1} \ldots \eta_n)$, der $\mathsf{P}(\xi_1 \ldots \xi_\mu)$ elementweise invariant läßt, und $\xi_{\mu+1} \ldots \xi_n$ in $\eta_{\mu+1} \ldots \eta_n$ überführt (**2**), woraus die Konjugiertheit von $\eta_{\mu+1} \ldots \eta_n$ mit $\xi_{\mu+1} \ldots \xi_n$ folgt (§ 2, **5**).

10. Folge: *Ist* $\mathfrak{p}$ *ein Ideal der Dimensionszahl* 0, *so sind alle (endlichvielen) Nullstellen im algebraisch-abgeschlossenen Erweiterungskörper* Ω *von* P *konjugiert in bezug auf* P. *Ist insbesondere* P *algebraisch-abgeschlossen, so gibt es demnach nur eine Nullstelle* $\{\xi_1 \ldots \xi_n\}$, *wo die* ξ_i *Elemente von* P *sind; das Ideal* $\mathfrak{p}$ *besteht in diesem Falle aus allen Polynomen, die an dieser Stelle verschwinden, hat mithin die Basis:*

$$(x_1 - \xi_1, \ldots, x_n - \xi_n).$$

11. Zum Schlusse sei bemerkt, daß ein Teil der Begriffe und Sätze dieses Paragraphen ihre Geltung beibehalten für Primideale in einem beliebigen kommutativen Ring R, der einen Körper P umfaßt, und der aus P durch Ringadjunktion von endlichvielen Elementen $x_1, \ldots, x_n$ entsteht, wobei aber diese Elemente noch durch Gleichungen verknüpft sein können. Alle Elemente des Rings sind dann als Polynome in $x_1, \ldots, x_n$ zu schreiben, aber nicht eindeutig, und damit versagt die Konstruktion **1**. Die umgekehrte Konstruktion **3**, die jedem Primideal einen Nullstellenkörper (dem Restklassenkörper isomorph) zuordnete, bleibt aber möglich, und damit wird der Dimensionsbegriff definierbar als Transzendenzgrad des

---

${}^{12})$ Für den Fall, daß P unendlich viele Elemente besitzt, ist dieser Satz von E. Noether bewiesen worden (Math. Ann. **90**, S. 250).

Nullstellenkörpers. Es gilt weiter Satz **6** samt dessen Beweis[12a]). Nimmt man noch allgemeiner an, daß der kommutative Ring R durch Ringadjunktion einer beliebigen Menge zu P entsteht, so wird die Dimensionszahl eines Primideals eine Mächtigkeit. Die erste Hälfte von Satz **6** bleibt samt ihrem Beweis uneingeschränkt bestehen, die zweite Hälfte aber gilt nur für Ideale endlicher Dimensionszahl.

§ 4.

Die Mannigfaltigkeit eines Primideals.

1. Unter dem (offenen, oder cartesischen) Raum $C_n(\mathsf{P})$ soll verstanden werden die Gesamtheit der geordneten Systeme von n Elementen $\xi_1, \ldots, \xi_n$ eines algebraisch-abgeschlossenen Körpers P[13]). Die Elemente des Raumes heißen Punkte, ihre Bestimmungszahlen Koordinaten.

2. Es genügt nun aber für die algebraische Geometrie nicht, sich auf die Betrachtung der Punkte in diesem Sinne zu beschränken, sondern es werden immer noch „unbestimmte Punkte" betrachtet, d. h. Punkte, deren Koordinaten entweder unabhängige Unbestimmte sind, oder doch algebraische Funktionen von Parametern, d. h. Elemente eines transzendenten Erweiterungskörpers Ω von P. Ein Elementsystem $\{\xi_1, \ldots, \xi_n\}$ eines solchen Körpers Ω (oder ein Punkt des Raumes $C_n(\Omega)$) soll *p-fach unbestimmter Punkt in* $C_n(\mathsf{P})$ heißen, wenn es den Transzendenzgrad p in bezug auf P hat, d. h. wenn der Punkt von p unabhängigen Parametern, aber nicht von weniger, algebraisch abhängt. Die in § 3 betrachteten „Nullstellen vom Transzendenzgrad p" waren solche p-fach unbestimmte Punkte.

3. Eine *algebraische Mannigfaltigkeit* M in $C_n(\mathsf{P})$ ist die Menge aller Nullstellen in $C_n(\mathsf{P})$ eines Ideals $\mathfrak{m}$ im Polynombereich $\mathsf{P}[x_1, \ldots x_n]$, vorausgesetzt, daß diese Menge nicht leer ist.

Verschiedene Ideale können die gleiche Mannigfaltigkeit definieren. Beispiel: Die drei Ideale $\mathfrak{p} = (x, y)$; $\mathfrak{q} = (x^2, y)$, $\mathfrak{r} = (x^2, xy, y^2, xz, yz)$ in $\mathsf{P}[x, y, z]$ definieren alle drei die Gerade $\xi = \eta = 0$ in $C_3(\mathsf{P})$. Bei den Polynomen von $\mathfrak{q}$ verschwindet in den Punkten dieser Geraden nicht nur das Polynom selbst, sondern auch die Ableitung nach x; bei denen

---

[12a]) **Zusatz bei der Korrektur.** Es gibt Ringe dieser Art, in denen Satz **8** nicht gilt. Also läßt sich die Dimensionszahl eines Primideals nicht allgemein durch Primteilerketten charakterisieren.

[13]) Diese Definition hält sich an die in der algebraischen Geometrie vorwaltende Richtung, die zum Raum immer die komplexen Punkte hinzunimmt. Unsere Definition umfaßt aber noch ganz andere Räume als den der gewöhnlichen Geometrie, z. B. solche, in denen der vierte harmonische Punkt immer mit dem dritten zusammenfällt. Man erhält einen solchen Raum nämlich, indem man für P einen solchen Körper nimmt, in dem $\varepsilon + \varepsilon = 0$ ist (Körper von der Charakteristik 2).

von $\mathfrak{r}$ verschwinden in einem Punkt der Geraden, nämlich im Punkt $\{0, 0, 0\}$, alle Ableitungen.

4. Ohne weiteres klar sind die folgenden beiden Sätze:

Die Mannigfaltigkeit eines K.G.V. von Idealen $[\mathfrak{m}_1, \ldots, \mathfrak{m}_r]$ *ist die Vereinigungsmenge der Mannigfaltigkeiten der Komponenten.*

Die Mannigfaltigkeit einer Idealsumme $(\mathfrak{m}_1, \ldots, \mathfrak{m}_r)$ *ist der Durchschnitt der Mannigfaltigkeiten der Summanden.*

5. Definition. Ein Polynom f *enthält* eine Mannigfaltigkeit M, wenn f verschwindet in allen Punkten von M.

6. Sei nun eine Mannigfaltigkeit M gegeben durch ein Ideal $\mathfrak{m}$. In der Gesamtheit der Ideale, welche die gleiche Mannigfaltigkeit definieren (s. obiges Beispiel), ist ein Ideal ausgezeichnet, nämlich die Gesamtheit aller Polynome, welche die Mannigfaltigkeit enthalten. Daß diese Gesamtheit ein Ideal ist, ist klar; daß sie in allen Punkten von M, und nur in diesen, Nullstellen hat, ist ebenfalls klar. Wir wollen dieses Ideal *das zu M gehörige Ideal* nennen.

7. Eine Mannigfaltigkeit heißt *irreduzibel*, wenn das zugehörige Ideal prim ist, d. h. wenn aus „fg enthält M" und „f enthält M nicht" folgt „g enthält M". Die *Dimensionszahl* einer irreduziblen Mannigfaltigkeit ist die Dimensionszahl des zugehörigen Primideals[13a]).

8. *Ist M irreduzibel, und sind* M_1, M_2 *beliebige Mannigfaltigkeiten, deren Vereinigungsmenge M enthält, ohne daß* M_1 *M enthält, so muß* M_2 *M enthalten.*

Denn gesetzt, weder M_1 noch M_2 würden M enthalten, so würde das heißen, daß in den Idealen $\mathfrak{m}_1, \mathfrak{m}_2$, die M_1 und M_2 definieren, Polynome f_1, f_2 vorhanden sein würden, die M nicht enthielten. Das Produkt $f_1 f_2$ aber würde sowohl M_1 wie M_2, also auch M enthalten. Das widerspricht der vorausgesetzten Irreduzibilität von M.

9. Sei eine Mannigfaltigkeit M gegeben, und sei $\mathfrak{m}$ das zugehörige Ideal. Die transzendenten Nullstellen des Ideals, also diejenigen Nullstellen, deren Koordinaten von Parametern algebraisch abhängen, werden als *unbestimmte Punkte der Mannigfaltigkeit* bezeichnet, weil sie zwar nicht der Mannigfaltigkeit angehören, aber doch allen algebraischen Gleichungen genügen, die in $C_n(\mathrm{P})$ die Mannigfaltigkeit definieren, und weil sie, wenn man die Parameter, von denen sie abhängen, regulär spezialisiert

---

[13a]) Zusatz bei der Korrektur. Ist eine Mannigfaltigkeit M in diesem Sinne irreduzibel, so ist sie nach 8 in der Tat unzerlegbar, d. h. nicht als Vereinigung von zwei echten algebraischen Teilmannigfaltigkeiten darstellbar. Wie leicht ersichtlich, gilt auch die Umkehrung.

($\S$ 2, **6**), in Punkte von $C_n(\mathsf{P})$ übergehen, die den nämlichen algebraischen Gleichungen genügen ($\S$ 2, **7**), mithin der Mannigfaltigkeit angehören.

Ist M irreduzibel, also $\mathfrak{m}$ prim, so heißt jede allgemeine Nullstelle des Ideals $\mathfrak{m}$ *allgemeiner Punkt der Mannigfaltigkeit* M. Diese Bezeichnung ist in Übereinstimmung mit der in der Geometrie geläufigen Bedeutung der Wörter allgemein und speziell. Man versteht doch meistens, wenn es auch nicht immer deutlich gesagt wird, unter einem allgemeinen Punkt einer Mannigfaltigkeit einen solchen Punkt, der keiner einzigen speziellen Gleichung genügt, außer denjenigen Gleichungen, die in allen Punkten erfüllt sind. Diese Forderung kann natürlich ein bestimmter Punkt von M niemals erfüllen, und so ist man genötigt, Punkte zu betrachten, die von hinreichend vielen Parametern abhängen, d. h. in einem Raum $C_n(\Omega)$ liegen, wo Ω eine transzendente Erweiterung von P ist. Fordert man aber von einem Punkt von $C_n(\Omega)$, daß er Nullstelle ist für alle die und nur die Polynome von $\mathsf{P}[x_1, \ldots, x_n]$, die in allen Punkten der Mannigfaltigkeit M verschwinden, so kommt man gerade auf unsere Definition eines allgemeinen Punktes der Mannigfaltigkeit M.

10. Da nach **7** zu jeder irreduziblen Mannigfaltigkeit ein Primideal gleicher Dimension gehört, so können die Sätze $\S$ 3 **2, 3, 7, 10** für Primideale unmittelbar auf irreduzible Mannigfaltigkeiten übertragen werden. Das ergibt:

Jede irreduzible Mannigfaltigkeit hat einen allgemeinen Punkt $\{\xi_1, \ldots, \xi_n\}$, *der von so vielen Parametern abhängt, wie die Dimensionszahl der Mannigfaltigkeit angibt* [14]), *und der Körper* $\mathsf{P}(\xi_1, \ldots, \xi_n)$ *ist dem Restklassenkörper des zugehörigen Primideals isomorph.*

Jeder spezielle Punkt einer μ-dimensionalen irreduziblen Mannigfaltigkeit hat einen Transzendenzgrad $< \mu$.

Eine nulldimensionale irreduzible Mannigfaltigkeit besteht aus einem Punkt.

11. Unter der *algebraischen Abschließung* einer Punktmenge M' in $C_n(\mathsf{P})$ verstehe ich die Menge M der gemeinsamen Nullstellen in $C_n(\mathsf{P})$ derjenigen Polynome aus $R = \mathsf{P}[x_1, \ldots, x_n]$, die in allen Punkten von M verschwinden. Da diese Polynome offenbar ein Ideal bilden, so ist M eine algebraische Mannigfaltigkeit [15]).

---

[14]) Diese Eigenschaft zeigt die Übereinstimmung unseres Dimensionsbegriffs mit dem aus der algebraischen Geometrie geläufigen.

[15]) Ist P der Körper der komplexen Zahlen, so umfaßt (weil jedes Polynom eine stetige Funktion darstellt) die algebraische Abschließung die topologische Abschließung. Z. B. hat in der Ebene $C_n(\mathsf{P})$ die Menge $\xi_1 = 0$, $|\xi_2| < 1$ die topologische Abschließung $\xi_1 = 0$, $|\xi_2| \leqq 1$, und die algebraische Abschließung $\xi_1 = 0$. Die algebraische Abschließung kann aber für die Algebra die Stelle der topologischen Abschließung voll-

12. Sind in einem algebraischen Erweiterungskörper Ω von $\mathsf{P}(\lambda_1, \ldots, \lambda_\mu)$, wo $\lambda_1, \ldots, \lambda_\mu$ als irreduzibles System angenommen ist, n algebraische Funktionen $\xi_1, \ldots, \xi_n$ von $\lambda_1, \ldots, \lambda_\mu$ gegeben, und ist M' die Menge der Funktionswertsysteme $\{\xi_1', \ldots, \xi_n'\}$, die zu regulären Argumentwerten $\lambda_1', \ldots, \lambda_\mu'$ gehören, so ist M' eine Punktmenge in $C_n(\mathsf{P})$; wenn nun M die algebraische Abschließung von M' ist, so sagen wir, daß M durch die Funktionen $\xi_1, \ldots, \xi_n$ in *Parameterdarstellung* gegeben ist. Die Parameterdarstellung ist nur regulär in den Punkten der Teilmenge M' von M, sie bestimmt aber M eindeutig.

Die algebraische Abschließung von M' wird nach Definition dadurch konstruiert, daß man das Ideal $\mathfrak{p}$ aller Polynome bildet, die in allen Punkten von M' verschwinden. Nach § 2, **7** kann man nun aber $\mathfrak{p}$ auch bestimmen als das Ideal aller Polynome f aus $\mathsf{P}[x_1, \ldots, x_n]$, für die $f(\xi_1, \ldots, \xi_n) = 0$, oder als dasjenige Primideal, das $\{\xi_1, \ldots, \xi_n\}$ zur allgemeinen Nullstelle hat (§ 3, **1**). Alle Polynome von $\mathfrak{p}$ verschwinden in allen Punkten von M, weil M ja die Mannigfaltigkeit von $\mathfrak{p}$ ist, und wenn umgekehrt ein Polynom verschwindet in allen Punkten von M, so verschwindet es auch in allen Punkten von M', gehört mithin zu $\mathfrak{p}$. Also ist $\mathfrak{p}$ das zu M gehörige Ideal (**3**). Damit ist bewiesen:

Jedes System von algebraischen Funktionen $\xi_1, \ldots, \xi_n$ von $\lambda_1, \ldots, \lambda_r$ bestimmt eine Mannigfaltigkeit M in Parameterdarstellung, und das zu M gehörige Primideal $\mathfrak{p}$ hat die allgemeine Nullstelle $\xi_1, \ldots, \xi_r$, die also zugleich allgemeiner Punkt der Mannigfaltigkeit ist.

13. Da auch jedes Primideal $\mathfrak{p}$ eine allgemeine Nullstelle $\{\xi_1, \ldots, \xi_n\}$ hat, wo die ξ_i algebraische Funktionen von Parametern $\lambda_1, \ldots, \lambda_\mu$ sind, so folgt:

Jedes Primideal $\neq R$ ist das zugehörige Ideal seiner Mannigfaltigkeit, die irreduzibel ist; und die allgemeine Nullstelle des Primideals ergibt eine Parameterdarstellung der Mannigfaltigkeit.

Folge: *Hat ein Primideal $\mathfrak{p}$ keine Nullstelle in $C_n(\mathsf{P})$, so ist $\mathfrak{p} = R$.*

14. Schließlich gilt, da auch zu jeder Mannigfaltigkeit ein Primideal $\neq R$ gehört:

Jede irreduzible algebraische Mannigfaltigkeit hat mindestens eine Parameterdarstellung. Die Dimensionszahl der Mannigfaltigkeit ist die kleinste Parameterzahl, die in eine Parameterdarstellung eingehen kann.

---

ständig vertreten. Definiert man z. B. eine Mannigfaltigkeit durch algebraische Parametergleichungen, die für gewisse Parameterwerte unbrauchbar (singulär) werden, so muß man zur Menge der regulären Punkte die Menge ihrer Grenzpunkte hinzunehmen; man kann aber auch ihre algebraische Abschließung bilden, wie wir sehen werden.

15. Aus § 3, **6** folgt: *Sind M_1, M_2 irreduzible Mannigfaltigkeiten der Dimensionszahlen μ_1, μ_2, und ist M_2 Untermenge von M_1, so ist $\mu_1 \geqq \mu_2$, und das Gleichheitszeichen gilt nur dann, wenn $M_1 = M_2$.*

§ 5.

Die Nullstellen beliebiger Ideale.

1. Zu einem Primärideal $\mathfrak{q}$ im Polynombereich R gehört, wie wir (§ 1, **13**) sahen, ein Exponent ϱ und ein Primideal $\mathfrak{p}$, so daß

$$\left\{ \begin{array}{l} \mathfrak{q} \equiv 0\,(\mathfrak{p}), \\ \mathfrak{p}^{\varrho} \equiv 0\,(\mathfrak{q}). \end{array} \right.$$

Aus der Definition von $\mathfrak{p}$ folgt: Die Mannigfaltigkeit von $\mathfrak{p}$ ist mit der von $\mathfrak{q}$ identisch. Daraus weiter: *Die Mannigfaltigkeit eines Primärideals $\mathfrak{q}$ ist irreduzibel.* Weiter: *Aus $fg \equiv 0\,(\mathfrak{q})$ folgt $g \equiv 0\,(\mathfrak{q})$, wenn f die Mannigfaltigkeit von $\mathfrak{q}$ nicht enthält.* Das letztere besagt nämlich nach § 4, **13**: $f \not\equiv 0\,(\mathfrak{p})$. Diese Eigenschaft des Primärideals benutzen Lasker[16]) und Macaulay[17]) als Definition. Die Äquivalenz mit unserer (E. Noetherschen) Definition folgt aus dem Hilbertschen Nullstellensatz (**9**).

2. Aus § 4, **13** und § 5, **1** folgt: *Hat ein Primärideal $\mathfrak{q}$ keine Nullstelle, so ist $\mathfrak{q} = R$.*

3. In denjenigen Ringen, für die der Hilbertsche Basissatz gilt, gilt auch, wie E. Noether[18]) gezeigt hat, der folgende Zerlegungssatz:

Jedes Ideal $\mathfrak{m}$ ist K.G.V. von endlich vielen Primäridealen: $\mathfrak{m} = [\mathfrak{q}_1, \ldots, \mathfrak{q}_r]$. Fordert man, daß dies größte primäre Komponenten sind, d. h. daß $[\mathfrak{q}_i, \mathfrak{q}_k]$ nicht mehr primär ist, und weiter, daß die Darstellung eine kürzeste ist, d. h. daß keine der $\mathfrak{q}_i$ überflüssig ist, so sind zwar nicht die Ideale $\mathfrak{q}_1, \ldots, \mathfrak{q}_r$ eindeutig bestimmt, wohl aber ihre Anzahl r und ihre zugehörigen Primideale $\mathfrak{p}_1, \ldots, \mathfrak{p}_r$.

4. Um zu zeigen, daß für den Polynombereich die gemachten Aussagen über Eindeutigkeit nicht verschärft werden können, und zugleich um die geometrische Natur der Primärideale zu erläutern, mögen die folgenden Beispiele gegeben werden, die sich auf den Polynombereich $\mathsf{P}\,[x, y]$ beziehen.

Beispiel 1. Das Ideal $\mathfrak{m} = (xy)$ besteht aus allen Polynomen, die

---

[16]) E. Lasker, Zur Theorie der Moduln und Ideale, Math. Ann. **60** (1905), S. 20—116.

[17]) F. S. Macaulay, Modular Systems, S. 33.

[18]) E. Noether, Math. Ann. **83**, S. 42. Der Beweis setzt den Wohlordnungssatz voraus. Einen etwas einfacheren Beweis gab W. Krull, Math. Ann. **90** (1923), S. 55—64. Für den Spezialfall des Polynombereichs: E. Lasker a. a. O.

auf der x-Achse und auf der y-Achse verschwinden, und hat u. a. die folgenden primären Teiler:

$q_1 = (x)$: Polynome, die auf der y-Achse verschwinden (Primideal).

$q_2 = (y)$: Entsprechend.

$q_3 = (x^2, xy, y^2)$: Polynome, die im Ursprung mindestens einen Doppelpunkt haben.

Daß q_1, q_2, q_3 primär sind, $\mathfrak{m}$ aber nicht, und daß $\mathfrak{m}$ Untermenge von q_1, q_2 und q_3 ist, folgt hier wie in allen folgenden Beispielen am einfachsten aus der eben gegebenen geometrischen Bestimmung dieser Ideale.

Offenbar ist $\mathfrak{m} = [q_1, q_2]$, aber auch $\mathfrak{m} = [q_1, q_2, q_3]$. Beide Zerlegungen sind Zerlegungen in größte primäre Komponenten, denn $[q_1, q_2]$, $[q_1, q_3]$ und $[q_2, q_3]$ sind nicht primär. Nur die erstere Darstellung $\mathfrak{m} = [q_1, q_2]$ ist eine kürzeste.

Beispiel 2. Das Ideal $\mathfrak{m} = (x^2, xy, y^2)$ (s. oben q_3) ist primär, und hat u. a. die folgenden primären Teile:

$q_2 = (x^2, y)$: Polynome, die die x-Achse im Ursprung zweifach schneiden oder ganz enthalten.

$q^2 = (x, y^2)$: Entsprechend.

Zu allen drei Idealen gehört als Primideal das zum Ursprung gehörige Primideal $\mathfrak{p} = (x, y)$. Weiter ist $\mathfrak{m} = \mathfrak{p}^2$, während q_1 und q_2 Ideale zwischen $\mathfrak{p}$ und $\mathfrak{p}^2$ sind[19]). Die Zerlegung $\mathfrak{m} = [q_1, q_2]$ ist eine kürzeste Darstellung, aber die q_i sind nicht größte primäre Ideale, denn $\mathfrak{m}$ ist selbst primär.

Beispiel 3. Das Ideal $\mathfrak{m} = (x^2, xy)$ besteht aus allen Polynomen, die die y-Achse enthalten und außerdem im Ursprung mindestens einen Doppelpunkt haben. $\mathfrak{m}$ ist nicht primär, und hat u. a. die folgenden primären Teiler:

$q_1 = \mathfrak{p}_1 = (x)$ (s. Beispiel 1).

$q_2 = (x^2, \mu x + y)$: Polynome, die die Gerade $\mu x + y = 0$ im Ursprung zweifach schneiden oder ganz enthalten. Zugehöriges Primideal: $\mathfrak{p}_2 = (x, y)$.

Die Zerlegung $\mathfrak{m} = [q_1, q_2]$ ist für jeden Wert von μ richtig, und immer eine kürzeste Darstellung durch größte primäre Komponenten. Nur die zugehörigen Primideale $\mathfrak{p}_1$, $\mathfrak{p}_2$ sind eindeutig bestimmt.

5. Definitionen. Die Mannigfaltigkeiten der primären Komponenten $q_1, \ldots, q_r$ eines Ideals $\mathfrak{m}$, oder, was dasselbe ist, die Mannigfaltigkeiten der zugehörigen Primideale $\mathfrak{p}_1, \ldots, \mathfrak{p}_r$, heißen die *wesentlichen Mannig-*

---

[19]) Daraus folgt nebenbei, daß für Ideale wie (x^2, y) eine Darstellung als Produkt von Primidealen, wie sie in der Theorie der Zahlkörper immer möglich ist, ausgeschlossen ist.

[20]) Nach Macaulay, zur Unterscheidung von den unwesentlichen Mannigfaltigkeiten, welche die Resolventenbildung nach Kronecker ergibt.

faltigkeiten des Ideals. Diejenigen unter ihnen, die in einer anderen enthalten sind, heißen *eingebettete Mannigfaltigkeiten*, die übrigen *isolierte Mannigfaltigkeiten*. Haben alle wesentlichen Mannigfaltigkeiten von $\mathfrak{m}$ die gleiche Dimensionszahl μ, so heißt das Ideal $\mathfrak{m}$ *ungemischt* und von der *Dimensionszahl μ*.

6. Ist wiederum $\mathfrak{m} = [q_1, \ldots, q_r]$, so ist die Mannigfaltigkeit von $\mathfrak{m}$ die Vereinigung der (irreduziblen) Mannigfaltigkeiten von $q_1, \ldots, q_r$:

$$M = \mathfrak{V}(M_1, \ldots, M_r).$$

Läßt man aus der Darstellung alle überflüssigen (eingebetteten) Mannigfaltigkeiten fort, so bleibt eine kürzeste Darstellung:

$$M = \mathfrak{V}(M_1, \ldots, M_s)$$

von M als Vereinigung von irreduziblen Mannigfaltigkeiten. Diese ist eindeutig, denn ist

$$M = \mathfrak{V}(M_1', \ldots, M_{s'}')$$

eine andere kürzeste Darstellung, so muß jede M_i in $\mathfrak{V}(M_1', \ldots, M_{s'}')$, mithin nach § 4, 8 in einer M_k' enthalten sein, und diese nach dem gleichen Schluß wiederum in einer M_j, welche dann notwendig gleich M_i sein muß, weil sonst M_i in M_j enthalten wäre, und die Darstellung keine kürzeste. Also ist jedes M_i einem M_k' gleich, und ebenso umgekehrt.

Damit ist gezeigt: *Jede algebraische Mannigfaltigkeit M läßt eindeutig eine kürzeste Darstellung als Vereinigung von endlichvielen irreduziblen Mannigfaltigkeiten zu.*

7. Aus **2** und **3** folgt: *Hat ein Ideal $\mathfrak{m}$ keine Nullstellen, so ist $\mathfrak{m} = R$.*

8. Aus **1** und **3** folgt: *Ist $fg \equiv 0\,(\mathfrak{m})$, und enthält f keine wesentliche Mannigfaltigkeit von $\mathfrak{m}$, so ist $g \equiv 0\,(\mathfrak{m})$.*

Für die Anwendung dieses äußerst wichtigen Satzes braucht man Kriterien, um zu entscheiden, ob ein Ideal eingebettete Mannigfaltigkeiten hat. Ohne Beweis führen wir zwei solche an, deren erstes sich unmittelbar aus Satz XI (S. 46) der E. Noetherschen Arbeit[21]) ergibt, während das zweite sich bei Macaulay[22]) findet:

Ein Ideal $\mathfrak{m}$ hat dann und nur dann eine in der Mannigfaltigkeit des anderen Ideals $\mathfrak{n} = (f_1, \ldots, f_r)$ enthaltene wesentliche Mannigfaltigkeit, wenn es ein Polynom f gibt, so daß $ff_i \equiv 0\,(\mathfrak{m})$ für jedes i, und dennoch $f \not\equiv 0\,(\mathfrak{m})$.

Hat ein Ideal $\mathfrak{m}$ von der Höchstdimension μ eine Basis aus $n - \mu$ Elementen, so ist es ungemischt, und jede seiner Potenzen ist ungemischt.

---

[21]) Math. Ann. **83**.
[22]) Modular Systems S. 49, 51.

9. Der Hilbertsche Nullstellensatz[23]) in der ursprünglichen Fassung lautet:

Verschwindet ein Polynom f, oder allgemeiner ein Ideal $\mathfrak{a}$, in allen Nullstellen eines Ideals $\mathfrak{m}$, so gibt es eine nur von $\mathfrak{m}$ abhängige Zahl ϱ, so daß $f^\varrho \equiv 0\,(\mathfrak{m})$ bzw. $\mathfrak{a}^\varrho \equiv 0\,(\mathfrak{m})$.

Beweis. Sei $\mathfrak{m} = [\mathfrak{q}_1, \ldots, \mathfrak{q}_r]$, und sei ϱ der größte unter den Exponenten von $\mathfrak{q}_1, \ldots, \mathfrak{q}_r$. Enthält f die Mannigfaltigkeit von M, so ist $f \equiv 0\,(\mathfrak{p}_i)$, wo $\mathfrak{p}_i$ das zu $\mathfrak{q}_i$ gehörige Primideal ist; daraus folgt $f^\varrho \equiv 0\,(\mathfrak{p}_i^\varrho)$, also $f^\varrho \equiv 0\,(\mathfrak{q}_i)$, also $f^\varrho \equiv 0\,(\mathfrak{m})$. Das gleiche gilt, wenn man $\mathfrak{a}$ statt f schreibt.

Genau so beweist sich eine etwas abweichende Fassung des Satzes, die für einige Anwendungen noch bequemer ist:

Verschwindet ein Ideal $\mathfrak{a}$ in den allgemeinen Nullstellen aller isolierten Primideale des Ideals $\mathfrak{m}$, so gibt es eine nur von $\mathfrak{m}$ abhängige Zahl ϱ, so daß $\mathfrak{a}^\varrho \equiv 0\,(\mathfrak{m})$.

10. *Ein ungemischtes $(n-1)$-dimensionales Ideal $\mathfrak{m}$ hat eine Basis aus einem Element („ist Hauptideal").*

Beweis. Sei zunächst $\mathfrak{m} = (f_1, \ldots, f_s)$ primär, $\mathfrak{p}$ das zugehörige Primideal, $\Omega = \mathsf{P}(\xi_1, \ldots, \xi_n)$ dessen Nullstellenkörper, und seien die Unbestimmten so numeriert, daß $\xi_1, \ldots, \xi_{n-1}$ ein dem Körper Ω äquivalentes irreduzibles System bilden.

Der größte gemeinsame Teiler f von $f_1, \ldots, f_s$ — Teiler im Polynomsinn — bleibt größter gemeinsamer Teiler, wenn man $f_1, \ldots, f_s$ als Polynome in x_n mit Koeffizienten aus $\mathsf{P}(x_1, \ldots, x_{n-1})$ betrachtet. Er ist also in der Form

$$f = a_1 f_1 + \ldots + a_s f_s$$

darstellbar, wo $a_i \,\varepsilon\, \mathsf{P}(x_1, \ldots, x_{n-1})[x_n]$. Multiplikation dieser Gleichung mit dem Hauptnenner $h(x_1, \ldots, x_{n-1})$ der rechten Seite ergibt:

$$f h \equiv 0\,(\mathfrak{m}),$$

also, da $h(\xi_1, \ldots, \xi_{n-1}) \neq 0$, mithin $h \not\equiv 0\,(\mathfrak{p})$:

$$f \equiv 0\,(\mathfrak{m}).$$

Auch ist

$$\mathfrak{m} = (f_1, \ldots, f_s) \equiv 0\,(f),$$

also folgt $\mathfrak{m} = (f)$, womit für primäre Ideale der Satz bewiesen ist.

Ist nun $\mathfrak{m} = [\mathfrak{q}_1, \ldots, \mathfrak{q}_r]$ ein beliebiges ungemischtes $(n-1)$-dimensionales Ideal, sind also $\mathfrak{q}_1, \ldots, \mathfrak{q}_r$ sämtlich $(n-1)$-dimensional, so ist nach dem eben bewiesenen $\mathfrak{q}_i = (f_i)$, also

$$\mathfrak{m} = [(f_1), \ldots, (f_r)].$$

---

[23]) D. Hilbert, Math. Ann. **42**, S. 320.

Ist nun f das K. G. V. (im Polynomsinn) von $f_1, \ldots, f_r$, so ist $\mathfrak{m} = (f)$, denn jedes Polynom, das durch $f_1, \ldots, f_r$ teilbar ist, ist durch f teilbar, und umgekehrt. Damit ist der Satz bewiesen.

Eine Zerlegung von f in Primfaktoren:

$$f = p_1^{\varrho_1} \ldots p_r^{\varrho_r}$$

ergibt eine Produktdarstellung für $\mathfrak{m}$:

$$\mathfrak{m} = (p_1)^{\varrho_1} \ldots (p_r)^{\varrho_r}.$$

Da die Ideale (p_i) prim sind, so folgt:

Jedes ungemischte $(n-1)$-dimensionale Ideal ist Produkt von Potenzen von Primidealen, deren jedes von einem Primelement erzeugt wird.

§ 6.

Der Hentzeltsche Nullstellensatz.

1. Der M. Noethersche Fundamentalsatz der Theorie der algebraischen Funktionen[24]) lautet bekanntlich folgendermaßen:

Ist P *ein algebraisch-abgeschlossener Körper,* $\mathfrak{m} = (f_1, f_2)$ *ein Ideal in* $R = \mathsf{P}[x_1, x_2]$, *wo* f_1 *und* f_2 *teilerfremd sind, und wo folglich das Ideal nur endlichviele Nullstellen* $\{\xi_1^{(i)}, \xi_2^{(i)}\}$ $[i = 1, \ldots, j]$ *in* $C_2(\mathsf{P})$ *hat[25]), und ist* f *ein Polynom in* R, *so daß in jeder dieser Nullstellen eine Gleichung besteht von der Form:*

$$(1) \qquad f = A_1^{(i)} f_1 + A_2^{(i)} f_2,$$

wo die A_1, A_2 *Potenzreihen nach* $(x_1 - \xi_1^{(i)}), (x_2 - \xi_2^{(i)})$ *sind, über deren Konvergenz nichts vorausgesetzt wird, so ist* $f \equiv 0\,(\mathfrak{m})$.

Dabei soll die Gleichung (1) in dem Sinne bestehen, daß, wenn beide Seiten rein formal nach Potenzen von $x_1 - \xi_1^{(i)}$, $x_2 - \xi_2^{(i)}$ geordnet werden, alle Koeffizienten übereinstimmen.

Verschärfung[26a]). *Es ist hinreichend, wenn beide Seiten von* (1) *bis auf Glieder von der Ordnung* ϱ *in* $x_1 - \xi_1^{(i)}$, $x_2 - \xi_2^{(i)}$ *übereinstimmen, wo* ϱ *eine nur von* $\mathfrak{m}$ *abhängige Zahl ist.*

Beweis. Sei $\mathfrak{q}_i$ eine primäre Komponente von $\mathfrak{m}$, $\mathfrak{p}_i$ ihr Primideal,

---

[24]) M. Noether, Math. Ann. 6 (1873), S. 351.

[25]) Sind nämlich f_1 und f_2 teilerfremd, so gibt es im Ideal $\mathfrak{m}$ sowohl ein von x_1, als ein von x_2 freies Polynom, die durch den Euklidischen Algorithmus gewonnen werden können. Diese beiden Polynome werden nur für endlichviele Werte von x_2 bzw. x_1 Null.

[26]) Sonst würde nämlich die Mannigfaltigkeit von $\mathfrak{p}_i$ aus mehreren getrennten Punkten bestehen, mithin reduzibel sein; vgl. § 3, 10.

[26a]) E. Bertini, Math. Ann. 34 (1889), S. 447; M. Noether, Math. Ann. 40 (1892), S. 140.

ϱ_i ihr Exponent, und $\varrho = \max(\varrho_i)$. Dann ist, da $\mathfrak{p}_i$ nur eine Nullstelle $\{\xi_1^{(i)}, \xi_2^{(i)}\}$ hat[26]:

$$\mathfrak{p}_i = (x_1 - \xi_1^{(i)}, x_2 - \xi_2^{(i)}).$$

Ersetzt man die Potenzreihen $A_1^{(i)}$, $A_2^{(i)}$ durch die Polynome $a_1^{(i)}$, $a_2^{(i)}$, die aus ihren Gliedern der Ordnung $< \varrho$ bestehen, so folgt aus (1):

$$f \equiv a_1^{(i)} f_1 + a_2^{(i)} f_2 \,(\mathfrak{p}_i^\varrho),$$

$$f \equiv 0 \,(\mathfrak{m}, \mathfrak{p}_i^\varrho).$$

$$f \equiv 0 \,(\mathfrak{q}_i)$$

für jedes i, mithin

$$f \equiv 0 \,(\mathfrak{m}) \qquad\qquad\qquad \text{q. e. d.}$$

2. Der Satz läßt sich nach verschiedenen Richtungen hin n-dimensional verallgemeinern.

Erstens gilt der Beweis offenbar für beliebige Ideale $\mathfrak{m} = (f_1, \ldots, f_r)$ von der Dimensionszahl 0 in $R = \mathsf{P}[x_1, \ldots, x_n]$. [27]

Zweitens kann man, wie wir zeigen werden, die Voraussetzung, der Körper P sei algebraisch-abgeschlossen, fallen lassen. Man hat dann die Nullstellen ξ_i und die Koeffizienten der Potenzreihen in einem algebraisch-abgeschlossenen Erweiterungskörper Ω von P anzunehmen. Bricht man von vornherein, was ja unwesentlich ist, die Potenzreihen mit den Gliedern $(\varrho - 1)$-ter Ordnung ab, und *schreibt man $\mathfrak{m}_\Omega$ für das von $\mathfrak{m}$ in* $\Omega[x_1, \ldots, x_n]$ *erzeugte Ideal* (§ 1, 4), so lautet der verallgemeinerte Satz mit „Verschärfung" zusammengefaßt:

Ist $\mathfrak{m}$ ein Ideal in $R = \mathsf{P}[x_1, \ldots, x_n]$, das die Dimension 0 hat (§ 5, **5**) *und folglich nur endlichviele Nullstellen $\xi_1^{(i)}, \ldots, \xi_n^{(i)}$ $[i = 1, \ldots, \gamma]$ in $C_n(\Omega)$, wo Ω ein algebraisch-abgeschlossener Erweiterungskörper von P ist, so gibt es eine nur von $\mathfrak{m}$ abhängige Zahl ϱ, so daß, wenn $f \varepsilon R$, aus*

$$f \equiv 0 \,(\mathfrak{m}_\Omega, (x_1 - \xi_1^{(i)}, \ldots, x_n - \xi_n^{(i)})^\varrho) \qquad [i = 1, \ldots, \gamma]$$

folgt

$$f \equiv 0 \,(\mathfrak{m}).$$

Der Beweis von vorhin ergibt nicht die gesuchte Gleichung, sondern die andere:

$$f \equiv 0 \,(\mathfrak{m}_\Omega).$$

Das heißt nach Definition von $\mathfrak{m}_\Omega$:

$$f = \textstyle\sum \alpha_k f_k, \qquad \text{wo} \quad \alpha_k \varepsilon \Omega, \quad f_k \varepsilon \mathfrak{m}.$$

Drückt man die Größen α_i durch endlichviele linear-unabhängige Elemente $\varepsilon, \omega_1, \omega_2, \ldots$ von Ω mit Koeffizienten aus P aus, so kommt:

$$f = g_0 + \omega_1 g_1 + \ldots, \qquad \text{wo} \quad g_i \varepsilon \mathfrak{m},$$

---

[27] Vgl. etwa Macaulay, Modular Systems, S. 60.

mithin, da die ω linear-unabhängig in bezug auf P sind:

$$f = g_0, \quad 0 = g_1, \ldots,$$

also

$$f \equiv 0 \ (\mathfrak{m}).$$

3. Läßt man nun drittens auch die Voraussetzung fallen, daß $\mathfrak{m}$ die Dimensionszahl 0 hat, so kann man zunächst, wie es Lasker[28] und Macaulay[29] versucht haben, unter Verzicht auf die Verschärfung daran festhalten, daß nur für eine endliche Anzahl Punkte $\{\xi_1, \ldots, \xi_n\}$ die Gleichungen (1) vorausgesetzt werden. Die Beweise von Lasker und Macaulay scheinen aber unvollständig[29a] und setzen außerdem voraus, daß P der Körper der gewöhnlichen komplexen Zahlen ist, und daß die Potenzreihen A in einem Gebiet konvergieren. Einen anderen Weg hat K. Hentzelt[30] eingeschlagen, indem er das Bestehen der Gleichung (1) in allen Punkten der Mannigfaltigkeit von $\mathfrak{m}$ fordert, und die angegebene Verschärfung beibehält. Sein Satz, der hier auf kürzerem Wege bewiesen werden soll, lautet:

(**Hentzeltscher Nullstellensatz, erste Fassung**) *Ist* $\mathfrak{m}$ *ein Ideal in* $R = \mathrm{P}[x_1, \ldots, x_n]$, *und* Ω *ein algebraisch-abgeschlossener Erweiterungskörper von* P, *so gibt es eine nur von* $\mathfrak{m}$ *abhängende Zahl* ϱ, *so daß, wenn* $f \,\varepsilon\, R$, *aus dem Bestehen der Kongruenz*

$$(2) \qquad f \equiv 0 \ (\mathfrak{m}_\Omega, (x_1 - \xi_1, \ldots, x_n - \xi_n)^\varrho)$$

für jede Nullstelle $\{\xi_1, \ldots, \xi_n\}$ *von* $\mathfrak{m}$ *in* $C_n(\Omega)$ *folgt*

$$f \equiv 0 \ (\mathfrak{m}).$$

Der Satz kann, genau so wie der Hilbertsche Nullstellensatz dahin modifiziert werden, daß die Kongruenz (2) nicht für alle algebraischen Nullstellen von $\mathfrak{m}$ gefordert wird, sondern nur für die allgemeine Nullstelle eines jeden zugehörigen Primideals $\mathfrak{p}_i$. So kommt man zum Satz[31]:

(**Hentzeltscher Nullstellensatz, zweite Fassung**) *Ist* $\mathfrak{m}$ *ein Ideal im* $R = \mathrm{P}[x_1, \ldots, x_n]$, *sind* $\mathfrak{p}_1, \ldots, \mathfrak{p}_r$ *die zugehörigen Primideale, und ist* $\Sigma_i = \mathrm{P}(\xi_1^{(i)}, \ldots, \xi_n^{(i)})$ *der Nullstellenkörper von* $\mathfrak{p}_i$, *so existiert eine Zahl* ϱ, *so daß, wenn* $f \,\varepsilon\, R$, *aus*

$$f \equiv 0 \ (\mathfrak{m}_{\Sigma_i}, (x_1 - \xi_1^{(i)}, \ldots, x_n - \xi_n^{(i)})^\varrho) \qquad [i = 1, \ldots, r]$$

---

[28] Math. Ann. **60**, S. 95.

[29] Modular Systems, S. 61.

[29a] **Zusatz bei der Korrektur.** Die Lücken des Macaulayschen Beweises sind in einem Briefwechsel zwischen Herrn Macaulay und dem Verfasser ausgefüllt worden. Die Konvergenzvoraussetzung blieb dabei aber aufrechterhalten. Ich besitze jetzt einen von dieser Voraussetzung freien Beweis, der bei beliebigem Koeffizientenkörper P gilt.

[30] Siehe Fußnote [6].

[31] Vgl. Grete Hermann, a. a. O. Die Bezeichnungen „erste" und „zweite Fassung" sind dort gerade umgekehrt wie hier.

folgt

$$f \equiv 0 \ (\mathfrak{m}).$$

Dabei bedeutet wiederum $\mathfrak{m}_{\Sigma_i}$ das von $\mathfrak{m}$ in $\Sigma_i[x_1, \ldots, x_n]$ erzeugte Ideal; die Forderung wird jeweils vorausgesetzt in $\Sigma_i[x_1, \ldots, x_n]$ für $i = 1, \ldots, r$.

Die zweite Fassung dürfte für Anwendungen wichtiger sein wie die erste; ihr Beweis ist einfacher.

4. Man braucht offenbar die beiden Sätze nur für Primärideale zu beweisen: Durch die Zerlegung $\mathfrak{m} = [\mathfrak{q}_1, \ldots, \mathfrak{q}_r]$ folgt dann der allgemeine Satz unmittelbar, wie im Spezialfall 1. Die zweite Fassung wird dadurch wesentlich vereinfacht; es ist nur zu beweisen:

Ist $\mathfrak{q}$ ein Primärideal in R, $\mathfrak{p}$ das zugehörige Primideal, $\Sigma = \mathsf{P}(\xi_1, \ldots, \xi_n)$, dessen Nullstellenkörper, so existiert eine Zahl ϱ, so daß, wenn $f \varepsilon R$, aus

$$(3) \qquad f \equiv 0 \ (\mathfrak{q}_\Sigma, (x_1 - \xi_1, \ldots, x_n - \xi_n)^\varrho) \quad \text{in} \quad \Sigma[x_1, \ldots, x_n]$$

folgt

$$f \equiv 0 \ (\mathfrak{q}).$$

Beweis. Wir wollen den allgemeinen Fall auf den schon erledigten Spezialfall, daß $\mathfrak{q}$ null-dimensional ist, zurückführen, indem wir gewisse x_i gleich Parametern setzen und so die Dimensionszahl erniedrigen.

Sei l die Dimensionszahl von $\mathfrak{q}$, und sei, evtl. nach Umnennung der Indizes, $\xi_1, \ldots, \xi_l$ ein irreduzibles System in Ω. Dann ist Σ algebraisch über $\Gamma = \mathsf{P}(\xi_1, \ldots, \xi_l)$. Sei Δ ein algebraisch-abgeschlossener Erweiterungskörper von Σ, und seien $\xi_{l+1}^{(k)}, \ldots, \xi_n^{(k)} \ [k = 1, \ldots, \gamma]$ die mit $\xi_{l+1}, \ldots, \xi_n = \xi_{l+1}^{(1)}, \ldots, \xi_n^{(1)}$ konjugierten Elementsysteme. Dann sind nach § 3, 9 die γ Punkte $\{\xi_1, \ldots, \xi_l, \xi_{l+1}^{(k)}, \ldots, \xi_n^{(k)}\}$ diejenigen Nullstellen von $\mathfrak{q}_\Delta$, in denen $x_1, \ldots, x_l$ die Werte $\xi_1, \ldots, \xi_l$ haben. Ersetzt man, in allen Polynomen von $\mathfrak{q}_\Delta$, die Unbestimmte $x_1, \ldots, x_l$ durch $\xi_1, \ldots, \xi_l$, so entsteht ein Ideal $\bar{\mathfrak{q}}_\Delta$ in $\Delta[x_{l+1}, \ldots, x_n]$, das nur die endlichvielen Nullstellen $\xi_{l+1}^{(k)}, \ldots, \xi_n^{(k)}$ in $C_{n+1}(\Delta)$ hat. Aus (3) folgt, wenn man zu Δ übergeht und $x_1, \ldots, x_l = \xi_1, \ldots, \xi_l$ setzt,

$$\bar{f} = f(\xi_1, \ldots, \xi_l, x_{l+1}, \ldots, x_n) \equiv 0 \ (\bar{\mathfrak{q}}_\Delta, (x_{l+1} - \xi_{l+1}, \ldots, x_n - \xi_n)^\varrho).$$

Ausübung der Automorphismen von Ω, die Γ elementweise invariant lassen, ergibt:

$$\bar{f} \equiv 0 \ (\bar{\mathfrak{q}}_\Delta, (x_{l+1} - \xi_{l+1}^{(k)}, \ldots, x_n - \xi_n^{(k)})^\varrho).$$

Mithin sind die Voraussetzungen des Hentzeltschen Satzes für das nulldimensionale Ideal $\bar{\mathfrak{q}}_\Gamma$ in $\Gamma[x_{l+1}, \ldots, x_n]$ erfüllt, und es folgt

$$(4) \qquad \bar{f} \equiv 0 \ (\bar{\mathfrak{q}}_\Gamma).$$

Das heißt, $\bar{f}$ entsteht aus einem Polynom aus $\mathfrak{q}_\Gamma$ dadurch, daß man $x_1, \ldots, x_l$ durch $\xi_1, \ldots, \xi_l$ ersetzt.

Ein Polynom aus $\mathfrak{q}_\Gamma$ hat die Form

$$\sum \frac{\varphi_i(\xi_1, \ldots, \xi_l)}{\psi_i(\xi_1, \ldots, \xi_l)}\, g_i(x_1, \ldots, x_n), \quad \text{wo} \quad g_i(x_1, \ldots, x_n) \equiv 0\ (\mathfrak{q}).$$

Mithin ist

$$\bar{f} = \sum \frac{\varphi_i(\xi_1, \ldots, \xi_l)}{\psi_i(\xi_1, \ldots, \xi_l)}\, g_i(\xi_1, \ldots, \xi_l, x_{l+1}, \ldots, x_n).$$

Multipliziert man beide Seiten mit dem Hauptnenner $\psi(\xi_1, \ldots, \xi_l)$, so kommt:

$$\psi(\xi_1, \ldots, \xi_l)\, f(\xi_1, \ldots, \xi_l, x_{l+1}, \ldots, x_n)$$
$$= \sum \chi_i(\xi_1, \ldots, \xi_l)\, g_i(\xi_1, \ldots, \xi_l, x_{l+1}, \ldots, x_n).$$

Da die $\xi_1, \ldots, \xi_l$ ein irreduzibles System bilden, kann man sie durch Unbestimmte $x_1, \ldots, x_l$ ersetzen. Die rechte Seite wird dann ein Polynom aus $\mathfrak{q}$, also:

$$\psi(x_1, \ldots, x_l)\, f(x_1, \ldots, x_n) \equiv 0\ (\mathfrak{q}).$$

Da aber $\psi(\xi_1, \ldots, \xi_l) \neq 0$, so ist $\psi(x_1, \ldots, x_l) \not\equiv 0\ (\mathfrak{p})$, mithin

$$f(x_1, \ldots, x_n) \equiv 0\ (\mathfrak{q}) \qquad\qquad \text{q. e. d.}$$

5. Zum Beweise des Hentzeltschen Satzes in der ersten Fassung für ein Primärideal $\mathfrak{q}$ können wir zunächst wieder voraussetzen, daß P algebraisch-abgeschlossen ist, mithin $\mathsf{P} = \Omega$. Die Voraussetzung ist hinterher wie in 2 leicht aufzuheben.

Der Beweis verläuft anfangs ungefähr wie der in 4 geführte bis zur Formel (4). Man numeriert wiederum die Unbestimmten so, daß im Nullstellenkörper die Elemente $\xi_1, \ldots, \xi_l$ ein irreduzibles System bilden, mithin $\xi_{l+1}, \ldots, \xi_n$ algebraische Funktionen von ihnen sind. Im Beweise 4 ist überall ξ_i durch ξ_i' zu ersetzen, wo $\xi_1', \ldots, \xi_l'$ reguläre Argumentwerte aus P, $\xi_{l+1}', \ldots, \xi_n'$ die zugehörigen Funktionswerte sind. Für diese $\xi_1', \ldots, \xi_n'$ gilt die Voraussetzung (2). An Stelle der in 4 benutzten Körper $\mathsf{P}, \Gamma, \Sigma, \Delta$ tritt der einzige Körper P. Man findet nach Analogie von (4):

$$(5) \qquad\qquad f' \equiv 0\ (\mathfrak{q}'),$$

wo der Strich überall bedeutet, daß $x_1, \ldots, x_l$ durch $\xi_1', \ldots, \xi_l'$ ersetzt sind.

Aus dieser Gleichung wollen wir die andere

$$(6) \qquad\qquad \bar{f} \equiv 0\ (\bar{\mathfrak{q}}_\Gamma)$$

[vgl. (4)] ableiten. Das geschieht, indem wir für $\mathfrak{q}$ eine Basis $f_1, \ldots, f_r$ annehmen. Die Polynome $f_1', \ldots, f_r'$ bilden dann offenbar eine Basis für $\mathfrak{q}'$, und $\bar{f}_1, \ldots, \bar{f}_r$ eine Basis für $\bar{\mathfrak{q}}_\Gamma$. Nach einer von Grete Hermann[32] nach Hentzelt angegebenen Methode kann man nun die Kongruenz (6) ersetzen durch die Forderung, daß ein gewisses lineares Gleichungssystem lösbar ist, oder daß ein gewisses Matrizenpaar m_1, m_2 gleichen Rang besitzt. Ebenso drückt sich (5) aus durch die Ranggleichheit der entsprechenden Matrizen m_1', m_2', die aus m_1, m_2 entstehen, indem $\xi_1, \ldots, \xi_l$ durch $\xi_1', \ldots, \xi_l'$ ersetzt werden. Für die Einzelheiten der Rechnung können wir auf die Arbeit von Fräulein Hermann verweisen. Aus den Sätzen § 2, 6, 7 folgt nun, daß man die Argumente $\xi_1, \ldots, \xi_l$ regulär so spezialisieren kann, daß m_1 und m_1' gleichen Rang haben, und ebenso m_2 und m_2'. Da die Voraussetzung (5) besagt, daß m_1' und m_2' gleichen Rang haben für alle regulären Argumentwerte $\xi_1', \ldots, \xi_l'$, so folgt die Ranggleichheit von m_1 und m_2, mithin die Kongruenz (6).

Der Beweis verläuft weiter wie oben von (4) an.

---

[32]) A. a. O. S. 760. Die dort für den Fall $t = n - 1$ gegebene Betrachtung läßt sich ohne weiteres auf den allgemeinen Fall übertragen.

(Eingegangen am 14. 8. 1925.)

3.

Der Multiplizitätsbegriff der algebraischen Geometrie

Mathematische Annalen 97, 4/5 (1927) 756–774

§ 1.
Einleitung.

In der algebraischen Geometrie kommt es oft vor, daß Lösungen eines Problems in einem Spezialfall mehrfach gezählt werden müssen, damit Anzahl und Eigenschaften der Lösungen beim Übergang vom allgemeinen Fall zum Spezialfall erhalten bleiben. Z. B. liegen auf der allgemeinen kubischen Fläche im projektiven Raum 27 Geraden, von denen jede durch 10 andere geschnitten wird; sollen diese Eigenschaften erhalten bleiben für eine kubische Fläche, die einen Doppelpunkt hat, im übrigen aber allgemein ist, so müssen die 6 Geraden der Fläche, welche durch diesen Doppelpunkt gehen, doppelt gezählt werden, die übrigen 15 einfach. D. h. man gibt jeder Geraden auf dieser Fläche eine *Multiplizität*. In gleicher Weise erteilt man den Schnittpunkten einer Geraden mit einer Fläche n-ten Grades eine Multiplizität, indem z. B. ein Berührungspunkt zweifach gezählt wird.

Es fragt sich nun: Ist dieser Multiplizitätsbegriff allgemein definierbar? Für welche Arten von Problemen hat es Sinn, von Multiplizitäten der Lösungen zu reden, und welchen Sinn hat es?

Im folgenden soll gezeigt werden, daß sich der Multiplizitätsbegriff definieren läßt für eine Klasse von Problemen, welche „Normalprobleme" genannt werden sollen. Ein Normalproblem ist ein geometrisches Problem, dessen Lösung auf die Lösung eines homogenen Gleichungssystems zurückgeführt werden kann (für die genaue Definition verweise ich auf § 2).

Die präzise Fassung des Multiplizitätsbegriffs gelingt nun mit Hilfe von zwei Sätzen über homogene Gleichungssysteme, die ich in den Amsterdamer Proceedings[1]) abgeleitet habe. Der erste Satz gibt ein algebraisches

---

[1]) Proc. Kon. Ac. Amsterdam **29** (1926), S. 142.

Kriterium für die Lösbarkeit eines homogenen Gleichungssystems; der zweite gibt einen algebraischen Ausdruck für die Lösungen, falls nur endlichviele wesentlich verschiedene vorhanden sind (die Sätze sind in § 2 genau angeführt). Diese Sätze gestatten es nämlich, das Verhalten der Lösungen eines homogenen Gleichungssystems bei Spezialisierung von in den Gleichungen vorkommenden unbestimmten Parametern genau zu verfolgen. Der wesentliche Begriff ist dabei der Begriff der *Relationstreue.* Es sei einerseits ein Gleichungssystem gegeben, das unbestimmte Parameter enthält, und dessen endlichviele wesentlich verschiedene Lösungen $X^{(1)}, \ldots, X^{(q)}$ aus algebraischen Funktionen der Parameter bestehen, und anderseits ein Gleichungssystem, das aus dem ersten entsteht, indem für die Parameter spezielle Werte aus irgendeinem Körper eingesetzt werden. Wenn nun eine Reihe von Lösungen $Y^{(1)}, \ldots, Y^{(q)}$ des spezialisierten Gleichungssystems die Eigenschaft hat, daß alle *homogenen* algebraischen Relationen, die zwischen den Lösungen $X^{(1)}, \ldots, X^{(q)}$ und den Parametern bestehen, bei der vorgenommenen Spezialisierung der Parameter und bei Ersetzung der X durch die Y bestehen bleiben, so heißt die Reihe $Y^{(1)}, \ldots, Y^{(q)}$ eine *relationstreue Spezialisierung* der Reihe $X^{(1)}, \ldots, X^{(q)}$.

Es kann nun mit Hilfe der eben genannten Sätze gezeigt werden, daß es eine und im wesentlichen nur eine relationstreue Spezialisierung der Reihe $X^{(1)}, \ldots, X^{(q)}$ bei gegebenen Parameterwerten gibt. Dabei ist angenommen, daß auch das spezialisierte Gleichungssystem nur endlichviele wesentlich verschiedene Lösungen hat. Beweis § 3 und § 4.

Auf Grund dieses Satzes kann man die Multiplizität einer gegebenen Lösung Y des spezialisierten Gleichungssystems eindeutig definieren als die Zahl, die angibt wie oft Y unter den Lösungen $Y^{(1)}, \ldots, Y^{(q)}$ vertreten ist (§ 5).

Die Multiplizität einer Lösung kann auch Null sein. Das heißt dann, daß eine Lösung Y des speziellen Gleichungssystems nicht durch Spezialisierung aus den Lösungen des allgemeinen Gleichungssystems entsteht, sondern neu hinzukommt (vgl. das in § 5 gegebene Beispiel, sowie auch die von G. Kohn[2]) gegebenen geometrischen Beispiele).

Die selbstverständliche Tatsache, daß die Summe der Multiplizitäten der Lösungen des spezialisierten Gleichungssystems gleich q ist, also gleich der Anzahl der Lösungen des allgemeinen Gleichungssystems, ergibt eine exakte Fassung des Schubertschen „Prinzips der Erhaltung der Anzahl", über dessen Tragweite und Anwendungsmöglichkeit (welche beide von der

---

[2]) G. Kohn, Archiv der Mathematik und Physik (3) **4** (1903), S. 312. Vgl. auch das dritte Beispiel von K. Rohn in seinem Zusatz zu E. Study, Das Prinzip der Erhaltung der Anzahl, Leipziger Ber. **68** (1916), S. 92.

„abzählenden Schule" weit überschätzt worden sind) § 6 einige Bemerkungen enthält.

Als Beispiele für die praktische Verwertung der allgemeinen Definitionen werden in den §§ 7, 8 einige geometrische Probleme behandelt. In § 7 wird ein Fall diskutiert, wo sich die Multiplizitäten direkt berechnen lassen; es handelt sich hier um die Schnittpunkte von n algebraischen Hyperflächen im R_n. Als spezieller Fall kommt der Schnitt einer Geraden mit einer Hyperfläche heraus; man wird hier auf die bekannten Formeln der Polarentheorie geführt. In § 8 wird gezeigt, daß auch die Bestimmung der Geraden auf einer kubischen Fläche ein Normalproblem ist. Hier konnte ich keine direkte Formel für die Multiplizität geben, konnte aber beweisen, daß die Multiplizitäten niemals Null werden können.

Bei einer späteren Gelegenheit hoffe ich zu zeigen, wie der Multiplizitätsbegriff sich anwenden läßt auf die Schnittpunkte zweier algebraischen Mannigfaltigkeiten der Dimensionen r und $n - r$ im projektiven Raum P_n.

§ 2.

Normalprobleme und Resultantensysteme.

Seien $f_1, \ldots, f_r$ Polynome in den Unbestimmten $x_0, \ldots, x_m$; $y_0, \ldots, y_n$; $\ldots$; $z_0, \ldots, z_p$; homogen in den x, in den y usw. (also *Formen*) mit Koeffizienten aus einem Körper P. Sei Ω ein algebraisch-abgeschlossener Erweiterungskörper von P. Wir betrachten diejenigen Elementsysteme $\xi_0, \ldots, \xi_m$; $\eta_0, \ldots$ aus Ω, die den Gleichungen

$$(1) \qquad \begin{cases} f_1(\xi_0, \ldots, \xi_m; \eta_0, \ldots, \eta_n; \ldots) = 0 \\ f_r(\xi_0, \ldots, \xi_m; \eta_0, \ldots, \eta_n; \ldots) = 0 \end{cases}$$

genügen. Sie zerfallen in *triviale Lösungen*, wo alle $\xi_i = 0$ oder alle $\eta_i = 0$ usw., und *nichttriviale Lösungen*. Ist $\xi_0, \ldots, \xi_m$; $\eta_0, \ldots, \eta_n$; $\ldots$ eine nichttriviale Lösung, so sind alle Elementsysteme $\lambda \xi_0, \ldots, \lambda \xi_m$; $\mu \eta_0, \ldots, \mu \eta_n$; $\ldots$ (wo $\lambda, \mu, \ldots$ Elemente von Ω sind) ebenfalls Lösungen; diese bilden zusammen eine *Lösungsklasse*, welche durch jede ihrer nichttrivialen Elemente bestimmt wird. Wir bezeichnen mit $X = \{\xi_0, \ldots, \xi_m; \ldots\}$ die durch das Elementsystem $\xi_0, \ldots, \xi_m$; $\ldots$ bestimmte Lösungsklasse.

Als *Normalproblem* bezeichnet man ein jedes geometrische Problem, das auf ein Gleichungssystem der Gestalt (1) zurückgeführt werden kann, in dem Sinne, daß jeder Lösungsklasse von (1) eindeutig eine Lösung des Normalproblems zugeordnet ist, und umgekehrt[3]).

In einer früheren Arbeit[4]) habe ich erstens eine algebraische Bedin-

---

[3]) Zum Beispiel sind die beiden in der Einleitung angeführten Probleme Normalprobleme, wie in §§ 7, 8 näher ausgeführt werden soll.

[4]) Proc. Kon. Ac. Amsterdam 29 (1926), S. 142.

gung hergeleitet für die Existenz mindestens einer nichttrivialen Lösung der Gleichungen (1) im Körper Ω, zweitens eine algebraische Formel gegeben, welche alle Lösungsklassen zu bestimmen gestattet, falls nur endlichviele solche vorhanden sind. Diese Ergebnisse sollen zunächst mitgeteilt werden.

1. Jedem Formensystem $F_1, \ldots, F_r$ mit unbestimmten Koeffizienten $a_1, \ldots, a_w$ ist ein *Resultantensystem* zugeordnet, dessen Elemente $c_1(a), \ldots, c_t(a)$ ganzzahlige Polynome in $a_1, \ldots, a_w$ sind. Unter dem Resultantensystem eines speziellen Formensystems $f_1, \ldots, f_r$ mit Koeffizienten $\alpha_1, \ldots, \alpha_w$ aus P wird verstanden das spezialisierte Resultantensystem $c_1(\alpha), \ldots, c_t(\alpha)$. Die notwendige und hinreichende Bedingung dafür, daß das Gleichungssystem (1) mindestens eine nichttriviale Lösung in Ω hat, ist gegeben im Verschwinden des Resultantensystems der Formen $f_1, \ldots, f_r$.

2. Ist

$$f_u = \sum_0 \ldots \sum_0^p u_{ij\ldots k}\, x_i\, y_j \ldots z_k$$

eine Multilinearform in den $x, y, \ldots, z$ mit unbestimmten Koeffizienten $u_{ij\ldots k}$ und $c_1(u), \ldots, c_t(u)$ das Resultantensystem der Formen $f_1, \ldots, f_r, f_u$, so heißt der größte gemeinsame Teiler $c(u)$ der Polynome $c_1(u), \ldots, c_t(u)$ die *u-Resultante* der Formen $f_1, \ldots, f_r$. Sie verschwindet identisch in den u, wenn das Gleichungssystem (1) unendlich viele Lösungsklassen in Ω hat; andernfalls zerfällt sie in Linearfaktoren:

$$(2) \qquad c(u) = \prod_1^q \{f^{(\alpha)}(u)\}^{\varrho_\alpha} \qquad\qquad (\varrho_\alpha > 0)$$

entsprechend den Lösungsklassen $X^{(\alpha)} = \{\xi_0^{(\alpha)}, \ldots, \xi_m^{(\alpha)}, \ldots\}$ $(\alpha = 1, \ldots, q)$, und von der Gestalt

$$(3) \qquad f^{(\alpha)}(u) = \sum \ldots \sum u_{ij\ldots k}\, \xi_i^{(\alpha)} \eta_j^{(\alpha)} \ldots \xi_k^{(\alpha)}.$$

Jeder dieser Linearfaktoren bestimmt die zugehörige Lösungsklasse eindeutig vermöge:

$$(4) \qquad \xi_0^{(\alpha)} : \xi_1^{(\alpha)} : \ldots : \xi_m^{(\alpha)} = \varphi_{0j\ldots k}^{(\alpha)} : \varphi_{1j\ldots k}^{(\alpha)} : \ldots : \varphi_{mj\ldots k}^{(\alpha)} \ \text{usw.,}$$

wo $\varphi_{ij\ldots k}^{(\alpha)}$ der Koeffizient von $u_{ij\ldots k}$ im Linearfaktor $f^{(\alpha)}(u)$ ist.

Von jetzt an soll der Einfachheit halber vorausgesetzt werden, daß nur *ein* System von Unbestimmten $x_0, \ldots, x_n$ in den Formen f_i auftritt. An Stelle von (2) und (3) treten dann die einfacheren Formeln:

$$(2\,\mathrm{a}) \qquad f^{(\alpha)}(u) = \xi_u^{(\alpha)} \quad \left(\text{Abkürzung für } \sum_0^n u_i \xi_i^{(\alpha)}\right)$$

$$(3\,\mathrm{a}) \qquad \xi_0^{(\alpha)} : \xi_1^{(\alpha)} : \ldots : \xi_n^{(\alpha)} = \varphi_0^{(\alpha)} : \ldots : \varphi_l^{(\alpha)}.$$

Man wird sich leicht überzeugen, daß alle folgenden Betrachtungen, die für den einfachen Fall angestellt werden, sich ohne weiteres auf den allgemeinen Fall übertragen lassen.

Um mich weiterhin kurz ausdrücken zu können, werde ich in Anschluß an die projektive Geometrie jede Menge von Elementsystemen $\{\lambda\,\xi_0, \ldots, \lambda\,\xi_n\}$, wo die ξ_i irgendwelche feste Elemente des Körpers Ω sind, nicht alle $= 0$, und λ alle Elemente des Körpers durchläuft, einen *Punkt* des projektiven Raumes $P_n(\Omega)$ nennen. Die Elemente ξ_i heißen die *Koordinaten* des Punktes. Wir sind in den „Lösungsklassen" schon solchen „Punkten" begegnet. Punkte werden, wie zuvor schon die Lösungsklassen, mit großen lateinischen Typen bezeichnet werden.

§ 3.

Relationstreue Spezialisierungen.

Einem beliebigen Körper P seien h unbestimmte Parameter $\lambda_1, \ldots, \lambda_h$ adjungiert, wodurch der rationale Funktionenkörper $P(\lambda_1, \ldots, \lambda_h)$ entsteht. Es sei Ω ein algebraisch-abgeschlossener Erweiterungskörper von $P(\lambda_1, \ldots, \lambda_h)$. Elemente von Ω sind also algebraische Funktionen von $\lambda_1, \ldots, \lambda_h$ und von eventuellen weiteren Unbestimmten.

Sind nun $X^{(1)}, \ldots, X^{(q)}$ Punkte im $P_n(\Omega)$, mit Koordinaten $\xi_i^{(a)}$, die algebraische Funktionen von $\lambda_1, \ldots, \lambda_h$ sind, und sind $\mu_1, \ldots, \mu_h$ irgendwelche Elemente von Ω (die insbesondere auch dem Grundkörper P angehören können), so verstehe ich unter einer *relationstreuen Spezialisierung* von $X^{(1)}, \ldots, X^{(q)}$ *für die Parameterwerte* $\mu_1, \ldots, \mu_h$ ein solches Punktsystem $Y^{(1)}, \ldots, Y^{(q)}$, daß alle richtigen Gleichungen der Gestalt

$$(5) \qquad h(\ldots, \xi_i^{(a)}, \ldots, \lambda_j, \ldots) = 0,$$

wo h jedesmal ein Polynom in $x_0^{(1)}, \ldots, x_n^{(q)}, \lambda_1, \ldots, \lambda_h$ mit Koeffizienten aus P, und homogen in jedem der Systeme $x_0^{(\alpha)}, \ldots, x_n^{(\alpha)}$ ist, richtig bleiben, wenn man jedes $\xi_i^{(a)}$ durch $\eta_i^{(a)}$, und jedes λ_j durch μ_j ersetzt[5]).

---

[5]) Die hier eingeführte „relationstreue Spezialisierung" unterscheidet sich von anderen derartigen Zuordnungen, wobei alle algebraischen Relationen erhalten bleiben (mehrstufige Isomorphismen oder Homomorphismen) dadurch, daß nicht Elemente, sondern Klassen proportionaler Elementsysteme (Punkte eines projektiven Raumes) zugeordnet werden. Unter Umständen werden dadurch Schwierigkeiten vermieden, die sonst entstehen, wenn ein Nenner verschwindet. Zum Beispiel: wenn ein endliches System von algebraischen Funktionen $y, z, \ldots$ von Unbestimmten $X_1, \ldots, X_n$ gegeben ist, so kann man definieren, was man unter den Werten $\eta, \zeta, \ldots$ dieser Funktionen für gegebene Argumentwerte $\xi_1, \ldots, \xi_n$ versteht, und kann die Erhaltung aller algebraischen Relationen bei dieser Spezialisierung zeigen (vgl. § 2 meiner Arbeit „Zur Nullstellentheorie der Polynomideale", Math. Annalen 96, S. 183); man muß dabei aber gewisse Argumentwerte ausschließen, für die die Definitionen

(Fortsetzung der Fußnote [5]) auf nächster Seite.)

41

Satz 1. *Es gibt immer mindestens eine relationstreue Spezialisierung des Punktsystems $X^{(1)}, \ldots, X^{(q)}$ für beliebige Parameterwerte μ_j.*

Beweis. In der Gesamtheit der homogenen Polynome h, welche der Gleichung (5) genügen, gibt es nach dem Hilbertschen Basissatz eine endliche Anzahl $h_1, \ldots, h_d$, derart, daß jedes andere Polynom dieser Gesamtheit im Polynombereich $\mathsf{P}\,[\ldots, x_i^{(a)}, \ldots, \lambda_j, \ldots]$ einer Kongruenz

$$h \equiv 0 \ (h_1, \ldots, h_d)$$

genügt. Es genügt, die Existenz eines Punktsystems $Y^{(1)}, \ldots, Y^{(q)}$ mit den Eigenschaften

$$h_k(\ldots, \eta_i^{(a)}, \ldots, \mu_j, \ldots) = 0 \qquad (k = 1, \ldots, d)$$

zu zeigen. Dazu hat man nur das Verschwinden des Resultantensystems der Polynome $h_1(x, \mu), \ldots, h_d(x, \mu)$, betrachtet als Formen in den q Unbestimmtensystemen $x_0^{(a)}, \ldots, x_n^{(a)}$, nachzuweisen. Dieses Resultantensystem entsteht durch die Substitution $\lambda \to \mu$ aus dem Resultantensystem der Formen $h_1(x, \lambda), \ldots, h_d(x, \lambda)$. Das letztere Resultantensystem verschwindet aber wegen (5). Also muß das erstere ebenfalls verschwinden, womit Satz 1 bewiesen ist.

Satz 2. *Notwendig und hinreichend, damit ein Punktsystem $Y^{(1)}, \ldots, Y^{(p)}$ $(p < q)$ sich zu einer relationstreuen Spezialisierung $Y^{(1)}, \ldots, Y^{(p)}, Y^{(p+1)}, \ldots, Y^{(q)}$ des Punktsystems $X^{(1)}, \ldots, X^{(q)}$ für gegebene Parameterwerte $\mu_1 \ldots \mu_h$ ergänzen läßt, ist, daß $Y^{(1)}, \ldots, Y^{(p)}$ eine relationstreue Spezialisierung von $X^{(1)}, \ldots, X^{(p)}$ darstellt.*

Beweis. Daß die Bedingung notwendig ist, ist klar, denn wenn $Y^{(1)}, \ldots, Y^{(p)}$ sich zu einer relationstreuen Spezialisierung $Y^{(1)}, \ldots, Y^{(q)}$ er-

---

sinnlos werden. Macht man aber vorher alle Relationen durch Einführung von Größen $y_0, z_0, \ldots$ homogen, und betrachtet statt der Elemente $\eta, \zeta, \ldots$ Klassen von proportionalen Elementpaaren, so ist die relationstreue Spezialisierung nach Satz 1 immer möglich.

Diese Definition der relationstreuen Spezialisierung bleibt sinnvoll für unendlich viele algebraische Funktionen (der Existenzbeweis ist mittels Satz 2 und Wohlordnung leicht zu erbringen); wendet man sie an auf *alle* Funktionen eines Körpers und nimmt die Argumentwerte $\mu_1, \ldots, \mu_h$ aus dem Grundkörper P, so kommt man zum Begriff eines *Punktes des zum Körper gehörigen algebraischen Gebildes.* Diese Definition ist (für algebraische Funktionen einer Veränderlichen) äquivalent mit der Dedekind-Weberschen Definition eines *Punktes der Riemannschen Fläche* (Crelle 92 (1882), S. 236), doch vermeidet sie die Einführung des Ausnahmswertes ∞, wodurch die Theorie um ein geringes vereinfacht wird.

Diese Begriffe sind allmählich in mündlichen Diskussionen mit E. Noether entstanden; die hier gegebene Definition des „Punktes des algebraischen Gebildes" ist eine Modifikation dessen, was ursprünglich E. Noether mit Homomorphismen definieren wollte.

gänzen läßt, so gelten alle homogenen Relationen $h(\ldots\xi_i^{(n)}, \ldots, \lambda_j, \ldots) = 0$ auch bei der Spezialisierung $X \to Y$, $\lambda \to \mu$; und das gilt insbesondere für die Relationen, die nur von den $\xi_i^{(1)}, \ldots, \xi_i^{(p)}$ abhängen.

Um die Hinreichendheit zu zeigen, nehmen wir an, es sei $Y^{(1)}, \ldots, Y^{(p)}$ eine relationstreue Spezialisierung von $X^{(1)}, \ldots, X^{(p)}$. Wir bilden wieder eine Basis $h_1, \ldots, h_d$ für die Polynome h mit der Eigenschaft (5). Es genügt, die Existenz von Punkten $Y^{(p+1)}, \ldots, Y^{(q)}$ mit den Eigenschaften

$$h_k(\eta_0^{(1)}, \ldots, \eta_n^{(p)}, \eta_0^{(p+1)}, \ldots, \eta_n^{(q)}, \mu_1, \ldots, \mu_h) = 0$$

zu zeigen. Dazu hat man das Verschwinden des Resultantensystems der Formen

$$h_k(\eta_0^{(1)}, \ldots, \eta_n^{(p)}, x_0^{(p+1)}, \ldots, x_n^{(q)}, \mu_1, \ldots, \mu_h)$$

nachzuweisen. Dieses entsteht durch die Substitution $\xi \to \eta$, $\lambda \to \mu$ aus dem Resultantensystem der Formen

$$h_k(\xi_0^{(1)}, \ldots, \xi_n^{(p)}, x_0^{(p+1)}, \ldots, x_n^{(q)}, \lambda_1, \ldots, \lambda_h).$$

Letzteres verschwindet wegen (5). Also muß wegen der vorausgesetzten Relationstreue auch das erstere verschwinden, womit Satz 2 bewiesen ist.

Satz 3. *Sind die Punkte $X^{(1)}, \ldots, X^{(q)}$ Lösungsklassen eines homogenen Gleichungssystems*

$$(6) \qquad\qquad f_k(\lambda_1, \ldots, \lambda_h, \xi_1, \ldots, \xi_n) = 0 \qquad\qquad (k = 1, \ldots, r),$$

so sind die relationstreu spezialisierten Punkte $Y^{(1)}, \ldots, Y^{(q)}$ Lösungsklassen des spezialisierten Gleichungssystems

$$(7) \qquad\qquad f_k(\mu_1, \ldots, \mu_h, \eta_0, \ldots, \eta_n) = 0 \qquad\qquad (k = 1, \ldots, r).$$

Beweis. Klar.

<h2 style="text-align:center">§ 4.</h2>

<h3 style="text-align:center">Die Eindeutigkeit der spezialisierten Lösungen $Y^{(1)}, \ldots, Y^{(q)}$.</h3>

Jetzt sei vorausgesetzt, daß das Gleichungssystem (6), sowie auch das spezialisierte Gleichungssystem (7), eine endliche Anzahl Lösungsklassen hat. Unter dieser Voraussetzung gilt:

Satz 4. *Alle relationstreuen Spezialisierungen $Y^{(1)}, \ldots, Y^{(q)}$ des Punktsystems $X^{(1)}, \ldots, X^{(q)}$ für feste Argumentwerte $\mu_1, \ldots, \mu_h$ sind bis auf die Reihenfolge der Punkte miteinander identisch.*

Beweis. Sei $Y^{(1)}, \ldots, Y^{(q)}$ eine solche Spezialisierung. Die u-Resultante $c(u)$ von $f_1, \ldots, f_r$ (§ 2, 2) ist ein Polynom in $u_1, \ldots, u_n, \lambda_1, \ldots, \lambda_h$ und läßt sich in der Gestalt

$$(8) \qquad\qquad c(\lambda, u) = \gamma \cdot \prod_1^q (\xi_u^{(\alpha)})^{\varrho_\alpha} \qquad\qquad (\varrho_\alpha > 0,\; \gamma \neq 0)$$

43

schreiben, wo γ von den u unabhängig ist und die $\xi_u^{(a)}$ die Bedeutung (2 a) haben. Die hier auftretenden Exponenten ϱ_a haben mit Multiplizitäten noch nichts zu tun; ihre Werte sind vollkommen gleichgültig.

Ein Automorphismus des Körpers Ω, der den Körper $\mathsf{P}(\lambda_1, \ldots, \lambda_h)$ elementweise invariant läßt, läßt auch das Gleichungssystem (6) invariant, vertauscht also seine verschiedenen Lösungsklassen.

Wählt man die Koordinaten einer jeden Lösungsklasse (die ja nur bis auf einen Proportionalitätsfaktor bestimmt sind) so, daß jeweils eine dieser Koordinaten gleich der Einheit ist (was immer möglich ist, da es eine Koordinate $\neq 0$ geben muß) und daß für konjugierte Lösungsklassen die entsprechende Koordinate gleich der Einheit ist, so vertauschen die ebengenannten Automorphismen nicht nur die Lösungsklassen, sondern auch ihre Koordinaten, mithin auch die Linearfaktoren $\xi_u^{(a)}$. Das Produkt $\prod\limits_1^q \xi_u^{(a)}$ bleibt also bei allen diesen Automorphismen invariant.

Im Fall, den wir zunächst betrachten wollen, daß der Körper $\mathsf{P}(\lambda_1, \ldots, \lambda_h)$ die Charakteristik Null hat[6]), folgt daraus, daß dieses Produkt dem Körper $\mathsf{P}(\lambda_1, \ldots, \lambda_h)$ angehört. Man kann also schreiben

$$(9) \qquad \prod\limits_1^q \xi_u^{(a)} = \frac{a(\lambda)}{b(\lambda)}\, e(\lambda, u),$$

wo $e(\lambda, u)$ ein Polynom ist, das keinen von u freien Faktor enthält.

Vergleicht man (8) mit (9), so folgt, daß $e(\lambda, u)$ in $c(\lambda, u)$ aufgeht.

Die aus (9) ersichtliche Tatsache, daß die Koeffizienten der entsprechenden Potenzprodukte der u in $\prod\limits_1^q \xi^{(a)}$ und $e(\lambda, u)$ proportional sind, kann dadurch ausgedrückt werden, daß alle zweireihigen Determinanten, aus entsprechenden Paaren dieser beiden Koeffizienten gebildet, verschwinden. Diese Gleichungen sind homogen in den Reihen $\xi^{(a)}$, da das Polynom $\prod\limits_1^q \xi_u^{(a)}$ homogen in den Reihen $\xi^{(a)}$ ist. Folglich bleiben sie bei der Spezialisierung $\xi \to \eta$, $\lambda \to \mu$ erhalten. Daraus folgt, daß die Polynome $\prod\limits_1^q \eta_u^{(a)}$ und $e(\mu, u)$ proportional sind. Da ersteres nicht identisch verschwindet, so folgt:

$$(10) \qquad e(\mu, u) = \delta \cdot \prod\limits_1^q \eta_u^{(a)}.$$

Es muß $\delta \neq 0$ sein. Nämlich $e(\lambda, u)$ geht auf in $c(\lambda, u)$, also in alle Formen des Resultantensystems von $f_1(\lambda, x), \ldots, f_2(\lambda, x), u_x$. Diese

---

[6]) E. Steinitz, Algebraische Theorie der Körper, Journ. f. Math. **137**, S. 167.

Polynomteilbarkeit bleibt bei der Spezialisierung $\lambda \to \mu$ bestehen, wobei das Resultantensystem in das von $f_1(\mu, x), \ldots, f_r(\mu, x), u_x$ übergeht. Das Polynom $e(\mu, u)$ geht also auf in dem größten gemeinsamen Teiler dieses Resultantensystems, d. h. in der u-Resultante. Diese u-Resultante verschwindet wegen der endlichvielen Lösungsklassen nicht identisch, also kann auch $e(\mu, u)$ nicht identisch verschwinden, was doch der Fall wäre, wenn $\delta = 0$.

Die Linearformen $\eta_u^{(\alpha)}$ sind durch (10) bis auf einen Proportionalitätsfaktor und bis auf die Reihenfolge eindeutig bestimmt, sobald $e(\mu, u)$ gegeben ist. Ihre Koeffizientenverhältnisse bestimmen die Lösungsklassen $Y^{(1)}, \ldots, Y^{(q)}$ eindeutig. Also sind diese bis auf die Reihenfolge eindeutig festgelegt.

Im Fall der Charakteristik p ist nicht $\prod\limits_1^q \xi_u^{(\alpha)}$ selbst, sondern erst eine p^f-te Potenz dieses Produktes rational in den λ. Diese p^f-te Potenz geht nach Ausscheidung eines Faktors $\dfrac{a(\lambda)}{b(\lambda)}$ in der p^f-ten Potenz von $c(\lambda, u)$ auf, und man findet durch die gleichen Betrachtungen wie zuvor:

$$e(\mu, u) = \delta \cdot \left\{ \prod\limits_1^q \xi_u^{(\alpha)} \right\}^{p^f} \quad \text{und} \quad \delta \neq 0.$$

Durch diese Gleichung sind wiederum die $Y^{(1)}, \ldots, Y^{(q)}$ bis auf die Reihenfolge eindeutig festgelegt.

Bemerkung. Die Gleichungen (9) und (10) geben zu gleicher Zeit ein Mittel, die Koordinatenverhältnisse der $Y^{(\alpha)}$ wirklich zu berechnen, wenn die der $X^{(\alpha)}$ gegeben sind.

§ 5.

Der Multiplizitätsbegriff und das Prinzip der Erhaltung der Anzahl.

Unter den Voraussetzungen des vorigen Paragraphen kann man definieren:

Die Multiplizität einer Lösungsklasse der Gleichungen (7), *betrachtet als Spezialisierung der Gleichungen* (6), *ist die Zahl, die angibt, wie oft diese Lösungsklasse unter den Lösungsklassen* $Y^{(1)}, \ldots, Y^{(q)}$ *vertreten ist.*

Die Definition ist eindeutig, denn die $Y^{(1)}, \ldots, Y^{(q)}$ sind nach § 4 bis auf die Reihenfolge eindeutig bestimmt.

Die Multiplizität einer Lösung eines *Normalproblems* definiert man naturgemäß als die Multiplizität der entsprechenden Lösung des entsprechenden Gleichungssystems.

Prinzip der Erhaltung der Anzahl. *Die Summe der Multiplizitäten der Lösungen eines spezialisierten Normalproblems ist gleich der Anzahl der Lösungen des allgemeineren Normalproblems, aus dem es ent-*

45

standen ist, vorausgesetzt, daß sowohl das allgemeine als das spezielle Problem eine endliche Anzahl Lösungen haben.

Beweis. Beide Anzahlen sind, mit den Bezeichnungen von § 4, gleich q.

Die Multiplizität einer Lösung eines Normalproblems kann auch Null sein, wie das folgende Beispiel zeigt.

P sei ein Körper der Charakteristik Null, λ eine Unbestimmte, Ω ein algebraisch-abgeschlossener Erweiterungskörper von $P(\lambda)$. Das allgemeine Normalproblem bestehe darin, die Lösungsklassen der Gleichungen

$$(11) \quad \begin{cases} (\xi_1 - \xi_0)(\xi_0^3 - \xi_1^2 \xi_2) = 0, \\ \xi_2(\xi_0^3 - \xi_1^2 \xi_2) = 0, \\ \xi_2 - \lambda \xi_1 = 0 \end{cases}$$

zu finden. Schließt man die triviale Lösung $\xi_1 = \xi_2 = \xi_3 = 0$ aus, so reduziert sich das Gleichungssystem leicht zu

$$\begin{cases} \xi_2 = \lambda \xi_1, \\ \xi_0^3 = \lambda \xi_1^3. \end{cases}$$

Sind also $\delta_1, \delta_2, \delta_3$ die dritten Wurzeln aus λ im Körper Ω, so sind

$$X^{(\alpha)} = \{\delta_\alpha, 1, \lambda\} \qquad (\alpha = 1, 2, 3)$$

die drei Lösungsklassen des Systems (11) in Ω. Nach (9) ist:

$$\prod_1^3 (\delta_\alpha u_0 + u_1 + \lambda u_2) = \lambda u_0^3 + (u_1 + \lambda u_2)^3 = e(\lambda, u).$$

Für $\lambda = 0$ gehen die Gleichungen (11) über in

$$(12) \quad \begin{cases} (\xi_1 - \xi_0)(\xi_0^3 - \xi_1^2 \xi_2) = 0, \\ \xi_2(\xi_0^3 - \xi_1^2 \xi_2) = 0, \\ \xi_2 = 0 \end{cases}$$

mit den beiden Lösungsklassen

$$Y' = \{0, 1, 0\},$$
$$Y'' = \{1, 1, 0\}.$$

Das Polynom $e(\lambda, u)$ geht für $\lambda = 0$ über in

$$e(0, u) = u_1^3,$$

ergibt also drei gleiche Linearfaktoren u_1, welche zu den drei zusammenfallenden Lösungsklassen

$$Y^{(\alpha)} = \{0, 1, 0\} \qquad (\alpha = 1, 2, 3)$$

gehören. Also hat die Lösung Y' die Multiplizität *drei*, die Lösung Y'' die Multiplizität *Null*.

Die obige Rechnung gilt auch für Körper P von Primzahlcharakteristik, ausgenommen die Charakteristik *drei*. In diesem Ausnahmefall gibt es in Ω nur eine dritte Wurzel δ aus λ, also nur eine Lösungsklasse

$$X^{(1)} = \{\delta, 1, \lambda\}$$

des Systems (11). Hier tritt der in § 5 erwähnte Fall auf, daß die Linearform $\delta u_0 + u_1 + \lambda u_2$ zwar mit allen ihren konjugierten identisch, aber nicht rational in λ ist; erst die dritte Potenz $\lambda u_0^3 + u_1^3 + \lambda^3 u_2^3$ ist rational. Für $\lambda = 0$ hat man wie vorhin zwei Lösungsklassen

$$\begin{cases} Y' = \{0, 1, 0\}, \\ Y'' = \{1, 1, 0\}, \end{cases}$$

während

$$e(0, u) = u_1^3,$$

also

$$Y^{(1)} = \{0, 1, 0\}.$$

Die Lösungsklasse Y' hat jetzt, da es nur ein $X^{(\alpha)}$ und ein $Y^{(\alpha)}$ gibt, die Multiplizität *eins*, die Lösung Y'' die Multiplizität *Null*.

Notwendig und hinreichend, damit die Multiplizität einer Lösung Y größer als Null ist, ist, daß sie in einer relationstreuen Spezialisierung $Y^{(1)}, \ldots, Y^{(q)}$ des Lösungssystems $X^{(1)}, \ldots, X^{(q)}$ vorkommt. Daraus folgt wegen Satz 2 das folgende Kriterium:

S a t z 5. *Eine Lösungsklasse Y eines spezialisierten Gleichungssystems hat dann und nur dann eine Multiplizität > 0, wenn sie eine relationstreue Spezialisierung einer Lösung X des allgemeinen Gleichungssystems ist.*

<h3 align="center">§ 6.</h3>

<h3 align="center">Einige Bemerkungen über die Tragweite des Multiplizitätsbegriffs und des Prinzips der Erhaltung der Anzahl.</h3>

Es ist sinnlos, von der Multiplizität einer Lösung eines Normalproblems zu reden, wenn man dieses Problem für sich allein betrachtet; man muß immer dabei angeben, aus welchem allgemeineren Problem man das gegebene durch Parameterspezialisierung abgeleitet denkt.

Gegen diesen Grundsatz ist oft verstoßen worden. Man redet z. B. ohne Definition von der Multiplizität eines Schnittpunktes zweier Mannigfaltigkeiten der Dimensionen r und $n - r$ im projektiven Raum P_n. Es wird dabei nicht angegeben, aus welchen allgemeineren Gebilden man die M_r und die M_{n-r} durch Spezialisierung entstanden denkt.

Wenn zwei Mannigfaltigkeiten der Dimensionen r und $n - r$ im P_n[7])

---

[7]) Unter der *Dimension einer Mannigfaltigkeit M*, gegeben durch k homogene Gleichungen $f_1(\xi_0, \ldots, \xi_n) = 0, \ldots, f_k(\xi_0, \ldots, \xi_n) = 0$ wird verstanden die um 1 ver-

(Fortsetzung der Fußnote [7]) auf nächster Seite.)

endlichviele Schnittpunkte haben, und wenn eine der Mannigfaltigkeiten, etwa die letztere, linear ist, d. h. durch r lineare Gleichungen gegeben sind, so kann man das Problem, die Schnittpunkte zu finden, als Spezialfall des allgemeineren Problems betrachten, das entsteht, indem man die gegebene r-dimensionale Mannigfaltigkeit mit einer allgemeinen linearen M_{n-r} schneidet, d. h. mit einer solchen, die durch r lineare Gleichungen mit unbestimmten Koeffizienten bestimmt wird. In diesem Fall hat es also einen guten Sinn, von der Multiplizität eines Schnittpunktes zu reden. Bei zwei nichtlinearen Mannigfaltigkeiten ist das aber ohne genauere Definition nicht der Fall.

Eine andere Bemerkung betrifft die Anwendungen, die in der „abzählenden Geometrie" von dem Prinzip der Erhaltung der Anzahl gemacht werden.

Um mittels der hier entwickelten Formen und Definitionen die Multiplizität einer Lösung eines spezialisierten Normalproblems zu bestimmen, ist es notwendig, daß man schon von vornherein eine vollständige Übersicht über alle Lösungen des allgemeinen Problems hat. Es ist also vorläufig noch unmöglich, auf Grund des Prinzips der Erhaltung der Anzahl durch eine geschickt gewählte Parameterspezialisierung die Lösungszahl eines allgemeinen Problems zu bestimmen, wenn dieses Problem nicht einer erschöpfenden algebraischen Behandlung fähig ist, was doch hauptsächlich das Ziel der abzählenden Geometrie ist. Vorbedingung aller einwandfreien abzählenden Geometrie ist demnach die Aufstellung von Methoden, die es gestatten, Multiplizitäten von Lösungen zu bestimmen, auch ohne die Lösungen des allgemeinen Problems zu kennen.

§ 7.

Schnittpunkte von n Hyperflächen.

In diesem Paragraphen soll ein Beispiel eines Normalproblems gegeben werden, wo die u-Resultante und das Polynom $e(\lambda, u)$, die in § 4 zur eindeutigen Festlegung des $Y^{(\alpha)}$ benutzt wurden, sich direkt berechnen lassen, und wo man dementsprechend explizite Multiplizitätsformeln angeben kann. Dieses Normalproblem besteht in der Bestimmung der Schnittpunkte von n Hyperflächen im $P_n(\mathsf{P})$, wo P ein algebraisch-abgeschlossener Körper ist.

---

ringerte Dimension des Ideals $(f_1, \ldots, f_k)$. Die Verringerung um 1 entspricht der Tatsache, daß die ξ_α nicht als inhomogene Koordinaten in einem Raum C_{n+1}, sondern als homogene Koordinaten in einem Raum P_n betrachtet werden. Vgl. für den algebraischen Dimensionsbegriff meine Arbeit „Zur Nullstellentheorie der Polynomideale", Math. Annalen **96**, S. 197.

Eine *Hauptmannigfaltigkeit*[8]) (oder Hyperfläche) im $P_n(\mathsf{P})$ ist die Gesamtheit der Punkte, deren Koordinaten einer einzigen Gleichung

$$\varphi(\xi_0, \ldots, \xi_n) = 0$$

genügen, wo φ eine Form mit Koeffizienten aus P ist.

Die Bestimmung der Schnittpunkte von n Hyperflächen ist demnach äquivalent mit der Lösung des Gleichungssystems

$$(13) \qquad \begin{cases} \varphi_1(\xi_0, \ldots, \xi_n) = 0 \\ \cdots \\ \varphi_n(\xi_0, \ldots, \xi_n) = 0. \end{cases}$$

Wir betrachten das Gleichungsproblem als Spezialisierung des allgemeineren

$$(14) \qquad \begin{cases} f_1(\xi_0, \ldots, \xi_n) = 0 \\ \cdots \\ f_n(\xi_0, \ldots, \xi_n) = 0, \end{cases}$$

wo $f_1, \ldots, f_n$ allgemeine Formen (mit unbestimmten Koeffizienten $\lambda_1, \ldots \lambda_h$) derselben Gradzahlen $q_1, \ldots, q_n$ sind. Gesucht werden zunächt die Lösungsklassen von (14) im algebraisch-abgeschlossenen Erweiterungskörper Ω von $\mathsf{P}(\lambda_1, \ldots, \lambda_h)$. Man findet sie durch Faktorzerlegung der u-Resultante $c(\lambda, u)$. Diese ist in diesem Fall einfach die Resultante der Formen $f_1, \ldots, f_n$ und der allgemeinen Linearform u_x. Sie ist bekanntlich irreduzibel im Polynombereich $\mathsf{P}[\lambda_1, \ldots, \lambda_h; u_0, \ldots, u_n]$; dagegen zerfällt sie in Linearfaktoren $\xi_u^{(1)}, \ldots, \xi_u^{(q)}$ $(q = \prod_1^n q_i)$ in $\Omega[u_0, \ldots, u_n]$. Wegen der erstgenannten Irreduzibilität müssen die Linearfaktoren $\xi_u^{(\alpha)}$ alle verschieden sein, es sei denn, daß P die Charakteristik p hat, und $c(\lambda, u)$ nur die p-ten Potenzen der u_i enthält. Wäre letzteres der Fall, so müßte es bei jeder Spezialisierung der Formen f_i so bleiben, insbesondere auch dann, wenn jedes f_i in ein Produkt von allgemeinen Linearfaktoren zerfällt. Die Resultante dieser zerfallenden Form ist aber ein Produkt von Resultanten der Faktoren, zu je n mit u_x zusammengenommen. Ein solches Produkt enthält offensichtlich nicht nur die p-ten Potenzen der u_i, also kann die ursprüngliche Determinante es auch nicht tun.

---

[8]) Der Name Hauptmannigfaltigkeit ist darum gewählt, weil das zugehörige Ideal (vgl. Math. Annalen 96, S. 196) einer solchen Mannigfaltigkeit Hauptideal ist. Jede algebraische Mannigfaltigkeit ist definitionsmäßig ein Durchschnitt von Hauptmannigfaltigkeiten. Man könnte sie auch ganz kurz und bequem „Häupter" nennen. Nach Einführung des algebraischen Dimensionsbegriffes für algebraische Mannigfaltigkeiten im P_n stellt sich heraus, daß alle rein $(n-1)$-dimensionalen Mannigfaltigkeiten im P_n Häupter sind, und umgekehrt.

Also sind die Linearformen $\xi_u^{(a)}$ alle verschieden. Daraus folgt wegen (9):

$$c(\lambda, u) = \gamma \cdot e(\lambda, u),$$

wo γ eine Konstante ist.

Um die spezialisierten Schnittpunkte $Y^{(1)}, \ldots, Y^{(q)}$ zu finden, hat man nach § 4 das spezialisierte Polynom $c(\lambda, u)$, oder, was jetzt auf dasselbe hinauskommt, die spezialisierte u-Resultante $e(\lambda, u)$ in Linearfaktoren zu zerlegen. Der Exponent eines Linearfaktors in dieser Zerlegung gibt die Multiplizität eines jeden Schnittpunktes an. Man erhält in dieser Weise nach der allgemeinen Resultantentheorie alle Schnittpunkte, also ist die Multiplizität eines Schnittpunktes niemals Null.

Bekanntlich[9]) ist der Exponent des zu einem Punkt Y gehörigen Linearfaktors in der Zerlegung der u-Resultante gleich der Anzahl der modulo q linear-unabhängigen Polynome, wo q die zum Punkt Y gehörige primäre Komponente des Ideals $(f_1, \ldots, f_n)_{x_0=1}$ ist.

Sind speziell die Formen $\varphi_2, \ldots, \varphi_n$ *linear*, so ist das Gleichungssystem (13) dem Problem äquivalent, die Schnittpunkte einer Hauptmannigfaltigkeit vom Grade $q = q$ und einer Geraden zu bestimmen. Die Gerade kann man auch festlegen durch zwei Punkte, von denen alle anderen Punkte linear abhängen:

$$(15) \qquad \xi_k = \sigma \eta_k + \tau \zeta_k.$$

Substitution von (15) in (13) ergibt eine binäre Gleichung:

$$(16) \qquad \varphi(\ldots, \sigma \eta_k + \tau \zeta_k, \ldots) = \alpha_0 \sigma^q + \alpha_1 \sigma^{q-1} \tau + \ldots + \alpha_q \tau^q = 0,$$

wo α_i eine Form i-ten Grades in den η und $(q-i)$-ten Grades in den ζ, eine „Polare" ist. Die Lösungsklassen von (16) ergeben, in (15) eingesetzt, die Schnittpunkte. Wir setzen natürlich voraus, daß (16) nicht identisch in σ, τ erfüllt ist.

Es ist naheliegend, zu vermuten, daß *die Multiplizität eines Schnittpunktes gleich dem Exponenten des entsprechenden Linearfaktors in der Faktorzerlegung von* (16) *ist.* Das beweist man so:

Bildet man die zu (16) analoge Gleichung für die allgemeine Form f_1 und die allgemeine Gerade $f_2 = \ldots = f_n = 0$, so kommt:

$$f(\ldots, \sigma y_k + \tau z_k, \ldots) = a_0 \sigma^q + a_1 \sigma^{q-1} \tau + \ldots + a_q \tau^q = 0.$$

Diese Gleichung hat in Ω n Lösungsklassen $\{\sigma^{(a)}, \tau^{(a)}\}$, die alle verschieden sein müssen, weil ja n verschiedene Schnittpunkte vorhanden sind. Diese Schnittpunkte werden gegeben durch

$$(17) \qquad \xi_k^{(a)} = \sigma^{(a)} x_k + \tau^{(a)} y_k.$$

---

[9]) F. S. Macaulay, Modular Systems, Cambridge Tracts 19 (1916), S. 77.

Es gibt nach § 3 eine relationstreue Spezialisierung $\{\overline{\sigma^{(a)}}, \overline{\tau^{(a)}}\}$ der Lösungsklassen $\{\sigma^{(a)}, \tau^{(a)}\}$ für die Parameterwerte ξ_k, η_k, usw. Die Zerlegung von $a_1\sigma^q + \ldots + a_q\tau^q$ in Linearfaktoren $\sigma\tau^{(a)} - \tau\sigma^{(a)}$ bleibt wegen der Relationstreue auch für $y \to \eta$, $z \leftarrow \zeta$, $\sigma^{(a)} \to \overline{\sigma^{(a)}}$, usw. bestehen [vgl. den analogen Schluß bei der Gleichung (9) § 4]; d. h. jede Lösungsklasse $\{\bar{\sigma}, \bar{\tau}\}$ kommt unter den $\{\overline{\sigma^{(a)}}, \overline{\tau^{(a)}}\}$ so oft vor, wie sein Exponent in der Zerlegung von (18) angibt. Setzt man nun:

$$(18) \qquad \overline{\xi_k^{(a)}} = \overline{\sigma^{(a)}}\eta_k + \overline{\tau^{(a)}}\zeta_k,$$

so bilden die dadurch definierten Punkte $\overline{X^{(a)}}$ eine relationstreue Spezialisierung des durch (17) definierten Punktsystems $X^{(1)}, \ldots, X^{(q)}$, denn eine jede algebraische Relation für die $X^{(a)}$ läßt sich mittels (14) unmittelbar als Relation für die $\sigma^{(a)}$, $\tau^{(a)}$ schreiben, und eine solche bleibt bei Spezialisierung erhalten. Also stimmen $\overline{X^{(1)}}, \ldots, \overline{X^{(q)}}$ mit $Y^{(1)}, \ldots, Y^{(q)}$ bis auf die Reihenfolge überein, und jeder Punkt Y kommt neben den $Y^{(i)}$ genau so oft vor wie unter den $X^{(i)}$, q. e. d.

Das Studium der Multiplizitäten ist durch diese Überlegung auf die Diskussion der Gleichung (16), d. h. auf die bekannte Polarentheorie, zurückgeführt.

An diesem Beispiel sieht man, wie sich die Multiplizitätsbestimmung vereinfacht, falls es möglich ist, ein gegebenes Normalproblem auf eine binäre Gleichung zurückzuführen.

§ 8.

Geraden auf der kubischen Fläche.

Es sei wieder P ein algebraisch-abgeschlossener Körper.

Eine kubische Fläche im Raum $P_3(\mathsf{P})$ besteht aus den Nullstellenklassen einer kubischen Form

$$\gamma(x) = \sum_{0 \leq i \leq k \leq l \leq 3} \gamma_{ikl}x_i x_k x_l.$$

Eine Gerade mit Plückerschen (Punkt-) Koordinaten π_{ik} liegt dann und nur dann auf der Fläche, wenn ihr Schnittpunkt mit irgendeiner Ebene $u_\xi = 0$ immer auf der Fläche liegt; dieser Schnittpunkt hat die Koordinaten $\sum u_i\pi_{ik}$; also ist die Bedingung, daß die Gerade auf der Fläche liegt:

$$(19) \qquad \gamma\left(\sum u_i\pi_{ik}\right) = 0 \quad \text{identisch in den } u.$$

Jedes Wertsystem π_{ik} (nicht alle $\pi_{ik} = 0$) stellt eine Gerade dar, falls

$$(20) \qquad \pi_{01}\pi_{23} + \pi_{02}\pi_{31} + \pi_{03}\pi_{12} = 0.$$

51

Die Geraden auf der Fläche werden mithin gegeben durch die Lösungsklassen des Gleichungssystems (19), (20), mithin:

Die Bestimmung der Geraden auf einer kubischen Fläche ist ein Normalproblem.

Die Gleichungen (19), (20) entstehen durch Spezialisierung aus den allgemeineren:

$$(21) \quad \begin{cases} c\left(\sum u_i \pi_{ik}\right) = 0 \qquad \text{identisch in den } u, \\ \pi_{01}\pi_{23} + \pi_{02}\pi_{31} + \pi_{03}\pi_{12} = 0, \end{cases}$$

wo

$$c(x) = \sum_{0 \leq i \leq k \leq l \leq 3} c_{ikl}\, x_i\, x_k\, x_l$$

eine *allgemeine* kubische Form mit unbestimmten Koeffizienten c_{ikl} ist.

Sei Ω der algebraisch-abgeschlossene Erweiterungskörper von $P(c_{ikl})$. Die Nullstellenklassen von $c(x)$ in Ω bilden eine „allgemeine kubische Fläche", und die darauf liegenden Geraden im $P_3(\Omega)$ werden durch die Lösungsklassen von (21) bestimmt.

Man kann die Beziehung zwischen Fläche und Geraden algebraisch noch anders darstellen. Ist π irgendeine Gerade der Fläche, so transformiere man zunächst die Koordinaten so, daß die Gerade nicht in einer Ebene $\varkappa_1 \xi_1 + \varkappa_0 \xi_0 = 0$ liegt. Dann kann man die Gleichungen der Geraden so schreiben:

$$(22) \quad \begin{cases} \xi_3 = \alpha_1 \xi_1 + \alpha_0 \xi_0, \\ \xi_2 = \beta_1 \xi_1 + \beta_0 \xi_0. \end{cases}$$

Soll die Fläche $\gamma(\xi) = 0$ durch die Gerade (22) gehen, so muß die Form $\gamma(x)$ nach Division durch $x_3 - \alpha_1 x_1 - \alpha_0 x_0$ einen von x_3 freien Rest lassen, der nach Division durch $x_2 - \beta_1 x_1 - \beta_0 x_0$ keinen Rest läßt. Also muß

$$(23) \quad \begin{aligned} \gamma(x) &= \sum \gamma_{ikl}\, x_i\, x_k\, x_l \\ &= \varphi(x) \cdot (x_3 - \alpha_1 x_1 - \alpha_0 x_0) + \psi(x) \cdot (x_2 - \beta_1 x_1 - \beta_0 x_0), \end{aligned}$$

wo $\varphi(x)$ und $\psi(x)$ quadratische Formen sind, und wo $\psi(x)$ von x_3 frei ist:

$$(24) \quad \begin{cases} \varphi(x) = \sum_{0 \leq i \leq k \leq 3} \alpha_{ik}\, x_i\, x_k, \\ \psi(x) = \sum_{0 \leq i \leq k \leq 2} \beta_{ik}\, x_i\, x_k. \end{cases}$$

Die Gleichung (23) ist notwendig und hinreichend, damit die Gerade (22) auf der Fläche $\gamma(\xi) = 0$ liegt.

Satz 5. 1. *Auf der allgemeinen kubischen Fläche im $P_3(\Omega)$ (und folglich nach Satz 3 auch auf jeder speziellen kubischen Fläche) liegt mindestens eine Gerade.*

2. *Alle Geraden auf der allgemeinen kubischen Fläche sind konjugiert in bezug auf* $\mathsf{P}(c_{ikl})$.

3. *Wenn auf einer speziellen kubischen Fläche im* $P_3(\mathsf{P})$ *nur endlich viele Gerade liegen, so hat keine die Multiplizität Null.*

Die beiden ersten Behauptungen sind (wenigstens für Körper der Charakteristik Null) bekannt; es ist mir aber hier darum zu tun, zu zeigen, daß eine und dieselbe Methode geeignet ist, die drei Behauptungen zu gleicher Zeit zu beweisen. Diese Methode, die auch in manchen anderen Fällen zum Ziel führt, besteht in einer Umkehrung des Problems: wir gehen aus von einer möglichst allgemeinen Geraden und suchen eine möglichst allgemeine kubische Fläche, die die Gerade enthält. Sodann wird gezeigt, daß diese Fläche durch einen Körperisomorphismus in jede allgemeine kubische Fläche übergeführt werden kann, womit 1. bewiesen sein wird. 2. und 3. folgen nachher leicht.

Beweis. Es seien a_i, b_i, a_{ik}, b_{ik} Unbestimmte, die, zu P adjungiert, einen Körper $\varDelta$ erzeugen. Eine (allgemeine) Gerade im $P_n(\varDelta)$ wird gegeben durch die Gleichungen

$$(25) \qquad \begin{cases} \xi_3 = a_1\,\xi_1 + a_0\,\xi_0, \\ \xi_2 = b_1\,\xi_1 + b_0\,\xi_0. \end{cases}$$

Eine kubische Fläche durch diese Gerade wird definiert durch die Form

$$(26) \qquad \delta(x) = \sum_{0 \leq i \leq k \leq l \leq 3} \delta_{ikl}\,x_i\,x_k\,x_l$$
$$= f(x) \cdot (x_3 - a_1\,x_1 - a_0\,x_0) + g(x) \cdot (x_2 - b_1\,x_1 - b_0\,x_0),$$

wo

$$(27) \qquad \begin{cases} f(x) = \sum_{0 \leq i \leq k \leq 3} a_{ik}\,x_i\,x_k, \\ g(x) = \sum_{0 \leq i \leq k \leq 2} b_{ik}\,x_i\,x_k. \end{cases}$$

Es handelt sich jetzt darum, zu zeigen, daß die durch (26) definierten δ_{ikl} algebraisch-unabhängig in bezug auf P sind.

Wären sie das nicht, so würde eine Gleichung

$$F(\delta_{ikl}) = 0$$

bestehen, während für unbestimmte x_{ikl}

$$F(x_{ikl}) \neq 0.$$

Ich werde das ad absurdum führen.

Ich gebe jedem a_{ik} und jedem b_{ik} als *Gewicht* die Anzahl der Zahlen 0 oder 1, die unter seinen Indizes vorkommen. Ausgenommen werden a_{00} und b_{11}, denen das Gewicht 3 gegeben wird.

Jedes δ_{ikl} ist ein Polynom der a_{ik}, b_{ik}, a_i, b_k. Wenn man von jedem dieser Polynome nur die Glieder vom höchsten Gewicht beibehält, so kommt:

$$\delta_{333} = a_{33} \qquad\qquad \delta_{222} = b_{22}$$
$$\delta_{332} = a_{32} \qquad\qquad \delta_{221} = b_{21} + \cdots$$
$$\delta_{331} = a_{31} + \cdots \qquad\qquad \delta_{220} = b_{20} + \cdots$$
$$\delta_{330} = a_{30} + \cdots \qquad\qquad \delta_{211} = b_{11} + \cdots$$
$$\delta_{322} = a_{22} + \cdots \qquad\qquad \delta_{210} = b_{10} + \cdots$$
$$\delta_{321} = a_{21} + \cdots \qquad\qquad \delta_{200} = b_{00} + \cdots$$
$$\delta_{320} = a_{20} + \cdots \qquad\qquad \delta_{111} = -\, b_{11}\, b_1 + \cdots$$
$$\delta_{311} = a_{11} + \cdots \qquad\qquad \delta_{110} = -\, b_{11}\, b_0 + \cdots$$
$$\delta_{310} = a_{10} + \cdots \qquad\qquad \delta_{100} = -\, a_{00}\, a_1 + \cdots$$
$$\delta_{300} = a_{00} + \cdots \qquad\qquad \delta_{000} = -\, a_{00}\, a_0 + \cdots$$

Wir geben jedem δ_{ikl} das Gewicht des eben hingeschriebenen jeweiligen Leitgliedes, das wir θ_{ikl} nennen, und geben jedem x_{ikl} dasselbe Gewicht wie das entsprechende δ_{ikl}.

Ist nun $F'(x_{ikl})$ der Bestandteil höchsten Gewichts in $F(x_{ikl})$, so ist $F'(\theta_{ikl})$ der Bestandteil vom selben Gewicht in $F(\delta_{ikl})$. Es ist folglich

$$F'(\theta_{ikl}) = 0\,,$$

aber

$$F'(x_{ikl}) \neq 0\,.$$

Da aber ersichtlich die θ_{ikl} algebraisch-unabhängig sind, so ist hier ein Widerspruch vorhanden.

Also sind die δ_{ikl} algebraisch-unabhängig. Also gibt es einen Isomorphismus der Körper $\mathsf{P}(\delta_{ikl})$ und $\mathsf{P}(c_{ikl})$, der δ_{ikl} in c_{ikl} überführt und P elementweise fest läßt. Diese Isomorphie läßt sich fortsetzen zu einer Isomorphie von $\varDelta$ mit einem Erweiterungskörper $\varDelta^*$ von c_{ikl}. Da $\varDelta$ und $\mathsf{P}(\delta_{ikl})$ beide den Transzendenzgrad 20[10]) haben, so ist $\varDelta$ algebraisch in bezug auf $\mathsf{P}(\delta_{ikl})$, also $\varDelta^*$ algebraisch in bezug auf $\mathsf{P}(c_{ikl})$. Also kann $\varDelta^*$ innerhalb $\varOmega$ gewählt werden. Die Elemente a_{ik}, b_{ik}, a_i, b_k mögen in a_{ik}^*, b_{ik}^*, a_i^*, b_k^* übergehen, die Polynome $f(x)$ und $g(x)$ in $f^*(x)$ und $g^*(x)$. Aus (26) folgt vermöge der Isomorphie:

$$(28) \qquad c(x) = f^*(x) \cdot (x_3 - a_1^* x_1 - a_0^* x_0) + g^*(x) \cdot (x_2 - b_1^* x_1 - b_0^* x_0),$$

---

[10]) Der Transzendenzgrad eines Körpers in bezug auf einen Unterkörper P ist die Maximalzahl der algebraisch-unabhängigen Größen (in bezug auf P), die im Körper vorhanden sind. Vgl. E. Steinitz, a. a. O.[6])

also liegt die Gerade

$$(29) \qquad \begin{cases} \xi_3 = a_1^* \xi_1 + a_0^* \xi_0 \\ \xi_2 = b_1^* \xi_1 + b_0^* \xi_0 \end{cases}$$

auf der allgemeinen kubischen Fläche $c(\xi) = 0$. Damit ist die Behauptung 1 bewiesen.

Um 2. zu beweisen, nehme ich an, es sei

$$\begin{cases} \xi_3 = \bar{a}_1 \xi_1 + \bar{a}_0 \xi_0 \\ \xi_2 = \bar{\beta}_1 \xi_1 + \bar{\beta}_0 \xi_0 \end{cases}$$

irgendeine Gerade auf der allgemeinen kubischen Fläche $c(\xi) = 0$. Man hat wie immer

$$(30) \qquad c(x) = \bar{\varphi}(x)(x_3 - \bar{a}_1 \xi_3 - \bar{a}_0 \xi_0) + \bar{\psi}(x)(x_2 - \bar{\beta}_1 \xi_1 - \bar{\beta}_0 \xi_0),$$

also hängen die 20 Größen c_{ikl} rational von den 20 Größen $\bar{a}_{ik}$, $\bar{a}_i$, $\bar{\beta}_{ik}$, $\bar{\beta}_i$ ab. Da die c_{ikl} algebraisch-unabhängig sind, so müssen die $\bar{a}_{ik}$, $\bar{a}_i$, $\bar{\beta}_{ik}$, $\bar{\beta}_k$ es auch sein. Also gibt es einen Isomorphismus der Körper $P(\bar{a}_{ik}, \bar{\beta}_{ik}, \bar{a}_i, \bar{\beta}_i)$ und $P(a_{ik}^*, b_{ik}^*, a_i^*, b_i^*)$, der $\bar{a}_{ik}$ in a_{ik}^*, usw. überführt und P elementweise fest läßt. Wegen (28) und (30) läßt dieser Isomorphismus die c_{ikl} fest; also sind die $\bar{a}_i$, $\bar{\beta}_i$ zu den a_i^*, b_i^* konjugiert in bezug auf $P(c_{ikl})$, womit 2. bewiesen ist.

Um schließlich 3. zu beweisen, verfahren wir genau so. Es sei diesmal

$$(31) \qquad \begin{cases} \xi_3 = \alpha_1 \xi_1 + \alpha_0 \xi_0 \\ \xi_2 = \beta_1 \xi_1 + \beta_0 \xi_0 \end{cases}$$

eine Gerade auf der speziellen kubischen Fläche $\gamma(\xi) = 0$. Die Plückerschen Koordinaten π_{ik}^* der Geraden (29) sind rationale Funktionen der a_i^* und b_i^*; die der Geraden (31) sind dieselben Funktionen der α_i und β_i. Die Koeffizienten c_{ikl} der allgemeinen kubischen Fläche sind nach (28) rationale Funktionen der algebraisch-unabhängigen Größen a_{ik}^*, b_{ik}^*, a_i^*, b_i^*; die γ_{ikl} sind nach (23) dieselben Funktionen der α_{ik}, β_{ik}, α_i, β_i. Daraus folgt, daß alle algebraischen Relationen, die zwischen den c_{ikl} und π_{ik}^* bestehen, auch gelten bei der Spezialisierung $c \to \gamma$, $\pi^* \to \pi$. Also bilden die π_{ik} eine relationstreue Spezialisierung der π_{ik}^* für die Argumentwerte γ_{ikl}, woraus nach Satz 5 folgt, daß die Gerade π auf der Fläche $\gamma(\xi) = 0$ eine Multiplizität > 0 hat.

(Eingegangen am 30. 6. 1926.)

55

4.

Eine Verallgemeinerung des Bézoutschen Theorems

Mathematische Annalen 99, 4/5 (1928) 497–541

§ 1.

Einleitung.

Um die Lösungsklassen (Klassen proportionaler Lösungen) eines Systems von n homogenen Gleichungen $f_1 = 0, \ldots, f_n = 0$ der Gradzahlen $m_1, \ldots, n_n$ mit $n + 1$ Unbekannten $x_0, \ldots, x_n$ zu finden, bildet man die „u-Resultante" des Gleichungssystems, d. h. die Resultante der Formen $f_1, \ldots, f_n$ und einer allgemeinen Linearform $\sum u_k x_k$. Die u-Resultante hat als Funktion der u_i den Grad $\prod_1^n m_i$ und zerfällt (in einem geeigneten Erweiterungskörper des Rationalitätsbereichs der Koeffizienten) in Linearfaktoren $\sum u_k \xi_k^{(a)}$. Im allgemeinen, das heißt wenn die Formenkoeffizienten Unbestimmte sind, sind diese Linearfaktoren alle verschieden; für spezielle Werte können aber vielfache Faktoren vorkommen; oder es kann vorkommen, daß die u-Resultante identisch verschwindet, nämlich dann, wenn unendlich viele Lösungsklassen vorhanden sind. Sehen wir von dem letzten Fall ab, so repräsentieren die Koeffizienten $\xi_0^{(a)}, \ldots, \xi_n^{(a)}$ eines jeden Linearfaktors eine Lösungsklasse. Ein μ-facher Linearfaktor gibt eine „Lösung der Multiplizität μ". Die Summe der Multiplizitäten der Lösungsklassen[1] ist somit $\prod_1^n m_i$; dies ist der *Satz von Bézout* in moderner Gestalt[2]. Wie man sieht, gibt hier die u-Resultante das Mittel, die Multiplizitäten so zu definieren, daß die „Erhaltung der Anzahl" gilt: die

---

[1] Man sagt oft ungenau „Anzahl" statt „Summe der Multiplizitäten".

[2] Vgl. etwa F. S. Macaulay, Algebraic Theory of Modular Systems, Cambridge Tract **19** (1916), p. 15—17. Im folgenden mit „Macaulay" zitiert.

Anzahl der Lösungen im „allgemeinen" Fall ist gleich der Summe der Multiplizitäten im Spezialfall.

Geometrisch heißt dieser Satz, daß die Summe der Multiplizitäten der Schnittpunkte von n algebraischen Hyperflächen im projektiven ^{n}Raum gleich dem Produkt der Gradzahlen ist, falls die Anzahl der Schnittpunkte endlich bleibt. Die Multiplizitäten sind dabei wie vorhin definiert durch die Exponenten der Linearfaktoren der u-Resultante.

Man hat oft in der Geometrie[3]) unter dem Namen „Bézoutschen Satz" einen viel weitergehenden Satz angeführt und meist ohne Beweis angewandt, der besagt, daß die Summe der Multiplizitäten einer r-dimensionalen und einer $(n-r)$-dimensionalen Mannigfaltigkeit im projektiven ^{n}Raum (oder speziell einer Kurve und einer Fläche im ^{3}Raum) gleich dem Produkt der Gradzahlen dieser Mannigfaltigkeiten ist, falls die Anzahl der Schnittpunkte endlich bleibt. Dabei versteht man unter dem Grad einer Mannigfaltigkeit von k komplexen Dimensionen die Anzahl der Schnittpunkte mit einem allgemeinen linearen $^{n-k}$Raum. Was aber unter der Multiplizität eines Schnittpunktes zu verstehen ist, hat man anzugeben versäumt.

Erst Hensel und Landsberg haben mit funktionentheoretischen Hilfsmitteln für den Fall $r = 1$ eine Defininition der Multiplizität und einen Beweis des Satzes gegeben. Die Multiplizität eines Schnittpunktes wird da definiert als Ordnung des Verschwindens einer gewissen Funktion (genauer: eines gewissen Divisors) in einem (oder mehreren) Punkten der zur gegebenen eindimensionalen Mannigfaltigkeit gehörigen Riemannschen Fläche. Der Beweis läßt sich aber nicht auf beliebige Werte von r übertragen.

Dasselbe gilt von einem Beweisansatz von Lasker, der die Multiplizität durch idealtheoretische Begriffe (Idealmultiplizität oder „Länge") präzisiert[4]).

---

[3]) Siehe z. B.: G. Salmon, Geometry of three dimensions (1874), S. 303 (Übers. von Fiedler, II. Teil, S. 86); H. Schubert, Kalkül der abzählenden Geometrie (1879), S. 47; E. Study, Über die Geometrie der Kegelschnitte (1885), S. 36, und Math. Annalen **40** (1892), S. 555. Der angebliche Beweis bei Halphen (Mémoire sur les courbes algébriques, Journal Ec. Pol. **52** (1882), p. 19) für den dreidimensionalen Fall kann nicht anerkannt werden, weil bei Halphen eine Multiplizitätsdefinition fehlt, also der Satz keinen klaren Sinn hat. Derselbe und noch andere Einwände lassen sich geltend machen auch gegen einen allgemeinen Beweis von M. Noether, Zur Eliminationstheorie, Math. Annalen **11** (1877), S. 571.

[4]) E. Lasker, Zur Theorie der Moduln und Ideale, Math. Annalen **60** (1905), S. 20; vgl. auch B. L. v. d. Waerden, On Hilbert's function etc., Proc. Kon. Ac. Amsterdam Sept. 1927, wo die Laskerschen Gedanken genauer ausgeführt sind.

Von geometrischer Seite her ist nur ein Beweis von Zeuthen zu verzeichnen, ebenfalls nur für den Fall $r = 1$ gültig[5]).

In dieser Arbeit wird das Problem für beliebige Werte von n und r in Angriff genommen. Ausgangspunkt ist die Forderung, daß die Multiplizitäten wie vorhin so definiert werden sollen, daß die Summe der Multiplizitäten der Schnittpunkte im Spezialfall gleich der Anzahl der Schnittpunkte im allgemeinen Fall ist. Was dabei als „allgemeiner Fall" anzusehen ist, ist nicht ohne weiteres klar[6]); es erweist sich als zweckmäßig, ihn folgendermaßen zu definieren. Man bringe die beiden gegebenen Mannigfaltigkeiten, deren Schnittpunkte gesucht werden, in eine allgemeine Lage zueinander, indem man die eine mit unbestimmten Transformationsparametern linear transformiert, und bringe sodann die Mannigfaltigkeiten zum Schnitt. Aus diesem allgemeineren Schnittproblem gewinnt man das ursprüngliche zurück, indem man die Transformationsmatrix U als Einheitsmatrix E spezialisiert.

Es handelt sich nun darum, eine Multiplizitätsdefinition zu finden, die die obige Forderung der „Erhaltung der Anzahl" erfüllt. Da hier die Bézoutsche Resultante ebenso wie die Riemannsche Fläche und der idealtheoretische Längenbegriff versagt (vgl. § 11, Schluß), so muß ein neuer Begriff an ihre Stelle treten. Dieser ist der von mir in der schon zitierten Arbeit W_2 eingeführte Begriff der „relationstreuen Spezialisierung". Es gibt, wie ich dort zeigte, zu dem vollständigen Schnittpunktssystem $X^{(1)}, \ldots, X^{(q)}$ der in allgemeine Lage gebrachten Mannigfaltigkeiten ein und im wesentlichen nur ein Schnittpunktssystem $Y^{(1)}, \ldots, Y^{(q)}$ der ursprünglichen Mannigfaltigkeiten, derart, daß alle zwischen den Koordinaten der Punkte $X^{(\alpha)}$ und den Transformationsparametern bestehenden algebraischen Relationen bei der Spezialisierunng $X^{(\alpha)} \to Y^{(\alpha)}$, $U \to E$ bestehen bleiben. Die *Multiplizität* ist die Zahl, die angibt, wie oft ein Schnittpunkt unter den Punkten $Y^{(\alpha)}$ vorkommt.

Auf Grund dieser Multiplizitätsdefinition, die selbstverständlich dem Prinzip der „Erhaltung der Anzahl" genügt, hat die zu beweisende Behauptung einen präzisen Sinn, und eben wegen dieser „Erhaltung der Anzahl" ist sie äquivalent mit der folgenden:

Sind M und M' algebraische Mannigfaltigkeiten von r und $n - r$-Dimensionen im projektiven ^{n}Raum, und unterwirft man eine der beiden, etwa M, einer linearen Transformation U mit unbestimmten

---

[5]) H. G. Zeuthen, Abzählende Methoden (1914), S. 30. Für beliebige r hat Zeuthen einige Andeutungen für einen Beweis gegeben, aber keinen vollständigen Beweis und keine Multiplizitätsdefinition.

[6]) Vgl. die Ausführung dazu in § 6 meiner Arbeit „Der Multiplizitätsbegriff der algebraischen Geometrie", Math. Annalen 97 (1927), S. 756 (im folgenden mit W_2 zitiert)

Koeffizienten, so ist die Anzahl der Schnittpunkte von M' mit der transformierten Mannigfaltigkeit MU gleich dem Produkt der Gradzahlen von M und M' [7]) *(Hauptsatz).*

Zum Beweis dieses Satzes wird eine Methode angewandt, deren Grundgedanke sich schon in der Literatur der abzählenden Geometrie vorfindet, und die systematisch (wenn auch nicht exakt) vor allem von G. Schaake [8]) angewandt worden ist, nämlich die folgende. Die Mannigfaltigkeit M wird einer ausgearteten linearen Transformation (vermittelt durch eine möglichst allgemeine Matrix vom Rang $n - r + 1$) unterworfen. Sie geht dabei in eine Mannigfaltigkeit MW über, die in so viele lineare Räume zerfällt, wie der Grad von M beträgt. Die Anzahl der Schnittpunkte dieser zerfallenden Mannigfaltigkeit MW mit M' ist gleich dem Produkt der Gradzahlen von M und M'. Dieses Schnittpunktssystem entsteht nun wie vorhin durch relationstreue Spezialisierungen aus dem Schnittpunktssystem von MU und M', und zwar entsteht, wie gezeigt wird, jeder Schnittpunkt genau einmal (m. a. W. alle Multiplizitäten sind eins). Also ist die Anzahl der Schnittpunkte von MU und M' auch gleich dem Produkt der Gradzahlen von M und M'.

Die nähere Ausführung des eben skizzierten Beweises geschieht mit den Hilfsmitteln der Idealtheorie [9]), die zu diesem Zwecke vorher in einigen Punkten ausgebaut werden muß, was in den einleitenden §§ 2—6 geschehen wird. In § 2 wird ein allgemeiner Prozeß behandelt, der aus Idealen wieder Ideale, und zwar isolierte Komponenten der ersteren erzeugt, und der eine Erweiterung der Hentzelt-Noetherschen „Grundidealbildung" darstellt. Dieser Prozeß wird gebraucht für die Untersuchung des Verhaltens eines Polynomideals in einem gegebenen Punkt, und wird in § 3 noch einmal verwendet, wo es sich darum handelt, den Übergang vom projektiven zum Cartesischen Raum (oder von homogenen zu inhomogenen Koordinaten) idealtheoretisch zu erfassen.

---

[7]) Es ist natürlich für die Anzahl der Schnittpunkte gleichgültig, ob man M allein, oder M' allein, oder beide unabhängig voneinander einer allgemeinen linearen Transformation unterwirft, da man immer durch eine simultane Transformation von M und M' die Transformation von M' rückgängig machen kann.

[8]) G. Schaake, Afbeelding van stelsels figuren op de punten eener lineaire ruimte, Diss. Amsterdam 1922, S. 147—188.

[9]) Für die Grundlagen der Idealtheorie in Polynombereichen und ihrer geometrischen Anwendungen, sowie für weitere Literatur darüber verweise ich auf meine zusammenfassende Arbeit „Zur Nullstellentheorie der Polynomideale", Math. Annalen 96 (1926), S. 183. Sie wird im folgenden mit W_1 zitiert. Für die körpertheoretischen Sätze, die zur Verwendung kommen werden, verweise ich auf die grundlegende Steinitzsche Arbeit „Algebraische Theorie der Körper", Journ. f. Math. 137 (1910). S. 167—309. Im folgenden mit „Steinitz" zitiert.

§ 4 ist eine Zusammenfassung der Ergebnisse von Hilbert und Lasker über die „charakteristische Funktion" eines von Formen erzeugten Ideals. Diese charakteristische Funktion eignet sich in einigen einfachen Fällen zur Bestimmung der Schnittpunktsmultiplizität, eignet sich aber nicht zu einer allgemeinen Definition derselben, weil sie nicht der „Erhaltung der Anzahl" genügt (vgl. § 11). Sie gibt aber immer eine obere Grenze für die Multiplizität (§ 10). Mit Hilfe der Sätze von § 5 über das Verhalten der Polynomideale bei Erweiterung des Grundkörpers (z. B. bei Adjunktion von Parametern) werden in § 6 die Schnittpunkte einer irreduziblen Mannigfaltigkeit mit allgemeinen linearen Räumen bestimmt.

§ 7 enthält den ersten Schritt zum Beweis des Hauptsatzes: den Nachweis, daß eine irreduzible Mannigfaltigkeit durch eine passend gewählte „ausgeartete" lineare Transformation in so viele lineare Räume zerfällt wie ihr Grad beträgt. Die definitive Formulierung des Hauptsatzes erfolgt in § 8; der Beweis wird in §§ 9—11 zu Ende geführt. § 9 bringt außerdem den Nachweis, daß die Schnittpunktsmultiplizitäten niemals Null (also immer positiv) sind.

In § 12 werden schließlich für den Fall $r = 1$, also wenn M eine Kurve ist, die Multiplizitäten durch charakteristische Funktionen von gewissen Idealen ausgedrückt, und damit der Anschluß an die Laskersche Multiplizitätsdefinition hergestellt. Der Nachweis, daß die gefundene Formel mit der Henselschen Multiplizitätsdefinition übereinstimmt, soll einer späteren Arbeit vorbehalten bleiben.

§ 2.

Die Komponenten der Ideale in einem Punkt.

Es seien einige allgemeine Betrachtungen vorausgeschickt, die auch später noch zur Anwendung kommen werden.

In einem Ring $\Re$ sei eine nichtleere Menge G gegeben, die zu je zwei Elementen a, b auch das Produkt ab enthält. Ist nun $\mathfrak{m}$ ein Ideal in $\Re$, so verstehe ich unter der *isolierten Komponente*[10]) $\mathfrak{m}_G$ von $\mathfrak{m}$ in bezug

---

[10]) Diese Bezeichnung rechtfertigt sich durch Satz 2. E. Noether versteht nämlich unter einem *isolierten Komponentenideal* von $\mathfrak{m}$ einen Durchschnitt von irgend ν Primäridealen $\mathfrak{q}_{i_1}, \ldots, \mathfrak{q}_{i_\nu}$, die in einer unverkürzbaren Darstellung von $\mathfrak{m}$ durch „größte Primärideale"

$$\mathfrak{m} = [\mathfrak{q}_1, \ldots, \mathfrak{q}_r]$$

auftreten, vorausgesetzt, daß, wenn $\mathfrak{p}_{i_\lambda}$ ein zu einem $\mathfrak{q}_{i_\lambda}$ gehöriges Primideal ist, und wenn irgendein zu einem $\mathfrak{q}_j$ gehöriges Primideal $\mathfrak{p}_j$ ein Vielfaches von $\mathfrak{p}_{i_\lambda}$ ist, dann auch $\mathfrak{q}_j$ unter den Idealen $\mathfrak{q}_{i_1}, \ldots, \mathfrak{q}_{i_\nu}$ vorkommt. Diese Voraussetzungen sind nach Satz 2 für $\mathfrak{m}_G$ alle erfüllt.

(Fortsetzung der Fußnote 10 auf nächster Seite.)

auf G die Menge aller Elemente f von $\Re$, zu denen ein Element g aus G existiert, so daß

$$fg \equiv 0\,(\mathfrak{m}).$$

Offenbar liegt $\mathfrak{m}$ in $\mathfrak{m}_G$. Weiter: wenn $\mathfrak{m}$ ganz in $\mathfrak{n}$ liegt, so liegt $\mathfrak{m}_G$ ganz in $\mathfrak{n}_G$. Außerdem gelten die folgenden Sätze:

Satz 1. $\mathfrak{m}_G$ *ist ein Ideal.*

Beweis. Wenn $fg \equiv 0\,(\mathfrak{m})$, so folgt daraus $rfg \equiv 0\,(\mathfrak{m})$, d. h. wenn f zu $\mathfrak{m}_G$ gehört, so gehört auch rf dazu. Und wenn $f_1 g_1 \equiv 0\,(\mathfrak{m})$ und $f_2 g_2 \equiv 0\,(\mathfrak{m})$, so folgt daraus

$$(f_1 - f_2)\,g_1 g_2 = (f_1 g_1)\,g_2 - (f_2 g_2)\,g_1 \equiv 0\,(\mathfrak{m}),$$

d. h. wenn f_1 und f_2 zu $\mathfrak{m}_G$ gehören, so gehört auch $f_1 - f_2$ dazu.

Satz 2. *Wenn* $\mathfrak{m}$ *als Durchschnitt von Primäridealen*[11]) *dargestellt ist*:

$$\mathfrak{m} = [\mathfrak{q}_1, \ldots, \mathfrak{q}_r],$$

und wenn von den zugehörigen Primäridealen[12]) $\mathfrak{p}_1, \ldots, \mathfrak{p}_r$ *nur* $\mathfrak{p}_1, \ldots, \mathfrak{p}_s$ $(s \leqq r)$ *ein Element von* G *enthalten, so ist*

$$\mathfrak{m}_G = [\mathfrak{q}_{s+1}, \ldots, \mathfrak{q}_r]$$

(bzw. $\mathfrak{m}_G = \Re$ *für den Fall* $s = r$).

Beweis. Erstens: Aus

$$f \equiv 0\,(\mathfrak{m}_G)$$

folgt

$$fg \equiv 0\,(\mathfrak{m}),$$
$$fg \equiv 0\,(\mathfrak{q}_i) \qquad\qquad (i = 1, \ldots, r).$$

Nun ist aber für $i = s + 1, \ldots, r$

$$g \not\equiv 0\,(\mathfrak{p}_i),$$

also folgt[13])

$$f \equiv 0\,(\mathfrak{q}_i) \qquad\qquad (i = s + 1, \ldots, r),$$
$$f \equiv 0\,[\mathfrak{q}_{s+1}, \ldots, \mathfrak{q}_r].$$

In der Hentzelt-Noetherschen Eliminationstheorie (Math. Annalen **88** (1923), S. 53) wird der Prozeß, der uns von $\mathfrak{m}$ auf $\mathfrak{m}_G$ führte, angegeben für den Fall, daß G besteht aus allen von $x_1, \ldots, x_{i-1}$ freien Polynomen. Das Ideal $\mathfrak{m}_G$ heißt dort „Grundideal $(i-1)$-ter Stufe". Auch bei Macaulay kommt es vor (Macaulay, S. 45, Nr. 43).

Es sei noch bemerkt, daß alle Sätze und Beweise ihre Geltung beibehalten, wenn man für G nicht eine Menge von Ringelementen, sondern eine Menge von Idealen nimmt, die zu $\mathfrak{a}$ und $\mathfrak{b}$ auch das Produkt $\mathfrak{a} \cdot \mathfrak{b}$ enthält. $\mathfrak{m}_G$ wird dann die Menge aller Elemente f von $\Re$, zu denen ein $\mathfrak{g}$ aus G existiert, so daß

$$\mathfrak{g}f \equiv 0\,(\mathfrak{m}).$$

[11]) W_1, § 1, **13**.
[12]) W_1, § 1, **13**.
[13]) W_1, § 1, **13**.

Zweitens: Aus

$$f \equiv 0\,[\mathfrak{q}_{s+1}, \ldots, \mathfrak{q}_r]$$

folgt

$$(1) \qquad fg \equiv 0\,[\mathfrak{q}_{s+1}, \ldots, \mathfrak{q}_r]$$

für beliebige g. Nun liegen in $\mathfrak{p}_1, \ldots, \mathfrak{p}_s$ Elemente $g_1, \ldots, g_s$ von G. Wenn g_i in $\mathfrak{p}_i$ liegt, so muß eine Potenz $g_i^{\varrho_i}$ in $\mathfrak{q}_i$ liegen. Setzen wir also

$$g = \prod_1^s g_i^{\varrho_i},$$

so folgt

$$g \equiv 0\,[\mathfrak{q}_1, \ldots, \mathfrak{q}_s],$$
$$(2) \qquad fg \equiv 0\,[\mathfrak{q}_1, \ldots, \mathfrak{q}_s].$$

Aus (1) und (2):

$$fg \equiv 0\,(\mathfrak{m}).$$

Satz 3. *Es ist* $\mathfrak{m}_{GG} = \mathfrak{m}_G$, *oder: aus* $fg \equiv 0\,(\mathfrak{m}_G)$ *folgt* $f \equiv 0\,(\mathfrak{m}_G)$.

Beweis. Aus $fg \equiv 0\,(\mathfrak{m}_G)$ folgt $fgg' \equiv 0\,(\mathfrak{m})$, also $f \equiv 0\,(\mathfrak{m}_G)$.

Satz 4. *Sind* $\mathfrak{m}$ *und* $\mathfrak{n}$ *Ideale in* $\mathfrak{R}$, *so gelten die Relationen* [14])

$$(3) \qquad (\mathfrak{m}, \mathfrak{n})_G = (\mathfrak{m}_G, \mathfrak{n})_G,$$
$$(4) \qquad (\mathfrak{m} \cdot \mathfrak{n})_G = (\mathfrak{m}_G \cdot \mathfrak{n})_G.$$

Beweis von (3). Erstens ist

$$(\mathfrak{m}, \mathfrak{n}) \equiv 0\,(\mathfrak{m}_G, \mathfrak{n}),$$

also

$$(\mathfrak{m}, \mathfrak{n})_G \equiv 0\,(\mathfrak{m}_G, \mathfrak{n})_G.$$

Zweitens: Aus

$$f \equiv 0\,(\mathfrak{m}_G, \mathfrak{n})_G$$

folgt

$$fg \equiv 0\,(\mathfrak{m}_G, \mathfrak{n}),$$
$$fg = a + b, \quad ag' \equiv 0\,(\mathfrak{m}), \quad b \equiv 0\,(\mathfrak{n}),$$
$$fgg' = ag' + bg' \equiv 0\,(\mathfrak{m}, \mathfrak{n}),$$
$$f \equiv 0\,(\mathfrak{m}, \mathfrak{n})_G.$$

Genau ebenso beweist man (4).

Es sei nun $\mathfrak{R}$ insbesondere ein Polynombereich $\mathsf{P}\,[x_1, \ldots, x_n]$, und G die Menge aller Polynome, die in einem festen Punkt $A = \{\alpha_1, \ldots, \alpha_n\}$ des Raumes $C_n(\mathsf{P})$ *nicht* verschwinden. Nach Satz 1 ist jedem Ideal $\mathfrak{m}$ ein Ideal $\mathfrak{m}_G$ zugeordnet, das in diesem Fall mit $\mathfrak{m}_A$ bezeichnet werden soll. Ist $\mathfrak{m} = [\mathfrak{q}_1, \ldots, \mathfrak{q}_r]$, so hat man nach Satz 2, um $\mathfrak{m}_A$ zu finden, von den $\mathfrak{q}_i$ nur diejenigen beizubehalten, deren zugehörige Primideale den

---

[14]) Für die Bedeutung von $(\mathfrak{m}, \mathfrak{n})$ und $\mathfrak{m}\mathfrak{n}$ siehe W_1, § 1, 19.

Punkt A als Nullstelle[15]) haben. Dementsprechend soll $\mathfrak{m}_A$ *die A-Komponente von* $\mathfrak{m}$ heißen.

Aus Satz 4 folgen weiter die Gleichungen

$$(5) \qquad (\mathfrak{m}, \mathfrak{n})_A = (\mathfrak{m}_A, \mathfrak{n})_A,$$

$$(6) \qquad (\mathfrak{m}\,\mathfrak{n})_A = (\mathfrak{m}_A\,\mathfrak{n})_A.$$

Das zum Punkt A gehörige Primideal

$$(x_1 - \alpha_1, \ldots, x_n - \alpha_n)$$

werde immer mit $\mathfrak{a}$ bezeichnet. Ein zu $\mathfrak{a}$ gehöriges Primärideal, oder ein Ideal mit nur einer Nullstelle A, heißt nach Macaulay[16]) *einfach*.

Satz 5. *Ist* $\mathfrak{m}_A$ *einfach, und* $\varrho \geqq$ *dem Exponenten*[17]) *von* $\mathfrak{m}_A$, *so ist*

$$\mathfrak{m}_A = (\mathfrak{m}, \mathfrak{a}^\varrho).$$

Beweis. Erstens:

$$\begin{cases} \mathfrak{m} \equiv 0\,(\mathfrak{m}_A) \\ \mathfrak{a}^\varrho \equiv 0\,(\mathfrak{m}_A), \end{cases}$$

also

$$(7) \qquad (\mathfrak{m}, \mathfrak{a}^\varrho) \equiv 0\,(\mathfrak{m}_A).$$

Zweitens: Ist

$$f \equiv 0\,(\mathfrak{m}_A),$$

so heißt das

$$fg \equiv 0\,(\mathfrak{m}), \qquad g(A) \neq 0.$$

Daraus

$$(8) \qquad fg \equiv 0\,(\mathfrak{m}, \mathfrak{a}^\varrho).$$

Das Ideal $(\mathfrak{m}, \mathfrak{a}^\varrho)$ hat nur die Nullstelle A, also enthält g keine Nullstelle des Ideals. Daraus und aus (8) folgt (vgl. W$_1$, § 5, 8).

$$f \equiv 0\,(\mathfrak{m}, \mathfrak{a}^\varrho),$$

also

$$(9) \qquad \mathfrak{m}_A \equiv 0\,(\mathfrak{m}, \mathfrak{a}^\varrho).$$

Aus (7) und (9) folgt die Behauptung.

Man kann diese Sätze noch verallgemeinern, indem man für G nimmt die Menge aller Polynome, die in endlichvielen Punkten $A, B, \ldots, D$ nicht verschwinden. Um jetzt $\mathfrak{m}_G$ zu erhalten, hat man von den Primärkomponenten von $\mathfrak{m}$ diejenigen beizubehalten, die in A oder $B \ldots$ oder D eine Nullstelle haben. Daraus folgt:

$$\mathfrak{m}_G = [\mathfrak{m}_A, \mathfrak{m}_B, \ldots, \mathfrak{m}_D].$$

---

[15]) W$_1$, § 3, 4.
[16]) Macauly, S. 36.
[17]) W$_1$, § 1, 13.

Genau so wie Satz 5 beweist man jetzt:

Satz 6. *Sind $A, B, \ldots, D$ isolierte Nullstellen eines Ideals* $\mathfrak{m}$, *d. h. sind* $\mathfrak{m}_A, \mathfrak{m}_B, \ldots, \mathfrak{m}_D$ *einfache Komponenten von* $\mathfrak{m}$, *und ist* $\varrho \geq$ *allen Exponenten dieser Komponenten, so ist*

$$[\mathfrak{m}_A, \mathfrak{m}_B, \ldots, \mathfrak{m}_D] = (\mathfrak{m}, \mathfrak{a}^\varrho \mathfrak{b}^\varrho \ldots \mathfrak{d}^\varrho).$$

§ 3.
Homogene Ideale und projektive Räume.

Sei P ein algebraisch-abgeschlossener Körper[18]), $x_0, \ldots, x_n$ Unbestimmte und $\mathfrak{R} = \mathsf{P}[x_0, \ldots, x_n]$.

Unter einem *Punkt des Cartesischen Raumes* $C_n(\mathsf{P})$[19]) wird verstanden ein Wertsystem $\{\xi_1, \ldots, \xi_n\}$ aus P. Unter einem *Punkt* $X = \{\xi_0, \ldots, \xi_n\}$ *des projektiven Raumes* $P_n(\mathsf{P})$[20]) wird verstanden die Gesamtheit der Wertsysteme $\{\lambda\xi_0, \ldots, \lambda\xi_n\}$ aus P, wo $\xi_0, \ldots, \xi_n$ feste Elemente sind, nicht alle $= 0$, und wo λ alle Werte aus P durchläuft. Ein Punkt von $P_n(\mathsf{P})$ ist also nichts anderes als eine Gerade durch den Koordinatenursprung im $C_{n+1}(\mathsf{P})$. Die $n+1$ Elemente $\xi_0, \ldots, \xi_n$ heißen die *Koordinaten* des Punktes. Die Elemente $\lambda\xi_0, \ldots, \lambda\xi_n$, wo $\lambda \neq 0$, können als Koordinaten des nämlichen Punktes verwertet werden.

Ist $\xi_0 \neq 0$, so kann man dem Punkt $X = \{\xi_0, \ldots, \xi_n\}$ des $P_n(\mathsf{P})$ den Punkt $\overline{X} = \left\{\dfrac{\xi_1}{\xi_0}, \ldots, \dfrac{\xi_n}{\xi_0}\right\}$ des $C_n(\mathsf{P})$ zuordnen. Die Punkte des $P_n(\mathsf{P})$ für die $\xi_0 \neq 0$, und die Punkte des $P_n(\mathsf{P})$ entsprechen sich so umkehrbar eindeutig. Die Elemente $\dfrac{\xi_i}{\xi_0}$ heißen die inhomogenen Koordinaten des Punktes X. Sind eine endliche Anzahl Punkte gegeben, so kann man durch eine lineare Koordinatentransformation immer erreichen, daß für alle diese Punkte zugleich $\xi_0 \neq 0$ ist, sodaß man für alle diese Punkte zugleich inhomogene Koordinaten einführen kann.

Ein *homogenes Ideal* oder *H-Ideal*[21]) in $\mathfrak{R} = \mathsf{P}[x_0, \ldots, x_n]$ ist ein solches Ideal, das mit einem Polynom f zugleich alle homogenen Bestandteile von f enthält. Ein homogenes Ideal hat offenbar eine Basis aus homogenen Polynomen oder *Formen*, denn die Polynome einer beliebigen Basis können in homogene Bestandteile aufgelöst werden.

Umgekehrt ist jedes Ideal, dessen Basis aus Formen besteht, homogen.

Sei $\mathfrak{m}$ ein *H*-Ideal in $\mathfrak{R}$. Vorausgesetzt werde, daß $\{0, \ldots, 0\}$ nicht die einzige Nullstelle des Ideals ist. Ist dann $\{\xi_0, \ldots, \xi_n\}$ eine andere

---

[18]) Steinitz S. 260.
[19]) W_1 § 4, 1.
[20]) W_2 § 2, Schluß.
[21]) Macaulay, S. 36.

Nullstelle von $\mathfrak{m}$ im $C_{n+1}(\mathsf{P})$, so sind auch alle Punkte $\{\lambda\,\xi_0, \ldots, \lambda\,\xi_n\}$ Nullstellen. Also kann man die Nullstellen zusammenfassen zu *Nullstellenklassen* $\{\lambda\,\xi, \ldots, \lambda\,\xi_n\}$, die *Punkte von* $P_n(\mathsf{P})$ sind. M. a. W. man kann die Mannigfaltigkeit von $\mathfrak{m}$ in $C_{n+1}(\mathsf{P})$ auch auffassen als Mannigfaltigkeit in $P_n(\mathsf{P})$.

Es soll gezeigt werden, daß zwischen den H-Idealen in $\mathfrak{R} = \mathsf{P}\,[x_0, \ldots, x_n]$ und den Idealen in $\overline{\mathfrak{R}} = \mathsf{P}\,[x_1, \ldots, x_n]$ ein Entsprechen stattfindet, ähnlich dem zwischen den Punkten X des $P_n(\mathsf{P})$ und den Punkten $\overline{X}$ des $C_n(\mathsf{P})$ (siehe oben).

Aus einem Polynom $f(x_0, \ldots, x_n)$ von $\mathfrak{R}$ entsteht, wenn man $x_0 = 1$ setzt, ein Polynom $\bar{f}(x_1, \ldots, x_n)$ von $\overline{\mathfrak{R}}$. Jedes Polynom $\bar{f}$ von $\overline{\mathfrak{R}}$ entsteht dabei mindestens einmal. Summen $f_1 + f_2$ und Produkte $f_1 f_2$ gehen dabei wieder in Summen $\bar{f}_1 + \bar{f}_2$ und Produkte $\bar{f}_1 \bar{f}_2$ über („Meromorphismus"). Daraus folgt, daß aus einem Ideal $\mathfrak{m}$ in $\mathfrak{R}$ wieder ein Ideal $\overline{\mathfrak{m}}$ in $\overline{\mathfrak{R}}$ entsteht.

Wählt man für $\mathfrak{m}$ insbesondere ein H-Ideal, so braucht man für die Berechnung der Polynome $\bar{f}$ von $\overline{\mathfrak{m}}$ nicht alle Polynome f von $\mathfrak{m}$ zu benutzen, sondern man kann sich auf die *Formen* von $\mathfrak{m}$ beschränken. Man kann nämlich jedes Polynom f von $\mathfrak{m}$ in homogene Bestandteile spalten, die ebenfalls zu $\mathfrak{m}$ gehören, und diese Bestandteile mit solchen Potenzen x_0^ϱ multiplizieren, daß sie den gleichen Grad bekommen. Die Summe ist dann eine Form von $\mathfrak{m}$ und ergibt für $x_0 = 1$ dasselbe Resultat wie das ursprüngliche Polynom.

Ist umgekehrt ein Ideal $\overline{\mathfrak{m}}$ in $\overline{\mathfrak{R}}$ gegeben, so erzeugen alle Formen in $\mathfrak{R}$, die durch die Substitution $x_0 = 1$ in Polynome von $\overline{\mathfrak{m}}$ übergehen, ein Ideal $\mathfrak{m}_0$, das (weil es von Formen erzeugt wird) H-Ideal ist, und das bei der Substitution $x_0 = 1$ offenbar wieder das Ideal $\overline{\mathfrak{m}}$ ergibt.

Das so konstruierte H-Ideal soll *das zu* $\mathfrak{m}$ *äquivalente H-Ideal* heißen[22]).

Satz 7. *Das zu* $\overline{\mathfrak{m}}$ *äquivalente H-Ideal* $\mathfrak{m}_0$ *ist das umfassendste H-Ideal, das bei der Substitution* $x_0 = 1$ *wieder* $\overline{\mathfrak{m}}$ *ergibt.*

Das heißt: Jedes H-Ideal $\mathfrak{m}$, das bei der Substitution $x_0 = 1$ wieder $\overline{\mathfrak{m}}$ ergibt, ist Untermenge (Vielfaches) von $\mathfrak{m}_0$.

Beweis. $\mathfrak{m}$ wird erzeugt von Formen, die bei der Substitution $x_0 = 1$ in Polynome von $\overline{\mathfrak{m}}$ übergehen; diese Formen liegen in $\mathfrak{m}_0$, also ist $\mathfrak{m}$ Untermenge von $\mathfrak{m}_0$, *q. e. d.*

Ist $\mathfrak{m}$ ein H-Ideal und $\overline{\mathfrak{m}}$ wie oben definiert, so soll das zu $\overline{\mathfrak{m}}$ äquivalente H-Ideal immer mit $\mathfrak{m}_0$ bezeichnet werden. Wir studieren die Beziehung zwischen $\mathfrak{m}$ und $\mathfrak{m}_0$.

---

[22]) Macaulay, S. 39.

Aus Satz 7 folgt zunächst

$$\mathfrak{m} \equiv 0\,(\mathfrak{m}_0).$$

Weiter gilt

Satz 8. $\mathfrak{m}_0$ *ist die Menge aller Polynome* f, *für die eine Potenz* x_0^ϱ *existiert so daß*

$$x_0^\varrho\, f \equiv 0\,(\mathfrak{m}).$$

Beweis. Erstens: Aus $x_0^\varrho\, f \equiv 0\,(\mathfrak{m})$ folgt, indem man f in homogene Bestandteile f_i spaltet, $x_0^\varrho\, f_i \equiv 0\,(\mathfrak{m})$; daraus weiter, indem man $x_0 = 1$ setzt, $\bar{f}_i \equiv 0\,(\overline{\mathfrak{m}})$, und daraus $f_i \equiv 0\,(\mathfrak{m}_0)$, mithin $f \equiv 0\,(\overline{\mathfrak{m}}_0)$.

Ist umgekehrt f ein Polynom von $\mathfrak{m}_0$, so ist f eine Summe von Formen f_i, die bei der Substitution $x_0 = 1$ in Polynome $\bar{f}_i$ von $\overline{\mathfrak{m}}$ übergehen. Diese $\bar{f}_i$ entstehen ihrerseits durch die Substitution $x_0 = 1$ aus Formen g_i von $\mathfrak{m}$. Wenn aber zwei Formen f_i, g_i bei der Substitution $x_0 = 1$ dasselbe Polynom $\bar{f}_i$ ergeben, so unterscheiden sie sich bloß um Faktoren x_0, d. h. es ist

$$x_0^{\varrho_i}\, f_i = x_0^{\sigma_i}\, g_i$$
$$\equiv 0\,(\mathfrak{m}).$$

Ist nun $\varrho = \max(\varrho_i)$, so folgt

$$x_0^\varrho\, f_i \equiv 0\,(\mathfrak{m}),$$

mithin wegen $f = \varSigma f_i$:

$$x_0^\varrho\, f \equiv 0\,(\mathfrak{m}),$$

womit der Satz bewiesen ist.

Aus diesem Satz folgt, daß man die Beziehung zwischen $\mathfrak{m}$ und $\mathfrak{m}_0$ der in § 2 gegebenen allgemeinen Theorie unterordnen kann, indem man unter G versteht die Menge aller Potenzen von x_0. Insbesondere gilt Satz 2:

Wenn $\mathfrak{m}$ *als Durchschnitt von Primäridealen dargestellt ist:*

$$\mathfrak{m} = [\mathfrak{q}_1, \ldots, \mathfrak{q}_r],$$

und wenn von den zugehörigen Primidealen $\mathfrak{p}_1, \ldots, \mathfrak{p}_r$ *nur* $\mathfrak{p}_1, \ldots, \mathfrak{p}_s\,(s \leq r)$ *eine Potenz von* x_0 *enthalten, so ist*

$$\mathfrak{m}_0 = [\mathfrak{q}_{s+1}, \ldots, \mathfrak{q}_r].$$

Soll ein Primideal eine Potenz von x_0, also x_0 selbst enthalten, so muß in der allgemeinen Nullstelle $\{\xi_0, \ldots, \xi_n\}$ des Primideals $\xi_0 = 0$ sein. Wir heben nun aus den $\mathfrak{q}_i$ besonders diejenigen hervor, die als einzige Nullstelle den Ursprung $\{0, \ldots, 0\}$ haben, d. h. deren zugehöriges Primideal ist $\mathfrak{u} = (x_0, \ldots, x_n)$, falls solche vorhanden sind. In den allgemeinen Nullstellen aller übrigen zugehörigen Primidealen $\mathfrak{p}_i$ ist mindestens eine Koordinate $\xi_i \neq 0$. Also läßt es sich durch eine lineare Koordinatentransformation erreichen, daß in allen diesen allgemeinen Nullstellen $\xi_0 \neq 0$ ist. Wir wollen uns diese Koordinatentransformation immer vorher ausgeführt denken. Unter dieser Voraussetzung vereinfacht sich der eben formulierte Satz zu:

Satz 9. *Aus einer Darstellung von* $\mathfrak{m}$ *als Durchschnitt von Primäridealen erhält man eine ebensolche Darstellung für* $\mathfrak{m}_0$, *indem man die Primärideale ausläßt, die nur die eine Nullstelle* $\{0, \ldots, 0\}$ *haben.*

Es sollen jetzt noch einige formale Relationen zwischen den Idealen $\mathfrak{m}, \overline{\mathfrak{m}}, \mathfrak{m}_0$ abgeleitet werden.

Satz 10. *Ist* $\mathfrak{m} = (f_1, \ldots, f_r)$, *so ist*

$$\overline{\mathfrak{m}} = (\bar{f}_1, \ldots, \bar{f}_r).$$

Beweis: Klar.

Folgen.   1. $\qquad\qquad (\overline{\mathfrak{m}, \mathfrak{n}}) = (\overline{\mathfrak{m}}, \overline{\mathfrak{n}}).$

           2. $\qquad\qquad \overline{\mathfrak{m}\,\mathfrak{n}} = \overline{\mathfrak{m}}.\overline{\mathfrak{n}}.$

Satz 11. *Sind* $\mathfrak{m}, \mathfrak{n}$ *zwei H-Ideale derart, daß nach vorheriger Koordinatentronsformation* (*siehe oben*)

$$\overline{\mathfrak{m}} = \overline{\mathfrak{n}},$$

so bestehen Relationen der Gestalt

$$\begin{cases} \mathfrak{u}^\varrho\, \mathfrak{m} \equiv 0\,(\mathfrak{n}) \\ \mathfrak{u}^\varrho\, \mathfrak{n} \equiv 0\,(\mathfrak{m}) \quad {}^{23)} \end{cases},$$

wo $\mathfrak{u} = (x_0, \ldots, x_n)$ *gesetzt ist.*

Beweis. Sei, wie in Satz 9,

$$\mathfrak{m} = [\mathfrak{q}_1, \ldots, \mathfrak{q}_r]$$

und

$$\mathfrak{m}_0 = [\mathfrak{q}_{s+1}, \ldots, \mathfrak{q}_r].$$

Da $\mathfrak{m}_0$ das zu $\overline{\mathfrak{m}}$ äquivalente und $\mathfrak{n}_0$ das zu $\overline{\mathfrak{n}}$ äquivalente H-Ideal ist, und da $\overline{\mathfrak{m}} = \overline{\mathfrak{n}}$, so muß $\mathfrak{m}_0 = \mathfrak{n}_0$. Ist nun ϱ der größte der Exponenten der (zum Primideal $\mathfrak{u}$ gehörigen) Primärideale $\mathfrak{q}_1, \ldots, \mathfrak{q}_s$, so ist

$$(1) \qquad\qquad \mathfrak{u}^\varrho \equiv 0\,([\mathfrak{q}_1, \ldots, \mathfrak{q}_s]).$$

Auch ist

$$\mathfrak{n} \equiv 0\,(\mathfrak{n}_0)$$

oder

$$(2) \qquad\qquad \mathfrak{n} \equiv 0\,([\mathfrak{q}_{s+1}, \ldots, \mathfrak{q}_r]).$$

---

[23]) Diese Relationen bilden den bequemsten algebraischen Ausdruck für die Tatsache, daß, sobald $\overline{\mathfrak{m}} = \overline{\mathfrak{n}}$, die Ideale $\mathfrak{m}, \mathfrak{n}$ sich bloß unterscheiden können um Primärkomponenten, die zum Primideal $\mathfrak{u}$ gehören, also die einzige Nullstelle $\{0, \ldots, 0\}$ haben.

Aus (1) und (2) folgt

$$\mathfrak{u}^{\varrho}\,\mathfrak{n} \equiv 0\,[\mathfrak{q}_1, \ldots, \mathfrak{q}_{s+1}, \ldots, \mathfrak{q}_r],$$

oder

$$\mathfrak{u}^{\varrho}\,\mathfrak{n} \equiv 0\,(\mathfrak{m}).$$

Genau so beweist man

$$\mathfrak{u}^{\sigma}\,\mathfrak{m}] \equiv 0\,(\mathfrak{n}); \qquad\qquad \text{q. e. d.}$$

Ist $X = \{\lambda\,\xi_0, \ldots, \lambda\,\xi_n\}$ eine Nullstellenklasse von $\mathfrak{m}$ im $P_n(\mathsf{P})$, und ist $\xi_0 \neq 0$, so kann man (durch Multiplikation mit einem passenden Proportionalitätsfaktor) erreichen, daß $\xi_0 = 1$ ist. Dann ist $\overline{X} = \{\xi_1, \ldots, \xi_n\}$ Nullstelle von $\overline{\mathfrak{m}}$.

Ist umgekehrt $\overline{X}$ Nullstelle von $\overline{\mathfrak{m}}$, so ist X Nullstellenklasse von $\mathfrak{m}$. Die Mannigfaltigkeit von $\overline{\mathfrak{m}}$ im $C_n(\mathsf{P})$ wird also aus der von $\mathfrak{m}$ im $P_n(\mathsf{P})$ dadurch konstruiert, daß man zu allen Punkten X der letzteren, soweit sie nicht in der Hyperebene $\xi_0 = 0$ liegen, die entsprechenden Punkte $\overline{X}$ des $C_n(\mathsf{P})$ aufsucht.

Ist insbesondere $\mathfrak{p}$ ein Primideal, und ist $\{0, \ldots, 0\}$ nicht die einzige Nullstelle von $\mathfrak{p}$, so ist nach Satz 9 $\mathfrak{p}_0 = \mathfrak{p}$. Es gilt weiter:

Satz 12. *Ist $\mathfrak{p}$ prim, und $\{0, \ldots, 0\}$ nicht seine einzige Nullstelle, so ist auch $\overline{\mathfrak{p}}$ prim, und jede allgemeine Nullstelle*[24]*) $\{\xi_1, \ldots, \xi_n\}$ von $\overline{\mathfrak{p}}$ ergibt, wenn λ eine neu adjungierte Unbestimmte ist, eine allgemeine Nullstelle $\{\lambda, \lambda\,\xi_1, \ldots, \lambda\,\xi_n\}$ von $\mathfrak{p}$.*

Beweis. 1. Ist $\overline{f}\,\overline{g} \equiv 0\,(\overline{\mathfrak{p}})$ und sind f, g die durch Homogenisieren aus $\overline{f}, \overline{g}$ gebildeten Formen, so folgt $fg \equiv 0\,(\mathfrak{p}_0)$, also entweder $f \equiv 0\,(\mathfrak{p}_0)$ oder $g \equiv 0\,(\mathfrak{p}_0)$, also entweder $\overline{f} \equiv 0\,(\overline{\mathfrak{p}})$ oder $\overline{g} \equiv 0\,(\overline{\mathfrak{p}})$. Also ist $\overline{\mathfrak{p}}$ prim.

2. Daß $\{\lambda, \lambda\,\xi_1, \ldots, \lambda\,\xi_n\}$ Nullstelle von $\mathfrak{p}$ ist, wurde oben schon bemerkt. Ist nun f ein Polynom in $\mathsf{P}[x_0, \ldots, x_n]$, und ist $f(\lambda, \lambda\,\xi_1, \ldots, \lambda\,\xi_n) = 0$, so kann man zunächst f in homogene Bestandteile zerlegen:

$$f = \sum_{1}^{s} f_i\,.$$

Ist γ_i der Grad von f_i, so folgt

$$f(\lambda, \lambda\,\xi_1, \ldots, \lambda\,\xi_n) = \sum \lambda^{\gamma_i} f_i(1, \xi_1, \ldots, \xi_n) = 0,$$

also

$$f_i(1, \xi_1, \ldots, \xi_n) = 0,$$

also

$$f_i(1, x_1, \ldots, x_n) \equiv 0\,(\overline{\mathfrak{p}}),$$

$$f_i(x_0, x_1, \ldots, x_n) \equiv 0\,(\mathfrak{p}_0) \equiv 0\,(\mathfrak{p}),$$

$$f = \sum_{1}^{s} f_i \equiv 0\,(\mathfrak{p}).$$

Mithin ist $\{\lambda, \lambda\,\xi_1, \ldots, \lambda\,\xi_n\}$ allgemeine Nullstelle von $\mathfrak{p}$.

------

[24]) W_1 § 3, 4.

Ist r die Dimensionszahl von $\bar{\mathfrak{p}}$ [25]), so ist offenbar $r+1$ die Dimensionszahl von $\mathfrak{p}$, weil durch die Adjunktion von λ der Transzendenzgrad der allgemeinen Nullstelle um eins vermehrt ist. Man ist gewohnt, bei der Berechnung der Dimensionszahl der Mannigfaltigkeit M von $\mathfrak{p}$ im $P_n(\mathsf{P})$ den Parameter λ als unwesentlich zu betrachten, also unter dieser Dimensionszahl die Zahl r zu verstehen. Die Dimensionszahl einer irreduziblen Mannigfaltigkeit im projektiven Raum unterscheidet sich demnach um eins von der Dimensionszahl des zugehörigen H-Ideals. Die Zahl r, die für H-Ideale im allgemeinen eine größere Rolle spielt als die wirkliche Dimensionszahl $r+1$, wollen wir die *reduzierte Dimension* des H-Ideals $\mathfrak{p}$ nennen. Die reduzierte Dimension eines Primärideals wird definiert als die des zugehörigen Primideals, und die eines beliebigen Ideals als die der höchstdimensionalen Primärkomponente.

Auch für H-Ideale kann man die Komponente in einem Punkt bilden (vgl. § 2). Ist $Y = \{\eta_0, \ldots, \eta_n\}$ ein Punkt im $P_n(\mathsf{P})$, $\mathfrak{m}$ ein H-Ideal, so wird die *Y-Komponente* $\mathfrak{m}_Y$ *von* $\mathfrak{m}$ definiert als Gesamtheit derjenigen Polynome f, zu denen eine Form g existiert, so daß

$$\begin{cases} fg \equiv 0\,(\mathfrak{m}), \\ g(Y) \neq 0. \end{cases}$$

Offenbar ist $\mathfrak{m}_Y$ wieder ein H-Ideal. Der Übergang $\mathfrak{m} \to \mathfrak{m}_Y$ läßt sich ohne weiteres in die allgemeine Theorie des § 2 einordnen, indem man unter G versteht die Menge aller Formen, die im Punkt Y nicht verschwinden. Demnach gelten auch hier die Gleichungen (5), (6) § 2, und kann man $\mathfrak{m}_0$ dadurch erhalten, daß man von den Primärkomponenten von $\mathfrak{m}$ nur diejenigen beibehält, deren zugehörige Primideale in Y eine Nullstelle haben.

Der Übergang zu inhomogenen Koordinaten wird ermöglicht durch

Satz 13. *Ist die Numerierung der Koordinaten so gewählt, daß* $\eta_0 \neq 0$ *ist, so ist*

$$\overline{\mathfrak{m}_Y} = \overline{\mathfrak{m}}_{\overline{Y}}.$$

Beweis. Sei $\bar{f}$ Element von $\overline{\mathfrak{m}_Y}$. Dann entsteht $\bar{f}$ aus einer Form f von $\mathfrak{m}_Y$ durch die Substitution $x_0 = 1$. Also ist

$$fg \equiv 0\,(\mathfrak{m}), \qquad g(Y) \neq 0.$$

Daraus

$$\bar{f}\bar{g} \equiv 0\,(\overline{\mathfrak{m}}), \qquad \bar{g}(\overline{Y}) \neq 0,$$

also gehört $\bar{f}$ zu $\overline{\mathfrak{m}}_{\overline{Y}}$. Gehört umgekehrt $\bar{f}$ zu $\overline{\mathfrak{m}}_{\overline{Y}}$, so folgt

---

[25]) W_1 § 3, 5.

$$\bar{f}\,\bar{g} \equiv 0\,(\overline{\mathfrak{m}}), \qquad \bar{g}\,(\overline{Y}) \doteq 0.$$

$$f\,g\,x_0^\varrho \equiv 0\,(\mathfrak{m}), \qquad g\,(Y) \doteq 0, \qquad \eta_0^\varrho \doteq 0,$$

$$f \equiv 0\,(\mathfrak{m}_Y),$$

$$\bar{f} \equiv 0\,(\overline{\mathfrak{m}_Y}).$$

§ 4.
Die Hilbertsche charakteristische Funktion.

Hilbert und Lasker haben einige Sätze entwickelt, die sich auf die Anzahl der modulo einem H-Ideal linear-unabhängigen Formen gegebener Ordnung beziehen. Sätze und Beweise sollen hier kurz rekapituliert werden, weil man die Sache noch etwas einfacher darstellen und allgemeiner fassen kann, als es in der (gegenüber Hilbert durch Vermeidung der Syzygienketten schon sehr vereinfachten) Laskerschen Darstellung geschehen ist.

Definitionen. Es sei P ein Körper und $\mathsf{P}\,[x_0, \ldots, x_n]$ der aus ihm abgeleitete Polynombereich. Mit $\varphi\,(\varrho)$ bezeichnen wir die Anzahl der Potenzprodukte der $x_0, \ldots, x_n$ vom Grad ϱ. Mit $\varphi\,(\varrho;\,\mathfrak{a})$ bezeichnen wir die Anzahl der linear-unabhängigen Formen vom Grad ϱ in einem H-Ideal $\mathfrak{a}$. Mit $\chi\,(\varrho;\,\mathfrak{a})$ schließlich bezeichnen wir die Anzahl der modulo dem Ideal $\mathfrak{a}$ linear-unabhängigen Formen vom Grad ϱ, oder die Anzahl der unabhängigen linearen Gleichungen, die die Koeffizienten einer Form ϱ-ten Grades zu erfüllen haben, damit die Form im Ideal liegt. Auf die χ-Funktion kommt es im folgenden hauptsächlich an.

Offenbar ist:

$$(1) \qquad\qquad \varphi\,(\varrho) = \varphi\,(\varrho;\,(1)),$$

$$(2) \qquad\qquad \chi\,(\varrho;\,\mathfrak{a}) = \varphi\,(\varrho) - \varphi\,(\varrho,\,\mathfrak{a}).$$

Satz 14. *Sind* $\mathfrak{a}, \mathfrak{b}$ *zwei* H-*Ideale,* $(\mathfrak{a}, \mathfrak{b})$ *ihre Summe,* $[\mathfrak{a}, \mathfrak{b}]$ *ihr Durchschnitt, so ist*

$$(3) \qquad \varphi\,(\varrho;\,(\mathfrak{a},\mathfrak{b})) = \varphi\,(\varrho;\,\mathfrak{a}) + \varphi\,(\varrho;\,\mathfrak{b}) - \varphi\,(\varrho;\,[\mathfrak{a},\mathfrak{b}]),$$

$$(4) \qquad \chi\,(\varrho;\,(\mathfrak{a},\mathfrak{b})) = \chi\,(\varrho;\,\mathfrak{a}) + \chi\,(\varrho;\,\mathfrak{b}) - \chi\,(\varrho;\,[\mathfrak{a},\mathfrak{b}]).$$

Beweis. (3) ist unmittelbar klar. (4) folgt aus (2) und (3).

Satz 15. *Ist* f *eine Form vom Grad* γ, *und relativ prim zu* $\mathfrak{a}$ (*das letztere heißt, daß aus* $fg \equiv 0\,(\mathfrak{a})$ *notwendig* $g \equiv 0\,(\mathfrak{a})$ *folgt), so ist*

$$(5) \qquad \chi\,(\varrho;\,(\mathfrak{a},f)) = \chi\,(\varrho;\,\mathfrak{a}) - \chi\,(\varrho - \gamma;\,\mathfrak{a}).$$

Beweis. Sei $\mathfrak{f}$ das von f erzeugte Ideal. Wir bestimmen zunächst die Funktionen $\varphi\,(\varrho;\,\mathfrak{f})$ und $\varphi\,(\varrho,\,[\mathfrak{a},\mathfrak{f}])$. Alle Formen vom Grad ϱ in $\mathfrak{f}$ haben die Gestalt $f \cdot g$, wo g den Grad $\varrho - \gamma$ hat. Also folgt

$$(6) \qquad\qquad \varphi\,(\varrho;\,\mathfrak{f}) = \varphi\,(\varrho - \gamma).$$

Die Formen vom Grad ϱ in $[\mathfrak{a}, \mathfrak{f}]$ müssen sowohl in $\mathfrak{f}$ liegen, also die Gestalt $f \cdot g$ haben, als auch in $\mathfrak{a}$ liegen:

$$f \cdot g \equiv 0\,(\mathfrak{a}).$$

Aus dieser Kongruenz folgt aber $g \equiv 0\,(\mathfrak{a})$, und umgekehrt. Also ist die Anzahl der linear-unabhängigen Formen $f \cdot g$ vom Grad ϱ in $[\mathfrak{a}, \mathfrak{f}]$ gleich der Anzahl der linear-unabhängigen Formen g vom Grad $\varrho - \gamma$ in $\mathfrak{a}$. Also:

$$(7) \qquad \varphi(\varrho; [\mathfrak{a}, \mathfrak{f}]) = \varphi(\varrho - \gamma; \mathfrak{a}).$$

Nunmehr hat man nach (4):

$$\chi(\varrho; (\mathfrak{a}, \mathfrak{f})) = \chi(\varrho; \mathfrak{a}) + \chi(\varrho; \mathfrak{f}) - \chi(\varrho; [\mathfrak{a}, \mathfrak{f}])$$

nach (2):

$$= \chi(\varrho; \mathfrak{a}) - \varphi(\varrho; \mathfrak{f}) - \varphi(\varrho; [\mathfrak{a}, \mathfrak{f}])$$

nach (6), (7):

$$= \chi(\varrho; \mathfrak{a}) - \varphi(\varrho - \gamma) + \varphi(\varrho - \gamma; \mathfrak{a}),$$

nach (2):

$$= \chi(\varrho; \mathfrak{a}) - \chi(\varrho - \gamma; \mathfrak{a}).$$

Satz 16. *Ist $\{0, \ldots, 0\}$ die einzige Nullstelle des Ideals $\mathfrak{a}$, so ist für hinreichend hohe ϱ:*

$$\chi(\varrho; \mathfrak{a}) = 0.$$

Beweis. Nach dem Hilbertschen Nullstellensatz[26] liegt jedes Potenzprodukt von hinreichend hohem Grad im Ideal.

Satz 17. *Ist a ein H-Ideal und d seine reduzierte Dimension, so wird die Funktion $\chi(\varrho; \mathfrak{a})$ für hinreichend hohe ϱ dargestellt durch ein Polynom in ϱ von der Gestalt*

$$(8) \qquad \chi(\varrho; \mathfrak{a}) = a_0 \binom{\varrho}{d} + a_1 \binom{\varrho}{d-1} + \ldots + a_d \qquad (a_0 \gneqq 0)$$

mit ganzzahligen Koeffizienten a_ν. Man nennt dieses Polynom die *charakteristische Funktion* des Ideals.

Beweis. Ist die reduzierte Dimension $d = -1$, also die gewöhnliche Dimension Null, so hat das Ideal $\mathfrak{a}$ nur die Nullstelle $\{0, \ldots, 0\}$ und nach Satz 16 ist $\chi(\varrho; \mathfrak{a}) = 0$. Wenn man also die Null als Funktion (-1)-ten Grades von ϱ ansieht, so stimmt der Satz für Ideale der reduzierten Dimension -1. Es sei also $d \geq 0$, und der Satz sei für alle reduzierten Dimensionen kleiner als d bewiesen. Das gegebene Ideal $\mathfrak{a}$ läßt sich darstellen als Durchschnitt von Primäridealen:

$$\mathfrak{a} = [\mathfrak{q}_1, \ldots, \mathfrak{q}_r],$$

wo alle $\mathfrak{q}_i$ höchstens die reduzierte Dimension d haben, und wo mindestens ein $\mathfrak{q}_i$ wirklich diese Dimension erreicht. Setzt man den Satz für Primär-

---

[26] W_1 § 5, 9.

ideale, sowie (falls $r > 1$) für Durchschnitte von weniger als r Primäridealen als bewiesen voraus, so folgt aus (4)

$$\chi(\varrho;\mathfrak{a}) = \chi(\varrho;[[\mathfrak{q}_1,\ldots,\mathfrak{q}_{r-1}],\mathfrak{q}_r])$$
$$= \chi(\varrho;[\mathfrak{q}_1,\ldots,\mathfrak{q}_{r-1}]) + \chi(\varrho;\mathfrak{q}_r) - \chi(\varrho;([\mathfrak{q}_1,\ldots,\mathfrak{q}_{r-1}],\mathfrak{q}_r)).$$

Das letzte Ideal $([\mathfrak{q}_1,\ldots,\mathfrak{q}_{r-1}]\,\mathfrak{q}_r)$ hat eine reduzierte Dimension $< d$, also hat seine χ-Funktion die gesuchte Gestalt mit $a_0 = 0$. Die ersten beiden Glieder in der vorigen Gleichung haben diese Gestalt nach Annahme auch (mit $a_0 \geqq 0$); also muß auch die linke Seite diese Gestalt haben. Somit bleibt nur, den Satz für Primärideale zu beweisen. Sei $\mathfrak{a}$ primär und $\mathfrak{p}$ das zugehörige Primideal. Wir wählen eine Linearform l, die nicht in $\mathfrak{p}$ liegt. Dann hat $(\mathfrak{a}, l)$ höchstens die reduzierte Dimension d, und l ist zu $\mathfrak{a}$ relativ prim. Also gilt (5):

$$(9) \qquad \chi(\varrho;(\mathfrak{a},l)) = \chi(\varrho;\mathfrak{a}) - \chi(\varrho-1;\mathfrak{a}).$$

Nach Induktionsvoraussetzung hat die linke Seite für $\varrho \geqq \varrho_0$ die Gestalt

$$a_0\binom{\varrho}{d-1} + a_1\binom{\varrho}{d-2} + \cdots + a_{d-1} \qquad\qquad (a_0 \geqq 0).$$

Aus der Rekursionsformel (9) findet man nun:

$$\chi(\varrho;\mathfrak{a}) - \chi(\varrho_0;\mathfrak{a}) = a_0\left\{\binom{\varrho+1}{d} - \binom{\varrho_0+1}{d}\right\}$$
$$+ a_1\left\{\binom{\varrho+1}{d-1} - \binom{\varrho_0+1}{d-1}\right\} + \cdots + a_{d-1}(\varrho-\varrho_0).$$

Wenn man alle Konstanten zusammensucht und a_d nennt, und beachtet, daß $\binom{\varrho+1}{k} = \binom{\varrho}{k} + \binom{\varrho}{k-1}$ ist, so steht die Behauptung (8) da.

Den (nichtnegativen) Koeffizienten a_0 von $\binom{\varrho}{d}$ kann man den *Grad* des Ideals $\mathfrak{a}$ nennen, weil er auf das engste mit der Anzahl der Schnittpunkte der zugehörigen Mannigfaltigkeit mit einem allgemeinen linearen Raum von $n-d$ Dimensionen zusammenhängt[27]). Aus dem eben gegebenen Beweis folgt noch:

Satz 18. *Der Grad eines H-Ideals ist die Summe der Grade der Primärkomponenten gleicher Dimension.*

Für $d = 0$ besteht die charakteristische Funktion nur aus einem Glied a_0. Das Ideal hat Nullstellen, die von $\{0,\ldots,0\}$ verschieden sind, und es gibt von beliebig hohem Grad Formen, die nicht in allen diesen

---

[27]) Für Primideale ist nämlich Grad des Ideals = Grad der Mannigfaltigkeit. Für Primärideale läßt sich zeigen: der Grad ist gleich l mal dem Grad des zugehörigen Primideals, wo l die „Länge" des Primärideals ist, d. h. die Länge einer maximalen Kette von Primäridealen $\mathfrak{q} = \mathfrak{q}_1, \mathfrak{q}_2, \ldots, \mathfrak{q}_l = \mathfrak{p}$ ($\mathfrak{q}_{i+1}$ echter Teiler von $\mathfrak{q}_i$), die zum selben Primideal gehören. Vgl. meine unter [4]) zitierte Arbeit.

Nullstellen verschwinden, also nicht dem Ideal angehören; demnach ist $a_0 > 0$. Dasselbe gilt nun, wie wir zeigen werden, für jede beliebige Dimension:

Satz 19. *Der Grad eines H-Ideals $\mathfrak{a}$ der reduzierten Dimension $d \geqq 0$ ist eine positive ganze Zahl.*

Es genügt, den Satz für Primärideale zu beweisen; wir können außerdem annehmen, es sei $d > 0$, und der Satz sei für Ideale niedrigerer Dimension schon bewiesen. Ist $\mathfrak{a}$ primär, $\mathfrak{p}$ das zugehörige Primideal, und $f \not\equiv 0\,(\mathfrak{p})$, so folgt aus (5) durch Vergleichung der Koeffizienten von ϱ^{d-1} auf beiden Seiten:

$$(10) \qquad\qquad \mathrm{Grad}\,(\mathfrak{a}, f) = \gamma \cdot \mathrm{Grad}\,\mathfrak{a},$$

falls $(\mathfrak{a}, f)$ genau die reduzierte Dimension $d - 1$ hat. Hat aber $(\mathfrak{a}, f)$ eine niedrigere Dimension, so hat man in (10) links *Null* zu schreiben. Die Behauptung wird nun bewiesen sein, sobald man eine Form f angeben kann, derart, daß $(\mathfrak{a}, f)$ genau die reduzierte Dimension $d - 1$ hat. Geht man zu inhomogenen Idealen über, so ist also zu zeigen, daß es ein Polynom $\bar{f}$ gibt, derart, daß $(\bar{a}, \bar{f})$ die Dimension $d - 1$ hat. Nun sei $\bar{\mathfrak{p}}$ das zu $\bar{a}$ gehörige Primideal, $\bar{\mathfrak{p}}_1$ ein Primteiler der Dimension $d - 1$ von $\bar{\mathfrak{p}}$ [28]), und $\bar{f}$ so gewählt, daß $\bar{f} \equiv 0\,(\bar{\mathfrak{p}}_1)$, aber $\bar{f} \not\equiv 0\,(\bar{\mathfrak{p}})$. Dann genügt $\bar{f}$ unseren Anforderungen und der Satz ist bewiesen. Nachträglich folgt nunmehr, daß in (10) rechts niemals Null stehen kann, wie auch $f \not\equiv 0\,(\mathfrak{p})$ gewählt werde, also daß $(\mathfrak{a}, f)$ *immer genau die reduzierte Dimension $d - 1$ hat.*

Ist $\mathfrak{m}$ ein H-Ideal und definiert man $\mathfrak{m}_0$ wie im § 3, unter Berücksichtigung der für Satz 9 nötigen Koordinatentransformation, so haben $\mathfrak{m}$ und $\mathfrak{m}_0$ dieselben Primärkomponenten bis auf solche Komponenten, die nur eine Nullstelle haben, also dieselbe charakteristische Funktion. Nach § 3 entspricht dem Ideal $\mathfrak{m}$ bei Übergang zu inhomogenen Koordinaten ein Ideal $\bar{\mathfrak{m}}$, derart, daß die Formen vom Grad ϱ in $\mathfrak{m}_0$ eineindeutig den Polynomen vom Grad $\leqq \varrho$ in $\bar{\mathfrak{m}}$ entsprechen, in die sie durch $x_0 = 1$ übergehen. Also ist die Anzahl der linear-unabhängigen Polynome vom Grad $\leqq \varrho$ in $\bar{\mathfrak{m}}$ gleich der Anzahl der linear-unabhängigen Formen vom Grad ϱ in $\mathfrak{m}_0$, und ebenso die Anzahl der modulo $\bar{\mathfrak{m}}$ linear-unabhängigen Polynome vom Grad $\leqq \varrho$ gleich der Anzahl der modulo $\mathfrak{m}_0$ linear-unabhängigen Formen vom Grad ϱ. Daraus und aus der Tatsache, daß $\mathfrak{m}$ und $\mathfrak{m}_0$ dieselbe charakteristische Funktion haben, folgt:

Satz 20. *Ist $\bar{\mathfrak{m}}$ ein d-dimensionales Ideal, so wird die Anzahl der modulo $\bar{\mathfrak{m}}$ linear-unabhängigen Polynome vom Grad $\leqq \varrho$ für große ϱ*

---

[28]) Einen solchen gibt es nach W_1 § 3, 8.

durch die charakteristische Funktion des H-Ideals $\mathfrak{m}$*, also durch ein Polynom d-ten Grades der Gestalt* (8) *dargestellt.*

Ist insbesondere $\overline{\mathfrak{m}}$ nulldimensional ($d = 0$), so ist die charakteristische Funktion von $\mathfrak{m}$ eine Konstante a_0, Summe der ebenfalls konstanten charakteristischen Funktionen der Primärkomponenten gleicher Dimension. Also ist auch die Anzahl der modulo $\mathfrak{m}$ linear-unabhängigen Polynome beliebig hohen Grades eine Konstante und gleich der Summe der entsprechend gebildeten Konstanten der Primärkomponenten. Man kann dieses auch so ausdrücken:

Satz 21. *Der Restklassenbereich* $\overline{\mathfrak{R}}/\overline{\mathfrak{m}}$ *ist, wenn* $\overline{\mathfrak{m}}$ *nulldimensional ist, ein* P*-Modul von endlichem Rang. Dieser Rang ist gleich der Summe der Rangzahlen der Restklassenbereiche nach den Primärkomponenten und gleich der charakteristischen Funktion des H-Ideals* $\mathfrak{m}$*.*

Für nulldimensionale Primideale $\overline{\mathfrak{p}}$ ist der Rang von $\overline{\mathfrak{R}}/\overline{\mathfrak{p}}$ wegen der Isomorphie $\overline{\mathfrak{R}}/\overline{\mathfrak{p}} \cong \mathsf{P}(\xi_1, \ldots, \xi_n)$ (W_1 § 3, 2) gleich dem Grad des Nullstellenkörpers $\mathsf{P}(\xi_1, \ldots, \xi_n)$ in bezug auf P. Ist insbesondere P algebraisch abgeschlossen, so ist dieser Grad 1. Also ist in diesem Fall auch die charakteristische Funktion von $\mathfrak{p}$ gleich 1.

§ 5.

Erweiterung des Grundkörpers.

Es sei Ω ein Erweiterungskörper von P; es sei

$$\mathfrak{R} = \mathsf{P}[x_1, \ldots, x_n]$$

und

$$\mathfrak{R}_\Omega = \Omega[x_1, \ldots, x_n].$$

Ein Ideal $\mathfrak{m}$ in $\mathfrak{R}$ erzeugt ein Ideal $\mathfrak{m}_\Omega$ in $\mathfrak{R}_\Omega$. Wir untersuchen die Beziehungen zwischen $\mathfrak{m}$ und $\mathfrak{m}_\Omega$.

Satz 22. *Der Durchschnitt von* $\mathfrak{m}_\Omega$ *mit* $\mathfrak{R}$ *ist wieder* $\mathfrak{m}$*.*

Beweis[29]). Es sei f ein gemeinsames Element von $\mathfrak{m}_\Omega$ und $\mathfrak{R}$. Zu beweisen ist:

$$f \varepsilon \mathfrak{m} \;\;^{30}).$$

Nach der Definition von $\mathfrak{m}_\Omega$ ist

$$(1) \qquad f = \sum_1^r \alpha_i f_i, \qquad f_i \varepsilon \mathfrak{m}, \qquad \alpha_i \varepsilon \Omega.$$

Drückt man die Größen α_i durch endlichviele linear-unabhängige Elemente $1, \omega_1, \omega_2, \ldots$ von Ω mit Koeffizienten aus P aus, so kommt:

$$f = g_0 + \omega_1 g_1 + \ldots, \qquad \text{wo } g_i \varepsilon \mathfrak{m},$$

---

[29]) Vgl. W_1 § 6, 2.

[30]) ε heißt: ist Element von.

mithin, da die ω_i linear-unabhängig in bezug auf P sind:

$$f = g_0, \qquad 0 = g_1, \ldots$$

also

$$f \,\varepsilon\, \mathfrak{m}.$$

Satz 23. *Die charakteristische Funktion eines H-Ideals ändert sich bei Erweiterung des Grundkörpers nicht.*

Beweis. Wir haben zu zeigen, daß

$$\chi(\varrho; \mathfrak{a}) = \chi(\varrho; \mathfrak{a}_\Omega)$$

oder, was nach (2) § 5 auf dasselbe hinauskommt, daß

$$\varphi(\varrho; \mathfrak{a}) = \varphi(\varrho; \mathfrak{a}_\Omega)$$

Sei nun $f_1, \ldots, f_\varphi$ ein Maximalsystem von linear-unabhängigen Formen vom Grad ϱ in $\mathfrak{a}$, so daß sich alle Formen desselben Grades in $\mathfrak{a}$ linear durch diese ausdrücken. Dann drücken sich auch alle Formen von $\mathfrak{a}_\Omega$ linear durch die f_i aus, und die lineare Unabhängigkeit bleibt bestehen, denn würde eine lineare Abhängigkeit im Körper Ω bestehen, so könnte man damit ebenso verfahren wie oben mit der Gleichung (1), und käme zu einer linearen Abhängigkeit in P. Damit ist $\varphi(\varrho; \mathfrak{a}) = \varphi(\varrho; \mathfrak{a}_\Omega)$ bewiesen.

Satz 24. Voraussetzung. *Ω sei Galoisch in bezug auf P. $\mathfrak{A}$ sei ein Ideal in $\mathfrak{R}_\Omega$, daß von lauter solchen Polynomen erzeugt wird, deren Koeffizienten von erster Art[30a] in bezug auf P sind. $\mathfrak{A}$ sei mit mit allen seinen konjugierten Idealen in bezug auf P identisch.*

Behauptung. *$\mathfrak{A}$ wird von einem Ideal $\mathfrak{a}$ aus $\mathfrak{R}$ erzeugt.*

Beweis. Sei $\mathfrak{a}$ der Durchschnitt von $\mathfrak{R}$ mit $\mathfrak{A}$, und $\mathfrak{a}_\Omega$ wie vorhin das von $\mathfrak{a}$ in Ω erzeugte Ideal. $\mathfrak{a}_\Omega$ ist sicher Untermenge von $\mathfrak{A}$; wenn wir zeigen können, daß alle erzeugenden Elemente von $\mathfrak{A}$ zu $\mathfrak{a}_\Omega$ gehören, so sind wir fertig.

Sei f eine solche Erzeugende. Der Unterkörper Γ von Ω, der durch Adjunktion der Koeffizienten von f und aller dazu konjugierten Größen entsteht, ist Galoisch, von erster Art und von endlichem Grad in bezug auf P. Eine Basis von Γ sei $(\omega_1, \ldots, \omega_m)$; die konjugierten Basen seien $(\omega_1^{(k)}, \ldots, \omega_m^{(k)})$ $(k = 1, \ldots, m)$. Das Polynom f drückt sich durch $\omega_1, \ldots, \omega_m$ linear aus:

$$(2) \qquad f = \textstyle\sum f_i \omega_i, \qquad f_i \,\varepsilon\, \mathfrak{R}.$$

Durch Anwendung desjenigen Automorphismus von Γ, der $(\omega_1, \ldots, \omega_m)$ in $(\omega_1^{(k)}, \ldots, \omega_m^{(k)})$ überführt, gehe f in $f^{(k)}$ über. Es folgt

$$f^{(k)} = \textstyle\sum f_i \omega_i^{(k)}.$$

[30a]) Steinitz § 13.

Aus diesem Gleichungssystem lassen sich wegen $|\omega_i^{(k)}| \neq 0$ die f_i ausrechnen:

$$f_i = \sum \alpha_i^{(k)} f^{(k)}.$$

Da die $f^{(k)}$, als Konjugierte von f, in $\mathfrak{A}$ liegen, liegen auch die f_i in $\mathfrak{A}$, also in $\mathfrak{a}$. Aus (2) folgt jetzt, daß f in $\mathfrak{a}_\Omega$ liegt, q. e. d.

Ist $\mathfrak{p}$ ein nulldimensionales Primideal in $\mathsf{P}\,[x_1, \ldots, x_n]$, so ist die allgemeine Nullstelle $\{\xi_1, \ldots, \xi_n\}$ algebraisch in bezug auf P. Man nennt $\mathfrak{p}$ *von erster Art*, wenn alle ξ_i von erster Art in bezug auf P sind; sonst *von zweiter Art.*

Eine wichtige Eigenschaft der Primideale erster Art ist:

Satz 25. *Es sei $\mathfrak{p}$ ein nulldimensionales Primideal von erster Art in $\mathfrak{R}$. Es sei Ω ein Galoischer Erweiterungskörper von P, der den Nullstellenkörper $\mathsf{P}(\xi_1, \ldots, \xi_n)$ umfaßt. Es seien $\{\xi_1^{(a)}, \ldots, \xi_n^{(a)}\}$ $(a = 1, \ldots, l)$ die zu $\xi_1, \ldots, \xi_n$ konjugierten Wertsysteme in Ω (oder die Nullstellen von $\mathfrak{p}$ in Ω). Dann zerfällt das von $\mathfrak{p}$ erzeugte Ideal $\mathfrak{p}_\Omega$ in $\mathfrak{R}_\Omega$ in lauter verschiedene Primideale:*

$$\mathfrak{p}_\Omega = [\mathfrak{p}^{(1)}, \ldots, \mathfrak{p}^{(l)}] = \mathfrak{p}^{(1)} \cdot \mathfrak{p}^{(2)} \ldots \mathfrak{p}^{(l)},$$

wo

$$\mathfrak{p}^{(a)} = (x_1 - \xi_1^{(a)}, \ldots, x_n - \xi_n^{(a)}).$$

Beweis. Da $\mathfrak{p}^{(1)}, \ldots, \mathfrak{p}^{(l)}$ paarweise teilerfremd sind, ist Durchschnitt = Produkt:

$$[\mathfrak{p}^{(1)}, \ldots, \mathfrak{p}^{(l)}] = \mathfrak{p}^{(1)} \cdot \mathfrak{p}^{(2)} \ldots \mathfrak{p}^{(l)}.$$

Das Produkt $\mathfrak{p}^{(1)} \ldots \mathfrak{p}^{(l)}$ ist mit allen seinen Konjugierten identisch, und wird erzeugt von Polynomen, deren Koeffizienten von erster Art in bezug auf P sind (sie hängen nämlich nur von den $\xi_i^{(a)}$ ab). Also wird es von P-Polynomen erzeugt (Satz 24). Ebenso wird $\mathfrak{p}_\Omega$ von P-Polynomen, nämlich von den Polynomen von $\mathfrak{p}$ erzeugt. Um also den Satz zu beweisen, genügt es, zu zeigen, daß jedes P-Polynom in $[\mathfrak{p}^{(1)}, \ldots, \mathfrak{p}^{(l)}]$ liegt sobald es in $\mathfrak{p}$ liegt, und umgekehrt.

Ein P-Polynom f gehört dann nur zu $\mathfrak{p}$, wenn

$$f(\xi_1, \ldots, \xi_n) = 0,$$

da ja $\{\xi_1, \ldots, \xi_n\}$ eine allgemeine Nullstelle von $\mathfrak{p}$ ist. Ist $f(\xi_1, \ldots, \xi_n) = 0$, so ist allgemeiner für jedes a

$$f(\xi_1^{(a)}, \ldots, \xi_n^{(a)}) = 0,$$

also

$$f \equiv 0 \, (\mathfrak{p}^{(a)})$$

für alle a, also

$$f \equiv ([\mathfrak{p}^{(1)}, \ldots, \mathfrak{p}^{(l)}]),$$

und derselbe Schluß gilt auch in umgekehrter Richtung. Damit ist die Behauptung bewiesen.

Bei Erweiterung des Grundkörpers P ändern sich nicht nur die Ideale, sondern auch ihre Mannigfaltigkeiten.

Sei $\mathfrak{m}$ ein Ideal und M seine Mannigfaltigkeit im $C_n(\mathsf{P})$, d. h. die Gesamtheit einer Nullstelle im Körper P. Sei wieder $\mathfrak{m}_\Omega$ das von $\mathfrak{m}$ in $\mathfrak{R}_\Omega$ erzeugte Ideal. Wir bezeichnen mit M_Ω seine Mannigfaltigkeit. Sie besteht aus den Nullstellen des Ideals $\mathfrak{m}_\Omega$ im Körper Ω, oder, was auf dasselbe hinauskommt, aus den Nullstellen der erzeugenden Polynome des Ideals, d. h. aus den Nullstellen von $\mathfrak{m}$ im Körper Ω.

Ist insbesondere P algebraisch-abgeschlossen, und ist $\mathfrak{n}$ das zu M gehörige Ideal (die Gesamtheit aller P-Polynome, die auf M verschwinden), so ist nach dem Hilbertschen Nullstellensatz

$$\mathfrak{n}^\varrho \equiv 0\,(\mathfrak{m}), \quad \mathfrak{m} \equiv 0\,(\mathfrak{n})$$

also auch

$$(\mathfrak{n}_\Omega)^\varrho \equiv 0\,(\mathfrak{m}_\Omega), \qquad \mathfrak{m}_\Omega \equiv 0\,(\mathfrak{n}_\Omega),$$

mithin stimmen nicht nur die Mannigfaltigkeiten von $\mathfrak{m}$ und $\mathfrak{n}$, sondern auch die von $\mathfrak{m}_\Omega$ und $\mathfrak{n}_\Omega$ überein. Da $\mathfrak{n}$ unabhängig von $\mathfrak{m}$ durch M allein bestimmt wird, so werden auch $\mathfrak{n}_\Omega$ und M_Ω durch M allein bestimmt: M_Ω *hängt nur von M, nicht von $\mathfrak{m}$ ab.*

§ 6.

Schnittpunkte einer irreduziblen Mannigfaltigkeit mit allgemeinen linearen Räumen.

Allen folgenden Untersuchungen werde ein fester algebraisch-abgeschlossener Körper P, für den man sich etwa den Körper der komplexen Zahlen denken kann, zugrunde gelegt.

Dem Körper P sollen abzählbar-unendlich-viele Unbestimmte adjungiert werden; zum so entstehenden rationalen Funktionenkörper sei Ω ein algebraisch-abgeschlossener Erweiterungskörper[31]). Ω hat dann die Eigenschaft, daß es immer, wenn im Laufe der Untersuchung endlichviele Größen aus Ω verwendet worden sind, noch beliebig viele neue, von diesen Größen unabhängige Unbestimmte in Ω gibt. Die Zugrundelegung des ein für allemal konstruierten Körpers Ω erspart uns also die immer erneute Adjunktion von Unbestimmten und alle Konstruktionen von algebraischen Erweiterungskörpern. Wenn im folgenden an irgendeiner Stelle „Unbestimmte aus Ω" eingeführt werden, so sind damit immer gemeint solche Unbestimmte von Ω, die von allen bis dahin verwendeten Größen aus Ω algebraisch-unabhängig sind.

---

[31]) Steinitz, § 21, Satz 9 (S. 287).

Außerdem hat Ω die folgende Eigenschaft: wenn irgendein Erweiterungskörper von P von endlichem Transzendenzgrad[32]) gegeben ist, so enthält Ω einen damit äquivalenten Unterkörper[33]). Insbesondere enthält Ω zu jedem Primideal aus einem Polynombereich P$[x_0, \ldots, x_n]$ einen Nullstellenkörper[34]).

Unter einem *allgemeinen linearen* $^{n-r}$*Raum* im $P_n(\Omega)$ werde verstanden eine solche $(n-r)$-dimensionale Mannigfaltigkeit in $P_n(\Omega)$, die aus den Nullstellen von r Linearformen

$$l_i = \sum u_{ik} x_k$$

mit *unbestimmten* Koeffizienten aus Ω besteht.

Es sei ein für allemal M eine irreduzible r-dimensionale Mannigfaltigkeit in $P_n(\mathsf{P})$, und $\mathfrak{p}$ das zugehörige Primideal in P $[x_0, \ldots, x_n]$. $\mathfrak{p}_\Omega$ und M_Ω seinen definiert wie im § 4. Wir werden in diesem Paragraphen sehen, daß M mit einem allgemeinen linearen $^{n-r-1}$Raum, keine Schnittpunkte, mit einem allgemeinen linearen $^{n-r}$Raum aber endlichviele Schnittpunkte hat. Die Anzahl dieser Schnittpunkte heißt der *Grad* von M.

Wie man diese Schnittpunkte zu finden hat, ist in der Eliminationstheorie schon wiederholt untersucht worden. Man verfährt dabei entweder so (Kronecker, Hentzelt), daß man den Unbestimmten $x_1, \ldots, x_n$ zuvor einer allgemeinen linearen Transformation unterwirft und dann diejenigen Punkte von M sucht, in denen etwa $x_{n-r+1}, \ldots, x_n$ gegebene (allgemeine) Werte haben; oder man sucht direkt durch Resultantenbildung (Mertens) die gemeinsamen Nullstellen von $\mathfrak{p}$ und $(n-r)$ allgemeinen Linearformen.

Die Sätze über die Schnittpunkte von M mit allgemeinen linearen Räumen, die wir hier brauchen werden, kommen daher in der Eliminationstheorie, vor allem in der Hentzeltschen, größtenteils schon vor, jedoch nicht in der idealtheoretischen Fassung, die für unsere Untersuchung nötig ist, und außerdem verquickt mit Eliminanten, Resolventen und Resultantenformen, die wir nicht brauchen.

Satz 26. *Ein allgemeiner linearer* $^{n-r-1}$*Raum in* $P_n(\Omega)$ *hat mit* M_Ω *keine Schnittpunkte.*

Beweis. Der $^{n-r-1}$Raum bestehe aus den Nullstellen der Formen

$$l_i = \sum_0^n u_{ik} x_k \qquad (i = 1, \ldots, r+1).$$

---

[32]) Steinitz, § 23 (S. 299).

[33]) Wenn man nämlich dem gegebenen Erweiterungskörper von endlichem Transzendenzgrad abzählbar unendlich viele Unbestimmte adjungiert und algebraisch abschließt, so entsteht ein zu Ω äquivalenter Erweiterungskörper von P (Steinitz § 22, Satz 5; § 21, Satz 9).

[34]) W$_1$, § 1, 4.

Gesetzt, es wäre $X = \{\xi_0, \ldots, \xi_n\}$ ein Schnittpunkt. Wir numerieren die Koordinaten so, daß $\xi_0 \neq 0$, und können dann annehmen $\xi_0 = 1$. Da X Nullstelle von $\mathfrak{p}$ ist, so ist $\overline{X} = \{\xi_1, \ldots, \xi_n\}$ Nullstelle von $\overline{\mathfrak{p}} = \mathfrak{p}_{(x_0 = 1)}$ (§ 3). Da $\overline{\mathfrak{p}}$ die Dimension r hat, so hat das System $\xi_1, \ldots, \xi_n$ höchstens den Transzendenzgrad r in bezug auf P, also um so mehr in bezug auf $\mathsf{P}(u_{11}, u_{12}, \ldots, u_{r+1,n})$.

Da nun vermöge

$$u_{i0} + \sum_1^n u_{ik}\xi_k = 0$$

die u_{i0} sich durch die ξ_k und die übrigen u_{ik} ausdrücken lassen, so hat auch das System $u_{10}, \ldots, u_{r+1,0}$ höchstens den Transzendenzgrad r in bezug auf $\mathsf{P}(u_{11}, u_{12}, \ldots, u_{r+1,n})$, entgegen der Voraussetzung, daß die u_{i0} Unbestimmte sind.

Hilfssatz 1. Voraussetzung. *In Ω sei $\xi_1, \ldots, \xi_n$ ein System vom Transzendenzgrad r in bezug auf P. Es seien u_{ik} $(i, k = 1, \ldots, n)$ Unbestimmte in Ω, unabhängig von den ξ, und es sei gesetzt*

$$(1) \qquad \xi_i^* = \sum_1^n u_{ik}\xi_k \qquad (i = 1, \ldots, n).$$

Behauptung 1. *$\xi_1^*, \ldots, \xi_r^*$ sind algebraisch-unabhängig in bezug auf $\mathsf{P}(u)$*[35].

2. $\xi_1^*, \ldots, \xi_n^*$ sind algebraisch und von erster Art[36] *in bezug auf $\mathsf{P}(u, \xi_1^*, \ldots, \xi_r^*)$*[37].

Beweis. Zunächst eine Vorbemerkung. Eine beliebige Permutation der ersten Indizes der u_{ik} erzeugt einen Automorphismus des Körpers $\mathsf{P}(u, \xi)$. Dabei werden die ξ_i^* ebenso permutiert wie die u_{ik}. Wenn wir also zeigen können, daß es überhaupt r linear-unbhängige $\xi_{\lambda_1}^*, \ldots, \xi_{\lambda_r}^*$ gibt, so folgt daraus vermöge dieser Automorphismen, daß auch $\xi_1^*, \ldots, \xi_r^*$ algebraisch-unabhängig sind (Behauptung 1); wenn wir weiter zeigen können, daß mindestens ein weiteres $\xi_{\lambda_{r+1}}^*$ algebraisch und von erster Art in bezug auf $\xi_{\lambda_1}^*, \ldots, \xi_{\lambda_r}^*$ ist, so folgt ebenso vermöge dieser Automorphismen, daß $\xi_{r+1}^*, \ldots, \xi_n^*$ alle algebraisch und von erster Art in bezug auf $\xi_1^*, \ldots, \xi_r^*$ sind (Behauptung 2).

Der Körper $\mathsf{P}(u, \xi^*) = \mathsf{P}(u, \xi)$ hat den Transzendenzgrad r in bezug auf $\mathsf{P}(u)$. Also gibt es ein irreduzibles System $\xi_{\lambda_1}^*, \ldots, \xi_{\lambda_r}^*$, von dem jedes weitere $\xi_{\lambda_{r+1}}^*$ algebraisch abhängt.

---

[35] Das Symbol $\mathsf{P}(u)$ soll eine Abkürzung für $\mathsf{P}(u_{11}, u_{12}, \ldots, u_{nn})$ sein.

[36] Steinitz, § 13.

[37] Der Satz ist enthalten im Hilfssatz V von E. Noether, Eliminationstheorie und allgemeine Idealtheorie, Math. Annalen **90** (1923), S. 245. Der dort gegebene Beweis ist aber fehlerhaft. Der hier gegebene Beweis gilt auch für jenen Hilfssatz V.

Für Charakteristik Null[38]) ist jede algebraische Größe zugleich von erster Art, also irgendein $\xi^*_{\lambda_{r+1}}$ algebraisch und von erster Art in bezug auf $P(u, \xi^*_{\lambda_1}, \ldots, \xi^*_{\lambda_r})$, womit nach der Vorbemerkung alles erledigt ist.

Für Charakteristik p bilde man die Gleichung niedrigsten Grades in $\xi^*_{\lambda_{r+1}}$, welche $\xi^*_{\lambda_{r+1}}$ mit $\xi^*_{\lambda_1}, \ldots, \xi^*_{\lambda_r}$ und den u_{ik} verknüpft. Auf Grund der Vorbemerkung ist nur zu beweisen, daß diese Gleichung in bezug auf mindestens eins der $\xi^*_{\lambda_i}$ von erster Art ist.

Wäre sie das nicht, d. h. enthielte sie $\xi^*_{\lambda_1}, \ldots, \xi^*_{\lambda_{r+1}}$ nur in der p-ten Potenz, so könnte man die linke Seite schreiben als Summe von Potenzprodukten

$$\Pi_\mu(u) = u_{11}^{\mu_{11}} u_{12}^{\mu_{12}}, \ldots, u_{nn}^{\mu_{nn}} \qquad (0 \leqq \mu_{ik} < p)$$

mit Koeffizienten, die nur von

$$u_{11}^p, u_{12}^p, \ldots, u_{nn}^p, \ldots, \xi^{*p}_{\lambda_1}, \ldots, \xi^{*p}_{\lambda_{r+1}}$$

abhängen. Die Gleichung würde also lauten:

$$(2) \qquad \sum_\mu \Pi_\mu(u) f_\mu(u_{11}^p, u_{12}^p, \ldots, u_{nn}^p, \xi^{*p}_{\lambda_1}, \ldots, \xi^{*p}_{\lambda_{r+1}}) = 0 .$$

Da die $\Pi_\mu(u)$ in bezug auf den Körper $P(u_{11}^p, u_{12}^p, \ldots, u_{nn}^p, \xi_1, \ldots, \xi_n)$, in dem auch die ξ_i^{*p} liegen, linear-unabhängig sind, so folgt aus (2):

$$f_\mu(u_{11}^p, u_{12}^p, \ldots, u_{nn}^p, \xi^{*p}_{\lambda_1}, \ldots, \xi^{*p}_{\lambda_{r+1}}) = 0$$

für alle μ. Mindestens eins der f_μ muß $\xi^*_{\lambda_{r+1}}$ wirklich enthalten. Nun ist, da P algebraisch-abgeschlossen ist, f_μ eine p-te Potenz eines Polynoms g_μ von niedrigerem Grad; also erhält man die Gleichung

$$g_\mu(u_{11}, u_{12}, \ldots, u_{nn}, \xi^*_{\lambda_1}, \ldots, \xi^*_{\lambda_{r+1}}) = 0 ,$$

die in $\xi^*_{\lambda_{r+1}}$ von niedrigerem Grad ist wie die ursprüngliche, entgegen der Voraussetzung.

Hilfssatz 2. *Es sei* $\mathfrak{p}$ *ein Primideal in* $P[x_1, \ldots, x_n]$, *und* $\{\xi_1, \ldots, \xi_n\}$ *seine allgemeine Nullstelle. Ist nun* $r \leqq n$, *so ist das Ideal*

$$\mathfrak{p}_r = (\mathfrak{p}, x_1 - \xi_1, \ldots, x_r - \xi_r)$$

in $P(\xi_1, \ldots, \xi_r)[x_1, \ldots, x_n]$ *ebenfalls prim, und es hat wiederum die allgemeine Nullstelle* $\{\xi_1, \ldots, \xi_n\}$.

Beweis. Daß $\{\xi_1, \ldots, \xi_n\}$ eine Nullstelle von $(\mathfrak{p}, x_1 - \xi_1, \ldots, x_r - \xi_r)$ ist, ist klar, denn dieser Punkt ist eine Nullstelle sowohl von $\mathfrak{p}$, wie von $x_1 - \xi_1, \ldots, x_r - \xi_r'$. Bleibt also zu zeigen, daß die Nullstelle allgemein ist, d. h. daß, wenn f ein Polynom aus $P(\xi_1, \ldots, \xi_r)[x_1, \ldots, x_n]$ ist, aus

$$f(\xi_1, \ldots, \xi_n) = 0$$

___

[38]) Steinitz, § 4.

folgt

$$f \equiv 0\,(\mathfrak{p}_r).$$

f hängt rational von $\xi_1, \ldots, \xi_r$ ab. Ohne Beschränkung der Allgemeinheit kann man f ganzrational in $\xi_1, \ldots, \xi_r$ annehmen:

$$f = F(\xi_1, \ldots, \xi_r, x_1, \ldots, x_n).$$

Ersetzt man in F überall ξ_i durch x_i, so entsteht das Polynom $F(x_1, \ldots, x_r, x_1, \ldots, x_n)$, das noch immer die Nullstelle $\xi_1, \ldots, \xi_n$ hat, also, da es nur von den x abhängt, in $\mathfrak{p}$ liegt. Weiter ist

$$F(\xi_1, \ldots, \xi_r, x_1, \ldots, x_n) \equiv F(x_1, \ldots, x_r, x_1, \ldots, x_n) \bmod (x_1 - \xi_1, \ldots, x_r - \xi_r),$$

also

$$f \equiv 0\,(\mathfrak{p}, x_1 - \xi_1, \ldots, x_r - \xi_r).$$

Also ist die Nullstelle allgemein, und das Ideal $\mathfrak{p}_r$ prim.

Eine kleine Spezialisierung der Voraussetzungen führt unmittelbar zu:

Hilfssatz 3. *Es sei* $\mathfrak{p}$ *ein r-dimensionales Primideal in* $\mathsf{P}[x_1, \ldots, x_n]$, *und* $\{\xi_1, \ldots, \xi_n\}$ *seine allgemeine Nullstelle. Sind nun* $\xi_{r+1}, \ldots, \xi_n$ *algebraisch und von erster Art in bezug auf* $\mathsf{P}(\xi_1, \ldots, \xi_r)$, *so ist für unbestimmte* $t_1, \ldots, t_r$ *das Ideal*

$$\mathfrak{c} = (\mathfrak{p}_{\mathsf{P}(t)}, x_1 - t_1, \ldots, x_r - t_r)$$

in $\mathsf{P}(t)[x_1, \ldots, x_n]$ *prim, nulldimensional und von erster Art.*

Beweis. Zunächst sind $\xi_1, \ldots, \xi_r$ algebraisch-unabhängig, also gilt die Isomorphie

$$\mathsf{P}(\xi_1, \ldots, \xi_r) \cong \mathsf{P}(t_1, \ldots, t_r).$$

Wir können demnach in der Behauptung alle t_i durch ξ_i ersetzen. Dann geht $\mathfrak{c}$ über in das Ideal $\mathfrak{p}_r$, von dem im Hilfssatz 2 nachgewiesen wurde, daß es prim ist und die allgemeine Nullstelle $\xi_1, \ldots, \xi_n$ hat. Diese allgemeine Nullstelle ist nach Voraussetzung algebraisch und von erster Art in bezug auf den Grundkörper $\mathsf{P}(\xi_1, \ldots, \xi_r)$; also ist $\mathfrak{p}_r$ nulldimensional und von erster Art. Das gleiche gilt vermöge der Isomorphie für $\mathfrak{c}$.

Auf Satz 25 und den Hilfssätzen 1 und 3 fußt das Hauptergebnis dieses Paragraphen:

Satz 27. Voraussetzung. *Die r-dimensionale Mannigfaltigkeit M in $P_n(\mathsf{P})$ gehöre zum homogenen Primideal $\mathfrak{p}$ in $\mathfrak{R} = \mathsf{P}[x_0, \ldots, x_n]$. Es seien u_{ik} $(i, k = 1, \ldots, n)$ und t_i $(i = 1, \ldots, r)$ Unbestimmte in Ω. Es sei L der allgemeine lineare $n-r$ Raum in $P_n(\Omega)$, bestehend aus den Nullstellen der r-Linearformen*

$$l_i = -\,t_i x_0 + \sum_1^n u_{ik} x_k \qquad (i = 1, \ldots, r).$$

Behauptung. 1. M_Ω *hat mit* L *nur endlichviele Schnittpunkte* $Y^{(1)}, \ldots, Y^{(s)}$ $(s > 0)$ *in* $P_n(\Omega)$ [39]). Wie schon bemerkt, nennt man die positive Zahl s den Grad von M.

2. *Die Koordinaten dieser Schnittpunkte sind, bei passender Wahl der Proportionalitätsfaktoren, konjugiert und von erster Art in bezug auf* $P(u, t) = P(u_{11}, \ldots u_{nn}, t_1, \ldots, t_r)$.

3. *Das Ideal* $\bar{\mathfrak{a}} = (\bar{\mathfrak{p}}_\Omega, \bar{l}_1, \ldots, \bar{l}_r)$[40]) *in* $\Omega[x_1, \ldots, x_n]$ *ist Produkt von lauter verschiedenen Primidealen, die je zu einem der Punkte* $\overline{Y^{(1)}}, \ldots, \overline{Y^{(s)}}$ *gehören.*

Beweis. Wir definieren neue Variable x_i^* durch

$$x_i^* = \sum_1^n u_{ik} x_k \qquad (i = 1, \ldots, n).$$

Dann ist

$$P(u)[x_1, \ldots, x_n] = P(u)[x_1^*, \ldots, x_n^*],$$
$$\Omega[x_1, \ldots, x_n] = \Omega[x_1^*, \ldots, x_n^*].$$

Ist nun $\xi_1, \ldots, \xi_n$ eine allgemeine Nullstelle von $\bar{\mathfrak{p}}_{P(u)}$, so ist, wenn man $x_1^*, \ldots, x_n^*$ als Variable desselben Polynombereichs $P(u)[x_1, \ldots, x_n]$ betrachtet, eine allgemeine Nullstelle von $\bar{\mathfrak{p}}_{P(u)}$ gegeben durch $\{\xi_1^*, \ldots, \xi_n^*\}$, wo

$$\xi_i^* = \sum u_{ik} \xi_k.$$

Da nun nach Hilfssatz 1 die Größen $\xi_{r+1}^*, \ldots, \xi_n^*$ algebraisch und von erster Art in bezug auf $P(\xi_1^*, \ldots, \xi_r^*)$ sind, so läßt sich Hilfssatz 3 anwenden, d. h. das Ideal

$$(\bar{\mathfrak{p}}_{P(u, t)}, x_1^* - t_1, \ldots, x_r^* - t_r)$$

in $P(u, t)[x_1^*, \ldots, x_n^*]$ ist prim, nulldimensional und von erster Art. Schreibt man für die x^* ihre Bedeutung, so sieht man, daß dieses Ideal dasselbe ist wie

$$(\bar{\mathfrak{p}}_{P(u, t)}, \bar{l}_1, \ldots, \bar{l}_r).$$

Durch Erweiterung des Körpers $P(u, t)$ zu Ω geht es also über in $\bar{\mathfrak{a}}$. Bei dieser Erweiterung bleibt es möglicherweise nicht mehr prim, aber nach Satz 25 zerfällt es in lauter verschiedene Primideale

$$\bar{\mathfrak{a}} = \prod_1^s \overline{\mathfrak{a}^{(\alpha)}}; \qquad \overline{\mathfrak{a}^{(\alpha)}} = (x_1 - \eta_1^{(\alpha)}, \ldots, x_n - \eta_n^{(\alpha)}),$$

---

[39]) Dieser Satz zeigt, daß der Mertenssche Begriff der Stufenzahl eines homogenen Gleichungssystems, nämlich die Anzahl der allgemeinen linearen Gleichungen, die man hinzufügen muß, um eine endliche Anzahl von Lösungsklassen zu erhalten, mit der Höchstdimension der zugehörigen Manigfaltigkeit übereinstimmt.

[40]) Die Bedeutung der Querstriche ist dieselbe wie in § 3.

wo die Punkte $\overline{Y^{(a)}} = \{\eta_1^{(a)}, \ldots, \eta_n^{(a)}\}$ konjugiert in bezug auf $\mathsf{P}(u, t)$ sind. Sie sind sicher die einzigen Schnittpunkte von M und L, bei denen die erste Koordinate $= 1$ gewählt werden kann, weil jedem solchen Schnittpunkt Y ein Punkt $\overline{Y}$ in $C_n(\Omega)$ zugeordnet werden kann, welcher eine gemeinsame Nullstelle von $\overline{\mathfrak{p}}_\Omega$ und $\overline{l}_1, \ldots, \overline{l}_r$, also unter den Punkten $\overline{Y^{(a)}}$ schon vertreten ist. Es fragt sich also nur, ob es nicht Schnittpunkte gibt, deren erste Koordinate verschwindet. Ein solcher Punkt wäre Nullstelle von $(\mathfrak{p}, x_0)$. Die Mannigfaltigkeit von $(\mathfrak{p}, x_0)$ ist aber, als echter Teil einer irreduziblen r-dimensionalen Mannigfaltigkeit, höchstens $(r - 1)$-dimensional, hat also nach Satz 26 mit L keine Punkte gemein. Also kann ein solcher Schnittpunkt nicht existieren.

Korrolar. Aus der hier abgeleiteten Gleichung

$$\bar{\mathfrak{a}} = \prod_1^s \overline{\mathfrak{a}^{(a)}}$$

kann man nach Satz 11 zwei Relationen herleiten zwischen den H-Idealen, definiert durch

$$(3) \qquad \begin{cases} \mathfrak{u} & = (x_0, \ldots, x_n), \\ \mathfrak{a} & = (\mathfrak{p}_\Omega, l_1, \ldots, l_r), \\ \mathfrak{a}^{(a)} & = (x_1 - \eta_1^{(a)} x_0, \ldots, x_n - \eta_n^{(a)} x_0), \end{cases}$$

nämlich die Relationen:

$$(4) \qquad \mathfrak{u}^\varrho\, \mathfrak{a} \equiv 0 \,\Big(\prod_1^s \mathfrak{a}^{(a)}\Big),$$

$$(5) \qquad \mathfrak{u}^\sigma \prod_1^r \mathfrak{a}^{(a)} \equiv 0\, (\mathfrak{a}).$$

Diese Gleichungen bilden den bequemsten algebraischen Ausdruck für die Tatsache, daß die Ideale $\mathfrak{a} = (\mathfrak{p}_\Omega, l_1, \ldots, l_r)$ und $\prod_1^s \mathfrak{a}^{(a)}$ in allen ihren Primärkomponenten, bis auf die zum Primideal $\mathfrak{u}$ gehörigen, übereinstimmen.

§ 7.

Das Verhalten der Ideale bei linearen Transformationen.

Sei wieder $R_\Omega = \Omega[x_0, \ldots, x_n]$, und sei $V = (v_{ik})$ $(i, k = 0, \ldots, n)$ eine Matrix mit Elementen aus Ω. Unter dem *Transformierten fV eines Polynoms f mit der Matrix V* soll verstanden werden das Polynom fV, das aus f entsteht nach Ersetzung von x_i durch $\sum v_{ik} x_k$. Unter dem *Transformierten $\mathfrak{m}V$ eines Ideals $\mathfrak{m}$ mit der Matrix V* soll verstanden werden das Ideal, das erzeugt wird von den transformierten Polynomen des Ideals $\mathfrak{m}$.

Ist $\mathfrak{m} = (f_1, \ldots, f_r)$, so lassen die Transformierten des Polynoms

$$f = \sum a_i f_i$$

von $\mathfrak{m}$ sich in die Gestalt

$$fV = \sum (a_i V \cdot f_i V)$$

schreiben, also wird das transformierte Ideal erzeugt von $f_1 V, \ldots, f_r V$:

$$\mathfrak{m} V = (f_1 V, \ldots, f_r V),$$

d. h.: *ein Ideal wird transformiert, indem man seine Basis transformiert.*
Daraus folgt direkt

$$(\mathfrak{a}, \mathfrak{b}) V = (\mathfrak{a} V, \mathfrak{b} V).$$
$$\mathfrak{a} \mathfrak{b} \cdot V = \mathfrak{a} V \cdot \mathfrak{b} V.$$

Weiter folgt aus $\mathfrak{a} \equiv 0 \, (\mathfrak{b})$ immer $\mathfrak{a} V \equiv 0 \, (\mathfrak{b} V)$.

Ist $\mathfrak{m}$ ein H-Ideal, also die Basiselemente $f_1, \ldots, f_r$ Formen, so sind auch $f_1 V, \ldots, f_r V$ Formen, also $\mathfrak{m} V$ ein H-Ideal.

Ist M eine Mannigfaltigkeit, $\mathfrak{m}$ das zugehörige Ideal, $\mathfrak{m} V$ das transformierte Ideal, so bezeichnen wir die Mannigfaltigkeit von $\mathfrak{m} V$ mit MV und nennen sie: *die Transformierte von M.*

Ist V nichtsingulär $(|v_{ik}| \neq 0)$, so kann man aus den transformierten Idealen die ursprünglichen zurückgewinnen, indem man mit der inversen Matrix V^{-1} transformiert.

Wir betrachten aber im folgenden insbesondere singuläre Matrizes (geometrisch: „ausgeartete Transformationen"), und zwar Matrizes vom Rang $n - r + 1$ $(0 \leqq r \leqq n)$.

Jede Matrix V vom Rang $n - r + 1$ läßt sich bekanntlich in der Gestalt

$$V = A^{-1} T B$$

darstellen, wo A und B nichtsinguläre Matrizes sind, und wo T eine Diagonalmatrix

$$\begin{cases} t_{ik} = 1 & \text{für} \quad i = k \leqq n - r \\ t_{ik} = 0 & \text{sonst.} \end{cases}$$

ist. Aus diesem Grund verstehen wir unter der *allgemeinen Matrix vom Rang $n - r + 1$* die Matrix

$$W = S^{-1} T U,$$

wo T wie oben definiert ist, und wo S und U Matrizes mit unbestimmten Koeffizienten sind.

Unser Ziel ist, das Verhalten des $(r + 1)$-dimensionalen H-Ideals $\mathfrak{p}_\Omega$ bei Transformation mit der allgemeinen Matrix vom Rang $(n - r + 1)$ zu untersuchen. Wozu diese Untersuchung nötig ist, wird sich aus dem nächsten Paragraphen ergeben.

Die Elemente der Matrizes S und U seien s_{ik}, u_{ik}, die in der inversen Matrizes s'_{ik}, u'_{ik}. Wir setzen

$$l_i = \sum s_{ik} x_k \qquad (i = n-r+1, \ldots, n).$$

Dann ist

$$l_i S^{-1} = \sum \sum s_{ik} s'_{kl} x_l = x_i$$

also,, wenn $\mathfrak{a} = (\mathfrak{p}_\Omega, l_{n-r+1}, \ldots, l_n)$ gesetzt wird,

$$\mathfrak{a} S^{-1} = (\mathfrak{p}_\Omega S^{-1}, x_{n-r+1}, \ldots, x_n)$$

Bei der Substitution $x_i \to \sum t_{ik} x_k$ gehen $x_{n-r+1}, \ldots, x_n$ in Null über. Also folgt

$$\mathfrak{a} S^{-1} T = (\mathfrak{p}_\Omega S^{-1} T, 0, \ldots, 0) = \mathfrak{p}_\Omega S^{-1} T$$

$$\mathfrak{a} S^{-1} T U = \mathfrak{p}_\Omega S^{-1} T U$$

$$(1) \qquad\qquad \mathfrak{a} W = \mathfrak{p}_\Omega W$$

Für das Ideal $\mathfrak{a}$ gelten die Gleichungen (4), (5) von § 6, da es natürlich nicht darauf ankommt ob die zur Definition von $\mathfrak{a}$ verwandten Linearformen mit $l_1, \ldots; l_r$ (vgl. (3) § 6) oder mit $l_{n-r+1}, \ldots, l_n$ bezeichnet werden. Man hat also

$$\begin{cases} \mathfrak{u}^\varrho \mathfrak{a} \equiv 0 \left(\prod_1^s \mathfrak{a}^{(\alpha)} \right), \\[2mm] \mathfrak{u}^\sigma \prod_1^s \mathfrak{a}^{(\alpha)} \equiv 0\,(\mathfrak{a}), \end{cases}$$

wo s der Grad von M ist, und $\mathfrak{u}$, $\mathfrak{a}^{(\alpha)}$ die Bedeutung (3) § 6 haben.

Durch Transformation mit W folgt daraus, wenn man (1) beachtet:

$$(2) \qquad (\mathfrak{u} W)^\varrho \cdot \mathfrak{p}_\Omega W \equiv 0 \left(\prod_1^s (\mathfrak{a}^{(\alpha)} W) \right),$$

$$(3) \qquad (\mathfrak{u} W)^\sigma \cdot \prod_1^s (\mathfrak{a}^{(\alpha)} W) \equiv 0\,(\mathfrak{p}_\Omega W).$$

Die Ideale $\mathfrak{u} W$, $\mathfrak{a}^{(\alpha)} W$ sind leicht explizite hinzuschreiben. Die untransformierten Ideale $\mathfrak{u}$, $\mathfrak{a}^{(\alpha)}$ haben, wie ihre Definitionen (3) § 6 zeigen, die allgemeinen Nullstellen

$$O = \{0, \ldots 0\},$$

bzw.

$$Y^{(\alpha)} = \{\mu, \mu \eta_1^{(\alpha)}, \ldots, \mu \eta_n^{(\alpha)}\}.$$

Die letztere kann man noch etwas symmetrischer schreiben, indem man $\eta_0^{(\alpha)} = 1$ setzt. Die transformierten Ideale $\mathfrak{u} S^{-1}$, $\mathfrak{a}^{(\alpha)} S^{-1}$ haben demnach allgemeine Nullstellen, die man durch Transformation mit S gewinnt:

$$\begin{cases} O S = \{0, \ldots, 0\} \\[1mm] Y^{(\alpha)} S = \{\mu \lambda_0^{(\alpha)}, \ldots, \mu \lambda_k^{(\alpha)}\}, \end{cases}$$

wo

$$\lambda_i^{(a)} = \sum_0^n s_{ik}\, \eta_k^{(a)}$$

gesetzt ist.

Also hat man für $\mathfrak{u}\, S^{-1}$ und $\mathfrak{a}^{(a)}\, S^{-1}$ die Basisdarstellungen

$$\begin{cases} \mathfrak{u}\, S^{-1} = (x_0, \ldots, x_n), \\ \mathfrak{a}^{(a)}\, S^{-1} = (\lambda_0^{(a)}\, x_1 - \lambda_1^{(a)}\, x_0; \ldots; \lambda_0^{(a)}\, x_n - \lambda_n^{(a)}\, x_0). \end{cases}$$

Nun waren die Punkte $Y^{(a)}$ ihrer Bedeutung nach (vgl. die Definition der $\eta_k^{(a)}$ in § 6) Nullstellen von $l_i = \sum s_{ik} x_k$ $(i > n - r)$. Also wird $\lambda_i^{(a)} = 0$ für $i > n - r$. Mithin ist

$$\mathfrak{a}^{(a)}\, S^{-1} = (\lambda_0^{(a)}\, x_1 - \lambda_1^{(a)}\, x_0; \ldots; \lambda_0^{(a)}\, x_{n-r} - \lambda_{n-r}^{(a)}\, x_0; x_{n-r+1}; \ldots; x_n).$$

Transformiert man jetzt weiter mit T und U, so folgt

$$\begin{cases} \mathfrak{u}\, S^{-1}\, T = (x_0, \ldots, x_{n-r}), \\ \mathfrak{a}^{(a)}\, S^{-1}\, T = (\lambda_0^{(a)}\, x_1 - \lambda_1^{(a)}\, x_0; \ldots; \lambda_0^{(a)}\, x_{n-r} - \lambda_{n-r}^{(a)}\, x_0), \end{cases}$$

$$(4) \qquad \mathfrak{u}\, W = \mathfrak{u}\, S^{-1}\, T U = \left(\sum_0^n u_{0k}\, x_k, \ldots, \sum_0^n u_{n-r,\, k}\, x_k \right).$$

$$\mathfrak{a}^{(a)}\, W = \mathfrak{a}^{(a)}\, S^{-1}\, T U$$

$$(5) \qquad = \left(\sum_0^n (\lambda_0^{(a)}\, u_{1k} - \lambda_1^{(a)}\, u_{0k})\, x_k; \ldots; \sum_0^n (\lambda_0^{(a)}\, u_{n-r,k} - \lambda_{n-r}^{(a)}\, u_{0k})\, x_k \right).$$

Aus (4) sieht man, daß $\mathfrak{u}\, W$ gehört zu einem allgemeinen $^{r-1}$Raum L in $P_n(\Omega)$, und aus (5) folgt, daß die Ideale $\mathfrak{a}^{(a)}\, W$ gehören zu linearen Räumen $L^{(a)}$ $(a = 1, \ldots, s)$, welche L enthalten. Die $L^{(a)}$ sind alle verschieden, denn da die Punkte $Y^{(a)} = \{\eta_0^{(a)}, \ldots, \eta_n^{(a)}\}$ verschieden sind, so sind die transformierten Punkte $Y^{(a)} S = \{\lambda_0^{(a)}, \ldots, \lambda_{n-r}^{(a)}, 0, \ldots, 0\}$ es auch, und sie sind die Schnittpunkte der Räume $L^{(a)}\, U^{-1}$ mit dem Raum $\xi_{n-r+1} = \ldots = \xi_n = 0$.

Die Mannigfaltigkeit von $\overset{\varrho}{\underset{1}{\prod}}(\mathfrak{a}^{(a)}\, W)$ ist die Vereinigung der linearen Räume $L^{(1)}, \ldots, L^{(s)}$. Sie enthält die Mannigfaltigkeit von $\mathfrak{u}\, W$. Also folgt aus (3), daß sie auch die Mannigfaltigkeit von $\mathfrak{p}_\Omega\, W$ enthält.

Umgekehrt enthält die Mannigfaltigkeit von $\mathfrak{p}_\Omega\, W$ die Räume $L^{(1)}, \ldots, L^{(s)}$, denn aus (2) folgt

$$(\mathfrak{u}\, W)^\varrho \cdot (\mathfrak{p}_\Omega\, W) \equiv 0\, (\mathfrak{a}^{(a)}\, W)$$

also, da $\mathfrak{a}^{(a)}\, W$ prim ist und $\mathfrak{u}\, W \not\equiv 0\, (\mathfrak{a}^{(a)}\, W)$:

$$(6) \qquad \mathfrak{p}_\Omega\, W \equiv 0\, (\mathfrak{a}^{(a)}\, W).$$

Damit ist bewiesen:

Satz 28. *Die Mannigfaltigkeit von $\mathfrak{p}_\Omega\, W$ ist die Vereinigung der linearen Räume $L^{(1)}, \ldots, L^{(s)}$.*

Dieser Satz enthält die für das folgende grundlegende, in der Einleitung erwähnte Tatsache, daß die singuläre Transformation W die Mannigfaltigkeit M_Ω in eine in lineare Räume zerfallende überführt.

Eine andere Folge der Gleichungen (2), (3) ist:

Satz 29. *Ist $Y = \{\eta_0, \ldots, \eta_n\}$ ein Punkt von $L^{(\beta)}$, der nicht in L liegt, so ist*

$$(\mathfrak{p}_\Omega W)_Y = \mathfrak{a}^{(\beta)} W \ {}^{41}).$$

Beweis. Aus (6) folgt:

$$(7) \qquad (\mathfrak{p}_\Omega W)_Y \equiv 0 \, (\mathfrak{a}^{(\beta)} W)_Y$$

und aus (3):

$$(8) \qquad \left((\mathfrak{u} \, W)^\varrho \prod_1^s (\mathfrak{a}^{(\alpha)} W) \right)_Y \equiv 0 \, (\mathfrak{p}_\Omega W)_Y .$$

Da Y nicht Nullstelle von $\mathfrak{u} \, W$ ist, und ebensowenig von $\mathfrak{a}^{(\alpha)} W \, (\alpha \neq \beta)$, so hat man

$$(\mathfrak{u} \, W)_Y = (1),$$
$$(\mathfrak{a}^{(\alpha)} W)_Y = (1) \qquad\qquad (\alpha \neq \beta).$$

Durch wiederholte Anwendung von (6) (§ 2) auf die linke Seite von (8) folgt jetzt

$$(9) \qquad (\mathfrak{a}^{(\beta)} W)_Y \equiv 0 \, (\mathfrak{p}_\Omega W)_Y$$

Aus (7) und (9):

$$(\mathfrak{p}_\Omega W)_Y = (\mathfrak{a}^{(\beta)} W)_Y = \alpha^\beta W,$$

weil Y Nullstelle des Primideals $\mathfrak{a}^{(\beta)} W$ ist.

§ 8
Der verallgemeinerte Bézoutsche Satz. Multiplizitäten.

Nach den Voruntersuchungen der vorangehenden Paragraphen sind wir imstande, den verallgemeinerten Bézoutschen Satz scharf zu formulieren und seinen Beweis in Angriff zu nehmen.

Hauptsatz. *Es sei P ein algebraisch-abgeschlossener Körper. Es seien M und M' irreduzible algebraische Mannigfaltigkeiten der Dimensionen r und $n-r$ im projektiven Raum $P_n(\mathsf{P})$. Es sei U eine Matrix mit unbestimmten (d. h. in bezug auf P algebraisch-unabhängigen) Koeffizienten u_{ik} im algebraisch-abgeschlossenen Erweiterungskörper Ω. Dann ist die Anzahl der Schnittpunkte von $M_\Omega U$ mit M'_Ω gleich dem Produkt der Gradzahlen von M und M'.*

Die Beschränkung auf irreduzible Mannigfaltigkeiten ist natürlich unwesentlich, da eine reduzible rein r-dimensionale bzw. rein $(n-r)$-dimen-

---

^{41}) Für die Bedeutung des Symbols $\mathfrak{m}_Y$ siehe § 3, Schluß.

sionale Mannigfaltigkeit sich aus ebensolchen irreduziblen aufbaut. Die irreduziblen Bestandteile von M haben miteinander nur Mannigfaltigkeiten von niedrigeren Dimensionszahlen gemein (W_1 § 4, **15**), welche, wie man leicht zeigt, nach Transformation mittels U keine Schnittpunkte mit M' haben (vgl. den analogen Satz 26). Also ist die Anzahl der Schnittpunkte der reduziblen Mannigfaltigkeit $M_\Omega U$ mit M_Ω'' gleich der Summe der Anzahlen der Schnittpunkte der irreduziblen Bestandteile von $M_\Omega U$ mit M_Ω'. Ebenso zerlegt man diese Anzahlen weiter nach den irreduziblen Bestandteilen von M'. Daraus folgt, daß der Satz für reduzible Mannigfaltigkeiten gilt sobald er für irreduzible bewiesen ist.

Sind $\mathfrak{p} = (f_1, \ldots, f_r)$ und $\mathfrak{p}' = (f_1', \ldots, f_l')$ die zu M und M' gehörigen Ideale, so sind die Schnittpunkte $X = \{\xi_0, \ldots, \xi_n\}$ von $M_\Omega U$ und M_Ω' die Lösungsklassen des homogenen Gleichungssystems:

$$(1) \quad \begin{cases} f_1\left(\sum u_{ik}\xi_k\right) = 0 \\ \cdots \cdots \cdots \\ f_r\left(\sum u_{ik}\xi_k\right) = 0 \\ f_1'(\xi) = 0 \\ \cdots \cdots \cdots \\ f_l'(\xi) = 0 \end{cases}$$

im Körper Ω. Ein geometrisches Problem, dessen Lösung mit der Lösung eines homogenen Gleichungssystems äquivalent ist, heißt *Normalproblem*.

Ist V irgendeine spezielle Matrix in Ω, so werden die Schnittpunkte von $M_\Omega V$ mit M_Ω' gegeben durch die Lösungsklassen des spezialisierten Gleichungssystems

$$(2) \quad \begin{cases} f_1\left(\sum v_{ik}\xi_k\right) = 0 \\ \cdots \cdots \cdots \\ f_r\left(\sum v_{ik}\xi_k\right) = 0 \\ f_1'(\xi) = 0 \\ \cdots \cdots \cdots \\ f_l'(\xi) = 0 \end{cases}$$

Insbesondere gilt das, wenn V die Einheitsmatrix ist: $V = E$; das Gleichungssystem (2) gilt dann für die Schnittpunkte von M und M'.

Wenn ein Normalproblem in dieser Weise durch eine Spezialisierung von Unbestimmten u_{ik} aus einem allgemeineren Normalproblem abgeleitet ist, so gelten die folgenden Sätze [42]):

I. *Sind $X^{(1)}, \ldots, X^{(q)}$ q verschiedene Lösungsklassen des allgemeinen Normalproblems* (1), *so gibt es q (nicht notwendig verschiedene) Lösungsklassen $Y^{(1)}, \ldots, Y^{(q)}$ des spezialisierten Normalproblems* (2) *derart, daß*

---

[42]) W_2 §§ 3—5.

alle in den Koordinaten der $X^{(i)}$ homogenen Relationen

$$f(u, X^{(1)}, \ldots, X^{(q)}) = 0$$

bei der Spezialisierung $u \to v$, $X \to Y$ bestehen bleiben.

Ein System $Y^{(1)}, \ldots, Y^{(q)}$ mit dieser Eigenschaft heißt eine *relationstreue Spezialisierung* des Systems $X^{(1)}, \ldots, X^{(q)}$.

II. *Hat das spezialisierte Normalproblem nur endlichviele Lösungsklassen, so hat das allgemeine auch nur endlichviele verschiedene Lösungsklassen, etwa $X^{(1)}, \ldots, X^{(q)}$. Alle relationstreuen Spezialisierungen $Y^{(1)}, \ldots, Y^{(q)}$ von $X^{(1)}, \ldots, X^{(q)}$ sind in diesem Fall bis auf die Reihenfolge miteinander identisch.*

Die Zahl, die angibt wie oft eine gegebene Lösung Y in der in Satz II definierten relationstreuen Spezialisierung vorkommt, heißt die *Multiplizität* der Lösung Y des spezialisierten Normalproblems. Sie kann positiv oder Null sein.

III. *Die Summe der Multiplizitäten aller Lösungen des spezialisierten Normalproblems ist gleich der Anzahl der Lösungen des allgemeinen Normalproblems* (Prinzip der Erhaltung der Anzahl).

Wir wollen diese Sätze auf die durch (1) und (2) gegebenen Normalprobleme anwenden und wählen dabei insbesondere für V die in § 8 definierte Matrix W, die allgemeine Matrix vom Rang $n - r + 1$. Die Mannigfaltigkeit $M_\Omega W$ zerfällt dann in lineare Räume $L^{(1)}, \ldots, L^{(s)}$, wo s der Grad von M ist. Die Schnittpunkte von $M_\Omega W$ und M'_Ω verteilen sich auf $L^{(1)}, \ldots, L^{(s)}$. Da die linearen Räume $L^{(1)}, \ldots, L^{(s)}$ allgemein sind, d. h. je durch r Linearformen mit algebraisch-unabhängigen Koeffizienten gegeben sind, so haben sie mit M'_Ω je so viele Schnittpunkte wie der Grad von M' beträgt. Ein solcher Schnittpunkt kann nur einem der Räume $L^{(\alpha)}$ zugleich angehören, denn der Raum L, in dem sich je zwei $L^{(\alpha)}, L^{(\beta)}$ schneiden, ist ein allgemeiner linearer $^{r-1}$Raum, hat also mit M'_Ω keine Punkte gemein (Satz 26). Also hat die zerfallende Mannigfaltigkeit $M_\Omega W$ mit M'_Ω so viele Schnittpunkte, wie das Produkt der Gradzahlen von M und M' beträgt. Wenn man noch zeigen kann, daß alle diese Schnittpunkte die Multiplizität *eins* haben, so wird nach dem Prinzip der Erhaltung der Anzahl die Anzahl der Schnittpunkte von $M_\Omega U$ und M'_Ω ebenfalls gleich dem Produkt der Gradzahlen von M und M' sein, also der Hauptsatz bewiesen sein.

Dem Beweis dieser Tatsache sind die beiden folgenden Paragraphen gewidmet: in § 9 wird gezeigt werden, daß die Multiplizitäten der Schnittpunkte von $M_\Omega W$ und $M'_\Omega > 0$ sind, in § 10, daß sie ≤ 1 sind. Es wird dabei zugleich noch mehr bewiesen werden: eine Zusammenfassung der erhaltenen Resultate folgt in § 11.

§ 9.
Die Multiplizitäten sind nicht Null.

Hilfssatz 4. *Sind* $f_1, \ldots, f_r$ *Polynome in* $\Sigma[y_1, \ldots, y_s]$, *und algebraisch-abhängig in bezug auf* $\mathsf{P}(y_1, \ldots, y_s)$, *wo* P *ein Unterkörper von* Σ *ist, so bleiben sie algebraisch-abhängig, wenn man* $y_s = 0$ *setzt.*

Beweis. Es sei

$$(1) \qquad F(y_1, \ldots, y_s, f_1, \ldots, f_r) = 0$$

die Relation mit Koeffizienten aus P, die nach Voraussetzung zwischen $f_1, \ldots, f_r$ besteht, wo F nicht identisch verschwindet:

$$(2) \qquad F(y_1, \ldots, y_s, z_1, \ldots, z_r) \neq 0 \quad \text{für unbestimmte } z_i.$$

Man kann annehmen, daß $F(y_1, \ldots, y_s, z_1, \ldots, z_r)$ nicht durch y_s teilbar ist, da man sonst den Faktor y_s weglassen könnte, ohne daß die Gültigkeit von (1) oder (2) gestört sein würde. In dieser Annahme ist

$$F(y_1, \ldots, y_{s-1}, 0, z_1, \ldots, z_r) \neq 0.$$

Bei der Substitution $y_s = 0$ mögen $f_1, \ldots, f_r$ in $f_1^0, \ldots, f_r^0$ übergehen. Aus (1) folgt dann durch die Substitution $y_s = 0$:

$$F(y_1, \ldots, y_{s-1}, 0, f_1^0, \ldots, f_r^0) = 0, \qquad \text{q. e. d.}$$

Bemerkung. Genau dasselbe gilt natürlich, wenn man, statt $y_s = 0$, $y_s = \alpha$ setzt, wo α irgendein Element von P ist, denn das heißt, daß man $y_s - \alpha$ als neue Unbestimmte auffaßt und durch Null ersetzt.

Wendet man den Hilfssatz mehreremal an, so folgt, daß man auch mehrere oder alle Unbestimmte y_i durch Konstante ersetzen kann, ohne daß die algebraische Abhängigkeit der f_i verloren geht. Diese Tatsache ist oft nützlich, um zu beweisen, daß gewisse Größen algebraisch-unabhängig sind, indem man durch eine passend gewählte Spezialisierung von vorkommenden Parametern die Annahme der algebraischen Abhängigkeit ad absurdum führt. In diesem Sinne wird der eben bewiesene Hilfssatz benutzt werden beim Beweis des folgenden Satzes.

Satz 30. Voraussetzung. *Es seien* M *und* M' *irreduzible algebraische Mannigfaltigkeiten der Dimensionen* r *und* $n - r$ *in* $P_n(\mathsf{P})$. *Es sei* U *eine Matrix mit unbestimmten Koeffizienten* u_{ik} $(i, k = 0, \ldots, n)$ *in* Ω.

Behauptung. 1. M_Ω *hat mit der transformierten* $M_\Omega U$ *mindestens einen Schnittpunkt.*

2. *Ist* V *irgendeine Matrix mit Elementen* v_{ik} *aus* Ω, *so daß* $M_\Omega V$ *und* M'_Ω *nur endlichviele Schnittpunkte haben, so ist die Multiplizität eines solchen Schnittpunkts* Y *niemals Null.*

Beweis. Der Grundgedanke des Beweises von 1. ist folgender: Wir kehren zunächst das Problem um, indem wir einen Punkt G auf M'_Ω (möglichst allgemein, also als allgemeine Nullstelle von $\mathfrak{p}'$) als gegeben annehmen, und eine möglichst allgemeine lineare Transformation Q so bestimmen, daß die transformierte Mannigfaltigkeit $M_\Omega Q$ auch den Punkt G enthält. Wenn es sich nun herausstellt, daß die so bestimmten Transformationskoeffizienten algebraisch-unabhängig sind, so folgt daraus, daß sie durch einen Körperisomorphismus in die Unbestimmte u_{ik} übergeführt werden können; der Punkt G geht dabei in einen Schnittpunkt X der Mannigfaltigkeiten $M_\Omega U$, M'_Ω über, womit die Existenz eines solchen dargetan ist.

Dieselbe Problemumkehrung führt nun aber auch zum Beweis der zweiten Behauptung. Es zeigt sich nämlich, daß man die allgemeine Nullstelle G und die weiteren beliebig angenommenen Größen relationstreu so spezialisieren kann, daß der Punkt G in jeden gegebenen Schnittpunkt Y von M'_Ω und $M_\Omega V$, und die Matrix Q in die gegebene Matrix V übergeht. Daraus folgt dann ohne weiteres, daß der gegebene Schnittpunkt Y auch eine relationstreue Spezialisierung von X ist, und daraus nach einem in W_2 bewiesenen Satz[43]), daß Y nicht die Multiplizität Null haben kann[44]).

Es seien also $\mathfrak{p}$ und $\mathfrak{p}'$ die zu M und M' gehörigen Primideale; es seien die Koordinaten so gewählt, daß in ihren allgemeinen Nullstellen die Koordinate ξ_0 nicht verschwindet. Dann kann man nach Satz 17 eine allgemeine Nullstelle von $\mathfrak{p}$ in der Gestalt

$$F = \{\lambda,\, \lambda\,\varphi_1,\, \ldots,\, \lambda\,\varphi_n\}$$

annehmen. Der Nullstellenkörper $\Phi = \mathsf{P}(\lambda, \varphi_1, \ldots, \varphi_n)$ kann als Unterkörper von Ω gedacht werden. Der Transzendenzgrad des Systems $\{\lambda, \varphi_1, \ldots, \varphi_n\}$ ist $r+1$; es seien etwa $\lambda, \varphi_1, \ldots, \varphi_r$ algebraisch unabhängig in bezug auf P. Mit Rücksicht auf den zweiten Teil des Beweises ist es zweckmäßig, $\lambda, \varphi_1, \ldots, \varphi_r$ in Ω so zu wählen, daß sie auch noch algebraisch unabhängig in bezug auf die Größen v_{ik} und η_i (Koordinaten von Y) sind. Die Nullstelle F bleibt dann eine allgemeine auch nach Adjunktion dieser Größen.

Eine allgemeine Nullstelle von $\mathfrak{p}'_\Phi$ kann ebenso in der Gestalt

$$G = \{\mu,\, \mu\,\psi_1,\, \ldots,\, \mu\,\psi_n\}$$

(wiederum in Ω) angenommen werden. Das Elementsystem $\psi_1, \ldots, \psi_n$ hat in bezug auf Φ den Transzendenzgrad $n-r$. Es seien etwa $\psi_{i_1}, \ldots, \psi_{i_{n-r}}$ algebraisch-unabhängig in bezug auf Φ.

---

[43]) W_2 Satz 5.

[44]) Ein ähnliches Beweisverfahren habe ich schon früher (W_2 § 8) benutzt, um zu zeigen, daß die Multiplizität einer Geraden einer kubischen Fläche (nicht Regelfläche) niemals Null ist.

Nun seien y_{ik} $(i = 0, \ldots, n;\ k = 1, \ldots, n)$ Unbestimmte in Ω, und es sei

$$q_{ik} = \mu\, y_{ik} \qquad (i = 0, \ldots, n;\ k = 1, \ldots, n),$$

$$q_{00} = \lambda - \sum_1^n y_{0k}\, \mu\, \psi_k,$$

$$q_{i0} = \lambda\, \varphi_i - \sum_1^n y_{ik}\, \mu\, \psi_k.$$

Dann wird die Transformation Q:

$$x_i = \sum_0^n q_{ik}\, x_k^*$$

den Punkt F von $P_n(\Omega)$ in $F^* = \{1, \psi_1, \ldots, \psi_n\} = G$ transformieren. Die Mannigfaltigkeiten $M_\Omega Q$ und M_Ω' haben also den Punkt G gemein. Wir werden zeigen, daß die Transformation Q eine allgemeine ist, d. h. daß die q_{ik} algebraisch-unabhängig in bezug auf P sind.

Wären die q_{ik} algebraisch-abhängig, so würde eine Relation

$$H(q) = 0$$

mit Koeffizienten aus P bestehen. In dieser Relation könnten $q_{00}, q_{10}, \ldots, q_{n0}$ sicher nicht fehlen, da die übrigen q_{ik} offensichtlich algebraisch-unabhängig sind. Also sind $q_{00}, q_{10}, \ldots, q_{n0}$ algebraisch-abhängig in bezug auf $P(\mu, y_{ik})$. Diese Abhängigkeit muß nach Hilfssatz 6 bestehen bleiben, wenn man spezialisiert:

$$\begin{cases} \mu = -1, \\ y_{r+1, i_1} = y_{r+2, i_2} = \ldots = y_{n, i_{n-r}} = 1, \\ \text{übrige}\quad y_{ik} = 0. \end{cases}$$

Dabei gehen $q_{00}, q_{10}, \ldots, q_{n0}$ über in

$$\lambda;\quad \lambda\, \varphi_1;\quad \ldots;\quad \lambda\, \varphi_r;\quad \lambda\, \varphi_{r+1} + \psi_{i_1};\quad \ldots;\quad \lambda\, \varphi_n + \psi_{i_{n-r}}.$$

Zwischen diesen Größen müßte nun eine algebraische Relation mit Koeffizienten aus P bestehen. Da $\psi_{i_1}, \ldots, \psi_{i_{n-r}}$ unabhängig in bezug auf $\Phi = P(\lambda, \varphi_1, \ldots, \varphi_r)$ sind, so können $\lambda\, \varphi_{r+1} + \psi_{i_1}, \ldots, \lambda\, \varphi_n + \psi_{i_{n-r}}$ zunächst in dieser Relation nicht vorkommen. Also bleibt eine Relation zwischen $\lambda, \lambda\, \varphi_1, \ldots, \lambda\, \varphi_r$. Diese sind aber algebraisch-unabhängig in bezug auf P. Der Widerspruch ist da und die Unabhängigkeit der q_{ik} bewiesen.

Da die Unbestimmte u_{ik} ebenfalls unabhängig sind, so folgt die Isomorphie:

$$P(q_{ik}) \simeq P(u_{ik}),$$

die sich zu einer Isomorphie des Erweiterungskörpers $P(q_{ik}, \lambda, \varphi_i)$ von $P(q_{ik})$ mit einem Erweiterungskörper von $P(u_{ik})$ innerhalb Ω erweitern läßt. Dabei geht der Punkt G, Schnittpunkt von $M_\Omega Q$ und M'_Ω, über in einen Punkt X, Schnittpunkt von $M_\Omega U$ und M'_Ω, womit die Behauptung 1. bewiesen ist.

Um 2. zu beweisen, nehmen wir an, $Y = \{\eta_0, \ldots, \eta_n\}$ sei ein Schnittpunkt von $M_\Omega V$ und M'_Ω, und es sei etwa $\eta_0 = 1$. Nach W_2 (Satz 5) wird 2. bewiesen sein, sobald man zeigen kann, daß Y eine relationstreue Spezialisierung des Schnittpunktes X von $M_\Omega U$ und M'_Ω ist. Es sei also

$$(2) \qquad f\left(\xi_0, \ldots, \xi_n; \begin{matrix} u_{00}, \ldots, u_{0n} \\ \ldots \ldots \ldots \\ u_{n0}, \ldots, u_{nn} \end{matrix}\right) = 0$$

eine in $\xi_0, \ldots, \xi_n$ homogene, richtige Gleichung mit Koeffizienten aus P. Wir haben zu zeigen, daß daraus folgt

$$(3) \qquad f\left(\eta_0, \ldots, \eta_n; \begin{matrix} v_{00}, \ldots, v_{0n} \\ \ldots \ldots \ldots \\ v_{n0}, \ldots, v_{nn} \end{matrix}\right) = 0.$$

Aus (2) folgt vermöge des Isomorphismus, der X in G und U in Q überführt:

$$(4) \quad f\left(\mu, \mu\psi_1, \ldots, \mu\psi_n; \begin{matrix} \lambda - \sum_1^n y_{0k}\mu\psi_k, & \mu y_{01}, & \ldots, & \mu y_{0n} \\ \lambda\varphi_1 - \sum_1^n y_{1k}\mu\psi_k, & \mu y_{11}, & \ldots, & \mu y_{1n} \\ \cdots & \cdots & \cdots & \cdots \\ \lambda\varphi_n - \sum_1^n y_{nk}\mu\psi_k, & \mu y_{n1}, & \ldots, & \mu y_{nn} \end{matrix}\right) = 0.$$

Die allgemeine Nullstelle $\{\mu, \mu\psi_1, \ldots, \mu\psi_k\}$ von $\mathfrak{p}'_\Phi$ kann durch irgendeine spezielle Nullstelle von $\mathfrak{p}'$ ersetzt werden, ohne daß die Gleichung (4) verloren geht; wir ersetzen sie durch $\{\eta_0, \ldots, \eta_n\}$, dabei $\eta_0 = 1$ beachtend. Zugleich ersetzen wir die Unbestimmten y_{ik} durch v_{ik}.

$$f\left(\eta_0, \ldots, \eta_n; \begin{matrix} \lambda - \sum_1^n v_{0k}\eta_k, & v_{01}, & \ldots, & v_{0n} \\ \lambda\varphi_1 - \sum_1^n v_{1k}\eta_k, & v_{11}, & \ldots, & v_{1n} \\ \cdots & \cdots & \cdots \\ \lambda\varphi_n - \sum_1^n v_{nk}\eta_k, & v_{n1}, & \ldots, & v_{nn} \end{matrix}\right) = 0.$$

Hier kann man weiter die allgemeine Nullstelle $\{\lambda, \lambda\varphi_1, \ldots, \lambda\varphi_n\}$ von $\mathfrak{p}_{P\ v_{ik},\eta_k}$ ersetzen durch irgendeine spezielle Nullstelle von $\mathfrak{p}_\Omega$. Nun ist $\{\eta_0, \ldots, \eta_n\}$ Nullstelle von $\mathfrak{p}_\Omega V$, also

$$\{\sum_0^n v_{0k}\,\eta_k, \ldots, \sum_0^n v_{nk}\,\eta_k\}$$

Nullstelle von $\mathfrak{p}_\Omega$. Man kann also $\{\lambda,\,\lambda\,\varphi_1,\,\ldots,\,\lambda\,\varphi_n\}$ dadurch ersetzen. Das ergibt gerade die zu beweisende Gleichung (3).

Satz 30 gilt insbesondere, wenn man für V die in den §§ 7 und 8 betrachtete Matrix W einsetzt. Wie gezeigt, hat $M_\Omega W$ mit M_Ω' nur endlichviele Schnittpunkte und es folgt, daß die Multiplizitäten dieser Schnittpunkte > 0 sind.

§ 10.
Eine obere Grenze für die Multiplizitäten.

Der folgende Satz gibt für beliebige Normalprobleme eine obere Schranke für die Multiplizitäten der Lösungen.

Satz 31. Voraussetzung. *Ein Normalproblem, gegeben durch die Gleichungen*

$$\begin{cases} f_1(\lambda,\,\xi) = 0 \\ \cdots \cdots \\ f_r(\lambda,\,\xi) = 0 \end{cases}$$

(wo $\lambda_1,\,\ldots,\,\lambda_h$ Unbestimmte in Ω sind), habe im Körper Ω endlichviele Lösungen $X^{(1)},\,\ldots,\,X^{(q)}$. Nach der Spezialisierung $\lambda \to \mu$ habe das Normalproblem immer noch endlichviele Lösungen. Eine relationstreue Spezialisierung der $X^{(1)},\,\ldots,\,X^{(q)}$ für $\lambda \to \mu$ sei $Y^{(1)},\,\ldots,\,Y^{(q)}$. Von diesen Punkten seien $Y^{(1)},\,\ldots,\,Y^{(m)}$ gleich Y (m ist demnach die Multiplizität von Y). In $\Omega[x_1,\,\ldots,\,x_n]$ betrachten wir die Ideale

$$\mathfrak{a} = (f_1(\lambda,\,x),\,\ldots,\,f_r(\lambda,\,x)),$$
$$\mathfrak{b} = (f_1(\mu,\,x),\,\ldots,\,f_r(\mu,\,x)),$$

deren Nullstellen gerade die Lösungen unserer Gleichungen sind. Es sei χ die charakteristische Funktion des Ideals $\mathfrak{b}_Y$, und χ_i die des Ideals $\mathfrak{a}_{X^{(i)}}$.

Behauptung.

$$\chi \geqq \sum_1^m \chi_i\,.$$

Eine Folge der Behauptung ist (weil notwendig jedes $\chi_i > 0$)

$$\chi \geqq m,$$

womit eine obere Schranke für die Multiplizität m gegeben ist.

Beweis. Durch eine Koordinatentransformation sei erreicht, daß x_0 in den Punkten $X^{(i)}, Y^{(i)}\,Y$ *nicht* verschwindet.

Die Summe $\overset{m}{\underset{1}{\sum}}\chi_i$ ist die charakteristische Funktion des Ideals $[\mathfrak{a}_{X^{(1)}}, \ldots, \mathfrak{a}_{X^{(m)}}]$ (Satz 18, § 4). Wir behaupten nun, daß, wenn σ eine hinreichend hohe Zahl ist, die beiden Ideale $[\mathfrak{a}_{X^{(1)}}, \ldots, \mathfrak{a}_{X^{(m)}}]$ und $\mathfrak{b}_Y$ dieselben charakteristischen Funktionen haben wie die beiden folgenden:

$$\mathfrak{v} = \left(\mathfrak{a}, \overset{m}{\underset{1}{\prod}}\mathfrak{m}_i^\sigma\right),$$

$$\mathfrak{w} = (\mathfrak{b}, \quad \mathfrak{n}^{m\sigma}).$$

wo

$$(1) \qquad \mathfrak{m}_i = (\xi_0^{(i)}x_1 - \xi_1^{(i)}x_0, \; \xi_0^{(i)}x_2 - \xi_2^{(i)}x_0, \ldots, \xi_0^{(i)}x_n - \xi_n^{(i)}x_0),$$

$$(2) \qquad \mathfrak{n} = (\eta_0 x_1 - \eta_1 x_0, \; \eta_1 x_2 - \eta_2 x_0, \ldots, \eta_0 x_n - \eta_n x_0)$$

gesetzt ist. Das wird nach Satz 21 bewiesen sein, sobald gezeigt ist, daß die zugehörigen inhomogenen Ideale übereinstimmen, also daß

$$[\bar{\mathfrak{a}}_{\overline{X^{(1)}}}, \ldots, \bar{\mathfrak{a}}_{\overline{X^{(m)}}}] = \left(\bar{\mathfrak{a}}, \overset{m}{\underset{1}{\prod}}\overline{\mathfrak{m}}_i^\sigma\right),$$

$$\bar{\mathfrak{b}}_{\overline{Y}} = (\bar{\mathfrak{b}}, \bar{\mathfrak{n}}^{m\sigma}).$$

Diese Gleichungen sind aber nach den Sätzen 5 und 6 (§ 2) richtig. Also sind $\overset{m}{\underset{1}{\sum}}\chi_i$ und χ in der Tat die charakteristischen Funktionen der Ideale $\mathfrak{v}, \mathfrak{w}$.

Sei nun $(r_1, \ldots, r_\tau)$ eine Basis für $\overset{m}{\underset{1}{\prod}}\mathfrak{m}_i^\sigma$, gebildet durch Multiplikation aus den Basen (1) der einzelnen Ideale $\mathfrak{m}_i$. Dieselbe gibt nach der Spezialisierung $\xi_k^{(i)} \rightarrow \eta_k^{(i)} = \eta_k$ eine Basis $(s_1, \ldots, s_\tau)$ für $\mathfrak{n}^{m\sigma}$. Man hat also:

$$\mathfrak{v} = \left(\mathfrak{a}, \overset{m}{\underset{1}{\prod}}\mathfrak{m}_i^\sigma\right) = (f_1(\lambda, x), \ldots, f_r(\lambda, x), r_1, \ldots, r_\tau),$$

$$\mathfrak{w} = (\mathfrak{b}, \mathfrak{n}^{m\sigma}) = (f_1(\mu, x), \ldots, f_r(\mu, x), s_1, \ldots, s_\tau).$$

Die Basis von $\mathfrak{w}$ ist demnach eine relationstreue Spezialisierung der Basis von $\mathfrak{v}$. Zu zeigen ist, daß für die charakteristischen Funktionen der Ideale $\mathfrak{v}, \mathfrak{w}$ die Beziehung besteht:

$$\chi(\varrho; \mathfrak{w}) \geqq \chi(\varrho; \mathfrak{v})$$

oder, was nach § 4, Gleichung (2), dasselbe ist,

$$(3) \qquad\qquad \varphi(\varrho; \mathfrak{w}) \leqq \varphi(\varrho; \mathfrak{v}).$$

Nun kann man eine Modulbasis für die Formen vom Grad ϱ von $\mathfrak{v}$ dadurch gewinnen, daß man die Basiselemente von $\mathfrak{v}$ multipliziert mit allen solchen Potenzprodukten der x_i, daß die Produkte jedesmal den Grad σ haben. Bildet man aus den Koeffizienten der so entstehenden Formen die Matrix (die „dialytische Matrix des Ideals $\mathfrak{v}$ für den Grad ϱ"), so gibt der Rang dieser Matrix den Wert von $\varphi(\varrho;\mathfrak{v})$ an. Ebenso kann man aber für das Ideal $\mathfrak{w}$ eine Matrix bilden, dessen Rang den Wert von $\varphi(\varrho;\mathfrak{w})$ angibt. Der Rang der $\mathfrak{w}$-Matrix ist (wegen der relationstreuen Spezialisierung) $\leq$ dem Rang der $\mathfrak{v}$-Matrix. Damit ist (3) bewiesen.

Dieser Satz werde nunmehr angewandt auf den Fall von § 8. Das Normalproblem bestand darin, die gemeinsamen Nullstellenklassen der Ideale $\mathfrak{p}_\Omega U$ und $\mathfrak{p}'_\Omega$ zu finden. Die Spezialisierung bestand darin, daß für U die singuläre Matrix W eingesetzt wird. Demnach hat man zu setzen:

$$\mathfrak{a} = (\mathfrak{p}_\Omega U, \mathfrak{p}'_\Omega),$$
$$\mathfrak{b} = (\mathfrak{p}_\Omega W, \mathfrak{p}'_\Omega),$$

$m =$ die Multiplizität einer Lösung Y des spezialisierten Normalproblems,

$\chi =$ die charakteristische Funktion von $\mathfrak{b}_Y$.

Der Satz ergibt

$$(4) \qquad\qquad \chi \geq m\,.$$

Um χ zu berechnen, beachte man, daß nach (5) § 2:

$$\mathfrak{b}_Y = (\mathfrak{p}_\Omega W, \mathfrak{p}'_\Omega)_Y = ((\mathfrak{p}_\Omega W)_Y, \mathfrak{p}'_\Omega)_Y\,.$$

Die Mannigfaltigkeit des Ideals $\mathfrak{p}_\Omega W$ zerfällt nach Satz 28 in lineare Räume $L^{(1)}, \ldots, L^{(s)}$, die einen linearen $^{r-1}$Raum L gemein haben. Y ist ein Punkt eines $L^{(\beta)}$, der, wie in § 8 schon bemerkt, nicht in L liegt. Daraus folgt nach Satz 29 (§ 7):

$$(\mathfrak{p}_\Omega W)_Y = \mathfrak{a}^{(\beta)} W,$$

wo $\mathfrak{a}^{(\beta)} W$ das zu $L^{(\beta)}$ gehörige Primideal

$$\mathfrak{a}^{(\beta)} W = \Big(\sum_k (\lambda_0^{(\beta)} u_{1k} - \lambda_1^{(\beta)} u_{0k}) x_k, \ldots, \sum_k (\lambda_0^{(\beta)} u_{n-r,k} - \lambda_{n-r}^{(\beta)} u_{0k}) x_k \Big)$$

ist. Setzt man

$$l_i = \sum_k (\lambda_0^{(\beta)} u_{ik} - \lambda_i^{(\beta)} u_{0k}) x_k \qquad (i = 1, \ldots, n-r),$$

so wird

$$\mathfrak{a}^{(\beta)} W = (l_1, \ldots, l_{n-r}),$$

mithin

$$\mathfrak{b}_Y = (\mathfrak{p}'_\Omega, (\mathfrak{p}_\Omega W)_Y)_Y = (\mathfrak{p}'_\Omega, \mathfrak{a}^{(\beta)} W)_Y = (\mathfrak{p}'_\Omega, l_1, \ldots, l_{n-r})_Y\,.$$

Die l_i sind allgemeine Linearformen. Also können wir auf das Ideal

$$(\overline{\mathfrak{p}'_\Omega}, \overline{l_1}, \ldots, \overline{l_{n-r}})$$

den dritten Teil von Satz 27 anwenden, der besagt, daß dieses Ideal ein Durchschnitt ist von lauter verschiedenen Primidealen, die je zu einem der Schnittpunkte von $\overline{M'_\Omega}$ und $\overline{L^{(\beta)}}$ gehören. Da Y einer dieser Schnittpunkte ist, so ist die $\overline{Y}$-Komponente

$$\overline{\mathfrak{b}_Y} = (\overline{\mathfrak{p}'_\Omega}, \overline{l_1}, \ldots, \overline{l_{n-r}})_{\overline{Y}}$$

das zu $\overline{Y}$ gehörige Primideal. Daraus folgt nach der Schlußbemerkung von § 4, daß die charakteristische Funktion χ von $\mathfrak{b}_Y$ gleich 1 ist.

Demnach kann man für (4) schreiben $m \leq 1$, womit bewiesen ist:

S a t z 32. *Die Schnittpunkte von $M_\Omega W$ und M'_Ω haben sämtlich eine Multiplizität ≤ 1.*

§ 11.
Zusammenfassung der Ergebnisse. Folgerungen.

In § 9 ist gezeigt, daß die Multiplizität der Schnittpunkte von $M_\Omega W$ und M'_Ω nicht Null sind, in § 10, daß sie ≤ 1 sind, also sind sie $= 1$, womit nach § 8 der Hauptsatz (verallgemeinerte Bézoutsche Satz) *bewiesen ist.*

Aus dem Hauptsatz folgt nun auf Grund des Prinzips der Erhaltung der Anzahl, daß *die Summe der Multiplizitäten der Schnittpunkte von $M_\Omega V$ mit M'_Ω immer gleich dem Produkt der Grundzahlen ist, wie auch die Matrix V beschaffen sein mag, vorausgesetzt, daß die Anzahl der Schnittpunkte endlich ist.*

Insbesondere gilt das, wenn man für V die Einheitsmatrix E einsetzt, also die Schnittpunkte von M_Ω mit M'_Ω betrachtet. Man muß dabei wieder voraussetzen, daß M_Ω und M'_Ω, also insbesondere auch M und M', nur endlichviele Schnittpunkte haben. Nun gilt:

S a t z 33. *Haben M und M' nur endlichviele Schnittpunkte, so haben M_Ω und M'_Ω auch nur endlichviele Schnittpunkte, und zwar genau dieselben wie M und M'.*

B e w e i s. Hätten M_Ω und M'_Ω einen Schnittpunkt, dessen Koordinatenverhältnisse nicht in P lägen, also transzendent in bezug auf P wären, so könnte man diese Koordinatenverhältnisse als algebraische Funktionen von gewissen Parametern auffassen, und unendlichviele reguläre Argumentwerte[45] für diese Funktionen finden, die zu unendlichvielen Schnittpunkten von M und M' führen würden, entgegen der Voraussetzung.

Nach diesem Satz braucht man nur über die Schnittpunkte von M und M', nicht auch über die von M_Ω und M'_Ω zu reden; jeder Schnittpunkt von M und M' hat eine gewisse Multiplizität, und *die Summe dieser Multiplizitäten ist gleich dem Produkt der Gradzahlen.* Dies ist eine zweite Fassung des verallgemeinerten Bézoutschen Theorems.

---

[45] W_1 § 2, 6.

Die Sätze der §§ 8, 9 ergeben aber noch mehr. Satz 23 besagt, daß die Multiplizität eines Schnittpunktes von $M_\Omega V$ und M'_Ω, insbesondere eines Schnittpunkts von M und M', niemals Null ist. Daraus folgt:

Satz 34. *Die Anzahl der Schnittpunkte von M und M' ist höchstens gleich dem Produkt der Gradzahlen.*

Weiter besagt Satz 31, daß die Multiplizität eines Schnittpunktes Y von $M_\Omega V$ und M'_Ω höchstens gleich der charakteristischen Funktion des Ideals

$$(\mathfrak{p}_\Omega V, \mathfrak{p}'_\Omega)_Y$$

ist. Insbesondere ist die Multiplizität m eines Schnittpunkts von M und M' höchstens gleich der charakteristischen Funktion χ des Ideals

$$(\mathfrak{p}_\Omega, \mathfrak{p}'_\Omega)_Y,$$

oder, was nach Satz 23 (§ 5) auf dasselbe herauskommt, des Ideals

$$(\mathfrak{p}, \mathfrak{p}')_Y.$$

Wir werden in § 12 sehen, daß im Fall $r = 1$, also wenn M eindimensional ist, die Ungleichung

$$\chi \leqq m$$

zu einer Gleichung

$$\chi = m$$

verschärft werden kann. Daß dies aber im allgemeinen nicht der Fall zu sein braucht, sieht man aus dem folgenden Beispiel, das im wesentlichen von Macaulay herrührt[46]).

Das H-Ideal

$$\mathfrak{p}' = (x_1 x_4 - x_2 x_3,\, x_2^3 - x_1^2 x_3,\, x_3^3 - x_2 x_4^2,\, x_2^2 x_4 - x_1 x_3^2)$$

in $P[x_0, \ldots, x_4]$ ist prim, weil es die allgemeine Nullstelle

$$\{\lambda,\, \mu^4,\, \mu^3 \nu,\, \mu \nu^3,\, \nu^4\}$$

hat. Die Punkte seiner Mannigfaltigkeit M' werden gefunden, indem man den Unbestimmten λ, μ, ν spezielle Werte λ', μ', ν' gibt.

Mit der linearen Mannigfaltigkeit M, gegeben durch das Primideal

$$\mathfrak{p} = (x_1, x_4)$$

hat M' offensichtlich nur einen Schnittpunkt, nämlich

$$Y = \{\lambda, 0, 0, 0, 0\}.$$

Mit der allgemein-transformierten Mannigfaltigkeit $M_\Omega U$, gegeben durch

$$\mathfrak{p}_\Omega U = (\textstyle\sum u_{1k} x_k,\, \sum u_{4k} x_k),$$

---

[46]) Macaulay, S. 98.

hat M'_Ω aber vier Schnittpunkte, wie man durch Auflösung der Gleichungen

$$\begin{cases} u_{10}\,\lambda' + u_{11}\,\mu'^4 + u_{12}\,\mu'^3\nu' + u_{13}\,\mu'\nu'^2 + u_{14}\,\nu'^4 = 0\,, \\ u_{40}\,\lambda' + u_{41}\,\mu'^4 + u_{42}\,\mu'^3\nu' + u_{43}\,\mu'\nu'^2 + u_{44}\,\nu'^2 = 0 \end{cases}$$

sieht. Also hat der eine Schnittpunkt Y von M und M' die Multiplizität 4:

$$m = 4\,.$$

χ ist die charakteristische Funktion des Ideals

$$(\mathfrak{p},\,\mathfrak{p}')_Y$$

in $\mathsf{P}[x_1,\,\ldots,\,x_4]$. Nun ist

$$(\mathfrak{p},\,\mathfrak{p}') = (x_1,\,x_4,\,x_2\,x_3,\,x_2^3,\,x_3^3)$$

ein Primärideal mit der Nullstelle Y, also folgt:

$$(\mathfrak{p},\,\mathfrak{p}')_Y = (x_1,\,x_4,\,x_2\,x_3,\,x_2^3,\,x_3^2)\,.$$

Geht man auf Grund von Satz 20 ($\S\,4$) auf inhomogene Ideale über durch die Substitution $x_0 = 1$, so wird ein Maximalsystem von modulo dem Ideal $(x_1,\,x_4,\,x_2\,x_3,\,x_2^3,\,x_3^3)$ linear-unabhängigen Polynomen gegeben durch

$$1,\,x_2,\,x_3,\,x_2^2,\,x_3^2\,.$$

Also ist $\chi = 5$. Damit ist die Gleichung $\chi = m$ widerlegt.

$$\S\ 12.$$

Der Fall $r = 1$.

Satz 35. *Es seien M und M' irreduzible Mannigfaltigkeiten der Dimensionen 1 und $n-1$ im $P_n(\mathsf{P})$, die sich in endlichvielen Punkten schneiden. Es seien $\mathfrak{p}$ und $\mathfrak{p}'$ die zugehörigen Primideale, Y ein Schnittpunkt von M und M', m seine Multiplizität, und χ die charakteristische Funktion des Ideals $(\mathfrak{p},\,\mathfrak{p}')_Y$. Dann ist $\chi = m$.*

Beweis. In $\S\,11$ wurde schon bemerkt, daß $\chi \leqq m$. Wenn wir also zeigen können, daß bei Summation über alle Schnittpunkte

$$\sum \chi \geqq \sum m$$

so wird der Satz bewiesen sein.

$\sum m$ ist nach $\S\,11$ gleich dem Produkt der Gradzahlen von M und M':

$$(1) \qquad\qquad \sum m = s \cdot s'\,.$$

$\sum \chi$ ist nach $\S\,4$, Gleichung (4), gleich der charakteristischen Funktion des Ideals $(\mathfrak{p},\,\mathfrak{p}')$. Nach $W_1\ \S\,5,\,\mathbf{10}$ ist $\mathfrak{p}'$ ein Hauptideal, das von einer irreduziblen Form f vom Grad γ erzeugt wird:

$$\mathfrak{p}' = (f)\,,$$
$$(\mathfrak{p},\,\mathfrak{p}') = (\mathfrak{p},\,f)\,,$$

wo f zu $\mathfrak{p}$ prim ist. Daraus folgt nach § 4, Gleichung (10):

$$\mathrm{Grad}\,(\mathfrak{p},\,f) = \gamma\cdot\mathrm{Grad}\,\mathfrak{p},$$

oder

$$(2)\qquad\qquad \sum\chi = \gamma\cdot\mathrm{Grad}\,\mathfrak{p}.$$

Dieselben Gleichungen gelten auch dann, wenn die Koeffizienten der auftretenden Polynome einem Erweiterungskörper Ω von P entnommen werden, denn bei dieser Erweiterung bleiben alle charakteristischen Funktionen ungeändert (Satz 23). Wählt man für f eine allgemeine Linearform, so wird die linke Seite von (2) mindestens gleich dem Grad von M, da schon die Anzahl der Summanden gleich diesem Grad ist, und die einzelnen Summanden positiv. Auf der rechten Seite wird $\gamma = 1$. Also kommt:

$$\mathrm{Grad}\,\mathfrak{p} \geqq \mathrm{Grad}\,M = s.$$

Die Formel (2) gilt aber auch dann, wenn M ein allgemeiner linearer [1]Raum ist, und M' beliebig. Dann ist aber die linke Seite mindestens gleich dem Grad von M', da schon die Anzahl der Summanden gleich diesem Grad ist, und die einzelnen Summanden positiv. Auf der rechten Seite ist $\mathrm{Grad}\,\mathfrak{p} = 1$. Also kommt

$$\gamma \geqq \mathrm{Grad}\ \text{von}\ M' = s',$$

und (2) ergibt

$$(3)\qquad\qquad \sum\chi \geqq s'\cdot s.$$

Aus (1) und (3) folgt die gesuchte Ungleichung

$$\sum\chi \geqq \sum m.$$

Hinterher schließt man, daß in allen Formeln dieses Paragraphen das Gleichheitszeichen gelten muß. Es ist also $s' = \gamma$ und $s = \mathrm{Grad}\,\mathfrak{p}$.

(Eingegangen am 3. 6. 1927.)

100

5.

Topologische Begründung des Kalküls der abzählenden Geometrie

Mathematische Annalen 102, 3 (1929) 337–362

§ 1.

Einleitung.

Eines der Pariser Probleme Hilberts[1] lautet: „Eine strenge Begründung des Schubertschen Abzählungskalküls".

In früheren Arbeiten[2] habe ich gesucht darzutun, daß das Kernproblem der abzählenden Geometrie besteht in der Aufstellung einer brauchbaren Definition der „Multiplizitäten" oder der Vielfachheiten, mit denen die Lösungen eines algebraisch-geometrischen Problems gezählt werden müssen, damit das „Prinzip der Erhaltung der Anzahl" für diese Lösungen bei jeder Spezialisierung der Daten des Problems gelte. In der Arbeit W_1 habe ich gezeigt, daß man bei jedem Problem, dessen Gleichungen homogen in den Unbekannten und rational in einigen Parametern sind, die Lösungen für jede spezielle Parameterzahl in einer und nur einer Weise mit solchen Vielfachheiten versehen kann, daß Anzahl und algebraische Eigenschaften der Lösungen bei diesen Parameterspezialisierungen erhalten bleiben, und daß bei allgemeiner Parameterwahl die Multiplizitäten gleich 1 sind. Damit war eine implizite Definition der Multiplizitäten gegeben, aber noch kein brauchbares Mittel, diese in vorliegenden Fällen (außer den allereinfachsten) wirklich zu bestimmen. Eine besondere Schwierigkeit bei der Anwendung war noch, daß mit der Möglichkeit von „Lösungen mit der

---

[1] D. Hilbert, Mathematische Probleme, Gött. Nachr. 1900, S. 253.

[2] B. L. v. d. Waerden, Diss. Amsterdam 1926. Der Multiplizitätsbegriff der algebraischen Geometrie, Math. Annalen **97** (1927), S. 756 (*zitiert* W_1). Eine Verallgemeinerung des Bézoutschen Theorems, Math. Annalen **99** (1928), S. 497 (*zitiert* W_2). On Hilbert's function etc., Proc. Kon. Ak. Amsterdam **31** (1928), S. 749.

Multiplizität Null", also von Lösungen, die bei spezieller Parameterwahl vorhanden sind, aber denen im allgemeinen Fall nichts entspricht, gerechnet werden mußte.

Die für den Schubertschen „Bedingungskalkül"[3] wichtigen Fälle sind die, wo es sich darum handelt, die Anzahl der gemeinsamen Elemente von zwei oder mehr algebraischen Varietäten[4] in einer festen singularitätenfreien Mannigfaltigkeit (z. B. im komplexen projektiven Raum) zu bestimmen. Für den Fall einer Kurve V_1 und einer Hyperfläche V_{n-1} im projektiven R_n sind die Bestimmungen der Vielfachheiten durch explizite Formeln und die Berechnung der Gesamtzahl sowohl auf funktionentheoretischem als auf idealtheoretischem Wege gelungen[5]. Beide Bestimmungsweisen scheiterten jedoch gänzlich im allgemeineren Fall der Schnittpunkte einer V_r und V_{n-r} im projektiven R_n[6]. Die oben dargestellte implizite Multiplizitätsdefinition führte aber in diesen Fällen noch zum Ziel[7]. Dadurch nämlich, daß die Varietäten V_r und V_s mittels einer projektiven Transformation mit unbestimmten Koeffizienten in allgemeine Lage zueinander gebracht wurden, war das Schnittpunktsproblem abhängig gemacht von den Transformationsparametern, und die obige Definition der Multiplizität wurde anwendbar. Es gelang durch eine äußerst mühevolle Analyse, für spezielle Werte der Parameter („ausgeartete Transformation") Anzahl und Multiplizitäten der Schnittpunkte algebraisch zu bestimmen, wobei sich das erwartete Ergebnis: Summe der Multiplizitäten = Produkt der Gradzahlen ergab, welches Ergebnis sich dann vermöge der „Erhaltung der Anzahl" auf die allgemeine Lage sowie auf alle überhaupt möglichen Spezialisierungen übertrug.

Soweit sie reichte, hatte die algebraische Methode eine größere Allgemeinheit als jede analytische, da sie auf beliebige abstrakte Geometrien (die zu abstrakten Körpern gehören) anwendbar war. Aber bei der Übertragung der Methode auf Varietäten von Geraden u. dgl. stieß die Durchführung der Beweise auf immer wachsende Schwierigkeiten, und für solche Gebilde, die nicht wie der projektive Raum eine transitive Gruppe von Transformationen in sich gestatten, ist die Übertragung der obigen Multiplizitätsdefinition ganz ausgeschlossen.

---

[3] H. Schubert, Kalkül der abzählenden Geometrie, Leipzig 1879.

[4] Da das Wort *Mannigfaltigkeit* in dieser Arbeit, dem topologischen Sprachgebrauch entsprechend, für *singularitätenfreie* Räume reserviert bleibt, werde ich für die durch algebraische Gleichungen definierten Punktmengen das französische Wort „*Varietäten*" gebrauchen.

[5] Für Literatur siehe W_2, Einleitung.

[6] W_2, Einleitung und § 11.

[7] W_2.

Auch die analytischen Methoden [Zeuthen [8]), Halphen [9]) u. a.] erreichen nur bestimmte Fälle.

Aber die Topologie besitzt einen Multiplizitätsbegriff: den Begriff des Index eines Schnittpunktes von zwei Komplexen [10]), der schon von Lefschetz [11]) mit Erfolg auf die Theorie der algebraischen Flächen sowie auf Korrespondenzen auf algebraischen Kurven angewandt wurde. Soll dieser Indexbegriff, angewandt auf algebraische Varietäten im komplexen Gebiet, sich als Multiplizitätsbegriff für die abzählende Geometrie eignen, so muß er die folgenden drei Eigenschaften besitzen:

1. Die Summe der Indizes soll dem „Prinzip der Erhaltung der Anzahl" bei stetigen Änderungen der schneidenden Varietäten genügen.

2. Die Indizes sollen für einfache Schnittpunkte (d. h. wenn die Tangentialräume der schneidenden Varietäten nur einen Punkt gemein haben) alle den Wert 1 haben.

3. Sie sollen keine negativen Werte annehmen.

Die erste Eigenschaft des topologischen Indexbegriffs folgt fast unmittelbar aus seiner Definition durch simpliziale Approximationen. Die zweite Eigenschaft wurde allgemein für analytische Varietäten von Lefschetz [12]) bewiesen. Und über die dritte Eigenschaft hinaus läßt sich sogar zeigen, daß die Indizes bei analytischen (also insbesondere algebraischen) Varietäten immer *positiv sind*, wodurch also zugleich das unangenehme Vorkommnis der „Multiplizität Null" ausgeschlossen wird. Dieser Satz über analytische Varietäten im komplexen Gebiet ist das Hauptergebnis dieser Arbeit. Aus den Eigenschaften 1. bis 3. folgt dann unschwer die Übereinstimmung des topologischen mit dem algebraischen Multiplizitätsbegriff für den projektiven Raum.

Die Topologie leistet aber noch mehr als die Ermöglichung einer brauchbaren Multiplizitätsdefinition. Sie verschafft zugleich eine Fülle von Mitteln, die Indexsumme aller Schnittpunkte oder „Schnittpunktszahl", deren Bestimmung das Ziel aller abzählenden Methoden ist, in einfacher Weise zu bestimmen, indem sie zeigt, daß diese Indexsumme nur von den

---

[8]) H. G. Zeuthen, Abzählende Methoden (Leipzig 1914); Enzyklopädie III, 3.

[9]) Halphen, Comptes Rendus (4. Sept. 1876).

[10]) S. Lefschetz, Trans. Am. Math. Soc. 28 (1926), S. 1 (*zitiert* L_1).

[11]) S. Lefschetz, L'analysis situs et la géométrie algébrique, Paris 1924.

[12]) loc. cit. [11]), S. 19. Lefschetz schließt aus den Eigenschaften 1., 2. auf die (zur Positivität verschärfte) Eigenschaft 3. mit der Begründung, man könne durch eine kleine Verschiebung immer die unter 2. vorausgesetzte Situation herstellen. Die Begründung scheint in dieser Form ungenügend, denn bei der Verschiebung könnten ja Schnittpunkte in Wegfall kommen. Erst die genauere Betrachtung von § 5 wird lehren, daß dieses tatsächlich nicht vorkommen kann.

Homologieklassen der zum Schnitt gebrachten Varietäten abhängt, und indem sie für die Bestimmung der Homologieklassen den ganzen Apparat der „kombinatorischen Topologie" zur Verfügung stellt. Zum Beispiel reduziert sich der algebraisch so mühevoll bewiesene Satz, daß im projektiven Raum die Schnittpunktsanzahl einer V_r und einer V_{n-r} gleich dem Produkt der Gradzahlen ist, von topologischem Gesichtspunkt auf die beiden leicht zu beweisenden Tatsachen, daß eine V_r (deren topologische Dimensionszahl im Komplexen gleich $2r$ ist) homolog dem g-fachen eines linearen L_r ist, wo g der Grad ist, und daß die Schnittpunktszahl zweier linearer Räume L_r und L_{n-r} gleich 1 ist.

Allgemein ergibt jede Homologierelation zwischen algebraischen Varietäten eine symbolische Gleichung im Schubertschen Sinn, und man darf diese Gleichungen unbeschränkt addieren und multiplizieren, wie es im Schubertschen Kalkül geschieht. Aus der Existenz einer endlichen Basis für die Homologien in jeder geschlossenen Mannigfaltigkeit ergibt sich weiter allgemein die Lösbarkeit der Schubertschen „Charakteristikenprobleme".

Durch vollständige Angabe aller benutzten topologischen, analytischen und algebraischen Definitionen und Zusammenstellung der wichtigsten Sätze hoffe ich, die Schwierigkeit, die darin besteht, daß eine gewisse Vertrautheit sowohl mit topologischen als auch mit abzählenden Methoden beim Leser notwendig vorausgesetzt werden mußte, zu einem Minimum reduziert zu haben. Der sachverständige Leser muß dafür einige Ausführungen über bekannte Tatsachen mit in den Kauf nehmen. Die Kenntnis der zitierten Arbeiten W_1 und W_2 ist für das Verständnis dieser Arbeit nicht erforderlich. Einige topologische Hilfsbetrachtungen fast trivialer Natur sind, um den Gedankengang nicht zu unterbrechen, in zwei Anhängen vereinigt.

Anwendungen der hier zu entwickelnden Methoden auf konkrete abzählende Probleme hoffe ich später zu geben.

§ 2.

Komplexe im euklidischen Raum.

In diesem Paragraphen sollen, um allen Zweifel beim Gebrauch von Worten mit schwankender Bedeutung auszuschließen, die nötigen topologischen Grundbegriffe ganz kurz zusammengestellt werden. Für genauere Erörterungen sei auf die Lehrbücher verwiesen[13]).

---

[13]) Etwa: O. Veblen, The Cambridge Colloquium Lectures, Cambridge (Mass.) 1922. Oder Hadamards Note zu J. Tannery, Introduction à la théorie des fonctions II, zu ergänzen durch J. W. Alexanders Proof of the Invariance of certain Numbers, Proc. Am. Math. Soc. 16 (1915), p. 148.

Ein *geradliniger k-dimensionaler Komplex* ist aus endlichvielen k-dimensionalen euklidischen Simplizes aufgebaut. Ein stetiges Bild eines geradlinigen k-dimensionalen Komplexes im euklidischen R^n (mit ganz beliebigen Singularitäten) heißt *Komplex* schlechthin und wird mit K^k bezeichnet. Der leere Komplex heißt *Null*. Ist von Rand, von Orientierung, Unterteilung oder Addition von Komplexen die Rede, so ist damit immer gemeint, daß man die genannten Operationen zunächst an den Urbildern vornimmt und dann auf das Bild überträgt. *Orientierung eines Simplex* (im Urbild) geschieht dadurch, daß man einer bestimmten Reihenfolge der Ecken und allen geraden Permutationen davon ein Vorzeichen $\varepsilon = \pm 1$, allen ungeraden Permutationen das entgegengesetzte Vorzeichen $-\varepsilon$ zuordnet. *Orientierung eines Komplexes K^k* geschieht durch (beliebige) Orientierung aller seiner k-dimensionalen Simplizes. Bei der *Addition von Komplexen* wird jedes Simplex, welches in der Summe zweimal mit entgegengesetzter Orientierung vorkommt, diese beiden Male weggelassen. Mit jeder Orientierung eines Simplex ist nach einer bestimmten Vorschrift eine bestimmte Orientierung seines Randes verknüpft. Wie diese Vorschrift lautet, ist gleichgültig; nur muß sie so eingerichtet werden, daß der orientierte Rand des orientierten Randes gleich Null wird. Unter dem Rand $R(K^k)$ eines orientierten Komplexes K^k ist zu verstehen die Summe der orientierten $(k-1)$-dimensionalen Ränder der Simplizes von K^k. Zwei Komplexe K_1^r, K_2^r heißen *homolog zueinander im Gebiet U* (Gebiet heißt in dieser Arbeit eine beliebige offene Menge des R^n), wenn es einen K^{r+1} in U gibt derart, daß $K_1^r = K_2 + R(K^{r+1})$ ist. Zeichen für Homologie: $K_1^r \sim K_2^r$.

Komplexe mit Rand Null heißen *Zyklen*.

Eine (geradlinige) δ-*Approximation* eines Komplexes K^s entsteht dadurch, daß man, von einer hinreichend feinen Simplexeinteilung des Urbildes von K^s ausgehend, die den Bildpunkt K^s definierenden Abbildungsfunktionen durch solche ersetzt, die in jedem Simplex des Urbildes (Rand eingeschlossen) linear sind, derart, daß jeder Bildpunkt nach dieser Approximation zu seiner ursprünglichen Lage eine Entfernung $< \delta$ hat, und daß solche Simplizes, die eine Seite gemein haben, auch nach der Approximation die entsprechende Seite gemein haben.

Verbindet man jeden Punkt von K^s geradlinig mit dem entsprechenden Punkt des approximierenden Komplexes K_1^s, so erhält man einen „Verbindungskomplex" K_v^{s+1} in der δ-Umgebung von K^s und ebenso einen Randverbindungskomplex K_v^s in der δ-Umgebung des Randes $R(K^s)$ mit folgenden Eigenschaften:

$$R(K_v^{s+1}) = K^s - K_1^s - K_v^s,$$

$$R(K_v^s) = R(K^s) - R(K_1^s),$$

$$K_v^s = 0, \text{ wenn } R(K^s) \text{ Null oder geradlinig ist.}$$

Eine bestimmte Reihenfolge $x_1, \ldots, x_n$ der Cartesischen Koordinaten des R^n induziert eine bestimmte Orientierung aller n-dimensionalen geradlinigen Simplizes dieses Raumes (oder, wie wir auch sagen wollen, eine *Orientierung des Raumes R^n*), nämlich diejenige, wobei einer bestimmten Reihenfolge der Ecken $x^0, \ldots, x^n$ eines solchen Simplex als Vorzeichen zugeordnet wird das Vorzeichen der Determinante

$$\left| 1\, x_1^\nu, \ldots, x_n^\nu \right| \quad \text{oder} \quad \left| x_1^\nu - x_1^0, \ldots, x_n^\nu - x_n^0 \right|.$$

§ 3.

Schnitte von Komplexen im euklidischen R^n.

Der Begriff des Schnittkomplexes zweier Komplexe ist in der hier gebrauchten Form vor allem von Lefschetz (L_1) entwickelt worden. Eine Übersicht über die benötigten Begriffsbildungen folgt in diesem Paragraphen.

Sind zwei (geradlinige) Simplizes X^r und X^s im R^n gegeben, so kann man durch beliebig kleine Verrückungen der Ecken eine „*allgemeine Lage*" der Simplizes zueinander erzwingen, welche darin bestehen soll, daß die Räume R^r, R^s, in denen sie liegen, einen linearen Raum R^k von genau $k = r + s - n$ Dimensionen gemein haben (bzw. zueinander fremd sind für $r + s - n < 0$), und daß in diesem Schnittraum die beiden Simplizes entweder fremd sind oder innere Punkte gemein haben. Im letzten Fall haben sie ein konvexes Polyeder P^k von der Dimension k gemein, das wir irgendwie in Simplizes eingeteilt denken.

Orientieren wir die Räume R^n, R^r, R^s (oder den Raum R^n und die Simplizes X^r, X^s), so ist dadurch eine Orientierung eines jeden Simplex von P^k mitbestimmt nach folgender *Vorschrift*: Man ergänze die Ecken $x^0, \ldots, x^k$ eines Simplex von P^k durch Punkte $y^1, \ldots, y^{r-k}$ zu einem Eckpunktssystem eines Simplex von R^r; dessen Vorzeichen in der Orientierung von R^r sei ε_1. Ebenso ergänzen wir $x^0, \ldots, x^k$ mit $z^1, \ldots, z^{s-k}$ in R^s und bestimmen das Vorzeichen ε_2. Dann bilden die $n + 1$ Punkte $x^0, \ldots, x^k, y^1, \ldots, y^{r-k}, z^1, \ldots, z^{s-k}$ die Ecken eines Simplex in R^n. Das zugehörige Vorzeichen in der Orientierung von R^n sei ε_3. Nunmehr ordnen wir der Eckenfolge der $x^0, \ldots, x^k$ das Vorzeichen

$$\varepsilon = \varepsilon_1\, \varepsilon_2\, \varepsilon_3$$

zu. Dadurch sind alle Simplizes von P^k und mithin auch P^k selbst orientiert.

Man nennt das so orientierte P^k den *orientierten Schnittkomplex* der Simplizes X^r und X^s und schreibt

$$P^k = X^r \cdot X^s.$$

106

Ist P^k leer, so schreibt man

$$X^r \cdot X^s = 0 \,.$$

Für zwei beliebige geradlinige Komplexe K^r, K^s definieren wir nun, falls alle Simplizes von K^r (sowie ihre Seiten) zu denen von K^s die oben präzisierte „allgemeine Lage" haben, den *orientierten Schnittkomplex* $K^r \cdot K^s$ als die Summe der orientierten Schnitte der Simplizes von K^r mit denen von K^s.

Die wichtigsten Rechnungsregeln für orientierte Schnitte sind:

(1a) $$K^s \cdot K^r = (-1)^{(n-r)(n-s)} K^r \cdot K^s \,;$$

(1b) $$- K^r \cdot K^s = (- K^r) \cdot K^s = K^r \cdot (- K^s) \,;$$

(1c) $$K^r \cdot (K^s \cdot K^t) = (K^r \cdot K^s) \cdot K^t \,;$$

(2) $$R(K^r \cdot K^s) = K^r \cdot R(K^s) + (-1)^{n-s} R(K^r) \cdot K^s$$

und die gewöhnlichen Distributivgesetze. Die ersten drei sind ziemlich trivialer Natur; die letzte ist die Zauberformel, die alle Existenz-, Invarianz- und Deformationssätze für Schnittpunktszahlen, Abbildungsgrade usw. in sich enthält.

Aus (2) folgt sofort:

In jedem Gebiet[14]), *das alle gemeinsamen Punkte von K^r und K^s enthält, gilt die Homologie:*

(3) $$K^r \cdot R(K^s) \sim (-1)^{n-s-1} R(K^r) \cdot K^s \,.$$

Ist speziell $R(K^r)$ zu K^s punktfremd, so wird die rechte Seite Null. Ersetzt man noch s durch $s+1$ und setzt man $R(K^{s+1}) = K_1^s - K_2^s$, so folgt:

Ist $K_1^s \sim K_2^s$ in einem Gebiet, das zum Rand von K^r fremd ist, so ist in diesem Gebiet auch

(4) $$K^r \cdot K_1^s \sim K^r \cdot K_2^s \,.$$

Sind nun zwei beliebige (nicht notwendig geradlinige) orientierte Komplexe K^r, K^s gegeben, und ist K^r zum Rand von K^s und K^s zum Rand von K^r fremd, ist weiter $\delta \leq$ dem Minimum der halben Abstände von K^r zum Rand von K^s und von K^s zum Rand von K^r, und δ-approximiert man K^r und K^s derart durch geradlinige Komplexe K_1^r und K_1^s, daß die Simplizes von K^r zu denen von K^s allgemeine Lage haben, so wird der orientierte Schnittkomplex $K_1^r \cdot K_1^s$ als ein *approximativer Schnittkomplex von K^r und K^s* bezeichnet.

---

[14]) Gebiet heißt hier eine jede offene Menge des Raumes R^n.

Inwieweit dieser approximative Schnittkomplex von der Wahl der Approximation unabhängig ist, lehrt der folgende Satz:

Ist U eine beliebige offene Umgebung der Menge aller gemeinsamen Punkte von K^r und K^s, die keine Randpunkte von K^r oder K^s enthält, und ist die Approximationsschranke $\delta \leq$ dem Minimum der halben Abstände des außerhalb U gelegenen Teils von K^r zu K^s und des außerhalb U gelegenen Teils von K^s zu K^r, so liegt der ganze approximative Schnittkomplex $K_1^{n-r-s} = K_1^r \cdot K_1^s$ innerhalb U, und je zwei zu verschiedenen Approximationen gehörige Schnittkomplexe sind innerhalb U homolog. (In dem Extremfall, wo U den ganzen Raum mit Ausnahme der Randpunkte von K^r und K^s ausfüllt, ist die im Satz gegebene Schranke für δ dieselbe wie oben, sonst möglicherweise kleiner.)

Da der Satz mit dieser genauen Schrankenangabe nicht bei Lefschetz steht, führe ich den Beweis an: Man bilde die verbindenden Komplexe von K^s mit K_1^s und K_2^s, deren Differenz ein verbindender Komplex von K_1^s mit K_2^s ist. Wenn dieser Komplex nicht geradlinig ist, mache man ihn geradlinig durch eine Approximation, ohne die schon von vornherein geradlinigen Simplizes von K_1^s und K_2^s zu verrücken und ohne die δ-Umgebung von K^s zu verlassen. Der so konstruierte Komplex K_v^{s+1} hat (bei passender Wahl der Approximation) allgemeine Lage zu K_1^r und hat als Rand

$$R(K_v^{s+1}) = K_1^s - K_2^s - K_v^s,$$

wobei K_v^s in der δ-Umgebung von $R(K^s)$ liegt. Aus (3), angewandt auf K_1^r und K_v^{s+1}, findet man wegen $R(K^r) \cdot K_v^{s+1} = 0$:

$$K_1^r \cdot (K_1^s - K_2^s + K_v^s) \sim 0 \quad (\text{in } U),$$

oder wegen $K_1^r \cdot K_v^s = 0$:

$$K_1^r \cdot K_1^s \sim K_1^r \cdot K_2^s,$$

q. e. d.

Auf Grund der bewiesenen Eindeutigkeit bis auf Homologie bezeichnet man den approximativen Schnittkomplex $K_1^r \cdot K_1^s$ einfach mit $K^r \cdot K^s$. Mit dieser Bezeichnung gelten auch für beliebige (nicht notwendig geradlinige) Komplexe die Gleichungen (1), (3), (4) unter denselben Voraussetzungen, während (2) im betrachteten Fall nur besagt, daß $K^r \cdot K^s$ ein Zyklus ist.

Wichtig ist insbesondere der Fall $r + s = n$, wo der Schnittkomplex aus endlichvielen Punkten besteht. Die eindeutige Bestimmtheit bis auf Homologie in U besagt in diesem Fall, daß bei jeder Zerlegung von U in getrennte Teilgebiete U' die algebraische Summe der Vorzeichen der Schnittpunkte in jedem Gebiet U' (kurz: die *Schnittpunktszahl in U'*) unabhängig von der Approximation ist. Auch die *Gesamtschnittpunktszahl* oder Schnittpunktszahl in U ist durch K^r und K^s allein eindeutig bestimmt: sie wird mit $\chi(K^r, K^s)$ oder einfach mit $(K^r \cdot K^s)$ bezeichnet. Ist speziell P ein isolierter Punkt des Durchschnitts von K^r und K^s, und wählt man für ein U' eine beliebig kleine Umgebung von P, deren abgeschlossene Hülle keine weiteren Schnittpunkte enthält, so heißt die Schnittpunktszahl in U' der *Index* des Schnittpunktes P. Er kann (wie alle Schnittpunktszahlen) positiv, negativ oder Null sein. Sind alle Schnittpunkte isoliert, so ist die Indexsumme die Gesamtschnittpunktszahl.

Ist insbesondere $r = 0$, $s = n$, K^0 ein Punkt P mit positiver Orientierung, so geht die Definition des Index von P als Schnittpunkt von K^n und K^0 in die des *Abbildungsgrades* der Abbildung des Urbildkomplexes im Raum R^n im Punkt P über [15]); dieser ist eindeutig definiert, sobald P kein Randpunkt von K^n ist. Die Grundeigenschaften des Abbildungsgrades sind aus den Formeln (3), (4) abzulesen. Eine weitere Eigenschaft wird im folgenden oft gebraucht: *Der Grad einer eineindeutigen Abbildung ist ± 1.* Das folgt aus der leicht zu beweisenden Tatsache, daß der Grad der aus der Abbildung und ihrer Inversen zusammengesetzten identischen Abbildung einerseits gleich dem Produkt der beiden Abbildungsgrade, andererseits gleich $+1$ sein muß. Die Abbildungen vom Grade $+1$ sind die, welche „die Orientierung erhalten".

Nach Lefschetz [16]) sind die Schnittpunktszahlen und Schnittkomplexe (bis auf Homologie) invariant bei topologischen Abbildungen dieses Gebietes mit Abbildungsgrad $+1$ (bei den anderen vom Grade -1 kehren die Schnittpunktszahlen ihr Vorzeichen um).

§ 4.

Schnittpunkte von differenzierbaren Komplexen.

Wir haben gesehen, daß eine Orientierung eines geradlinigen Simplex X^n des Raumes R^n dadurch festgelegt werden kann, daß man den Koordinaten eine bestimmte Reihenfolge zuerkennt. Dasselbe gilt nun, wenn im Simplex X^n nicht schiefwinklige, sondern beliebige (krummlinige) Koordinaten $u_1, \ldots, u_n$ eingeführt werden derart, daß das Simplex ein topologisches Bild eines Bereiches im u-Raum wird. Da nämlich der Abbildungsgrad einer eineindeutigen Abbildung stets ± 1 ist, so kann man (nach erfolgter Orientierung des u-Raums durch Wahl einer Reihenfolge $u_1, \ldots, u_n$) die Orientierung des Simplex in einer und nur einer Weise so wählen, daß der Grad der Abbildung des Simplex auf den u-Raum gleich $+1$ wird.

---

15) L. E. J. Brouwer, Über Abbildungen von Mannigfaltigkeiten, Math. Annalen **71** (1911), S. 97—115.

16) L_1 S. 21—26.

Ein Komplex K^r in R^n heißt (stückweise) *differenzierbar*, wenn für jedes Simplex des Urbildes Koordinaten $u_1, \ldots, u_r$ so eingeführt werden können, daß die Koordinaten des Bildpunktes $x_1, \ldots, x_r$ im Innern eines jeden Simplex differenzierbare, am Rande stetige Funktionen von $u_1, \ldots, u_r$ sind. Durch Wahl einer bestimmten Reihenfolge der Parameter u_j für jedes einzelne Simplex wird K^r orientiert; diese Orientierung ist noch ganz allgemein, da man sie durch Ersetzung eines u_j durch $-u_j$ für jedes einzelne Simplex in die entgegengesetzte überführen kann.

Ist P ein isolierter Schnittpunkt der beiden differenzierbaren Komplexe K^r, K^s $(r + s = n)$, ist weiter in beiden Komplexen P das Bild je eines einzelnen inneren Punktes eines Simplex mit Koordinaten $u_1, \ldots, u_r$ bzw. $v_1, \ldots, v_s$, ist schließlich an dieser Stelle die Determinante

$$D = \left| \frac{\partial x_j}{\partial u_1}, \ldots, \frac{\partial x_j}{\partial u_r}, \frac{\partial x_j}{\partial v_1}, \ldots, \frac{\partial x_j}{\partial v_s} \right| \neq 0,$$

so hat P als Schnittpunkt von K^r und K^s den Index $\varepsilon = \pm 1 = \operatorname{sign} D$.[17]

Beweis. Die x können linear so transformiert werden, daß die Determinante (im Punkt P) in der Hauptdiagonale lauter Einsen und überall sonst Nullen hat. Dabei kehrt sich die Orientierung des Raumes R^n, also auch der Index von P als Schnittpunkt, um oder nicht, je nachdem D positiv oder negativ war (§ 1). Auf Grund der Differenzierbarkeit der Funktionen $x(u)$ hat der Abstand des durch sie dargestellten Punktes vom Punkt $(u_1, \ldots, u_r, 0, \ldots, 0)$ für $|u_1| \leqq h, \ldots, |u_r| \leqq h$ die Größenordnung $o(h)$, mithin ist er bei passender Wahl von h kleiner als $\frac{h}{4}$. Ebenso haben die Punkte $x(v)$ von K^s von den Punkten $(0, \ldots, 0, v_1, \ldots, v_s)$ für $|v_j| \leqq h$ Entfernungen $< \frac{h}{4}$. Setzt man $\frac{h}{4} = \delta$, so kann man eine δ-Approximation der beiden durch $|u_j| \leqq h$, $|v_j| \leqq h$ gegebenen Ausschnitte von K^r und K^s dadurch bestimmen, daß man die Punkte x durch $(u_1, \ldots, u_r, 0, \ldots, 0)$ bzw. $(0, \ldots, 0, v_1, \ldots, v_s)$ ersetzt. Der Rand eines jeden dieser approximierenden Komplexausschnitte hat vom anderen eine Entfernung h, also hatte vor der Approximation der Rand eines jeden Ausschnittes vom anderen Ausschnitt eine Entfernung $> h - 2\delta = 2\delta$. Für die Bestimmung des Index von P als Schnittpunkt von K^r und K^s genügt es, sich auf die betrachteten Ausschnitte von K^r und K^s zu beschränken; für diese ist auf Grund der Randabstände eine δ-Approximation erlaubt (vgl. die Definition des Index in § 3); mithin hat man, um den Index von P zu bestimmen, nur das Vorzeichen von P als Schnitt-

---

[17] Entsprechendes gilt für isolierte Schnittpunkte von drei oder mehr differenzierbaren Komplexen.

punkt der beiden geradlinigen Komplexe

$$'K^r: \quad (x_1, \ldots, x_n) = (u_1, \ldots, u_r, 0, \ldots, 0) \qquad (|u_j| \leqq h),$$
$$'K^s: \quad (x_1, \ldots, x_n) = (0, \ldots, 0, v_1, \ldots, v_s) \qquad (|v_j| \leqq h)$$

zu bestimmen; dieses ist aber, wie man durch eine einfache Wahl zweier kleinen Simplizes in $'K^r$ und $'K^s$ und Kombination der Ecken zu einem Simplex in R^n erkennt, gleich $+1$. Damit ist alles bewiesen.

§ 5.

Schnittpunkte von analytischen Komplexen.

Für die Untersuchung der analytischen Komplexe brauchen wir zunächst einige funktionentheoretische Begriffe und Sätze.

Wenn für alle Wertsysteme der komplexen Variablen $z_1, \ldots, z_n$ in der Umgebung einer Stelle z^0 endlichviele Wertsysteme $w_1, \ldots, w_m$ so definiert sind, daß die symmetrischen Funktionen dieser k Wertsysteme reguläre analytische Funktionen der z sind, also die w selbst Wurzeln von algebraischen Gleichungen mit analytisch von den z abhängigen Koeffizienten, so folgt daraus bekanntlich, daß zu jedem passend gewählten Wertsystem z in dieser Umgebung die w in k eindeutige analytische „Funktionszweige" zerfallen. Nimmt man nun außerdem an, daß man in beliebig kleinen Umgebungen der Ausgangsstelle aus einem dieser Funktionszweige alle anderen durch analytische Fortsetzung erhalten kann (oder daß das durch die w definierte „analytische Gebilde" nicht „zerfällt"), so soll das durch einen dieser Funktionenzweige gegebene System $w_1, \ldots, w_m$ ein *System von m mehrdeutigen analytischen Funktionen in der Umgebung der Stelle z^0* heißen. Die eindeutigen analytischen Funktionen der w und z bilden dann einen *Funktionenkörper*. Die eindeutigen analytischen Zweige aller Funktionen w sind in der betrachteten Umgebung überall regulär bis auf einem „Verzweigungsgebilde" von höchstens $2n-2$ Dimensionen im n-dimensionalen Raum der $z_1, \ldots, z_n$.

Irgend r von den Funktionen $w_1, \ldots, w_m$ heißen *analytisch-unabhängig*, wenn keine als analytische Funktion der übrigen ausgedrückt werden kann. Unter je m Funktionen w_i, die nicht alle konstant sind, sind immer eine Anzahl analytisch-unabhängiger zu finden, von denen die übrigen analytisch abhängen. Ein Kriterium für die analytische Abhängigkeit von n analytischen Funktionen von n Variablen ist das identische Verschwinden der Funktionaldeterminante.

Es gilt weiter der folgende „Umkehrsatz": *Ist w eine mehrdeutige analytische Funktion von $z_1, \ldots, z_n$ in der Umgebung einer Stelle z^0 in einem Funktionenkörper und ist $\frac{\partial w}{\partial z_1}$ nicht identisch Null, so können um-*

111

gekehrt z_1 und daher weiter alle Funktionen des Körpers auch als mehr-deutige analytische Funktionen von w, z_2, ..., z_n in einer Umgebung der Stelle w^0, z_2^0, ... aufgefaßt werden. (Ist dagegen $\frac{\partial w}{\partial z_1} = 0$, so ist w eine Funktion von z_2, ..., z_n allein).

Durch wiederholte Anwendung folgt der „Austauschsatz": *Sind w_1, ..., w_m analytisch-unabhängige analytische Funktionen von z_1, ..., z_n in einem Funktionenkörper, so gibt es unter den z_j m Größen, die gegen die w ausgetauscht werden können, derart, daß alle Funktionen des Kör-pers als analytische Funktionen der w und der übrigbleibenden z aufge-faßt werden können.*

Ist insbesondere $m = n$, so folgt, daß alle z_j mehrdeutige analytische Funktionen der w in der Umgebung der Stelle w^0 werden.

n komplexe Veränderliche $z_j = x_j + i y_j$ können als komplexe Koor-dinaten in einem R^{2n} aufgefaßt werden. Für die reellen Koordinaten x_j, y_j setzen wir ein für allemal die Reihenfolge

$$x_1, y_1, x_2, y_2, \ldots, x_n, y_n$$

fest, wodurch nach § 2 eine bestimmte Orientierung des Raumes R^{2n} ge-geben ist.

Ein Komplex K^{2r} im R^{2r}, der aus eineindeutig stetigen Bildern von geradlinigen (2)-dimensionalen Simplizes aufgebaut ist, heißt ein *ana-lytischer Komplex von r komplexen Dimensionen*, wenn die Umgebung einer beliebigen Stelle von K^{2r} (Randstellen ausgenommen) sich aus endlich-vielen „analytischen Zweigen" zusammensetzt, auf denen jeweils die Ko-ordinaten z_1, ..., z_n als mehrdeutige analytische Funktionen von r unter ihnen, etwa von z_1, ..., z_r, aufgefaßt werden können.

Diese Struktur der Umgebungen einer Stelle besitzen nicht nur die algebraischen Varietäten, sondern nach O. Blumenthal[18] überhaupt alle durch analytische Gleichungen definierbaren Punktmengen. Sollen diese unter unsere Definition fallen, so müssen sie außerdem Komplexe, im obigen Sinn, sein, d. h. mit endlichvielen eineindeutig stetigen Bildern von Sim-plizes triangulierbar sein. Diese Bedingung kann man dadurch erfüllen, daß man sich auf einen beschränkten Bereich $|z_j| \leq M$ beschränkt, in dem die definierenden Gleichungen des untersuchten Gebildes überall regulär sind (vgl. Anhang 1). Richtet man die Triangulation so ein, daß die Ver-zweigungsgebilde der analytischen Funktionen z_j immer aus höchstens $(2r - 2)$-dimensionalen Seiten der zur Triangulation benutzten Simplizes bestehen, so kann man in den einzelnen Simplizes die unabhängigen

---

[18] O. Blumenthal, Zum Eliminationsproblem bei analytischen Funktionen mehrerer Veränderlicher, Math. Annalen **57** (1903), S. 356.

$z_1, \ldots, z_r$ als Koordinaten benutzen. Die übrigen Raumkoordinaten sind dann im Innern dieser Simplizes differenzierbare Funktionen dieser $2\,r$ Parameter, mithin:

Ein analytischer Komplex von r komplexen Dimensionen ist ein differenzierbarer Komplex von $2\,r$ Dimensionen.

Statt der speziellen Parameter $z_1, \ldots, z_r$ benutzen wir im folgenden beliebige komplexe Parameter $w_1, \ldots, w_r$, von denen die Koordinaten $z_1, \ldots, z_n$ im Innern eines Simplex regulär-analytisch abhängen. Setzt man $w_j = u_j + i v_j$, so ist durch die zu Anfang dieses Paragraphen gemachten Voraussetzungen über die Reihenfolge der Real- und Imaginärteile eine bestimmte Orientierung aller Simplizes des Komplexes K^{2r} festgelegt, unabhängig von der Reihenfolge der w.

Beim Übergang zu anderen Koordinaten $W_1, \ldots, W_r$, die von den w umkehrbar-analytisch abhängen, ist $\dfrac{\partial(W_1, \ldots, W_r)}{\partial(w_1, \ldots, w_r)} \neq 0$. Wir wollen, nach Aufspaltung in Real- und Imaginärteil $W_j = U_j + i V_j$, die Funktionaldeterminante $\dfrac{\partial(U_1, V_1, \ldots, U_r, V_r)}{\partial(u_1, v_1, \ldots, u_r, v_r)}$ berechnen. Sie ändert ihren Wert nicht, wenn wir in Zähler und Nenner formal Linearkombinationen $W = U + iV$, $\overline{W} = U - iV$ einführen. Also hat man

$$\frac{\partial(U_1, V_1, \ldots, U_r, V_r)}{\partial(u_1, v_1, \ldots, u_r, v_r)} = \frac{\partial(W_1, \overline{W}_1, \ldots, W_r, \overline{W}_r)}{\partial(w_1, \overline{w}_1, \ldots, w_r, \overline{w}_r)}$$

$$= \frac{\partial(W_1, \ldots, W_r)}{\partial(w_1, \ldots, w_r)} \cdot \frac{\partial(\overline{W}_1, \ldots, \overline{W}_r)}{\partial(\overline{w}_1, \ldots, \overline{w}_r)} = \left| \frac{\partial(W_1, \ldots, W_r)}{\partial(w_1, \ldots, w_r)} \right|^2 > 0.$$

Daraus folgt, daß die oben festgesetzte Orientierung sich bei Übergang zu den Parametern W nicht ändert.

Die Parameter in zwei längs einer $(2\,r-1)$-dimensionalen Seite aneinanderstoßenden Simplizes von K^{2r} hängen auf dieser Seite durch eine analytische Transformation miteinander zusammen, denn die Singularitäten der Parameterdarstellung liegen alle auf den höchstens $(2\,r-2)$-dimensionalen Seiten. Die Orientierungen in diesen anstoßenden Simplizes stimmen also überein, d. h. *der analytische Komplex K^{2r} ist orientierbar.*

Die Randpunkte analytischer Komplexe, wo die analytische Darstellung eventuell aufhören kann, mögen ein für allemal von der Untersuchung ausgeschlossen werden.

Nunmehr seien zwei analytische Komplexe K^{2r}, K^{2s} $(r + s = n)$ in der Umgebung eines isolierten (inneren) Schnittpunktes P gegeben durch Parameterfunktionen $z_j(w_1, \ldots, w_r)$ bzw. $z_j^*(w_1^*, \ldots, w_s^*)$. Im einfachsten Fall, daß P ein innerer Punkt eines Simplex von K^{2r} und eines Simplex von K^{2s} ist, und die komplexe Funktionaldeterminante $\dfrac{\partial z_j}{\partial w_1}, \ldots, \dfrac{\partial z_j}{\partial w_r}$,

$\dfrac{\partial z_j^*}{\partial w_1^*}, \ldots, \dfrac{\partial z_j^*}{\partial w_s^*}\bigg|$ in $P \neq 0$ ist, findet man durch dieselbe Rechnung wie oben für die reelle Determinante D von § 4 einen positiven Wert; *mithin hat in diesem Fall der Schnittpunktsindex den Wert* $+1$.

Wir wollen nun für einen beliebigen isolierten Schnittpunkt beweisen, daß der Schnittpunktsindex positiv sein muß, und zwar wollen wir durch eine kleine Parallelverschiebung den Fall auf den vorigen zurückführen.

Wir setzen zunächst $t_j = z_j - z_j^*$, und wollen zeigen, daß die n Funktionen t_j der n Veränderlichen w, w^* analytisch-unabhängig sind. Wären sie es nicht, und wären etwa nur $t_1, \ldots, t_m$ $(m < n)$ analytisch-unabhängig, so könnte man nach dem Austauschsatz $n - m$ von den Größen w und w^* so auswählen, daß alle z, z^*, w, w^* als mehrdeutige analytische Funktionen von $t_1, \ldots, t_m$ und den $n - m$ übriggebliebenen w, w^* aufgefaßt werden können. Läßt man nun, von den zum Schnittpunkt gehörigen Nullwerten dieser neuen Variablen ausgehend, ein w oder w^* sich stetig ändern, während man $t_1, \ldots, t_m$ konstant gleich' Null läßt (man kann sich dabei ganz beliebig für irgendeinen Zweig der mehrdeutigen Funktionen z, z^*, w, w^* entscheiden), so bleiben auch die übrigen t_j, die ja analytisch von $t_1, \ldots, t_m$ abhängen, konstant gleich Null, also $z_j = z_j^*$ $(j = 1, \ldots, n)$, mithin hat man es fortwährend mit einem Schnittpunkt zu tun, während doch ein w oder w^* variabel ist, entgegen der Voraussetzung, wir hätten es mit einem isolierten Schnittpunkt zu tun.

Also sind die t_j analytisch unabhängig. Daraus folgt erstens, daß ihre Funktionaldeterminante $D = \dfrac{\partial\,(t_1, \ldots, t_n)}{\partial\,(w_1, \ldots, w_s^*)}$ nicht identisch verschwindet, und zweitens, daß die w, w^* und z, z^* als endlichvieldeutige analytische Funktionen der t_j in der Umgebung der Null aufgefaßt werden können. Zu jeder Stelle t in dieser Umgebung gehören endlichviele Stellenpaare z, z^* mit $z_j - z_j^* = t_j$, und bei passend gewählten t (nämlich überall bis auf einem Gebilde von höchstens $2n - 2$ Dimensionen) sind an allen diesen Stellen die Funktionen $z(w)$, $z^*(w^*)$ unverzweigt und die Funktionaldeterminante D nicht Null.

Geometrisch bedeutet das: Wenn man einen der beiden gegebenen Komplexe, etwa K^{2r}, um beliebig wenig parallel verschiebt, indem man den Punkt z_j durch $z_j - t_j$ ersetzt, so hat der verschobene Komplex K^{2r} mit dem durch $z^*(w^*)$ gegebenen K^{2s} endlichviele Punkte in der Umgebung der betreffenden Stelle gemein, und bei passender Wahl der t_j sind alle diese Stellen innere Punkte der Simplizes von K^{2r} und K^{2s}, während die Funktionaldeterminante

$$D = \frac{\partial\,(z_1 - z_1^*, \ldots, z_n - z_n^*)}{\partial\,(w_1, \ldots, w_r, \ldots, w_s^*)} = (-1)^s \cdot \left| \frac{\partial z_j}{\partial w_1}, \ldots, \frac{\partial z_j}{\partial w_r}, \frac{\partial z_j^*}{\partial w_1^*}, \ldots, \frac{\partial z_j^*}{\partial w_s^*} \right|$$

an diesen Stellen nicht verschwindet.

Daraus folgt nach dem früheren, daß diese endlichvielen Schnittpunkte der verschobenen Komplexe die Indizes $+1$ haben. *Somit hat der ursprüngliche isolierte Schnittpunkt einen positiven Index, gleich der Anzahl der Schnittpunkte nach der Verschiebung.*

Der obige Beweis ist, bis auf die Ersetzung der algebraischen Abhängigkeiten durch analytische und einer kleinen Vereinfachung dadurch, daß eine Verschiebung statt einer projektiven Transformation benutzt wurde, wesentlich derselbe wie der Beweis für die Positivität der Schnittpunktsmultiplizitäten in W_2, § 9. Zugleich stellt er die Ausführung der Lefschetzschen Beweisandeutung (Fußnote [12])) dar. Er ist mühelos auf Schnittpunkte von mehr als zwei Komplexe auszudehnen.

§ 6.

Die Beziehung zum algebraischen Multiplizitätsbegriff.

Dieser Paragraph ist im logischen Zusammenhang dieser Arbeit entbehrlich und dient bloß dazu, den Anschluß der jetzigen Untersuchung an die frühere W_2 herzustellen.

Die durch algebraische Gleichungen definierten Punktmengen oder *algebraischen Varietäten* im komplexen projektiven Raum sind triangulierbare Komplexe (vgl. Anhang 1), die in irreduzible Bestandteile von verschiedener Dimension zerfallen, welche in der Umgebung einer jeden Stelle analytisch sind. Wir bezeichnen sie mit V_r oder V^{2r}, wenn r die algebraische oder komplexe Dimension und daher $2r$ die reelle Dimensionszahl ist. Orientierung nach der Vorschrift von § 5.

Die *algebraische Multiplizität* eines Schnittpunktes P zweier algebraischen Varietäten V_r, V_s $(r + s = n)$ kann folgendermaßen definiert werden (vgl. W_2, § 8): Man bringe zunächst die beiden Varietäten in „allgemeine Lage" zueinander, indem man die eine, etwa V_r, einer projektiven Transformation mit unbestimmten Koeffizienten unterzieht. Die Schnittpunkte der transformierten Varietäten (jeder einmal gezählt) ergeben, wenn man nachher die Transformationsmatrix zur Einheitsmatrix spezialisiert, durch „relationstreue Spezialisierung" ebenso viele (gleiche oder verschiedene) Schnittpunkte der untransformierten Varietäten, und die Zahl, die angibt, wie oft dabei ein jeder Schnittpunkt entsteht, ist die „Multiplizität" dieses Schnittpunkts. Dabei wird unter einer „relationstreuen Spezialisierung" eine solche verstanden, wobei alle durch homogene Gleichungen ausgedrückten algebraischen Relationen, die zwischen den Schnittpunkten und den Transformationskoeffizienten bestehen, erhalten bleiben (vgl. W_1, § 3).

Wählt man die uneigentliche Ebene des Raumes so, daß weder vor noch nach der Transformation Schnittpunkte in ihr liegen (das geht, wenn,

115

wie wir annehmen wollen, nur endlichviele Schnittpunkte vorhanden sind),
so kann man diese uneigentliche Ebene ganz weglassen, d. h. sich auf den
euklidischen Raum beschränken, und die homogenen Gleichungen, die in
der Definition der Relationstreue vorkamen, durch inhomogene ersetzen.
Weiter kann man die Matrix mit unbestimmten Koeffizienten ersetzen
durch eine von der Einheitsmatrix beliebig wenig verschiedene Matrix,
deren Elemente algebraisch-unabhängige Transzendente sind (unabhängig auch
von den Koeffizienten der definierenden Gleichungen der gegebenen Varietäten);
für die algebraischen Eigenschaften der Schnittpunkte macht das offenbar
nichts aus. Schließlich kann man die relationstreue Spezialisierung durch einen
Grenzübergang ersetzen, denn bei einem solchen bleiben tatsächlich alle
durch algebraische Gleichungen ausdrückbaren Relationen erhalten.

Bei einem Grenzübergang bleiben aber auch alle topologischen Schnittpunktszahlen im Sinne von § 3 [19]) erhalten, denn diese Anzahlen waren
durch eine Approximation definiert, welche natürlich für alle diejenigen
transformierten Varietäten, welche sich von der gegebenen um hinreichend
wenig unterscheiden, gültig bleibt. Also ist die Schnittpunktsanzahl in der
Umgebung einer Stelle P dieselbe für eine der approximierenden V_r mit
der festen V_s wie für die gegebene V_r mit derselben V_s.

Wie wir wissen, können wir in beliebiger Nähe von V_r parallel-ver-
schobene V_r finden, für die alle Schnittpunktsindizes mit V_s gleich Eins
werden. Die Matrix einer solchen Parallelverschiebung (als projektive Transformation aufgefaßt) sei B, und eine Matrix mit unabhängigen transzendenten Koeffizienten, die sich von B sehr wenig unterscheidet, sei A. Wir
können dann den Grenzübergang, der unsere relationstreue Spezialisierung
liefern soll, so ausführen, daß wir zuerst A zu B, dann B zur Einheitsmatrix E konvergieren lassen. Würden beim ersten Grenzübergang $A \to B$
mehrere Schnittpunkte zu einem konvergieren, so würde der Index dieses
einen Grenzpunktes größer als Eins werden, was unmöglich ist. Also konvergieren bei $A \to B$ alle Schnittpunkte wieder zu getrennten Schnittpunkten. Beim Übergang $B \to E$ rücken einige Schnittpunkte in P zusammen; ihre Anzahl ist erstens gleich dem Index von P (da die Indexsumme in der Umgebung erhalten bleibt), zweitens gleich der algebraischen
Multiplizität (da es sich um eine relationstreue Spezialisierung handelt).
Damit ist bewiesen:

Die algebraische Multiplizität eines Schnittpunktes zweier algebraischen
Mannigfaltigkeiten, die nur endlichviele Schnittpunkte besitzen, ist gleich
dem topologischen Index.

---

[19]) Um diese Definition anwenden zu können, beschränke man sich jeweils auf
einen ganz im Endlichen gelegenen Teil der Komplexe V_r bzw. V_s.

Die in W_2, § 9, auf algebraischem Wege bewiesene Tatsache, daß die Multiplizität eines jeden isolierten Schnittpunktes stets größer als Null ist, ergibt sich jetzt als Spezialfall des allgemeineren Ergebnisses von § 5 für analytische Komplexe.

§ 7.

Ausdehnung auf allgemeinere Mannigfaltigkeiten.

Unter einer *topologischen Mannigfaltigkeit* wird verstanden ein zusammenhängender topologischer Raum, in dem eine Umgebung eines jeden Punktes topologisch auf eine offene Menge eines euklidischen Parameterraums abbildbar ist. Ist der Raum kompakt (etwa ein Komplex), so spricht man von einer *geschlossenen Mannigfaltigkeit*. An jeder Stelle, wo zwei verschiedene euklidisch-abbildbare Umgebungen sich teilweise überdecken, hängen die beiden Parameterdarstellungen durch eine eineindeutige Parametertransformation mit Abbildungsgrad ± 1 miteinander zusammen. Können die Orientierungen der euklidischen Parameterräume so gewählt werden, daß diese Abbildungsgrade alle gleich $+1$ werden, so heißt die Mannigfaltigkeit *orientierbar*. Sind die genannten Parametertransformationen beiderseits differenzierbar, so spricht man von einer *differenzierbaren Mannigfaltigkeit*; auf ihr kann man in naheliegender Weise differenzierbare und nicht-differenzierbare Komplexe unterscheiden. Sind die Parametertransformationen sogar (nach Zusammenfassung je zweier reeller Parameter x, y zu einem komplexen $x + iy$) analytisch, so spricht man von einer (komplexen) *analytischen Mannigfaltigkeit* (vgl. die Weylsche Definition einer Riemannschen Fläche); eine solche ist von selbst orientierbar, da der Abbildungsgrad einer analytischen Abbildung im komplexen Gebiet stets positiv ist (§ 5).

Es sei M^n eine orientierbare und orientierte topologische Mannigfaltigkeit. Wir betrachten zwei Komplexe K^r, K^s ($r + s = n$) in M^n, bei denen der Rand von K^r zu K^s und der Rand von K^s zu K^r fremd ist. Der *Index* eines isolierten Schnittpunkts P von K^r und K^s wird definiert, indem die Umgebung von P vermöge einer der Parameterdarstellungen von M^n auf eine offene Menge des euklidischen R^n abgebildet wird und dort der Index im Sinne von § 2 bestimmt wird. Diese Indexdefinition ist invariant gegenüber Parametertransformationen mit Abbildungsgrad $+1$. Sind alle Schnittpunkte von K^r und K^s isoliert, so heißt die Indexsumme die *Schnittpunktszahl* von K^r und K^s.

Zur Definition der Schnittpunktszahlen und Schnittkomplexe bei nicht-isolierten Schnittpunkten und für $r + s > n$ braucht man simpliziale Approximationen, die nicht nur für die Umgebung einer Stelle, sondern in der ganzen Mannigfaltigkeit sinnvoll bleiben, und zwar braucht man verschiedene Approximationen für K^r und K^s derart, daß eine Zelle der Approximation von K^r mit einer Zelle der Approximation von K^s immer einen Schnittkomplex von der richtigen Dimension $r + s - n$ hat. Die Möglichkeit einer solchen Approximation ist bis heute nur bewiesen für *„Mannigfaltigkeiten im kombinatorischen Sinn"*, die derart in Simplizes eingeteilt werden können, daß die Simplizes um einen Punkt sich so aneinanderschließen wie Simplizes im euklidischen Raum, die eine Umgebung einer gemeinsamen Ecke ganz ausfüllen. Wir können dafür auf die schon mehrfach genannte Lefschetzsche Arbeit (L_1) verweisen. Für allgemeinere Mannigfaltigkeiten steht der Beweis noch aus (sogar wenn sie triangulierbar sind); wir setzen daher einstweilen voraus, daß wir es mit Mannigfaltigkeiten im kombinatorischen Sinn zu tun haben [20]).

Die Formeln (1), (3), (4) von § 3 können für beliebige Komplexe genau so bewiesen werden wie für den euklidischen Raum. Auch die Invarianz bei topologischen Abbildungen vom Grade $+1$ hat Lefschetz bewiesen; daraus folgt insbesondere die Übereinstimmung des mit diesen Methoden definierten Indexbegriffs mit dem vorhin durch direkte Übertragung aus dem euklidischen Raum definierten Indexbegriff im Fall eines isolierten Schnittpunkts.

Für analytische Komplexe in einer analytischen Mannigfaltigkeit sind, wenn man alle Orientierungen gemäß der Verabredung in § 5 normiert, die Indizes isolierter Schnittpunkte stets positiv [21]). Das folgt unmittelbar aus dem Satz von § 5.

Da außerdem die Dimensionen analytischer Komplexe gerade sind, gilt im kommutativen Gesetz (1a) für sie stets das $+$-Zeichen.

---

[20]) **Zusatz bei der Korrektur.** Inzwischen ist die Dissertation von E. R. van Kampen („Die kombinatorische Topologie und die Dualitätssätze", Leiden 1929) erschienen, in der die erwähnte Lücke ausgefüllt ist. Dort werden nämlich die Schnittpunktszahlen definiert und ihre Eigenschaften bewiesen für triangulierbare „Mannigfaltigkeiten" in einem sehr allgemeinen Sinn, unter die unter anderem alle oben definierten „topologischen Mannigfaltigkeiten" fallen, soweit sie triangulierbar sind. Für die algebraische Anwendung reicht das völlig hin, denn alle algebraischen Mannigfaltigkeiten sind triangulierbar (Anhang 1).

[21]) Bei nichtisolierten Schnittpunkten können aber die Schnittpunktszahlen sehr wohl negativ sein, wie das folgende Beispiel zeigt: Durch einen Kegelschnitt K im komplexen projektiven R_3 werde eine singularitätenfreie Fläche 5. Ordnung F gelegt. Die Ebene des Kegelschnittes schneidet die Fläche außerdem in einer kubischen Kurve C. Die Schnittpunktszahl von K mit jedem ebenen Schnitt von F auf F, also insbesondere mit $K + C$, ist $+2$, die von K mit C aber $+6$, also bleibt für die Schnittpunktszahl von K mit K nur -4 übrig.

Ist die Mannigfaltigkeit, in der wir uns befinden, der komplexe projektive Raum oder allgemeiner eine (singularitätenfreie) allgemeine Varietät V_n,[22] so sind alle in dieser Mannigfaltigkeit liegenden algebraischen Varietäten $V_r = V^{2r}$ erstens *triangulierbar* (vgl. Anhang 1), zweitens *orientierbar* (wie alle analytischen Komplexe) und drittens *unberandet* (weil die algebraischen Funktionen von r komplexen Variablen keine Singularitätengebilde von mehr als $2r - 2$ Dimensionen haben, mithin ein $(2r-1)$-dimensionaler Rand nicht auftreten kann), also *Zyklen*.

Für isolierte Schnittpunkte algebraischer V_r und V_s in einer Mannigfaltigkeit V_n $(r + s = n)$ ist noch keine allgemein anwendbare algebraische Multiplizitätsdefinition bekannt. Wir definieren daher (in Analogie zum projektiven Raum) als *Schnittpunktsmultiplizität* den topologischen Index des betreffenden Schnittpunktes.

Die Indexsumme genügt dem „Prinzip der Erhaltung der Anzahl", nicht nur bei stetigen Änderungen der schneidenden Varietäten, sondern sogar bei Ersetzung einer V_r durch einen dazu homologen Zykel. Nennt man zwei Zyklen *homolog mit erlaubter Division* (Zeichen $\approx$), wenn ein ganzzahliges Vielfaches ihrer Differenz homolog Null ist, so ist sofort einzusehen, daß man auch einen Zykel C^k durch einen anderen $D^k \approx C^k$ ersetzen kann, ohne seine Schnittpunktszahl mit irgendeinem Zykel der komplementären Dimension zu ändern. Denn eine Schnittpunktszahl, von der ein Vielfaches Null ist, ist selbst Null.

Jetzt sei allgemeiner $r + s \geq n$. Die Durchschnittsmenge V einer V_r und einer V_s ist eine algebraische Varietät, von der wir annehmen wollen, sie habe nicht mehr als $k = r + s - n$ komplexe Dimensionen. Der approximative Schnittkomplex S^{2k} von V_r und V_s kann in einer beliebig dünnen Umgebung von V gewählt werden. Legt man nun eine Triangulation des Raumes zugrunde, bei der V aus Seiten der Triangulation besteht, und approximiert man S^{2k} mit dieser Triangulation, so wird, wenn S^{2k} in einer genügend kleinen Umgebung von V liegt, der approximierende Zykel S_1^{2k} in V liegen, d. h. aus Zellen der höchsten Dimension $2k$ von V bestehen. Sind $V', V'', \ldots$ die $2k$-dimensionalen irreduziblen Bestandteile von V, so folgt:

$$(5) \qquad V_r \cdot V_s = S^{2k} \sim S_1^{2k} = \mu' V' + \mu'' V'' + \ldots.$$

Die Größen $\mu', \mu'', \ldots$ nennen wir die *Multiplizitäten* der Schnittbestandteile $V', V'', \ldots$[23]

---

[22] Der untere Index n soll die algebraische oder analytische Dimensionszahl darstellen; die topologische Dimensionszahl ist doppelt so groß und wird als oberer Index benutzt.

[23] Algebraisch kann man auch diese Multiplizitäten noch definieren im Fall des projektiven Raums, indem man die Varietäten V_r, V_s, V mit einem allgemeinen linearen Raum L_{n-k} schneidet und dort die Schnittpunktsmultiplizitäten bestimmt.

Bemerkung. Wenn ein irreduzibler Bestandteil von V eine Dimension $< 2k$ haben sollte, so kann er in (5) rechts nicht vorkommen. Man zeigt aber leicht, daß es solche Bestandteile gar nicht geben kann. Wäre nämlich P ein Punkt eines solchen, der nicht zugleich zu einem der $2k$-dimensionalen Bestandteile V', V'', ... gehört, so würde man V_r, V_s und V mit einem V_{n-k} durch P schneiden können, der in P einen isolierten Schnittpunkt mit V hätte. Die Schnittpunktszahl von $V_r \cdot V_s \cdot V_{n-k}$ in einer Umgebung von P müßte nach § 5 positiv sein, andererseits aber sich nicht ändern, wenn man für $V_r \cdot V_s$ die Approximation $\mu' V' + \mu'' V'' + \dots$ einsetzt, also (da P nicht zu $\mu' V' + \dots$ gehört) Null sein, was ein Widerspruch ist. Die hiermit bewiesene Tatsache, daß ein Schnittgebilde von analytischen Varietäten von r und s komplexen Dimensionen keine Komponenten von weniger als $k = r + s - n$ komplexen Dimensionen besitzen kann, ist eine Verallgemeinerung einer von Blumenthal [24]) bewiesenen Behauptung, die sich auf den Fall $s = n - 1$, $k = r - 1$ bezieht.

Hat die Durchschnittsvarietät V mehr als k komplexe Dimensionen, so versagt die Darstellung (5) von $V_r \cdot V_s$ als Summe von algebraischen Varietäten und man bleibt auf die approximativen Schnittkomplexe angewiesen [25]).

§ 8.

Der Schubertsche Kalkül.

Dem Schubertschen Bedingungskalkül [26]) liegt immer eine feste singularitätenfreie algebraische Varietät $M^{2n} = M_n$ zugrunde. Die Elemente von M_n können irgendwelche geometrische Gebilde sein (etwa alle Geraden des Raumes), aber wir verlangen, daß man M_n auch als Varietät von Punkten in einem mehrdimensionalen projektiven Raum auffassen kann. Wir verlangen außerdem, daß M^{2n} eine topologische Mannigfaltigkeit ist, in welcher Schnittkomplexe durch simpliziale Approximationen definiert werden können (§ 7).

Schubert verwendet Symbole für algebraische Bedingungen, die den Punkten von M_n aufgelegt werden: eine k-fache Bedingung ist eine solche, die eine algebraische Varietät V_{n-k} in M_n definiert, deren Punkte eben dieser Bedingung genügen. Es ist bequem, für diese Bedingung dasselbe Symbol zu benutzen wie für die Varietät V_{n-k} selber.

---

[24]) O. Blumenthal, loc. cit. [18]).

[25]) Man kann beweisen, daß man auch in diesem Fall noch $V_r \cdot V_s$ als Summe oder Differenz von algebraischen Varietäten darstellen kann. Für unsere Zwecke ist das aber nicht nötig.

[26]) H. Schubert, Kalkül der abzählenden Geometrie, Leipzig 1879.

Eine „symbolische Gleichung" zwischen Symbolen von k-fachen Bedingungen $U, V, \ldots$, etwa:

$$(6) \qquad U \doteq V + \frac{1}{2}\,W \quad {}^{27})$$

bedeutet nach Schubert die Aussage, daß für *jede* algebraische V_k der M_n die Anzahlen der Punkte von V_k, welche jeweils den Bedingungen U, V, W genügen, in derselben linearen Relation zueinander stehen wie die Symbole U, V, W in (6). Dabei sind natürlich die „Anzahlen" zu Indexsummen, also zu Schnittpunktszahlen χ zu präzisieren. In unseren Bezeichnungen würde die symbolische Gleichung (6) also heißen, daß für jede algebraische V_k in M_n die Gleichung

$$(7) \qquad \chi\,(U, V_k) = \chi\,(V, V_k) + \frac{1}{2}\,(W, V_k)$$

besteht. Es ist dabei bequem, die Gleichung (7) zu verlangen nicht nur für alle algebraischen V_k, sondern auch für die k-dimensionalen approximativen Schnittkomplexe von algebraischen Varietäten höherer Dimension [28].

In der Form (7) hat die Gleichung (6) einen ganz präzisen Sinn, der außerdem unabhängig davon ist, ob die Varietäten U und V_k, V und V_k usw. wirklich nur endlichviele Schnittpunkte haben.

Wir lassen mit Schubert in Ausdrücken wie (7) fortan die χ und die Klammer weg, verwenden also ein und dasselbe Symbol für den nulldimensionalen Komplex $U \cdot V_k$ und für die Anzahl (genauer: Summe der Vorzeichen) seiner Punkte.

Klar ist, daß man symbolische Gleichungen addieren, subtrahieren und mit rationalen Zahlen multiplizieren darf. Man darf aber auch alle Glieder mit dem Symbol einer Bedingung H multiplizieren (multiplizieren in der topologischen Bedeutung von Schnittbildung, welche in der Tat die richtige Präzisierung der Schubertschen Multiplikation von Bedingungen darstellt).

Ist nämlich H eine h-fache Bedingung, mit der man die Gleichung (6) zu multiplizieren wünscht, so kann man zunächst $h + k \geq n$ annehmen, da sonst alle Produkte $U \cdot H$ usw. Null werden. Die Behauptung

$$U \cdot H \doteq V \cdot H + \frac{1}{2}\,W \cdot H$$

bedeutet nun, daß für jede algebraische V_{h+k} gilt

$$U \cdot H \cdot V_{h+k} = V \cdot H \cdot V_{h+k} + \frac{1}{2}\,W \cdot H \cdot V_{h+k}.$$

---

[27] Schubert schreibt $=$, wo hier $\doteq$ steht.

[28] Nach dem in Fußnote [25] angeführten Satz macht es keinen wirklichen Unterschied, ob man dies verlangt oder nicht.

Auf Grund des Assoziativgesetzes (1c) erhält man aber dieselbe Gleichung, wenn man in (7) speziell $V_k = H \cdot V_{h+k}$ setzt. Damit ist die Behauptung bewiesen.

Man darf also auch symbolische Gleichungen miteinander multiplizieren. Für die Multiplikation gelten nach (1) die Assoziativ- und Kommutativgesetze. Damit ist der formale Teil des Bedingungskalküls exakt gerechtfertigt.

Wichtig für die Anwendungen ist insbesondere, daß *jede Homologie mit erlaubter Division zwischen algebraischen Varietäten zu einer symbolischen Gleichung im Schubertschen Sinne Anlaß gibt.* Ist nämlich $U \approx V$, so ist nach § 7 für jede V_k:

$$U \cdot V_k = V \cdot V_k,$$

mithin

$$U \doteq V.$$

Ob auch die Umkehrung gilt, ist mir nicht bekannt, aber ohne viel Interesse, denn man findet in praxi aus Stetigkeitsbetrachtungen immer zunächst Homologien, aus denen man dann sofort symbolische Gleichungen folgern kann.

Das *Charakteristikenproblem* für eine Mannigfaltigkeit M_n lautet in der Schubertschen Fassung: Es ist eine Basis $V^{(1)}, \ldots, V^{(m)}$ für die k-fachen Bedingungen aufzustellen derart, daß jede k-fache Bedingung V symbolisch gleich einer Linearkombination der Basiselemente wird:

$$(8) \qquad V \doteq \lambda_1 V^{(1)} + \ldots + \lambda_m V^{(m)}.$$

Die Lösung des Charakteristikenproblems konnte bis jetzt immer nur von Fall zu Fall gegeben werden. Die topologische Methode liefert zum erstenmal einen allgemeinen Beweis für die Existenz einer Basis. Denn es gibt in jeder triangulierbaren M^{2n} eine endliche Basis für die Klassen homologer Zyklen, also um so mehr eine Basis für die Homologieklassen mit erlaubter Division. Sucht man nun aus den letzteren Klassen diejenigen aus, die algebraische Varietäten enthalten, so haben diese wiederum eine endliche Basis. Die so erhaltene Basis in bezug auf die Relation $\approx$ ist um so mehr eine Basis für die Relation $\doteq$.

Die Anzahlen der nötigen Basiselemente für die Dimensionen k und $n - k$ sind gleich, und die Matrix der Schnittpunktszahlen ist regulär. Denn sonst würde es eine Linearkombination der Basiselemente für die Dimension k oder $n - k$ geben, welche mit jeder algebraischen Varietät von der komplementären Dimension die Schnittpunktszahl Null hätte, also symbolisch gleich Null wäre; damit wäre aber ein Basiselement überflüssig. — Folglich kann man die Basis für die Dimensionen k und $n - k$ auch so wählen, daß die Matrix eine Einheitsmatrix wird. Die Koeffizienten in (8) lassen sich dann deuten als Schnittpunktszahlen von V mit den Basiselementen der komplementären Dimension $n - k$.

Die wirkliche Aufstellung einer Homologiebasis ist in vielen Fällen nicht schwer. Hat man einmal die Basis für die Dimensionen k und $n - k$, sowie die Schnittpunktszahlen für die Basiselemente gefunden, so ist man im Stande, für je zwei (als Linearkombinationen der Basiselemente geschriebene) Varietäten V_k und V_{n-k} die Schnittpunktszahlen auszurechnen.

Für den Fall des projektiven Raumes soll das im nächsten Paragraphen ausgeführt werden.

<h2 style="text-align:center">§ 9.</h2>

<h2 style="text-align:center">Anwendung auf den projektiven Raum.</h2>

Im Anhang 2 wird bewiesen, daß die linearen Räume L_j eine Homologiebasis für alle Dimensionen im komplexen projektiven Raum $P_n = P^{2n}$ bilden. Führen wir mit Schubert für die einfache lineare Bedingung L_{n-1} das Symbol p ein, so stellt die Potenz p^h den L_{n-h} dar. Jede Varietät V_r ist homolog einem Vielfachen von L_r, also haben wir:

$$(9) \qquad V_r \doteq \alpha \cdot p^{n-r}.$$

Daß hier α den Grad von V_r (d. h. die Schnittpunktszahl mit einem L_{n-r}) darstellt, sieht man sofort, indem man die Gleichung (9) mit p^r multipliziert.

Ist für eine V_{n-r} ebenso

$$(10) \qquad V_{n-r} \doteq \beta \cdot p^r,$$

so erhält man durch Multiplikation von (9) mit (10) sofort den *„verallgemeinerten Bézoutschen Satz"*:

$$(11) \qquad V_r \cdot V_{n-r} = \alpha\beta.$$

Sind allgemeiner V_r und V_s mit $r + s \geqq n$ gegeben, und sind α und β wie vorhin die Gradzahlen, so erhält man ebenso:

$$V_r \cdot V_s \doteq \alpha\beta \cdot p^{2n-r-s},$$

d. h. der approximative Schnittkomplex von V^r und V^s hat als Grad das Produkt der Gradzahlen von V^r und V^s. Hat der wirkliche Durchschnitt von V^r und V^s die richtige Dimension $k = r + s - n$, und zerfällt er in irreduzible Komponenten $V', V'', \ldots$ von den Gradzahlen $g', g'', \ldots$ und Multiplizitäten $\mu', \mu'', \ldots$, so hat man

$$V_r \cdot V_s = \mu' V' + \mu'' V'' + \ldots = \mu' g' p^{n-k} + \mu'' g'' p^{n-k} + \ldots,$$

mithin

$$(12) \qquad \mu' g' + \mu'' g'' + \ldots = \alpha \cdot \beta.$$

123

Anhang 1.

Triangulierbarkeit der algebraischen Gebilde.

Satz. *Sind in einem beschränkten Teil (Simplex oder konvexen Polyeder) des reellen euklidischen R^n endlichviele algebraische Varietäten oder algebraisch-begrenzte Teilbereiche von algebraischen Varietäten gegeben, so gibt es eine Unterteilung dieses Raumteils derart, daß alle gegebenen algebraischen Varietäten aus Seiten der Zellen der Unterteilung bestehen.*

Beweis. Induktion nach n. (Für $n = 0$ ist nichts zu beweisen.)

Die Gleichungen der gegebenen Varietäten sind, eventuell nach einer linearen Koordinatentransformation, so zu schreiben, daß sie regulär in x_n sind (d. h. daß die höchste Potenz von x_n einen konstanten Koeffizienten $\neq 0$ hat). Die Projektionen dieser Varietäten (soweit ihre Dimension n ist) und ihrer Randgebilde auf den Raum $x_n = 0$ sind algebraische Gebilde von derselben Art, und zu jedem Punkt der Projektion gehören endlichviele Originalpunkte, algebraische Funktionen der Projektion. Wir verzeichnen im Raum $x_n = 0$ auch alle Verzweigungsgebilde dieser algebraischen Funktionen (wo zwei Wurzeln x_n zusammenrücken), wobei diejenigen Teilgebilde, wo noch mehr Wurzeln zusammenrücken als sonst auf dem Verzweigungsgebilde, noch besonders verzeichnet werden. Auf die Gesamtheit der so erhaltenen algebraischen Gebilde des Raumes $x_n = 0$ wird die Induktionsvoraussetzung angewandt. Über jeder Zelle oder Seite der damit erhaltenen Unterteilung des betreffenden Teils des Raumes $x_n = 0$ liegt eine zylindrische Punktmenge des R^n, und auf jeder erzeugenden Geraden ($x_1 = \text{konst.}, \ldots, x_{n-1} = \text{konst.}$) eines solchen Zylinders sind endlichviele zu den gegebenen algebraischen Varietäten gehörige x_n-Werte ausgezeichnet, welche die erzeugende Gerade also in endlichviele Strecken teilen. Ändert sich die erzeugende Gerade, so ändern sich diese Teilpunkte stetig, und es fallen nie (außer am Rande der Zelle) zwei Teilpunkte zusammen. Jede dieser Strecken durchläuft also eine Zelle des R^n. Damit ist die verlangte Zelleneinteilung des R^n konstruiert.

Durch eine baryzentrische Unterteilung erhält man, wenn man das wünscht, aus der Zelleneinteilung eine Triangulation oder Simplexeinteilung.

Mutatis mutandis läßt sich das alles auf analytische Varietäten übertragen. Wichtiger für uns ist aber die *Übertragung auf algebraische Varietäten im komplexen projektiven Raum.*

Es gibt, wenn man die komplexen Koordinaten $z_0, \ldots, z_n$ durch $\sum z_j \bar{z}_j = 1$ normiert, die folgende (bekannte) eineindeutige Abbildung des komplexen projektiven Raumes auf eine ganz im Endlichen liegende algebraische Varietät im reellen euklidischen Raum von n^2 Dimensionen:

$$z_j \bar{z}_k = x_{jk} + i y_{jk} \qquad (x_{jk} = x_{kj}, \; y_{jk} = - y_{kj}). \; {}^{29})$$

Bei dieser Abbildung gehen natürlich algebraische Varietäten von r komplexen Dimensionen in reelle algebraische Varietäten von $2r$ Dimensionen über. Der vorige Satz liefert daher die Möglichkeit einer Triangulation des komplexen projektiven Raumes, wobei beliebig vorgegebene algebraische Varietäten in ihm mit trianguliert werden können.

Anhang 2.
Die Topologie des reellen und des komplexen projektiven Raumes.

Zelleneinteilung. Die bequemste Zelleneinteilung des *reellen projektiven Raums* (homogene Koordinaten $z_0, \ldots, z_n$) erhält man so: Eine Zelle hat die Gleichungen

$$|z_1| \leqq |z_0|, \ldots, |z_n| \leqq |z_0|,$$

und die übrigen erhält man daraus durch Permutation der z_j. Verwandelt man einige der Ungleichungen in Gleichungen, so erhält man die Seiten der verschiedenen Dimensionen. Alle diese Zellen sind homöomorph Parallelotopen des euklidischen Raumes (man kann ja z. B. für alle Punkte der oben angeschriebenen Zelle die Größen $\frac{z_1}{z_0}, \ldots, \frac{z_n}{z_0}$ als inhomogene Koordinaten benutzen).

Im *komplexen projektiven Raum* hat man den obigen Ungleichungen noch andere von der Form $0 \leqq \arg\left(\frac{z_j}{z_0}\right) \leqq \pi$ für einige j, $\pi \leqq \arg\left(\frac{z_j}{z_0}\right) \leqq 2\pi$ für die übrigen j aus der Reihe $1, \ldots, n$ hinzuzufügen. Die Zellen werden topologische Produkte von Halbkreisen und Strecken. Bildet man eine baryzentrische Unterteilung, so sind alle Simplizes um einen Punkt so angeordnet wie Simplizes um einen Punkt im euklidischen Raum, wie man durch eine passende Abbildung leicht feststellt; der Raum ist mithin eine Mannigfaltigkeit im kombinatorischen Sinn [30]).

Die *Orientierbarkeit* des komplexen projektiven Raums wird genau so bewiesen wie die Orientierbarkeit aller algebraischen Varietäten im § 7:

---

[29]) Vgl. G. Mannoury, Surfaces-images, Nieuw Archief von Wiskunde (2) 4 (1898), p. 112.

[30]) Die etwas langwierige Verifikation dieser Behauptungen wird überflüssig gemacht durch die in Fußnote [20]) erwähnten neueren Untersuchungen. Denen zufolge genügt es ja, erstens nachzuweisen, daß es eine Triangulation überhaupt gibt, was in Anhang 1 schon geschah, und zweitens zu zeigen, daß der komplexe projektive Raum eine Mannigfaltigkeit im topologischen Sinn (vgl. § 7) ist. Letzteres ist klar, denn eine Umgebung eines jeden Punktes ist einem euklidischen Raum homöomorph: man kann für $z_0 \neq 0$ ja die Real- und Imaginärteile von $\frac{z_1}{z_0}, \ldots, \frac{z_n}{z_0}$ als Parameter benutzen.

362   B. L. van der Waerden. Abzählende Geometrie.

Für einen Raumteil $|z_j| \leq |z_0|$ $(j = 1, \ldots, n)$ kann man $\frac{z_1}{z_0}, \ldots, \frac{z_n}{z_0}$ als komplexe Parameter benutzen und den Raumteil entsprechend orientieren. Der Übergang zu einem angrenzenden Raumteil, wo etwa $\frac{z_0}{z_1}, \frac{z_2}{z_1}, \ldots, \frac{z_n}{z_1}$ als Parameter benutzt werden, vollzieht sich dann durch eine analytische Parametertransformation, und diese erhält die Orientierung.

Der reelle projektive Raum ist bekanntlich nur für die ungeraden Dimensionszahlen orientierbar.

Die Aufstellung einer *Homologiebasis* kann für den komplexen und den reellen Fall gemeinsam geschehen. Wir nehmen etwa den komplexen $P_n = P^{2n}$. Eine Basis für die Zyklen der Dimension $2n$ bildet natürlich der Raum P^{2n} selbst (im reellen Fall der Raum P^n, falls er orientierbar ist, also für ungerades n). Jeder Zykel C^k $(k < 2n)$ kann zunächst so deformiert werden, daß er den Punkt $(1, 0, \ldots, 0)$ nicht enthält. Deformiert man ihn jetzt weiter, indem man jeden Punkt $(x_0, \ldots, x_n)$ von C^k durch $(\lambda x_0, x_1, \ldots, x_n)$ ersetzt und die reelle Zahl λ von 1 nach 0 gehen läßt, so wird C^k schließlich ganz in den Unterraum $z = 0$ geschoben. Nimmt man nun als Induktionsvoraussetzung für den P_{n-1} an, daß in ihm eine Homologiebasis für alle Dimensionen durch die linearen Unterräume $L_j = L^{2j}$ $(j = 0, \ldots, n-1)$ gebildet wird, so folgt, daß unser Zykel Z^k für ungerades k homolog Null, für gerades $k = 2j$ homolog einem Vielfachen eines L_j im Unterraum $x_0 = 0$ ist. Damit ist die Induktionsvoraussetzung auch für den P_n bewiesen. Für P_0 ist sie trivial, mithin:

Im komplexen $P_n = P^{2n}$ bilden die linearen Unterräume $L_j = L^{2j}$ (inklusive P_n selbst) eine Homologiebasis für alle Dimensionen.

Daß die L_j sowie ihre Vielfache nicht homolog Null sind, sieht man daraus, daß die Schnittpunktszahl $L_j \cdot L_{n-j}$ den Wert 1 hat.

Im reellen projektiven P_n kommen die L_j als Zyklen nur für ungerades j in Betracht, da sie für gerades j ja nicht orientierbar sind; sie sind, doppelt gezählt, Rand eines L_{j+1}, mithin

$$2 L_j \sim 0.$$

Daß die L_j $(j = 2n+1)$ selbst nicht homolog Null und sogar nicht homolog Null modulo 2 sind, sieht man daraus, daß die „Schnittpunktszahl mod 2“ von L_j und L_{n-j} (mod 2 ist auch L_{n-j} ein Zykel) den Wert 1 hat.

Der Grenzübergang $\lambda \to 0$, der hier zur Bestimmung der Homologiebasis führte, entspricht der „ausgearteten Transformation“ von W_2, § 7. Was aber algebraisch sehr kompliziert durchzuführen war, wird topologisch äußerst einfach.

(Eingegangen am 23. 2. 1929.)

126

6.

Zur Begründung des Restsatzes mit dem Noetherschen Fundamentalsatz

Mathematische Annalen 104, 3 (1931) 472–475

In der seit Brill und Noether[1]) üblichen Begründung des Restsatzes mittels des „Noetherschen Fundamentalsatzes" ist eine Lücke, die bis jetzt unbemerkt geblieben zu sein scheint.

Es handelt sich nämlich beim Restsatz um folgenden Sachverhalt: $f = 0$ sei eine Kurve in der projektiven Ebene, von der man annehmen kann, daß sie keine anderen Singularitäten als s-fache Punkte mit getrennten Tangenten besitzt[2]). Die Kurven $\varphi = 0$ und $\varphi' = 0$ schneiden, außer einer beiden gemeinsamen Punktgruppe H, die Punktgruppen G und G' auf den Zweigen von $f = 0$ aus. Die Kurve $\psi = 0$ sei adjungiert, d. h. sie gehe $(s - 1)$-fach durch die s-fachen Punkte von $f = 0$ und sie schneide $f = 0$, außer in diesen auf jedem Zweig $(s - 1)$-fach gezählten Punkten — und außer einer Gruppe K —, genau nach der Gruppe G. Zu beweisen ist, daß es eine ebenfalls adjungierte Kurve $\psi' = 0$ gibt, welche in derselben Weise außer den vielfachen Punkten — und außer K — die Gruppe G' ausschneidet.

Zu dem Zweck genügt es offenbar, das Produkt $F = \varphi' \psi$ in der Form $Af + B\varphi$ darzustellen und $\psi' = B$ zu setzen[3]). Man muß also für F, f, φ die Voraussetzungen des Noetherschen Satzes als erfüllt nachweisen, d. h. man muß für jeden Schnittpunkt (a, b) die Noethersche Potenzreihenbedingung

$$F = Pf + R\varphi \quad (P \text{ und } R \text{ Potenzreihen in } x - a \text{ und } y - b)$$

---

[1]) A. Brill und M. Noether, Math. Annalen 7 (1874), S. 269.

[2]) Für die „Auflösung der Singularitäten" durch birationale Transformationen vgl. M. Noether, Gött. Nachr. 1871, S. 207 und Math. Annalen 9 (1875), S. 182, sowie E. Bertini, Rendiconti Ist. Lombardo (2) 21 (1888), S. 326.

[3]) Dabei sind F, f, φ als Formen in x_0, x_1, x_2 zu denken; es genügt aber bekanntlich, durch die Substitution $x_0 = 1$ zu inhomogenen Polynomen in (x_1, x_2) oder (x, y) überzugehen.

127

(oder die damit gleichlautende Polynomrelation mit Vernachlässigung von Gliedern, deren Grad eine gewisse Zahl ϱ überschreitet) nachweisen. Statt dessen beruft man sich meistens auf den „einfachen Fall" des Noetherschen Satzes, wo die Kurven $f = 0$ und $\varphi = 0$ in jedem Schnittpunkt getrennte Tangenten besitzen, während doch in Wirklichkeit die Kurve $\varphi = 0$ beliebig komplizierte Berührungen mit den Zweigen von $f = 0$ aufweisen kann.

Diese Lücke soll hier ausgefüllt werden, indem die Noethersche Potenzreihenbedingung als erfüllt nachgewiesen wird unter solchen Voraussetzungen, welche im obigen Fall der Kurven φ und $F = \varphi' \psi$ wirklich erfüllt sind

Ich schicke eine Erklärung der zum Verständnis des Satzes nötigen Begriffe voraus.

Ein *Zweig* einer ebenen algebraischen Kurve $f(x, y) = 0$ in einem Punkt (a, b) wird durch eine Puiseuxsche Reihenentwicklung

$$x - a = t^k,$$
$$y - b = b_1 t + b_2 t^2 + \cdots$$

definiert, welche diese Gleichung $f(x, y) = 0$ identisch befriedigt. Der Zweig heißt *einfach*, wenn in der Reihe für x oder y das Glied mit der ersten Potenz von t wirklich vorkommt. Läßt man alle Glieder höheren als ersten Grades weg, so erhält man die Parametergleichung der *Tangente* des einfachen Zweiges. Wählt man die Koordinatenrichtungen so, daß die Gerade $x = a$ nicht tangiert, so wird $k = 1$ und $x - a = t$.

Die *Multiplizität* des Schnittes eines Zweiges mit einer anderen Kurve $\varphi(x, y) = 0$ ist der niedrigste Exponent in der Reihenentwicklung von $\varphi(x, y)$ nach Potenzen von t.

Der zu beweisende Satz, der meines Wissens bis jetzt nur im Fall $s = 1$ ausdrücklich ausgesprochen und bewiesen worden ist[4]), lautet nun so:

Wenn der Punkt $(0, 0)$ *für die Kurve* $f = 0$ *ein s-facher Punkt mit s getrennten Tangenten ist und wenn jeder Zweig* Z_i *dieser Kurve in diesem Punkt von der Kurve* $\varphi = 0$ μ_i-*fach und von der Kurve* $F = 0$ *mindestens* $(s - 1 + \mu_i)$-*fach geschnitten wird, so ist*

$$F = P f + R \varphi$$

im Bereich der Potenzreihen nach x und y.

Beweis. Wir transformieren eventuell die Koordinaten so, daß keine der s Tangenten im betrachteten Punkt mit der Y-Achse ($x = 0$) zusammenfällt. Dann kann in der Puiseuxschen Reihenentwicklung der Para-

---

[4]) Siehe H. Kapferer, Sitzungsber. Heidelberg 1927, 8. Abhandlung, S. 79, sowie P. Dubreil, Thèse de Doctorat, Paris 1930, S. 73.

meter $t = x$ gewählt werden. Die Gleichung $f(x, y) = 0$ wird also befriedigt von s Puiseuxschen Reihen

$$y = P_i = \sum_1^\infty b_{i\nu}\, x^\nu.$$

Spaltet man die Faktoren $(y - P_1)(y - P_2) \ldots (y - P_s)$ von f ab, so ist der restliche Faktor eine Potenzreihe mit nichtverschwindendem Anfangsglied, also eine Einheit im Bereich der Potenzreihen. Also:

$$(1) \qquad f = (y - P_1)(y - P_2) \ldots (y - P_s) \cdot E.$$

Wir wollen nun unsere Behauptung durch Induktion nach s beweisen, müssen sie aber zu dem Zweck ausdehnen von Polynomen F, φ, f auf Potenzreihen F, φ, f, von denen f nach (1) zerfällt. Es genügt und ist bequem, F und φ und f als Polynome in y und Potenzreihen in x anzunehmen. Der Begriff Multiplizität bleibt derselbe: die Potenzreihe φ schneidet einen Zweig $y = P_i$ im Nullpunkt μ-fach, wenn die Potenzreihe φ^*, die aus φ durch die Substitution $y = P_i$ entsteht, durch x^μ teilbar ist.

Für $s = 1$ ist die Behauptung nun sehr leicht zu beweisen. Setzt man in F und φ für y die Potenzreihe P_1 ein, wodurch F^* und φ^* entstehen mögen, so wird φ^* gleich einem Einheitsfaktor mal x^{μ_1}, während F^* durch x^{μ_1}, also auch durch φ^* teilbar wird. Da man das Ergebnis der Substitution $y = P_1$ auch immer als Rest einer Division durch $y - P_1$ erhalten kann, so folgt

$$F = R\,\varphi + S \cdot (y - P_1),$$

woraus wegen (1) für $s = 1$ die Behauptung folgt.

Ist für $s - 1$ Zweige die Behauptung schon bewiesen, so erhält man im Fall von s Zweigen durch die Substitution $y = P_1$ aus F und $\varphi \cdot \prod_2^s (y - P_\nu)$ gewisse Potenzreihen F^* und $\varphi^* \cdot \prod_2^s (P_1 - P_\nu)$, von denen F^* durch $x^{\mu_1 + s - 1}$ teilbar ist, während das andere Produkt gleich einem Einheitsfaktor mal $x^{\mu_1 + s - 1}$ wird. Daraus folgt in derselben Weise wie vorhin

$$(2) \qquad F = A\,\varphi \prod_2^s (y - P_\nu) + B(y - P_1).$$

Aus dieser Gleichung ergibt sich, daß die Potenzreihe $B(y - P_1)$ beim Schnitt mit den Zweigen $y = P_k$ $(k = 2, \ldots, s)$ dieselben Multiplizitäten hat wie F, d. h. mindestens $\mu_k + s - 1$, mithin hat B als Multiplizität mindestens $\mu_k + s - 2$. Daraus folgt nach der Induktionsvoraussetzung (auf $f_1 = E \prod_2^s (y - P_\nu)$ angewandt)

$$(3) \qquad B = C\,E \prod_2^s (y - P_\nu) + D\,\varphi.$$

Setzt man (3) in (2) ein, so folgt wegen (1) die Behauptung.

129

Bemerkungen. Statt mit Potenzreihen kann man auch mit Kongruenzen nach beliebig hohen Potenzen des Ideals (x, y) operieren. Für den Beweis macht das gar nichts aus.

Dieselbe Beweismethode gestattet auch den Beweis des folgenden allgemeineren Satzes:

Wenn der Punkt $(0, 0)$ für die Kurve $f = 0$ ein s-facher Punkt ist, durch den s einfache Zweige hindurchgehen (die sich aber auch berühren dürfen), und wenn jeder Zweig Z_i in diesem Punkt von der Kurve $\varphi = 0$ μ_i-fach, von der Gesamtheit der anderen Zweige λ_i-fach und von der Kurve $F = 0$ mindestens $(\lambda_i + \mu_i)$-fach geschnitten wird, so ist

$$F = Pf + R\varphi$$

im Bereich der Potenzreihen nach x und y.

Wenn also die genannten Bedingungen in jedem Schnittpunkt der Kurven $f = 0$ und $\varphi = 0$ erfüllt sind, so ist $F = Af + B\varphi$ im Bereich der Polynome bzw. der ternären Formen.

Man kann die Relation $F = Af + B\varphi$ im Bereich der ternären Formen unter den angegebenen Voraussetzungen auch nach der Methode von Scott[5]) und Severi[6]) beweisen, indem man für Formen F von genügend hohem Grade die Anzahl der linear-unabhängigen Multiplizitätsbedingungen, die durch die Voraussetzungen des Satzes bedingt werden, vergleicht mit der Dimension der Schar $Af + B\varphi$ mit adjungiertem B.

---

[5]) C. A. Scott, Math. Annalen **52** (1899), S. 593.
[6]) F. Severi, Rendiconti Acc. Lincei (5) **11** (1902), S. 105.

(Eingegangen am 19. 11. 1930.)

7.

Zur algebraischen Geometrie I

Gradbestimmung von Schnittmannigfaltigkeiten einer beliebigen
Mannigfaltigkeit mit Hyperflächen

Mathematische Annalen 108, 1 (1933) 113–125

In drei früheren Annalenarbeiten[1]) habe ich einige algebraische und topologische Begriffe und Methoden entwickelt, die der mehrdimensionalen algebraischen Geometrie zugrunde gelegt werden können. Der Zweck der jetzigen Serie von Abhandlungen „Zur algebraischen Geometrie" ist, die Anwendbarkeit dieser Methoden auf verschiedene algebraisch-geometrische Probleme darzutun. Der Inhalt der ersten beiden unter [1]) zitierten Arbeiten, der in der Hauptsache auch in mein Buch „Moderne Algebra" II, Berlin 1931, Kapitel 13 übergegangen ist, wird immer als bekannt vorausgesetzt, der der dritten Arbeit nur dort, wo topologische Methoden zur Anwendung kommen.

Die Frage nach den Gradzahlen der Schnittmannigfaltigkeiten algebraischer Mannigfaltigkeiten beliebiger Dimension im projektiven Raum läßt sich mit topologischen Methoden erschöpfend beantworten (vgl. die dritte der unter [1]) zitierten Arbeiten). Es scheint mir jedoch methodisch von Wichtigkeit, daß der wichtigste Spezialfall, wo nur eine der zu schneidenden Mannigfaltigkeiten beliebige Dimension hat, die anderen dagegen Hyperflächen (d. h. von der Dimension $n - 1$) sind, sich mit relativ einfachen algebraischen Methoden behandeln läßt. Ich habe die Untersuchung dieses Falles mit Hilfe der Hilbertschen charakteristischen Funktion und den E. Noetherschen Kompositionsreihen von Idealen in einer früheren Arbeit[2]) durchgeführt. In dieser Arbeit soll nun eine einfachere Begründung mit einem Minimum an idealtheoretischem Aufwand gegeben werden[3]).

---

[1]) Math. Annalen **96**, S. 183—208; **97**, S. 756—774; **102**, S. 337—362. Diese Arbeiten werden im folgenden immer mit ihren abgekürzten Titeln: „Nullstellentheorie", „Multiplizitätsbegriff" und „Topologische Begründung" zitiert.

[2]) Proc. Kon. Ak. Amsterdam **31** (1927), S. 749—770.

[3]) Die Methode dieser Arbeit ist verwandt mit der in meiner „Verallgemeinerung des Bézoutschen Theorems", Math. Annalen 99 (1928), S. 497—541, angewandten, edoch sehr vereinfacht.

131

Man sieht zunächst leicht ein, daß die Berechnung einer Gradzahl auf die einer Schnittpunktszahl zurückgeführt werden kann, indem man noch eine gewisse Anzahl von allgemeinen linearen Gleichungen hinzunimmt. Um nun die Schnittpunktszahl zu bestimmen, bedient man sich der Methode der „relationstreuen Spezialisierung", die in der unter [1]) zitierten Arbeit „Multiplizitätsbegriff" erklärt wurde, durch die man aus den Lösungen eines allgemeinen Problems solche eines spezialisierten Problems erhält. Es handelt sich dabei um solche Probleme, die durch homogene Gleichungssysteme gegeben werden, in denen außer den Unbekannten noch Parameter vorkommen, die nachher spezialisiert werden. In unserem Fall wird das Gleichungssystem gegeben durch die Gleichungen einer r-dimensionalen Mannigfaltigkeit M_r und r weitere Gleichungen, $F_1 = 0, \ldots, F_r = 0$, welche r Hyperflächen definieren. Die Hyperflächen werden zunächst als „allgemeine" angenommen und dann so spezialisiert, daß sie je in so viele allgemeine Hyperebenen zerfallen, als ihr Grad beträgt. Bei dieser Spezialisierung ist die Anzahl der Schnittpunkte offenbar gleich dem Produkt der Gradzahlen von M_r und den Hyperflächen. Wenn man also noch nachweisen kann, daß bei der relationstreuen Spezialisierung keine Schnittpunkte zusammenrücken und keine neuen Schnittpunkte hinzukommen, kurz: daß die Multiplizitäten gleich Eins sind, so folgt, daß auch im allgemeinen Fall die Anzahl der Schnittpunkte gleich dem Produkt der Gradzahlen ist. Dieser Nachweis gelingt nun unter Zuhilfenahme einer typischen idealtheoretischen Betrachtung, welche zeigt, daß das zum Gleichungssystem gehörige Ideal $(\mathfrak{p}, F_1, \ldots, F_r)$, wo $\mathfrak{p}$ das zugehörige Ideal der Mannigfaltigkeit M_r ist, sowohl für allgemeine F_ν als auch nach der Spezialisierung der F_ν zu Produkten in lauter verschiedene Primideale zerfällt.

Der hier skizzierte Beweis wird in § 1 und 2 näher ausgeführt. In § 3 wird eine für die Anwendung sehr wichtige Verallgemeinerung vorgenommen, indem Mannigfaltigkeiten aus Punktepaaren, Punkttripeln usw. betrachtet werden: das führt zum Begriff eines „mehrfach-projektiven Raumes". Insbesondere erhält man hier eine Methode zur Bestimmung der Lösungsanzahlen von homogenen Gleichungssystemen in mehreren Variablenreihen. Für dieses Problem ist man auf die hier benutzten abzählenden Methoden angewiesen, weil eine Resultantentheorie für mehrfach-homogene Formen bisher fehlt. Die Nützlichkeit dieser Untersuchungen wird in § 4 an einigen Beispielen gezeigt.

Was die algebraischen Voraussetzungen der Untersuchungen betrifft: es wird ein ganz beliebiger Körper als Grundkörper angenommen, dem im Laufe der Untersuchungen verschiedene Unbestimmte adjungiert werden; die Koordinaten der betrachteten Punkte sind dann immer einem passenden algebraischen Erweiterungskörper des mit Unbestimmten erweiterten Grundkörpers zu entnehmen. Wir nehmen zunächst einmal an, daß alle vorzunehmenden algebraischen Erweiterungen separabel sind. Für Charakteristik Null trifft das natürlich immer zu, anderenfalls nicht immer. Wir werden aber sehen, daß im Fall eines vollkommenen Grundkörpers und des (einfach-) projektiven Raums die Voraussetzung der Separabilität, wo wir sie brauchen, auch erfüllt ist.

§ 1.

Schnittpunkte einer Kurve und einer Hyperfläche.

Die Anzahl der Schnittpunkte einer unzerlegbaren algebraischen Kurve C vom Grade l und einer Hyperfläche $F = 0$ vom Grade m im projektiven Raum P_n ist mit funktionentheoretischen Hilfsmitteln u. a. von Hensel und Landsberg[4], mit geometrisch-funktionentheoretischen Methoden von Zeuthen[5], mit idealtheoretischen von Lasker[6] und mir[2] bestimmt worden. Hier soll gezeigt werden, daß man mit dem Begriff der relationstreuen Spezialisierung und mit einem idealtheoretischen Hilfssatz allein auskommt.

Es sei zunächst F eine allgemeine Form m-ten Grades mit unbestimmten Koeffizienten $u_0, \ldots, u_h$, und die verschiedenen Schnittpunkte von $F = 0$ mit C seien $X^{(1)}, \ldots, X^{(q)}$. (Daß Schnittpunkte vorhanden sind, kommt bei den folgenden Beweisen als Nebenergebnis mit zum Vorschein.) Bei einer Spezialisierung der allgemeinen Form F zu irgendeiner Form G gibt es nun, wenn $G = 0$ die Kurve nicht ganz enthält, ein System von q Schnittpunkten $Y^{(1)}, \ldots, Y^{(q)}$ derart, daß alle algebraischen Relationen, welche zwischen den u_λ und $X^{(\alpha)}$ bestehen, bei der Spezialisierung $F \rightarrow G$, $X^{(\alpha)} \rightarrow Y^{(\alpha)}$ erhalten bleiben. Die Zahl, die angibt, wie oft irgendein Schnittpunkt Y in diesem System vertreten ist, heißt die *Multiplizität* dieses Schnittpunktes. Wir beweisen nun zunächst:

Die Multiplizität eines Schnittpunktes Y kann niemals Null sein.

Beweis. Nach Satz 5 der unter [1] zitierten Arbeit „Multiplizitätsbegriff" ist die Behauptung bewiesen, sobald gezeigt ist, daß es einen Schnittpunkt X der Kurve C mit der allgemeinen Hyperfläche $F = 0$ gibt derart, daß $X \rightarrow Y$, $F \rightarrow G$ eine relationstreue Spezialisierung ist. Sind $\eta_0, \ldots, \eta_n$ die Koordinaten von Y, so können wir z. B. $\eta_0 \neq 0$ annehmen und durch $\eta_0 = 1$ zu inhomogenen Koordinaten übergehen. Durch die Substitution $x_0 = 1$ mögen die Formen F und G in Polynome f und g übergehen. Ist u_0 das konstante Glied von f und v_0 das von g, so setzen wir

$$f = u_0 + f_0; \quad g = v_0 + g_0.$$

---

[4] Hensel und Landsberg, Theorie der algebraischen Funktionen einer Variablen, Leipzig 1902.

[5] H. G. Zeuthen, Abzählende Methoden der Geometrie, Leipzig 1914, S. 30.

[6] E. Lasker, Math. Annalen **60** (1905), S. 20—116.

Die Koeffizienten $u_1, \ldots, u_h$ von f_0 denken wir zum Grundkörper K adjungiert; der so erweiterte Körper heiße K^*.

Ist $\{\xi_1, \ldots, \xi_n\}$ ein allgemeiner Punkt der Kurve C, so ist $f_0(\xi_1, \ldots, \xi_n)$ sicher keine Konstante des Funktionenkörpers $\mathsf{K}^*(\xi_1, \ldots, \xi_n)$, weil f ein allgemeines Polynom war; also kann man $f_0(\xi)$ als unabhängige Veränderliche des Körpers annehmen, von der die übrigen Körperelemente algebraische Funktionen sind. Man kann dann auch $f_0(\xi) = -u_0$ setzen, da eine unabhängige Variable durch jede andere Unbestimmte ersetzt werden kann. So hat man einen allgemeinen Punkt der Kurve C erhalten, welcher zugleich eine Nullstelle von $f = u_0 + f_0$ ist.

Wenn nun eine Relation

$$\Phi(u_0, u_1, \ldots, u_h, \xi_1, \ldots, \xi_n) = 0$$

besteht, so bleibt diese Relation gültig, wenn man zunächst u_0 durch $-f_0(\xi)$, sodann den allgemeinen Punkt ξ durch den speziellen Kurvenpunkt η, schließlich die Unbestimmten $u_1, \ldots, u_h$ durch $v_1, \ldots, v_h$ ersetzt. Man erhält so:

$$\Phi(-g_0(\eta), v_1, \ldots, v_h, \eta_1, \ldots, \eta_n) = 0.$$

Nun ist $g(\eta) = 0$, also $v_0 + g_0(\eta) = 0$, daher

$$\Phi(v_0, v_1, \ldots, v_h, \eta_1, \ldots, \eta_n) = 0.$$

Also ist $X \to Y$, $F \to G$ in der Tat eine relationstreue Spezialisierung.

Aus dem Bewiesenen folgt, daß die Anzahl der Schnittpunkte im Spezialfall höchstens gleich der Anzahl q im allgemeinen Fall ist. Um nun die Anzahl im allgemeinen Fall zu bestimmen, spezialisieren wir zunächst die Form F so, daß sie in ein Produkt von m unabhängigen allgemeinen Linearformen zerfällt. Bei dieser Spezialisierung ist die Anzahl q' der Schnittpunkte offensichtlich gleich dem Produkt $l \cdot m$ der Gradzahlen von F und C. Also folgt

$$(1) \qquad\qquad q' = l \cdot m \leqq q.$$

Um nun andererseits $q' \geqq q$ zu beweisen, ziehen wir einen idealtheoretischen Hilfssatz heran, der so lautet:

$\mathfrak{p}$ sei ein eindimensionales Primideal in $\mathsf{K}[x_1, \ldots, x_n] = \mathsf{K}[x]$. f sci ein Polynom in den x, in welchem wenigstens das konstante Glied eine von den übrigen Koeffizienten unabhängige Unbestimmte u_0 ist. Der Körper, der aus K durch Adjunktion aller Koeffizienten von f entsteht, möge Ω heißen; $\mathfrak{p}_\Omega$ sei das von $\mathfrak{p}$ erzeugte Ideal in $\Omega[x]$. Wir setzen voraus, daß $\mathfrak{p}_\Omega$ wieder ein Primideal ist. Dann ist $(\mathfrak{p}_\Omega, f)$ ein Primideal in $\Omega[x]$, und zwar entweder das Einheitsideal oder ein nulldimensionales Primideal. Im letzteren Fall sind alle Nullstellen von $(\mathfrak{p}_\Omega, f)$ konjugiert in bezug auf Ω, und ihre Anzahl ist gleich dem Grad des Ideals $(\mathfrak{p}_\Omega, f)$, d. h. gleich der Anzahl der linear-unabhängigen Restklassen nach diesem Ideal.

Beweis. Wir setzen $f = u_0 + f_0$. Ein Produkt $a \cdot b$ von Polynomen aus $\Omega[x_1, \ldots, x_n]$ sei durch $(\mathfrak{p}_\Omega, f)$ teilbar. Wir können a und b als ganzrational in u_0 annehmen und schreiben, um die Abhängigkeit von u_0 hervorzuheben, $a(u_0, x) \cdot b(u_0, x)$. Ersetzen wir nun überall u_0 durch $- f_0$, so wird $f = 0$ und unser Produkt $a \cdot b$ wird durch $\mathfrak{p}_\Omega$ teilbar. Also muß einer der Faktoren es sein, etwa

$$a(- f_0, x) \equiv 0 \ (\mathfrak{p}_\Omega).$$

Daraus folgt

$$a(u_0, x) \equiv 0 \ (\mathfrak{p}_\Omega, f_0 + u_0).$$

Also ist $(\mathfrak{p}_\Omega, f)$ ein Primideal, echter Teiler von $\mathfrak{p}_\Omega$ und somit von kleinerer Dimension[7]. Wenn $(\mathfrak{p}_\Omega, f)$ nicht das Einheitsideal ist, so ist es nulldimensional. Seine Nullstellen sind dann konjugiert[7] zu einer einzigen Nullstelle η. Da wir hier annehmen, daß wir es nur mit separablen Körpererweiterungen zu tun haben, ist die Anzahl der konjugierten Nullstellen gleich dem Grad des Körpers $\Omega(\eta)$ über Ω. Dieser Körper ist[7] aber isomorph dem Restklassenring $\Omega[x]/(\mathfrak{p}_\Omega, f)$, somit ist die Anzahl der Nullstellen gleich der Anzahl der linear-unabhängigen Restklassen, womit der Hilfssatz bewiesen ist.

Die im Hilfssatz gemachte Voraussetzung, daß $\mathfrak{p}$ beim Übergang zu $\mathfrak{p}_\Omega$ Primideal bleibt, ist stets erfüllt, wenn Ω durch Adjunktion von Unbestimmten an K entsteht, also insbesondere, wenn f ein *allgemeines* Polynom vom Grade m ist, und zwar hat in diesem Fall, wie wir schon wissen, das Ideal $(\mathfrak{p}_\Omega, f)$ immer Nullstellen, deren Anzahl nach dem Hilfssatz gleich dem Grad des Ideals $(\mathfrak{p}_\Omega, f)$ ist. Dasselbe gilt nun aber auch, wie wir zeigen wollen, wenn f ein Produkt von m allgemeinen Linearfaktoren ist:

$$f = l_1 l_2 \ldots l_m.$$

Das Idealprodukt $(\mathfrak{p}_\Omega, l_1) \cdot (\mathfrak{p}_\Omega, l_2), \ldots, (\mathfrak{p}_\Omega, l_m)$ ist offenbar teilbar durch $(\mathfrak{p}_\Omega, f)$, welches Ideal seinerseits durch alle Ideale $(\mathfrak{p}_\Omega, l_\nu)$, also auch durch ihr K. G. V., welches $\mathfrak{d}$ heißen möge, teilbar. Die Ideale $(\mathfrak{p}_\Omega, l_\nu)$ sind aber paarweise teilerfremd, da die endlichvielen Nullstellen von $(\mathfrak{p}_\Omega, l)$ nicht gleichzeitig Nullstellen von $(\mathfrak{p}_\Omega, l)$ sein können; also ist Produkt $=$ K. G. V. und daher

$$\prod_1^m (\mathfrak{p}_\Omega, l_\nu) = (\mathfrak{p}_\Omega, f) = \mathfrak{d}.$$

Daraus folgt nun bekanntlich vermöge der additiven Zerlegung des Restklassenringes nach $(\mathfrak{p}_\Omega, f)$ in Ringe, die zu den Restklassenringen nach den einzelnen $(\mathfrak{p}_\Omega, l_\nu)$ isomorph sind[8], daß der lineare Rang jenes Restklassenringes gleich der Summe der linearen Ränge der Restklassenringe nach

---

[7]) Vgl. „Nullstellentheorie" § 3, oder „Moderne Algebra" II. § 90.

[8]) Siehe etwa „Moderne Algebra" II, § 85.

den Primidealen $(\mathfrak{p}_\Omega, l_\nu)$ ist[9]. Diese Ränge sind andererseits nach dem Hilfssatz gleich den Anzahlen der Nullstellen dieser Primideale, also ist der Grad des Ideals $(\mathfrak{p}_\Omega, f)$ gleich der Anzahl der Nullstellen von allen diesen Primidealen zusammengenommen, d. h. gleich der Anzahl der Nullstellen des Ideals $(\mathfrak{p}_\Omega, f)$, wie wir zeigen wollten.

Nun sei $\mathfrak{P}$ das zugehörige Primideal der Kurve C, erzeugt von allen Formen in $x_0, \ldots, x_n$, welche auf C Null werden, und es sei F eine allgemeine Form vom Grade m, die nachher zu einem Produkt von Linearformen spezialisiert wird. Die uneigentliche Ebene $x_0 = 0$ hat nur endlichviele Schnittpunkte mit der Kurve; diese sind natürlich nicht gleichzeitig Nullstellen der allgemeinen Form F; also können wir durch die Substitution $x_0 = 1$ zu inhomogenen Koordinaten übergehen: das Ideal $\mathfrak{P}$ möge dadurch in $\mathfrak{p}$ und F in f übergehen. Nach unserem Hilfssatz ist die Anzahl q der Schnittpunkte von Kurve und Hyperfläche, d. h. die Anzahl der Nullstellen des Ideals $(\mathfrak{p}_\Omega, f)$ gleich dem Grad dieses Ideals, d. h. gleich der Anzahl der mod $(\mathfrak{p}_\Omega, f)$ linearunabhängigen Polynomen. Macht man nun diese Polynome wieder homogen, so folgt leicht, daß dieselbe Zahl q auch gleich der Anzahl der modulo $(\mathfrak{P}_\Omega, F)$ linear-unabhängigen Formen eines genügend hohen Grades g ist. Ebenso ist, wenn man die Form F zu einem Produkt $L_1 L_2 \ldots L_m$ von allgemeinen Linearformen spezialisiert, die Anzahl q' der Schnittpunkte im Spezialfall gleich der Anzahl der mod $(\mathfrak{P}_\Omega, L_1 \ldots L_m)$ linear-unabhängigen Formen.

Wir wollen nun zeigen, daß die Anzahl der mod $(\mathfrak{P}_\Omega, F)$ linear-unabhängigen Formen bei einer beliebigen Spezialisierung der Koeffizienten von F niemals kleiner werden kann, woraus dann sofort die gewünschte Ungleichung

$$(2) \qquad\qquad q' \geqq q$$

folgt. Das ist aber sehr einfach: Wäre die genannte Anzahl von Formen g-ten Grades bei der Spezialisierung kleiner geworden, so wäre die Anzahl der linear-unabhängigen Formen g-ten Grades *im* Ideal größer geworden. Nun kann man aber ein volles System von Formen g-ten Grades im Ideal in der Weise erhalten, daß man die Basisformen des Ideals mit allen Potenzprodukten geeigneten Grades der x_ν multipliziert. Die Anzahl der linearunabhängigen unter diesen Formen ist dann der Rang ihrer Koeffizientenmatrix. Dieser Rang nun kann bei einer Spezialisierung von Unbestimmten niemals größer werden. Damit ist die Ungleichung (2) bewiesen.

Aus (1) und (2) folgt: *Die Anzahl q der Schnittpunkte einer allgemeinen Hyperfläche $F = 0$ mit einer irreduziblen Kurve C ist gleich dem Produkt $l \cdot m$ der Gradzahlen.*

---

[9]) Statt Anzahlen von Restklassen abzuzählen, kann man natürlich auch Anzahlen von unabhängigen Gleichungen abzählen, welche ein Polynom zu erfüllen hat, um dem Ideal $(\mathfrak{p}, f)$ anzugehören; das Endergebnis ist dasselbe. Vgl. F. S. Macaulay, Cambridge Tracts 19, Kap. IV.

Die Summe der Multiplizitäten der Schnittpunkte in jedem Spezialfall ist natürlich auch gleich dem Produkt der Gradzahlen. Daß die Multiplizitäten niemals Null sein können, sahen wir schon.

Die Voraussetzung, daß ausschließlich separable Körpererweiterungen vorkommen sollten, wurde im vorangehenden zweimal benutzt; nämlich um für die Primideale $(\mathfrak{p}_\Omega, l_\nu)$ und $(\mathfrak{p}_\Omega, f)$ zu schließen, daß der Grad des Nullstellenkörpers $\Omega(\eta)$ gleich der Anzahl der zu η konjugierten Nullstellen ist. Wäre nur im Fall $(\mathfrak{p}_\Omega, f)$ der Körper $\Omega(\eta)$ über Ω inseparabel, so wäre das nicht schlimm: Die Anzahl q der Nullstellen würde kleiner werden und die Ungleichung (2) würde um so mehr gelten. Wesentlich geht aber in den Beweis von (2) die Voraussetzung ein, daß im Fall $(\mathfrak{p}_\Omega, l_\nu)$ der Körper $\Omega(\eta)$ über Ω separabel ist; dabei war η eine allgemeine Nullstelle von $\mathfrak{p}$, es waren $u_1, \ldots, u_n$ Unbestimmte und $u_0 = -u_1\eta_1 - \ldots - u_n\eta_n$. Daß diese Voraussetzung bei vollkommenem Grundkörper K wirklich zutrifft, habe ich früher[10]) bewiesen. Die Ergebnisse dieses Paragraphen sind also allgemein gültig. Dasselbe gilt von § 2, nicht aber von § 3, da dort im Fall einer Primzahlcharakteristik wirklich inseparable Erweiterungen auftreten können.

§ 2.

Schnitt einer Mannigfaltigkeit mit Hyperflächen.

M_r sei eine r-dimensionale Mannigfaltigkeit in P_n vom Grade l, und $F_1 = 0$, $F_2 = 0$, $\ldots$, $F_r = 0$ seien r allgemeine Hyperflächen der Grade $m_1, \ldots, m_r$. Dann kann man fast wörtlich wie in § 1 beweisen, daß die Anzahl der Schnittpunkte von M_r mit diesen Hyperflächen endlich und gleich dem Produkt $l\,m_1 \ldots m_r$ der Gradzahlen ist. Ich ziehe es aber vor, das vorliegende Problem durch das folgende sukzessive Verfahren auf den in § 1 erledigten Spezialfall zurückzuführen: Man schneidet zuerst M_r mit der (allgemeinen) Hyperfläche $F_1 = 0$; das ergibt eine Mannigfaltigkeit M_{r-1}, von der wir beweisen werden, daß sie (in bezug auf den Körper Ω, der aus K durch Adjunktion aller Koeffizienten von F_1 entsteht) irreduzibel und $(r-1)$-dimensional ist und den Grad $l \cdot m_1$ hat. Schneidet man dann in derselben Weise M_{r-1} mit $F_2 = 0$, usw., so findet man ohne weiteres die gesuchte Zahl $l\,m_2 \ldots m_r$.

Ist $\mathfrak{P}$ das zugehörige Primideal von M_r und nimmt man in $\mathfrak{P}$ und F die Substitution $x_0 = 1$ vor, wodurch sie in $\mathfrak{p}$ und f übergehen, so beweist man, wie in § 1 (Hilfssatz), daß $(\mathfrak{p}_\Omega, f_1)$ ein Primideal ist. Daraus folgt die Irreduzibilität desjenigen Teils von M_{r-1}, der sich nicht in der uneigentlichen Ebene $x_0 = 0$ befindet. Da aber diese Ebene $x_0 = 0$ durch jede andere Hyperebene ersetzt werden kann, so folgt die Irreduzibilität von M_{r-1}.

Diese gilt natürlich auch dann, wenn statt F_1 eine allgemeine Linearform L_1 genommen wird. Schneidet man nun M_r sukzessiv mit den linearen Räumen $L_1 = 0$, $L_2 = 0$, $\ldots$, $L_r = 0$, so erhält man lauter (relativ-)

---

[10]) Math. Annalen 99 (1928), S. 520, Hilfssatz 1. Vgl. auch Satz 27, 2 ebenda.

irreduzible Mannigfaltigkeiten von abnehmender Dimension. Zum Schluß ist aber die Schnittpunktszahl endlich und gleich dem Grad l von M_r nach Definition dieses Grades[11]), also war nach $r-1$ Schritten eine irreduzible Kurve vom Grade l da. Diese schneidet die allgemeine Hyperfläche $F_1 = 0$ nach § 1 in $l\,m_1$ Punkten. Also ist der Schnitt M_{r-1} von M_r mit $F_1 = 0$ eine irreduzible Mannigfaltigkeit, die von einem allgemeinen linearen Raum $L_1 = 0, \ldots, L_{r-1} = 0$ in genau $l \cdot m_1$ Punkten geschnitten wird, woraus folgt, daß M_{r-1} die Dimension $r-1$ und den Grad $l m_1$ hat, wie wir beweisen wollten.

Werden die Formen $F_1, \ldots, F_r$ spezialisiert, so erhält, falls nicht der unangenehme Ausnahmefall unendlichvieler Schnittpunkte eintritt, jeder Schnittpunkt eine gewisse Multiplizität, und die Summe dieser Multiplizitäten ist $l m_1 \ldots m_r$. Daß die Multiplizitäten nicht Null sind, folgt genau so wie in § 1 (vgl. auch den analogen Beweis im Fall linearer F_j an der unter [8]) zitierten Stelle).

Wir gehen jetzt zum Fall über, daß eine Mannigfaltigkeit M_r mit s ($< r$) Hyperflächen $F_1 = 0, \ldots, F_s = 0$ geschnitten wird, wobei wir annehmen, daß bei jeder Hinzufügung einer neuen Form F_j die Dimension der Schnittmannigfaltigkeit sich wirklich verringert, so daß die Dimension der Schnittmannigfaltigkeit höchstens $r-s$ beträgt. Fügt man noch $r-s$ allgemeine lineare Gleichungen $L_1 = 0, \ldots, L_{r-s} = 0$ hinzu, so erhält man endlichviele Schnittpunkte, also war die Dimension auch wirklich $r-s$. Schnittbestandteile von einer Dimension $< r-s$ können nicht vorkommen, denn sie würden bei der Hinzunahme von allgemeinen linearen Gleichungen keine Schnittpunkte, dagegen bei passend gewählten speziellen linearen Gleichungen wohl Schnittpunkte ergeben, welche dann, wie leicht ersichtlich, die Multipliziät 0 hätten, was nicht sein kann.

Also zerfällt die Schnittmannigfaltigkeit M_{r-s} von M_r mit $F_1 = 0, \ldots,$ $F_s = 0$ in irreduzible Bestandteile $M^{(1)}, \ldots, M^{(h)}$ von der Dimension $r-s$, mit Gradzahlen $g_1, \ldots, g_h$. Beim Schnitt von $M^{(1)}$ mit dem allgemeinen linearen Raum $L_1 = 0, \ldots, L_{r-s} = 0$ ergibt sich ein System von g_1 konjugierten Schnittpunkten, welche als Schnittpunkte von M_r mit $F_1 = 0, \ldots, F_s = 0, L_1 = 0, \ldots, L_{r-s} = 0$ je eine gewisse Multiplizität μ_1 haben: dieselbe für alle auf Grund der Konjugiertheit. Ebenso bestimmt man für $M^{(2)}, \ldots, M^{(h)}$ die Multiplizitäten $\mu_2, \ldots, \mu_h$. Die Summe aller dieser Multiplizitäten ist

$$(1) \qquad \mu_1 g_1 + \mu_2 g_2 + \ldots + \mu_h g_h = l m_1 \ldots m_r.$$

Wir bezeichnen nun die Zahlen $\mu_1, \ldots, \mu_h$ als *die Multiplizitäten der irreduziblen Bestandteile* $M^{(1)}, \ldots, M^{(h)}$ *der Schnittmannigfaltigkeit von* M_r *mit* $F_1 = 0, \ldots, F_s = 0$. Dann gilt also der Satz:

---

[11]) Vgl. „Moderne Algebra" II, § 96.

Eine irreduzible Mannigfaltigkeit M_r vom Grade l wird von s Hyperflächen der Grade $m_1, \ldots, m_s$ geschnitten nach einer Mannigfaltigkeit M_{r-s}, deren irreduzible Bestandteile gewisse Gradzahlen g_j und Multiplizitäten μ_j haben, für welche die Formel (1) gilt, es sei denn, die Dimensionszahl des Schnittes wäre größer als $r - s$.

§ 3.

Verallgemeinerung: die mehrfach-projektiven Räume.

Ein Punktepaar X, Y mit Koordinaten $\xi_0, \ldots, \xi_m; \eta_0, \ldots, \eta_n$, wo der Punkt X einem projektiven Raum P_m und der Punkt Y einem projektiven Raum P_n angehört, kann als *Punkt* eines *zweifach-projektiven Raumes $P_{m,n}$* von $m + n$ Dimensionen aufgefaßt werden. Algebraische Mannigfaltigkeiten, Hyperflächen usw. werden in derselben Weise definiert wie im projektiven Raum; die dabei benutzten Formen müssen natürlich homogen in zwei Variablenreihen $x_0, \ldots, x_m; y_0, \ldots, y_n$ sein. Genau entsprechend kann man auch dreifach-homogene Räume, usw. definieren. Wir beschränken uns der Einfachheit halber auf zweifach-projektive Räume.

Eine Kurve C des Raumes $P_{m,n}$ hat zwei Gradzahlen: die Anzahl p der Kurvenpunkte, die einer allgemeinen linearen Gleichung in den ξ, und die Anzahl q derjenigen, die einer allgemeinen linearen Gleichung in den η genügen[12]). Ist nun F eine allgemeine Form der Gradzahlen k, l in den x und y, so ist die Anzahl der Schnittpunkte der Kurve C mit der Hyperfläche $F(\xi, \eta) = 0$ gleich

$$(1) \qquad\qquad k\,p + l\,q.$$

Das beweist man genau so wie in § 1 durch Spezialisierung der Form F zu einem Produkt von k Linearformen in den x und l Linearfaktoren in den y. Die einzige Stelle, wo der in § 1 gegebene Beweis versagen kann, ist der beim Beweis des ersten Satzes benutzte Schluß: „Ist nun $\{\xi_1, \ldots, \xi_n\}$ ein allgemeiner Punkt der Kurve C, so ist $f_0(\xi_1, \ldots, \xi_n)$ sicher keine Konstante des Funktionenkörpers, weil f ein allgemeines Polynom war". Dieser Schluß versagt genau in dem Fall, wo f nur von den x (bzw. nur von den y) abhängt und die ξ- (bzw. η-) Koordinaten des allgemeinen Punktes von C Konstante sind. In diesem Fall ist die Anzahl der Schnittpunkte von C mit der allgemeinen Hyperfläche $F = 0$ in Übereinstimmung mit der Formel (1) Null. Die spezialisierte Gleichung $G = 0$ (oder inhomogen $g = 0$) hängt dann ebenfalls nur von den ξ ab; daher ist die Anzahl der Schnittpunkte von C mit $G = 0$ Null oder unendlich, je nachdem, ob die konstanten ξ-Koordinaten

---

[12]) Eine dieser beiden Gradzahlen kann Null sein, nämlich dann, wenn für einen allgemeinen Punkt (ξ, η) der Kurve C die Verhältnisse der ξ oder die der η konstant sind.

des allgemeinen Punktes von C die Gleichung $g = 0$ erfüllen oder nicht. Die Behauptung über die Multiplizität eines Schnittpunktes im Fall endlichvieler Schnittpunkte ist in diesem Fall inhaltslos, also richtig.

In derselben Weise überträgt man die Sätze des § 2 auf mehrfachprojektive Räume. Eine irreduzible Mannigfaltigkeit M von r Dimensionen in $P_{m,n}$ hat eine ganze Reihe von Gradzahlen g_{ij} $(i + j = r, \ 0 \leq i \leq m, \ 0 \leq j \leq n)$, welche die Anzahlen der Punkte von M bedeuten, welche i allgemeinen linearen Gleichungen in den ξ und j allgemeinen linearen Gleichungen in den η genügen. Einige von diesen Gradzahlen können Null sein; dieser Fall tritt z. B. dann ein, wenn mehr Gleichungen in den ξ allein verlangt werden, als der Transzendenzgrad des Systems der Verhältnisse der ξ-Koordinaten eines allgemeinen Punktes von M_r beträgt.

Die Anzahl der Schnittpunkte einer M_r mit r allgemeinen Hyperflächen von gegebenen Gradzahlen ergibt sich auf Grund der Schlußweise von § 2 *ebenso groß wie die Anzahl der Schnittpunkte, die man erhält, wenn man die betreffenden Formen zu Produkten von allgemeinen Linearformen spezialisiert.* Ist diese Anzahl > 0, so sind auch bei einer beliebigen Spezialisierung der Hyperflächen die Multiplizitäten der Schnittpunkte von Null verschieden und die Summe dieser Multiplizitäten ist gleich der Anzahl im allgemeinen Fall, falls nicht nach der Spezialisierung unendlichviele Schnittpunkte da sind.

Für die Berechnung dieser Anzahl ist folgender Algorithmus zweckmäßig, der dem Schubertschen Kalkül der abzählenden Geometrie[13]) nachgebildet ist. Man ordnet jeder Form $F(x, y)$ mit den Gradzahlen k und l ein „Bedingungssymbol" $kp + lq$ zu, in dem p und q zunächst Unbestimmte sind. Wenn dann eine Mannigfaltigkeit M_r mit r Hyperflächen $F_1 = 0, \ldots,$ $F_r = 0$ geschnitten werden soll, so multipliziert man die Bedingungssymbole der Formen $F_1, \ldots, F_r$ miteinander und ersetzt im Ergebnis die Potenzprodukte $p^i q^j$ durch die Gradzahlen g_{ij} der Mannigfaltigkeit M_r. Solche Potenzprodukte, die den Faktor p^{m+1} oder q^{n+1} enthalten, sind dabei $= 0$ zu setzen. Das Ergebnis dieser Einsetzung gibt in Übereinstimmung mit der oben cursivierten Regel die gesuchte Schnittpunktsanzahl an.

Man kann den obigen Algorithmus noch dadurch modifizieren, daß man auch der Mannigfaltigkeit M_r ein Bedingungssymbol

$$\sum_{i+j=r} g_{ij} \, p^{m-i} \, q^{n-j}$$

zuordnet und dieses mit den Bedingungssymbolen der Formen $F_1, \ldots, F_r$ multipliziert. Der Koeffizient von $p^m q^n$ im entstehenden Produkt gibt dann die Schnittpunktszahl an.

---

[13]) Vgl. „Topologische Begründung", § 8 und § 9.

Wird die Mannigfaltigkeit M_r mit s $(< r)$ Hyperflächen geschnitten und hat die Schnittmannigfaltigkeit M_{r-s} nicht mehr als $r - s$ Dimensionen, so zerfällt sie in irreduzible Bestandteile $M^{(1)}$, ..., $M^{(h)}$ von der Dimension $r - s$. Die Multiplizitäten, mit denen diese Schnittbestandteile zu zählen sind, werden zunächst einmal dadurch definiert, daß man $r - s$ allgemeine Bilinearformen in x und y hinzunimmt und wie in § 2 die Systeme konjugierter Schnittpunkte bestimmt. Läßt man nun hinterher die Bilinearformen in Produkte von allgemeinen Linearformen zerfallen, so bleibt die Anzahl der Schnittpunkte dieselbe, also sind bei dieser Spezialisierung keine Schnittpunkte zusammengerückt, also sind die Multiplizitäten ungeändert geblieben. D. h. die durch den Schnitt mit den allgemeinen Bilinearformen definierten Multiplizitäten sind dieselben, die man auch beim Schnitt mit ebenso vielen Linearformen in x oder y findet. Durch passende Kombination dieser Linearformen kann man alle Gradzahlen der Schnittmannigfaltigkeit M_{r-s} bestimmen, wobei unter der Gradzahl g_{ij} der gesamten Schnittmannigfaltigkeit M_{r-s} die Summe der entsprechend bezeichneten Gradzahlen der Bestandteile $M^{(1)}$, ..., $M^{(h)}$, multipliziert mit ihren Multiplizitäten, zu verstehen ist.

Nach dem obigen Algorithmus wird das Bedingungssymbol $\sum g_{ij}\,p^i\,q^j$ der Schnittmannigfaltigkeit M_{r-s} gefunden durch Multiplikation der Bedingungssymbole von M_r und der s Hyperflächen, mit denen man M_r geschnitten hat.

§ 4.

Beispiele und Anwendungen.

1. In der Literatur begegnet man vielfach der Frage nach dem Grad derjenigen Mannigfaltigkeit, die durch das Verschwinden aller s-reihigen Unterdeterminanten in einer Matrix

$$\begin{pmatrix} a_{11} & \cdots & a_{1s} \\ a_{21} & \cdots & a_{2s} \\ \vdots & & \vdots \\ a_{r1} & \cdots & a_{rs} \end{pmatrix}$$

definiert wird, wo die a_{ij} Formen k-ten Grades in $x_0, \ldots, x_n$ sind. Die Frage wird gewöhnlich mit einer nicht ganz einwandfreien Differenzenmethode beantwortet, indem aus den $\binom{r}{s}$ vorgelegten Gleichungen ein System von unabhängigen ausgewählt wird, von deren Lösungen dann diejenigen auszuscheiden sind, welche gewisse kleinere Unterdeterminanten annullieren. Die Methode ist darum ungenügend, weil den eventuell auftretenden Vielfachkeiten der auszuscheidenden Mannigfaltigkeiten nicht Rechnung getragen wird.

Die Anwendung der mehrfach-projektiven Räume führt aber zu einer sehr einfachen einwandfreien Lösung des gestellten Problems. Das Verschwinden der s-reihigen Unterdeterminanten für die Werte ξ_ν der Unbestimmten x_ν ist nämlich gleichbedeutend mit der Lösbarkeit des folgenden linearen Gleichungssystems:

$$a_{\lambda 1}(\xi)\,\eta_1 + a_{\lambda 2}(\xi)\,\eta_2 + \ldots + a_{\lambda s}(\xi)\,\eta_s = 0 \quad (\lambda = 1, 2, \ldots, r).$$

Dieses Gleichungssystem definiert im Raum $P_{n,\,s-1}$ der Punktepaare (ξ, η) eine Mannigfaltigkeit von mindestens $n + s - 1 - r$ Dimensionen; im „allgemeinen Fall", wo die $a_{\lambda\mu}$ allgemeine Formen sind, ist die Dimension genau $n + s - 1 - r$. Nehmen wir noch $n + s - 1 - r$ allgemeine lineare Gleichungen in den ξ allein hinzu, so erhalten wir eine endliche Schnittpunktszahl, welche offenbar die gesuchte Gradzahl darstellt. Nach dem in § 3 entwickelten Algorithmus ist diese Gradzahl im „allgemeinen" Fall gleich dem Koeffizienten von $p^n q^{s-1}$ im Bedingungssymbol

$$(k\,p + q)^r\,p^{n+s-1-r}.$$

Dieser Koeffizient ist $\binom{r}{s-1} k^{r-s-1}$. *Also ist die gesuchte Gradzahl im allgemeinen und höchstens gleich* $\binom{r}{s-1} k^{r-s-1}$.

2. In Hilberts Abhandlung „Über die Singularitäten der Diskriminantenfläche", Math. Annalen **27** (1886), S. 158—161, kommt das folgende Gleichungssystem vor:

$$\left\{\begin{array}{l} X_1\,(t_1,\ t_2,\ \ldots,\ t_\varkappa) = 0 \\ \ldots\ldots\ldots\ldots\ldots, \\ X_\varkappa\,(t_1,\ t_2,\ \ldots,\ t_\varkappa) = 0 \end{array}\right.$$

wo die linken Seiten Polynome vom Grad μ_1 in t_1, vom Grad μ_2 in t_2, usw. sind, welche durch Polarenbildung aus *allgemeinen* Polynomen n-ten Grades:

$$a_0^{(\nu)} - n\,a_1^{(\nu)}\,t + \binom{n}{2}\,a_2^{(\nu)}\,t^2 - \ldots + (-1)^n\,a_n^{(\nu)}\,t^n$$

entstehen. Hilbert gibt an, daß die Anzahl der Lösungen dieses Gleichungssystems $n!\ \mu_1\mu_2\ldots\mu_\varkappa$ beträgt[14]), ohne mitzuteilen, wie er zu diesem Ergebnis gekommen ist. Macht man (wie es naturgemäß ist) die Gleichungen homogen durch Einführung von Verhältnisgrößen $u_1 : v_1$ statt t_1, usw., so läßt sich die Theorie der $\varkappa$-fach projektiven Räume anwenden, welche lehrt, daß die Anzahl

---

[14]) Hilbert dividiert noch durch $\varPi(i_\varkappa!)$, wo $i_\varkappa$ die Anzahlen der untereinander gleichen μ_ν bedeuten, weil er solche Lösungen, die durch Vertauschung der t_ν auseinander hervorgehen, als nichtverschieden betrachtet.

der Lösungen eines allgemeinen Gleichungssystems der angegebenen Gradzahlen gleich dem Koeffizienten von $p_1 p_2 \ldots p_\varkappa$ im Bedingungssymbol

$$(\mu_1 p_1 + \mu_2 p_2 + \ldots + \mu_\varkappa p_\varkappa)^\varkappa$$

ist; das ergibt genau die von Hilbert angegebene Zahl.

Nun ist aber unser Gleichungssystem kein allgemeines von den gegebenen Gradzahlen, denn es hat gewisse (durch die Polarenbildung bedingte) Symmetrien. Trotzdem ist die Anzahl der Lösungen dieselbe wie im allgemeinen Fall. Da nämlich die konstanten Glieder der Polynome $X_1, \ldots, X_\varkappa$ lauter unabhängige Unbestimmte $a_0^{(\nu)}$ sind, so ist nach dem Hilfssatz von § 1 (auf die inhomogenen Polynome angewandt) das Ideal $(X_1, \ldots, X_\varkappa)$ prim und die Anzahl seiner Nullstellen gleich dem Grad des Ideals. Daraus folgt wie im § 1, daß die Anzahl q' der Lösungen im betrachteten Fall nicht kleiner ist als die im allgemeinen Fall.

(Eingegangen am 12. 7. 1932.)

8.

Zur algebraischen Geometrie II
Die geraden Linien auf den Hyperflächen des P_n

Mathematische Annalen 108, 2 (1933) 253–259

Im dreidimensionalen Raum liegen auf einer quadratischen Fläche ∞^1 Geraden, auf einer kubischen Fläche F^3 im allgemeinen 27, während eine allgemeine Fläche höheren Grades keine Geraden enthält.

Diese Tatsachen lassen sich leicht n-dimensional verallgemeinern. Man findet (vgl. § 1), daß für $m \leq 2\,n - 3$ auf einer allgemeinen Hyperfläche F^m m-ten Grades des projektiven Raumes P_n

$$\infty^{(2\,n-3)-m}$$

Geraden liegen, insbesondere für $m = 2\,n - 3$ endlich viele, aber für $m > 2\,n - 3$ keine. Es erhebt sich nun vor allem die Frage: *Wieviele Geraden gibt es im Fall $m = 2\,n - 3$, und wie liegen sie?*

Die Anzahlbestimmung wird hier durchgeführt. Das Ergebnis ist das folgende: *die gesuchte Anzahl h ist gleich dem Koeffizienten von $p^n q^{n-1}$ in der Form*

$$(p - q) \cdot m\,p\,\{(m-1)\,p + q\}\,\{(m-2)\,p + 2\,q\} \ldots \{p + (m-1)\,q\}\,m\,q.$$

Man findet also für:

$$n = 2,\ m = 1:\ \ h = 1$$
$$n = 3,\ m = 3:\ \ h = 27 = 3^3$$
$$n = 4,\ m = 5:\ \ h = 2875 = 5^3 \cdot 23$$
$$n = 5,\ m = 7:\ \ h = 698005 = 7^3 \cdot 2035.$$

Die Methode, mit der dieses Ergebnis gefunden wird, ist die folgende. Das Problem führt auf ein System von $2\,n - 2$ homogenen Gleichungen in zwei Reihen von $n + 1$ Unbekannten: den Koordinaten von zwei Punkten, welche die gesuchten Geraden bestimmen. Dieses Gleichungssystem hat nun aber außer den gesuchten Lösungen noch „falsche" Nebenlösungen, bei denen die beiden Punkte zusammenfallen und keine Gerade bestimmen. (Die Dimension der falschen Lösungsmannigfaltigkeit ist dabei für $n > 3$ noch größer als die der richtigen Lösungen.) Nun werden die Gleichungen des Problems sukzessiv herangezogen und auf jeder Stufe werden die „falschen‘‘ Lösungen ausgeschieden und die Gradzahlen der übrigbleibenden Mannigfaltigkeiten bestimmt. Zum Schluß bleibt eine Anzahl übrig, welche die Summe der Multiplizitäten der richtigen Lösungen darstellt[1]). Nun stellt sich heraus, daß die Multiplizitäten gleich Eins sind, mithin stellt die gefundene Zahl zugleich die Anzahl der Geraden auf der allgemeinen F^{2n-3} dar.

Was die Lage der gefundenen Geraden betrifft, ergibt sich weiter, daß sie alle zueinander windschief sind, daß kein Tripel in einem Raum P_3 liegt und kein Quadrupel eine gemeinsame Transversale besitzt, überhaupt daß zwischen weniger als 7 Geraden keinerlei algebraische Relation besteht. Die Zahl 7 läßt sich übrigens durch verfeinerte Betrachtungen noch mindestens verdoppeln.

§ 1.

Die Existenz der Geraden.

Sind ξ und η zwei Punkte des P_n und $f = 0$ eine Hyperfläche m-ten Grades, so findet man die Bedingungen dafür, daß die Gerade $\lambda_1 \xi + \lambda_2 \eta$ der Hyperfläche angehört, indem man

$$f(\lambda_1 \xi + \lambda_2 \eta) = \lambda_1^m f_0 + \lambda_1^{m-1} \lambda_2 f_1 + \ldots + \lambda_2^m f_m$$

identisch in λ_1 und λ_2 gleich Null setzt. Das ergibt $m + 1$ Gleichungen:

$$(1) \qquad f_0 = 0;\ f_1 = 0;\ \ldots;\ f_m = 0,$$

zu denen man, da die Punkte ξ und η willkürlich auf der Geraden gewählt werden können, wenn erwünscht, noch eine lineare Bedingung für die Koordinaten $\xi_0, \ldots, \xi_n$ und eine für $\eta_0, \ldots, \eta_n$ hinzunehmen kann, etwa:

$$\xi_0 = 0, \qquad \eta_1 = 0.$$

Schließt man den Fall aus, daß die Gerade $\xi\eta$ die Ebene $\xi_0 = \xi_1 = 0$ schneidet (bei einer allgemeinen Hyperfläche spielen ja die besonderen Geraden, welche diese Ebene schneiden, keine Rolle: sie bilden eine Mannigfaltigkeit von geringerer Dimension), so kann man $\xi_1 \neq 0$ und $\eta_0 \neq 0$, etwa $\xi_1 = \eta_0 = 1$ annehmen. Die Gleichungen (1) werden dann inhomogen; ihre konstanten Glieder auf der linken Seite lauten:

$$a_0,\ a_1,\ \ldots,\ a_m,$$

wenn a_i der Koeffizient von $x_0^i x_1^{m-i}$ im Polynom f ist. Es handelt sich also bei einer allgemeinen Form f um solche Gleichungen, deren konstante

---

[1]) Die Methode ist dieselbe, die ich in meiner Arbeit „Die Alternative bei nichtlinearen Gleichungen‘‘, Nachr. Ges. Wiss. Göttingen 1928, S. 1—11, angewandt habe; sie wird auch sonst in der abzählenden Geometrie häufig benutzt.

Glieder lauter unabhängige Unbestimmte sind, welche sonst in diesen Gleichungen nicht mehr vorkommen.

Ein solches Gleichungssystem ist offensichtlich unlösbar, wenn die Anzahl der Gleichungen größer als die Anzahl der Unbekannten ist, also für

$$m + 1 > 2\,n - 2,$$
$$m > 2\,n - 3.$$

Für $m \leqq 2\,n - 3$ ist das Gleichungssystem lösbar, falls die linken Seiten, nach Weglassung ihrer konstanten Glieder, algebraisch-unabhängige Polynome in $\xi_2, \ldots, \xi_n, \eta_2, \ldots, \eta_n$ darstellen [2]), und die Dimensionszahl der Lösungsmannigfaltigkeit ist in diesem Fall $2\,n - 3 - m$.

Wir setzen also

$$f_i\,(\xi,\,\eta) = a_i + g_i\,(\xi,\,\eta)$$

und haben noch die algebraische Unabhängigkeit der Polynome $g_i(\xi,\,\eta)$ für Unbestimmte $\xi,\,\eta$ zu beweisen. Wären die g_i algebraisch-abhängig, so würden sie es bleiben bei jeder Spezialisierung der unbestimmten Koeffizienten der Form f. Spezialisieren wir nun f folgendermaßen:

$$f = \sum_0^n a_i\,x_0^i\,x_1^{m-i} + \sum_0^q x_0^{2i}\,x_1^{m-2i-1}\,x_{2+i},$$

wo $q = \left[\dfrac{m-1}{2}\right] \leqq n - 2$ (wegen $m \leqq 2\,n - 3$) ist. Dann wird

$$g_0 = \xi_2, \; g_1 = \eta_2, \; g_3 = \xi_3, \; g_4 = \eta_3, \; \ldots, \; g_{2q} = \xi_{2+q}, \; g_{2q+1} = \eta_{2+q}.$$

Diese Größen sind aber offensichtlich algebraisch-unabhängig. Also sind die g_i es auch und die aufgestellten Behauptungen über die Lösbarkeit und die Dimensionszahl sind bewiesen.

Gleichzeitig folgt, daß die Lösungen eine irreduzible algebraische Mannigfaltigkeit bilden (irreduzibel in bezug auf den Rationalitätsbereich Ω der Unbestimmten a_i), was insbesondere im Fall $m = 2\,n - 3$ bedeutet, daß alle Lösungen algebraisch-konjugiert (in bezug auf den genannten Rationalitätsbereich) sind.

Schließlich folgt noch [vgl. Fußnote [2])], daß das Ideal $(f_0, \ldots, f_m)$ im Polynombereich $\Omega\,[\xi_2, \ldots, \xi_n, \eta_2, \ldots, \eta_n]$ ein Primideal ist. Im Fall $m = 2\,n - 3$, wo das Primideal nulldimensional ist, ist also sein Grad $(=$ Restklassenzahl$)$ gleich seiner Nullstellenzahl.

Diese Relation bleibt erhalten, wenn die Nullstellen des Ideals zum Körper Ω adjungiert werden. Also zerfällt das Ideal dann in so viele Primideale vom Grad Eins, als es Nullstellen gibt.

Daraus folgt, daß die (idealtheoretische) Multiplizität der Nullstellen gleich Eins ist. Diese Erkenntnis ist für die Abzählung in § 2 von Wichtigkeit.

------

[2]) Vgl. § 2 der unter [1]) zitierten Arbeit.

§ 2.

Die Anzahl der Geraden.

Wir kehren nun zum Gebrauch homogener Koordinaten zurück und fassen ein Punktepaar (ξ, η) auf als einen Punkt eines zweifach-projektiven Raumes $P_{n,n}$. In diesem Raum definieren die Gleichungen (1) eine algebraische Mannigfaltigkeit M. Sie besteht aus der zweidimensionalen Mannigfaltigkeit M_2' der Punktepaare der Geraden auf der Hyperfläche $f = 0$ und der $(n-1)$-dimensionalen Mannigfaltigkeit M_{n-1}'' der „falschen" Lösungen $\xi = \eta$, $f(\xi) = 0$.

Die Ausscheidung dieser falschen Lösungen geschieht folgendermaßen: Die $n+1$ Gleichungen $f_0 = 0$, $f_1 = 0$, $\ldots$, $f_n = 0$ definieren eine Mannigfaltigkeit M_{n-1}, welche in den „falschen" Teil M_{n-1}'' ($\xi = \eta$, $f_0 = 0$) und einen Rest M_{n-1}' zerfällt. Wir schneiden M_{n-1}' mit $f_{n+1} = 0$ und erhalten eine M_{n-2}, welche sich wieder in einen „falschen" Teil M_{n-2}'' mit $\xi = \eta$ und einen Rest M_{n-2}' spaltet.

So fortfahrend, findet man schließlich als Schnitt von M_3' mit $f_m = 0$ ($m = 2n - 3$) eine Mannigfaltigkeit $M_2 = M_2' + M_2''$. Die Mannigfaltigkeit M_2' besteht aus den Punktepaaren der gesuchten Geraden.

Die Formen f_i $(i = 0, 1, \ldots, n)$ haben in ξ und η die Gradzahlen $m - i$ und i. Daraus folgt nach § 4 der ersten Arbeit dieser Serie[3]), daß die Gradzahlen g_{ij} der Mannigfaltigkeit M_{n-1}, wenn ihre Bestandteile mit der richtigen Vielfachheit gezählt werden, durch die Koeffizienten des Polynoms

$$(2) \qquad \sum g_{ij} p^{n-i} q^{n-j} = \prod_{i=0}^{n} ((m-i)p + iq)$$

gegeben werden. Auf der rechten Seite sind übrigens p^{n+1} und q^{n+1} durch Null zu ersetzen. Die Gradzahlen der Teilmannigfaltigkeit M_{n-1}'' sind alle gleich, da auf dieser Mannigfaltigkeit $\xi = \eta$ ist. Dasselbe gilt von $M_{n-2}'', \ldots, M_2''$. Wir haben also, um die Gradzahlen des Restes M_{n-1}' zu erhalten, von allen Koeffizienten des Polynoms (2) ein und dieselbe Zahl γ zu subtrahieren. Wie groß γ ist, wissen wir natürlich nicht, da die Multiplizität, mit der M_{n-1}'' zu zählen ist, nicht bekannt ist. Die Gradzahlen von M_{n-1}' werden also durch das Polynom

$$(3) \qquad \prod_0^{n} ((m-i)p + iq) - \gamma \sum_0^{n-1} p^{n-\lambda} q^{\lambda+1}$$

gegeben.

Wenn nun diese Mannigfaltigkeit M_{n-1}' mit der nächsten Gleichung $f_{n+1} = 0$ geschnitten wird, so kommt ein weiterer Faktor

$$(m - n - 1)p + (n+1)q$$

---

[3]) Zur algebraischen Geometrie. I. Math. Annalen **108**, S. 113—125.

auf beiden Seiten hinzu. Das ergibt, wenn wieder die Faktoren p^{n+1} und q^{n+1} durch Null ersetzt werden:

$$(4) \qquad \prod_{1}^{n+1} ((m-i)\, p + i\, q) - \gamma\, m \sum_{0}^{n-2} p^{n-\lambda}\, q^{\lambda+2}.$$

Die Gradzahlen von M_{n-2} sind die Koeffizienten dieses Polynoms. M''_{n-2} hat wieder lauter gleiche Gradzahlen, also hat man, um die Gradzahlen von M'_{n-2} zu erhalten, von allen Koeffizienten von (4) die gleiche (unbekannte) Zahl δ zu subtrahieren. Zum Glück treten die Unbekannten γ und δ nur in der Kombination $\gamma' = \gamma m + \delta$ auf und wir erhalten:

$$\prod_{1}^{n+1} ((m-i)\, p + i\, q) - \gamma' \sum_{0}^{n-2} p^{n-\lambda}\, q^{\lambda+2}.$$

Da der Ausdruck dieselbe Gestalt hat wie (3), können wir das Verfahren in derselben Weise fortsetzen bis zur Gleichung $f_m = 0$. Die unbekannten Multiplizitäten kombinieren sich in höchst angenehmer Weise zu einer einzigen Unbekannten γ'' und man erhält für die Gradzahlen von M'_2 die Formel:

$$\prod_{0}^{m} ((m-i)\, p + i\, q) - \gamma'' \, (p^n\, q^{n-2} + p^{n-1}\, q^{n-1} + p^{n-2}\, q^n).$$

Die Ausrechnung des Produktes $\prod_{1}^{m}$ (mit der schon erwähnten Vernachlässigung von p^{n+1} und q^{n+1}) ergibt einen Ausdruck der Gestalt

$$\alpha_{02}\, p^n\, q^{n-2} + \alpha_{11}\, p^{n-1}\, q^{n-1} + \alpha_{20}\, p^{n-2}\, q^n,$$

die Gradzahlen von M'_2 sind daher

$$\alpha_{02} - \gamma'', \quad \alpha_{11} - \gamma'', \quad \gamma_{20} - \gamma''.$$

Zur Bestimmung von γ'' genügt nun die Bemerkung, daß *die beiden äußersten Gradzahlen von M'_2 gleich Null ausfallen müssen*. Da es nämlich nur endlichviele Geraden $\xi\eta$ auf der Hyperfläche gibt, so können zwei allgemeine lineare Gleichungen in den ξ allein oder in den η allein durch die Punktepaare $\xi\eta$ nicht befriedigt werden. Also ist

$$\gamma'' = \alpha_{02} = \alpha_{20}$$

und die Gradzahlen von M'_2 sind

$$0, \quad \alpha_{11} - \alpha_{02}, \quad 0.$$

Die mittlere Gradzahl gibt die Anzahl der Geraden an, jede mit der Multiplizität gezählt, mit der sie bei Hinzunahme einer linearen Gleichung in ξ und einer ebensolchen in η, z. B. der Gleichungen $\xi_0 = 0$, $\eta_1 = 0$ erscheint. Diese Multiplizitäten sind aber nach § 1 gleich Eins; also *stellt die mittlere Gradzahl $\alpha_{11} - \alpha_{02}$ zugleich die Anzahl h der gesuchten Geraden dar.*

Multipliziert man das Polynom

$$\overset{m}{\underset{0}{\varPi}}\,((m-i)\,p + i\,q) = \alpha_{02}\,p^n\,q^{n-2} + \alpha_{11}\,p^{n-1}\,q^{n-1} + \alpha_{20}\,p^{n-2}\,q^n$$

noch mit $p - q$, so wird der Koeffizient von $p^n\,q^{n-1}$ gerade $\alpha_{11} - \alpha_{02}$: das ergibt genau die in der Einleitung erwähnte Regel zur Berechnung von h.

Mit derselben Methode kann man übrigens auch die Mannigfaltigkeit der geraden Linien auf Hyperflächen F^m vom Grad $m < 2\,n - 3$ untersuchen. Man findet z. B. für $n = 4$, daß auf einer F^3 des P_4 eine absolut-irreduzible Mannigfaltigkeit von ∞^2 Geraden liegt, von denen durch jeden Punkt der Hyperfläche 6 gehen, während in jeder Hyperebene 27 von diesen Geraden liegen und jede zwei ebene Schnittkurven der Hyperfläche durch 45 Geraden verbunden werden. Ebenso bilden die ∞^1 Geraden auf einer Fläche F^4 des P_4, eine absolut-irreduzible Mannigfaltigkeit, die eine Fläche vom Grade 160 einfach überdeckt.

<h3 style="text-align:center">§ 3.</h3>

<h3 style="text-align:center">Die Lage der Geraden.</h3>

Wir zeigen jetzt, daß für $n > 3$ kein Paar schneidender Geraden auf einer allgemeinen F^{2n-3} liegt.

Ein Paar schneidender Geraden hängt von $3\,n - 2$ Parametern ab. Soll eine Hyperfläche F^m das Paar enthalten, so müssen ihre $\binom{m+n}{n}$ Koeffizienten den $2\,m + 1$ linearen Bedingungen genügen, welche ausdrücken, daß die Hyperfläche den Schnittpunkt S und noch m beliebig gewählte Punkte der einen und m der anderen Geraden enthalten soll. Diese linearen Bedingungen sind unabhängig, denn man kann leicht zerfallende Hyperflächen angeben, welche alle diese Bedingungen erfüllen bis auf eine nicht erfüllte. Also bleiben nach Wahl des Geradenpaares nur $\binom{m+n}{n} - (2\,m + 1)$ Koeffizienten willkürlich verfügbar. Zählt man zu dieser Parameterzahl noch die $3\,n - 2$ des Geradenpaares und setzt $m = 2\,n - 3$, so erhält man

$$\binom{m+n}{n} - (2\,m + 1) - (3\,n - 2) = \binom{m+n}{n} - n + 3$$

Parameter, also um $n - 3$ weniger als die Konstantenzahl einer allgemeinen Gleichung m-ten Grades. Also ist eine F^{2n-3}, welche ein schneidendes Geradenpaar enthält, keine allgemeine.

Mit derselben Methode beweisen wir, daß die allgemeine F^{2n-3} $(n > 3)$ nicht 6 Geraden enthält, zwischen denen irgendeine algebraische Relation besteht. Es genügt, den Fall $n = 4$ zu betrachten. 6 windschiefe Gerade hängen von 36 Parametern ab. Wenn zwischen diesen eine algebraische Ab-

149

hängigkeit besteht, so hängen sie von höchstens 35 unter ihnen algebraisch ab. Wenn eine Fläche F^2 die 6 Geraden enthalten soll, hat sie aber 36 unabhängige lineare Bedingungen zu erfüllen, welche ausdrücken, daß sie von jeder dieser Geraden 6 Punkte enthält. Die Unabhängigkeit der 36 Bedingungen folgt wie oben durch Angabe von vollständig zerfallenden F^5, welche alle Bedingungen bis auf eine erfüllen. Die Parameterzahl der Hyperflächengleichung ist also höchstens

$$\binom{4+5}{4} - 36 + 35 < \binom{4+5}{4},$$

also ist die Fläche keine allgemeine.

Insbesondere folgt daraus, daß keine drei Geraden des F^5 in einem Raum P_3 liegen, daß keine vier eine gemeinsame Transversale besitzen, usw.

(Eingegangen am 27. 7. 1932.)

9.

Zur algebraischen Geometrie III
Über irreduzible algebraische Mannigfaltigkeiten

Mathematische Annalen 108, 5 (1933) 694-698

Jede irreduzible algebraische Mannigfaltigkeit besitzt eine Parameterdarstellung, bei der die Koordinaten $\xi_1, \ldots, \xi_n$ eines allgemeinen Punktes algebraische Funktionen von r algebraisch-unabhängigen Größen, etwa von $\xi_1, \ldots, \xi_r$ sind. Die algebraischen Funktionen sind nicht überall sinnvoll: sie sind es im allgemeinen nur für solche Werte der Unbestimmten $\xi_1, \ldots, \xi_r$, für die ein gewisses, als Nenner auftretendes Polynom $V(\xi_1, \ldots, \xi_r)$ nicht verschwindet[1]. *Wie findet man nun die Punkte η der Mannigfaltigkeit, für die $V(\eta) = 0$ ist?*

Die Definition der irreduziblen Mannigfaltigkeit, die zu einer gegebenen Parameterdarstellung gehört, gibt auf diese Frage nur die folgende Antwort: Man bilde *alle* Polynome $f(x_1, \ldots, x_n)$, für die $f(\xi) = 0$ ist, und suche dann alle Punkte η, für die alle $f(\eta) = 0$ sind. Die Antwort ist darum nicht restlos befriedigend, weil man kein Mittel hat, *alle* Polynome f mit der erwähnten Eigenschaft $f(\xi) = 0$ wirklich aufzufinden. Wir werden im folgenden zeigen, daß eine in der Hentzeltschen Eliminationstheorie[2] schon vorkommende Normbildung eine wirklich brauchbare Antwort auf die gestellte Frage gibt.

Diese Antwort ist nicht nur rechnerisch, sondern auch theoretisch von Interesse. Aus ihr ergibt sich nämlich gleichzeitig ein einfacher Beweis eines Satzes von J. F. Ritt[3], der so lautet: *Ist der Grundkörper der Körper der komplexen Zahlen, und ist $g(x_1, \ldots, x_n)$ ein Polynom, welches*

---

[1] Vgl. meine „Nullstellentheorie", § 2, oder meine Moderne Algebra II, § 88.

[2] K. Hentzelt - E. Noether, Zur Theorie der Polynomideale und Resultanten, Math. Annalen **88** (1923), S. 53—70.

[3] J. F. Ritt, Differential equations, Am. Math. Soc. Colloq. Publ. **14**, New-York 1932, S. 91.

151

nicht in allen Punkten der irreduziblen Mannigfaltigkeit M verschwindet, so sind alle die Punkte von M, in denen $g = 0$ ist, Limespunkte von solchen Punkten von M, in denen $g \neq 0$ ist. Dieser Satz zeigt, daß man die anfangs gestellte Frage im Fall des komplexen Zahlkörpers auch mit Hilfe der Limesrelation beantworten kann. Aus ihm folgt sehr leicht, daß eine irreduzible Mannigfaltigkeit M im Komplexen topologisch zusammenhängend ist.

Schließlich wird das gefundene Ergebnis noch benutzt, um eine Lücke in meiner Arbeit „Topologische Begründung des Kalküls der abzählenden Geometrie"[4]) auszufüllen. Im Anhang I zu dieser Arbeit wurde bewiesen, daß eine r-dimensionale (komplexe) algebraische Mannigfaltigkeit ein $2\,r$-dimensionaler Komplex im Sinne der Topologie ist. Dabei wurde die Möglichkeit noch offen gelassen, daß dieser Komplex in Teilkomplexe verschiedener Dimension zerfällt. In § 5 derselben Arbeit werden aber *rein* r-dimensionale algebraische Mannigfaltigkeiten betrachtet, und es wird ohne Beweis angenommen, daß solche Mannigfaltigkeiten auch topologisch rein $2\,r$-dimensional sind, d. h. aus lauter $2\,r$-dimensionalen Zellen samt ihren Randzellen bestehen. Um das Fehlende zu ergänzen, beweisen wir im folgenden, daß *eine irreduzible algebraische Mannigfaltigkeit M_r im komplexen Gebiet ein rein $2\,r$-dimensionaler Komplex ist.*

§ 1.

Definition der Norm.

Es sei $\{\xi_1, \ldots, \xi_n\}$ ein allgemeiner Punkt der irreduziblen Mannigfaltigkeit M_r. Dann besteht zwischen $\xi_1, \ldots, \xi_r, \xi_n$ eine algebraische Abhängigkeit. Durch eine lineare Koordinatentransformation kann man immer erreichen, daß ξ_n *ganz*-algebraisch von $\xi_1, \ldots, \xi_r$ abhängt[5]). Ebenso besteht zwischen $\xi_1, \ldots, \xi_r, \xi_{n-1}$ eine Abhängigkeit, und durch eine Transformation dieser Variablen erreicht man, daß ξ_{n-1} ganz von $\xi_1, \ldots, \xi_r$ abhängt. Was man für ξ_n schon erreicht hat, wird bei dieser zweiten Transformation nicht zerstört, denn ξ_n hängt nachher jedenfalls ganz-algebraisch von $\xi_1, \ldots, \xi_r$ und ξ_{n-1} ab, also wegen der Transitivität der ganzen Abhängigkeit auch von $\xi_1, \ldots, \xi_r$ allein. So fortfahrend, erreicht man schließlich, daß $\xi_{r+1}, \ldots, \xi_n$ ganze algebraische Funktionen von $\xi_1, \ldots, \xi_r$ sind.

Sind nun $u, v, \ldots, w$ Unbestimmte, die dem Grundkörper P adjungiert werden, so ist $u\,\xi_{r+1} + v\,\xi_{r+2} + \ldots + w\,\xi_n$ eine ganze algebraische

---

[4]) Math. Annalen **102** (1929), S. 337—362.

[5]) Der Grundkörper P werde (was nicht wesentlich ist) als algebraisch-abgeschlossen vorausgesetzt; dann hat er sicher unendlich viele Elemente.

Funktion von $\xi_1, \ldots, \xi_r, u, \ldots, w$. Wir bilden nun mit einer weiteren Unbestimmten z die Norm[6])

$$(1) \quad N(z - u\,\xi_{r+1} - v\,\xi_{r+2} - \ldots - w\,\xi_n) = F(\xi_1, \ldots, \xi_r;\, u, \ldots, w, z)$$

im Körper $\mathsf{P}(\xi_1, \ldots, \xi_n, u, \ldots, w, z)$ in bezug auf den Körper $\mathsf{P}(\xi_1, \ldots, \xi_r, u, \ldots, w, z)$. Diese Norm ist ein Polynom in $\xi_1, \ldots, \xi_r;\, u, \ldots, w, z$.

Behauptung: *Setzt man für* $\xi_1, \ldots, \xi_r$ *irgendwelche Werte* $\eta_1, \ldots, \eta_r$ *aus dem Grundkörper* P *ein, so zerfällt* $F(\eta_1, \ldots, \eta_r;\, u, \ldots, z)$ *ganz in Linearfaktoren* $(z - u\,\eta_{r+1} - v\,\eta_{r+2} - \ldots - w\,\eta_n)$, *welche zu den Punkten* $\{\eta_1, \ldots, \eta_{r+1}, \ldots, \eta_n\}$ *der Mannigfaltigkeit* M_r *gehören, und so erhält man alle Punkte der Mannigfaltigkeit.*

Der letzte Teil der Behauptung (der übrigens für die beiden hier beabsichtigten Anwendungen allein schon hinreicht) ist am leichtesten zu beweisen. Das Polynom (1) wird nämlich Null für $z = u\,\xi_{r+1} + \ldots + w\,\xi_n$. Wenn eine algebraische Relation $F(\xi_1, \ldots, \xi_r;\, u, \ldots, w, u\,\xi_{r+1} + \ldots + w\,\xi_n) = 0$ aber für den allgemeinen Punkt ξ von M_r gilt, so gilt sie auch für jeden speziellen Punkt η. Also hat das Polynom $F(\eta_1, \ldots, \eta_r;\, u, \ldots, w, z)$ die Nullstelle $z = u\,\eta_{r+1} + \ldots + w\,\eta_n$, es enthält daher den Linearfaktor $z - u\,\eta_{r+1} - \ldots - w\,\eta_n$.

Zum Beweis des ersten Teiles bemerken wir, daß die Norm (1) nach Definition ein Produkt von konjugierten Linearfaktoren ist:

$$(2) \quad F(\xi_1, \ldots, \xi_r;\, u, \ldots, z) = \prod_\nu (z - u\,\xi_{r+1}^{(\nu)} - \ldots - w\,\xi_n^{(\nu)}).$$

Die Mannigfaltigkeit M_r sei durch ein Gleichungssystem

$$(3) \quad f_\lambda(\xi_1, \ldots, \xi_n) = 0$$

gegeben. Wir nehmen noch eine Größe ζ und eine Gleichung

$$(4) \quad \zeta = u\,\xi_{r+1} + v\,\xi_{r+2} + \ldots + w\,\xi_n$$

hinzu. Aus den Gleichungen (3) und (4) eliminieren wir sukzessiv $\xi_n, \xi_{n-1}, \ldots, \xi_{r+1}$, was ohne weiteres geht, da unter den Gleichungen (3) auch solche der Form

$$\xi_k^g + a_1(\xi_1, \ldots, \xi_r)\,\xi_k^{g-1} + \ldots + a_r(\xi_1, \ldots, \xi_r) = 0$$

vorkommen, welche zum Ausdruck bringen, daß $\xi_k\ (k = r+1, \ldots, n)$ ganz-algebraisch von $\xi_1, \ldots, \xi_r$ abhängt. Man erhält als Eliminationsergebnis ein Gleichungssystem

$$(5) \quad h_\lambda(\xi_1, \ldots, \xi_r, \zeta) = 0.$$

---

[6]) Im Fall eines separablen Erweiterungskörpers wird die Norm einfach als Produkt von konjugierten Linearfaktoren definiert. Im inseparablen Fall kann man die Norm als eine passende Potenz dieses Produktes definieren; noch einfacher jedoch ist es, in (1) die Norm durch das Minimalpolynom (Polynom kleinsten Grades mit der Nullstelle $\zeta = u\,\xi_{r+1} + \ldots + w\,\xi_n$) zu ersetzen, welches in einem passenden Erweiterungskörper jedenfalls in konjugierte Linearfaktoren, die nicht alle verschieden sein müssen, zerfällt.

Da nun für unbestimmte $\xi_1, \ldots, \xi_r$ das Gleichungssystem (3), (4), also auch (5), die Lösungen

$$\zeta = u\,\xi_{r+1}^{(\nu)} + v\,\xi_{r+2}^{(\nu)} + \ldots + w\,\xi_n^{(\nu)}$$

besitzt, so sind die Polynome h_λ nach Ersetzung von ζ durch eine Unbestimmte z durch alle Linearfaktoren von (2) teilbar, d. h. eine Potenz eines jeden $h_\lambda(\xi_1, \ldots, \xi_r, z)$ ist durch $F(\xi_1, \ldots, \xi_r; u, \ldots, z)$ teilbar.

Setzt man nun für $\xi_1, \ldots, \xi_r$ irgendwelche Werte $\eta_1, \ldots, \eta_r$ ein und bestimmt ζ in einem passenden Erweiterungskörper derart, daß $F(\eta_1, \ldots, \eta_r; u, \ldots, w, \zeta) = 0$ wird, so werden auch alle $h_\lambda(\eta_1, \ldots, \eta_r, \zeta)$ Null, also wird das Gleichungssystem (3), (4) lösbar, d. h. wir erhalten einen Punkt η der Mannigfaltigkeit M_r, für den $\zeta = u\,\eta_{r+1} + \ldots + w\,\eta_n$ ist. Mithin haben alle Nullstellen ζ von $F(\eta_1, \ldots, \eta_r; u, \ldots, w, z)$ die Gestalt $\zeta = u\,\eta_{r+1} + \ldots + w\,\eta_n$; d. h. F zerfällt vollständig in Linearfaktoren $z - u\,\eta_{r+1} - \ldots - w\,\eta_n$, was zu beweisen war.

<h3 style="text-align:center">§ 2.</h3>

<h3 style="text-align:center">Beweis des Satzes von J. F. Ritt.</h3>

Indem man aus $g(\xi_1, \ldots, \xi_n)$ die Norm in bezug auf $\mathsf{P}(\xi_1, \ldots, \xi_r)$ bildet, erhält man ein Polynom $G(\xi_1, \ldots, \xi_r)$, das nur von $\xi_1, \ldots, \xi_r$ abhängt und durch $g(\xi_1, \ldots, \xi_n)$ teilbar ist, also Null wird in allen den Punkten von M, wo $g = 0$ wird. Es genügt also, zu zeigen, daß alle Punkte η von M, in denen $G = 0$ ist, Limespunkte von solchen Punkten von M sind, in denen $G \neq 0$ ist.

Ein Wertsystem $\{\eta_1, \ldots, \eta_r\}$, für welches $G(\eta) = 0$ ist, ist jedenfalls Limes von solchen Wertsystemen $\{\dot\eta_1, \ldots, \dot\eta_r\}$, für die $G \neq 0$ ist. Die zugehörigen $\dot\eta_{r+1}, \ldots, \dot\eta_n$ werden nach dem Obigen durch die Nullstellen $\zeta = u\,\dot\eta_{r+1} + v\,\dot\eta_{r+2} + \ldots + w\,\dot\eta_n$ des Polynoms $F(\dot\eta_1, \ldots, \dot\eta_r; u, \ldots, w, z)$ bestimmt. Nach dem Satz von der Stetigkeit der Wurzeln algebraischer Gleichungen hängen $\dot\eta_{r+1}, \ldots, \dot\eta_n$ stetig von $\dot\eta_1, \ldots, \dot\eta_r$ ab[7]); daraus folgt die Behauptung.

---

[7]) Der Satz von der Stetigkeit der Wurzeln wird für gewöhnlich nur bewiesen für den Fall, daß die Wurzeln Zahlen sind, nicht für den hier vorliegenden, daß die Wurzeln Linearformen in den Unbestimmten $u, v, \ldots, w$ sind. Der folgende Beweis gilt allgemein und läßt sich außerdem zu einem alternativen Beweis des ersten (schwierigeren) Teils der Behauptung des § 1 im Fall des komplexen Zahlkörpers verwenden:

Es sei $F(\eta, z)$ ein von Variablen η stetig abhängiges Polynom in z. Für eine Folge von Werten $\overset{1}{\eta}, \overset{2}{\eta}, \ldots$, die gegen einen Wert η konvergieren, möge $F(\eta, z)$ folgendermaßen zerfallen:

$$F\left(\overset{\lambda}{\eta}, z\right) = \overset{m}{\underset{r=1}{\Pi}} \left(z - \overset{\lambda}{\zeta_\nu}\right).$$

Die $\overset{\lambda}{\zeta_\nu}$ bleiben, als Wurzeln einer Gleichung mit Anfangskoeffizienten 1 und beschränkten Koeffizienten, beschränkt. Also kann man aus der Folge der Wert-

§ 3.

Die Mannigfaltigkeit M_r als $2\,r$-dimensionaler Komplex.

Nach Anhang 1 der erwähnten Arbeit[4]) gibt es eine Zelleneinteilung der Mannigfaltigkeit M_r, bei der die Zellen aller Dimensionen in dem Sinne algebraisch sind, daß sie durch algebraische Gleichungen und Ungleichungen in den Real- und Imaginärteilen der Variablen $\eta_1, \ldots, \eta_n$ gegeben werden. Projizieren wir die Zellen der Dimensionen $0, 1, \ldots, 2r - 1$ in den Raum der Veränderlichen $\eta_1, \ldots, \eta_r$ hinein, indem die übrigen Variablen durch Null ersetzt werden, so bleibt die Dimension dieser Zellen kleiner als $2\,r$. Ein Punkt $\eta_1, \ldots, \eta_r$ der Projektion einer solchen Zelle Z^k $(k < 2\,r)$ ist Limes von solchen Punkten $(\dot{\eta}_1, \ldots, \dot{\eta}_r)$, welche außerhalb der weniger als $2\,r$-dimensionalen Zellenprojektionen liegen. Die zugehörigen Punkte $(\dot{\eta}_1, \ldots, \dot{\eta}_r, \dot{\eta}_{r-1}, \ldots, \dot{\eta}_n)$ der Mannigfaltigkeit können nur den $2\,r$-dimensionalen Zellen angehören. Jeder Punkt $(\eta_1, \ldots, \eta_n)$ von M_r ist daher (wieder nach dem Satz von der Stetigkeit der Wurzeln algebraischer Gleichungen) Limes von inneren Punkten $\dot{\eta}$ der $2\,r$-dimensionalen Zellen von M_r. Das war aber die zu beweisende Behauptung.

---

systeme $\left\{ \overset{\lambda}{\zeta}_1, \ldots, \overset{\lambda}{\zeta}_m \right\}$ $(\lambda = 1, 2, \ldots)$ eine konvergente Teilfolge auswählen. Ist $\{\zeta_1, \ldots, \zeta_m\}$ der Limes dieser Folge, so folgt:

$$F(\eta, z) = \overset{m}{\underset{1}{\varPi}}(z - \zeta_\nu).$$

Die Konvergenz einer Teilfolge der $\overset{\lambda}{\zeta}_\nu$ zu ζ_ν genügt für die meisten Anwendungen schon. Will man aber die Konvergenz der ganzen Folge beweisen, so schließt man folgendermaßen: Hätte eine andere Teilfolge einen wesentlich anderen (d. h. nicht nur durch die Reihenfolge der ζ_ν verschiedenen) Limes, so würde man durch Grenzübergang eine von der ersten verschiedene Faktorzerlegung von $F(\eta, z)$ erhalten, was unmöglich ist. Also ist (bei passender Anordnung der $\overset{\lambda}{\zeta}_\nu$ für jedes λ):

$$\zeta_\nu = \lim \overset{\lambda}{\zeta}_\nu.$$

(Eingegangen am 27. 10. 1932.)

10.

Zur algebraischen Geometrie IV
Die Homologiezahlen der Quadriken und die Formeln von Halphen
der Liniengeometrie

Mathematische Annalen 109, 1 (1933) 7–12

In meiner Arbeit „Topologische Begründung des Kalküls der abzählenden Geometrie"[1]) habe ich gezeigt, daß das „Charakteristikenproblem": die Bestimmung der „Anzahl" der gemeinsamen Punkte zweier beliebiger Teilmannigfaltigkeiten einer singularitätenfreien algebraischen Mannigfaltigkeit M, zurückgeführt werden kann auf die Bestimmung der Homologiegruppen von M als topologische Mannigfaltigkeit. Diese Bestimmung soll nun durchgeführt werden für den Fall, daß M eine Quadrik Q_{2n} oder Q_{2n+1}, d. h. eine quadratische Hyperfläche im projektiven Raum ist. Als Spezialfall erhält man daraus, wenn M die Mannigfaltigkeit der Geraden des dreidimensionalen Raumes ist, die Halphenschen Formeln für die Anzahlen der gemeinsamen Strahlen zweier Geradensysteme.

Die Beweismethode ist im wesentlichen dieselbe, die in der „Topologischen Begründung", Anhang II, zur Bestimmung der Homologiegruppe des projektiven Raumes benutzt wurde: Durch eine Deformation werden alle Zyklen in einen Teilraum von geringerer Dimension „hineingeschoben". Während aber im Fall des projektiven Raumes die Deformation stetig war für alle Zyklen, die einen Punkt nicht enthalten, ist sie es hier nur für die Zyklen, die eine gewisse Teilmannigfaltigkeit P_n^* von der Dimension n nicht treffen, was eine Schwierigkeit mit sich bringt, die aber durch Anwendung eines Dualitätssatzes der kombinatorischen Topologie behoben werden kann. Dieser Dualitätssatz gibt nämlich ein Kriterium dafür, wann ein gegebener Zykel homolog einem solchen ist, der P_n^* nicht trifft: das Kriterium besteht in dem Nullsein aller Schnittpunktszahlen des gegebenen Zykels mit den Zyklen von P_n^*.

Die eben skizzierte Deformationsmethode ist im wesentlichen nur eine Übertragung ins Topologische einer algebraisch-geometrischen Methode,

---

[1]) Math. Annalen **102** (1929), S. 337—362.

156

welche G. Schaake[2]) in vielen Fällen mit Erfolg zur Auffindung von Charakteristikenformeln angewandt hat. Die Umwandlung ins Topologische hat nur den Vorteil, eine bessere Beherrschung aller Multiplizitätsfragen zu ermöglichen. Es ist zu erwarten, daß mit einer ähnlichen Deformationsmethode noch eine Reihe von Charakteristikenproblemen exakt gelöst werden können. Als typisches Beispiel nenne ich das Charakteristikenproblem für die Mannigfaltigkeit aller linearen Unterräume P_m des Raumes P_n [2a]).

§ 1.
Die Quadriken gerader Dimensionszahl.

Im folgenden bezeichnet ein Index rechts unten immer die algebraische, ein Index rechts oben die (doppelt so große) topologische Dimensionszahl.

Die Gleichung einer allgemeinen Quadrik Q_{2n} im komplexen projektiven Raum P_{2n+1} kann auf die Form

$$x_0 y_0 + x_1 y_1 + \ldots + x_n y_n = 0$$

gebracht werden. Auf der Fläche liegen zwei Scharen von linearen Räumen P_n. Die Räume P_n der einen Schar werden durch Gleichungen der Form

$$(1) \qquad\qquad x_i = \sum a_{ik} y_k \qquad\qquad (a_{ik} = - a_{ki})$$

gegeben, während die Gleichungen der Räume P'_n der anderen Schar daraus durch Vertauschung eines x_i mit einem y_i hervorgehen. Vertauscht man noch ein x mit einem y, so erhält man wieder einen Raum der ersten Schar usw. Zwei Räume P_n oder P'_n derselben Schar schneiden sich im allgemeinen in einem Punkt, wenn n gerade, und schneiden sich nicht, wenn n ungerade ist. Zwei Räume P_n und P'_n aus verschiedenen Scharen schneiden sich umgekehrt nicht, wenn n gerade, und in einem Punkt, wenn n ungerade ist.

Die Räume $P_n = P^{2n}$ sind $(2n)$-dimensionale Zyklen im $(4n)$-dimensionalen Raum $Q_{2n} = Q^{4n}$. Der Schnittpunktindex zweier sich in einem Punkt schneidender Räume P_n oder P'_n ist ± 1. Um das einzusehen, genügt es auf Grund der topologischen Invarianz des Schnittpunktsindex,

---

[2]) G. Schaake, Afbeeldingen van figuren op de punten eener lineaire ruimte. Diss. Amsterdam 1922, Kap. VI.

[2a]) **Zusatz bei der Korrektur.** Dieses Problem ist inzwischen für den Fall $m = 1$, also für die Mannigfaltigkeit der Geraden des P_m, von Herrn Ehresmann (C. R. Acad. Sci. Paris **196** (1932), p. 152) auf Grund einer ähnlichen, aber etwas einfacheren topologischen Methode gelöst worden. Damit ist insbesondere die Halphensche Regel neu bewiesen.

die Umgebung der betreffenden Stelle von Q_{2n} durch stereographische Projektion topologisch auf ein Stück des euklidischen Raumes E_{2n} abzubilden, wobei die Räume P_n in lineare Räume E_n übergehen, für welche definitionsmäßig ein isolierter Schnittpunkt immer den Index ± 1 hat. Bei passender Orientierung ist der Index sogar $+1$.

Die Matrix der Schnittpunktszahlen zweier Räume P_n, P_n' aus verschiedenen Scharen mit zwei ebensolchen lautet demnach:

$$\text{für } n \text{ gerade:} \quad \begin{pmatrix} 1 & 0 \\ 0 & 1 \end{pmatrix},$$

$$\text{für } n \text{ ungerade:} \quad \begin{pmatrix} 0 & 1 \\ 1 & 0 \end{pmatrix}.$$

Daraus folgt, daß P_n und P_n' im Sinn der Homologie linear-unabhängig sind.

In P_n liegen natürlich lineare Räume von allen kleineren Dimensionen: P_0, P_1, ..., P_{n-1}. Unter Q_{n+1}, Q_{n+2}, ..., Q_{2n-1} verstehen wir schließlich die ebenen Schnitte von Q_{2n} mit linearen Räumen der Dimensionen $n+2$, $n+3$, ..., $2n$. Jedes $P_\nu = P^{2\nu}$ ($\nu = 0, 1, ..., n-1$) schneidet $Q_{2n-\nu} = Q^{2(2n-\nu)}$ in einem Punkt vom Index Eins, also ist kein Vielfaches von P_ν oder $Q_{2n-\nu}$ homolog Null.

Die Bestimmung der Homologiegruppe von Q_{2n} geschieht nun auf Grund der folgenden Überlegung.

Ist C^k ein Zykel von einer Dimension $k < 2n$ auf $Q^{2 \cdot 2n}$, so kann man C^k durch einen Homologen ersetzen, der den Raum

$$P_n^*: \; y_0 = y_1 = \dots = y_n = 0$$

nicht trifft. Dieser Raum P_n^* wird von einem Raum P_n' der zweiten Schar in genau einem Punkte geschnitten. Ist C^{2n} ein Zykel von der Dimension $2n$, also von derselben Dimension wie P_n', so kann man die Zahl j so bestimmen, daß

$$'C^{2n} = C^{2n} - j \cdot P_n'$$

mit dem Raum P_n^* die Schnittpunktszahl Null hat. Nach einem Dualitätssatz von van Kampen und Pontrjagin[3]) ist dann $'C^{2n}$ einem Zykel $''C^{2n}$, der P_n^* nicht trifft, homolog mit erlaubter Division. In dieser Weise werden alle Zyklen mit $k \leqq 2n$ auf solche zurückgeführt, welche P_n^* nicht treffen. Das gleiche könnte man übrigens auch für die C^k mit $k > 2n$ machen, indem man passende Vielfache von Q_{n+1}, ..., Q_{2n-1} von ihnen subtrahiert, jedoch brauchen wir das für das folgende nicht.

[3]) E. R. van Kampen, Die kombinatorische Topologie und die Dualitätssätze, Diss. Leiden 1929, S. 64 (Satz 4) und S. 72 (Schluß von § 4). L. Pontrjagin, Math. Ann. 105 (1931), S. 190, Formel (3).

Jetzt definieren wir eine Deformation, welche den Punkt

$$x_0, \ldots, x_n, y_0, \ldots, y_n$$

in

$$\lambda x_0, \ldots, \lambda x_n, y_0, \ldots, y_n$$

überführt, wobei der reelle Parameter λ von 1 nach 0 geht. Für alle nicht zu P_n^* gehörigen Punkte bleibt diese Deformation eindeutig und stetig auch bei $\lambda = 0$, also werden unsere Zyklen C^k bzw. $''C^{2n}$ stetig in solche übergeführt, welche dem Raum P_n:

$$x_0 = x_1 = \ldots = x_n = 0$$

angehören. Für diesen projektiven Raum P_n bilden nun bekanntlich die Unterräume $P_\nu = P^{2\nu}$ ($\nu = 0, 1, \ldots, n$) eine vollständige Homologiebasis.

Also bilden diese Zyklen $P^{2\nu}$ ($\nu = 0, \ldots, n-1$) auch eine Homologiebasis für $Q^{2 \cdot 2n}$ für alle Dimensionen $k < 2n$, während für $k = 2n$ die Zyklen P_n und P_n' zusammen eine Basis im Sinne der Homologie mit erlaubter Division bilden.

Damit sind die Homologiezahlen $p_0 = 1$, $p_1 = 0$, $p_2 = 1$, $\ldots$, $p_{2n-1} = 0$, $p_{2n} = 2$ bestimmt, während sich gleichzeitig ergibt, daß für die Dimensionen $k < 2n$ keine Torsion existiert. Aus dem Poincaréschen Reziprozitätsgesetz für die Homologiezahlen und Torsionskoeffizienten:

$$p_{4n-\nu} = p_\nu; \quad \tau_{4n-\nu} = \tau_{\nu-1}$$

bestimmen sich nun alle übrigen Homologiezahlen:

$$p_{2n+1} = 0, \quad p_{2n+2} = 1, \quad \ldots, \quad p_{4n-1} = 0, \quad p_{4n} = 1,$$

während sich andererseits die völlige Torsionsfreiheit ergibt. Zieht man schließlich noch die Tatsache heran, daß die Determinanten der oben aufgestellten Schnittzahlmatrizen gleich ± 1 sind, so folgt, *daß die Zyklen*

$$P_0, P_1, \ldots, P_{n-1}; \quad P_n, P_n'; \quad Q_{n+1}, \ldots, Q_{2n}$$

eine vollständige Homologiebasis für Q_{2n} bilden.

§ 2.

Die Quadriken ungerader Dimensionszahlen.

Die Gleichung einer allgemeinen Quadrik $Q_{2n+1} = Q^{2(2n+1)}$ im komplexen projektiven Raum P_{2n+2} kann auf die Form

$$x_0 y_0 + \ldots + x_n y_n + z^2 = 0$$

gebracht werden. Die Quadrik enthält lineare Räume P_n, z. B. die Räume

$$P_n^*: \quad y_0 = y_1 = \ldots = y_n = 0, \quad z = 0,$$
$$P_n: \quad x_0 = x_1 = \ldots = x_n = 0, \quad z = 0,$$

und die Teilräume P_ν ($\nu = 0, 1, \ldots, n-1$) von P_n.

Jeder Zykel C^k von einer Dimension $< 2(n+1)$ kann durch einen homologen ersetzt werden, der den Raum P_n^* nicht trifft. Sodann deformieren wir den Zykel innerhalb Q_{2n+1} dadurch, daß der Punkt

$$x_0, \ldots, x_n, y_0, \ldots, y_n, z$$

in

$$\lambda^2 x_0, \ldots, \lambda^2 x_n, y_0, \ldots, y_n, \lambda z$$

übergeführt wird, wobei λ von 1 nach 0 geht. Der Zykel C^k kommt dadurch innerhalb P_n zu liegen.

Also bilden die Zyklen

$$P_\nu = P^{2\nu} \qquad\qquad (\nu = 0, 1, \ldots, n)$$

eine Homologiebasis für alle Zyklen der Dimensionen $k < 2(n+1)$. Daß diese Zyklen nicht homolog Null sind, ergibt sich daraus, daß jeder Raum P_ν mit einem ebenen Schnitt $Q_{2n+1-\nu}$ genau einen Schnittpunkt vom Index 1 bestimmt. Die Homologiezahlen und Homologiebasen für die Dimensionen $k \geq 2(n+1)$ bestimmen sich wie in § 1 auf Grund des Reziprozitätsprinzips. Die Homologiezahlen $p_0, p_1, \ldots, p_{2(2n+1)}$ sind abwechselnd 1 und 0, und die Homologiebasis besteht aus den Zyklen

$$P^0, P^2, \ldots, P^{2n}, Q^{2(n+1)}, \ldots, Q^{2(2n+1)}.$$

Bemerkung. Die in § 1 und § 2 angewandte Methode kann auch auf reelle Quadriken in reellen projektiven Räumen angewandt werden. Die Gleichungen dieser Quadriken können nämlich immer auf die Form

$$x_0 y_0 + \ldots + x_l y_l + z_1^2 + \ldots + z_m^2 = 0$$

gebracht werden; die zu benutzende Deformation führt den Punkt

$$\{x_0, \ldots, x_l, y_0, \ldots, y_l, z_1, \ldots, z_m\}$$

in

$$\{\lambda^2 x_0, \ldots, \lambda^2 x_l, y_0, \ldots, y_l, \lambda z_1, \ldots, \lambda z_m\}$$

über und ist stetig außer für die Punkte des Teilraums

$$P_{l-1}^*; \quad y_0 = \ldots = y_l = 0, z_1, \ldots, z_m = 0.$$

Diese Deformation führt zu einer vollständigen Bestimmung aller Homologiegruppen der Dimensionen $k < l + m - 1$. Falls die Quadrik Q_{2l+m} orientierbar ist, was aber nur für gerade Werte von m (und im trivialen Fall der Sphäre: $l = 1$) der Fall ist, kann man weiter aus dem Reziprozitätsprinzip die fehlenden Homologiegruppen bestimmen.

§ 3.

Die Formel von Halphen.

Nachdem die Homologiebasen der Quadriken Q_{2n} und Q_{2n+1} bestimmt und ihre Schnittpunktmatrizen berechnet sind, kann man die „Schnittpunktszahl" zweier algebraischer Teilmannigfaltigkeiten komplementärer

Dimensionszahl, d. h. im Fall isolierter Schnittpunkte die Summe ihrer Schnittpunktsindizes, ohne weiteres berechnen („Topologische Begründung" § 8). Das Ergebnis läßt sich für beide Fälle Q_{2n} und Q_{2n+1} gemeinsam so formulieren:

Jede algebraische Teilmannigfaltigkeit M_k von Q_m hat eine Gradzahl, nämlich die Schnittpunktszahl von M mit dem Basiselement P_{m-k} oder Q_{m-k} der Homologiegruppe; nur im Fall $m = 2n$, $k = n$ hat M_k zwei Gradzahlen b und f, nämlich die Schnittpunktszahlen von M_n mit P_n und P'_n. Die Schnittpunktszahl zweier Mannigfaltigkeiten M_k und M_{m-k} ist gleich dem Produkt der Gradzahlen; nur im Fall $m = 2n$, $k = n$ wird die Schnittpunktszahl gegeben durch die Ausdrücke:

$$b\,b' + f\,f' \quad \text{für } n \text{ ungerade,}$$

$$b\,f' + f\,b' \quad \text{für } n \text{ gerade,}$$

wobei b, f die Grade von M_k und b', f' die von M_{m-k} sind.

Ein interessanter Fall ist der, wo Q_m die vierdimensionale Mannigfaltigkeit aller Geraden des Raumes P_3 ist, deren Gleichung (in Plückerschen Linienkoordinaten) lautet:

$$p_{01}\,p_{23} + p_{02}\,p_{31} + p_{03}\,p_{12} = 0.$$

Man erhält das Ergebnis, daß jede Regelschar und jeder Geradenkomplex eine Gradzahl besitzt, nämlich die Schnittzahl mit einem (eventuell speziellen) linearen Komplex bzw. mit einem Geradenbüschel, während eine Geradenkongruenz zwei Gradzahlen, nämlich den „Bündelgrad" b (Schnittzahl mit einem Gradenbündel) und den „Feldgrad" f (Schnittzahl mit einem ebenen Geradenfeld) hat. Die Schnittzahl einer Regelschar und eines Komplexes ist gleich dem Produkt der Gradzahlen, während die Schnittzahl zweier Kongruenzen mit den Gradzahlen b, f und b', f' durch die von Halphen[4]) gefundene Formel

$$b\,b' + f\,f'$$

geliefert wird.

---

[4]) G. Halphen, C. R. Ac. Paris 1872, S. 41. Vgl. auch H. Schubert, Math. Annalen 10, S. 96, sowie Kalkül der abzählenden Geometrie, S. 62; H. G. Zeuthen, Lehrbuch der abzählenden Methoden der Geometrie, S. 268 und 275. Alle zitierten Beweise sind, da sie sich nicht auf eine klare Multiplizitätsdefinition beziehen, als ungenügend zu bezeichnen.

(Eingegangen am 27. 10. 1932.)

11.

Zur algebraischen Geometrie V
Ein Kriterium für die Einfachheit von Schnittpunkten

Mathematische Annalen 110, 1 (1934) 128–133

In meiner Abhandlung über die Alternative bei nichtlinearen Gleichungen[1]) habe ich folgenden Satz bewiesen:

1. *Die Lösungen eines Systems von n Gleichungen mit n Unbekannten, deren konstante Glieder unabhängige Unbestimmte sind, sind sämtlich einfach.*

Dieser Satz ist ein Spezialfall des folgenden, in der italienischen Literatur häufig benutzten:

2. *Schneidet man eine irreduzible d-dimensionale Mannigfaltigkeit $\mathfrak{M}$ des projektiven Raumes P_n mit d Hyperflächen*

$$(1) \quad \begin{cases} \lambda_0\, F_0 + \lambda_1\, F_1 + \ldots + \lambda_r\, F_r = 0, \\ \mu_0\, G_0 + \mu_1\, G_1 + \ldots + \mu_s\, G_s = 0, \\ \cdots\cdots\cdots\cdots\cdots\cdots\cdots\cdots \\ \nu_0\, H_0 + \nu_1\, H_1 + \ldots + \nu_s\, H_s = 0, \end{cases}$$

die je eine lineare Schar durchlaufen, so haben für unbestimmte $\lambda, \mu, \ldots, \nu$ diejenigen Schnittpunkte, die nicht Basispunkte einer linearen Schar sind, stets die Multiplizität Eins.

Dieser Satz folgt wiederum leicht aus dem bekannten Spezialfall $d = 1,\ r = 1$:

3. *Schneidet man eine irreduzible Kurve C mit den Hyperflächen eines Büschels*

$$(2) \qquad \lambda_0\, F_0 + F_1 = 0,$$

so sind für unbestimmte λ diejenigen Schnittpunkte, die nicht Basispunkte des Büschels sind, stets einfach.

Die Sätze **2** und **3** sollen nun auf Grund der in der ersten Abhandlung dieser Serie (ZAG I, Math. Annalen **108**, S. 115) gegebenen Definition der Schnittpunktsmultiplizität möglichst elementar bewiesen werden.

---

[1]) Nachr. Ges. Wiss. Göttingen 1928.

Meine früheren Beweise des Satzes **1** und ähnlicher Sätze, welche behaupten, daß gewisse Schnittpunkts- oder Spezialisierungsmultiplizitäten gleich Eins sind, beruhten immer darauf, daß gewisse Ideale als Primideale nachgewiesen wurden, und dann die allgemeine Ungleichung

$$(3) \qquad \text{Multiplizität} \leq \text{Ideallänge}$$

angewandt wurde. Die bekannten Beweise des Satzes **3** sowie des allgemeineren Satzes von Bertini in der italienischen Lehrbuchliteratur beruhen auf einem anderen Prinzip, nämlich auf der Untersuchung des differentialgeometrischen Verhaltens der untersuchten Mannigfaltigkeiten: im Fall des Satzes **3** wird z. B. nachgewiesen, daß die allgemeine Hyperfläche des Büschels die Kurve nicht berühren kann. Der Übergang von dieser differentialgeometrischen Einfachheit zur Einfachheit im Sinne einer algebraischen Multiplizitätsdefinition ist aber nicht immer trivial. Man kann sich zur Rechtfertigung dieses Übergangs auf die idealtheoretische Relation (3) oder auch auf die „Zeuthensche Regel" stützen, aber beide Wege erscheinen etwas künstlich. Der wahre Grund für die Möglichkeit des besprochenen Übergangs scheint mir vielmehr in folgendem allgemeinem Satze zu liegen:

4. *Wenn bei einer relationstreuen Spezialisierung der Lösungen $\xi^{(1)}$, $\xi^{(2)}$, ... eines homogenen Gleichungssystems*

$$(4) \qquad G_j(\lambda; \xi_0, \xi_1, \ldots, \xi_n) = 0 \qquad (j = 1, 2, \ldots, r)$$

für $\lambda \to \mu$ zwei Lösungen $\xi^{(1)}$ und $\xi^{(2)}$ in eine Lösung η des spezialisierten Gleichungssystems

$$G_j(\mu; \eta_0, \ldots, \eta_n) = 0$$

hineinrücken, so gibt es eine Gerade durch den Punkt η, die alle Hyperflächen $G_j = 0$ im Punkt η berührt, d. h. alle Berührungs- (oder Polar-) hyperebenen

$$(5) \qquad \zeta_0 \frac{\partial G_j}{\partial \eta_0} + \zeta_1 \frac{\partial G_j}{\partial \eta_1} + \cdots + \zeta_n \frac{\partial G_j}{\partial \eta_n} = 0$$

haben eine Gerade durch η gemeinsam. (Einige oder alle Gleichungen (5) können natürlich auch identisch erfüllt sein.)

Aus **4** folgt offenbar das Kriterium: Wenn der Rang der Matrix $(\partial_\nu G_j)$[2]) an der Stelle η gleich n ist, so hat der Punkt η bei der Spezialisierung $\lambda \to \mu$ höchstens die Spezialisierungsmultiplizität Eins. Ist $\eta_0 \neq 0$, so kann man die Spalte $(\partial_0 G_j)$ der Matrix weglassen: *Hat die dann übrig bleibende Matrix den maximalen Rang n, oder, was dasselbe*

---

2) $\partial_\nu G$ bedeutet Differentiation der Form $G(x_0, x_1, \ldots, x_n)$ nach x_ν.

ist, hat das Gleichungssystem (5) *mit* $\zeta_0 = 0$ *nur die Nullösung, so ist die Spezialisierungsmultiplizität gleich Eins.* In dieser letzten Form ist das Kriterium auch bei Benutzung inhomogener Koordinaten anwendbar.

Mit Hilfe der Kriterien **2** und **4** lassen sich eine Reihe von Beweisen, die ich früher mit idealtheoretischen Methoden geführt habe, ohne Benutzung der Idealtheorie durchführen. So läßt sich auf Grund von **2** (Spezialfall $d = 1$) die Ungleichung (1) aus Z A G I[3]), § 1 sofort zur Gleichung $q' = q$ verschärfen, ohne daß man den dortigen idealtheoretischen Hilfssatz dazu braucht. Dieser idealtheoretische Hilfssatz wird in Z A G I nur noch einmal benutzt, nämlich in § 2, um die (relative) Irreduzibilität des Durchschnittes einer irreduziblen Mannigfaltigkeit M_r mit einer allgemeinen Hyperfläche $F_1 = 0$ zu zeigen. Aus § 5 der nächstfolgenden Arbeit Z A G VI dieser Serie wird man sehen, daß auch diese Irreduzibilität sich ohne Idealtheorie sehr leicht ergibt. Bei einer anderen Gelegenheit werde ich zeigen, daß man auch die Beweise meiner Arbeit „Verallgemeinerung des Bézoutschen Theorems"[4]) mit Hilfe des Kriteriums **4** von der Idealtheorie befreien und dabei sehr erheblich vereinfachen kann.

Beweis von 4.

Wir können $\eta_0 \neq 0$ annehmen; dann liegen auch $\xi^{(1)}$ und $\xi^{(2)}$ nicht in der Hyperebene $\xi_0 = 0$ und wir können $\eta_0 = \xi_0^{(1)} = \xi_0^{(2)} = 1$ normieren. Die Verbindungslinie von $\xi^{(1)}$ und $\xi^{(2)}$ schneidet diese Hyperebene in einem Punkt τ mit $\tau_0 = 0$ und man hat

$$(6) \qquad \xi_\nu^{(2)} = \xi_\nu^{(1)} + \gamma\, \tau_\nu \qquad\qquad (\nu = 0, 1, \ldots, n).$$

Die Gleichungen $G_j = 0$ haben die Lösungen $\xi^{(1)}$ und $\xi^{(2)}$, also haben die Gleichungen

$$G_j(\xi^{(1)} + z\,\tau_\nu) = 0$$

die Lösungen $z = 0$ und $z = \gamma$. Entwickelt man die linke Seite nach Potenzen von z, dividiert durch z und setzt $z = \gamma$, so kommt

$$(7) \qquad \sum_\nu \tau_\nu\, \partial_\nu G_j(\xi^{(1)}) + \gamma\, H_2(\xi^{(1)}, \tau) + \gamma^2 H_3(\xi^{(1)}, \tau) + \ldots = 0.$$

Zu der vorgegebenen relationstreuen Spezialisierung $\lambda \to \mu$, $\xi^{(1)} \to \eta$, $\xi^{(2)} \to \eta$ läßt sich nun eine relationstreue Spezialisierung $\tau \to \omega$ finden, wobei wieder $\omega_0 = 0$, aber nicht alle $\omega_\nu = 0$ sind. Es sei etwa $\omega_1 \neq 0$, dann ist auch $\tau_1 \neq 0$ und wir können $\tau_1 = \omega_1 = 1$ setzen. Dann folgt aus (6):

$$(8) \qquad \gamma = \xi_1^{(2)} - \xi_1^{(1)}.$$

---

[3]) Math. Annalen **108** (1933), S. 118.
[4]) Ebenda **99** (1928), S. 497—541.

Das denken wir uns in (7) eingesetzt. Die so entstehende Gleichung machen wir zunächst durch Einführung von $\xi_0^{(1)}$, $\xi_0^{(2)}$ und τ_1 homogen, machen dann den Übergang $\xi^{(1)} \to \eta$, $\xi^{(2)} \to \eta$, $\tau \to \omega$ und setzen dann wieder $\eta_0 = 1$, $\omega_1 = 1$. Dasselbe Ergebnis hätten wir natürlich auch ohne Homogenisierung erhalten: die Homogenisierung war nur nötig, um die Definition der relationstreuen Spezialisierung anwenden zu können. Bei der Spezialisierung geht γ in 0 über wegen (8), also folgt aus (7):

$$\sum_\nu \omega_\nu \cdot \partial_\nu G_j(\eta) = 0.$$

Ebenso ist (nach Euler) wegen $G_j(\eta) = 0$

$$\sum_\nu \eta_\nu \cdot \partial_\nu G_j(\eta) = 0,$$

also liegt die Verbindungslinie von η mit ω ganz in den Polarhyperebenen (5), w. z. b. w.

Beim Beweis wurde der folgende, auch sonst nützliche **Hilfssatz** benutzt:

Sind $\xi_0, \ldots, \xi_n$; $\tau_0, \ldots, \tau_m$ Elemente eines Funktionenkörpers, zwischen denen gewisse in den ξ und τ einzeln homogene algebraische Relationen $F(\xi, \tau) = 0$ bestehen, und ist $\xi \to \eta$ eine relationstreue Spezialisierung der ξ, so läßt sich diese zu einer relationstreuen Spezialisierung $\xi \to \eta, \tau \to \omega$ ergänzen.

Zum Beweis des Hilfssatzes bilde man aus den Gleichungen $F(\xi, \tau) = 0$ durch Elimination der τ das Resultantensystem $R(\xi) = 0$. Bei der Spezialisierung $\xi \to \eta$ bleiben die Relationen $R(\xi) = 0$ erhalten; es ist also $R(\eta) = 0$ und daraus folgt, daß die Gleichungen $F(\eta, \omega) = 0$ eine nicht triviale Lösung ω besitzen.

Beweis von 3.

ξ sei ein variabler Schnittpunkt der Hyperfläche

$$G_1 = \lambda F_0 + F_1 = 0$$

mit der Kurve C. Wir werden zeigen, daß es möglich ist, aus den Gleichungen der Kurve $n - 1$ solche $G_2 = 0, \ldots, G_n = 0$ auszuwählen, daß die Berührungshyperebenen dieser Hyperflächen im Punkt ξ nur eine Gerade (nämlich die Kurventangente) gemeinsam haben, und daß die Berührungshyperebene der Hyperfläche $G_1 = 0$ diese Gerade nicht enthält. Daraus folgt dann auf Grund von **4**, daß der Punkt ξ ein einfacher Schnittpunkt von C und $G_1 = 0$ ist.

Die inhomogenen Koordinaten $\xi_1, \ldots, \xi_n$ des Punktes ξ sind algebraische Funktionen von λ und als solche differenzierbar[5]) nach λ. Man kann auch alle ξ_ν als algebraische Funktionen von einer unter ihnen, etwa von ξ_1 auffassen; es gibt also Gleichungen

$$G_2(\xi_1, \xi_2) = 0; \quad G_3(\xi_1, \xi_3) = 0; \quad \ldots; \quad G_n(\xi_1, \xi_n) = 0.$$

Differentiation ergibt

$$(\partial_1 G_\nu) \frac{d\xi_1}{d\lambda} + (\partial_\nu G_\nu) \frac{d\xi_\nu}{d\lambda} = 0 \qquad\qquad [\partial_\nu G_\nu \neq 0].$$

Da durch diese Formeln die Verhältnisse der Größen $\dfrac{d\xi_n}{d\lambda}$ eindeutig bestimmt sind, so ist jede Lösung ζ der Gleichungen

$$(9) \qquad\qquad (\partial_1 G_\nu) \zeta_1 + (\partial_\nu G_\nu) \zeta_\nu = 0$$

zu diesen Größen proportional:

$$(10) \qquad\qquad \zeta_1 : \zeta_2 : \ldots : \zeta_n = \frac{d\xi_1}{d\lambda} : \frac{d\xi_2}{d\lambda} : \ldots : \frac{d\xi_n}{d\lambda}.$$

D. h. geometrisch: Es gibt nur einen Punkt ζ der Hyperebene $\zeta_0 = 0$ und somit auch nur eine Gerade durch den Punkt ξ, welche allen Berührungsebenen der Hyperflächen $G_2 = 0, \ldots, G_n = 0$ im Punkt ξ angehört. Wäre nun ξ ein mehrfacher Schnittpunkt von C mit $G_1 = 0$ (d. h. ein solcher, in welchen bei der Spezialisierung einer allgemeinen Hyperfläche $G_1^* = 0$ zu der speziellen $G_1 = 0$ mehrere verschiedene Schnittpunkte hineinrücken), so müßte nach Satz **4** derselbe Punkt ζ auch in der Berührungshyperebene der Hyperfläche $G_1 = 0$ in ξ liegen, d. h. es müßte sein

$$\sum_1^n (\partial_\nu G_1) \zeta_\nu = 0$$

oder nach (10)

$$(11) \qquad\qquad \sum_1^n (\partial_\nu G_1) \frac{d\xi_\nu}{d\lambda} = 0.$$

Das ist aber nicht der Fall. Denn aus $G_1(\xi) = 0$ folgt durch Differentiation nach λ

$$(12) \qquad\qquad \sum_1^n (\partial_\nu G_1) \frac{d\xi_\nu}{d\lambda} + F_0(\xi) = 0.$$

Aus (11) und (12) würde folgen $F_0(\xi) = 0$, also wegen $G_1(\xi) = \lambda F_0(\xi) + F_1(\xi) = 0$ auch $F_1(\xi) = 0$, d. h. ξ wäre ein Basispunkt des Büschels (2), entgegen der Voraussetzung.

Bemerkung 1. In genau analoger Weise zeigt man, daß unter den Gleichungen einer r-dimensionalen Mannigfaltigkeit $n - r$ solche

$$G_{r+1} = 0, \ldots, G_n = 0$$

---

[5]) Siehe meine algebraische Theorie der Differentiation, N. Archief v. Wisk. **15** (1926), S. 111.

stets so ausgewählt werden können, daß die Berührungshyperebenen dieser Hyperflächen in einem allgemeinen Punkt ξ von $\mathfrak{M}$ nur einen r-dimensionalen Raum (den Tangentialraum von $\mathfrak{M}$ im *P*unkt ξ) gemeinsam haben.

Bemerkung 2. Der von Enriques verallgemeinerte Satz von Bertini[6]), der besagt, daß eine allgemeine Hyperfläche eines Büschels (2) aus einer algebraischen Fläche eine Kurve ausschneidet, welche außerhalb der Basispunkte des Büschels und außerhalb der Doppelpunkte der Fläche keine mehrfachen Punkte besitzt, kann auf den eben bewiesenen Satz **3** zurückgeführt werden. Wäre nämlich ein von λ abhängiger mehrfacher Punkt vorhanden, so würde dieser bei Änderung von λ eine Kurve durchlaufen und die Hyperflächen (1) würden mit dieser Kurve einen veränderlichen mehrfachen Schnittpunkt haben, was nach **3** unmöglich ist.

Bemerkung 3. In den Beweis geht ganz wesentlich die Voraussetzung ein, daß $\xi_1, \ldots, \xi_n$ *separable* algebraische Funktionen von λ sind; ohne diese Voraussetzung sind Satz **3** und ebenso der allgemeinere Satz **4** falsch. Man kann demnach die Sätze **3** und **4** nur dann ohne Kautelen anwenden, wenn der Grundkörper die Charakteristik Null hat.

Beweis von 2.

Wir zeigen zunächst, daß der Durchschnitt einer irreduziblen d-dimensionalen Mannigfaltigkeit $\mathfrak{M}_d$ mit einer allgemeinen Hyperfläche der Schar $\lambda_0 F_0 + \lambda_1 F_1 + \ldots + \lambda_r F_r = 0$ außerhalb der Basismannigfaltigkeit der Schar aus lauter einfach zu zählenden $\mathfrak{M}_{d-1}$ besteht.

Schneidet man $\mathfrak{M}_d$ und alle Hyperflächen mit einem allgemeinen linearen Raum P_{n-d+1}, so erhält man eine Kurve C und eine lineare Schar von Hyperflächen $F(\lambda) = 0$ des Raumes P_{n-d+1}. Ist ξ ein Schnittpunkt von C mit $F(\lambda) = 0$ außerhalb der Basispunkte, so gibt es unter den Hyperflächen eine, etwa F_0, welche den Punkt ξ nicht enthält. Schreibt man nun $F(\lambda)$ in der Form

$$\lambda_0 F_0 + F_1'; \quad F_1' = \lambda_1 F_1 + \ldots + \lambda_r F_r,$$

so folgt aus **3**, daß ξ ein einfacher Schnittpunkt ist. Also sind auch alle Bestandteile $\mathfrak{M}_{d-1}$ der Schnittmannigfaltigkeit in P_n einfach gewesen.

Wendet man das eben Bewiesene d-mal hintereinander (auf $\mathfrak{M}_d$, $\mathfrak{M}_{d-1}, \ldots, \mathfrak{M}_1$) an, so folgt Satz **2**.

---

[6]) Siehe etwa F. Enriques, Introduzione alla Geometria sopra le Superficie Algebriche, Mem. Soc. It. Sci. (3) 10 (1896), S. 22.

(Eingegangen am 8. 10. 1933.)

12.

Zur algebraischen Geometrie VI
Algebraische Korrespondenzen und rationale Abbildungen

Mathematische Annalen 110, 1 (1934) 134–160

Das Ziel der Serie meiner Abhandlungen „Zur algebraischen Geometrie" (ZAG) ist nicht nur, neue Sätze aufzustellen, sondern auch, die
weitreichenden Methoden und Begriffsbildungen der italienischen geometrischen Schule in exakter algebraischer Begründung dem Leserkreis
der Math. Annalen näherzubringen. Wenn ich dabei vielleicht einiges,
was schon mehr oder weniger einwandfrei bewiesen vorliegt, hier wieder
beweise, so hat das einen doppelten Grund. Erstens setzen die italienischen Geometer in ihren Beweisen meistens eine ganze Begriffswelt,
eine Art geometrischen Denkens voraus, mit der z. B. der Deutsche von
heute nicht von vornherein vertraut ist. Zweitens aber ist es mir unmöglich, bei jedem einzelnen Satz alle in der Literatur vorhandenen
Beweise dahin nachzuprüfen, ob sich ein völlig einwandfreier darunter
befindet, sondern ich ziehe es vor, die Sätze in meiner eigenen Art zu
formulieren und zu beweisen. Wenn ich also hin und wieder einmal auf
Unzulänglichkeiten in den verbreitetsten Darstellungen hinweisen werde,
so erhebe ich damit keineswegs den Anspruch, der erste zu sein, der die
Sachen nun wirklich exakt darstellt.

Algebraische Korrespondenzen und insbesondere rationale Abbildungen
spielen in der algebraischen Geometrie eine grundlegende Rolle. Wenn
man aber die Erklärungen dieser Begriffe in den verbreitetsten Lehrbüchern näher ansieht, fallen eine Reihe von logischen Unzulänglichkeiten auf.

Es wird etwa eine rationale Abbildung definiert durch Formeln, die
für einen allgemeinen Punkt der abzubildenden Mannigfaltigkeit sinnvoll
sind, aber für spezielle Punkte unbestimmt werden. Es fragt sich nun:
Welche Punkte werden als Bildpunkte dieser speziellen Punkte angesehen?
Man könnte die Frage durch einen Grenzübergang beantworten. Es zeigt
sich aber, daß die Auffassung einer rationalen Abbildung als irreduzible
Mannigfaltigkeit von Punktpaaren und die Anwendung des Begriffs eines
„allgemeinen Punktpaares" dieser Mannigfaltigkeit eine der Stetigkeits-
betrachtung äquivalente, aber rein algebraische Beantwortung derselben
Frage gestattet, welche kurz so lautet: Alle die Punktpaare gehören zur
Abbildung, die durch relationstreue Spezialisierung aus dem allgemeinen
Punktpaar der Abbildung (oder allgemeiner der irreduziblen Korrespondenz)

entstehen. Aus dieser algebraischen Auffassung folgt weiter leicht die (bei der Limesauffassung gar nicht selbstverständliche, aber wohl meistens als selbstverständlich angenommene) Tatsache, daß irgendeinem Punkt der abgebildeten Mannigfaltigkeit $\mathfrak{M}$ in einer algebraischen Abbildung oder Korrespondenz eine nicht leere *algebraische* Mannigfaltigkeit von Punkten der Bildmannigfaltigkeit $\mathfrak{N}$ entspricht.

Genauer gilt (§ 2): Wenn einem allgemeinen Punkt von $\mathfrak{M}$ ∞ Punkte von $\mathfrak{N}$ entsprechen, so entsprechen jedem Punkt von $\mathfrak{M}$ mindestens ∞^b Punkte, und die Mannigfaltigkeit dieser Punkte enthält keine Bestandteile von einer Dimension $< b$. Aus dieser Tatsache leiten wir in § 2 einen neuen Beweis des von Severi[1]) zuerst bewiesenen und verschärften „Prinzips von Plücker-Clebsch" her.

In § 3 gehen wir auf das „Prinzip der Erhaltung der Anzahl" in der von Severi[2]) gegebenen präzisen Fassung ein. Dieses Prinzip heißt so: Wenn in einer irreduziblen Korrespondenz einem allgemeinen Punkt ξ von $\mathfrak{M}$ genau α Punkte $\eta_1, \ldots, \eta_\alpha$ einer Bildmannigfaltigkeit $\mathfrak{N}$ entsprechen, so entsprechen einem speziellen Punkt ξ' entweder unendlich viele, oder höchstens α Punkte, welche dann zusammen ebenfalls die Anzahl α ergeben, wenn sie mit denjenigen Multiplizitäten gezählt werden, mit denen sie bei einem Grenzübergang aus den Punkten $\eta_1, \ldots, \eta_\alpha$ entstehen, wenn ξ gegen ξ' rückt.

Aus einer späteren Präzisierung, die Severi[3]) dem Beweis gegeben hat, geht hervor, daß die Multiplizitäten nur dann eindeutig bestimmt und positiv sind, wenn die Mannigfaltigkeit $\mathfrak{M}$ als singularitätenfrei vorausgesetzt wird. Ich werde auf Grund meiner rein algebraischen Multiplizitätsdefinition einen neuen (von Grenzübergängen freien) Beweis der Eindeutigkeit der Multiplizitäten und damit auch einen neuen Beweis des Prinzips der Erhaltung der Anzahl geben. Der Beweis beruht darauf, daß der allgemeine Fall einer singularitätenfreien Mannigfaltigkeit durch Projektion auf den von mir schon früher[4]) behandelten Spezialfall zurückgeführt wird, wo $\mathfrak{M}$ ein linearer Raum ist.

Der Multiplizitätsbegriff tritt in der Literatur noch in einer weitergehenden Verallgemeinerung auf: Er wird nämlich auch dann angewandt,

---

[1]) F. Severi, Sulla compatibilità dei sistemi di equazioni, Rend. Accad. naz. Lincei (4) **17** (1933), S. 3—10.

[2]) F. Severi, Il Principio della Conservazione del Numero, Rend. Circ. Mat. Palermo **33** (1912), S. 313.

[3]) F. Severi, Abh. Math. Sem. Hamburg **9** (1933), S. 335—364, insbesondere Nr. 13.

[4]) „Multiplizitätsbegriff", Math. Annalen **97**, S. 756—774.

wenn einem Punkt ξ nicht ein endliches System von Punkten $\eta_1, \ldots, \eta_a$, sondern eine Kurve $\mathfrak{C}$ oder eine höherdimensionale Mannigfaltigkeit entspricht, welche bei der Spezialisierung $\xi \to \xi'$ in eine Mannigfaltigkeit $\mathfrak{C}'$ derselben Dimension übergeht, deren Bestandteile dann mit gewissen Multiplizitäten zu zählen sind. Um nun diese Multiplizitäten zu definieren, schneiden wir nach Severi[3]) die d-dimensionalen Mannigfaltigkeiten $\mathfrak{C}$ und $\mathfrak{C}'$ mit einem System von d allgemeinen Hyperflächen, welches $\mathfrak{C}$ und $\mathfrak{C}'$ je in endlichvielen Punkten schneidet, wodurch der Fall auf den schon erledigten zurückgeführt wird. Es zeigt sich (§ 4), daß diese Definition von dem Grad der gewählten Hyperflächen unabhängig und in einem gewissen Sinn auch birational-invariant ist; weiter zeigt sich, daß die gleichen Multiplizitätszahlen auch für das Verhalten der Schnittpunkte von $\mathfrak{C}$ mit nichtallgemeinen Hyperflächen maßgebend sind.

Der wichtigste Spezialfall der vorstehenden Begriffsbildung entsteht, wenn die Mannigfaltigkeit $\mathfrak{C}$ eine lineare Schar durchläuft. In § 5 werden die Grundbegriffe der Theorie der linearen Scharen entwickelt und werden zwei Sätze bewiesen, welche die Multiplizitäten der Bestandteile spezieller Elemente $\mathfrak{C}'$ der Schar, wie sie nach dem obigen durch die Spezialisierung $\mathfrak{C} \to \mathfrak{C}'$ definiert werden, mit gewissen Schnittpunktsmultiplizitäten in Beziehung bringen.

Die Theorie der linearen Scharen ist wiederum aufs engste verknüpft mit der Theorie der rationalen Abbildungen, denn jede rationale Abbildung einer Mannigfaltigkeit auf ein „projektives Modell" wird bekanntlich durch eine lineare Schar vermittelt. Mit Hilfe dieses Zusammenhangs werden in § 6 die Fundamentalpunkte der Abbildung (d. h. diejenigen Punkte von $\mathfrak{M}$, denen mindestens ∞^1 Bildpunkte auf $\mathfrak{N}$ entsprechen) studiert. Es wird bewiesen, daß die „singulären Punktepaare" (ξ', η') der Abbildung, bei denen ξ' ein Fundamentalpunkt ist, eine rein $(a-1)$-dimensionale Teilmannigfaltigkeit $\mathfrak{S}$ der a-dimensionalen Mannigfaltigkeit aller Punktepaare der Abbildung bilden.

In § 7 werden birationale Abbildungen betrachtet, die in *einer* Richtung ausnahmslos eindeutig sind. Es wird eine Formel bewiesen, die ich schon in meiner Dissertation (Amsterdam 1926) aufgestellt habe, welche die Schnittpunktzahl einer Kurve mit einer $(a-1)$-dimensionalen Mannigfaltigkeit auf $\mathfrak{M}$ ausdrückt durch die Schnittpunktzahlen der Bildkurve mit der Bildmannigfaltigkeit auf $\mathfrak{N}$ und mit den Bestandteilen der singulären Mannigfaltigkeit $\mathfrak{S}$. Diese Formel gestattet es, das „Charakteristikenproblem" für die Kurven und Hypermannigfaltigkeiten von $\mathfrak{N}$ auf das entsprechende Problem für $\mathfrak{M}$ zurückzuführen. In dieser Weise wird in § 8 das Charakteristikenproblem für die Mannigfaltigkeit „verbundener Punktepaare" (Punktepaare mit Verbindungslinie) gelöst.

Auch das berühmte Charakteristikenproblem für Kegelschnitte und das entsprechende für quadratische Flächen kann man nach Studys Vorbild in dieser Weise behandeln: ich hoffe darauf später zurückzukommen.

Die Beweismethoden der vorliegenden Untersuchung bestehen erstens in einer immer wiederholten Anwendung der „relationstreuen Spezialisierung"[4]) und zweitens der Ergänzung beliebiger Teilmannigfaltigkeiten einer Mannigfaltigkeit $\mathfrak{M}$ zu vollständigen Schnitten von $\mathfrak{M}$ durch Hinzunahme von Restschnitten, welche einen vorgegebenen Punkt nicht enthalten. Die zweite Methode habe ich von Severi[3]) übernommen.

§ 1.

Die Grundbegriffe. Das Prinzip der Konstantenzählung.

Eine algebraische Mannigfaltigkeit von Punktepaaren zweier affiner Räume E_m, E_n heißt eine *algebraische Korrespondenz* $\mathfrak{K}$. Die Gleichungen der Korrespondenz mögen lauten

$$(1) \qquad f_\nu (\xi'_1, \ldots, \xi'_m;\ \eta'_1, \ldots, \eta'_n) = 0.$$

Macht man die Gleichungen homogen durch Einführung von ξ'_0 und η'_0, so erhält man eine Korrespondenz zwischen zwei projektiven Räumen P_m und P_n:

$$(2) \qquad F_\nu (\xi'_0, \ldots, \xi'_m;\ \eta'_0, \ldots, \eta'_n) = 0.$$

Aus den homogenen Gleichungen (2) kann man bekanntlich[5]) die η' oder die ξ' eliminieren und erhält als „Resultantensystem" ein Gleichungssystem

$$(3) \qquad G_\lambda (\xi'_0, \ldots, \xi'_m) = 0,$$
$$(4) \qquad H_\mu (\eta'_0, \ldots, \eta'_m) = 0$$

mit der Eigenschaft, daß zu jeder Lösung ξ' von (3) oder η' von (4) mindestens ein Punktepaar (ξ', η') der Korrespondenz (2) gehört. Die Eigenschaft gilt im allgemeinen nicht für die inhomogenen Gleichungen (1). Die Gleichungen (3), (4) definieren zwei Mannigfaltigkeiten $\mathfrak{M}$ in P_m und $\mathfrak{N}$ in P_n: die *Urmannigfaltigkeit* und *Bildmannigfaltigkeit* der Korrespondenz (2). Man spricht auch von einer *Korrespondenz $\mathfrak{K}$ zwischen $\mathfrak{M}$ und $\mathfrak{N}$*.

Hält man den Punkt ξ' fest, so definieren die Gleichungen (2) eine algebraische Mannigfaltigkeit im Raum $\mathfrak{S}_n$, und zwar eine Teilmannigfaltigkeit $\mathfrak{N}_{\xi'}$ von $\mathfrak{N}$. Ebenso entspricht jedem Punkt η' von $\mathfrak{N}$ eine Teilmannigfaltigkeit $\mathfrak{M}_{\eta'}$ von $\mathfrak{M}$. Sind $\mathfrak{M}$ und $\mathfrak{N}$ irreduzibel und entsprechen jedem allgemeinen Punkt ξ von $\mathfrak{M}$ α Punkte von $\mathfrak{N}$ und jedem

---

[5]) Siehe etwa meine Moderne Algebra II (Berlin 1931), § 76.

171

allgemeinen Punkt η von $\mathfrak{N}$ β Punkte von $\mathfrak{M}$, so spricht man von einer (α, β)-*Korrespondenz zwischen* $\mathfrak{M}$ *und* $\mathfrak{N}$.

Die Elemente ξ' und η', die durch die Korrespondenz (2) verknüpft werden, müssen nicht notwendig als Punkte gedeutet werden: es können irgendwelche Gebilde sein, welche durch *homogene* Koordinaten bestimmt werden. Beispiele davon sind: ebene Kurven oder Hyperflächen, welche durch ihre Gleichungskoeffizienten festgelegt werden; Geraden oder allgemeiner lineare Räume P_d eines P_s, bestimmt durch ihre Plückerschen Koordinaten; Raumkurven, welche durch die Koeffizienten ihrer Gleichung in Plückerschen Linienkoordinaten (Gleichung des Treffgeradenkomplexes) oder in Ebenenkoordinaten (Gleichung des Systems der berührenden Ebenen) bestimmt werden; Kollineationen, bestimmt durch ihre Koeffizientenmatrix; usw. Wir werden im folgenden trotzdem immer von „Punktepaaren (ξ', η')" reden.

Ist die Mannigfaltigkeit $\mathfrak{K}$ irreduzibel, so sprechen wir von einer *irreduziblen Korrespondenz.* In diesem Fall sind auch $\mathfrak{M}$ und $\mathfrak{N}$ irreduzibel; denn wenn ein Produkt von zwei Formen $F(\xi') \cdot G(\xi')$ Null wird für alle Punkte ξ' von $\mathfrak{M}$, so wird es Null für alle Punktepaare (ξ', η') von $\mathfrak{K}$, also wird ein Faktor F oder G Null für alle Punktepaare (ξ', η') von $\mathfrak{K}$, also für alle Punkte ξ' von $\mathfrak{M}$.

Eine irreduzible Korrespondenz ist (wie jede irreduzible Mannigfaltigkeit) bestimmt durch ihr *allgemeines Punktepaar* (ξ, η). Die charakteristische Eigenschaft dieses allgemeinen Punktepaares ist, daß alle homogenen algebraischen Relationen $F(\xi, \eta) = 0$, welche für das allgemeine Punktepaar gelten, überhaupt für alle Punktepaare (ξ', η') gelten (und umgekehrt); und durch diese Eigenschaft sind auch alle zur Korrespondenz gehörigen Punktepaare (ξ', η') charakterisiert.

Die Koordinaten des allgemeinen Punktepaars (ξ, η) sind algebraische Funktionen von q Parametern, wenn q die Dimension der Korrespondenz ist. Wenn etwa $\xi_0 \neq 0$ und $\eta_0 \neq 0$ ist, so können wir $\xi_0 = \eta_0 = 1$ annehmen; die Zahl q läßt sich dann definieren als die Anzahl der algebraisch-unabhängigen unter den Größen $\xi_1, \ldots, \xi_m, \eta_1, \ldots, \eta_n$. Ist a die Anzahl der algebraisch-unabhängigen unter den ξ_ν, und b die Anzahl der algebraisch-unabhängigen unter den η_μ, nachdem die ξ als gegeben angenommen werden (d. h. dem Grundkörper K adjungiert werden), so ist offenbar

$$(5) \qquad q = a + b.$$

Ebenso ist, wenn c die Anzahl der algebraisch-unabhängigen unter den η ist und d die der algebraisch-unabhängigen unter den ξ nach Adjunktion der η:

$$(6) \qquad q = c + d.$$

Die Zahlen a, b, c, d lassen sich auch als Dimensionen von Mannig-faltigkeiten deuten. Die ξ definieren offenbar einen allgemeinen Punkt von $\mathfrak{M}$, also ist a die Dimension von $\mathfrak{M}$ und ebenso c die von $\mathfrak{N}$. Sucht man nun, nachdem man ξ einmal zu K adjungiert hat, die dem Punkt ξ entsprechende Teilmannigfaltigkeit $\mathfrak{N}_\xi$ von $\mathfrak{N}$, so besteht diese aus allen Punkten η', derart, daß das Punktepaar (ξ, η') der Korrespondenz an-gehört, d. h. daß alle algebraischen Relationen, die für (ξ, η) gelten, auch für (ξ, η') gelten. Das besagt aber, daß η der allgemeine Punkt der Mannigfaltigkeit $\mathfrak{N}_\xi$ ist. $\mathfrak{N}_\xi$ ist also (relativ zum Körper $\mathsf{K}\,(\xi)$) irreduzibel und von der Dimension b. Bei einer Erweiterung (z. B. bei algebraischem Abschluß) des Körpers $\mathsf{K}\,(\xi)$ kann $\mathfrak{N}_\xi$ zerfallen in algebraisch-konjugierte Teilmannigfaltigkeiten von der gleichen Dimension b. Ebenso ist d die Dimension der Mannigfaltigkeit $\mathfrak{M}_\eta$, die dem allgemeinen Punkt η von $\mathfrak{N}$ in der Korrespondenz entspricht.

Aus (5) und (6) folgt das *Prinzip der Konstantenzählung:*

Wenn in einer q-dimensionalen irreduziblen Korrespondenz zwischen $\mathfrak{M}$ und $\mathfrak{N}$ einem allgemeinen Punkt ξ der a-dimensionalen Urmannigfaltigkeit $\mathfrak{M}$ eine b-dimensionale Mannigfaltigkeit von Punkten auf $\mathfrak{N}$ und einem allgemeinen Punkt η der c-dimensionalen Bildmannigfaltigkeit $\mathfrak{N}$ eine d-dimensionale Mannigfaltigkeit von Punkten auf $\mathfrak{M}$ entspricht, so ist

$$(7) \qquad q = a + b = c + d.$$

In den meisten Anwendungen benutzt man die Formel (7) zur Be-stimmung der Dimension c der Bildmannigfaltigkeit $\mathfrak{N}$, wenn a, b und d gegeben sind (vgl. das Beispiel am Schluß von § 1).

Ist $b = 0$, so entspricht einem allgemeinen Punkt ξ ein System von endlichvielen in bezug auf $\mathsf{K}\,(\xi)$ konjugierten Punkten. Besteht das System aus nur einem Punkt η, so ist η mit allen seinen konjugierten identisch und daher (falls η_ν nicht inseparabel über $\mathsf{K}\,(\xi)$ sind) rational von ξ ab-hängig:

$$(8) \qquad \eta_\nu = \varphi_\nu\,(\xi_1, \ldots, \xi_m) \qquad (\nu = 1, \ldots, n),$$

was man in ganzrationaler Form auch so schreiben kann:

$$\eta_0 : \eta_1 : \ldots : \eta_n = \varPhi_0\,(\xi) : \varPhi_1\,(\xi) : \ldots : \varPhi_n\,(\xi).$$

Die Korrespondenz heißt in diesem Fall eine *rationale Abbildung*. Ist auch umgekehrt ξ rational abhängig von η, so heißt die Abbildung *birational*.

Eine rationale Abbildung ist durch Angabe des Urbildes $\mathfrak{M}$ und der Abbildungsformel (8) vollständig bestimmt; aber aus dieser Formel läßt sich für spezielle Punkte ξ' nicht ohne weiteres entnehmen, welche Punkte η' ihnen entsprechen, da die Funktionen φ_ν sehr wohl unbestimmt werden können. Zur Bestimmung der zulässigen η' hat man die schon erwähnte allgemeine Regel: Alle homogenen algebraischen Relationen, die für das allgemeine Punktepaar (ξ, η) gelten, sollen auch für (ξ', η') gelten.

Man kann die so charakterisierten Punktepaare (ξ', η') auch durch Grenzübergang aus solchen erhalten, welche direkt durch (8) geliefert werden. Daß man so alle Punktepaare der Abbildung erhält, folgt aus dem in § 2 der Abhandlung ZAG III[6]) bewiesenen Satz von Ritt. Andere Mittel zur Erhaltung der η' sind: Darstellung von ξ und η als Funktionen anderer Variablen, welche an der betreffenden Stelle sinnvoll bleiben, oder Reihenentwicklung von ξ und η auf den Zweigen algebraischer Kurven durch ξ'. Bei dem letzteren Verfahren stellt sich häufig heraus, daß der Grenzwert η' von η nur von der Richtung des Kurvenzweiges an der Stelle ξ', also von dem zu ξ' unendlich-benachbarten Punkt des Kurvenzweiges abhängt; in anderen Fällen muß man noch die unendlich-benachbarten Punkte höherer Ordnung heranziehen. Es steht zu vermuten, daß die Betrachtung der benachbarten Punkte einer endlichen Ordnung in jedem Fall zum Ziel führt. So verhält es sich jedenfalls dann, wenn $\mathfrak{M}$ eine singularitätenfreie algebraische Fläche ist.

Beispiele und Anwendungen zu § 1.

Ein typisches Beispiel für die Anwendung des Prinzips der Konstantenzählung möge hier angeführt werden. Für weitere Anwendungen verweisen wir auf die Lehrbuchliteratur, insbesondere auf das bekannte Werk von Enriques-Chisini.

Nach Aronhold[7]) kann man in der Ebene zu 7 gegebenen Geraden in allgemeiner Lage immer eine Kurve vierter Ordnung konstruieren, die diese 7 Geraden und noch 21 weitere, rational aus ihnen konstruierbare, zu Doppeltangenten hat. Durch diese Konstruktion ist eine rationale Abbildung der 14-dimensionalen Mannigfaltigkeit aller Punktseptupel der Ebene auf eine gewisse Bildmannigfaltigkeit $\mathfrak{N}$ im Raum der Kurven vierter Ordnung definiert. Da eine ebene Kurve vierter Ordnung nur endlichviele Doppeltangenten hat, so entsprechen jedem Punkt η' von $\mathfrak{N}$ nur endlichviele Punktseptupel. In der Formel (7) ist daher $b = d = 0$, $a = 14$ zu setzen; es folgt $c = 14$, also füllt die Bildmannigfaltigkeit $\mathfrak{N}$ den ganzen 14-dimensionalen Raum aus. Daraus folgt, daß es zu *jeder* Kurve vierter Ordnung ein System von 7 Doppeltangenten gibt, derart, daß alle algebraischen Relationen, die bei allgemeiner Lage der 7 Geraden zwischen ihnen und der Kurve bestehen, auch für die besondere Lage der Kurve und der 7 Doppeltangenten gelten. Insbesondere gilt die Aronholdsche Konstruktion der 21 restlichen Doppeltangenten aus 7 passend gewählten ganz allgemein für jede Kurve vierter Ordnung, solange die Konstruktionsschritte nicht unbestimmt werden.

Eine andere Anwendung der hier bewiesenen Sätze ist die folgende. Wir wollen zeigen, daß der Schnitt einer irreduziblen Mannigfaltigkeit $\mathfrak{N}$ mit einer allgemeinen Hyperebene $u_0 \eta_0' + u_1 \eta_1' + \cdots + u_n \eta_n' = 0$ in bezug auf den Koeffizientenkörper $\mathsf{K}\,(u_0, \ldots, u_n)$ stets irreduzibel ist.

---

6) Math. Annalen **108** (1933), S. 694.

7) Aronhold, Monatsber. preuß. Akad. Wiss. Berlin 1864, S. 499—522.

Die Gleichung

$$(9) \qquad u_0' \eta_0' + u_1' \eta_1' + \ldots + u_n' \eta_n' = 0$$

definiert zusammen mit den Gleichungen von $\mathfrak{N}$ eine Korrespondenz zwischen dem u'-Raum und $\mathfrak{N}$. Diese ist irreduzibel, denn in ihr entspricht jedem Punkt η' von $\mathfrak{N}$ nach (1) eine Hyperebene des u'-Raums, also eine irreduzible Mannigfaltigkeit, während in einer zerfallenden Korrespondenz offenbar mindestens einem Punkt η' eine zerfallende Mannigfaltigkeit des u'-Raums entsprechen müßte. Nach unseren allgemeinen Sätzen folgt daraus, daß einem allgemeinen Punkt u des u'-Raums in der Korrespondenz eine relativ-irreduzible Teilmannigfaltigkeit $\mathfrak{N}_u$ von $\mathfrak{N}$ entspricht, das heißt der Schnitt von $\mathfrak{N}$ mit der allgemeinen Hyperebene $u_0 \eta_0' + \ldots + u_n \eta_n' = 0$ ist irreduzibel.

Genau so beweist man, daß der Schnitt von $\mathfrak{N}$ mit einer allgemeinen Hyperfläche h-ten Grades relativ zum Körper der Koeffizienten irreduzibel ist. Wir werden diese Sätze in § 5 noch weitgehend verschärfen.

§ 2.

Die Dimension der $\mathfrak{M}_\xi$. Das Prinzip von Plücker-Clebsch.

Wir brauchen in diesem Paragraph mehrmals den folgenden

Hilfssatz. *Ein Durchschnitt einer irreduziblen r-dimensionalen Mannigfaltigkeit $\mathfrak{M}_r$ in einem affinen oder projektiven Raum mit k Hyperflächen $g_1 = 0,\ \ldots,\ g_k = 0$ enthält keine isolierten irreduziblen Bestandteile von einer Dimension $< r - k$.* Dabei wird unter einem isolierten irreduziblen Bestandteil ein solcher verstanden, der nicht in einem irreduziblen Bestandteil von höherer Dimension enthalten ist.

Der Hilfssatz folgt durch k-malige Anwendung sofort aus folgendem: *Ein Durchschnitt einer r-dimensionalen irreduziblen Mannigfaltigkeit $\mathfrak{M}_r$ mit einer Hyperfläche $g = 0$ ist entweder $\mathfrak{M}_r$ selbst oder zerfällt in lauter $(r-1)$-dimensionale irreduzible Bestandteile.* Für den Beweis siehe etwa W. Krull[8]) oder ZAG I (Math. Annalen **108**) § 2.

Es sei nun eine beliebige Korrespondenz $\mathfrak{K}$ zwischen $\mathfrak{M}$ und $\mathfrak{N}$ gegeben. Dann gilt der folgende

Satz. *Diejenigen Punkte ξ' von $\mathfrak{M}$, denen eine mindestens h-dimensionale Mannigfaltigkeit von Punkten η' auf $\mathfrak{N}$ entspricht, bilden für jedes h eine (eventuell leere) algebraische Mannigfaltigkeit.*

Beweis. Es seien $L_1 = 0, L_2 = 0, \ldots, L_h = 0$ die Gleichungen von h allgemeinen Hyperebenen im Raum der η' mit unbestimmten Koeffizienten $u_{10}, \ldots, u_{1n}; u_{20}, \ldots; \ldots, u_{hn}$. Die Korrespondenz $\mathfrak{K}$ sei wieder durch die Gleichungen (2) gegeben. Damit einem Punkt ξ' mindestens ∞^h Punkte η' entsprechen, muß das Gleichungssystem

$$(10) \qquad \begin{cases} L_1(\eta') = 0, \ \ldots, \ L_h(\eta') = 0 \\ F_\nu(\xi', \eta') = 0 \end{cases}$$

---

8) W. Krull, Primidealketten in allgemeinen Ringbereichen, S.-B. Heidelberger Akad. Wiss. 1928, 7. Abhandlung, § 3, Hauptidealsatz.

mindestens eine Lösung η' besitzen. Das Resultantensystem

$$(11) \qquad\qquad \mathfrak{R}_\mu\,(u,\,\xi') = 0$$

muß also identisch in u erfüllt sein. Umgekehrt, wenn (11) identisch in u erfüllt ist, hat die dem Punkt ξ' entsprechende Mannigfaltigkeit $\mathfrak{N}_{\xi'}$ mit einem System von h allgemeinen Hyperebenen einen Punkt gemeinsam, mithin ist $\mathfrak{N}_{\xi'}$ mindestens h-dimensional. Durch Nullsetzen der Koeffizienten aller Potenzprodukte der u im Resultantensystem (11) erhält man somit die Gleichungen der gesuchten Mannigfaltigkeit.

Aus dem Satz folgt, wenn $\mathfrak{M}$ irreduzibel ist und einem allgemeinen Punkt von $\mathfrak{M}$ ∞^b Punkte von $\mathfrak{N}$ entsprechen, daß einem speziellen Punkt ξ^0 von $\mathfrak{M}$ auch mindestens eine b-dimensionale Mannigfaltigkeit $\mathfrak{M}_{\xi_0}$ entspricht. Für irreduzible Korrespondenzen kann man aber noch Genaueres aussagen, nämlich:

Die Mannigfaltigkeit $\mathfrak{N}_{\xi^0}$, die einem Punkt ξ^0 von $\mathfrak{M}$ entspricht, enthält keine Bestandteile von niedrigerer Dimension als b.

B e w e i s. Gesetzt, $\mathfrak{N}^0$ wäre ein isolierter Bestandteil von $\mathfrak{N}_{\xi^0}$ von einer Dimension $< b$. Es sei η^0 ein allgemeiner Punkt von $\mathfrak{N}^0$. Diejenigen Punkte von $\mathfrak{M}$, denen in der Korrespondenz der Punkt η^0 entspricht, bilden eine echte Teilmannigfaltigkeit von $\mathfrak{M}$ von einer Dimension $< a$. Der Kegel, der diese Teilmannigfaltigkeit von ξ^0 aus projiziert, hat höchstens die Dimension a. Es gibt also durch ξ^0 einen linearen Raum P_{n-a}, der diesen Kegel außer in ξ^0 nicht trifft und der $\mathfrak{M}$ außer in ξ^0 nur noch in endlichvielen Punkten $\xi^1,\,\ldots,\,\xi^{\gamma-1}$ trifft. Die Gleichungen dieses linearen Raumes seien

$$(12) \qquad\qquad L_1\,(\xi') = 0,\ \ldots,\ L_a\,(\xi') = 0.$$

Den Punkten $\xi^1,\,\ldots,\,\xi^{\gamma-1}$ entsprechen in der Korrespondenz Mannigfaltigkeiten, welche den Punkt η^0 nicht enthalten, also auch die Mannigfaltigkeit $\mathfrak{N}^0$ nicht enthalten. Insgesamt bilden die Punkte $\xi^0,\,\xi^1,\,\ldots,\,\xi^{\gamma-1}$ zusammen mit den ihnen entsprechenden Punkten η' von $\mathfrak{N}$ eine Mannigfaltigkeit $\mathfrak{K}'$ von Punktepaaren, welche als isolierten Bestandteil die Mannigfaltigkeit $(\xi^0,\,\mathfrak{N}^0)$ enthält, bestehend aus den Punktepaaren $(\xi^0,\,\eta')$ mit η' in $\mathfrak{N}^0$. Die Dimension dieses Bestandteils ist gleich der von $\mathfrak{N}^0$,

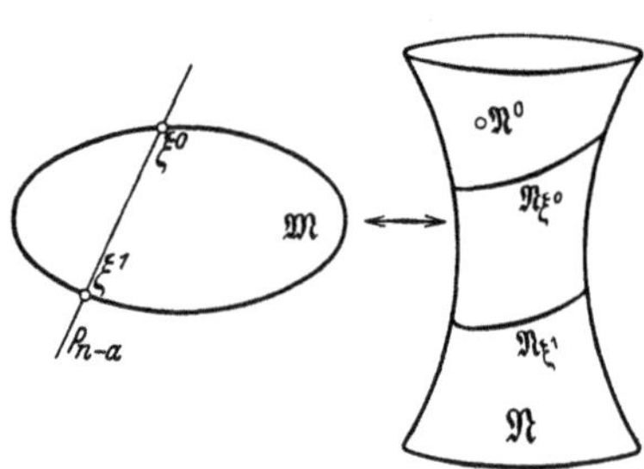

also $< b$. Andererseits wird die Mannigfaltigkeit $\mathfrak{K}'$ durch die Gleichungen (2) und (12) definiert, wobei (2) eine irreduzible Mannigfaltigkeit von der Dimension $q = a + b$ und (12) ein System von a Hyperflächen im Raum der Punktepaare $(\xi',\,\eta')$ bedeutet. Nach dem Hilfssatz zu

Anfang dieses Paragraphen enthält ein solcher Durchschnitt niemals isolierte Bestandteile von einer Dimension $< b$. Damit sind wir auf einen Widerspruch gestoßen.

Der eben bewiesene Satz gilt offenbar nicht nur im projektiven, sondern auch im affinen Raum. Eine wichtige Anwendung ist das *Prinzip von Plücker-Clebsch*, das in einer von Severi verschärften Form so lautet:

Wenn ein System von n Gleichungen in n Unbekannten $\eta_1', \ldots, \eta_n$, welches noch von Parametern $t_1, \ldots, t_m$ abhängt:

$$(13) \qquad f_\nu(t_1, \ldots, t_m, \eta_1', \ldots, \eta_n') = 0$$

für irgendwelche speziellen Werte t^0 der Parameter t eine isolierte Lösung besitzt, so ist das System für allgemeine Werte von t stets lösbar.

Beweis. Wir deuten (13) als das Gleichungssystem einer Korrespondenz zwischen den Punkten t^0 und den Punkten η'. Die Korrespondenz zerfällt in irreduzible, von denen eine die Eigenschaft haben muß, dem Punkt t^0 eine Mannigfaltigkeit zuzuordnen, die einen isolierten Punkt besitzt. Diese irreduzible Korrespondenz sei $\mathfrak{K}$; ihre Urmannigfaltigkeit und Bildmannigfaltigkeit seien $\mathfrak{M}$ und $\mathfrak{N}$. Da die Korrespondenz (13) durch n Gleichungen im $(m+n)$-dimensionalen Raum der Punktepaare (t, η') definiert wird, hat jede ihrer irreduziblen Bestandteile mindestens die Dimension m. In der Formel (5) ist also $q \geqq m$. Weiter ist $b = 0$, denn sonst könnte (nach dem zuletzt bewiesenen Satz) die Bildmannigfaltigkeit $\mathfrak{N}_{\tau^0}$ von τ^0 keinen isolierten Punkt enthalten. Aus (5) folgt also $a \geqq m$, d. h. die Urmannigfaltigkeit $\mathfrak{M}$ ist der ganze t-Raum; mithin entspricht einem allgemeinen Punkt dieses Raumes mindestens ein Punkt η', was zu beweisen war.

Bemerkungen. 1. Hätten wir es mit homogenen Gleichungen zu tun, so könnten wir weiter schließen, daß *jedem* Wert der t mindestens ∞^{n-r} Lösungen von (11) entsprechen. Bei inhomogenen Gleichungen geht das aber nicht.

2. Genau so beweist man allgemeiner: *Wenn das System (13) aus $n-r$ Gleichungen besteht und für $t = t^0$ eine isolierte r-dimensionale Lösungsmannigfaltigkeit besitzt, so hat das System für allgemeine t mindestens ∞^r Lösungen.*

3. Aus dem Beweis geht hervor, daß es nicht nötig ist, daß der Punkt t einen linearen Raum durchläuft, sondern daß das Prinzip von Plücker-Clebsch genau so gilt, wenn t ein allgemeiner Punkt einer irreduziblen Mannigfaltigkeit $\mathfrak{M}$ ist, welche den Punkt t^0 enthält.

§ 3.

Das Prinzip der Erhaltung der Anzahl.

In einer Korrespondenz $\Re$ zwischen $\mathfrak{M}$ und $\mathfrak{N}$, wo $\mathfrak{M}$ als irreduzibel vorausgesetzt wird, möge einem allgemeinen Punkt ξ von $\mathfrak{M}$ eine endliche Anzahl α von Punkten $\eta^{(1)}, \ldots, \eta^{(\alpha)}$ von $\mathfrak{N}$ entsprechen. ξ' sei ein spezieller Punkt von $\mathfrak{M}$; dann gibt es auf Grund der schon in § 1 angewandten Resultantenschlußweise mindestens ein System von Punkten $'\eta^{(1)}, \ldots, '\eta^{(\alpha)}$, welches eine relationstreue Spezialisierung des Systems $\eta^{(1)}, \ldots, \eta^{(\alpha)}$ darstellt, in dem Sinne, daß alle homogenen Relationen $F(\xi, \eta^{(1)}, \ldots, \eta^{(\alpha)}) = 0$ auch für $\xi', '\eta^{(1)}, \ldots, '\eta^{(\alpha)}$ gültig bleiben. Die Zahl μ, die angibt, wie oft ein Punkt η' in der Reihe $'\eta^{(1)}, \ldots, '\eta^{(\alpha)}$ vorkommt, heißt die *Multiplizität* von η bei der gewählten Spezialisierung. Nun gilt:

1. *Wenn ξ' ein einfacher Punkt von $\mathfrak{M}$ ist und wenn ihm nur endlichviele Punkte η' von $\mathfrak{N}$ entsprechen, so ist die Multiplizität eines jeden dieser Punkte, unabhängig von der Art, wie die Spezialisierung $\eta^{(\nu)} \to '\eta^{(\nu)}$ gewählt wird, eindeutig bestimmt.*

2. *Wenn außerdem die Korrespondenz $\Re$ irreduzibel ist (oder sich aus solchen irreduziblen zusammensetzt, die alle die ganze Mannigfaltigkeit $\mathfrak{M}$ als Urmannigfaltigkeit haben), so sind alle Multiplizitäten positiv, d. h. jeder Punkt η' kommt mindestens einmal unter den $'\eta^{(\nu)}$ vor.*

Beweis 1. Es möge ein Punkt η', der ξ' entspricht, fixiert werden. Wir projizieren die Punkte von $\mathfrak{M}$ aus einem allgemeinen linearen Raum P_{m-a-1} des P_m auf einen P_a. Die Projektion des allgemeinen Punktes ξ

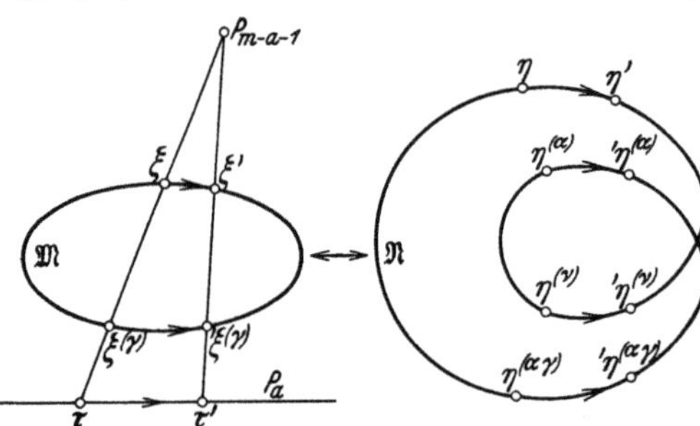

sei τ, die von ξ' sei τ'. Der projizierende Raum von ξ hat mit $\mathfrak{M}$ endlichviele Punkte $\xi^{(1)}, \xi^{(2)}, \ldots, \xi^{(\gamma)}$ gemein. Zu $\xi^{(1)} = \xi$ gehören in der Korrespondenz die Punkte $\eta^{(1)}, \ldots, \eta^{(\alpha)}$, zu $\xi^{(2)}$ ebenso $\eta^{(\alpha+1)}, \ldots, \eta^{(2\alpha)}$, usw. bis $\eta^{(\gamma\alpha)}$. Ist nun eine relationstreue Spezialisierung $\xi \to \xi', \eta^{(\nu)} \to '\eta^{(\nu)}$ $(\nu = 1, \ldots, \alpha)$ gegeben, so läßt sich diese zu einer relationstreuen Spezialisierung aller Punkte $\xi^{(\lambda)}$ und $\eta^{(\nu)}$ erweitern:

$$\xi^{(\lambda)} \to '\xi^{(\lambda)}, \quad \tau \to \tau', \quad \eta^{(\nu)} \to '\eta^{(\nu)} \qquad (\lambda = 1, \ldots, \alpha; \ \nu = 1, \ldots, \alpha\gamma).$$

Dabei sind $'\xi^{(\lambda)}$ die Schnittpunkte des projizierenden Raumes von ξ' mit $\mathfrak{M}$, jeder mit seiner Multiplizität gezählt. Der Punkt $'\xi$ kommt aber nur einmal unter ihnen vor, weil er ein einfacher Punkt von $\mathfrak{M}$ ist und die Projektionsrichtung allgemein gewählt wurde. Den übrigen $'\xi^{(\nu)}$ $(\nu = 2, \ldots, \gamma)$

entspricht in der Korrespondenz sicher nicht der Punkt η'; denn die Punkte von $\mathfrak{M}$, denen η' entspricht, bilden eine Teilmannigfaltigkeit von $\mathfrak{M}$ von höchstens $a-1$ Dimensionen, und diese wird von dem linearen Raum P_{m-a}, der ξ' mit dem allgemeinen P_{m-a-1} verbindet, nicht zum zweiten Male getroffen. Also ist die Multiplizität μ von η' gleich der Zahl, die angibt, wie oft η' unter den Punkten $\eta^{(1)}, \ldots, \eta^{(\alpha\gamma)}$ vorkommt. Ebenso sieht man ein, daß den Punkten $'\xi^{(\nu)}$ je nur endlichviele Punkte in der Korrespondenz entsprechen; denn die Punkte von $\mathfrak{M}$, denen ∞^1 oder mehr Punkte von $\mathfrak{N}$ entsprechen, bilden wieder eine echte Teilmannigfaltigkeit von $\mathfrak{M}$, die von dem Verbindungsraum P_{m-a} nicht getroffen wird.

Nunmehr kann man die $\xi^{(\nu)}$ ganz eliminieren und alles auf den Fall eines linearen Raums zurückführen; denn τ durchläuft einen linearen Raum und alle algebraischen Relationen, die für $\tau, \eta^{(1)}, \ldots, \eta^{(\alpha\gamma)}$ gelten, bleiben bei der Spezialisierung $\tau \to \tau'$, $\eta^{(\nu)} \to '\eta^{(\nu)}$ bestehen. Wir können demnach den Satz aus § 4 meiner unter [4]) zitierten Arbeit „Multiplizitätsbegriff" anwenden, wenn wir noch beweisen können, daß die $\eta^{(\nu)}$ Lösungen eines von τ ganz rational abhängigen Gleichungssystems sind, welches auch für $\tau = \tau'$ nur endlichviele Lösungen hat. Dieses Gleichungssystem ist aber ohne weiteres angebbar: es entsteht aus den Gleichungen (2) und den Relationen, welche ausdrücken, daß ξ (bzw. ξ') in dem Verbindungsraum von τ (bzw. τ') mit P_{m-a-1} liegt, durch Elimination der ξ (bzw. ξ'). Die Lösungen dieses Gleichungssystems sind genau die Punkte von $\mathfrak{N}$, denen solche Punkte von $\mathfrak{M}$ in der Korrespondenz entsprechen, welche in dem erwähnten Verbindungsraum liegen; das sind aber für unbestimmte τ genau die Punkte $\eta^{(1)}, \ldots, \eta^{(\alpha\gamma)}$ und für τ' die endlichvielen Punkte von $\mathfrak{N}$, die den Punkten $'\xi^{(1)}, \ldots, '\xi^{(\gamma)}$ entsprechen. Nach dem erwähnten Satz folgt daraus, daß die Multiplizität μ von der gewählten Spezialisierung unabhängig ist.

Beweis 2. Wenn $\mathfrak{K}$ irreduzibel ist, so ist jedes Punktepaar (ξ', η') von $\mathfrak{K}$ eine relationstreue Spezialisierung des allgemeinen Punktepaares (ξ, η). Daraus folgt genau so wie in „Multiplizitätsbegriff" § 3 (Satz 2), daß η' mit einer positiven Multiplizität unter den $'\eta^{(\nu)}$ vorkommt. — Wenn $\mathfrak{K}$ in mehrere irreduzible Korrespondenzen zerfällt, die alle $\mathfrak{M}$ als Urmannigfaltigkeit haben, so muß eine dieser irreduziblen Korrespondenzen das Punktepaar (ξ', η') enthalten; dieser Fall führt somit auf den einer irreduziblen Korrespondenz zurück.

Durch das Vorangehende ist der Begriff der Spezialisierungsmultiplizität eindeutig fixiert. Die Summe der Multiplizitäten aller Punkte η' ist offenbar gleich der Anzahl α der Punkte $\eta^{(\nu)}$ vor der Spezialisierung: das ist das *Prinzip der Erhaltung der Anzahl.*

§ 4.

Verallgemeinerung des Multiplizitätsbegriffs: Mehrfach-gezählte Mannigfaltigkeiten.

Wir nehmen zunächst an, daß in einer irreduziblen Korrespondenz $\Re$ einem allgemeinen Punkt ξ der irreduziblen Mannigfaltigkeit $\mathfrak{M}$ eine Kurve $\mathfrak{C}$ auf $\Re$ entspricht und einem speziellen Punkt ξ' ebenfalls eine Kurve $\mathfrak{C}'$, die in $\mathfrak{C}'_1, \mathfrak{C}'_2, \ldots$ zerfallen möge. Wir wollen nun die Multiplizitäten definieren, mit denen $\mathfrak{C}'_1, \mathfrak{C}'_2, \ldots$ zu zählen sind, wenn ξ gegen ξ' und $\mathfrak{C}$ gegen $\mathfrak{C}'$ rückt.

Zu diesem Zweck schneiden wir alle betrachteten Kurven mit einer allgemeinen Hyperfläche $G(\eta') = 0$ eines festen Grades g (z. B. mit einer Hyperebene: $g = 1$). Diese schneidet $\mathfrak{C}$ in einem System konjugierter Punkte $\eta^{(1)}, \ldots, \eta^{(g\,\nu)}$ und $\mathfrak{C}'_1, \mathfrak{C}'_2, \ldots$ je in einem System konjugierter Punkte $'\eta_1^{(1)}, \ldots, '\eta_1^{(g\,\nu_1)};\ '\eta_2^{(1)}, \ldots, '\eta_2^{(g\,\nu_2)};\ \ldots$ Durch die Hinzunahme der Gleichung $G(\eta') = 0$ zu den Gleichungen der Korrespondenz $\Re$ erhält man nach § 1, Schluß eine relativ-irreduzible Korrespondenz $\Re'$, in welcher dem Punkt ξ genau die Punkte $\eta^{(\nu)}$ und dem Punkt ξ' die Punkte $'\eta_j^{(\nu)}$ entsprechen. Wenn daher ξ gegen ξ' rückt, so rücken die $\eta^{(\nu)}$ gegen die $'\eta_j^{(\nu)}$, welche dabei auf Grund von § 3 mit ganz bestimmten, positiven Multiplizitäten erscheinen. Da algebraisch-konjugierte Punkte in allen algebraischen Eigenschaften übereinstimmen, so sind die Multiplizitäten der konjugierten Punkte $'\eta_1^{(1)}, \ldots, '\eta_1^{(g\,\nu_1)}$ alle untereinander gleich, etwa gleich μ_1. Diese Zahl μ_1 definieren wir nun als *die Multiplizität des Kurvenbestandteils* $\mathfrak{C}_1$ *bei der Spezialisierung* $\xi \to \xi'$, $\mathfrak{C} \to \mathfrak{C}'$.

Wir wollen nun zeigen, daß die so definierten Multiplizitäten μ_j der Kurvenbestandteile $\mathfrak{C}_j$ von dem Grad g der benutzten Hyperflächen G unabhängig sind und daß man auch beim Schnitt mit einer nicht-allgemeinen Hyperfläche stets dieselben Multiplizitäten wiederfindet.

Es sei $\overline{G} = 0$ eine spezielle Hyperfläche desselben Grades g, welche die Kurve $\mathfrak{C}$ in endlichvielen Punkten $\overline{\eta}^{(\nu)}$ mit Schnittmultiplizitäten σ_ν und die Kurven $\mathfrak{C}_j$ ebenfalls in endlichvielen Punkten $'\overline{\eta}_j^{(\nu)}$ schneidet. Der Sachverhalt ist nun folgender: Gehen wir von der allgemeinen Hyperfläche $G = 0$ und der allgemeinen Kurve $\mathfrak{C}$ aus, welche sich in $\eta^{(1)}, \ldots, \eta^{(g\,\nu)}$ schneiden, und spezialisieren wir G zu $\overline{G}$ und $\mathfrak{C}$ zu $\overline{\mathfrak{C}}'$, so gehen die $\eta^{(\nu)}$ relationstreu in gewisse $'\overline{\eta}_j^{(\nu)}$ über, wobei ein beliebiger dieser Punkte mit einer gewissen Spezialisierungsmultiplizität erscheint, welche wir „Gesamtmultiplizität" nennen wollen. Diese Gesamtmultiplizitäten sind eindeutig bestimmt und man findet für sie daher dieselben Werte, wenn man *zuerst* G zu $\overline{G}$ und *dann* $\mathfrak{C}$ zu $\mathfrak{C}'$ spezialisiert, als wenn man *zuerst* $\mathfrak{C}$ in $\mathfrak{C}'$ und *dann* G in $\overline{G}$ übergehen läßt. Spezialisiert man zuerst G zu $\overline{G}$, so

180

gehen die $\eta^{(\nu)}$ in die $\overline{\eta}^{(\nu)}$ über, jedes σ_ν mal gezählt nach der Definition der σ_ν. Spezialisiert man dann $\mathfrak{C}$ zu $\mathfrak{C}'$, so können gewisse $\overline{\eta}^{(\nu)}$ in einen Punkt η' hineinrücken, und die Summe $\Sigma\,\sigma_\nu$ der Multiplizitäten aller dieser Punkte ist die Gesamtmultiplizität von η' bei dieser Reihenfolge der Spezialisierungen. Spezialisiert man aber umgekehrt zuerst $\mathfrak{C}$ zu $\mathfrak{C}'$, so gehen nach dem Obigen die $\eta^{(\nu)}$ in $'\eta_j^{(\nu)}$ mit den Multiplizitäten μ_j über. Rückt dann G gegen $\overline{G}$, so rücken auf jeder einzelnen Kurve $\mathfrak{C}_j'$ eine Anzahl σ_j' Punkte $'\eta_j^{(\nu)}$ in η' hinein, wo σ_j' die Schnittmultiplizität von $\overline{G}$ mit $\mathfrak{C}_j'$ im Punkt η' ist. Die Gesamtmultiplizität von η' bei dieser Reihenfolge der Spezialisierungen ist $\sum\limits_j \mu_j\,\sigma_j'$, wobei die Summation über j nur dann nötig ist, wenn der Punkt η' mehreren Kurven $\mathfrak{C}_j'$ angehört. Gleichsetzung der beiden Werte der Gesamtmultiplizität ergibt den folgenden Satz:

Wenn bei dem Übergang $\mathfrak{C} \to \mathfrak{C}'$ eine gewisse Anzahl von Schnittpunkten $\overline{\eta}^{(\nu)}$ von $\mathfrak{C}$ mit $\overline{G} = 0$ in einen Punkt η' hineinrücken, so ist die Summe der Schnittpunktsmultiplizitäten σ_ν dieser Schnittpunkte gleich

$$(15) \qquad \sum \sigma_\nu = \sum\limits_j \mu_j\,\sigma_j',$$

wobei μ_j die Multiplizität des Kurvenbestandteils $\mathfrak{C}_j'$ bei der Spezialisierung $\mathfrak{C} \to \mathfrak{C}'$ und σ_j' die Schnittmultiplizität von $\overline{G}$ mit $\mathfrak{C}_j'$ im Punkt η' ist.

Wichtig für die Anwendungen ist insbesondere der Fall, wo alle Multiplizitäten σ_ν und σ_j' in (15) gleich Eins werden. Dieser Fall tritt nach ZAG V, Satz 2 dann ein, wenn $\overline{G} = 0$ das allgemeine Element einer linearen Schar von Hyperflächen und η' ein Schnittpunkt von $\overline{G} = 0$ mit $\mathfrak{C}'$ außerhalb der Basispunkte der Schar ist. Die Summation über j in (15) fällt in diesem Fall weg, da η' ein mit $\overline{G}$ variabler Punkt von $\mathfrak{C}_j'$ ist und daher nicht gleichzeitig einer anderen Kurve $\mathfrak{C}_k'$ angehören kann. Die Formel (15) ergibt dann $\Sigma\,1 = \mu_j$, d. h.:

Ist $\overline{G} = 0$ die allgemeine Hyperfläche einer linearen Schar, welche die Kurve $\mathfrak{C}_j'$ außerhalb der Basispunkte in einem Punkt η' trifft, so rücken bei der Spezialisierung $\mathfrak{C} \to \mathfrak{C}'$ genau μ_j Schnittpunkte von $\overline{G} = 0$ mit $\mathfrak{C}$ in den Punkt η' hinein.

Nach diesem Satz kann man zur Definition der μ_j statt der linearen Schar *aller* Hyperflächen vom Grade g auch jede andere Teilschar von positiver Dimension benutzen. Wählt man dafür insbesondere die Schar aller Hyperebenen des Raumes, ergänzt durch $g - 1$ feste Hyperebenen, so folgt, daß man die μ_j statt durch Hyperflächen g-ten Grades auch durch Hyperebenen definieren kann, d. h.: *Die μ_j sind von der Zahl g unabhängig.*

Wenn das betrachtete Kurvensystem $|\mathfrak{C}|$ auf einer singularitätenfreien r-dimensionalen Mannigfaltigkeit $\mathfrak{N}_r$ gelegen ist $(c \leqq r \leqq n)$, so kann man die Kurven $\mathfrak{C}$ und $\mathfrak{C}'$ auch mit einer $\mathfrak{N}_{r-1}$ des $\mathfrak{N}_r$ schneiden. Die Schnittmultiplizität des $\mathfrak{N}_{r-1}$ mit $\mathfrak{C}$ oder $\mathfrak{C}'$ in einem Punkt η' wird nach Severi[3]) definiert als die Schnittmultiplizität der Projektion $\mathfrak{N}_{n-1}$ von $\mathfrak{N}_{r-1}$ aus einem allgemeinen P_{n-r} des P_n mit $\mathfrak{C}$ oder $\mathfrak{C}'$. Da für $\mathfrak{N}_{n-1}$ die Formel (15) gilt, gilt sie auch für $\mathfrak{N}_{r-1}$. Auch die Folgerungen gelten: Man kann die Multiplizitäten μ_j der Kurvenbestandteile $\mathfrak{C}'_j$ definieren, indem man $\mathfrak{N}_{r-1}$ als allgemeines Element einer linearen Schar (vgl. § 5) wählt. Daraus folgt weiter:

Die Multiplizitäten μ_j sind invariant bei einer birationalen Abbildung der singularitätenfreien Trägermannigfaltigkeit $\mathfrak{M}_r$ des Kurvensystems $|\mathfrak{C}|$, vorausgesetzt, daß diese Abbildung einen allgemeinen Punkt einer jeden Kurve $\mathfrak{C}'_j$ eineindeutig (in einen allgemeinen Punkt der Bildkurve) transformiert.

Es ändert sich nichts Wesentliches, wenn $\mathfrak{C}$ und $\mathfrak{C}'$ keine Kurven, sondern b-dimensionale Mannigfaltigkeiten sind. Man bringt sie dann nicht mit *einer* allgemeinen Hyperfläche, sondern mit b allgemeinen Hyperflächen zum Schnitt und beweist genau so wie oben, daß die Multiplizitäten von den Gradzahlen der benutzten Hyperflächen unabhängig sind und auch mit Hilfe beliebiger linearer Scharen von Hyperflächen definiert werden können, vorausgesetzt, daß ein System von allgemeinen Hyperflächen dieser Scharen die Mannigfaltigkeiten $\mathfrak{C}'_j$ außer in den Basispunkten je mindestens einmal (und dann nach ZAG V, Satz 2 notwendig einfach) schneidet.

§ 5.

Anwendung auf die Theorie der linearen Scharen.

Einen wichtigen Spezialfall der vorstehenden Begriffsbildungen liefern die linearen Kurvenscharen auf Flächen, allgemeiner die linearen Scharen von $(c-1)$-dimensionalen Mannigfaltigkeiten auf einer c-dimensionalen Mannigfaltigkeit $\mathfrak{N}$. Eine solche lineare Schar wird erzeugt von einer Schar von Hyperflächen, von denen keine $\mathfrak{N}$ enthält[9]):

$$(16) \qquad F(\lambda) = \lambda_0 F_0(\eta') + \lambda_1 F_1(\eta') + \ldots + \lambda_n F_n(\eta') = 0.$$

Die allgemeine Hyperfläche dieser Schar schneidet aus $\mathfrak{N}$ eine $(b-1)$-dimensionale Mannigfaltigkeit $\mathfrak{C}(\lambda)$, welche aus einem festen, d. h. von λ

---

[9]) Sollte eine gewisse Teilschar von (16) $\mathfrak{N}$ enthalten, so kann man in bekannter Weise aus der ganzen Schar eine andere Teilschar herausgreifen, deren Hyperflächen $\mathfrak{N}$ nicht enthalten, und die auf $\mathfrak{N}$ dasselbe Kurvensystem ausschneidet wie die ursprünglich gegebene Schar.

unabhängigen Bestandteil $\mathfrak{C}_0$ und einem von λ abhängigen Bestandteil $\mathfrak{C}'(\lambda)$ bestehen möge. Dann ist $\mathfrak{C}'(\lambda)$ das allgemeine Element einer *linearen Schar von* $(c-1)$-*dimensionalen Mannigfaltigkeiten auf* $\mathfrak{N}$. Die Schar selbst wird mit $|\mathfrak{C}'(\lambda)|$ bezeichnet.

Durch die obige Definition ist nur das *allgemeine* Element $\mathfrak{C}'(\lambda)$ der Schar bestimmt, nicht die spezielle Kurve $\mathfrak{C}'(\lambda')$, die zu einem speziellen Wert von λ gehört, und ebensowenig die Vielfachheiten, mit denen die Bestandteile von $\mathfrak{C}'(\lambda')$ zu zählen sind. Wir führen nun diese Fragen auf die in § 4 gelösten zurück, indem wir eine Korrespondenz $\mathfrak{R}'$ definieren, welche einem allgemeinen Wert λ alle Punkte von $\mathfrak{C}'(\lambda)$ zuordnet.

Für spezielle Werte λ' des Unbestimmten λ geht $F(\lambda)$ über in eine Hyperfläche $F(\lambda')$ mit der Gleichung

$$(17) \qquad F(\lambda') = \lambda_0' F_0(\eta') + \lambda_1' F_1(\eta') + \ldots + \lambda_n' F_n(\eta') = 0.$$

Die Gleichung (17) bestimmt, zusammen mit den Gleichungen von $\mathfrak{N}$, eine Korrespondenz $\mathfrak{R}$ zwischen dem λ'-Raum und der Mannigfaltigkeit $\mathfrak{N}$, wobei einem Punkt λ' alle Punkte η' der Schnittmannigfaltigkeit $\mathfrak{C}(\lambda')$ von $F(\lambda')$ mit $\mathfrak{N}$ zugeordnet sind. Diese Korrespondenz $\mathfrak{R}$ zerfällt in folgende Bestandteile:

1. Eine irreduzible Korrespondenz $\mathfrak{R}'$, deren allgemeines Punktepaar (λ, η) dadurch gegeben ist, daß einem allgemeinen Punkt η von $\mathfrak{N}$ ein allgemeiner Punkt λ der Hyperebene

$$(18) \qquad \lambda_0 F_0(\eta) + \lambda_1 F_1(\eta) + \ldots + \lambda_n F_n(\eta)$$

des λ'-Raumes zugeordnet wird. Aus dem Prinzip der Konstantenzählung (7) folgt wegen $d = n-1$, $b \leqq c-1$, daß $a = n$, also die Urbildmannigfaltigkeit von $\mathfrak{R}$ der *ganze* λ'-Raum ist. Der Punkt η des allgemeinen Punktepaares (λ, η) gehört wegen (18) zu $\mathfrak{C}(\lambda)$, aber als allgemeiner Punkt von $\mathfrak{N}$ nicht zu $\mathfrak{C}_0$, also zu $\mathfrak{C}'(\lambda)$.

2. Eine Korrespondenz $\mathfrak{R}_0$, in welcher jedem Punkt von $\mathfrak{C}_0$ der ganze λ'-Raum zugeordnet ist, und welche offenbar in irreduzible Korrespondenzen zerfällt, die den irreduziblen Bestandteilen von $\mathfrak{C}_0$ entsprechen.

Ich behaupte, daß die Korrespondenzen $\mathfrak{R}_0$ und $\mathfrak{R}'$ zusammen die ganze Korrespondenz $\mathfrak{R}$ bilden. Dazu ist zu zeigen, daß jedes Punktepaar (λ', η') von $\mathfrak{R}$, bei dem η' nicht zu $\mathfrak{C}_0$ gehört, eine relationstreue Spezialisierung des allgemeinen Punktepaares (λ, η) von $\mathfrak{R}'$ darstellt, also zu $\mathfrak{R}'$ gehört. Wir unterscheiden:

Fall 1. Der Punkt η' gehört nicht allen Mannigfaltigkeiten $\mathfrak{C}(\lambda)$ an; dann ist mindestens ein $F_\nu(\eta') \neq 0$, etwa $F_0(\eta') \neq 0$, und die für das allgemeine Punktepaar (λ, η) gültige rationale Darstellung

$$\lambda_0 = -\frac{\lambda_1 F_1(\eta) + \ldots + \lambda_n F_n(\eta)}{F_0(\eta)}$$

bleibt bei Ersetzung der η durch η' und λ durch λ' sinnvoll; also ist (λ', η') in diesem Fall eine relationstreue Spezialisierung von (λ, η).

Fall 2. Der Punkt η' gehört allen $\mathfrak{C}(\lambda)$, also, da er nicht zu $\mathfrak{C}_0$ gehört, allen $\mathfrak{C}'(\lambda)$ an. Wenn η' einem irreduziblen Bestandteil $\mathfrak{C}^*(\lambda)$ von $\mathfrak{C}'(\lambda)$ angehört, so befindet sich ein allgemeiner Punkt η^* dieses Bestandteils im Fall 1, also ist (λ, η^*) eine relationstreue Spezialisierung von (λ, η). Weiter ist (λ, η') eine relationstreue Spezialisierung von (λ, η^*), da η' ein Punkt von $\mathfrak{C}^*(\lambda)$ ist. Schließlich ist (λ', η') eine relationstreue Spezialisierung von (λ, η'), da η' ein fester, von λ unabhängiger Punkt und λ ein allgemeiner Punkt des λ'-Raumes ist. Also ist auch in diesem Fall (λ', η') eine relationstreue Spezialisierung von (λ, η).

Die Korrespondenz $\mathfrak{K}'$ ordnet einem allgemeinen Punkt λ die Mannigfaltigkeit $\mathfrak{C}'(\lambda)$ zu. Da $\mathfrak{K}'$ irreduzibel ist, ist $\mathfrak{C}'(\lambda)$ auch (relativ zum Körper $\mathsf{K}(\lambda)$) irreduzibel. In bezug auf einen Erweiterungskörper von $\mathsf{K}(\lambda)$ kann $\mathfrak{C}'(\lambda)$ natürlich zerfallen: In diesem Fall lehrt uns ein schöner Beweis von Bertini, daß die absolut irreduziblen Bestandteile von $\mathfrak{C}'(\lambda)$ alle demselben irreduziblen System von ∞^1 Mannigfaltigkeiten angehören, wobei durch jeden allgemeinen Punkt von $\mathfrak{N}$ genau eine Mannigfaltigkeit des Systems hindurchgeht [10]).

Die Korrespondenzen $\mathfrak{K}$, $\mathfrak{K}'$ und $\mathfrak{K}_0$ ordnen einem speziellen Punkt λ' des λ'-Raumes die $(c-1)$-dimensionalen Mannigfaltigkeiten $\mathfrak{C}(\lambda')$, $\mathfrak{C}'(\lambda')$ und $\mathfrak{C}_0$ zu. Da $\mathfrak{K}$ in $\mathfrak{K}'$ und $\mathfrak{K}_0$ zerfällt, so zerfällt $\mathfrak{C}(\lambda')$ in $\mathfrak{C}'(\lambda')$ und $\mathfrak{C}_0$. Damit ist nun $\mathfrak{C}'(\lambda')$ endgültig definiert. Wendet man auf die irreduzible Korrespondenz $\mathfrak{K}'$ die Betrachtungen von § 5 an, so folgt, daß die Bestandteile $\mathfrak{C}'_1, \mathfrak{C}'_2, \ldots$ von $\mathfrak{C}'(\lambda')$ bei der Spezialisierung $\lambda \to (\lambda')$, $\mathfrak{C}'(\lambda) \to \mathfrak{C}'(\lambda)'$ je mit einer gewissen Multiplizität behaftet erscheinen. Über diese Multiplizitäten gelten folgende Sätze:

1. *Die Multiplizität eines Bestandteils $\mathfrak{C}'_1$ einer Mannigfaltigkeit $\mathfrak{C}'(\lambda')$ bei der Spezialisierung $\lambda \to \lambda'$ ist gleich der Multiplizität von $\mathfrak{C}'_1$ als Teil des Schnittes von $\mathfrak{N}$ mit der Hyperfläche $F(\lambda') = 0$, eventuell vermindert um die Multiplizität von $\mathfrak{C}'_1$ als Bestandteil des Schnittes von $\mathfrak{N}$ mit $F(\lambda) = 0$, falls $\mathfrak{C}'_1$ zur Mannigfaltigkeit $\mathfrak{C}_0$ gehört* [11]).

Beweis. Die Spezialisierungsmultiplizität des Bestandteils $\mathfrak{C}'_1$ wird nach § 4 dadurch definiert, daß man die Mannigfaltigkeiten $\mathfrak{N}, \mathfrak{C}'(\lambda), \mathfrak{C}'_1$ mit einem allgemeinen linearen P_{n-c-1} zum Schnitt bringt, wodurch $\mathfrak{N}$

---

[10]) Bertini, Atti Ist. Lomb. Rend. (2) **15** (1882), S. 24. Vgl. Enriques, Soc. ital. dei XI Mem. (3) **10** (1896) und Severi, Atti R. Acc. Torino **41** (1906), S. 207.

[11]) Satz 1 gilt nur für *lineare* Scharen, ist also keineswegs trivial. Wir hätten ihn natürlich als Definition der Kurve $C'(\lambda')$ und der Vielfachheiten ihrer Bestandteile benutzen können, halten aber die obige Definition, die auch für nichtlineare Scharen gilt, für natürlicher.

in eine Kurve, $\mathfrak{C}(\lambda)$ in ein System konjugierter Punkte $\eta^{(\nu)}(\lambda)$ und $\mathfrak{C}_1'$ ebenfalls in ein solches Punktsystem $'\eta^{(\nu)}$ übergeht, und daß man in diesem Raum P_{n-c-1} die Spezialisierungsmultiplizitäten beim Übergang $\lambda \to \lambda'$ betrachtet. Die Schnittmultiplizität von $\mathfrak{C}_1'$ ist aber ebenfalls in dem allgemeinen linearen Raum P_{n-c-1} definiert. Wir können uns also vollständig beschränken auf den Fall $b = 1$, wo $\mathfrak{N}$ eine Kurve ist und $\mathfrak{C}(\lambda)$ und $\mathfrak{C}_1'$ Systeme konjugierter Punkte $\eta^{(\nu)}(\lambda)$ bzw. $'\eta^{(\nu)}$ sind. Die Multiplizität von $\mathfrak{C}_1'$ als Schnittpunkt von $\mathfrak{N}$ mit der Hyperfläche $F(\lambda') = 0$ wird nun so definiert, daß man diese als Spezialfall einer allgemeinen Hyperfläche $G = 0$ desselben Grades betrachtet und dann die relationstreue Spezialisierung $G \to F(\lambda')$ vornimmt. Diese Spezialisierung kann man in zwei Schritten vornehmen: zuerst läßt man G in $F(\lambda)$ und dann $F(\lambda)$ in $F(\lambda')$ übergehen. Beim ersten Schritt erhält man aus den Schnittpunkten von $G = 0$ mit $\mathfrak{N}$ die von $F(\lambda) = 0$ mit $\mathfrak{N}$, also die Punkte $\eta^{(\nu)}(\lambda)$ von $\mathfrak{C}(\lambda)$ und die Punkte ξ^0 von $\mathfrak{C}_0$, die ersteren mit der Multiplizität 1 auf Grund von ZAG V, Satz 3, und die letzteren mit ihrer eigenen Schnittmultiplizität μ^0. Bei der zweiten Spezialisierung $F(\lambda) \to F(\lambda')$ bleiben die Punkte ξ^0 fest und die Punkte $\eta^{(\nu)}(\lambda)$ rücken gegen die Punkte $'\eta^{(\nu)}$, wobei etwa μ Punkte $\eta^{(\nu)}(\lambda)$ gegen einen Punkt η' rücken mögen. Falls der Punkt η' gleichzeitig ein Punkt ξ^0 ist, möge er als Schnittpunkt von $\mathfrak{N}$ mit $F(\lambda) = 0$ die Multiplizität μ^0 haben. Dann gehen also insgesamt $\mu + \mu^0$ Schnittpunkte von $G = 0$ mit $\mathfrak{N}$ bei der Spezialisierung $G \to F(\lambda) \to F(\lambda')$ in η' über, d. h. die Schnittmultiplizität von η' ist $\mu + \mu^0$. Diese Summe, vermindert um μ^0, ergibt die Spezialisierungsmultiplizität μ, wie behauptet.

Wenn die Multiplizitäten der Bestandteile $\mathfrak{C}_j'$ von $\mathfrak{C}'(\lambda')$ ausdrücklich hervorgehoben werden sollen, so schreibt man

$$\mathfrak{C}'(\lambda') = \mu_1\mathfrak{C}_1' + \mu_2\mathfrak{C}_2' + \ldots = \Sigma\,\mu_j\mathfrak{C}_j'.$$

2. *Wenn ein lineares System $|K|$ eine reduzible Mannigfaltigkeit $K' = \Sigma\,\mu_j K_j'$ enthält und wenn eine Kurve $\mathfrak{D}$ auf $\mathfrak{N}$ die allgemeine Mannigfaltigkeit K in gewissen Punkten $\eta^{(\nu)}$ und die Mannigfaltigkeit K' außerhalb der Basispunkte in η' trifft, so ist die Anzahl μ der Punkte $\eta^{(\nu)}$, die bei der Spezialisierung $K \to K'$ in η' hineinrücken, gleich $\Sigma\,\mu_j\sigma_j$, wo σ_j die Schnittmultiplizitäten von $\mathfrak{D}$ mit K_j' in η' sind, vorausgesetzt, daß η' ein einfacher Punkt von $\mathfrak{N}$ ist.*

Beweis. Nach Severi[3]) wird σ_j dadurch definiert, daß man K_j' durch einen Rest R_j ergänzt, der η' nicht enthält, derart, daß $K_j' + R_j$ je einer linearen Schar $|L_j|$ ohne Basispunkte angehören. Die Mannigfaltigkeit $K' + R = \Sigma\,\mu_j K_j' + \Sigma\,\mu_j R_j = \Sigma\,\mu_j(K_j' + R_j)$ gehört dann dem System $|\Sigma\,\mu_j L_j|$ an und dieses ist wieder in einer *vollständigen* linearen Schar $|H|$ ohne Basispunkte enthalten [12]). Da $K' + R$ in $|H|$ enthalten ist und K' zum System $|K|$ gehört, ist auch $K + R$ in $|H|$ enthalten. Man kann nun die Spezialisierung $H \to K' + R$ in zwei Weisen vornehmen:

$$1. \quad H \to K + R \to K' + R,$$
$$2. \quad H \to \Sigma \mu_j L_j \to \Sigma \mu_j (K_j + R_j).$$

Berechnet man für diese beiden Arten der Spezialisierung die Anzahl der Schnittpunkte von H mit C, die gegen η' rücken, so findet man einmal μ und das andere Mal $\Sigma \mu_j \sigma_j$. Mithin ist, wie behauptet,

$$\mu = \Sigma \mu_j \sigma_j.$$

§ 6.

Rationale Abbildungen und ihre Fundamentalpunkte.

Eine rationale Abbildung $\mathfrak{T}$ von $\mathfrak{M}$ auf $\mathfrak{N}$ sei gegeben durch die für das allgemeine Punktsystem der Abbildung gültigen Formeln

$$(19) \qquad \eta_0 : \eta_1 : \ldots : \eta_n = \Phi_0(\xi) : \Phi_1(\xi) : \ldots : \Phi_n(\xi),$$

oder auch

$$(20) \qquad \left\| \begin{array}{cccc} \eta_0 & \eta_1 & \cdots & \eta_n \\ \Phi_0(\xi) & \Phi_1(\xi) & \cdots & \Phi_n(\xi) \end{array} \right\| = 0.$$

Die Gleichungen (20) müssen auch für alle speziellen Punktepaare der Abbildung gelten. Daraus folgt: Durchläuft η' einen (allgemeinen) ebenen Schnitt $E(\lambda)$ von $\mathfrak{N}$ mit der Gleichung

$$(21) \qquad \lambda_0 \eta_0' + \lambda_1 \eta_1' + \ldots + \lambda_n \eta_n' = 0,$$

so liegen die zugehörigen Punkte ξ' auf der Mannigfaltigkeit $\mathfrak{M}(\lambda)$ mit der Gleichung

$$(22) \qquad \lambda_0 \Phi_0(\xi) + \lambda_1 \Phi_1(\xi) + \ldots + \lambda_n \Phi_n(\xi) = 0.$$

Wie in § 5 spalten wir $\mathfrak{M}(\lambda)$ in einen von λ unabhängigen Teil $\mathfrak{M}_0$ und einen Rest $\mathfrak{M}'(\lambda)$, welcher eine lineare Schar auf $\mathfrak{M}$ durchläuft, und ebenso die durch (22) definierte Korrespondenz $\mathfrak{K}$ zwischen $\mathfrak{M}$ und dem λ-Raum in eine irreduzible Korrespondenz $\mathfrak{K}'$, in der dem allgemeinen Punkt λ alle Punkte von $\mathfrak{M}'(\lambda)$ entsprechen, und einen Rest $\mathfrak{K}_0$, in welchem alle Punkte des λ'-Raumes allen Punkten von $\mathfrak{M}_0$ entsprechen. Die Gleichungen der irreduziblen Korrespondenz $\mathfrak{K}'$ seien

$$(23) \qquad G_1(\lambda', \xi') = 0, \ldots, G_h(\lambda', \xi') = 0.$$

Dann sind (23) bei gegebenem λ' auch die Gleichungen der zugehörigen Mannigfaltigkeit $\mathfrak{M}'(\lambda')$.

---

[12]) Vgl. etwa F. Enriques, Acc. Torino Memorie (2) 44 (1893), S. 196.

Bekanntlich bestimmt das lineare System $|\mathfrak{M}'(\lambda)|$ oder, was dasselbe ist, die Korrespondenz $\mathfrak{K}'$ auch umgekehrt die Abbildung $\mathfrak{T}$ eindeutig. Einem allgemeinen Punkt ξ entspricht nämlich in der Korrespondenz $\mathfrak{K}'$ eine Hyperebene des λ'-Raumes und die Koeffizienten der Gleichung dieser Hyperebene sind die Koordinaten η_ν des Bildpunktes η.

Wir beweisen nun:

1. *Wenn (ξ', η') ein Punktepaar der Abbildung $\mathfrak{T}$ ist und η' der Hyperebene $E(\lambda')$ angehört, so gehört ξ' zu $\mathfrak{M}'(\lambda')$.*

Beweis. Es sei (ξ, η) ein allgemeines Punktepaar von $\mathfrak{T}$; wir verbinden η mit $n-1$ allgemeinen Punktepaaren $A_1, \ldots, A_n$ von P_n zu einem Raum $E(\lambda)$, wobei λ rational von η und den A_ν abhängt. Da η zu $E(\lambda)$ gehört, gehört ξ zu $\mathfrak{M}(\lambda)$. Da ξ als allgemeiner Punkt von $\mathfrak{M}$ nicht zu $\mathfrak{M}_0$ gehört, gehört ξ zu $\mathfrak{M}'(\lambda)$, d. h. es ist

$$(24) \qquad G_1(\lambda, \xi) = 0, \ \ldots, \ G_h(\lambda, \xi) = 0.$$

Spezialisiert man alle (ξ, η) zu (ξ', η') und die Punkte A_ν zu $n-1$ Punkten, welche zusammen mit η' die Hyperebene $E(\lambda')$ aufspannen, so bleiben die Gleichungen (24) erfüllt, d. h. ξ' gehört zu $\mathfrak{M}'(\lambda')$.

2. *Einem Punkt ξ' von $\mathfrak{M}'(\lambda')$ entspricht in der Abbildung $\mathfrak{T}$ mindestens ein Punkt η' von $E(\lambda')$.*

Beweis. Wir gehen aus von einem allgemeinen Punktepaar (ξ, λ) der irreduziblen Korrespondenz $\mathfrak{K}'$. Da ξ zu $\mathfrak{M}'(\lambda)$ gehört, gehört ξ auch zu $\mathfrak{M}(\lambda)$, d. h. es ist

$$\lambda_0 \Phi_0(\xi) + \lambda_1 \Phi_1(\xi) + \ldots + \lambda_n \Phi_n(\xi) = 0$$

und daraus wegen (19)

$$(25) \qquad \lambda_0 \eta_0 + \lambda_1 \eta_1 + \ldots + \lambda_n \eta_n = 0.$$

Spezialisiert man nun das allgemeine Punktepaar (λ, ξ) relationstreu zu einem speziellen Punktepaar (λ', ξ') von $\mathfrak{K}'$, so gehört dazu mindestens eine Spezialisierung $\eta \to \eta'$ derart, daß die Gleichung (25) und die Gleichungen der Abbildung $\mathfrak{T}$ erfüllt bleiben. Das heißt, es gibt einen Punkt η' von $E(\lambda')$, welcher dem Punkt ξ' in der Abbildung $\mathfrak{T}$ zugeordnet ist.

Die Sätze 1. und 2. bilden nur die genaue Präzisierung der bekannten Aussage, daß die ebenen Schnitte $E(\lambda')$ von $\mathfrak{N}$ den Mannigfaltigkeiten $\mathfrak{M}'(\lambda')$ der Schar $|\mathfrak{M}'(\lambda)|$ entsprechen.

Wir untersuchen nun die *Fundamentalpunkte* der Abbildung, d. h. diejenigen Punkte ξ' von $\mathfrak{M}$, denen mindestens ∞^1 Punkte η' entsprechen.

3. *Die Fundamentalpunkte der Abbildung $\mathfrak{T}$ sind genau die Punkte, die allen $\mathfrak{M}'(\lambda')$ gemeinsam sind.*

Beweis. Ist ξ' ein Fundamentalpunkt, so bilden die entsprechenden Punkte η' mindestens eine Kurve, also liegt in jeder Hyperebene $E(\lambda')$

mindestens ein Punkt η'. Daraus folgt nach 1., daß ξ' allen $\mathfrak{M}'(\lambda')$ an-gehört. Gehört umgekehrt ξ' allen $\mathfrak{M}'(\lambda')$ an, so enthält nach 2. jede Hyperebene $E(\lambda')$ mindestens einen Bildpunkt η', was unmöglich wäre, wenn der Punkt ξ' nur endlichviele Bildpunkte η' hätte; also ist ξ' ein Fundamentalpunkt.

Einem Nichtfundamentalpunkt ξ' entsprechen nach Definition nur endlichviele Punkte η'. Darüber hinaus gilt:

4. *Einem einfachen Punkt ξ' von $\mathfrak{M}$, der nicht Fundamentalpunkt ist, entspricht nur ein Punkt η'.*

B e w e i s. Nach § 3 hat jeder von den endlichvielen Punkten η', die dem Punkt ξ' entsprechen, eine bestimmte positive Multiplizität, und die Summe dieser Multiplizitäten ist gleich der Anzahl der Punkte η', die einem allgemeinen Punkt ξ entsprechen, also gleich Eins. Also kann es nur einen Punkt η' geben.

Ist $\mathfrak{M}$ singularitätenfrei, so kann man demnach die Fundamental-punkte auch definieren als diejenigen Punkte ξ', denen mehr als ein Punkt η' entspricht.

Aus 3. oder aus § 1 folgt, daß die Fundamentalpunkte von $\mathfrak{T}$ (wenn es sie gibt) eine algebraische Mannigfaltigkeit bilden. Wir nennen ein Punktepaar (ξ', η') der Abbildung *singulär*, wenn ξ' Fundamentalpunkt ist. Die singulären Punktepaare bilden eine algebraische Mannigfaltig-keit $\mathfrak{S}$, deren Gleichungen gefunden werden, indem man die Gleichungen der Korrespondenz $\mathfrak{T}$ mit den Gleichungen der Mannigfaltigkeit der Fundamentalpunkte zusammennimmt. Ist nun $\mathfrak{M}$ singularitätenfrei, so gilt:

5. *Die singulären Punktepaare bilden (wenn es sie gibt) eine rein $(a-1)$-dimensionale Mannigfaltigkeit $\mathfrak{S}$, wobei a die Dimension von $\mathfrak{M}$ und daher auch von $\mathfrak{T}$ ist.*

B e w e i s. (ξ^0, η^0) sei ein singuläres Punktepaar; wir wollen zeigen, daß (ξ^0, η^0) einem irreduziblen $(a-1)$-dimensionalen Bestandteil von $\mathfrak{S}$ angehört. Wir wählen eine Ebene $E(\lambda')$, die η^0 nicht enthält. Die zugehörige Mannigfaltigkeit $\mathfrak{M}'(\lambda')$ enthält nach 3. den Punkt ξ^0. Wir projizieren $\mathfrak{M}'(\lambda')$ aus einem allgemeinen P_{m-a-1} und erhalten eine projizierende Hyperfläche $H = 0$, die $\mathfrak{M}$ schneidet nach $\mathfrak{M}'(\lambda')$ und einem Restschnitt $\mathfrak{M}''$, der den Punkt ξ^0 nicht enthält. Die Gleichung $H = 0$, kombiniert mit den Gleichungen von $\mathfrak{T}$, ergibt eine rein $(a-1)$-dimensionale Mannig-faltigkeit von Punktepaaren, zu der das Paar (ξ^0, η^0) gehört. Ein gewisser irreduzibler Bestandteil $\mathfrak{S}'$ dieser Mannigfaltigkeit enthält (ξ^0, η^0). Das allgemeine Punktepaar (ξ', η') von $\mathfrak{S}'$ muß singulär sein; denn wenn ξ' kein Fundamentalpunkt wäre, so müßte der nach 4. eindeutig bestimmte Bildpunkt η' in $E(\lambda')$ liegen; mithin müßte, da (ξ^0, η^0) alle Relationen erfüllt, die für (ξ', η') gelten, auch η^0 in $E(\lambda')$ liegen, was nicht der Fall ist. Also gehört $\mathfrak{S}'$ ganz der Mannigfaltigkeit der singulären Punktepaare an, was wir beweisen wollten.

Auf Grund von 5. zerfällt $\mathfrak{S}$ in irreduzible Mannigfaltigkeiten $\mathfrak{S}_1, \mathfrak{S}_2, \ldots$ von der Dimension $a - 1$. Jede Mannigfaltigkeit $\mathfrak{S}_j$ stellt eine gewisse irreduzible Korrespondenz zwischen einer Teilmannigfaltigkeit $\mathfrak{M}_j$ von $\mathfrak{M}$ und einer Teilmannigfaltigkeit $\mathfrak{N}_j$ von $\mathfrak{N}$ dar. Ist p_j die Dimension von $\mathfrak{M}_j$ und entsprechen einem allgemeinen Punkt von $\mathfrak{M}_j$ ∞^j Punkte η' von $\mathfrak{N}_j$, so ist nach dem Prinzip der Konstantenzählung

$$p_j + q_j = a - 1.$$

Dabei ist $q_j \geqq 1$, also $p_j \leqq a - 2$. Alle Fundamentalpunkte der Abbildung $\mathfrak{T}$ verteilen sich auf die Mannigfaltigkeiten $\mathfrak{M}_1, \mathfrak{M}_2, \ldots$, von denen einige auch ineinander liegen können.

Bemerkung. Die Beweise der Sätze 1., 2., 3. gelten mit geringfügigen Modifikationen auch dann, wenn $|E(\lambda)|$ nicht das System der Hyperebenen, sondern das System der Hyperflächen h-ten Grades des Raumes P_n ist. Den Beweis von Satz 1 hat man dann so zu modifizieren, daß nicht n, sondern $\binom{n+h}{h} - 1$ Punkte die Hyperfläche $E(\lambda)$ bestimmen. Man könnte die Sätze 1. und 2. noch weiter verallgemeinern, jedoch brauchen wir das für das Folgende nicht.

<h3 style="text-align:center">§ 7.</h3>

<h2 style="text-align:center">Das Verhalten
der Schnittpunktszahlen bei gewissen birationalen Abbildungen.</h2>

Gegeben sei eine birationale Abbildung $\mathfrak{T}$ zwischen zwei singularitätenfreien a-dimensionalen Mannigfaltigkeiten $\mathfrak{M}$ und $\mathfrak{N}$. Wir nehmen an, daß *die inverse Abbildung ausnahmslos eindeutig* ist, also keine Fundamentalpunkte auf $\mathfrak{N}$ besitzt. Dann wird nach § 6 das System der hyperebenen Schnitte von $\mathfrak{M}$ oder allgemeiner das System $|\mathfrak{H}|$ der Schnitte $\mathfrak{H}$ von $\mathfrak{M}$ mit den Hyperflächen h-ten Grades des Raumes P_m in ein lineares System $|\mathfrak{K}|$ ohne Basispunkte auf $\mathfrak{N}$ übergeführt, wobei jede Mannigfaltigkeit $\mathfrak{K}$ des Systems $|\mathfrak{K}|$ besteht aus den Punkten η', deren Urbilder ξ' der entsprechenden Mannigfaltigkeit $\overline{\mathfrak{H}}$ des Systems $|\mathfrak{H}|$ angehören.

Die Fundamentalpunkte ξ' auf $\mathfrak{M}$ verteilen sich nach § 6 auf gewisse Fundamentalmannigfaltigkeiten $\mathfrak{M}_j$, welche den „singulären Mannigfaltigkeiten" $\mathfrak{S}_j$ von der Dimension $a - 1$ auf $\mathfrak{N}$ entsprechen. Jedem Nichtfundamentalpunkt ξ' entspricht eineindeutig ein Punkt η', der zu keinem $\mathfrak{S}_j$ gehört. Dieses eineindeutige Entsprechen gilt auch dann, wenn ξ' der allgemeine Punkt einer Teilmannigfaltigkeit $\mathfrak{A}$ von $\mathfrak{M}$ ist. η' wird dann der allgemeine Punkt einer gleichdimensionalen Bildmannigfaltigkeit $\mathfrak{B}$ von $\mathfrak{N}$. Daraus folgt:

Es besteht ein eineindeutiges Entsprechen zwischen den d-dimensionalen irreduziblen Mannigfaltigkeiten $\mathfrak{A}$ auf $\mathfrak{M}$, die nicht aus lauter Fundamentalpunkten bestehen, und den irreduziblen d-dimensionalen Mannigfaltigkeiten $\mathfrak{B}$ auf $\mathfrak{N}$, die keinem $\mathfrak{S}_j$ angehören.

Geht man zu den speziellen Punkten von $\mathfrak{A}$ und $\mathfrak{B}$ über, so entspricht jedem speziellen Punkt η' von $\mathfrak{B}$ immer ein bestimmter Punkt ξ', und die Gesamtheit dieser Punkte ξ' ist genau $\mathfrak{A}$. Sucht man aber alle Punkte η' von $\mathfrak{N}$, deren Urbilder ξ' zu $\mathfrak{A}$ gehören, so findet man nicht nur die Punkte von $\mathfrak{B}$, sondern auch alle diejenigen Punkte eines $\mathfrak{S}_j$, die in der inversen Abbildung dasselbe Bild haben wie ein Punkt von $\mathfrak{B}$.

Die Menge dieser Punkte η' läßt sich besonders leicht übersehen in dem Fall, wo $\mathfrak{A}$ und $\mathfrak{B}$ die Dimension $a-1$ haben. Der projizierende Hyperkegel $H = 0$ von $\mathfrak{A}$ aus einem allgemeinen P_{n-a-1} des P_n schneidet $\mathfrak{M}$ nach $\mathfrak{A}$ und einem Restschnitt $\mathfrak{A}'$, der einen beliebig vorgegebenen Punkt ξ^0 nicht enthält und der außerdem keine der Mannigfaltigkeiten $\mathfrak{M}_j$ enthält. Die Mannigfaltigkeit $\overline{\mathfrak{H}} = \mathfrak{A} + \mathfrak{A}'$ gehört dann dem System $|\mathfrak{H}|$ an und ihr entspricht eine Mannigfaltigkeit $\overline{\mathfrak{K}}$ des Systems $|\mathfrak{K}|$, bestehend aus allen Punkten η', deren Urbilder ξ' zu $\overline{\mathfrak{H}}$ gehören. Diese Punkte η' sind erstens die Punkte der Bildmannigfaltigkeiten $\mathfrak{B}$ und $\mathfrak{B}'$ von $\mathfrak{A}$ und $\mathfrak{A}'$, zweitens eventuell noch Punkte gewisser $\mathfrak{S}_j$. Da $\overline{\mathfrak{K}}$ aber rein $(a-1)$-dimensional ist, kann $\overline{\mathfrak{K}}$ außer $\mathfrak{B}$ und $\mathfrak{B}'$ nur noch aus *vollen* $(a-1)$-dimensionalen Mannigfaltigkeiten $\mathfrak{S}_j$ (nicht aus Teilen davon) bestehen, und zwar gehört $\mathfrak{S}_j$ dann zu $\overline{\mathfrak{K}}$, wenn das Urbild $\mathfrak{M}_j$ von $\mathfrak{S}_j$ zu $\mathfrak{A} + \mathfrak{A}'$, also zu $\mathfrak{A}$ gehört.

Bei der Spezialisierung $\mathfrak{K} \rightarrow \overline{\mathfrak{K}}$ erhalten die Bestandteile $\mathfrak{B}$, $\mathfrak{B}'$, $\mathfrak{S}_j$ von $\overline{\mathfrak{K}}$ gewisse Multiplizitäten μ, μ', μ_j; wir schreiben also

$$\text{(27)} \qquad \overline{\mathfrak{K}} = \mu\,\mathfrak{B} + \mu'\,\mathfrak{B}' + \sum_j \mu_j\,\mathfrak{S}_j.$$

Der Bestandteil $\mu'\,\mathfrak{B}'$ ist für alles folgende unwesentlich, da $\mathfrak{B}'$ einen beliebigen fest gewählten Punkt η^0 nicht enthält; es ist also auch unwesentlich, ob $\mathfrak{B}'$ etwa reduzibel ist und ob in (27) eventuell verschiedene Teile von $\mathfrak{B}'$ mit verschiedenen Koeffizienten stehen müßten. Wesentlich sind die Bestandteile $\mu\,\mathfrak{B}$ und $\mu_j\,\mathfrak{S}_j$, denn sie umfassen genau die Punkte, deren Bilder in der inversen Abbildung zu $\mathfrak{A}$ gehören. Nachher wird sich zeigen, daß $\mu = 1$ ist. Ein μ_j ist genau dann Null, wenn $\mathfrak{A}$ die Fundamentalmannigfaltigkeit $\mathfrak{M}_j$ nicht enthält.

Nun sei $\mathfrak{C}$ eine irreduzible, nicht aus lauter Fundamentalpunkten bestehende Kurve auf $\mathfrak{M}$ und $\mathfrak{D}$ ihre Bildkurve auf $\mathfrak{N}$. Wir wollen zeigen:

Die Schnittmultiplizität von $\mathfrak{C}$ mit $\mathfrak{A}$ in einem Punkte ξ^0 ist gleich

$$\text{(28)} \qquad \sum_{\eta^0} \left(k\,\mu + \sum_j k_j\,\mu_j \right),$$

wobei die Summation sich über diejenigen Punkte η_0 der Bildkurve $\mathfrak{D}$ erstreckt, deren Urbild ξ^0 ist, und wo k und k_j die Schnittmultiplizitäten von $\mathfrak{D}$ mit $\mathfrak{B}$ und mit $\mathfrak{S}_j$ im Punkt η^0 sind.

Beweis. Die Anzahl der Schnittpunkte $\eta^{(\nu)}$ von $\mathfrak{K}$ mit $\mathfrak{D}$, die bei der Spezialisierung $\mathfrak{K} \to \overline{\mathfrak{K}}$ in einen Punkt η^0 hineinrücken, ist nach § 5, Satz 2 gleich

$$k\,\mu + \sum_j k_j\,\mu_j.$$

Den Schnittpunkten $\eta^{(\nu)}$ von $\mathfrak{K}$ mit $\mathfrak{D}$ entsprechen eineindeutig die Schnittpunkte $\xi^{(\nu)}$ von $\mathfrak{H}$ mit $\mathfrak{C}$, und wenn die $\eta^{(\nu)}$ in ein η^0 übergehen, so gehen die entsprechenden $\xi^{(\nu)}$ in ξ^0 über. Die Summe (28) gibt also an, wie viele $\xi^{(\nu)}$ bei der Spezialisierung $\mathfrak{H} \to \overline{\mathfrak{H}} = \mathfrak{A} + \mathfrak{A}'$ in ξ^0 hineinrücken. Diese Anzahl ist aber nach Severis Definition genau die Schnittmultiplizität von $\mathfrak{A}$ und $\mathfrak{C}$ in ξ^0.

Um nun zu zeigen, daß $\mu = 1$ ist, wählen wir für ξ^0 einen solchen Punkt von $\mathfrak{A}$, der nicht Fundamentalpunkt der Abbildung $\mathfrak{T}$ ist. Ihm entspricht ein einziger Punkt η^0, der nicht auf $\mathfrak{S}$ liegt: in (28) sind also alle $k_j\,\mu_j$ Null und die Summation über η_0 fällt weg. Wählt man nun die Kurve $\mathfrak{C}$ so, daß der Punkt ξ^0 ein einfacher Schnittpunkt wird, so wird nach (28)

$$1 = k\,\mu,$$

mithin, da k eine ganze Zahl ist, $\mu = 1$.

Summiert man die Multiplizitäten (28) über alle Schnittpunkte ξ^0 von $\mathfrak{C}$ mit $\mathfrak{A}_x$ und beachtet $\mu = 1$, so folgt die *Regel für die Schnittpunktszahlen:*

$$(29) \qquad [\mathfrak{C} \cdot \mathfrak{A}] = [\mathfrak{D} \cdot \mathfrak{B}] + \sum_j \mu_j\,[\mathfrak{D} \cdot \mathfrak{S}_j],$$

wo das Symbol $[\mathfrak{X} \cdot \mathfrak{Y}]$ die Schnittpunktszahl von $\mathfrak{X}$ mit $\mathfrak{Y}$, d. h. die Summe der Multiplizitäten aller Schnittpunkte bedeutet.

Die Bedeutung der Formel (29) beruht darauf, daß man mit ihrer Hilfe die Berechnung von Schnittpunktszahlen in $\mathfrak{N}$ zurückführen kann auf die entsprechende Berechnung in $\mathfrak{M}$ (oder umgekehrt). Gesetzt z. B., man habe für $\mathfrak{M}$ schon eine „Charakteristikenformel" [13]) gefunden, welche die Schnittpunktszahl $[\mathfrak{C}, \mathfrak{A}]$ für beliebige Kurven $\mathfrak{C}$ und Mannigfaltigkeiten $\mathfrak{A}$ linear durch gewisse spezielle Schnittpunktszahlen $[\mathfrak{C} \cdot \mathfrak{A}_\lambda]$ ausdrückt:

$$(30) \qquad [\mathfrak{C} \cdot \mathfrak{A}] = \sum_\lambda \alpha_\lambda\,[\mathfrak{C} \cdot \mathfrak{A}_\lambda],$$

---

[13]) Vgl. H. Schubert, Calcül der abzählenden Geometrie, Leipzig 1879, Abschn. 6. In meiner „Topologischen Begründung", Math. Ann. 102, § 8, habe ich bewiesen, daß eine Charakteristikenformel der Gestalt (30) für jedes singularitätenfreie $\mathfrak{M}$ existiert.

wobei die Koeffizienten α_λ nur von $\mathfrak{A}$ abhängen. Von den $\mathfrak{A}_\lambda$ nehmen wir noch an, daß sie die Fundamentalmannigfaltigkeiten $\mathfrak{M}_j$ nicht enthalten [14]). Formeln dieser Art gelten nach ZAG I z. B. dann, wenn $\mathfrak{M}$ ein projektiver oder mehrfach-projektiver Raum ist: die $\mathfrak{A}_\lambda$ sind dann Hyperebenen und die α_λ die Gradzahlen der Hyperfläche $\mathfrak{A}$. Geht man nun in (30) auf beiden Seiten mittels (29) zu $\mathfrak{N}$ über, so folgt

$$(31) \qquad [\mathfrak{D} \cdot \mathfrak{B}] = \sum_\lambda \alpha_\lambda \, [\mathfrak{D} \cdot \mathfrak{B}_\lambda] - \sum_j \mu_j \, [\mathfrak{D} \cdot \mathfrak{S}_j],$$

was man unter Weglassung des „Faktors" $\mathfrak{D}$ auch in Gestalt einer „symbolischen Gleichung" im Sinne Schuberts

$$(32) \qquad \mathfrak{B} \doteq \sum \alpha_\lambda \, \mathfrak{B}_\lambda - \sum \mu_j \, \mathfrak{S}_j$$

schreiben kann. (32) bedeutet nichts anderes, als daß die Hinzufügung einer beliebigen Kurve $\mathfrak{D}$ als Faktor immer eine richtige Zahlenrelation, nämlich eben (31) ergibt.

Die Formel (31) gilt auch für solche Kurven $\mathfrak{D}$, die auf einem $\mathfrak{B}_\lambda$ oder $\mathfrak{S}_j$ gelegen sind, vorausgesetzt, daß man das Symbol $[\mathfrak{D} \cdot \mathfrak{S}_j]$ in bekannter Weise als „virtuelle Schnittpunktszahl" definiert, d. h. als Differenz $[\mathfrak{D}_1 \cdot \mathfrak{S}_j] - [\mathfrak{D}_2 \cdot \mathfrak{S}_j]$, wobei $\mathfrak{D}_1$ eine Kurve ist, die ein irreduzibles System durchläuft, welches auch die Kurve $\mathfrak{D}_2 + \mathfrak{D}$ enthält [15]).

§ 8.

Die Charakteristikenformel für verbundene Punktepaare.

Ein einfaches und lehrreiches Beispiel zu § 6 ist das Folgende.

$\mathfrak{M}$ sei der zweifach-projektive Raum, bestehend aus den (geordneten) Punktepaaren (p, q) eines P_n. $\mathfrak{N}$ sei die Mannigfaltigkeit der „verbundenen Punktepaare" (p, q, g), wobei ein verbundenes Punktepaar besteht aus einem Punktepaar und einer die beiden Punkte enthaltenden Geraden g [16]). Offenbar ist $\mathfrak{M}$ auf $\mathfrak{N}$ birational abgebildet, denn ein allgemeines Punktepaar (p, q) hat nur eine, rational bestimmbare Verbindungsgerade g. Fundamental für die Abbildung sind die koinzidenten Punktepaare, deren Verbindungsgeraden unbestimmt sind. Die entsprechende singuläre Mannigfaltigkeit $\mathfrak{S}$ besteht aus den koinzidenten Punktepaaren $p = q$ mit Geraden g durch p. $\mathfrak{S}$ ist irreduzibel und hat, wie es sein muß, die Dimension -1. Es gibt also nur ein $\mathfrak{S}_j$, nämlich $\mathfrak{S}_1 = \mathfrak{S}$.

---

[14]) Die Annahme ist nicht wesentlich; sie vereinfacht aber die Rechnungen und läßt sich in allen praktisch wichtigen Fällen leicht erfüllen.

[15]) Vgl. etwa die unter [3]) zitierte Arbeit von Severi.

[16]) Will man $\mathfrak{M}$ und $\mathfrak{N}$ als Mannigfaltigkeiten in einem projektiven Raum ansehen, so hat man die Produkte der Koordinaten von p und q bzw. von p, q und g als homogene Koordinaten in einem projektiven Raum zu deuten.

Sind $\mathfrak{P}$ und $\mathfrak{Q}$ die Mannigfaltigkeiten der Punktepaare, bei denen p oder q in einer festen (allgemein gewählten) Hyperebene des P_n liegt, und bezeichnen wir ihre Bildmannigfaltigkeiten in $\mathfrak{N}$ der Einfachheit halber ebenfalls mit $\mathfrak{P}$ und $\mathfrak{Q}$, so gilt in $\mathfrak{M}$ die Charakteristikenformel (vgl. ZAG I, § 3):

$$[\mathfrak{C}\cdot\mathfrak{A}] = \alpha_1\,[\mathfrak{C}\cdot\mathfrak{P}] + \alpha_2\,[\mathfrak{C}\cdot\mathfrak{Q}].$$

Daraus folgt nach § 6 für $\mathfrak{N}$ die Charakteristikenformel

$$(33) \qquad [\mathfrak{D}\cdot\mathfrak{B}] = \alpha_1\,[\mathfrak{D}\cdot\mathfrak{P}] + \alpha_2\,[\mathfrak{D}\cdot\mathfrak{Q}] - \mu_1\,[\mathfrak{D}\cdot\mathfrak{S}]$$

oder kürzer

$$(34) \qquad \mathfrak{B} \doteq \alpha_1\,\mathfrak{P} + \alpha_2\,\mathfrak{Q} - \mu_1\,\mathfrak{S}.$$

Schubert[13]) verwendet statt $\mathfrak{P}$, $\mathfrak{Q}$, $\mathfrak{S}$ die Symbole p, q, ε ($p =$ die Bedingung, daß p einer festen Hyperebene angehört; $\varepsilon =$ Koinzidenzbedingung). In seiner Bezeichnung ist also *jede einfache Bedingung für ein verbundenes Punktepaar linear durch p, q und ε ausdrückbar.*

Die geometrische Bedeutung der Koeffizienten α_1, α_2, μ_1 in (34) wird klar, wenn man (33) auf spezielle Kurven $\mathfrak{D}_1$, $\mathfrak{D}_2$. $\mathfrak{D}_3$ anwendet, und zwar mögen diese Kurven folgendermaßen definiert werden:

$\mathfrak{D}_1 = p^{n-1}q^n$: Gesamtheit der Punktepaare mit gegebenem q in allgemeiner Lage, wobei p einer gegebenen Geraden angehört;

$\mathfrak{D}_2 = p^n q^{n-1}$: Gesamtheit der Punktepaare mit gegebenem p, wobei q einer gegebenen Geraden angehört.

$\mathfrak{D}_3 = G\,p$: Gesamtheit der verbundenen Punktepaare, bei denen g gegeben ist und p einer gegebenen Hyperebene angehört (also ebenfalls fest liegt).

Zur Durchführung der Rechnungen haben wir die Schnittpunktszahlen $[\mathfrak{D}_\nu\cdot\mathfrak{P}]$, $[\mathfrak{D}_\nu\cdot\mathfrak{Q}]$ und $[\mathfrak{D}_\nu\cdot\mathfrak{S}]$ zu bestimmen. Die Multiplizitäten der Schnittpunkte von $\mathfrak{D}_1$ und $\mathfrak{P}$, von $\mathfrak{D}_2$ und $\mathfrak{Q}$ usw. ergeben sich auf Grund der Kriterien ZAG V, 2 und 4 alle gleich Eins. Daher findet man: $[\mathfrak{D}_1\cdot\mathfrak{P}] = [\mathfrak{D}_2\cdot\mathfrak{Q}] = [\mathfrak{D}_3\cdot\mathfrak{Q}] = [\mathfrak{D}_3\cdot\mathfrak{S}] = 1$, alle übrigen $= 0$. Setzt man also $\mathfrak{D}_1$, $\mathfrak{D}_2$ und $\mathfrak{D}_3$ statt $\mathfrak{D}$ in (33) ein, so folgt

$$(35) \qquad \begin{cases} [\mathfrak{D}_1\cdot\mathfrak{B}] = \alpha_1, \\ [\mathfrak{D}_2\cdot\mathfrak{B}] = \alpha_2, \\ [\mathfrak{D}_3\cdot\mathfrak{B}] = \alpha_2 - \mu_1. \end{cases}$$

Zum Beispiel sei $\mathfrak{B} = \mathfrak{G}$ die Gesamtheit der Geraden, die einen allgemeinen P_{n-2} schneidet (Schuberts Bedingung g). Man findet $\alpha_1 = \alpha_2 = \mu_1 = 1$, mithin

$$\mathfrak{G} \doteq \mathfrak{P} + \mathfrak{Q} - \mathfrak{S}$$

oder

$$(36) \qquad \mathfrak{S} \doteq \mathfrak{P} + \mathfrak{Q} - \mathfrak{G}.$$

Das ist Schuberts berühmte Koinzidenzformel $\varepsilon = p + q - g$, welche besagt: *Die (virtuelle) Anzahl der Koinzidenzen in einem einfach-unendlichen System von verbundenen Punktepaaren ist gleich der Summe der Anzahlen der Punktepaare des Systems, bei denen p oder q einer allgemein gegebenen Hyperebene angehören, vermindert um die Anzahl der Punktepaare des Systems, bei denen die Verbindungsgerade g einen allgemein gegebenen P_{n-2} schneidet.*

Mit Hilfe von (36) können wir die virtuelle Schnittpunktzahl $[\mathfrak{D}_4 \cdot \mathfrak{S}]$ berechnen, wo $\mathfrak{D}_4 = \varepsilon\, p^n\, g^{n-2}$ die Gesamtheit der koinzidenten verbundenen Punktepaare mit gegebenem Punkt p ist, bei denen g eine gegebene Gerade trifft. $\mathfrak{D}_4$ liegt auf $\mathfrak{S}$. Man findet mittels (36):

$$[\mathfrak{D}_4 \cdot \mathfrak{S}] = [\mathfrak{D}_4 \cdot \mathfrak{P}] + [\mathfrak{D}_4 \cdot \mathfrak{Q}] - [\mathfrak{D}_4 \cdot \mathfrak{G}] = 0 + 0 - 1 = -1.$$

Setzt man nun $\mathfrak{D}_4$ in (33) ein, so folgt

$$(37) \qquad\qquad [\mathfrak{D}_4 \cdot \mathfrak{B}] = \mu_1.$$

Damit haben wir auch die geometrische Bedeutung von μ_1 gefunden. Die Formel (37) kann statt der letzten Formel von (35) zur Bestimmung von μ_1 benutzt werden.

Bemerkung. Die Formel (34) kann, wie ihr Beweis lehrt, auch als Äquivalenz im Sinne der Theorie der linearen Scharen gedeutet werden: sie bedeutet, daß die Mannigfaltigkeiten $\mathfrak{B} + \mu_1 \mathfrak{S}$ und $\alpha_1 \mathfrak{P} + \alpha_2 \mathfrak{Q}$ einer linearen Schar angehören. Diese Schar besteht nämlich aus allen Mannigfaltigkeiten, die durch eine homogene Bedingungsgleichung vom Grade α_1 in den Koordinaten des Punktes p und vom Grade α_2 in denen von q definiert sind. Eine spezielle derartige Gleichung definiert auf $\mathfrak{M}$ die Mannigfaltigkeit $\mathfrak{A}$ und auf $\mathfrak{N}$ ihr Bild $\mathfrak{B} + \mu_1 \mathfrak{S}$. Spezialisiert man andererseits die Gleichung so, daß sie zerfällt in α_1 Linearfaktoren in p und β_1 Linearfaktoren in q, so definiert sie die Mannigfaltigkeit $\alpha_1 \mathfrak{P} + \alpha_2 \mathfrak{Q}$. Diese Betrachtungsweise rückt die Bedeutung der Formel (34) in ein helles Licht [17]).

---

[17]) Nicht immer hat eine symbolische Gleichung im Sinne Schuberts ($\doteq$) die Bedeutung einer Äquivalenz ($\equiv$). Die Erzeugenden einer nicht-rationalen Regelschar sind im Schubertschen Sinne gleich, aber nicht äquivalent, da sie keiner linearen Schar angehören.

(Eingegangen am 11. 10. 1933.)

13.

Zur algebraischen Geometrie VII
Ein neuer Beweis des Restsatzes

Mathematische Annalen 111, 3 (1935) 432–437

Der Brill-Noethersche Restsatz folgt, wie ich früher[1]) bemerkt habe, fast unmittelbar aus dem folgenden Satz, den ich aus nachher zu erwähnenden Gründen den *Satz vom Doppelpunktsdivisor* nennen will:

S a t z D. *Wenn alle Schnittpunkte der Kurven $f = 0$ und $\varphi = 0$ gewöhnliche Punkte oder gewöhnliche Singularitäten der Kurve $f = 0$ sind, d. h. wenn alle Zweige Z dieser Kurve in diesen Punkten einfach sind und getrennte Tangenten haben, und wenn jeder Zweig Z eines solchen, etwa s-fachen Punktes, der von der Kurve $\varphi = 0$ etwa μ-fach geschnitten wird, von einer weiteren Kurve $F = 0$ mindestens $(\mu + s - 1)$-fach geschnitten wird, so gilt für die ternären Formen F, φ, f die Identität*

$$(1) \qquad\qquad F = A\,f + B\,\varphi;$$

überdies hat die Kurve $B = 0$ in jedem solchen Schnittpunkt einen mindestens $(s - 1)$-fachen Punkt.

Gelten umgekehrt (1) und die zuletzt formulierte Adjungiertheitsbedingung für B, so erfüllen die Kurven $F = 0$ und $\varphi = 0$ die im Satz angegebenen Schnittmultiplizitätsbedingungen für jeden Zweig Z.

Ich habe in meiner früheren Note[1]) den Satz D aus dem Noetherschen Fundamentalsatz (in der ursprünglichen Noetherschen Potenzreihenformulierung) hergeleitet. Ich hatte damals nicht bemerkt, daß der Satz im Grunde nicht neu, sondern einem bekannten Satz über den Doppelpunktsdivisor aus der arithmetischen Theorie der algebraischen Funktionen[2]) im wesentlichen gleichwertig ist. Die Voraussetzung unseres Satzes D besagt nämlich, falls sie für alle vielfachen Punkte der Kurve $f = 0$ zutrifft, daß der Divisor $\dfrac{F}{\varphi}$ durch den Doppelpunktsdivisor der Kurve $f = 0$ (der

---

[1]) B. L. van der Waerden, Zur Begründung des Restsatzes mit dem Noetherschen Fundamentalsatz, Math. Annalen **104** (1931), S. 472–475.

[2]) Siehe K. Hensel und G. Landsberg, Theorie der algebraischen Funktionen einer Variabeln, 1902, 25. Vorlesung, § 3, S. 430. Dort auf S. 435 auch die Herleitung des Restsatzes.

195

aus allen s-fachen Punkten der Kurve besteht, je $(s-1)$-fach gezählt auf jedem Zweig Z_i) teilbar ist. Die Behauptung (1) läßt sich andererseits auch so lesen:

$$F = B\,\varphi \quad \text{oder} \quad \frac{F}{\varphi} = B \quad \text{auf der Kurve } f = 0.$$

Satz D ist also in diesem Fall gleichwertig mit der Aussage, daß ein Divisor der Klasse A^d (wo A eine Linearform in den drei homogenen Koordinaten ist), der durch den Doppelpunktsdivisor teilbar ist, gleich einer Form d-ten Grades in den homogenen Koordinaten ist. Aus diesem Grunde habe ich oben die Bezeichnung „Satz vom Doppelpunktsdivisor" vorgeschlagen. Allerdings ist Satz D insofern etwas spezieller als der allgemeine Satz vom Doppelpunktsdivisor [2]), als in Satz D die Singularitäten der Kurve $f = 0$ als gewöhnliche vielfache Punkte mit getrennten Tangenten vorausgesetzt wurden, was in der allgemeinen Theorie der algebraischen Funktionen nicht nötig ist. Für die Anwendungen ist das nicht von ausschlaggebender Wichtigkeit, da man die komplizierteren Singularitäten immer durch Cremonatransformationen auflösen kann.

Ich werde jetzt zeigen, daß der Satz D besonders einfach mit derselben Methode bewiesen werden kann, mit der Herr van der Woude in der unmittelbar vorangehenden Arbeit den „einfachen Fall" des Noetherschen Fundamentalsatzes behandelt hat. Ich glaube nicht, daß diese Methode zum Beweis des allgemeinen Falles des Noetherschen Fundamentalsatzes hinreicht. Für die Geometrie auf der Kurve ist aber der Satz D, aus dem der Restsatz folgt, wichtiger als der allgemeine Fall des Noetherschen Fundamentalsatzes.

Der Vollständigkeit halber gebe ich am Schluß noch einmal die Herleitung des Restsatzes aus dem Satz D an und füge dann noch einige Bemerkungen über andere Beweise des Restsatzes hinzu.

Wie in der vorangehenden Arbeit von W. van der Woude gehen wir von den Formen F, f und φ durch die Substitution $z = 1$ zu inhomogenen Polynomen in x und y über, wobei wir annehmen können, 1. daß das Glied x^n in f wirklich vorkommt, wo n der Grad von f ist; 2. daß die Formen, die von den Gliedern höchsten Grades in f und φ gebildet werden, teilerfremd sind; 3. daß keine zwei Schnittpunkte der Kurven $f = 0$ und $g = 0$ auf einer Geraden $y = $ konst. liegen; 4. daß keine Tangente in einem von diesen Schnittpunkten parallel zur x-Achse ist. Ist dann $\varrho\,(y)$ die Resultante von f und φ nach x, so gilt bekanntlich eine Identität

$$\varrho\,(y) = P\,f + Q\,\varphi$$

aus der unmittelbar folgt

$$(2) \qquad \varrho(y)\,F = R\,f + S\,\varphi.$$

Eine solche Identität bleibt bestehen, wenn man S durch seinen Rest bei Division durch f ersetzt, dessen Grad in x kleiner als n ist.

Nun sei $Q\,(a,b)$ ein Schnittpunkt der Kurven $f = 0$ und $\varphi = 0$, und zwar ein s-facher Punkt von $f = 0$, durch den s einfache Zweige $Z_1,\ldots,Z_s$ dieser Kurve hindurchgehen. Setzt man die Potenzreihenentwicklung eines solchen Zweiges Z_i in (2) ein, so wird $f = 0$, also $\varrho(y)\,F = S\,\varphi$. Die Funktionen $\varrho(y)\,F$ und $S\,\varphi$ haben also an der Stelle Q (als Funktionen der Ortsuniformisierenden t) eine Nullstelle von derselben Ordnung. Hat φ die Ordnung μ_i, so hat F nach Voraussetzung mindestens die Ordnung $\mu_i + s - 1$, und da $\varrho(y)$ auch Null wird an der Stelle Q, so hat $\varrho(y)\,F$ mindestens die Ordnung $\mu_i + s$. Somit hat die Funktion S eine mindestens s-fache Nullstelle in Q auf jedem Zweig $Z_i\,(i = 1,\ldots,s)$.

Hieraus folgt aber, daß die Kurve $S = 0$ einen mindestens s-fachen Punkt in Q hat. Hätte nämlich $S = 0$ nur einen k-fachen Punkt $(k < s)$, so hätten die Zweige dieser Kurve in Q insgesamt höchstens k Tangenten, also käme unter den s Zweigen Z_i mindestens einer vor, der keinen Zweig von $S = 0$ berührte. Dieser Zweig Z_i würde dann die Kurve $S = 0$ genau k-fach schneiden, was wegen $k < s$ dem vorhin Gesagten widerspricht.

Von den n Schnittpunkten der Geraden $y = b$ mit $f = 0$ wird der Punkt Q s-fach gezählt; die anderen, $n - s$ an der Zahl, liegen wegen (2) notwendig auf $S = 0$. Die Kurve $S = 0$ schneidet also die Gerade $y = 0$ in dem s-fachen Punkt Q und außerdem noch in $n - s$ Punkten. Da S aber in x einen Grad $< n$ hat, so muß S für $y = b$ identisch verschwinden, d. h. S ist durch $y - b$ teilbar.

In (2) sind nun $\varrho(y)\,F$ und $S\,\varphi$ durch $y - b$ teilbar; also ist auch $R\,f$ und somit R durch $y - b$ teilbar. Man kann also mit $y - b$ kürzen. Durch Wiederholung derselben Schlußweise kann man alle Linearfaktoren von $\varrho(y)$ in (2) der Reihe nach zum Verschwinden bringen. Es folgt

$$(3) \qquad F = A\,f + B\,\varphi.$$

Das ganze Verfahren kann so eingerichtet werden, daß die beiden Glieder rechter Hand in (2) und daher auch in (3) denselben Grad haben wie die linke Seite. Sollten daran noch Zweifel bestehen, so kann man durch eine bekannte Umformung (vgl. § 6 der Arbeit von W. v. d. Woude) auch nachträglich erreichen, daß in (3) die beiden Glieder $A\,f$ und $B\,\varphi$ denselben Grad wie F haben. Sodann kann die Identität (3) durch Einführung von z wieder homogen gemacht werden. Damit ist die Identität (1) bewiesen.

Dieselbe Überlegung, durch die wir oben fanden, daß die Kurve $S = 0$ in jedem s-fachen Punkt von $f = 0$ ebenfalls einen s-fachen Punkt haben muß, zeigt jetzt, auf (3) angewandt, daß die Kurve $B = 0$ in jedem s-fachen Punkt von $f = 0$, Schnittpunkt von $f = 0$ und $\varphi = 0$, einen mindestens $(s - 1)$-fachen Punkt besitzt. Damit ist Satz D in allen Teilen bewiesen.

Die Umkehrung des Satzes heißt so: Aus (1) und den Adjungiertheitsbedingungen für B in den Schnittpunkten von $f = 0$ und $\varphi = 0$ folgt, daß jeder Zweig Z_i eines s-fachen Punktes von $f = 0$, der von $\varphi = 0$ genau μ_i-fach geschnitten wird, von $F = 0$ mindestens $(\mu_i + s - 1)$-fach geschnitten wird. Zum Beweis braucht man nur in (1) $f = 0$ zu setzen und die Ordnungen der Nullstellen links und rechts zu vergleichen.

Der *Restsatz* besagt folgendes. Die Kurve $f = 0$ besitze keine anderen als gewöhnliche Singularitäten. Die Kurven $\varphi = 0$ und $\varphi' = 0$ mögen auf den verschiedenen Zweigen der Grundkurve $f = 0$, außer einer beiden gemeinsamen Punktgruppe H, die Punktgruppen G und G' ausschneiden. Die Kurve $\psi = 0$ sei adjungiert, d. h. sie habe einen $(s - 1)$-fachen Punkt in jedem s-fachen Punkt der Grundkurve und sie schneide diese, außer in diesem auf jedem Zweig $(s - 1)$-fach gezählten mehrfachen Punkten und einer Gruppe K genau nach der Gruppe G. Dann gibt es eine ebenfalls adjungierte Kurve $\psi' = 0$, welche in derselben Weise außer den vielfachen Punkten und außer K die Gruppe G' ausschneidet.

Zu der Formulierung ist noch zu bemerken, daß mit dem Wort „Punktgruppe“ nicht einfach eine Menge von Punkten der Kurve $f = 0$ gemeint ist, sondern eine endliche Menge von eventuell mit Vielfachheiten versehenen Punkten auf den verschiedenen Zweigen der Kurve.

Beweis des Restsatzes. Man setze $F = \varphi' \psi$. Dann erfüllt F die Voraussetzungen des Satzes D. Daher gilt (1). Setzt man $\psi' = B$, so schneiden $F = \varphi' \psi$ und $B \varphi = \psi' \varphi$ auf der Grundkurve dieselbe Punktgruppe aus; denn für $f = 0$ wird $F = B \varphi$. Daraus folgt die Behauptung.

Der hier dargestellte Beweis scheint mir an Einfachheit und Natürlichkeit alle bekannten Beweise des Restsatzes zu übertreffen. Er beruht auf denselben Grundgedanken wie der ursprüngliche Noethersche, nämlich auf der Untersuchung der Bedingungen „im Kleinen“, welche für die Identität „im Großen“ (1) hinreichend sind. Auch die Methode des sukzessiven Weghebens der Resultantenfaktoren wurde von Noether[3]) schon früher angewandt. Zum Unterschied von den anderen Beweisen

---

[3]) M. Noether, Math. Annalen 40 (1892), S. 140—144.

werden bei uns bloß Schnittpunktsmultiplizitäten herangezogen und wird die Diskussion der Vielfachheiten der betrachteten Punkte als Punkte der Kurven $\varphi = 0$ und $F = 0$, welche ja auch mit dem eigentlichen Problem nichts zu tun haben, vermieden.

Man hat mich nach der Publikation meines ersten Beweises des Satzes vom Doppelpunktsdivisor[1]) verschiedentlich gefragt, ob mir die Beweismethode des Restsatzes nicht bekannt sei, die ebenfalls von M. Noether[4]) herrührt und auch in verschiedenen Lehrbüchern[5]) dargestellt ist, welche sich auf den folgenden (einwandfrei richtigen) Satz stützt:

Satz E. Hat die Kurve $F = 0$ in allen gemeinsamen Punkten der Kurven $\varphi = 0$ und $f = 0$, welche für diese Kurven bzw. r-fache und s-fache Punkte sind, jeweils einen mindestens $(r + s - 1)$-fachen Punkt und ist dieselbe Bedingung auch für alle zu diesen Schnittpunkten unendlich benachbarten Schnittpunkte erfüllt, so gilt (1).

Mir war dieser Beweis natürlich bekannt, aber die Herleitung des Restsatzes aus dem Satz E scheitert zunächst daran, daß die Voraussetzungen des Satzes E unter den Voraussetzungen des Restsatzes nicht immer erfüllt sind. Es sei etwa $f = x$, $\varphi = x^2 - y^2$ und $F = \varphi' = x - y^2$. Die Kurven $f = 0$ und $\varphi = 0$ haben in $Q\,(0,0)$ einen Schnittpunkt, der zweifach für φ und einfach für f ist ($r = 2$, $s = 1$). Die Kurve $F = 0$ geht nur einfach durch diesen Schnittpunkt, nicht $(r + s - 1)$-fach. Satz E ist also nicht anwendbar. — Man kann auch umgekehrt $\varphi = x - y^2$ und $F = x^2 - y^2$ setzen. Es gibt dann einen Schnittpunkt von f und φ in Q und einen dazu unendlich benachbarten Schnittpunkt Q_1. Die Kurve $F = 0$ geht durch Q, aber nicht durch Q_1. Sie erfüllt somit die Bedingungen des Satzes E nicht.

Es ist leicht, kompliziertere Beispiele zu geben, in denen die Behauptung des Restsatzes nicht so trivial ist wie oben. Das Prinzip ist aber immer das gleiche: Die Voraussetzungen des Satzes D stimmen für $F = \varphi'\,\psi$ genau mit denen des Restsatzes überein und sind auch notwendig und hinreichend für (1) mit adjungiertem B; die Voraussetzungen von E aber verlangen etwas mehr als die von D und sind daher wohl hinreichend, aber nicht notwendig.

Diese Schwierigkeit ist übrigens wohlbekannt. Sie ist es, die Severi[6]) veranlaßte, den Begriff der scheinbaren Multiplizität einzuführen. Mittels

---

4) M. Noether, Math. Annalen 23 (1884), S. 348—351.

5) Siehe z. B. F. Enriques - O. Chisini, Teoria geometrica delle equazioni algebriche, I, Bologna 1915, S. 131.

6) F. Severi-E. Löffler, Vorlesungen über algebraische Geometrie, Leipzig 1921, S. 117 und 120—121.

dieses Begriffes kann der Beweis des Restsatzes in Ordnung gebracht werden, in der Weise, wie es bei Severi angedeutet ist. Vom Standpunkt der Einfachheit aber scheint der oben dargestellte Beweis, in welchem weder unendlich benachbarte Schnittpunkte noch Umdeutungen des Multiplizitätsbegriffs herangezogen zu werden brauchen, vorzuziehen zu sein.

Es war nicht, um die Max Noether gebührende Ehre als Begründer der geometrischen Theorie der algebraischen Funktionen zu verkleinern, daß ich in meiner Note[1]) auf eine kleine Lücke in Noethers Beweisführung hingewiesen habe. Seine grundlegenden Ideen haben ihn nicht nur zu Entdeckung der fruchtbarsten und schönsten Methoden der algebraisch-geometrischen Theorie geführt, sondern sie reichen auch zur exakten Begründung aller Sätze dieser Theorie hin. Das beweist auch wieder der hier dargestellte Beweis des Restsatzes, in welchem (ebenso wie in meinem ersten Beweis) im Grunde ausschließlich Noethersche Schlußweisen benutzt werden.

(Eingegangen am 29. 3. 1935.)

14.

Zur algebraischen Geometrie
Berichtigung und Ergänzungen

Mathematische Annalen 113, 1 (1936) 36–39

Durch Zuschriften von verschiedenen Seiten, für die ich sehr dankbar bin, bin ich auf einige Punkte in meiner Aufsatzreihe „Zur algebraischen Geometrie" aufmerksam gemacht worden, die einer näheren Erläuterung bedürfen, sowie auf einen Beweis, der nicht ganz in Ordnung ist. Ich bringe hier die erforderlichen Ergänzungen der Beweise. Die Ergebnisse bleiben davon unberührt.

1. *Das Verhalten eines Primideals bei rein transzendenten Erweiterungen des Grundkörpers.*

In ZAG I (Math. Annalen **108**, S. 117) wird von dem Satze Gebrauch gemacht, daß ein Primideal $\mathfrak{p}$ eines Polynombereichs $\mathsf{K}[x_1, \ldots, x_n]$ Primideal bleibt, wenn man durch Adjunktion von Unbestimmten $u_1, \ldots, u_h$ den Körper K zu $\Omega = \mathsf{K}(u)$ erweitert und das Erweiterungsideal $\mathfrak{p}_\Omega$ bildet. Der Beweis lautet so:

Es genügt offenbar, sich auf den Fall zu beschränken, wo nur eine Unbestimmte u an K adjungiert wird. Gesetzt, ein Produkt $f(u, x) \cdot g(u, x)$ wäre durch $\mathfrak{p}_\Omega$ teilbar, ohne daß $f(u, x)$ oder $g(u, x)$ es wären. Nach Definition des Ideals $\mathfrak{p}_\Omega$ wäre dann

$$(1) \qquad f(u, x) \cdot g(u, x) = \sum_\nu a_\nu(u)\, p_\nu(x); \qquad p_\nu(x) \text{ in } \mathfrak{p}.$$

Multipliziert man alle a_ν sowie $f(u, x)$ und $g(u, x)$ mit passenden Polynomen in u allein, so kann man erreichen, daß f, g und alle a_ν ganzrational auch in u sind. Man ordne nun $f(u, x)$ und $g(u, x)$ nach Potenzen von u und streiche in dieser Entwicklung alle die Glieder, deren Koeffizienten durch $\mathfrak{p}$ teilbar sind, ganz weg. Die Gültigkeit von (1) wird durch diese Weglassung nicht gestört. Der Anfangskoeffizient von $f(u, x)$, d. h. der Koeffizient der höchsten Potenz von u, ist nunmehr nicht durch $\mathfrak{p}$ teilbar; dasselbe gilt von $g(u, x)$. Da $\mathfrak{p}$ prim ist, ist das Produkt dieser Anfangskoeffizienten, also der Anfangskoeffizient von $f(u, x) \cdot g(u, x)$, auch nicht durch $\mathfrak{p}$ teilbar. Der Vergleich der Koeffizienten der Potenzen von u links und rechts in (1) lehrt aber das Gegenteil.

2. Die Koordinaten der Schnittpunkte einer Kurve C mit einer allgemeinen Hyperfläche eines Büschels als algebraische Funktionen des Büschelparameters.

In ZAG V (Math. Annalen **110**, S. 132, oben) bin ich den Beweis schuldig geblieben, daß die inhomogenen Koordinaten $\xi_1, \ldots, \xi_n$ des Punktes ξ algebraische Funktionen von λ sind. Dabei war ξ ein Schnittpunkt einer algebraischen Kurve C mit der allgemeinen Hyperfläche eines Büschels $\lambda F_0 + F_1 = 0$ (λ eine Unbestimmte), von welchem vorausgesetzt war, daß er kein Basispunkt des Büschels ist. Man schließt so: Wäre $F_0(\xi) = 0$, so würde aus $\lambda F_0(\xi) + F_1(\xi) = 0$ folgen $F_1(\xi) = 0$, also wäre ξ ein Basispunkt des Büschels. Also ist $F_0(\xi) \neq 0$ und

$$\lambda = -\frac{F_1(\xi)}{F_0(\xi)}.$$

Das System $(\xi_1, \ldots, \xi_n, \lambda)$ hat höchstens den Transzendenzgrad Eins, da ξ ein Punkt einer algebraischen Kurve und λ eine rationale Funktion von den ξ_ν ist. Da nun λ eine Unbestimmte ist, hängt das ganze System $(\xi_1, \ldots, \xi_n, \lambda)$ von λ algebraisch ab. Damit ist die Behauptung bewiesen.

3. Der Satz von Bertini über die mehrfachen Punkte der Kurven einer linearen Kurvenschar auf einer Fläche.

Es sei λ wieder eine Unbestimmte. Die Hyperfläche

$$(2) \qquad\qquad \lambda F_0 + F_1 = 0$$

schneidet auf einer algebraischen Fläche eine Kurve aus. Der Satz von Bertini besagt, daß diese Kurve außerhalb der Doppelpunkte des Büschels und der mehrfachen Punkte der Fläche keine mehrfachen Punkte besitzt. Der Beweis wurde in ZAG V (Math. Annalen **110**, S. 133) nur angedeutet. Ich führe ihn hier vollständig aus.

Gesetzt, ξ wäre ein Doppelpunkt der Schnittkurve. Dann ist wieder $F_0(\xi) \neq 0$ und

$$\lambda = -\frac{F_1(\xi)}{F_0(\xi)}.$$

Das System $(\xi_1, \ldots, \xi_n, \lambda)$ hat den Transzendenzgrad 1 oder 2. Im Fall des Transzendenzgrades 2 sind alle ξ_ν algebraische Funktionen von λ und einem ξ_μ. Dieses ξ_μ kann man so spezialisieren, daß die algebraischen Funktionen sinnvoll bleiben und die übrigen wesentlichen Eigenschaften des Punktes ξ (ξ nicht auf der Doppelkurve der Fläche und nicht auf der Basiskurve des Büschels) erhalten bleiben. So wird dieser Fall auf den Fall des Transzendenzgrades 1 zurückgeführt. Wenn nun der Punkt $(\xi_1, \ldots, \xi_n)$ den Transzendenzgrad 1 hat, so ist er allgemeiner Punkt einer irreduziblen Kurve C, und zwar Schnittpunkt dieser Kurve mit der Hyperfläche (2). Nach Satz 3 der zitierten Arbeit ZAG V ist dieser Schnittpunkt notwendig einfach. Legt man nun durch die Kurve C irgendeine weitere Hyperfläche H, welche die Fläche M nicht berührt, z. B. die projizierende Hyperfläche der Kurve aus einem allgemeinen P_{n-3}, so ist C eine einfach zählende Schnittkurve von H mit M und ξ ein einfacher Schnittpunkt von C mit der Hyperfläche (2), also ξ ein einfacher Schnittpunkt von M, H und (2). Wäre nun ξ ein Doppelpunkt der Schnittkurve von M und (2), so würde ξ auch als Schnittpunkt von M, H und (2) mindestens doppelt zählen. Das widerspricht aber dem eben Bewiesenen.

4. *Ein Dimensionssatz bei irreduziblen Korrespondenzen.*

In ZAG VI (Math. Annalen **110**, S. 142) wurde der folgende Satz ausgesprochen:

Wenn in einer irreduziblen Korrespondenz $\mathfrak{K}$ zwischen $\mathfrak{M}$ und $\mathfrak{N}$ dem allgemeinen Punkt ξ von $\mathfrak{M}$ eine b-dimensionale Mannigfaltigkeit von Punkten η entspricht, so entspricht einem speziellen Punkt ξ' eine Mannigfaltigkeit von Punkten η', welche keine Bestandteile von niedrigerer Dimension als b enthält.

Zum Beweise wurde angenommen, es gäbe wohl solche Bestandteile, und η^0 sei ein allgemeiner Punkt eines solchen. Der weitere Beweis ging nun von der Annahme aus, daß die Punkte von $\mathfrak{M}$, die dem Punkte η^0 entsprechen, nicht die ganze Mannigfaltigkeit $\mathfrak{M}$ ausfüllen können.

Herr D. Pedoe (Princeton) hat mich nun aufmerksam gemacht, daß diese Annahme keineswegs immer zutreffen muß. Der Beweis enthält also eine Lücke, die dadurch geschlossen werden soll, daß der Fall, wo einem Punkt η^0 alle Punkte von $\mathfrak{M}$ entsprechen, auf den anderen Fall zurückgeführt wird.

Wir können $\mathfrak{M}$ und $\mathfrak{N}$ als Mannigfaltigkeiten in affinen Räumen E_m und E_n annehmen. Wir erweitern nun E_m und E_n zu E_{m+1} und E_{n+1}, indem wir zu den Koordinaten $\xi_1', \ldots, \xi_m'$ und $\eta_1', \ldots, \eta_n'$ je eine neue Koordinate ξ_{m+1}' bzw. η_{n+1}' hinzunehmen. In E_{m+1} und E_{n+1} mögen die Mannigfaltigkeiten $\mathfrak{M}^*$ und $\mathfrak{N}^*$ durch dieselben Gleichungen definiert werden, welche vorhin $\mathfrak{M}$ und $\mathfrak{N}$ definierten. Die Dimensionen von $\mathfrak{M}$ und $\mathfrak{N}$ erhöhen sich dabei offenbar um Eins: ist z. B. $\mathfrak{M}$ eine ebene Kurve, so ist $\mathfrak{M}^*$ ein Zylinder. Zu den Gleichungen der gegebenen Korrespondenz $\mathfrak{K}$ nehmen wir nun eine neue Gleichung

$$(3) \qquad \xi_{m+1}' = \eta_{n+1}'$$

hinzu, wodurch wir eine neue irreduzible Korrespondenz $\mathfrak{K}^*$ zwischen $\mathfrak{M}^*$ und $\mathfrak{N}^*$ erhalten. Wenn einem Punkt $(\xi_1', \ldots, \xi_m')$ von $\mathfrak{M}$ in $\mathfrak{K}$ eine

Mannigfaltigkeit $\mathfrak{N}_{\xi'}$ mit Gleichungen $f_\nu(\eta'_1, \ldots, \eta'_n) = 0$ entspricht, so entspricht dem Punkt $(\xi'_1, \ldots, \xi'_m, 0)$ von $\mathfrak{M}^*$ in $\mathfrak{R}^*$ eine Mannigfaltigkeit $\mathfrak{N}^*_{\xi'}$ mit den Gleichungen $f_\nu(\eta'_1, \ldots, \eta'_n) = 0$ und $\eta'_{n+1} = 0$. Offenbar ist $\mathfrak{N}^*_{\xi'}$ genau dieselbe Mannigfaltigkeit wie $\mathfrak{N}_{\xi'}$, nur daß der Raum E_n, in welchem $\mathfrak{N}_{\xi'}$ lag, jetzt in einem E_{n+1} eingebettet erscheint. Die Dimensionen der irreduziblen Bestandteile von $\mathfrak{N}_{\xi'}$ ändern sich also beim Übergang von $\mathfrak{R}$ zu $\mathfrak{R}^*$ nicht. Daher genügt es, den Satz für $\mathfrak{R}^*$ zu beweisen. Hier entspricht aber einem Punkt η^0 wegen der Bedingung (3) niemals die ganze Mannigfaltigkeit $\mathfrak{M}^*$. Mithin ist der in ZAG VI gegebene Beweis für $\mathfrak{R}^*$ gültig.

(Eingegangen am **13. 4. 1936**.)

15.

Zur algebraischen Geometrie VIII
Der Grad der Graßmannschen Mannigfaltigkeit
der linearen Räume S_m in S_n

Mathematische Annalen 113, 2 (1936) 199–205

Das Hauptziel der vorliegenden Note ist, eine exakte Anwendung des Prinzips der Erhaltung der Anzahl zu geben, welche für viele ähnliche Anwendungen als Vorbild genommen werden kann.

§ 1.
Problemstellung und Methode.

Die m-dimensionalen linearen Teilräume $[m]$ des projektiven n-dimensionalen Raumes $[n]$ lassen sich vermöge ihrer Plückerschen Koordinaten auf Punkte eines Raumes $[N]$ von der Dimension

$$N = \binom{n+1}{m+1} - 1$$

abbilden. Die Bildpunkte bilden eine rationale algebraische Mannigfaltigkeit $\mathfrak{G}$ von der Dimension

$$M = (m+1)(n-m)$$

in $[N]$, die *Graßmannsche Mannigfaltigkeit* der $[m]$ in $[n]$. Wir wollen den Grad g dieser Mannigfaltigkeit bestimmen.

Ein obere Schranke für g hat Th. Vahlen[1]) angegeben.

Der Grad g ist die Anzahl der Schnittpunkte von $\mathfrak{G}$ mit M allgemeinen Hyperebenen ω des Bildraumes $[N]$, oder anders ausgedrückt: *g ist die Anzahl der $[m]$ in $[n]$, deren Plückersche Koordinaten einem System von M allgemeinen linearen Gleichungen genügen.*

Man kann, wie wir in § 3 zeigen werden, die linearen Gleichungen ohne Verringerung der Zahl g so spezialisieren, daß sie die Bedingungen bedeuten, daß der gesuchte Raum $[m]$ mit M allgemein gegebenen Räumen $[n-m-1]$ je einen Punkt gemeinsam hat.

Gleichzeitig mit dem Grad g bestimmen wir auch die Gradzahlen einer Reihe von Teilmannigfaltigkeiten von $\mathfrak{G}$, die von Schubert[2]) angegeben worden sind und die nach Ehresmann[3]) gleichzeitig eine Homo-

---

[1]) K. Th. Vahlen, Journ. f. reine u. angew. Math. **112** (1893), S. 306.

[2]) H. Schubert, Acta Math. **8** (1886) S. 97.

[3]) C. Ehresmann, Ann. of Math. (2) **35** (1934), S. 396—443.

logiebasis für $\mathfrak{G}$ abgeben. Schubert bezeichnet nämlich mit $[i_0, i_1, \ldots, i_m]$ die Gesamtheit der $[m]$, die mit einem gegebenen $[i_0]$ mindestens einen Punkt gemeinsam haben, weiter mit einem gegebenen, $[i_0]$ enthaltenden $[i_1]$ mindestens eine Gerade $[1]$ gemeinsam haben, usw., und die schließlich ganz in einem gegebenen, $[i_{m-1}]$ enthaltenden $[i_m]$ liegen. Dabei ist

$$(1) \qquad\qquad 0 \leqq i_0 < i_1 < \ldots i_m \leqq n$$

vorausgesetzt. Die Bildmannigfaltigkeit von $[i_0, i_1, \ldots, i_m]$ in $\mathfrak{G}$ wird ebenfalls mit $[i_0, i_1, \ldots, i_m]$ bezeichnet. Die Dimension von $[i_0, i_1, \ldots, i_m]$ ist

$$(2) \qquad\qquad \varrho = i_0 + (i_1 - 1) + \ldots + (i_m - m) = \sum_{\lambda} (i_\lambda - \lambda).$$

Der größte Wert von ϱ ist

$$\varrho = M \text{ für } [i_0, i_1, \ldots, i_m] = [n - m, n - m + 1, \ldots, n] = \mathfrak{G}.$$

Der kleinste Wert ist $\varrho = 0$ für die aus einem einzigen Element $[m]$ bestehende Mannigfaltigkeit $[0, 1, \ldots, m]$.

Die Gradbestimmung von $[i_0, i_1, \ldots, i_m]$ beruht auf der im wesentlichen von Schubert[2]) schon angegebenen Formel

$$(3) \qquad \omega \cdot [i_0, i_1, \ldots, i_m]$$
$$= [i_0 - 1, i_1, \ldots, i_m] + [i_0, i_1 - 1, \ldots, i_m] + \ldots + [i_0, i_1, \ldots, i_m - 1].$$

Dabei bedeutet die linke Seite den Schnitt von $[i_0, i_1, \ldots, i_m]$ mit einer allgemeinen Hyperebene ω, während das Gleichheitszeichen bedeutet, daß die Schnittpunktszahl der Mannigfaltigkeit $\mathfrak{M}_{\varrho-1}$ linker Hand mit einer beliebigen Mannigfaltigkeit $\mathfrak{M}'$ von der Dimension $\mathfrak{M} - (\varrho - 1)$ in $\mathfrak{G}$ gleich der Summe der Schnittpunktszahlen der Mannigfaltigkeiten rechter Hand mit $\mathfrak{M}'$ ist. Dabei sind auf der rechten Seite diejenigen Summanden, welche nicht der Bedingung (1) genügen, wegzulassen.

Wie der Beweis der Formel (3) zeigen wird, kann man die symbolische Gleichung (3) sogar zu einer *Äquivalenz* im Sinne der Theorie der linearen Scharen verschärfen, welche besagt, daß die Schnittmannigfaltigkeit $\mathfrak{M}_{\varrho-1}$ linker Hand, wenn ω variiert, auf $[i_0, i_1, \ldots, i_m]$ eine lineare Schar durchläuft, zu welcher auch die zerfallende Mannigfaltigkeit rechter Hand gehört.

Aus dieser Äquivalenz oder auch schon aus der symbolischen Gleichung (3) folgt, daß der Grad der Mannigfaltigkeit $\mathfrak{M}_{\varrho-1}$ links gleich dem Grad der zerfallenden Mannigfaltigkeit rechts, also gleich der Summe der Grade ihrer Bestandteile ist[4]). Bezeichnet also $g(i_0, i_1, \ldots, i_m)$ den Grad der Bildmannigfaltigkeit $[i_0, i_1, \ldots, i_m]$, so ergibt sich

$$(4) \qquad g(i_0, i_1, \ldots, i_m)$$
$$= g(i_0 - 1, i_1, \ldots, i_m) + g(i_0, i_1 - 1 \ldots, i_m) + \ldots + g(i_0, i_1, \ldots, i_m - 1).$$

---

[4]) In die Schubertsche Sprache übersetzt, heißt diese Überlegung so: Man multipliziere beide Seiten der symbolischen Gleichung (3) mit $\omega^{\varrho-1}$ und erhält (4).

Aus dieser Rekursionsformel und der Anfangsbedingung

(5) $$g(0, 1, \ldots, m) = 1$$

folgen die Werte der Gradzahlen

(6) $$g(i_0, i_1, \ldots, i_m) = \frac{\varrho!}{i_0!\, i_1! \ldots i_m} \prod_{\mu > \nu} (i_\mu - i_\nu); \quad \varrho = \sum_\lambda (i_\lambda - \lambda)$$

in Übereinstimmung mit Schubert. Insbesondere ist der Grad von $\mathfrak{G}$ gleich

(7) $$\begin{aligned} g &= g(n-m,\, n-m+1,\, \ldots,\, n) \\ &= \frac{M!\, 1!\, 2! \ldots m!}{n!\,(n-1)! \ldots (n-m)!}; \qquad M = (m+1)(n-m). \end{aligned}$$

Es handelt sich also — abgesehen von der rein arithmetischen Frage der Lösung der Rekursionsformel (4) durch (6) — nur um den Beweis der Schubertschen Formel (3).

Schubert beweist (3) mittels des Prinzips der Erhaltung der Anzahl, indem er die allgemeine Hyperebene ω so spezialisiert, daß die Schnittmannigfaltigkeit linker Hand in die zerfallende Mannigfaltigkeit rechter Hand übergeht. Diese Überlegung muß nur ergänzt werden durch eine exakte Multiplizitätsbetrachtung, welche zeigt, daß die Bestandteile dieser zerfallenden Schnittmannigfaltigkeit rechts alle mit der Multiplizität Eins zu zählen sind. Diese Multiplizitätsbetrachtung — das einzige Neue der vorliegenden Untersuchung — wird in § 2 gegeben.

§ 2.

Beweis der Schubertschen Formel (3).

Die Mannigfaltigkeit $\mathfrak{M}_\varrho = [i_0, i_1, \ldots, i_m]$ wird durch $m+1$ ineinander geschachtelte Räume $[i_0], [i_1], \ldots, [i_m]$ gegeben, deren Dimensionen die Bedingung (1) erfüllen. Die Dimension ϱ von $\mathfrak{M}_\varrho$ wird durch (2) gegeben. Die Hyperebenen ω schneiden auf $\mathfrak{M}_\varrho$ eine lineare Schar von $(\varrho - 1)$-dimensionalen Teilmannigfaltigkeiten $\mathfrak{M}_{\varrho - 1}$ aus. Die Gleichung der Hyperebene in den (Plückerschen) Koordinaten $p_{(k)}$ möge lauten

(8) $$\sum_{(k)} \omega^{k_0\, k_1 \ldots k_m}\, p_{k_0\, k_1 \ldots k_m} = 0.$$

Wir spezialisieren nun die Koeffizienten $\omega^{(k)}$ so, daß sie selbst die dualen Plückerschen Koordinaten eines Raumes $[n-m-1]$ sind; dann bedeutet (8) die Bedingung, daß die Räume $[n-m-1]$ und $[m]$ einen Punkt gemeinsam haben. Wir spezialisieren den Raum $[n-m-1]$ weiter so, daß er mit dem gegebenen i_0 einen $[i_0 - 1]$, mit $[i_1]$ einen $[i_1 - 2]$, allgemein mit $[i_\lambda]$ einen $[i_\lambda - (\lambda + 1)]$ gemeinsam hat. Diese

207

Räume $[i_\lambda - (\lambda + 1)]$ sind als gegebene Räume zu betrachten; jeder vorangehende Raum $[i_{\lambda-1} - \lambda]$ liegt in dem folgenden $[i_\lambda - (\lambda + 1)]$ und ist nur dann mit ihm identisch, wenn $i_\lambda - 1 = i_{\lambda-1}$ ist.

Der Raum $[m]$ soll nun die Bedingung (8) erfüllen, d. h. er soll mit dem gegebenen $[n - m - 1]$ einen Punkt P gemeinsam haben. Dieser Punkt P liegt in $[i_m]$, da der ganze Raum $[m]$ in $[i_m]$ liegt. Wir unterscheiden nun $m + 1$ Fälle:

Fall 1. P liegt in $[i_0]$.

Fall 2. P liegt in $[i_1]$, aber nicht in $[i_0]$.

. . . .

Fall $m + 1$. P liegt in $[i_m]$, aber nicht in $[i_{m-1}]$.

Im Falle 1 liegt P in $[i_0]$ und in $[n - m - 1]$, also in deren Durchschnitt $[i_0 - 1]$. Der Raum $[m]$ hat also mit $[i_0 - 1]$ den Punkt P gemeinsam, genügt also einer Bedingung $[i_0 - 1, i_1, \ldots, i_m]$.

Im Falle $\lambda + 1$ liegt P in $[i_\lambda]$ und $[n - m - 1]$, also in deren Durchschnitt $[i_\lambda - (\lambda + 1)]$, aber nicht in $[i_{\lambda-1}]$. Der Raum $[m]$ hat, wenn er der Bedingung $[i_0, i_1, \ldots, i_m]$ genügt, mit $[i_{\lambda-1}]$ einen $[\lambda - 1]$ gemeinsam, der nicht durch P geht, also mit P zusammen einen Raum $[\lambda]$ bestimmt, der im Verbindungsraum $[i_\lambda - 1]$ von $[i_{\lambda-1}]$ mit $[i_\lambda - (\lambda + 1)]$ liegt. Die Dimension $i_\lambda - 1$ dieses Verbindungsraumes ergibt sich aus der bekannten Formel

$$(9) \qquad\qquad v + d = p + q,$$

in welcher p und q die Dimensionen der zu verbindenden Räume, v die Dimension des Verbindungsraumes und d die des Durchschnittes bedeutet. In unserem Falle wird die Formel (9) zu

$$(i_\lambda - 1) + (i_{\lambda-1} - \lambda) = i_{\lambda-1} + (i_\lambda - \lambda - 1).$$

Der Raum $[m]$ hat also mit $[i_\lambda - 1]$ einen $[\lambda]$ gemeinsam, genügt also einer Bedingung $[i_0, i_1, \ldots, i_\lambda - 1, \ldots, i_m]$.

Umgekehrt, wenn $[m]$ der Bedingung $[i_0, i_1, \ldots, i_\lambda - 1, \ldots, i_m]$ genügt, so hat $[m]$ mit $[i_\lambda - 1]$ einen $[\lambda]$ gemeinsam, der mit $[i_\lambda - (\lambda + 1)]$ zusammen in $[i_\lambda - 1]$ liegt und der daher mit $[i_\lambda - (\lambda + 1)]$ mindestens einen Punkt P gemeinsam hat. Der Raum $[m]$ hat dann also mit $[n - m]$ einen Punkt P gemeinsam, d. h. er genügt der Bedingung ω und außerdem offenbar der Bedingung $[i_0, i_1, \ldots, i_\lambda, \ldots, i_m]$.

Aus dieser Betrachtung folgt, daß die vollständige Schnittmannigfaltigkeit $\omega \cdot [i_0, i_1, \ldots, i_m]$ für das speziell gewählte ω in die (irreduziblen) Bestandteile $[i_0, i_1, \ldots, i_\lambda - 1, \ldots, i_m]$ zerfällt. Erfüllt eines von diesen Symbolen $[i_0, i_1, \ldots, i_\lambda - 1, \ldots, i_m]$ nicht die Bedingungen (1), ist also einmal $i_\lambda - 1 = i_{\lambda-1}$, so wird der Fall $\lambda + 1$ ganz von selbst unmöglich, da in diesem Falle der Raum $[i_\lambda - (\lambda + 1)]$, in dem der Punkt P liegen soll, mit dem Raum $[i_{\lambda-1} - \lambda]$, in dem der Punkt P nicht liegen soll, zusammenfällt. Die Symbole $[i_0, i_1, \ldots, i_\lambda - 1, \ldots, i_m]$ auf der rechten Seite in (3), welche (1) nicht erfüllen, sind also einfach wegzulassen.

Soweit sind wir im wesentlichen Schubert gefolgt. Wir müssen nun noch zeigen, daß alle irreduziblen Bestandteile $[i_0, i_1, \ldots, i_\lambda - 1, \ldots, i_m]$ in der Schnittmannigfaltigkeit (3) einfach zählen. Dazu genügt es nach ZAG. V[5]), zu beweisen, daß in einem gemeinsamen Punkt von $\mathfrak{M}_\varrho$ und ω der Tangentialraum von $\mathfrak{M}_\varrho$ nicht in der Hyperebene ω liegt.

Ein Raum $[m]$ des Systems $\mathfrak{M}_\varrho = [i_0, i_1, \ldots, i_m]$ wird bestimmt durch einen Punkt Q_0 von $[i_0]$, einen davon linear-unabhängigen Punkt Q_1 von $[i_1]$, usw. bis Q_m von $[i_m]$. Soll nun der Raum $[m]$ etwa dem System $[i_0, \ldots, i_\lambda - 1, \ldots, i_m]$ angehören, so muß man den Punkt Q_λ in $[i_\lambda - 1]$ statt in $[i_\lambda]$ wählen. Diesen Punkt $Q_\lambda = z$ lasse man nun eine lineare Punktreihe

$$z = \gamma_1 x + \gamma_2 y$$

in $[i_\lambda]$ durchlaufen, von welcher nur ein Punkt z' in $[i_\lambda - 1]$ liegt. Die übrigen Punkte $Q_0, \ldots, Q_{\lambda-1}, Q_{\lambda+1}, \ldots, Q_m$ werden beliebig in den Räumen $[i_0], \ldots, [i_{\lambda-1}], [i_{\lambda+1}], \ldots, [i_m]$ gewählt, aber so, daß Q_ν nicht dem Teilraum $[i_\nu - 1]$ angehört ($\nu = 0, \ldots, \lambda - 1, \lambda + 1, \ldots, m$). Durchläuft nun $Q_\nu = z$ die lineare Punktreihe, während die übrigen Punkte Q_ν fest bleiben, so hängen die Plückerschen Koordinaten des Raumes $[m]$,

$$p_{k_0\,k_1\ldots k_m} = \Sigma \pm (Q_0)_{k_0}\,(Q_1)_{k_1} \cdots (Q_\lambda)_{k_\lambda} \cdots (Q_m)_{k_m}$$

linear von den Parametern γ_1, γ_2 der Punktreihe ab. Der Bildpunkt von $[m]$ im Raum $[N]$ durchläuft also ebenfalls eine lineare Punktreihe, welche ganz auf der Bildmannigfaltigkeit $\mathfrak{M}_\varrho$ liegt und welche mit der Hyperebene ω nur einen einzigen Punkt gemeinsam hat, weil eben nur der Punkt z' in $[i_\lambda - 1]$ liegt. Der Träger dieser Punktreihe ist eine Tangente von $\mathfrak{M}_\varrho$, weil er ganz auf $\mathfrak{M}_\varrho$ liegt. Es gibt also in einem Schnittpunkt von $\mathfrak{M}_\varrho$ und ω eine Tangente von $\mathfrak{M}_\varrho$, welche nicht in der Hyperebene ω liegt, was wir beweisen wollten.

§ 3.

Der Grad von $\mathfrak{G}$ und seine geometrische Deutung.

Durch die oben bewiesenen Rekursionsformel (4) mit der selbstverständlichen Anfangsbedingung (5) sind die Gradzahlen $g\,(i_0, i_1, \ldots, i_m)$ offenbar festgelegt. Daß nun die in § 1 angegebenen Werte (6) tatsächlich die Rekursionsformeln (4), (5) erfüllen und somit die gesuchten Gradzahlen darstellen, ergibt sich aus der unter [2]) zitierten Abhandlung von

---

[5]) B. L. van der Waerden, Math. Annalen 110, S. 128.

Schubert[6]) oder auch aus der Abhandlung von Frobenius[7]) über die Charaktere der symmetrischen Gruppe $\mathfrak{S}_\varrho$. Die durch (6) gegebenen Zahlen $g\,(i_0, i_1, \ldots, i_m)$ stimmen nämlich merkwürdigerweise mit den Graden der irreduziblen Darstellungen der Gruppe $\mathfrak{S}_\varrho$ überein, welche ebenfalls die Rekursionsformeln (4), (5) erfüllen.

Die folgende Tabelle gibt die Werte des Gesamtgrades g der Mannigfaltigkeit $\mathfrak{G}$ für $n \leqq 6$ an. Für jeden Wert von n wird der Reihe nach $m = 0, 1, 2, \ldots, n - 1$ gesetzt.

$$
\begin{array}{llccccccc}
n = 1: & & & & & 1 & & & \\
n = 2: & & & & 1 & & 1 & & \\
n = 3: & & & 1 & & 2 & & 1 & \\
n = 4: & & 1 & & 5 & & 5 & & 1 \\
n = 5: & 1 & & 14 & & 42 & & 14 & & 1 \\
n = 6: & 1 & 42 & & 462 & & 462 & & 42 & & 1 \\
\end{array}
$$

Die geometrische Bedeutung dieser Anzahlen ergibt sich, wenn man die Hyperebenen ω, die zur Definition des Grades von $\mathfrak{M}_\varrho$ verwendet wurden, nicht, wie bisher, als allgemeine Hyperebenen annimmt, sondern sie (wie es auch Schubert getan hat) von vornherein so spezialisiert, daß die Koeffizienten $\omega^{(k)}$ in der Gleichung (8) einer solchen Hyperebene die Plückerschen Koordinaten eines allgemeinen $[n - m - 1]$ sind. Die Gleichung (8) bedeutet dann, wie schon bemerkt, die Bedingung, daß der Raum $[m]$ mit einem allgemein gegebenen $[n - m - 1]$ einen Punkt gemeinsam hat. Wir dürfen dann also nicht mehr, wie bisher, von dem Grad von $\mathfrak{M}_\varrho$ reden (denn Grad bedeutet Anzahl der Schnittpunkte mit einem allgemeinen System von Hyperebenen), sondern wir müssen unter $g\,(i_0, i_1, \ldots, i_m)$ die Anzahl der Punkte von $\mathfrak{M}_\varrho$ verstehen, welche einem System von ϱ linearen Gleichungen (8) genügen, deren Koeffizienten die Koordinaten von ϱ allgemein gewählten Räumen $[n - m - 1]$ sind, oder, was dasselbe ist, die Anzahl der Räume $[m]$ des Systems $[i_0, i_1, \ldots, i_m]$, welche allgemein gegebene Räume $[n - m - 1]$ schneiden. Für diese neu definierten Zahlen $g\,(i_0, i, \ldots, i_m)$ gelten nun alle Überlegungen von § 2 nahezu ungeändert, da ja die spezialisierten Hyperebenen ω, die in dem Beweis gebraucht wurden, schon die hier vorausgesetzten speziellen Eigenschaften haben. Eines besonderen Beweises bedarf es lediglich, daß der „Grad" (im neuen Sinn) einer zerfallenden Mannigfaltigkeit, wie sie rechts in (3) auftritt, gleich der Summe der „Grade" der Bestandteile ist. Dazu muß man nachweisen, daß die Mannigfaltigkeiten $\mathfrak{M}_{\varrho-2}$ der Dimension

---

[6]) Vgl. dazu auch H. Schubert, Math. Annalen 38 (1891), S. 589.

[7]) G. Frobenius, Sitz.-Ber. Akad. Berlin 1900, S. 516.

$\varrho - 2$, welche je zwei Bestandteilen der zerfallenden Mannigfaltigkeit $\mathfrak{M}_{\varrho-1}$ gemeinsam sind, *keine* Schnittpunkte mit $\varrho - 1$ Hyperebenen ω der hier betrachteten Art aufweisen. Dieses folgt nun wiederum daraus, daß die Dimension einer Mannigfaltigkeit $\mathfrak{M}_{\varrho-2}$ sich bei Schnitt mit einer genügend allgemein gewählten Hyperebene der betrachteten Art immer um mindestens Eins erniedrigt, so daß also beim Schnitt mit $\varrho - 1$ Hyperebenen schließlich nichts mehr übrig bleibt.

Die Zahlen $g\,(i_0, \ldots, i_m)$ nach der neuen Definition genügen also genau denselben Rekursionsformeln wie die alten; sie haben daher auch dieselben Werte (6). Insbesondere wird die Anzahl der Räume $[m]$, welche $M = (m + 1)\,(n - m)$ allgemein gegebene Räume $[n - m - 1]$ schneiden, durch (7) gegeben. An Hand der obigen Tabelle findet man z. B.:

Es gibt zwei Geraden im dreidimensionalen Raum, welche vier allgemein gegebene Geraden schneiden (übrigens allgemein bekannt). Es gibt fünf Geraden im vierdimensionalen Raum S_4, welche sechs allgemein gegebene Ebenen schneiden[8]). Es gibt (dual dazu) fünf Ebenen im S_4, welche sechs allgemein gegebene Geraden schneiden. Es gibt 14 Geraden im S_5, welche acht allgemein gegebene Teilräume S_3 schneiden, usw.

---

[8]) Vgl. C. Stephanos, C. R. Acad. Paris **93** (1881), S. 578, 633; C. Segre, Atti di Torino **22** (1887); R. Weitzenböck, Sitz.-Ber. Akad. Wien **121** (1912), S. 2553.

(Eingegangen am 26. 3. 1936.)

16.

Zur algebraischen Geometrie IX
Über zugeordnete Formen und algebraische Systeme von algebraischen Mannigfaltigkeiten
(mit Wei-Liang Chow)

Mathematische Annalen 113, 5 (1937) 692–704

Es ist prinzipiell wichtig, geometrische Gebilde durch Koordinaten darstellen zu können. Ist das nämlich für eine bestimmte Art von Gebilden G einmal geschehen, so hat es einen Sinn, von einer algebraischen Mannigfaltigkeit oder einem algebraischen System von Gebilden G zu sprechen und die gesamte Theorie der algebraischen Mannigfaltigkeiten (Zerlegung in irreduzible, Dimensionsbegriff, Begriff der allgemeinen Elemente einer irreduziblen Mannigfaltigkeit) darauf anzuwenden. Erwünscht ist dabei, daß die Gesamtheit aller Gebilde G der betrachteten Art (eventuell nach Hinzufügung von geeigneten Grenzgebilden) eine algebraische Mannigfaltigkeit darstellt, also durch algebraische Gleichungen in den Koordinaten charakterisiert werden kann.

Die *Punkte* des projektiven Raumes S_n werden durch $n + 1$ homogene Koordinaten, die *Teilräume* S_r durch ihre Plückerschen Koordinaten, die *Hyperflächen* vom Grade g durch die Koeffizienten ihrer Gleichungen gegeben. In allen diesen Fällen ist für die Gesamtheit aller dargestellten Gebilde die eben gestellte Bedingung erfüllt.

Es soll nun ein Mittel angegeben werden, die *r-dimensionalen Mannigfaltigkeiten M* eines festen Grades g in S_n durch Koordinaten darzustellen. Wir werden dabei so verfahren: Wir stellen die Bedingung dafür auf, daß $r + 1$ Hyperebenen $u^{(0)}, u^{(1)}, \ldots, u^{(r)}$ einen Punkt mit M gemeinsam haben. Diese Bedingung wird durch eine einzige Gleichung $F(u) = 0$ vom Grade g in jeder der Variablenreihen $u^{(0)}, \ldots, u^{(r)}$ gegeben, welche in ebensoviele irreduzible Faktoren zerfällt, als M irreduzible Bestandteile besitzt. $F(u)$ heißt die *zugeordnete Form* der Mannigfaltigkeit. Die Grenzfälle, in denen einige von diesen irreduziblen Bestandteilen von M und dementsprechend auch einige Faktoren der Form $F(u)$ mehrfach gezählt werden, sind mit eingeschlossen. Die Koeffizienten der zugeordneten Form $F(u)$ werden nunmehr als Koordinaten von M genommen.

Man sieht leicht ein, daß die Mannigfaltigkeit M durch ihre zugeordnete Form, also durch ihre Koordinaten eindeutig bestimmt ist. Nicht so leicht ist aber der Nachweis, daß *die Gesamtheit aller M vom Grade g und von der Dimension r im Koordinatenraum eine algebraische Mannig-*

212

faltigkeit darstellt. Dieser Nachweis, der dem ersten Verfasser gelungen ist, wird in § 1 dargestellt. In § 2 wird der Begriff eines algebraischen Systems von Mannigfaltigkeiten M näher erörtert und zu der Theorie der algebraischen Korrespondenzen in Beziehung gesetzt.

§ 1.

Die zugeordnete Form einer Mannigfaltigkeit M.

Eine r-dimensionale irreduzible algebraische Mannigfaltigkeit M im n-dimensionalen projektiven Raum S_n sei durch die Formen f_μ aus $\mathsf{K}[x] = \mathsf{K}[x_0, x_1, \ldots, x_n]$ definiert. Ein allgemeiner linearer Unterraum S_{n-r}, der durch die r allgemeinen Linearformen

$$l_i = \sum_{j=0}^{n} u_j^{(i)} x_j \qquad (i = 1, \ldots, r)$$

definiert sein möge, schneidet M in einer nulldimensionalen, in bezug auf $\mathsf{K}(u^{(1)}, \ldots, u^{(r)})$ irreduziblen Mannigfaltigkeit[1]), die aus endlichvielen in bezug auf $\mathsf{K}(u^{(1)}, \ldots, u^{(r)})$ konjugierten, in bezug auf K allgemeinen Punkten von M besteht. Diese Punkte seien mit

$$p^{(i)} = (p_0^{(i)}, p_1^{(i)}, \ldots, p_n^{(i)}) \qquad (i = 1, \ldots, q)$$

bezeichnet. Bildet man die g Linearformen

$$L_i(u^{(0)}) = \sum_{j=1}^{n} p_j^{(i)} u_j^{(0)},$$

wo die $u_j^{(0)}$ Unbestimmte sind, so ist ihr Produkt

$$G(u^{(0)}) = \prod_{i=1}^{g} L_i(u^{(0)})$$

eine Form in $u^{(0)}$, die mit allen ihren Konjugierten in bezug auf $\mathsf{K}(u^{(1)}, \ldots, u^{(r)})$ übereinstimmt. Daraus folgt, daß eine gewisse Potenz von $G(u^{(0)})$ eine Form mit Koeffizienten aus $\mathsf{K}(u^{(1)}, \ldots, u^{(r)})$ ist, und wenn man diese noch mit einem geeigneten Polynom aus $\mathsf{K}[u^{(1)}, \ldots, u^{(r)}]$ multipliziert, so bekommt man eine Form in $u^{(0)}$ mit Koeffizienten aus $\mathsf{K}[u^{(1)}, \ldots, u^{(r)}]$, die nur durch Potenzprodukte der $L_i(u^{(0)})$ teilbar ist. Da je zwei solche Formen einen gemeinsamen Faktor haben, der wieder eine solche Form ist, so gibt es eine bis auf einen konstanten Faktor aus K eindeutig bestimmte Form $F(u^{(0)})$, die diese Eigenschaft besitzt und als Polynom aus $\mathsf{K}[u^{(0)}, u^{(1)}, \ldots, u^{(r)}]$ irreduzibel ist. Diese Form wird die *zugeordnete Form* von M genannt, und ihr Grad heißt der *Grad* von M[2]).

---

[1]) Für den Beweis siehe B. L. v. d. Waerden, ZAG. V, Math. Annalen 110, S. 140.

[2]) Die Anzahl der verschiedenen Schnittpunkte $p^{(i)}$, die bisher immer als Grad von M bezeichnet wurde, sollte man besser den reduzierten Grad nennen. Im Fall eines vollkommenen Grundkörpers K ist $F(u^{(0)}) = G(u^{(0)})$ und der Grad gleich dem reduzierten Grad.

213

Es ist klar, daß zwei verschiedene irreduzible Mannigfaltigkeiten nicht dieselbe zugeordnete Form besitzen können. Denn man kann durch Faktorzerlegung aus der zugeordneten Form einen allgemeinen Punkt der irreduziblen Mannigfaltigkeit erhalten, und zwei irreduzible Mannigfaltigkeiten müssen gleich sein, wenn sie einen gemeinsamen allgemeinen Punkt haben.

Die zugeordnete Form $F(u^{(0)}) = F(u^{(0)}, u^{(1)}, \ldots, u^{(r)})$ einer irreduziblen Mannigfaltigkeit, die definitionsgemäß homogen in $u^{(0)}$ ist, ist aber auch homogen in den übrigen $u^{(i)}$, und zwar von demselben Grad wie in $u^{(0)}$. Das folgt daraus, daß die u-Resultante $D(u)$ von $f_\mu, l_1, l_2, \ldots, l_r$ mit der hinzugenommenen Linearform $l_0 = \sum_{j=0}^{n} u_j^{(0)} x_j$ auch eine Form in $u^{(0)}$ mit Koeffizienten aus $\mathsf{K}[u^{(1)}, \ldots u^{(r)}]$ ist, die nur durch Potenzprodukte der $L_i(u^{(0)})$ teilbar ist. Diese u-Resultante $D(u)$ ist nun aber der größte gemeinsame Teiler des Resultantensystems von $f_\mu, l_0, l_1, \ldots, l_r$, und muß daher von derselben Beschaffenheit in bezug auf alle $u^{(i)}$ sein; genauer: D wird bei Vertauschung von irgend zwei $u^{(i)}$ höchstens um einen konstanten Faktor aus K geändert. Wegen der Irreduzibilität von F muß D bis auf einen Faktor aus $\mathsf{K}[u^{(1)}, \ldots, u^{(r)}]$ eine Potenz von F sein. Dieser Faktor kann aber nur eine Konstante aus K sein. Denn wenn $D(u)$ einen von $u^{(0)}$ freien, aber etwa von $u^{(1)}$ abhängigen Faktor besäße, so würde $D(u)$ wegen der Symmetrie auch einen von $u^{(1)}$ freien, aber von $u^{(0)}$ abhängigen Faktor besitzen. Dieser Faktor könnte wieder nur eine Potenz von F sein, also wäre F von $u^{(1)}$ und ebenso von $u^{(2)}, \ldots, u^{(r)}$ unabhängig, was offenbar absurd ist. Also ist $D(u)$ eine Potenz von F. Folglich ist F auch von derselben Beschaffenheit in bezug auf alle $u^{(i)}$, insbesondere homogen von demselben Grad in bezug auf alle $u^{(i)}$.

Eine beliebige rein r-dimensionale algebraische Mannigfaltigkeit M ist eine Summe von endlichvielen r-dimensionalen irreduziblen Mannigfaltigkeiten $M_1, M_2, \ldots, M_h$, wobei jede Mannigfaltigkeit M_i mit einer beliebigen Vielfachheit oder Multiplizität n_i versehen werden möge. Die zugeordnete Form von M_i sei $F_i(u^{(0)}, u^{(1)}, \ldots, u^{(r)})$ vom Grade g_i. Dann heißt

$$F(u^{(0)}, u^{(1)}, \ldots, u^{(r)}) = \prod_{i=1}^{h} F_i(u^{(0)}, u^{(1)}, \ldots, u^{(r)})^{n_i}$$

die zugeordnete Form von M und ihr Grad $g = \sum n_i g_i$ der Grad von M. Es ist klar daß auch hier, wie bei irreduziblen Mannigfaltigkeiten, eine Mannigfaltigkeit durch ihre zugeordnete Form eindeutig festgelegt ist. Die Form ist homogen vom selben Grad in bezug auf alle $u^{(i)}$. Im folgenden werden Formen, die homogen von demselben Grade g in bezug auf alle $u^{(i)}$ sind, kurz als Formen vom Grade g aus $\mathsf{K}[u^{(0)}, u^{(1)}, \ldots, u^{(r)}]$ bezeichnet.

Jetzt wird gefragt, welche die Bedingungen sind, denen eine beliebig vorgegebene Form vom Grade g aus $\mathsf{K}\,[u^{(0)},\, u^{(1)},\, \ldots,\, u^{(r)}]$ genügen muß, damit sie die zugeordnete Form einer r-dimensionalen Mannigfaltigkeit ist. Die Antwort wird durch den folgenden Satz gegeben.

Satz 1. *Damit eine Form* $F(u^{(0)},\, u^{(1)},\, \ldots,\, u^{(r)})$ *vom Grade* g *aus* $\mathsf{K}\,[u^{(0)},\, u^{(1)},\, \ldots,\, u^{(r)}]$ *die zugeordnete Form einer r-dimensionalen algebraischen Mannigfaltigkeit ist, ist notwendig und hinreichend, daß die folgenden Bedingungen erfüllt sind:*

1. $F(u^{(0)}) = F(u^{(0)},\, u^{(1)},\, \ldots,\, u^{(r)})$, *betrachtet als Form in* $u^{(0)}$, *zerfällt in einem Erweiterungskörper von* $\mathsf{K}\,(u^{(1)},\, \ldots,\, u^{(r)})$ *vollständig in Linearfaktoren:*

$$F(u^{(0)}) = \prod_{i=1}^{g} L_i(u^{(0)}) = \prod_{i=1}^{g} \left(\sum_{j=0}^{n} p_j^{(i)}\, u_j^{(0)} \right).$$

2. $L_i(u^{(k)}) = 0$ *für* $i = 1, 2, \ldots, g;\; k = 1, \ldots, r$.

3. Sind $v_j^{(1)},\, \ldots,\, v_j^{(r)}$ $(j = 0, 1, \ldots, n)$ *Elemente irgend eines Erweiterungskörpers von* $\mathsf{K}\,(p^{(i)})$ *und ist* $L_i(v^{(k)}) = 0$ *für* $k = 1, \ldots, r$ *(alles für ein festes* i), *so ist* $F(u^{(0)},\, v^{(1)},\, \ldots,\, v^{(r)})$ *als Form in* $u^{(0)}$ *durch* $L_i(u^{(0)})$ *teilbar.*

Beweis. Zunächst ist zu bemerken, daß man sich auf den Fall beschränken kann, wo die Form F irreduzibel ist. Der allgemeine Fall läßt sich nämlich auf diesen Fall zurückführen, indem man die Form F in ihre irreduziblen Bestandteile zerlegt. Denn gelten die obigen drei Bedingungen für eine Form F, so gelten sie für jeden ihrer Bestandteile, und umgekehrt. Für 1. und 2. ist dies ohne weiteres klar. Für 3. sieht man es so. F_1 sei ein irreduzibler Faktor von F, dessen Zerlegung in Linearfaktoren laute

$$F_1 = \prod_{i=1}^{g'} L_i(u^{(0)}).$$

Bedingung 3. sei für F erfüllt; wir wollen zeigen, daß 3. auch für F_1 erfüllt ist. Die $v^{(k)}$ seien also Lösungen des linearen Gleichungssystems

$$(1) \qquad\qquad L_i(v^{(k)}) = 0 \qquad\qquad (k = 1, \ldots, r)$$

für ein festes i ($1 \leq i \leq g'$). Aus der Theorie der linearen Gleichungen entnehmen wir, daß alle Lösungen dieses Gleichungssystems durch Parameterspezialisierung aus einer allgemeinen Lösung entstehen, in welche gewisse unbestimmte Parameter linear eingehen. Eine spezielle Lösung dieses Gleichungssystems ist nach 2. $v^{(k)} = u^{(k)}$. Wir haben nun zu beweisen, daß $F_1(u^{(0)},\, v^{(1)},\, \ldots,\, v^{(r)})$ durch $L_i(u^{(0)})$ teilbar ist, wenn $v^{(k)}$ die allgemeine Lösung des erwähnten linearen Gleichungssystems ist. Nach Voraussetzung ist $F(u^{(0)},\, v^{(1)},\, \ldots,\, v^{(r)})$ durch $L_i(u^{(0)})$ teilbar; folglich muß wenigstens ein irreduzibler Faktor von F durch $L_i(u^{(0)})$ teilbar sein. Dieser Faktor sei F_2. Die Teilbarkeit bleibt erhalten, wenn man die in der allgemeinen Lösung $v^{(k)}$ steckenden unbestimmten Parameter so spezialisiert, daß $v^{(k)}$ in $u^{(k)}$ übergeht. Also hat $F_2(u^{(0)}, u^{(1)}, \ldots, u^{(r)})$ mit $F_1(u^{(0)}, u^{(1)}, \ldots, u^{(r)})$ einen Faktor $L_i(u^{(0)})$ gemeinsam. Da aber F_1 und F_2 beide irreduzibel sind, so folgt $F_1 = F_2$ (bis auf einen konstanten Faktor). Also ist $F_1(u^{(0)}, v^{(1)}, \ldots, v^{(r)})$ durch $L_i(u^{(0)})$ teilbar, womit 3. für F_1 bewiesen ist.

Jetzt sei F irreduzibel. Die g Punkte $p^{(1)}, \ldots, p^{(g)}$ sind dann in bezug auf $\mathsf{K}(u^{(1)}, \ldots, u^{(r)})$ zueinander konjugiert. Folglich gibt es dann eine irreduzible Mannigfaltigkeit M, die alle $p^{(i)}$ als allgemeine Punkte besitzt. Wegen 2. schneidet ein allgemeiner $(n-r)$-dimensionaler linearer Unterraum S_{n-r} mit den Gleichungen

$$\sum_{j=0}^{n} u_j^{(k)} x_j = 0 \qquad\qquad (k = 1, \ldots, r)$$

die Mannigfaltigkeit M in mindestens g allgemeinen Punkten. Die Bedingung 3. besagt nun, daß S_{n-r} keinen weiteren allgemeinen Punkt von M enthält. Denn wenn S_{n-r} einen allgemeinen Punkt $q = (q_0, q_1, \ldots, q_n)$ von M enthält, so gibt es wegen der algebraischen Äquivalenz der allgemeinen Punkte einen Isomorphismus, der den Körper $\mathsf{K}(q)$ in $\mathsf{K}(p^{(i)})$ überführt. Dieser läßt sich zu einem Isomorphismus von $\mathsf{K}(q, u^{(1)}, \ldots, u^{(r)})$ mit einem Körper $\mathsf{K}(p^{(i)}, v^{(1)}, \ldots, v^{(r)})$ fortsetzen. Die Relationen $\sum u_j^{(k)} q_j = 0$, die aussagen, daß der Punkt q in S_{n-r} liegt, bleiben beim Isomorphismus erhalten; also gilt für $k = 1, 2, \ldots, r$

$$\sum_{j=0}^{n} v_j^{(k)} p_j^{(i)} \qquad \text{oder} \qquad L_i(v^{(k)}) = 0.$$

Daraus folgt wegen 3., daß $F(u^{(0)}, v^{(1)}, \ldots, v^{(r)})$ durch $L_i(u^{(0)})$ teilbar ist. Wendet man nun den obigen Isomorphismus in umgekehrter Richtung an, so folgt, daß $F(u^{(0)}, u^{(1)}, \ldots, u^{(r)})$ durch $\sum u_j^{(0)} q_j$ teilbar ist, d. h. wegen 1., daß q mit einem der Punkte $p^{(i)}$ zusammenfällt.

Da nun ein allgemeiner $(n-r)$-dimensionaler linearer Unterraum nur endlichviele allgemeine Punkte aus M ausschneidet, muß M r-dimensional sein. Daß F die zugeordnete Form von M ist, ist jetzt leicht zu sehen. Denn die g Punkte $p^{(1)}, \ldots, p^{(g)}$ sind wegen 2. nichts anderes als die g Schnittpunkte von M mit dem durch $u^{(1)}, \ldots, u^{(r)}$ definierten allgemeinen S_{n-r}. F ist also eine Form in $u^{(0)}$ mit Koeffizienten aus $\mathsf{K}[u^{(1)}, \ldots, u^{(r)}]$, die nur durch Potenzprodukte der L_i teilbar und als Polynom aus $\mathsf{K}[u^{(0)}, u^{(1)}, \ldots, u^{(r)}]$ irreduzibel ist, d. h. nach Definition die zugeordnete Form von M.

Daß die Bedingungen 1. und 2. für die zugeordnete Form einer Mannigfaltigkeit M auch notwendig sind, folgt sofort aus der Definition der zugeordneten Form. Daß 3. auch notwendig ist, sieht man so. Es genügt offenbar, 3. zu beweisen für den Fall, daß $v_j^{(k)}$ die *allgemeine*

Lösung des Gleichungssystems (1) bilden. In dem Fall sind die $v_j^{(k)}$ algebraisch-unabhängige Größen. Denn man kann, wie oben schon bemerkt, aus der allgemeinen Lösung $v^{(k)}$ durch Parameterspezialisierung die speziellere Lösung $u^{(k)}$ bilden, und sogar die $u_j^{(k)}$ sind noch algebraisch-unabhängig, weil unbestimmt. Also gibt es einen Isomorphismus $\mathsf{K}(v^{(1)}, \ldots, v^{(r)}) \simeq \mathsf{K}(u^{(1)}, \ldots, u^{(r)})$, welcher $v^{(k)}$ in $u^{(k)}$ überführt. Dieser Isomorphismus läßt sich zu einem Isomorphismus $\mathsf{K}(v^{(1)}, \ldots, v^{(r)}, p^{(i)}) \simeq \mathsf{K}(u^{(1)}, \ldots, u^{(r)}, q)$ erweitern. Der Punkt q liegt auf Grund des Isomorphismus in allen Ebenen $u^{(1)}, \ldots, u^{(r)}$ und auf M; daher ist q einer der Punkte $p^{(l)}$ und $F(u^{(0)}, u^{(1)}, \ldots, u^{(r)})$ enthält den Faktor $\sum q_j u_j^{(0)}$. Macht man nun den Isomorphismus wieder rückgängig, so folgt die Behauptung, daß $F(u^{(0)}, v^{(1)}, \ldots, v^{(r)})$ durch $L_i(u^{(0)})$ teilbar ist.

Jetzt werden wir zeigen, daß die im obigen Satz aufgestellten Bedingungen sich durch homogene algebraische Relationen zwischen den Koeffizienten a_λ der Form ausdrücken lassen. Der Gedankengang ist dabei folgender: Zunächst zeigen wir, daß die Bedingungen 1., 2., 3. sich durch homogene algebraische Relationen zwischen den a_λ, den $u^{(k)}$ und den Koeffizienten $p_j^{(i)}$ der Linearfaktoren L_i von F ausdrücken lassen, sodann eliminieren wir die $p_j^{(i)}$ aus diesen Bedingungen, und schließlich verlangen wir, daß die entstehenden Bedingungen identisch in den $u_j^{(k)}$ erfüllt seien.

Um die Bedingung 1. durch homogene algebraische Relationen auszudrücken, setzen wir an

$$F(u^{(0)}, u^{(1)}, \ldots, u^{(r)}) = \varrho \prod_{i=1}^{g} \left(\sum_{j=0}^{n} p_j^{(i)} u_j^{(0)} \right),$$

vergleichen die Koeffizienten der Potenzprodukte der $u_j^{(0)}$ links und rechts:

$$\varphi_k(u^{(1)}, \ldots, u^{(r)}) = \varrho \, \psi_k(p^{(1)}, \ldots, p^{(g)})$$

und eliminieren schließlich den Faktor ϱ, wodurch die Gleichungen homogen werden:

$$(2) \qquad \varphi_k \psi_l - \varphi_l \psi_k = 0.$$

Die Bedingung 2. hat von selbst schon die Gestalt eines homogenen Relationensystems:

$$(3) \qquad \sum_{j=0}^{n} p_j^{(i)} u_j^{(k)} = 0 \qquad (i = 1, \ldots, g; \; k = 1, \ldots, r).$$

Die Bedingung 3. muß vorher auf eine etwas andere Form gebracht werden. Die Aussage, daß $F(u^{(0)}, v^{(1)}, \ldots, v^{(r)})$ durch die Linearform $L_i(u^{(0)})$ teilbar ist, ist gleichbedeutend mit der anderen, daß alle Nullstellen der Linearform zugleich Nullstellen von $F(u^{(0)}, v^{(1)}, \ldots, v^{(r)})$ sind, d. h. daß aus $L_i(v^{(0)}) = 0$ folgt $F(v^{(0)}, v^{(1)}, \ldots, v^{(r)}) = 0$. Demnach ist die Bedingung 3 gleichwertig mit der folgenden Bedingung: Sind $v^{(0)}, v^{(1)}, \ldots, v^{(r)}$ aus einem Erweiterungskörper von $\mathsf{K}(p^{(i)})$ entnommen und ist $L_i(v^{(k)}) = 0$ für $k = 0, 1, \ldots, r$, so ist $F(v^{(0)}, v^{(1)}, \ldots, v^{(r)}) = 0$.

Die allgemeine Lösung des linearen Gleichungssystems

$$L_i(v) = \sum_{j=0}^{n} p_j^{(i)} v_j = 0$$

lautet

$$v_j = \sum_{l=0}^{n} s_{jl} p_l^{(i)}, \quad s_{jl} = - s_{lj}.$$

Also ist die Bedingung 3 auch gleichbedeutend mit der folgenden Forderung: Sind $s_{jl}^{(k)} = - s_{lj}^{(k)}$ $(k = 0, \ldots, r;\ j, l = 0, \ldots, n)$ lauter neue Unbestimmte und setzt man $v_j^{(k)} = \sum s_{jl}^{(k)} p_l^{(i)}$ $(k = 0, 1, \ldots, r)$, so wird

(4) $$F(v^{(0)}, v^{(1)}, \ldots, v^{(r)}) = 0.$$

Setzt man nun $v_j^{(k)} = \sum s_{jl}^{(k)} p_l^{(i)}$ in diese Gleichung wirklich ein und vergleicht die Koeffizienten der neuen Unbestimmten auf beiden Seiten, so erhält man ein System von homogenen Bedingungsgleichungen

(5) $$\chi_h(a_\lambda, p_j^{(i)}) = 0.$$

Demnach sind die Bedingungen 1., 2., 3. mit den Gleichungen (2), (3), (5) gleichbedeutend. Betrachtet man diese nun als Gleichungen zur Bestimmung der $p_j^{(i)}$ und stellt die Bedingungen für ihre Lösbarkeit auf, so erhält man ein System von homogenen Gleichungen in $a_\lambda, u^{(1)}, \ldots, u^{(r)}$. Ordnet man schließlich diese nach Potenzprodukten der Unbestimmten $u_j^{(k)}$ und setzt die Koeffizienten der einzelnen Potenzprodukte Null, so erhält man ein System von homogenen Bedingungsgleichungen für die a_λ allein.

Damit ist bewiesen:

Satz 2: *Notwendig und hinreichend dafür, daß eine Form F vom Grade g aus $\mathsf{K}[u^{(0)}, u^{(1)}, \ldots, u^{(r)}]$ die zugeordnete Form einer Mannigfaltigkeit M vom Grade g und der Dimension r ist, ist ein System von homogenen Bedingungsgleichungen für die Koeffizienten der Form F.*

In dem einfachsten Spezialfall, wo die Mannigfaltigkeit M eine gerade Linie des Raumes S_n ist, lautet die Form F

$$F = \Sigma \Sigma p_{ij} u_i^{(0)} u_j^{(1)}.$$

Ihre Koeffizienten sind die Plückerschen Koordinaten p_{ij} der Geraden M. Die Umformung der Bedingungen 1., 2., 3. in der oben angegebenen Weise ergibt ein System von kubischen Relationen zwischen den p_{ij}, welche natürlich den bekannten linearen und quadratischen Relationen

$$p_{ij} = - p_{ji}, \quad p_{ij} p_{kl} + p_{ik} p_{lj} + p_{il} p_{jk} = 0$$

äquivalent sein müssen.

Wie man aus den Gleichungen einer Mannigfaltigkeit die zugeordnete Form erhält, nämlich durch Bildung des Resultantensystems aus diesen Gleichungen und $r + 1$ allgemeinen linearen Gleichungen, haben wir anfangs schon erörtert. Wir untersuchen nun, wie man umgekehrt aus der zugeordneten Form die Gleichungen der Mannigfaltigkeit erhalten kann.

S a t z 3: *Die Gleichungen $f_\mu (y) = 0$ einer r-dimensionalen Mannigfaltigkeit M werden erhalten, indem man in der zugeordneten Form $F(u^{(0)}, \ldots, u^{(r)})$ für die $u_l^{(k)}$ die Größen*

$$v_l^{(k)} = \sum_{j=0}^{n} y_j \, s_{jl}^{(k)} \quad (s_{jl}^{(k)} = - s_{lj}^{(k)})$$

einsetzt und die erhaltene Form $G(y, s)$ identisch in den s_{jl} gleich Null setzt.

B e w e i s : Es genügt offenbar, den Fall einer irreduziblen Mannigfaltigkeit M zu betrachten. Daß die Gleichung $G(y, s) = 0$ für einen allgemeinen Punkt $y = p^{(i)}$ der Mannigfaltigkeit tatsächlich erfüllt ist, das besagt gerade die in (4) umgeformte Bedingung 3. Ist sie aber für einen allgemeinen Punkt erfüllt, so auch für jeden speziellen Punkt y.

Nun sei umgekehrt $G(y, s) = 0$, d. h. $F(v^{(0)}, v^{(1)}, \ldots v^{(r)}) = 0$ für $v_l^{(k)} = \sum y_j s_{jl}^{(k)}$. Wir können die $s_{jl}^{(k)}$ beliebig spezialisieren, also für $v^{(k)}$ irgend $(r + 1)$ Hyperebenen durch den Punkt y wählen. Wir können diese $r + 1$ Hyperebenen so wählen, daß sie mit der r-dimensionalen Mannigfaltigkeit M keinen Punkt außer y gemeinsam haben. Denn: man kann die erste Ebene $v^{(0)}$ so wählen, daß die einen beliebigen Punkt von M nicht enthält, also mit M nur einen $(r - 1)$-dimensionalen Durchschnitt M' hat; sodann kann man die zweite Hyperebene $v^{(1)}$ so wählen, daß sie je einen beliebigen Punkt in jedem irreduziblen Bestandteil von M' nicht enthält, also mit M' nur einen $(r - 2)$-dimensionalen Durchschnitt hat, usw. Nun ist $F(v^{(0)}, v^{(1)}, \ldots, v^{(r)}) = 0$, und $F(v^{(0)}, v^{(1)}, \ldots, v^{(r)})$ ist ein Teiler des Resultantensystems der Gleichungen von M und der Gleichungen der Hyperebenen $v^{(0)}, \ldots, v^{(r)}$; also ist dieses Resultantensystem Null. Es gibt somit einen gemeinsamen Punkt von M und den Hyperebenen $v^{(0)}, \ldots, v^{(r)}$, welcher auf Grund der Wahl dieser Hyperebenen nur der Punkt y sein kann. Folglich ist y ein Punkt von M.

<h2 style="text-align:center">§ 2.</h2>

<h2 style="text-align:center">Algebraische Systeme von Mannigfaltigkeiten.</h2>

Die zugeordnete Form einer Mannigfaltigkeit M wird durch die Gesamtheit ihrer Koeffizienten a_λ gegeben. Faßt man diese als Koordinaten eines Punktes a in einem Bildraum $\mathfrak{B}$ auf, so entspricht jeder Mannigfaltigkeit M genau ein Bildpunkt a und umgekehrt. Unter einem *alge-*

braischen System von Mannigfaltigkeiten M verstehen wir nun eine solche Menge von Mannigfaltigkeiten M, deren Bildmenge in $\mathfrak{B}$ eine algebraische Mannigfaltigkeit ist.

Nach Satz 2 bilden alle Mannigfaltigkeiten M (von gegebener Dimension und gegebenem Grad) ein algebraisches System. In derselben Weise kann man beweisen, daß alle Mannigfaltigkeiten M, die auf einer gegebenen Mannigfaltigkeit M_0 in S_n liegen, ein algebraisches System bilden. Man braucht zu dem Zweck nur in dem Beweis von Satz 2 zu den Gleichungen (2), (3), (5) diejenigen Gleichungen hinzuzufügen, die aussagen, daß die Punkte $p^{(1)}, \ldots, p^{(g)}$ sämtlich auf M_0 liegen. Sind die Koordinaten von M_0 bekannt, so kann man diese Gleichungen für $p^{(1)}, \ldots, p^{(g)}$ nach Satz 3 ohne weiteres aufstellen. Eliminiert man dann wie im Beweis von Satz 2 aus dem gesamten Gleichungssystem die Koordinaten von $p^{(1)}, \ldots, p^{(g)}$, so erhält man ein System von algebraischen Relationen zwischen den Koordinaten von M und denen von M_0, welche ausdrücken, daß M auf M_0 liegt.

Ist $\mathfrak{S}$ ein algebraisches System von Mannigfaltigkeiten M in S_n, so gibt es stets eine algebraische Korrespondenz zwischen einer algebraischen Mannigfaltigkeit $\mathfrak{L}$ und dem Raum S_n, in welcher jedem Punkte x von $\mathfrak{L}$ alle Punkte einer Mannigfaltigkeit $M(x)$ von S entsprechen, derart, daß, wenn x die Mannigfaltigkeit $\mathfrak{L}$ durchläuft, $M(x)$ das ganze System $\mathfrak{S}$ durchläuft. Man kann für $\mathfrak{L}$ nämlich die Bildmannigfaltigkeit des Systems $\mathfrak{S}$ im Bildraum $\mathfrak{B}$ wählen; der Punkt x ist dann der Bildpunkt a von M und die Gleichungen der Korrespondenz sind diejenigen Gleichungen, welche nach Satz 3 einen Punkt y von M mit den Koordinaten a_λ verbinden.

Eliminiert man die Koordinaten a_λ aus diesen Gleichungen, so folgt, daß jedes algebraische System von Mannigfaltigkeiten M eine *Trägermannigfaltigkeit* $\mathfrak{T}$ besitzt, welche von den Mannigfaltigkeiten M ganz überdeckt wird. $\mathfrak{T}$ ist die Bildmannigfaltigkeit von $\mathfrak{L}$ in der obigen Korrespondenz.

Wir fragen nun, inwieweit auch umgekehrt jede Korrespondenz $\mathfrak{K}$ zwischen zwei Mannigfaltigkeiten $\mathfrak{L}$ und $\mathfrak{T}$ ein algebraisches System von Mannigfaltigkeiten $\mathfrak{M}$ auf der Trägermannigfaltigkeit $\mathfrak{T}$ definiert, derart, daß den einzelnen Punkten von $\mathfrak{L}$ in der Korrespondenz gerade die einzelnen Mannigfaltigkeiten M des Systems entsprechen. Wir können uns dabei auf irreduzible Korrespondenzen $\mathfrak{K}$ beschränken. $\mathfrak{L}$ und $\mathfrak{T}$ sind dann auch irreduzibel. Nach der allgemeinen Theorie der Korrespondenzen entspricht jedenfalls jedem Punkt x von $\mathfrak{L}$ eine algebraische Mannigfaltigkeit $\mathfrak{T}_x$ auf $\mathfrak{T}$. Diese Mannigfaltigkeiten $\mathfrak{T}_x$ bilden aber keineswegs immer ein algebraisches System, denn es können unter ihnen bekanntlich

Mannigfaltigkeiten von verschiedenen Dimensionen vorkommen (vgl. das Beispiel am Schluß). Aber auch wenn alle $\mathfrak{X}_x$ dieselbe Dimension haben, bilden sie noch nicht immer ein algebraisches System, denn es kann noch vorkommen, daß einzelne $\mathfrak{X}_x$ einen höheren Grad haben als andere. Verbindet man z. B. alle Punkte einer ebenen Kurve dritter Ordnung C^3, die einen Knotenpunkt O besitzt, mit dem Punkt O, so erhält man eine Korrespondenz, in welcher einem allgemeinen Punkt von C^3 eine Gerade, dem Knotenpunkt O aber zwei Geraden (die Doppelpunktstangenten) entsprechen. Man muß also einschränkende Voraussetzungen machen, damit die $\mathfrak{X}_x$ ein algebraisches System bilden. Und zwar gilt der folgende Satz:

Satz 4: *Wenn jedem Punkt x von $\mathfrak{L}$ in einer irreduziblen Korrespondenz $\mathfrak{R}$ eine r-dimensionale Mannigfaltigkeit $\mathfrak{X}_x$ von Punkten auf $\mathfrak{X}$ entspricht und wenn $\mathfrak{L}$ keine mehrfachen Punkte enthält, so bilden die Bildmannigfaltigkeiten $\mathfrak{X}_x$ ein algebraisches System, vorausgesetzt, daß man ihre irreduziblen Bestandteile jeweils mit den in ZAG. VI, § 4 eben für diesen Fall definierten Multiplizitäten zählt.*

Dabei wird der Grundkörper K, wie übrigens auch in ZAG. VI, als *vollkommen* vorausgesetzt.

Beweis: Wir gehen von einem allgemeinen Punkt ξ von $\mathfrak{L}$ und der zugeordneten Mannigfaltigkeit $\mathfrak{X}_\xi$ aus. Um ihre zugeordnete Form zu erhalten, hat man die Schnittpunkte $p^{(1)}, \ldots, p^{(g)}$ von $\mathfrak{X}_\xi$ mit r allgemeinen Hyperebenen $u^{(1)}, \ldots, u^{(r)}$ zu bestimmen und das Produkt

$$(1) \qquad F(u^{(0)}) = \varrho \prod_{i=1}^{g} L_i(u^{(0)}) = \varrho \prod_{i=1}^{g} \left(\sum_{j=0}^{n} p_j^{(i)} u_j^{(0)} \right)$$

zu bilden. Die Beziehungen zwischen den a_λ, den $u^{(k)}$ und $p^{(i)}$ werden nach § 1, Gleichung (2), durch ein System von homogenen Relationen

$$(2) \qquad H(u^{(1)}, \ldots, u^{(r)}, a_\lambda, p^{(1)}, \ldots, p^{(g)}) = 0$$

ausgedrückt, welche der Gleichung (1) äquivalent sind.

Geht man nun durch relationstreue Spezialisierung von dem allgemeinen Punkt ξ zu einem speziellen Punkt x und von $\mathfrak{X}_\xi$ zu $\mathfrak{X}_x$ über, so werden die Vielfachheiten der irreduziblen Bestandteile von $\mathfrak{X}_x$ nach ZAG. VI, § 4 dadurch gefunden, daß man $\mathfrak{X}_x$ wieder mit den allgemeinen Hyperebenen $u^{(1)}, \ldots, u^{(r)}$ schneidet und zusieht, wie die $p^{(i)}$ relationstreu in gewisse Schnittpunkte $q^{(i)}$ übergehen. Die Vielfachheit eines Bestandteils von $\mathfrak{X}_x$ ist dann die Zahl, die angibt, wie oft ein Schnittpunkt dieses Bestandteils mit den Hyperebenen unter den Punkten $q^{(i)}$ vorkommt. Erweitert man nun diese relationstreue Spezialisierung durch eine dazu passende Spezialisierung $a_\lambda \to a'_\lambda$, so bleiben bei dieser Spezialisierung die Relationen (2), also auch die Zerlegung (1) erhalten. Die Vielfachheiten der Bestandteile von $\mathfrak{T}_x$ sind also gleich den Vielfachheiten, mit denen ihre zugeordneten Formen in der Zerlegung der zugeordneten Form $F'(u^{(0)})$ mit Koeffizienten a'_λ vorkommen. Das heißt aber, die spezialisierten a'_λ sind genau die Koordinaten der spezialisierten Mannigfaltigkeit $\mathfrak{T}_x$.

Nun wird aber durch das allgemeine Punktepaar (ξ, a) eine irreduzible Korrespondenz zwischen $\mathfrak{L}$ und dem Bildraum $\mathfrak{B}$ erzeugt. Die einzelnen Punktepaare (x, a') dieser Korrespondenz sind genau diejenigen, welche durch relationstreue Spezialisierung aus dem allgemeinen Punktepaar (ξ, a) entstehen. Die Punkte a', die in dieser Korrespondenz vorkommen, bilden eine algebraische Mannigfaltigkeit $\mathfrak{A}$ im Bildraum $\mathfrak{B}$. Also bilden die Mannigfaltigkeiten $\mathfrak{T}_x$ ein algebraisches System in unserem Sinne.

Läßt man die Voraussetzungen, daß $\mathfrak{L}$ keine mehrfachen Punkte enthält und daß jedem einzelnen Punkte x von $\mathfrak{L}$ eine genau r-dimensionale Mannigfaltigkeit $\mathfrak{T}_x$ entspricht, fallen, so bleibt nur der letzte Teil des Beweises in Kraft. Es gibt demnach auch dann eine irreduzible Korrespondenz, deren allgemeines Punktepaar (ξ, a) ist und in welcher jedem speziellen Punkt x von $\mathfrak{L}$ eine oder mehrere, eventuell auch unendlichviele Punkte a' entsprechen. Zu jedem solchen Punkt a' gehört eine Mannigfaltigkeit $M(a')$ von der Dimension r, und wir können zeigen, daß die Vereinigungsmenge dieser Mannigfaltigkeiten $M(a')$ genau die Mannigfaltigkeit $\mathfrak{T}_x$ ist. Ist nämlich y ein Punkt von $\mathfrak{T}_x$, so ist (x, y) eine relationstreue Spezialisierung des allgemeinen Punktepaares (ξ, η) der Korrespondenz $\mathfrak{K}$ und diese läßt sich ergänzen zu einer relationstreuen Spezialisierung $(\xi, \eta, a) \to (x, y, a')$, wobei a' der Bildmannigfaltigkeit $\mathfrak{A}$ angehört und wobei insbesondere diejenigen algebraischen Relationen erhalten bleiben, die ausdrücken, daß η auf $M(a)$ liegt. Also liegt y auf der Mannigfaltigkeit $M(a')$. Umgekehrt liegen alle Punkte von $M(a')$ auch auf $\mathfrak{T}_x$, denn diejenigen algebraischen Relationen zwischen ξ und a, welche ausdrücken, daß $M(a)$ auf $\mathfrak{T}_\xi$ liegt, bleiben bei der Spezialisierung $\xi \to x$, $a \to a'$ erhalten.

Damit ist bewiesen:

Satz 5. *Wenn in einer irreduziblen Korrespondenz $\mathfrak{K}$ zwischen $\mathfrak{L}$ und $\mathfrak{T}$ einem allgemeinen Punkt ξ von $\mathfrak{L}$ eine r-dimensionale Mannigfaltigkeit $M(a)$ auf $\mathfrak{T}$ entspricht, deren Bildpunkt a sei, so definiert das allgemeine Punktepaar (ξ, a) eine irreduzible Korrespondenz zwischen $\mathfrak{L}$ und einem Bildraum $\mathfrak{A}$ in $\mathfrak{B}$, welche jedem speziellen Punkt x einen oder mehrere Punkte a' zuordnet, zu denen Mannigfaltigkeiten $M(a')$ gehören, deren Vereinigungsmenge gerade diejenige Mannigfaltigkeit $\mathfrak{T}_x$ ist, die dem Punkt x*

in der Korrespondenz $\Re$ *entspricht. Die Mannigfaltigkeiten M(a') bilden ein irreduzibles algebraisches System* $\mathfrak{S}$.

Wenn durch irgendeine Vorschrift dem allgemeinen Punkt ξ einer Mannigfaltigkeit $\mathfrak{L}$ ein Punkt η mit Koordinaten aus $\mathsf{K}(\xi)$ zugeordnet wird, und wenn η der allgemeine Punkt einer Mannigfaltigkeit $\mathfrak{N}$ ist, so werden wir des geschmeidigen Ausdrucks wegen folgende Ausdrucksweise benutzen: *„Wenn ξ die ganze Mannigfaltigkeit $\mathfrak{L}$ durchläuft, so durchläuft η die Bildmannigfaltigkeit $\mathfrak{N}$".* Gemeint ist damit der folgende präzise Sachverhalt: (ξ, η) ist das allgemeine Punktepaar einer irreduziblen Korrespondenz zwischen $\mathfrak{L}$ und $\mathfrak{N}$, und diese Korrespondenz ordnet jedem Punkt von $\mathfrak{L}$ mindestens einen Punkt von $\mathfrak{N}$ und umgekehrt jedem Punkt von $\mathfrak{N}$ mindestens einen Punkt von $\mathfrak{L}$ zu.

Allgemeiner: Wenn durch irgendeine Vorschrift dem allgemeinen Punkt ξ einer Mannigfaltigkeit $\mathfrak{L}$ eine r-dimensionale Mannigfaltigkeit M_ξ zugeordnet ist, deren Gleichungen dem Körper $\mathsf{K}(\xi)$ angehören, und wenn M_ξ das allgemeine Element eines algebraischen Systems $\mathfrak{S}$ von Mannigfaltigkeiten ist, so werden wir folgende Ausdrucksweise benutzen: *„Wenn ξ die ganze Mannigfaltigkeit $\mathfrak{L}$ durchläuft, so durchläuft M_ξ das System $\mathfrak{S}$".* Ist $\mathfrak{T}$ die Trägermannigfaltigkeit von $\mathfrak{S}$, so werden wir auch sagen: *„M_ξ durchläuft $\mathfrak{T}$".* Mit der ersten Aussage ist gemeint, daß es eine irreduzible Korrespondenz zwischen $\mathfrak{L}$ und $\mathfrak{S}$ gibt, welche dem allgemeinen Punkt ξ die Mannigfaltigkeit M_ξ und jedem speziellen Punkt von $\mathfrak{L}$ mindestens eine Mannigfaltigkeit M des Systems $\mathfrak{S}$ zuordnet. Ordnet man weiter jeder Mannigfaltigkeit M alle Punkte von M zu, so erhält man eine Korrespondenz zwischen $\mathfrak{S}$ und der Trägermannigfaltigkeit $\mathfrak{T}$. Die beiden Korrespondenzen werden, wenn a_λ die Koordinaten einer Mannigfaltigkeit M von $\mathfrak{S}$, x_k die Koordinaten eines Punktes von $\mathfrak{L}$ und y_k die eines Punktes von $\mathfrak{T}$ sind, durch Gleichungen

$$f_\mu(x, a) = 0 \quad \text{bzw.} \quad g_\nu(a, y) = 0$$

gegeben. Eliminiert man die a_λ aus diesen Gleichungen, so erhält man eine Korrespondenz $\Re$ zwischen $\mathfrak{L}$ und $\mathfrak{T}$, welche einem Punkte x alle Punkte y aller zugeordneten Mannigfaltigkeiten des Systems $\mathfrak{S}$ zuordnet. Diese Korrespondenz $\Re$ steht somit zum System $\mathfrak{S}$ genau in der durch Satz 5 dargelegten Beziehung.

Das folgende Beispiel möge die eben eingeführte Sprechweise sowie den Satz 5 erläutern. Gegeben seien im Raum S_3 zwei windschiefe Geraden g, h und eine Ebene $\mathfrak{L}$, welche g und h in G und H schneidet. Durch einen allgemeinen Punkt ξ von $\mathfrak{L}$ geht eine und nur eine Gerade M_ξ, welche g und h schneidet. Durchläuft ξ nun die ganze Ebene, so durchläuft M_ξ die ganze Geradenkongruenz mit den Leitstrahlen g, h. Das heißt, M_ξ ist das allgemeine Element dieser Kongruenz. Jedem Punkte x von $\mathfrak{L}$ entspricht eine einzige Gerade der Kongruenz mit Ausnahme der Punkte G und H, denen je ein ganzes Geradenbüschel entspricht. Ordnet man nun jedem Punkt x von $\mathfrak{L}$ alle Punkte y zu, die mit x zusammen auf irgend einer g und h treffenden Geraden liegen, so erhält man eine Korrespondenz $\mathfrak{K}$ zwischen $\mathfrak{L}$ und S_3, in welcher die einem willkürlichen Punkt x von $\mathfrak{L}$ entsprechenden Punkte y im allgemeinen eine Gerade bilden. Die dem Punkte G oder H entsprechenden Punkte y bilden aber keine Gerade, sondern je eine Ebene, welche die Vereinigungsmenge der durch G bzw. H gehenden Kongruenzstrahlen ist, wie es nach Satz 5 auch sein muß.

(Eingegangen am 24. 5. 1936.)

17.

Zur algebraischen Geometrie X
Über lineare Scharen von reduziblen Mannigfaltigkeiten

Mathematische Annalen 113, 5 (1937) 705–712

Ein bekannter Satz von Bertini[1]) über lineare Scharen von reduziblen ebenen Kurven, der von Enriques[2]) auf lineare Scharen von Kurven auf algebraischen Flächen übertragen wurde, besagt, daß die Kurven einer linearen Schar, deren allgemeine Kurve reduzibel ist (oder auch: deren sämtliche Kurven reduzibel sind) entweder eine feste Kurve als Bestandteil enthalten, oder aus Kurven eines (nicht notwendig linearen) Büschels zusammengesetzt sind.

Analysiert man den Beweis dieses Satzes, so zeigt sich, daß in ihm von folgender Tatsache Gebrauch gemacht wird: Hängt eine reduzible Kurve C rational von irgendwelchen Parametern ab, so hängen ihre irreduziblen Bestandteile algebraisch von diesen Parametern ab.

So selbstverständlich dieser Satz erscheinen möge, so ist doch ein Beweis erforderlich. Ich bin Herrn Finsler zu Dank verpflichtet, daß er mich darauf aufmerksam gemacht hat. Ich werde hier (§ 1) diesen Beweis dadurch erbringen, daß ich allgemein zeige, daß die absolut-irreduziblen Bestandteile einer algebraischen Mannigfaltigkeit algebraisch von den Koeffizienten der Gleichungen dieser Mannigfaltigkeit abhängen (Satz 1). Diese Tatsache wird zurückgeführt auf die entsprechende Tatsache für reduzible Polynome, welche ihrerseits direkt aus einem Satz von Kronecker[3]) folgt. Ich werde den Beweis aber so formulieren, daß die Kenntnis des Satzes von Kronecker zum Verständnis des Beweises nicht erforderlich ist.

In § 2 werden einige allgemeine Begriffe und Sätze erörtert, die sich auf algebraische Kurvensysteme auf einer irreduziblen Fläche beziehen, insbesondere der Begriff des Grades einer linearen Kurvenschar.

In § 3 wird dann der Satz von Bertini-Enriques bewiesen. In § 4 gehe ich auf die Verallgemeinerung dieses Satzes auf lineare Scharen von M_{d-1} auf M_d ein, welche Verallgemeinerung wahrscheinlich nicht neu ist.

---

[1]) E. Bertini, Rendiconti R. Ist. Lombardo 15 (1882), S. 24—28.

[2]) F. Enriques, Introduzione alla Geometria sopra le Superficie Algebriche, Mem. Soc. It. Sci. (3) 10 (1896), S. 22.

[3]) L. Kronecker, Sitz.-Ber. Akad. Berlin 37 (1883), S. 957.

225

§ 1.

Zerlegung von Mannigfaltigkeiten in absolut-irreduzible.

Satz 1. *Wenn eine algebraische Mannigfaltigkeit M, deren Gleichungen Koeffizienten aus dem Körper K besitzen, bei einer Erweiterung von K in irreduzible Bestandteile $M_1, \ldots, M_s$ zerfällt, so kann man die Gleichungen von $M_1, \ldots, M_s$ so wählen, daß ihre Koeffizienten algebraisch in bezug auf K sind. Eine endliche Erweiterung von K genügt, um M in absolut-irreduzible Mannigfaltigkeiten zu zerlegen, welche bei keiner Erweiterung mehr zerfallen.*

Beweis. Wir können uns auf solche Mannigfaltigkeiten beschränken, die in K irreduzibel sind. Die Zerlegung solcher Mannigfaltigkeiten in irreduzible in irgendeinem Erweiterungskörper kommt nach der voranstehenden Abhandlung ZAG. IX, § 1 auf die Zerlegung der zugeordneten Form $F(u^{(0)}, \ldots, u^{(r)})$ der Mannigfaltigkeit M hinaus. Aus der Faktorenzerlegung von F erhält man nämlich durch rationale Prozesse die Gleichungen der irreduziblen Bestandteile von M. Alles kommt also darauf hinaus, die Behauptungen des Satzes 1 für Formen $F(u_1, \ldots, u_m)$ zu beweisen, welche bei Erweiterung des Grundkörpers in Faktoren $F_1 F_2 \ldots F_s$ zerfallen.

Durch die Kroneckersche Substitution

$$u_k = t^{g^{k-1}} \qquad (k = 0, 1, 2, \ldots, m),$$

wobei g größer als der Grad der Form F gewählt wird, werden F und $F_1, \ldots, F_s$ in Polynome $f, f_1, \ldots, f_s$ einer Veränderlichen t verwandelt, welche genau dieselben Koeffizienten wie $F, F_1, \ldots, F_s$ haben und für welche die Zerlegung

$$f(t) = f_1(t) \ldots f_s(t)$$

gilt. Nun wird aber unsere Behauptung selbstverständlich, denn $f(t)$ zerfällt in einem endlichen Erweiterungskörper von K vollständig in Linearfaktoren, aus denen sich dann die Faktoren $f_1, \ldots, f_s$ rational zusammensetzen.

Der Begriff eines irreduziblen algebraischen Systems von Mannigfaltigkeiten M wurde in der vorangehenden Arbeit ZAG. IX erklärt. Für solche Systeme gilt nun der folgende Satz:

Wenn das allgemeine Element eines irreduziblen Systems von Mannigfaltigkeiten M zerfällt, so zerfallen alle Mannigfaltigkeiten des Systems. Wenn aber das allgemeine Element M absolut-irreduzibel ist, so sind alle Mannigfaltigkeiten des Systems absolut-irreduzibel, höchstens mit Ausnahme der Elemente eines Teilsystems von niedrigerer Dimensionszahl.

226

Beweis. Das Zerfallen einer Mannigfaltigkeit M kommt auf das Zerfallen der zugeordneten Form $F(u)$ hinaus. Zerfällt nun $F(u)$ in Bestandteile von gegebenen Gradzahlen:

$$F(u) = \varrho \cdot G(u) \cdot H(u),$$

so ergibt der Vergleich der Koeffizienten der Potenzprodukte der u rechts und links ein System von Bedingungen für die Koeffizienten a von $F(u)$, b von $G(u)$ und c von $H(u)$:

$$a_\lambda = \varrho \, \varphi_\lambda (b, c).$$

Elimination von ϱ ergibt die homogenen Gleichungen

$$a_\lambda \, \varphi_\mu (b, c) - a_\mu \, \varphi_\lambda (b, c) = 0.$$

Elimination der homogenen Variablenreihen b und c ergibt ein Gleichungssystem für die a allein [4]):

$$\psi_\nu (a) = 0.$$

Ist dieses Gleichungssystem für das allgemeine Element einer irreduziblen Mannigfaltigkeit im a-Raum erfüllt, so für jedes Element dieser Mannigfaltigkeit. Ist es aber für das allgemeine Element nicht erfüllt, so kann es nur auf einer Teilmannigfaltigkeit von geringerer Dimension erfüllt sein.

§ 2.

Algebraische Systeme von Kurven auf einer Fläche.

Ist auf einer festen irreduziblen Fläche Φ eine Kurve C gegeben, in deren Gleichungen algebraische Funktionen von r unbestimmten Parametern eingehen, so ist dadurch ein irreduzibles, höchstens r-dimensionales Kurvensystem $|C|$ gegeben, dessen allgemeine Kurve C ist (vgl. ZAG. IX, § 2). Lassen wir den Fall, wo das System nur aus endlichvielen konjugierten Kurven besteht, außer Betracht, so kann die Trägermannigfaltigkeit des Kurvensystems nur die ganze Fläche Φ sein. Durch jeden Punkt von Φ geht also mindestens eine Kurve des Systems. Handelt es sich insbesondere um ein System von ∞^1 Kurven C (also um ein eindimensionales Kurvensystem), so geht durch einen allgemeinen Punkt von C eine endliche Anzahl h von Kurven C. Ist $h = 1$, so heißt das System ∞^1 ein *Büschel*. Ist das Büschel zugleich eine lineare Schar (vgl. ZAG VI, § 5), so spricht man von einem linearen Büschel.

Ein irreduzibles algebraisches Kurvensystem (z. B. eine lineare Schar) heißt *zusammengesetzt aus einem Büschel*, wenn die allgemeine Kurve C des Systems aus ν Kurven des Büschels besteht.

---

[4]) Vgl. E. Noether, Ein algebraisches Kriterium für absolute Irreduzibilität, Math. Annalen **85** (1922), S. 26, sowie E. Fischer, Math. Annalen **94** (1925), S. 163, und B. L. van der Waerden, Proc. Kon. Acad. Amsterdam **29** (1926), S. 142.

Zerlegt man die allgemeine Kurve C eines irreduziblen algebraischen Kurvensystems in irreduzible Bestandteile und wählt aus einem irreduziblen Bestandteil C' einen allgemeinen Punkt P, so können zwei Fälle eintreten. P kann den Transzendenzgrad 2 haben, d. h. von zwei Parametern abhängen; dann ist P ein allgemeiner Punkt der Fläche Φ. Oder P kann den Transzendenzgrad 1 (über dem Grundkörper K) haben, also allgemeiner Punkt einer irreduziblen Kurve D sein. Da alle Punkte von C' relationstreue Spezialisierungen von P sind, gehören sie alle zu D, und da D irreduzibel ist, ist $C' = D$. Die allgemeine Kurve C des Systems enthält in diesem Falle also einen festen (d. h. von den Systemparametern unabhängigen) Bestandteil D. Wir sehen also:

Wenn die allgemeine Kurve C eines irreduziblen algebraischen Kurvensystems $|C|$ auf Φ keine festen Bestandteile hat, so ist jeder allgemeine Punkt eines Bestandteils der allgemeinen Kurve C zugleich ein allgemeiner Punkt der Fläche Φ.

Unter dem *Grad* einer linearen Kurvenschar (oder allgemeiner eines irreduziblen Kurvensystems) auf einer Fläche versteht man die Anzahl der Schnittpunkte von zwei unabhängig voneinander gewählten allgemeinen Kurven der Schar außerhalb der Basispunkte.

Sind λ, μ die unbestimmten Parameter der beiden allgemeinen Kurven und sind $\overset{1}{x}, \ldots, \overset{g}{x}$ die Schnittpunkte (also g der Grad), so haben zwei spezielle Kurven $C_{\lambda'}$ und $C_{\mu'}$ der Schar entweder einen Bestandteil gemeinsam, oder bei der Spezialisierung $\lambda \to \lambda'$, $\mu \to \mu'$, $\overset{v}{x} \to \overset{v}{y}$ erscheinen die Schnittpunkte y von $C_{\lambda'}$ und $C_{\mu'}$ mit ganz bestimmten Multiplizitäten (Spezialisierungsmultiplizitäten) behaftet. Vollführt man die Spezialisierung in zwei Schritten, indem man zuerst $\lambda \to \lambda'$ und dann $\mu \to \mu'$ spezialisiert, und wendet man bei jedem Schritt den Satz 2 aus § 5 der Arbeit ZAG. VI (Math. Annalen **110**, S. 151) an, so folgt: Die Spezialisierungsmultiplizität eines Schnittpunktes y von $C_{\lambda'}$ und $C_{\mu'}$ ist gleich

$$\sum_i \sum_j \alpha_i \beta_j \sigma_{ij},$$

wobei α_i die Vielfachheiten der Bestandteile A_i der Kurve $C_{\lambda'}$ bei der Spezialisierung $\lambda \to \lambda'$, β_i die der Bestandteile B_j der Kurve C bei der Spezialisierung $\mu \to \mu'$ bedeuten, während σ_{ij} die Schnittmultiplizität von y als Schnittpunkt von A_i und B_j ist und die Summation sich über diejenigen Bestandteile A_i und B_j erstreckt, welche den Punkt y enthalten. Also grob gesagt: Die Spezialisierungsmultiplizität ist gleich der Schnittmultiplizität der spezialisierten Kurven. Insbesondere folgt daraus, daß die Spezialisierungsmultiplizitäten aller Schnittpunkte y stets *positiv* sind. Diese Eigenschaft ist die einzige, die wir im folgenden brauchen; sie läßt sich auch leicht direkt beweisen (vgl. § 4).

§ 3.

Der Satz von Bertini-Enriques.

Der Beweis des verallgemeinerten Bertinischen Satzes über die linearen Scharen von reduziblen Kurven wird wie bei Enriques[5]) in zwei Schritten geführt.

Satz 2. *Eine lineare Schar* $|C|$ *vom Grade Null ohne feste Bestandteile ist aus einem Büschel (von irreduziblen Kurven) zusammengesetzt.*

Beweis. Die Kurven der Schar $|C|$, die durch einen allgemeinen Punkt P von Φ gehen, bilden eine lineare Teilschar von der Dimension $r - 1$, wenn die ursprüngliche Schar die Dimension r hat. Je zwei Kurven C, C' dieser Teilschar müssen eine Kurve K durch P gemeinsam haben, denn sonst hätte P als isolierter Schnittpunkt eine positive Multiplizität; was der Voraussetzung des Grades Null widerspricht. Nimmt man C' fest, aber C als allgemeines Element der Teilschar an, so ist K als Bestandteil von C' eine feste Kurve, unabhängig von den Parametern von C. Also ist K in allen Kurven der Teilschar als Bestandteil enthalten.

Ist nun P' ein anderer Punkt von K (außerhalb der Basispunkte der Schar $|C|$), so bilden die durch P' gehenden Kurven der Schar $|C|$ wieder eine lineare Schar von der Dimension $r - 1$, welche die vorige umfaßt, also mit ihr identisch sein muß. Die durch P' gehenden Kurven der Schar $|C|$ haben also wieder die Kurve K miteinander gemeinsam.

Die Kurve K möge in absolut-irreduzible Bestandteile K_1, K_2, ... zerfallen. Wir wollen zeigen, daß jede dieser Kurven ein irreduzibles System K_1, und zwar ein Büschel durchläuft.

Wählen wir aus der linearen Schar $|C|$ ein lineares Büschel $C_1 + \lambda C_2$ ohne feste Kurve aus, so geht durch den allgemeinen Punkt P von Φ genau eine Kurve des Büschels. Die Gleichungen der absolut-irreduziblen Bestandteile, in die $C_1 + \lambda C_2$ zerfällt, also insbesondere die Gleichungen von K_1, K_2, ... gehören nach Satz 1 einem algebraischen Erweiterungskörper von $\mathsf{K}(\lambda)$ an, hängen also algebraisch von nur einem Parameter λ ab. Also durchlaufen K_1, K_2, ... algebraische Systeme $|K_1|$, $|K_2|$, ... von je höchstens ∞^1 Kurven.

Da das Büschel $C_1 + \lambda C_2$ keine feste Kurve enthält, so enthält auch das System $|K_1|$ keine festen Bestandteile. Daraus folgt erstens, daß das System $|K_1|$ wirklich eindimensional ist, und zweitens, daß die Trägermannigfaltigkeit des Systems die ganze Fläche Φ ist. Durch jeden allgemeinen Punkt P von Φ geht also mindestens eine Kurve K_1 des

---

[5]) Enriques-Campedelli, Lezioni sulla teoria delle superficie algebriche, Padova 1932, § 9.

Systems $|K_1|$, ebenso mindestens eine Kurve K_2 von $|K_2|$ usw., insgesamt etwa h verschiedene Kurven K_ν. Dabei ist K_1 eine allgemeine Kurve des Systems $|K_1|$ (usw.), denn eine spezielle Kurve würde nicht durch den allgemeinen Punkt P gehen. Alle durch P gehenden Kurven C von $|C|$ müssen (nach dem zweiten Absatz dieses Beweises) alle h Kurven K_ν enthalten. Das heißt, es gehen mindestens h verschiedene Bestandteile jeder Kurve C durch den Punkt P.

Nach § 2 kann man nun den allgemeinen Punkt P insbesondere so bestimmen, daß man zuerst eine allgemeine Kurve C und auf einem beliebigen Bestandteil von C einen allgemeinen Punkt P wählt. Dieser Punkt P liegt natürlich nur auf einem irreduziblen Bestandteil von C. Also muß $h = 1$ sein. Das heißt, es gibt nur ein einziges System $|K_1|$ und von diesem geht durch einen allgemeinen Punkt P nur eine Kurve. Demnach ist $|K_1|$ ein Büschel. Da weiter, wie wir sahen, die allgemeinen Punkte aller irreduziblen Bestandteile von C je einer in C enthaltenen Kurve K_1 angehören, so ist C aus lauter Kurven K_1 zusammengesetzt.

Satz 3. *Eine lineare Schar $|C|$, deren allgemeine Kurve C reduzibel ist, ohne aber feste Bestandteile zu enthalten, hat den Grad Null und ist aus einem Büschel zusammengesetzt.*

Beweis. Es genügt nach Satz 2, zu zeigen, daß die Schar den Grad Null hat. Zwei unabhängig gewählte allgemeine Kurven C_1 und C_2 der Schar definieren ein lineares Büschel $C_1 + \lambda C_2$ und wir haben zu zeigen, daß jeder möglicherweise vorhandene Schnittpunkt von C_1 und C_2 ein Basispunkt der ganzen Schar $|C|$ ist.

Das lineare Büschel $C_1 + \lambda C_2$ hat (wie jedes lineare Büschel) den Grad Null und ist daher nach Satz 2 aus einem Büschel $|K_1|$ von irreduziblen Kurven K_1 zusammengesetzt. Ist nun P' ein Schnittpunkt von C_1 und C_2, so gehen durch P' alle ∞^1 Kurven $C_1 + \lambda C_2$, also auch ∞^1 Kurven K_1, also alle Kurven des Systems $|K_1|$, welches ja irreduzibel ist. Daraus folgt, daß alle Bestandteile der Kurve $C_1 + \lambda C_2$ durch den Punkt P' gehen, welcher daher ein mehrfacher Punkt dieser Kurve ist. Da nun C_1 und C_2 beide allgemeine Kurven von $|C|$ waren, ist auch $C_1 + \lambda C_2$ eine allgemeine Kurve von $|C|$, und eine solche hat nach einem bekannten Satz von Bertini [6] außerhalb der Basispunkte des Systems $|C|$ und außerhalb der mehrfachen Punkte der Fläche Φ keine mehrfachen Punkte. Mithin ist P' entweder ein Basispunkt von $|C|$ oder ein mehrfacher Punkt der Fläche.

Die Kurve C_1 schneidet aber die Doppelkurve der Fläche Φ höchstens in endlichvielen Punkten, und diese liegen nicht auf einer zweiten all-

---

[6] Siehe etwa ZAG. V. Math. Annalen **110**, S. 133, sowie Math. Annalen **113**, S. 37.

gemeinen Kurve C_2, es sei denn, sie wären Basispunkte der Schar $|C|$. Demnach ist P' in jedem Fall ein Basispunkt der Schar $|C|$, w. z. b. w.

§ 4.

Lineare Scharen von M_{d-1} auf M_d.

Geht man von der Fläche Φ zu einer d-dimensionalen Mannigfaltigkeit M_d über und betrachtet auf M_d eine lineare Schar von Mannigfaltigkeiten C von der Dimension $d-1$, so wird zunächst der Begriff des Grades für unsere Zwecke unbrauchbar. Die Unterscheidung der beiden Fälle: Grad Null und Grad ungleich Null ist vielmehr durch eine andere zu ersetzen, auf die man folgendermaßen geführt wird.

Die Paare von Mannigfaltigkeiten C_1, C_2 aus $|C|$ werden durch zweimal $r+1$ homogene Parameter $\lambda_0, \ldots, \lambda_r$; $\mu_0, \ldots, \mu_r$, also durch Punkte eines zweifach-projektiven Raumes $S_{r,\,r}$ dargestellt. Ordnet man nun einem Punkte P von M_d, der nicht Basispunkt von $|C|$ sei, alle die Paare C_1, C_2 zu, welche beide durch P gehen, so erhält man eine Korrespondenz zwischen M_d und $S_{r,\,r}$, in der jedem Punkte P eine irreduzible Teilmannigfaltigkeit $S_{r-1,\,r-1}$ von der Dimension $2(r-1)$ entspricht. Die Korrespondenz ist demnach irreduzibel und von der Dimension $d+2(r-1)$. Die Bildmannigfaltigkeit der Korrespondenz, also die Gesamtheit der Punkte von $S_{r,\,r}$, die in der Korrespondenz überhaupt vorkommen, habe die Dimension a, und einem allgemeinen Punkt der Bildmannigfaltigkeit mögen ∞^b Punkte P entsprechen. Dann ist nach dem Prinzip der Konstantenzählung

$$a + b = d + 2(r-1).$$

Offenbar ist $b \leq d-1$, da die einem gegebenen Punkt der Bildmannigfaltigkeit zugeordneten Punkte P immer auf zwei gegebenen $(d-1)$-dimensionalen Mannigfaltigkeiten C_1 und C_2 liegen müssen. Also ist $a \geq 2r-1$. Es gibt nun zwei Fälle:

Fall 1. $a = 2r$, $b = d-2$. Die Bildmannigfaltigkeit ist der gesamte $S_{r,\,r}$, und jedem Punkt von $S_{r,\,r}$ entsprechen ∞^{d-2} Punkte P. Das heißt: Je zwei allgemeine Elemente C_1, C_2 der Schar $|C|$ schneiden sich außerhalb der Basispunkte der Schar in einer Mannigfaltigkeit M_{d-2}.

Fall 2. $a = 2r-1$, $b = d-1$. Zwei allgemeine Elemente C_1, C_2 der Schar $|C|$ schneiden sich außerhalb der Basispunkte der Schar nicht, aber wenn zwei Elemente C_1, C_2 einmal einen Punkt P außerhalb der Basispunkte gemeinsam haben, so haben sie gleich einen $(d-1)$-dimensionalen Bestandteil K gemeinsam.

Der Fall 2 ist offenbar die d-dimensionale Verallgemeinerung einer linearen Kurvenschar vom Grade Null. Im Fall 2 gilt der ganze Beweis des Satzes 2 wörtlich. Also gilt

S a t z 4. *Eine lineare Schar $|C|$ von M_{d-1} auf M_d, in welcher ein allgemeines Paar C_1, C_2 sich außerhalb der Basispunkte der Schar nicht schneidet, ist aus einem Büschel zusammengesetzt.*

Ebenso entspricht dem Satz 3 der folgende

S a t z 5. *Eine lineare Schar $|C|$ von M_{d-1} auf M_d, ohne feste Teilmannigfaltigkeit M_{d-1}, deren allgemeines Element C reduzibel ist, gehört in der obigen Fallunterscheidung unter Fall 2 und ist aus einem Büschel zusammengesetzt.*

B e w e i s. Es genügt, zu zeigen, daß der Fall 1 nicht eintreten kann. Die Mannigfaltigkeit C_1 hat mit der Mannigfaltigkeit der Doppelpunkte von M_d höchstens eine M_{d-2} gemeinsam. Diese hat mit einer weiteren allgemeinen C_2 außerhalb der Basispunkte der Schar höchstens eine M_{d-3} gemeinsam. Würde nun der Fall 1 eintreten, so hätten C_1 und C_2 außerhalb der Basispunkte eine M_{d-2} gemeinsam, welche nicht aus lauter mehrfachen Punkten von M_d bestehen kann. Ist nun P' ein Punkt von M_{d-2} außerhalb der Basispunkte der Schar und außerhalb der mehrfachen Punkte von M_d, so kann man genau so wie beim Beweis von Satz 3 schließen, daß P' doch ein Basispunkt der Schar oder ein mehrfacher Punkt von M_d sein muß, was einen Widerspruch ergibt.

(Eingegangen am 29. 5. 1936.)

18.

Zur algebraischen Geometrie XI

Projektive und birationale Äquivalenz und Moduln von ebenen Kurven

Mathematische Annalen 114, 5 (1937) 683–699

Man möchte meinen, daß die projektive Äquivalenz von zwei ebenen irreduziblen Kurven n-ter Ordnung eine algebraische Eigenschaft sei, die sich durch algebraische Gleichungen in den Koordinaten (Koeffizienten) der Kurven ausdrückt. Dem ist aber nicht so. In § 1 bestimmen wir die kleinste algebraische Mannigfaltigkeit von Kurvenpaaren, welche alle projektiv äquivalenten Paare umfaßt. Diese Mannigfaltigkeit ist irreduzibel und ihr allgemeines Element ist ein projektiv äquivalentes Kurvenpaar; aber es gibt auch Kurvenpaare in der Mannigfaltigkeit, die zwar Grenzfälle von projektiv äquivalenten Kurvenpaaren, aber selbst nicht projektiv äquivalent sind. Alle solche „uneigentlich projektiven‟ Kurvenpaare werden durch Satz 1 gegeben. Z. B. sind alle Kurven vierter Ordnung mit einem Berührungsknoten untereinander uneigentlich projektiv, aber sie sind noch nicht einmal birational ineinander transformierbar.

Aus dieser letzten Bemerkung folgt, daß auch die birationale Äquivalenz von zwei Kurven keine algebraische Eigenschaft dieser Kurven ist. Daraus ergibt sich eine wesentliche Schwierigkeit für die Frage der „Moduln‟, d. h. der Konstanten, von denen eine Klasse birational äquivalenter Kurven abhängt.

In einem Gespräch, das den Anlaß zu der vorliegenden Untersuchung bildete, formulierte M. Deuring die Frage nach den Moduln so: Gibt es eine algebraische Korrespondenz, welche jeder ebenen Kurve vom Geschlechte p einen Punkt einer $(3p - 3 + \varrho_p)$-dimensionalen ($\varrho_0 = 3$, $\varrho_1 = 1$, $\varrho_p = 0$ für $p > 1$) „Modulmannigfaltigkeit‟ zuordnet, derart, daß zwei Kurven genau dann der gleiche Punkt zugeordnet wird, wenn sie birational äquivalent sind? Diese Frage muß jetzt verneint werden, denn wenn es eine Korrespondenz der verlangten Art gäbe, und wenn sie auch nur stetig (nicht einmal algebraisch) wäre, so würden zwei Kurven, die sich als Grenzfälle von birational äquivalenten Kurven darstellen lassen, die gleiche Moduln haben und daher selber birational äquivalent sein müssen.

Wir werden aber in § 3 zeigen, daß die Frage nach den Moduln bejaht werden kann, falls man sich auf Kurven von genügend hohem Grad

233

mit nur gewöhnlichen Knotenpunkten beschränkt. Für diese „regulären Kurven" gibt es eine rationale Abbildung von der obigen Art auf eine Modulmannigfaltigkeit von der Dimension $3p - 3 + \varrho_p$.

Zu diesem Ergebnis führt ein etwas mühsamer Weg, bei dem die Sätze des § 1 über projektive Äquivalenz wesentlich gebraucht werden. In § 2 wird zunächst in Anlehnung an Severi[1]) bewiesen, daß die Kurven mit einer gegebenen Anzahl von Knotenpunkten sich auf endlich viele irreduzible Mannigfaltigkeiten von der Dimension $3n + p - 1$ verteilen, und daß die zu einer gegebenen Kurve birational äquivalenten Kurven einer irreduziblen Mannigfaltigkeit von der Dimension $3n - 2p + 2$ angehören. In § 3 wird die kleinste algebraische Mannigfaltigkeit untersucht, welche alle birational äquivalenten Kurvenpaare umfaßt, und es wird gezeigt, daß die *regulären Kurvenpaare* dieser Mannigfaltigkeit auch wirklich birational äquivalent sind. Zieht man dann noch die Sätze aus ZAG IX[2]) über algebraische Systeme von algebraischen Mannigfaltigkeiten heran, so folgt schließlich (§ 4), daß die Gesamtheit der regulären Kurven vom Geschlechte p mit einem algebraischen System von algebraischen Mannigfaltigkeiten, deren jede nur birational äquivalente Kurven enthält, einfach überdeckt werden kann, und daß diese Mannigfaltigkeiten sich auf Punkte einer Bildmannigfaltigkeit von der Dimension $3p - 3 + \varrho_p$ eineindeutig abbilden lassen.

§ 1.
Projektive Äquivalenz von ebenen Kurven.

Die ebenen Kurven n-ter Ordnung Γ bilden einen projektiven Raum S_r, $r = \frac{1}{2} n (n + 3)$. Die Paare (Γ, Δ) von ebenen Kurven n-ter Ordnung bilden einen zweifach projektiven Raum $S_{r,\,r}$. Ist Γ eine allgemeine Kurve (mit unbestimmten Koeffizienten) und Δ die allgemeinste projektiv transformierte Kurve von Γ, so hängt das Paar (Γ, Δ) von $r + 8$ wesentlichen Parametern ab. Das Paar (Γ, Δ) ist also das allgemeine Element einer irreduziblen Mannigfaltigkeit $\mathfrak{M}$ der Dimension $r + 8$. Zu $\mathfrak{M}$ gehören alle Paare von projektiv äquivalenten Kurven, da sie alle durch Parameterspezialisierung aus dem allgemeinen Paar (Γ, Δ) hervorgehen. Zu $\mathfrak{M}$ gehören aber, wie wir sehen werden, noch andere Kurvenpaare, die wir *uneigentlich projektive Kurvenpaare* nennen. Wir wollen diese alle bestimmen.

Hilfssatz. *Jeder Punkt einer irreduziblen Mannigfaltigkeit $\mathfrak{M}$ kann mit einem allgemeinen Punkt ξ der Mannigfaltigkeit durch eine irreduzible, auf $\mathfrak{M}$ verlaufende Kurve verbunden werden.*

---

[1]) F. Severi-E. Löffler, Vorlesungen über algebraische Geometrie, Leipzig 1921, Anhang F.

[2]) W.-L. Chow und B. L. van der Waerden, Math. Annalen 113 (1936), S. 692—704.

Beweis. Wir können $\mathfrak{M}$ als q-dimensionale Mannigfaltigkeit im affinen Raum A_n annehmen. Nach einer Koordinatentransformation können wir annehmen, daß $\xi_{q+1}, \ldots, \xi_n$ ganze algebraische Funktionen von $\xi_1, \ldots, \xi_q$ sind. Sind dann $z, u_{q+1}, \ldots, u_n$ Unbestimmte, so ist

$$N(z - u_{q+1}\xi_{q+1} - \cdots - u_n\xi_n) = F(z, u_{q+1}, \ldots, u_n, \xi_{q+1}, \ldots, \xi_n)$$

ein Polynom in z, dessen Faktorzerlegung nicht nur für allgemeine $\xi_1, \ldots, \xi_q$ sondern auch für alle speziellen Werte dieser Größen die zugehörigen Punkte von $\mathfrak{M}$ liefert[3]). Ist nun $(\eta_1, \ldots, \eta_n)$ irgend ein Punkt der Mannigfaltigkeit und setzt man für $j = 1, \ldots, q$

$$\xi_j = \eta_j + v_j \lambda,$$

wobei v_j und λ neue Unbestimmte sind, so werden in der Faktorzerlegung

$$F(z, u_{q+1}, \ldots, u_n, \xi_{q+1}, \ldots, \xi_n) = \prod_v (z - u_{q+1}\xi_{q+1}^{(v)} - \cdots - u_n\xi_n^{(v)})$$

alle ξ_j algebraische Funktionen von λ.

Mindestens ein Faktor von F nimmt den Wert $z - u_{q+1}\eta_{q+1} - \cdots - u_n\eta_n$ für $\lambda = 0$ an, also durchläuft mindestens ein Punkt $\xi^{(v)}$ bei variablem λ eine algebraische Kurve, die den Punkt η enthält. Für $\lambda = 1$ aber ist $\xi^{(v)}$ ein allgemeiner Punkt der Mannigfaltigkeit, da die v_j unabhängige Unbestimmte sind. Dasselbe gilt übrigens für jeden von 0 verschiedenen konstanten Wert von λ, sowie auch für unbestimmte λ. Damit ist der Hilfssatz bewiesen.

Die Koordinaten eines Punktes einer algebraischen Kurve lassen sich in der Umgebung einer jeden Stelle in Potenzreihen nach einer Ortsuniformisierenden entwickeln. Aus dem Hilfssatz folgt also, daß es zu jedem Punkte η einer algebraischen Mannigfaltigkeit einen allgemeinen Punkt ξ derselben Mannigfaltigkeit gibt, dessen Koordinaten Potenzreihen in einer Veränderlichen t sind, die für $t = 0$ in die Koordinaten von η übergehen.

Dieses Ergebnis wenden wir nun zur Bestimmung der Kurvenpaare unserer Mannigfaltigkeit $\mathfrak{M}$ an. Es seien also $\Gamma(t)$ und $\Omega(t)$ zwei Kurven, deren Koordinaten Potenzreihen in t sind, und die durch eine projektive Transformation $T(t)$ ineinander übergehen. Für $t = 0$ mögen $\Gamma(t)$ und $\Omega(t)$ in zwei Kurven Γ und Ω übergehen, die wir untersuchen wollen.

Die Matrixelemente der projektiven Transformation $T(t)$, die $\Gamma(t)$ in $\Omega(t)$ überführt, sind natürlich algebraische Funktionen von t und können daher (eventuell nach Wahl einer neuen Ortsuniformisierenden t) auch als Potenzreihen in t angenommen werden. Wir bezeichnen die Matrix ebenfalls mit $T(t)$. Bekanntlich läßt sich eine solche Matrix durch Multi-

---

[3]) **B. L. van der Waerden**, ZAG III, Math. Annalen **108** (1933), S. 696.

plikation von vorn und hinten mit zwei Matrizes, die für $t = 0$ regulär bleiben, auf Diagonalform bringen[4])

$$Q^{-1} T(t) P = \begin{pmatrix} \alpha_0 & 0 & 0 \\ 0 & \alpha_1 & 0 \\ 0 & 0 & \alpha_2 \end{pmatrix}.$$

Dabei können α_0, α_1, α_2 sogar als Potenzen von t angenommen werden:

$$\alpha_0 = t^{-\varrho},$$
$$\alpha_1 = t^{-\sigma},$$
$$\alpha_2 = t^{-\tau}.$$

Statt der Kurven $\Gamma(t)$ und $\Omega(t)$ können wir ebensogut die Kurven $P^{-1} \Gamma(t)$ und $Q^{-1} \Omega(t)$ betrachten, welche zu $\Gamma(t)$ und $\Omega(t)$ auch für $t = 0$ projektiv äquivalent sind. Diese neuen Kurven $P^{-1} \Gamma(t)$ und $Q^{-1} \Omega(t)$ gehen durch die Transformation $Q^{-1} T(t) P$ oder

$$x_0' = t^{-\varrho} x_0,$$
$$x_1' = t^{-\sigma} x_1,$$
$$x_2' = t^{-\tau} x_2$$

auseinander hervor. Die definierenden Formen dieser Kurven seien

$$f_t = \Sigma\, a_{ikl}\, x_0^i\, x_1^k\, x_2^l,$$
$$h_t = \Sigma\, a_{ikl}\, x_0^i\, x_1^k\, x_2^l\, t^{\varrho i + \sigma k + \tau l}.$$

Der Übergang $t \to 0$ wird so gemacht, daß aus f_t und h_t die niedrigste wirklich vorkommende Potenz von t als Faktor abgespalten und dann (im anderen Faktor) $t = 0$ gesetzt wird. Es bleiben also nur die niedrigsten Potenzen übrig. Die so entstandenen Formen seien f_0 und h_0; sie definieren (bis auf zwei eigentliche projektive Transformationen P und Q) die gesuchten Kurven Γ und Ω.

Das Anfangsglied in a_{ikl} sei $\alpha_{ikl} t^{b_{ikl}}$. Wir können voraussetzen, indem wir nötigenfalls f_t durch $t^{-\alpha} f_t$ ersetzen, daß alle $b_{ikl} \geqq 0$, und daß einige $b_{ikl} = 0$ sind. Für die Kurve Γ sind dann nur die Glieder mit $b_{ikl} = 0$ von Interesse. Weiter sei ω so gewählt, daß

$$\text{alle } b_{ikl} + \varrho\, i + \sigma\, k + \tau\, l \geqq \omega, \quad \text{einige} = \omega$$

sind. Für die Kurve Ω sind nur die Glieder mit

$$b_{ikl} + \varrho\, i + \sigma\, k + \tau\, l = \omega$$

von Interesse.

---

[4]) Vgl. etwa B. L. v. d. Waerden, Moderne Algebra II, § 106.

Wir haben also drei Arten von interessanten Koeffizienten b_{ikl}:

1. die mit $b_{ikl} = 0$, $b_{ikl} + \varrho\,i + \sigma\,k + \tau\,l - \omega > 0$,
2. die mit $b_{ikl} = 0$, $b_{ikl} + \varrho\,i + \sigma\,k + \tau\,l - \omega = 0$,
3. die mit $b_{ikl} > 0$, $b_{ikl} + \varrho\,i + \sigma\,k + \tau\,l - \omega = 0$.

Für die Glieder unter 1. gilt

$$(1) \qquad \varrho\,i + \sigma\,k + \tau\,l - \omega > 0,$$

für die unter 2.

$$(2) \qquad \varrho\,i + \sigma\,k + \tau\,l - \omega = 0,$$

für die unter 3.

$$(3) \qquad \varrho\,i + \sigma\,k + \tau\,l - \omega < 0.$$

Wir lassen nun die Glieder mit höheren Exponenten von t, die auf f_0 und h_0 keinen Einfluß haben, aus f_t und h_t einfach fort. Verstehen wir unter $\sum\limits_{(1)}$, $\sum\limits_{(2)}$, $\sum\limits_{(3)}$ Summen über alle i, k, l, welche die Bedingungen (1) bzw. (2) bzw. (3) erfüllen, so bleiben folgende Formen übrig:

$$(4) \quad \begin{cases} f_t = \sum\limits_{(1)} \alpha_{ikl}\, x_0^i\, x_1^k\, x_2^l + \sum\limits_{(2)} \alpha_{ikl}\, x_0^i\, x_1^k\, x_2^l + \sum\limits_{(3)} \alpha_{ikl}\, t^{-\varrho\,i - \sigma k - \tau l + \omega}\, x_0^i\, x_1^k\, x_2^l \\[2ex] h_t = \{ \sum\limits_{(1)} \alpha_{ikl}\, t^{\varrho\,i + \sigma k + \tau l - \omega}\, x_0^i\, x_1^k\, x_2^l + \sum\limits_{(2)} \alpha_{ikl}\, x_0^i\, x_1^k\, x_2^l \\[2ex] \qquad\qquad\qquad\qquad + \sum\limits_{(3)} \alpha_{ikl}\, x_0^i\, x_1^k\, x_2^l \} \, t^\omega, \end{cases}$$

$$(5) \quad \begin{cases} f_0 = \sum\limits_{(1)} \alpha_{ikl}\, x_0^i\, x_1^k\, x_2^l + \sum\limits_{(2)} \alpha_{ikl}\, x_0^i\, x_1^k\, x_2^l \\[2ex] g_0 = \sum\limits_{(2)} \alpha_{ikl}\, x_0^i\, x_1^k\, x_2^l + \sum\limits_{(3)} \alpha_{ikl}\, x_0^i\, x_1^k\, x_2^l. \end{cases}$$

Die Formel (5) gibt in expliziter Form alle uneigentlich-projektiven Formenpaare f_0, h_0. Die Koeffizienten α_{ikl} sind beliebige Konstanten. Die Formel (4) setzt gleichzeitig in Evidenz, daß f_0 und h_0 tatsächlich Grenzelemente von projektiv äquivalenten Formen sind, daß sie also uneigentlich projektiv sind. Wir haben also

S a t z 1. *Notwendig und hinreichend dafür, daß zwei ebene Kurven uneigentlich-projektiv sind, ist, daß ihre definierenden Formen f_0 und h_0 in bezug auf zwei geeignet gewählte Koordinatensysteme die Gestalt* (5) *annehmen, wobei die ganzen Zahlen ϱ, σ, τ, ω, welche die Klasseneinteilung* (1), (2), (3) *definieren, willkürlich gewählt werden können.*

Deutet man i, k, l als baryzentrische Koordinaten (oder, was auf das gleiche hinauskommt, i und k als schiefwinklige Koordinaten) in einer Ebene, so entspricht jedem $x_0^i\, x_1^k\, x_2^l$ mit $i + k + l = n$ ein Gitterpunkt im Inneren oder auf dem Rande des Fundamentaldreieckes (vgl. Fig. 1). Die Gitterpunkte (2) liegen auf einer Geraden G, die Gitterpunkte (1) und (3) liegen auf den beiden Seiten dieser Geraden. Hat man die Lage der Geraden G, so kann man die Formen f_0 und h_0 nach (5) anschreiben.

Will man sich auf Paare von irreduziblen Kurven beschränken, so muß man die Gerade G durch eine Ecke des Fundamentaldreiecks legen; denn sonst liegen immer zwei Ecken auf der einen Seite von G und eine der Formen f_0 oder h_0 spaltet einen Faktor x_0 oder x_1 oder x_2 ab.

Beispiel 1. Kurven 3. Ordnung. Die möglichen Lagen der Geraden G, die zu Paaren von irreduziblen Kurven führen, sind in Fig. 1

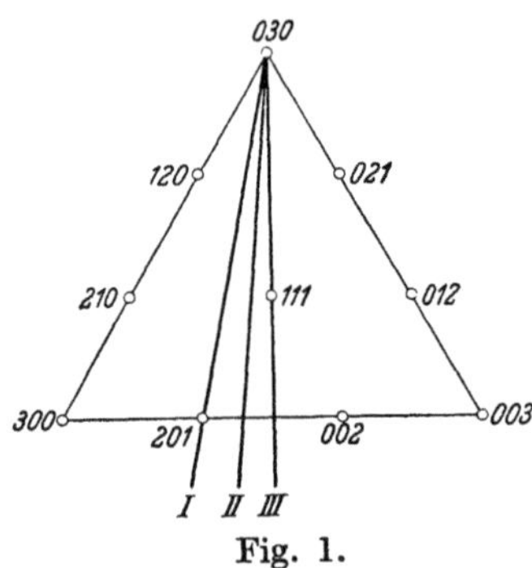

Fig. 1.

mit I, II, III angegeben. Wir nehmen etwa an, daß die Gitterpunkte links von G die Bedingung (1), die rechts von G die Bedingung (3) erfüllen. Dann definiert

im Fall I f_0 eine Kurve mit einer Spitze im Punkte (0, 0, 1), h_0 eine Kurve mit einem Wendepunkt im Punkte (1, 0, 0);

im Fall II f_0 eine Kurve mit einer Spitze im Punkte (0, 0, 1), h_0 eine Kurve mit einem Doppelpunkt im Punkte (1, 0, 0);

im Fall III f_0 eine Kurve mit einem Doppelpunkt im Punkte (1, 0, 0), h_0 eine Kurve mit einem Doppelpunkt im Punkte (0, 0, 1).

Im Fall III sind die Kurven f_0 und h_0 sogar eigentlich projektiv. Aus der Betrachtung der Fälle I und II folgt: *Eine Kurve 3. Ordnung mit einer Spitze ist zu jeder anderen Kurve 3. Ordnung uneigentlich projektiv.*

Beispiel 2. Kurven 4. Ordnung. Die Gerade G möge die Gleichung $i = l$ haben. Die Bedingung (1) besagt $i > l$, die Bedingung (2) $i = l$, die Bedingung (3) $i < l$. Nach (5) ist also

$$f_0 = \sum_{i \geq l} \alpha_{ikl}\, x_0^i\, x_1^k\, x_2^l,$$

$$h_0 = \sum_{i \leq l} \alpha_{ikl}\, x_0^i\, x_1^k\, x_2^l$$

oder voll ausgeschrieben

$$f_0 = \alpha\, x_0^4 + \beta\, x_0^3 x_1 + \gamma\, x_0^3 x_2 + \delta\, x_0^2 x_1^2 + \varepsilon\, x_0^2 x_1 x_2 + \zeta\, x_0 x_1^3$$
$$+ \eta\, x_0^2 x_2^2 + \vartheta\, x_0 x_1^2 x_2 + \iota\, x_1^4,$$

$$h_0 = \eta\, x_0^2 x_2^2 + \vartheta\, x_0 x_1^2 x_2 + \iota\, x_1^4 + \varkappa\, x_0 x_1 x_2^2 + \lambda\, x_0 x_2^3$$
$$+ \mu\, x_1^3 x_2 + \nu\, x_1^2 x_2^2 + o\, x_1 x_2^3 + \pi\, x_2^4.$$

Nehmen wir $\vartheta^2 - 4\,\iota\,\eta \neq 0$ an, so hat f_0 einen Berührungsknoten (Doppelpunkt mit zwei sich zweipunktig berührenden Zweigen) in (0, 0, 1) und h_0 einen Berührungsknoten in $(1, 0, 0)$. Sonst sind beide Kurven völlig beliebig. Wir schließen:

Zwei Kurven 4. Ordnung mit Berührungsknoten sind immer uneigentlich projektiv.

Wir bemerken noch, daß zwei solche Kurven nicht immer projektiv, ja nicht einmal birational äquivalent sein müssen. Aus dem Doppelpunkt kann man nämlich, falls keine weiteren Doppelpunkte vorhanden sind, außer der Doppelpunktstangente vier verschiedene Tangenten an die Kurve ziehen. Deren Doppelverhältnis ist eine birationale Invariante der Kurve und kann natürlich für verschiedene Kurven verschieden ausfallen.

Es sei Γ eine irreduzible Kurve mit nur gewöhnlichen Knotenpunkten, d. h. Doppelpunkten mit getrennten Tangenten. Wir wollen untersuchen, welche Kurven zu ihr uneigentlich projektiv sind.

Diejenige Seite der Geraden G, auf der die Ungleichung (1) gilt, nennen wir die linke Seite. Auf der linken Seite von G oder auf G müssen mindestens zwei Ecken des Fundamentaldreiecks liegen, da sonst f_0 reduzibel sein würde. Es seien etwa die Ecken $(n, 0, 0)$ und $(0, n, 0)$. Liegt die dritte Ecke auf G, so liegen rechts von G keine Gitterpunkte und auf G höchstens die einer Seite des Dreiecks $\varDelta$; also ist dann

$$h_0 = \alpha_0 x_1^n + \alpha_1 x_1^{n-1} x_2 + \ldots + \alpha_n x_2^n,$$

d. h. die Kurve Ω zerfällt in lauter Geraden durch einen Punkt.

Wir nehmen nun an, die dritte Ecke $(0, 0, n)$ liege rechts von G. G schneide die beiden durch diese Ecken gehenden Seiten in den Punkten $(i_1, 0, l_1)$ und $(0, k_2, l_2)$ mit $i_1 + l_1 = k_2 + l_2 = n$.
Die Gleichung von G lautet dann

$$\frac{i}{i_1} + \frac{k}{k_2} = 1.$$

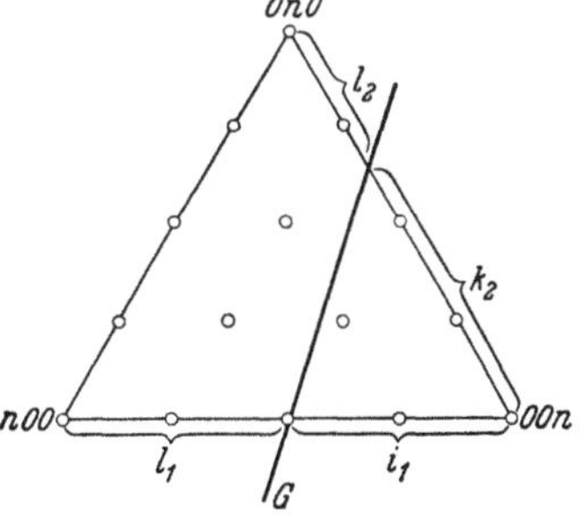

Fig. 2.

Links von G gilt $>$ statt $=$, rechts von G $<$. Für die Glieder von f_0 gilt also $\geqq$ und für die von h_0 gilt $\leqq$. Wir können $i_1 \leqq k_2$ annehmen. Dann ist $i_1 \leqq 2$, denn sonst würden in f_0 die Glieder mit x_3^n, x_3^{n-1} und x_3^{n-2} fehlen und Γ hätte einen dreifachen Punkt. Wir unter scheiden nun drei Fälle:

Fall 1. $i_1 = 2$, $k_2 > 2$. Dann fehlen in f_0 die Glieder mit x_3^n und x_3^{n-1}, sowie die mit x_3^{n-2} mit Ausnahme von $x_0^2 x_3^{n-2}$. Also hat Γ eine Spitze oder eine höhere Singularität in $(0, 0, 1)$, entgegen der Voraussetzung.

Fall 2. $i_1 = 2$, $k_2 = 2$. Dann kommen in h_0 nur die Glieder mit x_2^{n-2}, x_2^{n-1} und x_2^n vor; also besteht Ω aus einer $(n-2)$-fachen Geraden und einem Kegelschnitt.

Fall 3. $i_2 < 2$. In h_0 fehlen alle durch x_0^2 teilbaren Glieder. Die Kurve Ω hat also einen $(n-1)$-fachen oder n-fachen Punkt in $(1, 0, 0)$. Sie besteht daher aus Geraden durch diesen Punkt O und (oder) einer rationalen Kurve, welche von jeder Geraden durch O außer in O nur in einem Punkte geschnitten wird. Diese letzte Beschreibung paßt auch auf den Fall 2. Wir haben also

Satz 2. *Ist eine irreduzible Kurve Γ, die keine Singularitäten außer Knotenpunkten hat, zu einer Kurve Ω uneigentlich projektiv, so besteht Ω ausschließlich aus Geraden durch einen festen Punkt O und (oder) einer rationalen Kurve, die von jeder Geraden durch O außer in O nur in einem Punkte geschnitten wird.*

§ 2.

Die regulären Kurven vom Geschlechte p und ihre birationalen Abbildungen.

Die Brill-Noethersche geometrische Theorie der algebraischen Funktionen einer Veränderlichen wird in diesem Paragraphen als bekannt vorausgesetzt. Für eine moderne, strenge Begründung dieser Theorie siehe etwa die Dissertation von W.-L. Chow in diesem Bande der Math. Annalen (S. 655—682).

Wir nennen eine Kurve *regulär*, wenn sie keine anderen Singularitäten als gewöhnliche Knotenpunkte hat.

Satz 3. *Die regulären Kurven vom Grade n und vom Geschlechte p verteilen sich auf einige irreduzible Mannigfaltigkeiten I_ν von der Dimension $3n + p - 1$, deren allgemeine Elemente tatsächlich regulär und vom Geschlechte p sind.*

Beweis. Die Anzahl d der Knotenpunkte einer regulären Kurve n-ten Grades vom Geschlechte p ist

$$d = \frac{(n-1)(n-2)}{2} - p.$$

Wir bezeichnen mit Σ immer ein System von d verschiedenen Punkten $Q_1, \ldots, Q_d$ der Ebene, und mit Γ eine Kurve $f = 0$, welche in $Q_1, \ldots, Q_d$ Knotenpunkte hat. Die Gesamtheit aller Paare (Σ, Γ) wird durch die Gleichungen

$$(6) \qquad \frac{\partial}{\partial x_\nu} f(Q_k) = 0 \qquad (\nu = 0, 1, 2; \; k = 1, 2, \ldots, d)$$

und durch einige Ungleichungen gegeben, die ausdrücken, daß $Q_j \neq Q_k$ und daß Q_k nur ein gewöhnlicher Knotenpunkt von f ist. Wir lassen die Ungleichungen weg und betrachten die durch die Gleichungen (6) definierte algebraische Mannigfaltigkeit $\Re$, verabreden dabei aber gleichzeitig, daß diejenigen irreduziblen Teile von $\Re$, welche gar keine regulären, d. h. die Ungleichungen erfüllenden Paare (Σ, Γ) enthalten, außer Betracht zu lassen sind. Wir wollen nun beweisen, daß alle übrig bleibenden irreduziblen Teile der Mannigfaltigkeit $\Re$ genau die Dimension $N - d$ haben, wobei d die Anzahl der vorgeschriebenen Knotenpunkte Q_k und

$$N = \frac{n\,(n+3)}{2}$$

ist. Mit anderen Worten, wir beweisen, daß die durch (6) definierte Mannigfaltigkeit in jedem regulären Punkte (Σ, Γ) die Dimension $N - d$ hat. Darüber hinaus zeigt sich, daß die regulären Punkte von $\mathfrak{M}$ auch *einfache* Punkte von $\mathfrak{M}$ sind.

Die Elemente (Σ, Γ) gehören einem $(2\,d + N)$-dimensionalen $(d+1)$-fach projektiven Raum an. Wenn es sich aber nur um die Umgebung einer einzelnen Stelle (Σ, Γ) handelt, so kann man inhomogene Koordinaten einführen und zum affinen $(2\,d + N)$-dimensionalen Raum übergehen. Wenn wir nun zeigen, daß die Tangentialhyperebenen der $3\,d$ Hyperflächen (6) an jeder regulären Stelle (Σ, Γ) linear-unabhängig sind, so folgt daraus, daß ihr Durchschnitt an dieser Stelle keine höhere Dimension als $(2\,d + N) - 3\,d = N - d = 3\,n + p - 1$ haben kann. Daß die Dimension auch nicht niedriger sein kann, folgt aus einem allgemeinen Satz über Durchschnitte von Mannigfaltigkeiten mit Hyperflächen[5]).

In inhomogenen Koordinaten lauten die Gleichungen (6)

$$(7) \qquad \begin{cases} f\,(x_k,\,y_k) = 0, \\[2mm] \dfrac{\partial}{\partial x_k}\,f\,(x_k,\,y_k) = 0, \\[2mm] \dfrac{\partial}{\partial y_k}\,f\,(x_k,\,y_k) = 0. \end{cases}$$

Ist $f + df$ eine Nachbarform von f und sind $(x_k + d\,x_k,\ y_k + d\,y_k)$ Nachbarpunkte zu $(x_k,\,y_k)$, so daß $(d\,f,\ d\,x_k,\ d\,y_k)$ eine Tangentialrichtung an die Hyperflächen (7) bildet, so folgt aus (7) durch Differentiation

$$(8) \qquad \begin{cases} d\,f\,(x_k,\,y_k) + \dfrac{\partial f}{\partial x_k}\,d\,x_k + \dfrac{\partial f}{\partial y_k}\,d\,y_k = 0, \\[3mm] \dfrac{\partial}{\partial x_k}\,d\,f\,(x_k,\,y_k) + \dfrac{\partial^2 f}{\partial x_k^2}\,d\,x_k + \dfrac{\partial^2 f}{\partial x_k\,\partial y_k}\,d\,x_k = 0, \\[3mm] \dfrac{\partial}{\partial y_k}\,d\,f\,(x_k,\,y_k) + \dfrac{\partial^2 f}{\partial x_k\,\partial y_k}\,d\,x_k + \dfrac{\partial^2 f}{\partial y_k^2}\,d\,y_k = 0. \end{cases}$$

Wir haben zu zeigen, daß diese linearen Gleichungen für die $d\,x_k,\ d\,y_k$ und die Koeffizienten der Form $d\,f$ untereinander linear-unabhängig sind.

---

[5]) B. L. v. d. Waerden, ZAG I, § 2, Math. Annalen **108**, S. 120.

Im Fall eines gewöhnlichen Knotenpunktes ist $f_{xx}f_{yy} - f^2_{xy} \neq 0$, also kann man aus den letzten beiden Gleichungen (8) die Differentiale $d\,x_k$ und $d\,y_k$ berechnen. Die erste Gleichung (8) ergibt wegen (7), wenn $df = g$ gesetzt wird,

$$(9) \qquad\qquad g\,(x_k,\, y_k) = 0,$$

d. h., die Kurve $g = 0$ geht durch den Punkt Q_k hindurch. Wir haben noch zu zeigen, daß die linearen Gleichungen (9) für die Koeffizienten der Form g linear unabhängig sind.

Dieser Nachweis wird wie bei Severi[1]) durch eine einfache Abzählung geführt. Die Bedingungen (9) drücken aus, daß die Kurve n-ter Ordnung g zu der Kurve n-ter Ordnung f adjungiert ist. Die lineare Schar dieser adjungierten Kurven schneidet aus f eine Vollschar von Punktgruppen von der Ordnung $n^2 - 2\,d = n^2 - (n-1)\,(n-2) + 2\,p = 3\,n + 2\,p - 2$ aus, welche keine Spezialschar ist und daher nach dem Riemann-Rochschen Satz die Dimension $3\,n + p - 2$ hat. Die Dimension der Schar der Kurven g ist, da die Kurve f selbst auch noch zu dieser Schar gehört, $3\,n + p - 1$, d. h. es gibt $3\,n + p$ linear-unabhängige Formen g, welche die Adjungiertheitsbedingungen (9) erfüllen. Da es insgesamt $\frac{(n+1)\,(n+2)}{2}$ linear-unabhängige Formen n-ten Grades gibt, so gibt es unter den linearen Gleichungen (9) genau

$$\frac{(n+1)\,(n+2)}{2} - 3\,n - p = \frac{(n-1)\,(n-2)}{2} - p = d$$

linear-unabhängige, was zu beweisen war[6]).

Die Mannigfaltigkeit $\mathfrak{K}$ der Paare $(\Sigma,\, \Gamma)$ besteht also aus lauter Bestandteilen der Dimension $N - d$. Es sei $\mathfrak{K}_\nu$ ein solcher Bestandteil. Jeder Kurve Γ entsprechen nur endlich viele Punktsysteme Σ, also bilden die Kurven Γ, die zu den Paaren $(\Sigma,\, \Gamma)$ von $\mathfrak{K}_\nu$ gehören, eine irreduzible Mannigfaltigkeit I_ν von der Dimension $N - d = 3\,n + p - 1$. Die allgemeine Kurve von I_ν kann außer den d Doppelpunkten nicht noch weitere Singularitäten haben, denn sonst wäre die Dimension des Systems höchstens $N - d - 1$. Also hat die allgemeine Kurve von I_ν wirklich das Geschlecht

$$p = \frac{(n-1)\,(n-2)}{2} - d$$

und ist regulär.

Damit ist Satz 3 in allen Teilen bewiesen.

---

[6]) Dieser Beweis gilt bei sinngemäßer Anwendung des Riemann-Rochschen Satzes auch für reduzible Kurven. Man kann aber auch, wie Severi es tut, mittels des Noetherschen Fundamentalsatzes den Fall der reduziblen Kurven auf den der irreduziblen zurückführen.

Genau wie bei Severi[1]) beweist man weiter mit Hilfe vollständig zerfallender Kurven, daß es für $d \leqq \frac{1}{2} n (n - 1)$ tatsächlich reguläre Kurven vom Geschlecht $p = \binom{n-1}{2} - d$ gibt, und daß es für $d \leqq \frac{1}{2} (n - 1) (n - 2)$ sogar irreduzible reguläre Kurven vom Geschlecht p gibt, daß also unsere Mannigfaltigkeiten nicht leer sind.

Severi hat weiter bewiesen, daß die irreduziblen Kurven n-ten Grades mit d Knotenpunkte für $d \leqq \frac{1}{2} (n - 1) (n - 2)$ eine einzige irreduzible Mannigfaltigkeit bilden, d. h. also, daß es nicht mehrere Systeme I_ν, sondern nur ein System I_1 gibt. Da wir dieses Ergebnis hier aber nicht brauchen, so werden wir weiterhin von den Systemen I_ν reden.

S a t z 4. *Die zu einer gegebenen irreduziblen Kurve Γ des Geschlechtes p und der Ordnung $n > 2p - 2$ birational äquivalenten Kurven der gleichen Ordnung gehören einer irreduziblen Mannigfaltigkeit von der Dimension $3n - 2p + 2 - \varrho$ an, deren allgemeines Element tatsächlich irreduzibel und zu Γ birational äquivalent ist. Dabei ist $\varrho = 0$ für $p > 1$, $\varrho = 1$ für $p = 1$ und $\varrho = 3$ für $p = 0$ (es gibt ∞^ϱ birationale Transformationen von Γ in sich).*

B e w e i s. Jede birationale Abbildung wird durch eine Schar g_n^2 vermittelt. Diese ist in einer Vollschar g_n^r enthalten, welche durch eine einzelne Punktgruppe $P_1, \ldots, P_n$ bestimmt wird und durch adjungierte Kurven ausgeschnitten wird. Die Schar ist wegen $n > 2p - 2$ nicht spezial, also ist $r = n - p$, und $P_1, \ldots, P_n$ können alle verschieden und außerhalb der vielfachen Punkte von Γ gewählt werden. Durch $P_1, \ldots, P_n$ kann man immer eine adjungierte Kurve von der Ordnung $n - 1$ legen, denn die Schar der adjungierten Kurven von der Ordnung $n - 1$ hat nach bekannten Rechnungen die Dimension $p + 2n - 2 \geqq n$. Welche adjungierte Kurve man wählt, ist für die Vollschar g_n^r vollkommen gleichgültig, da diese allein durch die Punktgruppe $P_1, \ldots, P_n$ bestimmt wird. Man kann also von vornherein aus der Schar der adjungierten Kurven der Ordnung $n - 1$ eine Teilschar der Dimension n auswählen, von welcher durch $P_1, \ldots, P_n$ nur eine einzige Kurve geht. Diese adjungierte Kurve $g = 0$ schneidet die Kurve außer in $P_1, \ldots, P_n$ und außer in den Doppelpunkten $Q_1, \ldots, Q_d$ noch in Punkten $R_1, \ldots, R_s$. Drei willkürliche adjungierte Kurven $g = 0$, $g_1 = 0$, $g_2 = 0$ durch $R_1, \ldots, R_s$ schneiden aus der Kurve Γ drei Punktgruppen der Schar g_n^2 aus. Die gesuchte birationale Abbildung wird dann durch

$$(10) \qquad \begin{cases} y_0 = g_0 (x), \\ y_1 = g_1 (x), \\ y_2 = g_2 (x) \end{cases}$$

gegeben.

Die Punktgruppen $P_1, \ldots, P_n$ auf Γ bilden eine irreduzible n-dimensionale Mannigfaltigkeit: alle sind relationstreue Spezialisierungen einer allgemeinen Punktgruppe. Durch $P_1, \ldots, P_n$ ist die Form g rational bestimmt. Die Koordinaten von $R_1, \ldots, R_s$ sind algebraische Funktionen, ihre symmetrischen Funktionen rationale Funktionen der Koeffizienten von g und von der Gleichung $f = 0$ von Γ. g_0, g_1, g_2 sind Lösungen von linearen Gleichungen: sie müssen zu Γ adjungiert sein und durch $R_1, \ldots, R_s$ gehen. Sie gehen also durch relationstreue Spezialisierung aus der allgemeinen Lösung dieses Gleichungssystems hervor. Die allgemeine Lösung ist eine Linearkombination mit unbestimmten Koeffizienten von gewissen speziellen Lösungen, in denen gewisse Koordinaten $= 0$ und andere $= 1$ sind. Diese speziellen Lösungen hängen symmetrisch von den Koordinaten von $R_1, \ldots, R_s$ und daher rational von den Koeffizienten von g ab. Die Anzahl der unbestimmten Parameter in der allgemeinen Lösung ist $r + 1 = n - p + 1$, wobei $r = n - p$ die Dimension der Vollschar g_n^r ist. In den drei Lösungen g_0, g_1, g_2 stecken also $3(n - p + 1)$ Parameter, von denen ein willkürlicher Proportionalitätsfaktor, der für (10) nichts ausmacht, abzuziehen ist. Zählt man noch die n Parameter hinzu, von der die Punktgruppe $P_1, \ldots, P_n$ abhängt, so erhält man $4n - 3p + 2$ Parameter. Für allgemeine Parameter ist die Schar g_n^2 einfach und frei von festen Punkten, also die Abbildung (10) birational. Die Bildkurve hat den Grad n und ihre Gleichung $h = 0$ ist durch die Bedingung

$$(11) \qquad \gamma\, h\big(g_0(x),\ g_1(x),\ g_2(x)\big) = q(x)\, f(x)$$

eindeutig und rational bestimmt. Eine Mehrdeutigkeit in der Bestimmung der Kurve h, die nach (11) alle Punkte der Bildkurve enthalten muß, könnte nämlich nur dann eintreten, wenn die Bildkurve einen Grad $< n$ hätte. Man erhält ein lineares Gleichungssystem für die Koeffizienten von h, indem man aus (11) die Koeffizienten von $q(x)$ und den Faktor γ, die alle linear homogen vorkommen, eliminiert.

Das Gebilde, das aus n Punkten $P_1, \ldots, P_n$, einer Abbildung (10) und einer Bildkurve Ω besteht, hängt also von $4n - 3p + 2$ wesentlichen Parametern derart ab, daß man für unbestimmte Werte der Parameter ein „allgemeines Gebilde" erhält, aus welchem alle speziellen birationalen Abbildungen durch relationstreue Spezialisierung entstehen. Die Gebilde dieser Art bilden also eine irreduzible algebraische Mannigfaltigkeit von der Dimension $4n - 3p + 2$.

Daher bilden auch die Bildkurven Ω eine irreduzible Mannigfaltigkeit. Ihre Dimension wird nach dem Prinzip der Konstantenzählung gefunden, indem man von der Zahl $4n - 3p + 2$ die Zahl der Freiheitsgrade subtrahiert, die für die Punkte $P_1, \ldots, P_n$ und für die Abbildung (10) noch zur Verfügung stehen, wenn die Bildkurve Ω *vorgegeben* ist.

Die Kurve Γ hat bekanntlich, wenn $\varrho = 0$ für $p > 1$, $\varrho = 1$ für $p = 1$ und $\varrho = 3$ für $p = 0$ gesetzt wird, ∞^ϱ birationale Transformationen in sich. Bei gegebenen birational äquivalenten Kurven Γ, Ω gibt es also ∞^ϱ birationale Abbildungen von Γ auf Ω. Bei einer gegebenen Abbildung kann die Punktgruppe $(P_1, \ldots, P_n)$ noch willkürlich innerhalb einer Vollschar von der Dimension $n - p$ gewählt werden. Die gesuchte Anzahl der Freiheitsgrade ist also $n - p + \varrho$. Für die Dimension der Mannigfaltigkeit der Kurven Ω erhält man somit

$$(4\,n - 3\,p + 2) - (n - p + \varrho) = 3\,n - 2\,p + 2 - \varrho.$$

Damit ist Satz 4 vollständig bewiesen.

Es sei noch bemerkt, daß die algebraische Abhängigkeit der Bildkurve Ω von den angegebenen Parametern auch dann noch bestehen bleibt, wenn man außerdem noch die Koordinaten der Kurve Γ veränderlich macht. Man beschränkt sich dabei am besten auf reguläre Kurven und läßt die Kurve Γ und das System ihrer Doppelpunkte eines der im Beweis von Satz 3 erwähnten irreduziblen Systeme $\mathfrak{R}_\nu$ durchlaufen. Die Berechnung der Formen g, g_1, g_2, g_3, h verläuft dann reibungslos nach dem obigen Schema, und man findet *eine irreduzible Mannigfaltigkeit von Kurvenpaaren* (Γ, Ω), *deren allgemeines Element ein birational äquivalentes Kurvenpaar ist, und die alle birational äquivalenten Kurvenpaare umfaßt, bei denen Γ regulär ist und dem System I_ν angehört.*

Bis hierher sind wir im wesentlichen Severi gefolgt. Aus den Sätzen 3 und 4 würde nach dem Prinzip der Konstantenzählung ohne weiteres folgen, daß die Klassen birational äquivalenter Kurven von $3\,p - 3 + \varrho$ wesentlichen Konstanten abhängen, wenn man eine algebraische Korrespondenz hätte, welche diese Klassen eineindeutig auf Punkte einer Bildmannigfaltigkeit abbilden würde. In der Herstellung dieser Korrespondenz liegt aber erst die eigentliche Schwierigkeit, die wir in § 3 zu lösen haben.

§ 3.

Grenzfälle birational äquivalenter Kurvenpaare.

Wir untersuchen die Grenzfälle birationaler Äquivalenz in derselben Weise, wie wir in § 1 die Grenzfälle algebraischer Äquivalenz untersucht haben. Dabei sprechen wir von einem „Grenzübergang" $t \to 0$, wobei in Wirklichkeit ein rein algebraischer Prozeß gemeint ist: Man setzt in einem System von Potenzreihen, welche algebraische Funktionen darstellen, $t = 0$. Handelt es sich dabei um homogene Koordinaten eines Punktes (oder einer Kurve, einer Ebene usw.), so sind die fraglichen Potenzreihen vorher mit einer passenden Potenz zu multiplizieren, damit für $t = 0$ keine Koordinaten unendlich und nicht alle Null werden können. In diesem Sinne hat jeder von t abhängige Punkt, dessen Koordinaten Potenzreihen in t sind, einen Grenzpunkt für $t \to 0$.

Es seien $T(t)$ und $\Omega(t)$ zwei birational äquivalente, von t abhängige Kurven vom Grade n. Die rationale Abbildung von $\Gamma(t)$ auf $\Omega(t)$ sei durch (10) gegeben. Die Formen g_0, g_1, g_2 können wieder vom Grade $n - 1$ gewählt werden.

Die Formen g_0, g_1, g_2 sind für allgemeines t linear unabhängig, brauchen es aber für $t = 0$ nicht zu bleiben. Die lineare Schar $\lambda_0 g_0 + \lambda_1 g_1 + \lambda_2 g_2$ hat jedoch für $t \to 0$ eine bestimmte Grenzlage. Um das einzusehen, bilde man die Kurven $(n - 1)$-ter Ordnung auf die Punkte eines projektiven Raumes S_N ab; einer linearen Schar $\lambda_0 g_0 + \lambda_1 g_1 + \lambda_2 g_2$ entspricht dann eine Ebene $\varepsilon(t)$ in S_N, die durch ihre Plückerschen Koordinaten gegeben wird. Für $t = 0$ hat diese Ebene eine bestimmte Grenzlage $\varepsilon(0)$. Es ist leicht möglich, in der Ebene $\varepsilon(t)$ drei Punkte auszuwählen, deren Koordinaten Potenzreihen in t sind, und welche für $t = 0$ drei linear unabhängige Punkte von $\varepsilon(0)$ ergeben. (Es genügt z. B., $\varepsilon(t)$ mit drei passend gewählten Räumen S_{N-2} in S_N zum Schnitt zu bringen.) Diesen drei Punkten entsprechen nun wieder Formen $g_0^*(x, t)$, $g_1^*(x, t)$, $g_2^*(x, t)$, welche sowohl für allgemeine t als auch für $t = 0$ linear unabhängig sind und von denen g_0, g_1, g_2 drei Linearkombinationen sind.

Wir nehmen nun Γ als *irreduzible* Kurve an. Die Abbildung

$$(12) \quad \begin{aligned} y_0^* &= g_0^*(x, t), \\ y_1^* &= g_1^*(x, t), \\ y_2^* &= g_2^*(x, t) \end{aligned}$$

bildet $\Gamma(t)$ auf eine Kurve $\Delta(t)$ ab, die zu $\Omega(t)$ projektiv äquivalent ist. Durch Einschaltung dieser Kurve $\Delta(t)$ haben wir das Problem der Bestimmung der Kurvenpaare Γ, Ω in zwei Teilprobleme gespalten; Γ, Δ bilden einen Grenzfall eines birational äquivalenten Kurvenpaares $\Gamma(t)$, $\Delta(t)$, wobei die Abbildungsfunktionen g_ν^* noch für $t = 0$ linear unabhängig bleiben, während Δ, Ω einen Grenzfall eines projektiv äquivalenten Kurvenpaares, also ein uneigentlich projektives Kurvenpaar im Sinne von § 1 bilden.

Die Abbildung (12) ergibt für $t = 0$ wegen der linearen Unabhängigkeit der $g_\nu^*(0)$ jedenfalls eine rationale Abbildung der Kurve Γ auf eine Kurve Δ^*, welche keine Gerade ist. Die Kurve Δ^* kann natürlich einen niedrigeren Grad als n haben und die Abbildung braucht nicht birational zu sein. Es gilt aber folgendes: Eine beliebige Kurve $\Sigma \lambda_\nu g_\nu^*(x, t)$ schneidet die Kurve $\Gamma(t)$ außer in gewissen festen Punkten in n Punkten $\xi^{(1)}, \ldots, \xi^{(n)}$, welche durch die Abbildung (12) in die n Schnittpunkte der Kurve $\Delta(t)$ mit der Geraden $\Sigma \lambda_\nu y_\nu^* = 0$ übergeführt werden. Für $t \to 0$ streben $\xi^{(1)}, \ldots, \xi^{(n)}$ gegen gewisse Grenzpunkte $\xi_0^{(1)}, \ldots, \xi_0^{(n)}$. Die Tatsache, daß die Bildpunkte $g(\xi^{(1)}), \ldots, g(\xi^{(n)})$ auf $\Delta(t)$ und auf $\Sigma \lambda_\nu y_\nu^* = 0$ liegen, bleibt natürlich für $t = 0$ bestehen, soweit diese Bildpunkte nicht unbestimmt werden. Also ist Δ^* als Bestandteil in Δ enthalten. Ist nun die Abbildung für $t = 0$ nicht birational, so ist die durch $\Sigma \lambda_\nu g_\nu^*(x, 0)$ aus Γ ausgeschnittene Schar nicht mehr einfach, also ergibt eine gewisse Anzahl von Schnittpunkten, etwa $\xi^{(1)}, \ldots, \xi^{(\mu)}$, denselben Bildpunkt $g^*(\xi)$. Das heißt, für $t \to 0$ rücken die Punkte $g^*(\xi^{(1)}), \ldots, g^*(\xi^{(\mu)})$, Schnittpunkte von Δ mit $\Sigma \lambda_\nu y_\nu^* = 0$, in einen Punkt zusammen. Dieser Punkt muß dann ein μ-facher Schnittpunkt sein, und da die Gerade $\Sigma \lambda_\nu y_\nu^* = 0$ ganz beliebig ist, so ist der Bestandteil notwendigerweise μ-mal in Δ enthalten. *Die Kurve Δ enthält also als Bestandteil entweder eine zu Γ birational-äquivalente Kurve Δ^* oder eine mehrfach gezählte Kurve Δ^*, die keine Gerade ist.*

Nun setzen wir voraus, daß Γ und Ω beide irreduzibel sind, daß Γ nicht rational ist und daß Ω keine anderen Singularitäten als Knotenpunkte hat. Dann enthält auf Grund des letzten Satzes Δ entweder einen nicht rationalen oder einen mehrfach gezählten nicht linearen Bestandteil. Demnach treffen für Δ die Behauptungen von Satz 2 nicht zu, also kann Δ zu Ω nicht uneigentlich projektiv sein. Daher sind Δ und Ω eigentlich projektiv, d. h. Δ ist irreduzibel und vom Grade n. Dann ist aber Γ zu Δ, also auch zu Ω birational äquivalent.

Damit ist bewiesen:

Satz 5. *Wenn zwei Kurven n-ten Grades Γ und Δ beide irreduzibel sind und wenn Γ nicht rational und Ω regulär ist (d. h. nur gewöhnliche Knotenpunkte hat), und wenn das Paar (Γ, Ω) einen Grenzfall eines birational äquivalenten Kurvenpaares bildet, so sind Γ und Ω selber birational äquivalent.*

Betrachten wir jetzt die in § 2 zuletzt konstruierte irreduzible Mannigfaltigkeit von Kurvenpaaren, deren allgemeines Element (Γ, Ω) birational äquivalent ist. Aus Satz 5 folgt, daß die speziellen Kurvenpaare (Γ, Ω) dieser Mannigfaltigkeit, soweit Γ regulär ist und Γ, Ω beide irreduzibel sind, ebenfalls birational äquivalent sind. Wir haben also

Satz 6. *Die Paare (Γ, Ω) von irreduziblen birational äquivalenten Kurven n-ten Grades vom Geschlechte $p > 1$, bei denen mindestens eine Kurve* *regulär ist, verteilen sich auf endlich viele irreduzible Mannigfaltigkeiten, derart, daß auch umgekehrt jedes Paar (Γ, Ω) einer solchen Mannigfaltigkeit birational äquivalent ist, sofern nur die Kurven Γ und Ω beide irreduzibel und von positivem Geschlecht sind, und mindestens eine von ihnen regulär ist.*

Die Mannigfaltigkeit dieser Paare (Γ, Ω) heißt im folgenden die „Korrespondenz $\mathfrak{B}$", ihre irreduziblen Bestandteile, bei denen Γ und Ω zu I_ν gehören, heißen $\mathfrak{B}_\nu$.

§ 4.

Die Modulmannigfaltigkeit.

Wir beschränken uns auf reguläre irreduzible Kurven vom Geschlechte p eines Systems I_ν (§ 2). Die Korrespondenz $\mathfrak{B}_\nu$ ordnet jeder solchen Kurve Γ eine $(3\,n - 2\,p + 2 - \varrho)$-dimensionale irreduzible Mannigfaltigkeit von Kurven Ω zu, die (wieder unter Beschränkung auf reguläre irreduzible Kurven) zu Γ birational äquivalent sind. Diese Mannigfaltigkeit bezeichnen wir mit $\mathfrak{P}_\Gamma$ und nehmen dabei Γ zunächst als allgemeines Element von I_ν an. Die Mannigfaltigkeit $\mathfrak{P}_\Gamma$ kann nach ZAG IX[2]) durch einen Punkt P_Γ eines Raumes S_h dargestellt werden. Durchläuft Γ die Mannigfaltigkeit I_ν, so durchläuft $\mathfrak{P}_\Gamma$ nach ZAG IX, Satz 5, ein irreduzibles algebraisches System von Mannigfaltigkeiten, d. h. der Bildpunkt P_Γ durchläuft eine irreduzible algebraische Mannigfaltigkeit $\mathfrak{M}_\nu$. Nach Satz 4 (s. o. § 2) ist einem speziellen Punkt Γ ebenfalls eine genau $(3\,n - 2\,p + 2 - \varrho)$-dimensionale irreduzible Mannigfaltigkeit $\mathfrak{P}_\Gamma$ zugeordnet. Zählt man diese mit der richtigen Vielfachheit, so gehört sie (nach Satz 4 aus ZAG IX) ebenfalls dem algebraischen System von Mannigfaltigkeiten an, d. h. ihr Bildpunkt P_Γ gehört zu $\mathfrak{M}_\nu$. Wir erhalten so eine irreduzible Korrespondenz zwischen den Kurven Γ und den Bildpunkten P_Γ, welche jeder Kurve einen Bildpunkt P_Γ zuordnet, während allen zu Γ birational äquivalenten Kurven, soweit sie wieder regulär sind, derselbe Bildpunkt entspricht. Wir nennen $\mathfrak{M}_\nu$ die *Modulmannigfaltigkeit* des irreduziblen Kurvensystems I_ν. Wenden wir auf die eben erwähnte Korrespondenz das Prinzip der Konstantenzählung an, so folgt, da I_ν die Dimension $3\,n + p - 1$ hat, für die Dimension von $\mathfrak{M}_\nu$ der Wert

$$(3\,n + p - 1) - (3\,n - 2\,p + 2 - \varrho) = 3\,p - 3 + \varrho.$$

Damit haben wir den Hauptsatz dieser Arbeit bewiesen:

Satz 7. *Es gibt eine algebraische Korrespondenz, welche jeder regulären Kurve Γ vom Geschlechte $p \geqq 1$ des irreduziblen Kurvensystems I_ν einen Punkt P_Γ einer $(3\,p - 3 + \varrho)$-dimensionalen irreduziblen Modul-*

mannigfaltigkeit $\mathfrak{M}$, *eindeutig zuordnet, derart, daß zwei reguläre Kurven dann und nur dann birational äquivalent sind, wenn ihnen derselbe Punkt von* $\mathfrak{M}_\nu$ *zugeordnet wird. Dabei ist* $\varrho = 1$ *für* $p = 1$ *und* $\varrho = 0$ *für* $p > 1$.

In Wirklichkeit gibt es, wie wir in § 2 schon bemerkten, nur *ein* irreduzibles Kurvensystem I_1 und daher auch nur *eine* irreduzible Modulmannigfaltigkeit $\mathfrak{M}_1$. Die obige schwächere Formulierung wurde nur gewählt in dem Bestreben, in dieser Arbeit nur solche Sätze aufzustellen, deren vollständiger Beweis aus der Arbeit selbst entnommen werden kann.

Da die Korrespondenz in einer Richtung eindeutig ist, so ist sie rational. Die Moduln einer allgemeinen Kurve des irreduziblen Systems I_ν sind also rationale Funktionen der Koeffizienten der Gleichung dieser Kurve.

(Eingegangen am 19. 3. 1937.)

19.

Zur algebraischen Geometrie XII
Ein Satz über Korrespondenzen und die Dimension einer Schnittmannigfaltigkeit

Mathematische Annalen 115 (1938) 330–332

In ZAG VI, § 2 findet sich[1]) der folgende Satz:

I. *Wenn in einer irreduziblen Korrespondenz K zwischen M und N einem allgemeinen Punkt ξ von M eine b-dimensionale Mannigfaltigkeit von Punkten η entspricht, so entspricht jedem speziellen Punkt ξ' von M eine Mannigfaltigkeit von Punkten η', welche keine irreduziblen Bestandteile von einer Dimension $< b$ enthält.*

Der Beweis dieses Satzes enthielt eine Lücke, die später ausgefüllt werden konnte[2]). Ich werde jetzt einen allgemeingültigen Beweis geben, der auf der von Chow und mir[3]) entwickelten Theorie der algebraischen Systeme von Mannigfaltigkeiten beruht.

Aus Satz I wird sodann folgender Satz hergeleitet, der eine mir von J. F. Ritt gestellte Frage beantwortet:

II. *Der Durchschnitt D zweier Mannigfaltigkeiten M_r und M_s von den Dimensionen r und s im projektiven Raum S_n besitzt keine irreduziblen Bestandteile von einer Dimension $< r + s - n$.*

Beweis von I. Wir bezeichnen mit N_ξ die Mannigfaltigkeit von der Dimension b und vom Grade g, die dem Punkte ξ in der Korrespondenz entspricht, und ebenso mit $N_{\xi'}$ die dem Punkte ξ' entsprechende Mannigfaltigkeit. Die Gleichungen der Korrespondenz ergeben, wenn man den Punkt ξ bzw. ξ' in ihnen einsetzt, unmittelbar die Gleichungen der Mannigfaltigkeiten N_ξ und $N_{\xi'}$. Wir bilden nach ZAG IX[3]) die Mannigfaltigkeit N_ξ auf einen Bildpunkt ζ in einem Bildraum B ab und bezeichnen allgemein die einem Bildpunkt ζ' entsprechende Mannigfaltigkeit von der Dimension b und vom Grade g mit $M(\zeta')$. Dann ist also $N_\xi = M(\zeta)$.

---

[1]) B. L. v. d. Waerden, Math. Annalen **110** (1934), S. 142.

[2]) Zur algebraischen Geometrie, Berichtigung und Ergänzungen, Math. Annalen **113** (1936), S. 38.

[3]) W.-L. Chow und B. L. v. d. Waerden, ZAG IX, Math. Annalen **113** (1937), S. 699.

Ist nun η' irgendein Punkt von $N_{\xi'}$, so ist (ξ', η') ein Paar der Korrespondenz, also eine relationstreue Spezialisierung von (ξ, η). Diese läßt sich zu einer relationstreuen Spezialisierung (ξ', η', ζ') von (ξ, η, ζ) fortsetzen. Die algebraischen Relationen, die ausdrücken, daß

  a) ζ Bildpunkt einer Mannigfaltigkeit $M(\zeta)$ ist,

  b) η zu $M(\zeta)$ gehört,

  c) $M(\zeta)$ in $N(\xi)$ enthalten ist,

bleiben bei der Spezialisierung $\xi \to \xi'$, $\eta \to \eta'$, $\zeta \to \zeta'$ erhalten. Also ist ζ' Bildpunkt einer rein b-dimensionalen Mannigfaltigkeit $M(\zeta')$, die den Punkt η'

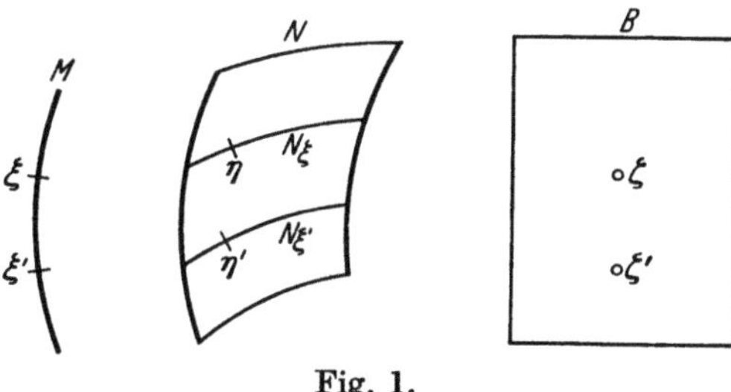

Fig. 1.

enthält und in $N(\xi')$ enthalten ist. Also ist η' in einem mindestens b-dimensionalen irreduziblen Bestandteil von $N(\xi')$ enthalten, und das gilt für jeden Punkt η' von $N(\xi')$.

Beweis von II. Wir ordnen jedem Punkte ξ von M_r alle projektiven Transformationen T zu, welche ξ in Punkte von M_s überführen. So erhält man eine Korrespondenz K zwischen ξ und T. Die Gleichungen von K erhält man, indem man zum Ausdruck bringt, daß ξ in M_r und

$$(1) \qquad \eta_j = \sum \tau_{jk}\,\xi_k$$

in M_s liegt. Ein allgemeines Elementepaar der Korrespondenz erhält man, indem man für ξ einen allgemeinen Punkt von M_r wählt, unabhängig davon einen allgemeinen Punkt η auf M_s annimmt und für τ_{jk} die allgemeine Lösung des linearen Gleichungssystems (1) nimmt. Alle speziellen Paare (ξ', T') der Korrespondenz entstehen offensichtlich aus dem angegebenen Paar (ξ, T) durch relationstreue Spezialisierung. Die Korrespondenz ist also irreduzibel.

Fragt man nun, welche Punkte ξ einer allgemeinen Projektivität T in der Korrespondenz entsprechen, so sind es die Punkte, die sowohl zu M_r als zu $T^{-1} M_s$ gehören. Es ist bekannt[4]), daß für $r + s = n$ und somit um

---

[4]) Eine Verallgemeinerung des Bézoutschen Satzes, Math. Annalen **99** (1928), Satz 30, S. 531. Ein einfacherer Beweis wird in der übernächsten Abhandlung ZAG XIV gegeben werden.

so mehr für $r + s \geqq n$ die Mannigfaltigkeiten M_r und $T^{-1} M_s$ wirklich gemeinsame Punkte haben. Wendet man nun auf die Korrespondenz K das Prinzip der Konstantenzählung[5]) an, so folgt sofort, daß der Durchschnitt von M_r und $T^{-1} M_s$ für allgemeine T die Dimension $r + s - n$ hat. Nach Satz I folgt daraus, daß für irgendein T, insbesondere für $T = I$ (die Identität), der Durchschnitt von M_r und $T^{-1} M_s$ keine Bestandteile von einer Dimension $< r + s - n$ besitzt.

------

[5]) Vgl. ZAG VI, Math. Annalen **110**, § 1, S. 139. In der Gleichung $a + b = c + d$, die das Prinzip der Konstantenzählung ausdrückt, ist in unserem Fall

$$a = r$$
$$b = s + n^2 + n$$
$$c = n^2 + 2 n$$

also

$$d = r + s - n.$$

(Eingegangen am 5. 8. 1937.)

20.

Zur algebraischen Geometrie XIII
Vereinfachte Grundlagen der algebraischen Geometrie

Mathematische Annalen 115, 3 (1938) 359–378

In einer Reihe von Arbeiten[1]) habe ich versucht, eine rein algebraische
und strenge Grundlage der algebraischen Geometrie zu schaffen. Da diese
Arbeiten sich über mehr als zehn Jahre erstrecken, war es unvermeidlich,
daß die in früheren Arbeiten entwickelten Begriffe später etwas verallgemeinert und abgewandelt wurden, daß der Schwerpunkt sich manchmal
verschoben hat, daß Begriffsbildungen durch andere ersetzt wurden. Auch
ergaben sich starke Vereinfachungen vor allem dadurch, daß die schwierigen
idealtheoretischen Hilfsmittel, die anfangs herangezogen wurden, sich als
unnötig herausgestellt haben. Schließlich bin ich durch Zuschriften und eigene
Vorlesungserfahrung auf die schwierige, manchmal etwas zu knappe Darstellung in einigen der erwähnten Arbeiten aufmerksam geworden. Aus allen
diesen Gründen erschien eine neue, vereinfachte und verbesserte Darstellung
der Grundlagen der algebraischen Geometrie angebracht. In dieser Arbeit
beschränke ich mich in der Hauptsache darauf, die Ergebnisse der Arbeiten
„Nullstellentheorie“, „Multiplizitätsbegriff“ und ZAG I sowie den Anfang
von ZAG VI neu darzustellen; in einer anschließenden Arbeit (ZAG XIV)
sollen die von „Bézout“ neu bewiesen und erweitert werden. Dabei wird aus
der Idealtheorie ausschließlich der Hilbertsche Basissatz (vgl. meine Moderne
Algebra II, § 80), ferner aus der Eliminationstheorie der Hilbertsche Nullstellensatz und die Existenz des Resultantensystems für homogene Formen
benutzt werden. Aus der Algebra wird sonst nur die Körpertheorie (Moderne
Algebra I, Kap. 3, 4 und 10) als bekannt vorausgesetzt, aus der Geometrie
der Begriff des projektiven Raumes und die einfachsten Sätze über lineare
Unterräume.

---

[1]) Zur Nullstellentheorie der Polynomideale, Math. Annalen **96** (1926), S. 183—208.
Zitiert „Nullstellentheorie“. — Der Multiplizitätsbegriff der algebraischen Geometrie,
ebenda **97** (1927), S. 756—774. Zitiert „Multiplizitätsbegriff“. — Eine Verallgemeinerung des Bézoutschen Theorems, ebenda **99** (1928), S. 497—541. Zitiert „Bézout“.
— Zur algebraischen Geometrie I, ebenda **108** (1933), S. 113—125. Zitiert ZAG I. —
Vgl. auch die weiteren Arbeiten ZAG III (ebenda **108**), ZAG V und VI (ebenda **110**)
und vor allem ZAG IX (ebenda **113**, gemeinsam mit W.-L. Chow).

Die vorliegende Arbeit ZAG XIII sowie die vorangehenden ZAG V, VI und IX und die nachfolgende ZAG XIV werden die Grundlage meiner weiteren Untersuchungen zur algebraischen Geometrie bilden.

§ 1.

Algebraische Mannigfaltigkeiten. Zerlegung in irreduzible.

Die Definitionen der algebraischen Mannigfaltigkeit im projektiven Raum S_n, der Summe, des Durchschnitts, der Irreduzibilität von Mannigfaltigkeit in bezug auf einen Grundkörper K entnimmt man am besten aus der Dissertation von W.-L. Chow[2]). Die von Chow entwickelten Definitionen zeichnen sich dadurch aus, daß die Koeffizienten der Gleichungen der algebraischen Mannigfaltigkeiten immer aus einem beliebigen, aber festen Grundkörper K, die Koordinaten der Punkte der Mannigfaltigkeit dagegen aus beliebigen algebraischen oder transzendenten Erweiterungskörpern von K entnommen werden. Die Möglichkeit der Darstellung einer algebraischen Mannigfaltigkeit als Vereinigung von endlich vielen irreduziblen folgt entweder aus der Eliminationstheorie[3]) oder aus der Idealtheorie[4]); man kann sie aber auch sehr einfach direkt aus dem Hilbertschen Basissatz folgern[5]). Die Eindeutigkeit beweist man besser direkt, auf folgendem Wege.

---

[2]) W.-L. Chow, Die geometrische Theorie der algebraischen Funktionen für beliebige vollkommene Körper, Math. Annalen **115** (1937), S. 655—682, § 1.

[3]) F. S. Macaulay, Algebraic Theory of Modular Systems, Cambridge Tracts 19 (1916), Chapter I.

[4]) B. L. van der Waerden, Moderne Algebra II, Kap. 13.

[5]) Der Beweis verläuft etwa so:

Hilfssatz. *Eine Folge von algebraischen Mannigfaltigkeiten* $M_1 \supset M_2 \supset \cdots$, *in denen jede folgende eine echte Teilmannigfaltigkeit der vorangehenden ist, bricht nach endlich vielen Schritten ab.*

Beweis. Man vereinige alle definierenden Polynome aller Mannigfaltigkeiten $M_1, M_2, \ldots$ zu einer Menge Ω. Nach dem Hilbertschen Basissatz gibt es endlich viele Polynome $f_1, \ldots, f_r$ in Ω, durch die sich alle Polynome von Ω linear mit Polynomkoeffizienten darstellen lassen. Diese endlich vielen Polynome gehören zu endlich vielen Mannigfaltigkeiten $M_1, \ldots, M_n$. Wegen $M_1 \supset M_2 \supset \cdots \supset M_n$ werden $f_1, \ldots, f_r$ alle Null auf M_n. Also werden auch alle weiteren Polynome von Ω, da sie die Gestalt $A_1 f_1 + \cdots + A_r f_r$ haben, Null auf M_n. Das gilt insbesondere für die definierenden Gleichungen von M_{n+1}. Das widerspricht der Voraussetzung, daß M_{n+1} eine echte Teilmannigfaltigkeit von M_n ist. Also bricht die Folge bei M_n ab.

Zerlegungssatz. *Jede algebraische Mannigfaltigkeit ist Summe von endlich vielen irreduziblen.*

Beweis. Gesetzt, ein M_1 wäre nicht Summe von endlich vielen irreduziblen. Dann ist M_1 zerlegbar: $M_1 = M_2 + M_2'$. Einer der beiden Summanden, etwa M_2,

Hilfssatz. *Ist eine irreduzible Mannigfaltigkeit M in einer Summe $M_1 + M_2$ enthalten, so ist M in M_1 oder M_2 enthalten.*

Beweis. Wären $M \cap M_1$ und $M \cap M_2$ echte Teilmannigfaltigkeiten von M, so hätte man eine Zerlegung

$$M = (M \cap M_1) + (M \cap M_2).$$

Spezialfall. *Ein Produkt $f_1 f_2$ von zwei Formen wird dann und nur dann Null in allen Punkten der irreduziblen Mannigfaltigkeit M, wenn f_1 oder f_2 dieselbe Eigenschaft hat.* Diese Eigenschaft ist, wie leicht ersichtlich [6]), für die irreduziblen Mannigfaltigkeiten charakteristisch.

Der Hilfssatz läßt sich sofort auf mehrgliedrige Summen

$$M_1 + M_2 + \ldots + M_r$$

ausdehnen.

Eindeutigkeitssatz. *Die Darstellung einer Mannigfaltigkeit M als unverkürzbare Summe[7]) von irreduziblen ist, abgesehen von der Reihenfolge der Bestandteile, eindeutig.*

Beweis. Es seien $M = M_1 + \ldots + M_r = M_1' + \ldots + M_{r'}'$ zwei unverkürzbare Darstellungen. Aus dem Hilfssatz folgt, daß M_1 in einer der Mannigfaltigkeiten M_{ν}' enthalten ist. Nach Änderung der Reihenfolge der M_{ν}' können wir annehmen. daß M_1 in M_1' enthalten ist. Ebenso ist M_1' in einer M_{μ} enthalten. Wäre $\mu \neq 1$, so wäre $M_1 \subseteq M_1' \subseteq M_{\mu}$, also die Summe $M_1 + \ldots + M_{\mu} + \ldots$ verkürzbar; also ist $\mu = 1$ und $M_1 = M_1'$. Ebenso erreicht man, daß $M_2 = M_2', \ldots, M_r = M_r'$ ist. Weitere Summanden $M_{r+1}', \ldots$ können dann in der zweiten Summe nicht mehr vorkommen, da diese sonst verkürzbar wäre.

Alle Betrachtungen dieses (und der nächsten) Paragraphen gelten unverändert für algebraische Mannigfaltigkeiten in mehrfach-projektiven Räumen. Der *zweifach-projektive Raum* $S_{m, n}$ besteht aus allen Punktepaaren ξ, η, wobei ξ ein Punkt von S_m und η ein Punkt von S_n ist. Analog werden die drei- und mehrfach-projektiven Räume definiert. Eine algebraische Mannigfaltigkeit in $S_{m, n}$ wird durch Gleichungen gegeben, welche homogen sowohl in den ξ als in den η sind. Eine solche Mannigfaltigkeit von Punktepaaren heißt auch eine *algebraische Korrespondenz* zwischen den Punkten ξ und den Punkten η.

---

ist wiederum nicht als Summe von irreduziblen darstellbar und ist eine echte Teilmannigfaltigkeit von M_2. Wiederholung derselben Schlußweise ergibt eine unendliche Kette $M_1 \supset M_2 \supset M_3 \supset \ldots$, was nach dem Hilfssatz unmöglich ist.

[6]) Siehe etwa Moderne Algebra II, § 87.

[7]) Verkürzbar ist eine Summe, wenn ein Summand in der Summe der übrigen enthalten ist, also weggelassen werden kann.

§ 2.

Der allgemeine Punkt
und die Dimension einer irreduziblen Mannigfaltigkeit.

Ein Punkt ξ heißt ein *allgemeiner Punkt* einer Mannigfaltigkeit M, wenn ξ zu M gehört, und wenn alle homogenen algebraischen Gleichungen mit Koeffizienten aus K, die für den Punkt ξ gelten, für alle Punkte von M gelten.

Irreduzibilitätskriterium. *Wenn eine Mannigfaltigkeit M einen allgemeinen Punkt ξ besitzt, so ist sie irreduzibel.*

Beweis. Wäre M zerlegbar, so gäbe es ein Produkt fg von zwei Formen, das auf M überall verschwindet, ohne daß die Faktoren es tun. Es folgt

$$f(\xi) \cdot g(\xi) = 0,$$

also, da $f(\xi)$ und $g(\xi)$ einem Körper angehören,

$$f(\xi) = 0 \quad \text{oder} \quad g(\xi) = 0,$$

folglich ist $f = 0$ in allen Punkten von M oder $g = 0$ in allen Punkten von M, entgegen der Annahme.

Existenzsatz. *Jede nicht leere irreduzible Mannigfaltigkeit M besitzt einen allgemeinen Punkt ξ.*

Beweis. Jeder Quotient von zwei Formen gleichen Grades

$$\frac{f(x_0, x_1, \ldots, x_n)}{g(x_0, x_1, \ldots, x_n)}$$

definiert eine rationale Funktion auf der Mannigfaltigkeit M, sofern man annimmt, daß der Nenner nicht in allen Punkten von M Null ist. Zwei solche Funktionen heißen gleich,

$$\frac{f}{g} = \frac{f'}{g'}, \quad \text{wenn} \quad fg' = f'g \quad \text{auf } M.$$

Addition, Subtraktion, Multiplikation und Division von rationalen Funktionen auf M ergeben wieder rationale Funktionen auf M. Die rationalen Funktionen auf M bilden also einen Körper, der den Konstantenkörper K umfaßt.

Wir können annehmen, daß x_0 nicht in allen Punkten von M gleich Null ist. Die rationalen Funktionen

$$\frac{x_1}{x_0}, \quad \frac{x_2}{x_0}, \quad \ldots, \quad \frac{x_n}{x_0}$$

bezeichnen wir mit $\xi_1, \xi_2, \ldots, \xi_n$. Weiter setzen wir $\xi_0 = 1$. Dann ist $(\xi_0, \xi_1, \ldots, \xi_n)$ ein allgemeiner Punkt von M. Denn aus

$$f(\xi_0, \xi_1, \ldots, \xi_n) = 0$$

oder, was dasselbe ist,

$$f\left(1, \frac{x_1}{x_0}, \ldots, \frac{x_n}{x_0}\right) = 0 \quad \text{auf } M$$

folgt, da f homogen ist,

$$f(x_0, x_1, \ldots x_n) = 0 \quad \text{auf } M,$$

und umgekehrt. Also gelten alle homogenen Gleichungen, die für den Punkt ξ gelten, für alle Punkte von M, und umgekehrt.

Ein Punkt heißt *normiert*, wenn seine erste von Null verschiedene Koordinate gleich Eins ist. Jeder Punkt kann so normiert werden. Durch Umnumerierung der Koordinaten kann man sogar $\xi_0 \neq 0$, also $\xi_0 = 1$ annehmen. $\xi_1, \ldots, \xi_n$ heißen dann die *inhomogenen Koordinaten* von ξ.

Eindeutigkeitssatz. *Je zwei normierte allgemeine Punkte ξ, η einer Mannigfaltigkeit M können durch einen Körperisomorphismus $K(\xi) \simeq K(\eta)$, der die Elemente von K festläßt, ineinander übergeführt werden. Die algebraischen Eigenschaften von ξ und η stimmen also genau überein.*

Beweis. Nach Definition des allgemeinen Punktes gelten alle homogenen algebraischen Gleichungen, die für ξ gelten, auch für η, und umgekehrt. Ist also $\xi_0 = 0$, so auch $\eta_0 = 0$, und umgekehrt; ist ξ_i die erste von Null verschiedene Koordinate von ξ, so gilt dasselbe von η_i. Durch Umnumerierung der Koordinaten erreichen wir, daß $\xi_0 \neq 0$ und $\eta_0 \neq 0$, also wegen der Normierung $\xi_0 = \eta_0 = 1$ ist. Jedes Polynom in $\xi_1, \ldots, \xi_n$ kann durch Hinzufügung von Faktoren ξ_0 zu den einzelnen Gliedern homogen gemacht werden. Wir ordnen nun jedem solchen Polynom $f(\xi_1, \ldots, \xi_n)$ dasselbe Polynom in $\eta_1, \ldots, \eta_n$ zu. Ist $f(\xi_1, \ldots, \xi_n) = g(\xi_1, \ldots, \xi_n)$, so ist $f(\xi) - g(\xi) = 0$, und diese Relation, homogen gemacht, gilt nach dem eingangs bemerkten auch für η:

$$f(\eta) - g(\eta) = 0, \quad \text{also} \quad f(\eta) = g(\eta).$$

Unsere Zuordnung $f(\xi) \to f(\eta)$ ist also eindeutig. Offensichtlich ist sie eineindeutig und isomorph. Sie führt weiter ξ_ν in η_ν über. Der so erhaltene Isomorphismus der Ringe $K[\xi_1, \ldots, \xi_n]$ und $K[\eta_1, \ldots, \eta_n]$ läßt sich ohne weiteres zu einem Isomorphismus der Quotientenkörper $K(\xi_1, \ldots, \xi_n)$ und $K(\eta_1, \ldots, \eta_n)$ erweitern. Damit ist alles bewiesen.

Umkehrsatz. *Zu jedem Punkt ξ (dessen Koordinaten irgendeinem Erweiterungskörper von K angehören, z. B. algebraische Funktionen von unbestimmten Parametern sind) gehört eine (irreduzible) algebraische Mannigfaltigkeit M, deren allgemeiner Punkt ξ ist.*

Beweis. Aus der Gesamtheit aller Formen $f(x_0, \ldots, x_n)$ mit konstanten Koeffizienten und mit der Eigenschaft $f(\xi) = 0$ läßt sich nach dem Hilbertschen Basissatz eine endliche Basis $(f_1, \ldots, f_r)$ herausheben. Die Gleichungen $f_1 = 0, \ldots, f_r = 0$ definieren eine algebraische Mannigfaltigkeit M.

Offensichtlich ist ξ ein allgemeiner Punkt von M; denn ξ gehört zu M, und alle homogenen Gleichungen, die für ξ gelten, sind Folgen der Gleichungen $f_1 = 0, \ldots, f_r = 0$ und gelten daher für alle Punkte von M.

Auf Grund des Existenz- und Eindeutigkeitssatzes können wir die *Dimension* einer irreduziblen Mannigfaltigkeit definieren als die Anzahl der algebraisch-unabhängigen unter den Koordinaten eines normierten allgemeinen Punktes ξ von M. Nach Chow kann man diese Zahl auch die *Dimension* des allgemeinen Punktes ξ nennen. Die Dimension einer zerfallenden Mannigfaltigkeit M ist die höchste Dimension eines irreduziblen Bestandteils, oder, was dasselbe ist, die höchste Dimension eines Punktes von M.

Dimensionssatz. *Ist M irreduzibel und r ihre Dimension, so hat jeder normierte Punkt von M, der nicht allgemeiner Punkt ist, einen kleineren Transzendenzgrad als r.*

Beweis. Siehe „Nullstellentheorie" [1]), § 3, oder Moderne Algebra II, § 90.

Aus dem Dimensionssatz folgt zunächst, daß bei einer nulldimensionalen irreduziblen Mannigfaltigkeit alle Punkte allgemein sind. Sie sind alle algebraisch über K und sind auf Grund des Eindeutigkeitssatzes untereinander konjugiert. Da eine algebraische Gleichung nur endlich viele Wurzeln hat, so gibt es (in jedem K umfassenden Körper) nur endlich viele Punkte der Mannigfaltigkeit. Letzteres gilt für alle nulldimensionalen Mannigfaltigkeiten und nur für diese.

Aus dem Dimensionssatz folgt weiter: *Ist $N \subset M$ und M irreduzibel, so ist die Dimension von N kleiner als die von M.* Denn wenn J ein irreduzibler Bestandteil von N ist, so ist der allgemeine Punkt von J kein allgemeiner Punkt von M, also seine Dimension kleiner als die von M.

Eine Mannigfaltigkeit, deren irreduzible Bestandteile alle die Dimension r haben, heißt *rein r-dimensional*.

Die rein $(n-1)$-dimensionalen Mannigfaltigkeiten in S_n heißen *Hyperflächen*. Ist ξ ein allgemeiner Punkt einer irreduziblen Hyperfläche, so kann man

$$\frac{\xi_1}{\xi_0} = u_1, \ldots, \frac{\xi_{n-1}}{\xi_0} = u_{n-1}$$

als algebraisch unabhängige Größen annehmen. Es gilt dann bekanntlich eine irreduzible Gleichung

$$f\left(\frac{\xi_1}{\xi_0}, \ldots, \frac{\xi_{n-1}}{\xi_0}, \frac{\xi_n}{\xi_0}\right) = 0$$

derart, daß jedes andere Polynom g mit der Eigenschaft

$$g\left(\frac{\xi_1}{\xi_0}, \ldots, \frac{\xi_{n-1}}{\xi_0}, \frac{\xi_n}{\xi_0}\right) = 0$$

durch f teilbar ist. Macht man f und g homogen, so folgt die Existenz einer irreduziblen Form f mit $f(\xi_0, \xi_1, \ldots, \xi_{n-1}, \xi_n) = 0$ derart, daß jede Form g mit der Eigenschaft $g(\xi_0, \ldots, \xi_n) = 0$ durch f teilbar ist. Die Betrachtung läßt sich umkehren, indem man von einer beliebigen irreduziblen Form f ausgeht, etwa $\xi_0 = 1$ setzt, $\xi_1, \ldots, \xi_{n-1}$ gleich Unbestimmten $u_1, \ldots, u_{n-1}$ setzt und ξ_n als algebraische Funktion dieser Unbestimmten durch die irreduzible Gleichung $f(\xi_0, \xi_1, \ldots, \xi_{n-1}, \xi_n) = 0$ definiert. In dieser Weise sieht man, daß *jede irreduzible Hyperfläche durch eine einzige homogene irreduzible Gleichung $f = 0$ gegeben wird* und daß umgekehrt *jede solche Gleichung eine irreduzible Hyperfläche definiert*, sofern die Form f keine Konstante ist.

Eine reduzible Gleichung $f = 0$ definiert nun offenbar eine reduzible Hyperfläche. Ihre irreduziblen Bestandteile werden durch Zerlegung der Form f in Primfaktoren gefunden. Man pflegt diese Bestandteile mit derselben Vielfachheit zu zählen, welche die Primfaktoren in der Zerlegung von f besitzen.

§ 3.

Erweiterung des Grundkörpers. Absolute Irreduzibilität.

In § 1 haben wir schon erwähnt, daß der Theorie der algebraischen Mannigfaltigkeiten ein beliebiger Grundkörper K, der also weder algebraisch abgeschlossen noch vollkommen zu sein braucht, zugrunde liegt, daß aber die Koordinaten der betrachteten Punkte aus ganz beliebigen Erweiterungskörpern entnommen werden dürfen. Man könnte darin eine unnötige Erschwerung erblicken und versuchen, die Sachlage dadurch zu vereinfachen, daß man ein für allemal einen algebraisch-abgeschlossenen Grundkörper, etwa den der komplexen Zahlen, zugrunde legt, aus dem sowohl die Koeffizienten der Gleichungen als auch die Koordinaten der Punkte entnommen werden. Diese Auffassung erweist sich aber als unzweckmäßig: Man würde mit ihr die für einen weiteren Ausbau der Theorie unentbehrliche Bewegungsfreiheit verlieren. Man würde z. B. die in § 2 erklärten allgemeinen Punkte einer Mannigfaltigkeit nicht als Punkte des Raumes betrachten können, denn ihre Koordinaten gehören ja im allgemeinen nicht dem Grundkörper K an.

Aber warum nimmt man nicht wenigstens den Grundkörper als algebraisch abgeschlossen an? so könnte man fragen. Man würde dadurch gewisse Komplikationen, wie z. B. die Unterscheidung zwischen absoluter und relativer Irreduzibilität (siehe unten) vermeiden. Bei näherer Betrachtung erweist sich aber auch diese Annahme als unzweckmäßig; denn es kommt häufig vor, daß man während einer Untersuchung den Grundkörper erweitern muß, z. B. durch Adjunktion von Unbestimmten, wodurch die algebraische Abgeschlossenheit wieder verloren geht. Gesetzt z. B., man will eine algebraische Fläche mit einer allgemeinen Hyperebene zum Schnitt bringen. Die Koeffizienten $u_0, u_1, \ldots, u_n$ der linearen Gleichung der Hyperebene seien Unbestimmte. Damit nun das Schnittgebilde wieder als algebraische Mannigfaltigkeit betrachtet werden kann, muß man diese Unbestimmten dem Grundkörper adjungieren. Auch erweist es sich häufig als nötig, allgemeine Punkte von Mannigfaltigkeiten dem Grundkörper zu adjungieren.

Eine völlige Bewegungsfreiheit erfordert somit, daß der Grundkörper K völlig frei bleibt und jederzeit durch Adjunktion erweitert werden kann, und daß die Koordinaten der betrachteten Punkte aus beliebigen Erweiterungskörpern entnommen werden können. Diese Adjunktion wird häufig sogar stillschweigend geschehen: Wenn wir sagen, daß eine Mannigfaltigkeit mit einer allgemeinen Hyperebene geschnitten wird, so ist in dieser Ausdrucksweise die Adjunktion der Koeffizienten der Hyperebene zum Grundkörper schon mit enthalten. Wenn mehrere allgemeine Punkte, Hyperebenen usw. gleichzeitig betrachtet werden, so werden diese, wenn nicht das Gegenteil ausdrücklich gesagt wird, immer als unabhängig voneinander angenommen. Das heißt, es sollen jedesmal neue Unbestimmte genommen werden, und ein neu angenommener allgemeiner Punkt einer Mannigfaltigkeit soll *nach Adjunktion aller früher betrachteten Größen* ein allgemeiner Punkt der Mannigfaltigkeit sein. Dabei muß natürlich vorausgesetzt werden, daß die Mannigfaltigkeit bei diesen Adjunktionen irreduzibel bleibt.

Eine irreduzible algebraische Mannigfaltigkeit kann bei Erweiterung des Grundkörpers zerlegbar werden. Eine Mannigfaltigkeit aber, die bei jeder Erweiterung des Grundkörpers K irreduzibel bleibt, heißt *absolut irreduzibel*. In ZAG X [8]) wurde bewiesen, daß eine *endliche algebraische* Erweiterung von K genügt, um jede in bezug auf K irreduzible Mannigfaltigkeit in absolut irreduzible Mannigfaltigkeiten zu zerlegen, welche *dieselbe Dimension* haben wie die ursprüngliche und untereinander *konjugiert* sind in bezug auf K. Daraus folgt weiter: Wenn eine Mannigfaltigkeit M bei algebraischer Erweiterung von K nicht zerfällt, so ist sie absolut irreduzibel.

Daß die Dimension einer Mannigfaltigkeit bei algebraischer Erweiterung des Grundkörpers sich nicht ändert, hätte man auch direkt beweisen können (z. B. mittels der Eliminationstheorie, welche es gestattet, die Dimension zu bestimmen, ohne dabei den Grundkörper zu verlassen).

S a t z. *Die Relationen $M \subseteq N$, $M \supseteq N$ und $M = N$ bleiben bei Erweiterung des Grundkörpers erhalten.*

Denn wenn alle Punkte mit Koordinaten aus allen Erweiterungskörpern von K, welche die Gleichungen von M erfüllen, auch die von N erfüllen, so

---

[8]) Math. Annalen **113** (1937), S. 705—712, § 1.

gilt das insbesondere für diejenigen Punkte, deren Koordinaten den Erweiterungskörpern eines Erweiterungskörpers von K angehören.

Satz. *Damit M in N enthalten ist, genügt es, daß alle über K algebraischen Punkte von M zu N gehören.*

Beweis. Sind $f_\lambda = 0$ die definierenden Gleichungen von M und $g_\mu = 0$ die von N, so folgt aus dem Hilbertschen Nullstellensatz, wenn alle algebraischen Punkte von M auch die Gleichungen $g_\mu = 0$ erfüllen:

$$g_\mu^\varrho = h_1 f_1 + \ldots + h_r f_r.$$

Daraus folgt, daß alle gemeinsamen Nullstellen von $f_1, \ldots, f_r$ auch Nullstellen von g_μ sind.

§ 4.
Relationstreue Spezialisierungen.

Wenn alle homogenen algebraischen Relationen $f(\xi) = 0$ mit Koeffizienten aus K, die für einen Punkt ξ gelten, auch für einen Punkt ξ' gelten, so heißt ξ' eine *relationstreue Spezialisierung* von ξ. Ebenso definiert man eine gleichzeitige relationstreue Spezialisierung $\xi \to \xi'$, $\eta \to \eta'$, ... von mehreren Punkten. Die Koordinaten von ξ können auch von einigen unbestimmten Parametern t abhängen; bleiben dann alle *in den ξ homogenen* Relationen $f(t, \xi) = 0$ bei der Ersetzung der t durch irgendwelche spezielle Werte t' und von ξ durch irgend einen Punkt ξ' aus einem Erweiterungskörper von $K(t')$ erhalten, so heißt $\xi \to \xi'$ eine relationstreue Spezialisierung zu der Parameterspezialisierung $t \to t'$.

Der Übergang von einem allgemeinen Punkt ξ einer Mannigfaltigkeit zu irgendeinem Punkt ξ' der Mannigfaltigkeit ist (nach Definition des allgemeinen Punktes) eine relationstreue Spezialisierung.

Existenzsatz. *Hängt der Punkt ξ (oder hängen mehrere Punkte $\xi, \eta, \ldots$) irgendwie von den Parametern t ab, so gibt es zu jeder Parameterspezialisierung $t \to t'$ eine relationstreue Spezialisierung $\xi \to \xi'$.*

Beweis. Aus den unendlich vielen homogenen Relationen $f(t, \xi) = 0$ lassen sich nach dem Hilbertschen Basissatz endlich viele auswählen, die eine Basis für alle bilden. Aus diesen endlich vielen Relationen $f_\nu(t, \xi) = 0$ bilde man durch Elimination der ξ das Resultantensystem [9]. Dieses muß (identisch in den t) verschwinden, denn es verschwindet dann und nur dann, wenn das Gleichungssystem eine Lösung besitzt. Die formalen Prozesse aber, die zur Definition des Resultantensystems führen, bleiben nach Ersetzung der t durch die t' und der ξ durch die (noch unbekannten) ξ' erhalten. Also ver-

---

[9]) B. L. van der Waerden, Moderne Algebra II, § 76.

schwindet das Resultantensystem der Gleichungen $f_\nu(t', \xi') = 0$ für beliebige t'. Also ist das Gleichungssystem lösbar, d. h. es gibt einen Punkt ξ' mit den verlangten Eigenschaften.

Genau so beweist man den

Fortsetzungssatz. *Sind ξ und η Punkte mit Koordinaten aus einem K umfassenden Körper, so läßt sich jede relationstreue Spezialisierung $\xi \to \xi'$ zu einer ebensolchen $\xi \to \xi'$, $\eta \to \eta'$ fortsetzen.*

Zum Beweis braucht man nur zu bemerken, daß das aus den homogenen Gleichungen $f_\nu(\xi, \eta) = 0$ durch Elimination der η entstehende Resultantensystem homogen in den ξ ist, also bei einer relationstreuen Spezialisierung $\xi \to \xi'$ Null bleibt.

Der Fortsetzungssatz gilt offensichtlich auch bei der Anwesenheit von Parametern t, sowie auch dann, wenn mehrere Punkte $\xi^{(1)}, \xi^{(2)}, \ldots$ im Spiel sind.

Zu dem Existenzsatz tritt ein Eindeutigkeitssatz, der aber nur unter stark einschränkenden Bedingungen gilt.

Eindeutigkeitssatz. *Es seien $\xi^{(1)}, \ldots, \xi^{(h)}$ alle Lösungen[10]) eines in den ξ homogenen algebraischen Gleichungssystems*

$$\text{(1)} \qquad\qquad f_\nu(t, \xi) = 0,$$

jede nur einmal gezählt. Für spezielle Werte t' der t möge das Gleichungssystem

$$\text{(2)} \qquad\qquad f_\nu(t', \eta) = 0$$

ebenfalls nur endlich viele Lösungen haben. Dann sind in einer relationstreuen Spezialisierung $t \to t'$, $\xi^{(1)} \to \eta^{(1)}, \ldots, \xi^{(h)} \to \eta^{(h)}$ die Punkte $\eta^{(1)}, \ldots, \eta^{(h)}$ bis auf ihre Reihenfolge eindeutig bestimmt.

Beweis. Die Gleichungen (1) definieren nach Voraussetzung eine nulldimensionale Mannigfaltigkeit. Ihre Punkte $\xi^{(1)}, \ldots, \xi^{(h)}$ zerfallen also in Systeme konjugierter Punkte. Es genügt, ein solches System $\xi^{(1)}, \ldots, \xi^{(k)}$ zu betrachten und diese Punkte durch $\xi_0^{(\nu)} = 1$ zu normieren. Die Koordinaten $\xi_k^{(\nu)}$ sind dann algebraisch über dem Grundkörper K.

Sind $u_0, u_1, \ldots, u_n$ Unbestimmte, so ist

$$\vartheta = -(u_1 \xi_1^{(1)} + \ldots + u_n \xi_n^{(1)})$$

algebraisch über $K(t, u_1, \ldots, u_n)$ und daher Nullstelle eines in diesem Körper irreduziblen Polynoms $G(u_0)$. In einem passenden Erweiterungskörper zerfällt $G(u_0)$ ganz in Linearfaktoren, die alle zu $u_0 - \vartheta$ konjugiert sind und daher die Gestalt $u_0 + u_1 \xi_1^{(\nu)} + \ldots + u_n \xi_n^{(\nu)}$ haben:

$$\text{(3)} \qquad G(u_0) = h(t) \cdot \Pi\, (u_0 + u_1 \xi_1^{(\nu)} + \ldots + u_n \xi_n^{(\nu)}).$$

---

[10]) „Alle Lösungen" heißt: Alle Lösungen in einem algebraischen Erweiterungskörper Ω von K, der so umfassend ist, daß bei Erweiterung von Ω keine Lösungen mehr hinzukommen.

Den willkürlichen Faktor $h\,(t)$ denken wir uns so normiert, daß das Polynom $G\,(u_0)$ nicht nur rational, sondern ganz rational in den t wird und keinen von den t allein abhängigen Faktor enthält. Wir nennen es dann $G\,(t,\,u)$. $G\,(t,\,u)$ ist ein unzerlegbares Polynom in den t und u und heißt die *zugeordnete Form* des Punktsystems $\xi^{(1)},\,\ldots,\,\xi^{(k)}$.

Entwickeln wir beide Seiten von (3) nach Potenzprodukten der u und vergleichen die Koeffizienten dieser Potenzprodukte, so folgt

$$a_\lambda\,(t) = h\,(t)\,b_\lambda\,(\xi).$$

Elimination von $h\,(t)$ ergibt die homogenen Relationen

$$a_\lambda\,(t)\,b_\mu\,(\xi) - a_\mu\,(t)\,b_\lambda\,(\xi) = 0.$$

Diese müssen bei der relationstreuen Spezialisierung $t \to t',\ \xi^{(\nu)} \to \eta^{(\nu)}$ erhalten bleiben:

$$(4) \qquad a_\lambda\,(t')\,b_\mu\,(\eta) - a_\mu\,(t')\,b_\lambda\,(\eta) = 0.$$

Die $b_\lambda\,(\eta)$ sind die Koeffizienten der Form

$$\Pi\,(u_0 + u_1\eta_1^{(\nu)} + \ldots + u_n\eta_n^{(\nu)});$$

sie verschwinden somit nicht alle. Also folgt aus (4)

$$a_\lambda\,(t') = \varrho\,b_\lambda\,(\eta).$$

Das heißt aber, da $a_\lambda\,(t')$ die Koeffizienten der Form $G\,(t',\,u)$ sind,

$$(5) \qquad G\,(t',\,u) = \varrho\,\Pi_\nu\,(u_0 + u_1\eta_1^{(\nu)} + \ldots + u_n\eta_n^{(\nu)}).$$

Damit ist gezeigt, daß die Relation (3) bei der relationstreuen Spezialisierung $t \to t',\ \xi^{(\nu)} \to \eta^{(\nu)}$ im wesentlichen erhalten bleibt. Wenn wir noch zeigen können, daß die Form $G\,(t',\,u)$ nicht identisch verschwindet, so folgt $\varrho \neq 0$. Wegen der eindeutigen Faktorenzerlegung sind dann die Linearfaktoren rechter Hand in (5) und damit auch die Punkte $\eta^{(1)},\,\ldots,\,\eta^{(k)}$ bis auf ihre Reihenfolge eindeutig bestimmt.

Zum Nachweis des Nichtverschwindens der Form $G\,(t',\,u)$ bilden wir das Resultantensystem der Formen $f_\nu\,(t,\,x)$ und der Linearform $u_0x_0 + \ldots + u_nx_n$ nach den x. Die Formen $R\,(t,\,u)$ dieses Resultantsystems werden für spezielle Werte der u nur dann Null, wenn die Ebene u durch einen von den Punkten $\xi^{(\nu)}$ geht. Also sind die Formen $R\,(t,\,u)$ durch die Linearformen $u_0\xi_0^{(\nu)} + \ldots + u_n\xi_n^{(\nu)}$ und daher auch durch die Form (3) teilbar. Da die Form (3) ganzrational und primitiv in den t ist, gilt die erwähnte Teilbarkeit schon im Bereich der Polynome in den t und u:

$$(6) \qquad R_\lambda\,(t,\,u) = A_\lambda\,(t,\,u)\,G\,(t,\,u).$$

Wäre nun $G\,(t',\,u) = 0$, so würde aus (6) folgen: $R_\lambda\,(t',\,u) = 0$. Das ist aber nicht der Fall, denn die $R_\lambda\,(t',\,u)$ bilden das Resultantensystem der Formen

$f_\nu(t', x)$ und der Linearform $u_0 x_0 + \ldots + u_n x_n$, und dieses verschwindet für spezielle u nur dann, wenn die Ebene u durch einen von den endlich vielen Punkten η geht, welche die Gleichungen (2) befriedigen. Also ist in der Tat $G(t', u) \neq 0$, womit der Beweis beendet ist.

Auf Grund des Eindeutigkeitssatzes kann man für jede Lösung des spezialisierten Gleichungssystems $f_\nu(t', \eta) = 0$ eine *Vielfachheit* oder *Multiplizität* definieren, die angibt, wie oft der Punkt η unter den spezialisierten Lösungen $\eta^{(1)}, \ldots, \eta^{(h)}$ vorkommt. Die Multiplizität ist eine ganze Zahl $\geqq 0$ und hängt wesentlich davon ab, aus welchem allgemeineren Gleichungssystem (1) das System (2) entstanden ist. Die Summe der Vielfachheiten aller Lösungen η ist offensichtlich gleich h, also gleich der Anzahl der Lösungen des nicht spezialisierten Gleichungssystems (1). Das ist das *Prinzip der Erhaltung der Anzahl.*

Es kann vorkommen, daß eine Lösung η des spezialisierten Problems die Vielfachheit Null hat, also nicht unter den spezialisierten Lösungen $\eta^{(\nu)}$ vorkommt, sondern neu auftaucht. Der folgende Satz gibt Auskunft darüber, wann dieser unangenehme Fall *nicht* eintritt.

Kriterium für positive Vielfachheit. *Wenn eine Lösung η des Gleichungssystems (2) eine relationstreue Spezialisierung einer Lösung ξ des Gleichungssystems (1) ist, so hat sie eine positive Vielfachheit.*

Beweis. Es sei etwa $\xi = \xi^{(1)}$. Die relationstreue Spezialisierung $t \to \tau$, $\xi \to \eta$ läßt sich zu einer ebensolchen Spezialisierung $\xi^{(1)} \to \eta$, $\xi^{(2)} \to \eta^{(2)}, \ldots, \xi^{(h)} \to \eta^{(h)}$ fortsetzen. Also kommt η unter den spezialisierten Lösungen $\eta^{(1)}, \ldots, \eta^{(h)}$ mindestens einmal vor.

Aus dem Kriterium folgt insbesondere, daß die Vielfachheit einer jeden Lösung η dann positiv ist, wenn die Gleichungen (2) in den t' homogen sind und eine *irreduzible Korrespondenz* zwischen den Punkten t' und den Lösungen η vermitteln. Denn dann ist jedes Punktepaar (t', η) der Korrespondenz eine relationstreue Spezialisierung des allgemeinen Punktepaares (t, ξ).

In ZAG V[1]) wurde auch ein Kriterium dafür angegeben, wann die Vielfachheit einer Lösung nicht größer als Eins ist. Es bedeute $\partial_k f(t, x)$ die partielle Ableitung von $f(t, x)$ nach x_k. Dann lautet das Kriterium:

Wenn eine Lösung η des Gleichungssystems (2) eine Multiplizität > 1 hat, so haben die Polarhyperebenen

$$(7) \qquad \zeta_0 \partial_0 f_\nu(t', \eta) + \zeta_1 \partial_1 f_\nu(t', \eta) + \ldots + \zeta_n \partial_n f_\nu(t', \eta) = 0$$

des Punktes η in bezug auf die Gleichungen (2) mindestens eine Gerade durch den Punkt η miteinander gemeinsam. Haben sie nur den einen Punkt η gemeinsam, so kann daher die Multiplizität von η nicht größer als Eins sein.

Der Beweis, der hier nicht ausführlich wiederholt werden soll, beruht auf folgendem Gedanken. Wenn bei der Spezialisierung $t \to t'$ zwei Lösungen

$\xi^{(1)}$, $\xi^{(2)}$ in eine Lösung η hineinrücken, so geht die Verbindungslinie $\xi^{(1)} \xi^{(2)}$, als Sehne der Hyperfläche $f_\nu(t, \xi) = 0$, in eine Tangente der Hyperfläche $f_\nu(t', \eta) = 0$ über. Diese Tangenten liegen aber alle in der Tangentialhyperebene (7).

Die beiden letzten Kriterien, zusammen genommen, ermöglichen in vielen Fällen die Feststellung, daß die bei einer bestimmten Spezialisierung auftretenden Vielfachheiten alle gleich Eins sind. In diesen Fällen ist also die Anzahl der verschiedenen Lösungen η im Spezialfall gleich der Anzahl der Lösungen ξ im allgemeinen Fall.

§ 5.
Algebraische Korrespondenzen.

Wie in § 2 schon erwähnt, versteht man unter einer *algebraischen Korrespondenz* eine algebraische Mannigfaltigkeit von Punktepaaren (ξ', η'), wobei ξ' einem Raum S_m, η' einem Raum S_n angehört. Eliminiert man aus den Gleichungen der Korrespondenz die Koordinaten ξ' oder η', so erhält man ein System von Gleichungen in den η' bzw. ξ' allein, welche eine algebraische Mannigfaltigkeit $\mathfrak{N}$ in S_n bzw. $\mathfrak{M}$ in S_m definieren: die *Bildmannigfaltigkeit* bzw. *Urmannigfaltigkeit* der Korrespondenz. Zu jedem Punkt ξ' der Urmannigfaltigkeit (bzw. zu jedem Punkt η' der Bildmannigfaltigkeit) gibt es mindestens ein Punktepaar (ξ', η') der Korrespondenz.

Ist die Korrespondenz $\mathfrak{K}$ irreduzibel, so entstehen alle ihre Punktepaare (ξ', η') durch relationstreue Spezialisierung aus einem allgemeinen Punktepaar (ξ, η). Der Punkt ξ ist dann offensichtlich ein allgemeiner Punkt der Urmannigfaltigkeit $\mathfrak{M}$ und η ist ein allgemeiner Punkt von $\mathfrak{N}$. $\mathfrak{M}$ und $\mathfrak{N}$ sind also auch irreduzibel. Ihre Dimensionen seien a und c.

Jedem Punkt ξ' der Urmannigfaltigkeit $\mathfrak{M}$ entspricht eine algebraische Mannigfaltigkeit $\mathfrak{N}_{\xi'}$ von Punkten der Bildmannigfaltigkeit $\mathfrak{N}$. *Insbesondere entspricht in einer irreduziblen Korrespondenz $\mathfrak{K}$ einem allgemeinen Punkt ξ von $\mathfrak{M}$ eine über dem Körper $K(\xi)$ irreduzible Bildmannigfaltigkeit $\mathfrak{N}_\xi$.*

Beweis. (ξ, η) sei ein allgemeines Punktepaar der Korrespondenz, η' irgend ein Punkt von $\mathfrak{N}_\xi$, dann ist (ξ, η') eine relationstreue Spezialisierung von (ξ, η), also η' eine relationstreue Spezialisierung von η in bezug auf den Körper $K(\xi)$. Ist umgekehrt η' eine solche relationstreue Spezialisierung von η, so ist η' ein Punkt von $\mathfrak{N}_\xi$. Also ist η ein allgemeiner Punkt von $\mathfrak{N}_\xi$ in bezug auf den Grundkörper $K(\xi)$. Also ist $\mathfrak{N}_\xi$ irreduzibel.

Die Dimension von $\mathfrak{N}_\xi$ sei b. Ebenso entspricht umgekehrt einem allgemeinen Punkt η von $\mathfrak{N}$ eine irreduzible Mannigfaltigkeit $\mathfrak{M}_\eta$ auf $\mathfrak{M}$. Ihre Dimension sei d. *Dann ist die Dimension von $\mathfrak{K}$ durch*

$$(1) \qquad e = a + b = c + d$$

gegeben.

Beweis. Die Zahl der algebraisch unabhängigen unter den normierten Koordinaten ξ, η kann entweder erhalten werden, indem man die algebraisch unabhängigen ξ zählt und dazu die Anzahl der von ihnen algebraisch unabhängigen η addiert, oder umgekehrt, indem man zunächst die unabhängigen η und dann die von ihnen unabhängigen ξ zählt.

Die Formel (1) nennt man *das Prinzip der Konstantenzählung*. Sie dient häufig dazu, die Dimension c der Bildmannigfaltigkeit $\mathfrak{N}$ auszurechnen, wenn a, b und d bekannt sind.

Zum Nachweis der Irreduzibilität einer Korrespondenz bedient man sich vorzugsweise des Irreduzibilitätskriteriums von § 2, indem man ein allgemeines Punktepaar (ξ, η) zu konstruieren versucht. Dabei kann man entweder von einem allgemeinen Punkt ξ von $\mathfrak{M}$ oder von einem allgemeinen Punkt η von $\mathfrak{N}$ ausgehen. Das allgemeine Punktepaar (ξ, η) hat, wie es auch gewonnen wird, immer dieselben algebraischen Eigenschaften.

Die gesamte Theorie der Korrespondenzen gilt ungeändert auch dann, wenn die Symbole ξ und η nicht Punkte, sondern irgendwelche andere geometrische Gebilde bezeichnen, sofern diese durch eine oder mehrere homogene Koordinatenreihen gegeben werden. Man erhält z. B. eine Korrespondenz, indem man einem Kegelschnitt ξ alle seine Tangenten η zuordnet (und dann relationstreu $\xi \to \xi'$, $\eta \to \eta'$ spezialisiert); oder es kann ξ ein Punktepaar, η die Verbindungslinie dieses Punktepaares bedeuten, usw. Jede irgendwie geartete algebraische Verknüpfung von geometrischen Objekten mit anderen geometrischen Objekten führt zu einer Korrespondenz.

Die Theorie der algebraischen Korrespondenzen wurde in ZAG VI ausführlicher entwickelt. Bei der Lektüre von ZAG VI ist zu beachten, daß der Beweis des Satzes auf S. 142 nicht in allen Fällen stichhaltig ist; er wurde in Math. Annalen **113** (S. 38) berichtigt und in ZAG XII (Math. Annalen **115**) durch einen eleganteren Beweis ersetzt. Die übrigen Beweise der Arbeit ZAG VI setzen den Inhalt des nächsten Paragraphen (§ 6) der vorliegenden Arbeit voraus.

§ 6.

**Schnitt von Mannigfaltigkeiten mit allgemeinen Hyperflächen.
Der Grad einer Mannigfaltigkeit.**

Wir wollen zunächst eine algebraische Mannigfaltigkeit M (mit Koeffizienten aus dem Grundkörper K) in S_n mit einer allgemeinen Hyperebene

$$u_0\xi_0 + u_1\xi_1 + \ldots + u_n\xi_n = 0,$$

deren Koeffizienten $u_0, \ldots, u_n$ also Unbestimmte sind, zum Schnitt bringen. Wir werden zeigen:

Der Durchschnitt einer irreduziblen a-dimensionalen Mannigfaltigkeit M mit einer allgemeinen Hyperebene u ist eine relativ zum Körper

$$K(u) = K(u_0, \ldots, u_n)$$

irreduzible Mannigfaltigkeit von der Dimension $a - 1$.

Beweis. Ordnen wir den Punkten ξ' der Mannigfaltigkeit M die durch ξ' gehenden Hyperebenen η' zu, so erhalten wir eine algebraische Korrespondenz. Ein allgemeines Elementepaar der Korrespondenz erhalten wir, indem wir durch einen allgemeinen Punkt ξ von M die allgemeinste Hyperebene η legen. Die Korrespondenz ist also irreduzibel. Das Prinzip der Konstantenzählung (§ 4) ergibt, wenn c die Dimension der Bildmannigfaltigkeit der Korrespondenz und d die des Durchschnittes von M mit der Hyperebene η ist:

$$(1) \qquad\qquad a + (n - 1) = c + d.$$

Da die durch den allgemeinen Punkt ξ gelegte allgemeinste Hyperebene η einen zweiten beliebig, aber fest gewählten Punkt ξ' nicht enthält, so ist ihr Durchschnitt mit der Mannigfaltigkeit M höchstens $(a - 1)$-dimensional. Daher ist $d \leq a - 1$. Aus (1) folgt nun $c \geq n$, also ist die Bildmannigfaltigkeit N der Korrespondenz die ganze n-dimensionale Mannigfaltigkeit aller Hyperebenen des Raumes S_n. Es ist somit $c = n$ und $d = a - 1$. Also entspricht einer allgemeinen Hyperebene u in der Korrespondenz eine relativ zum Körper $K(u)$ irreduzible Mannigfaltigkeit von der Dimension $a - 1$.

Genau so beweist man allgemeiner:

Der Durchschnitt einer irreduziblen a-dimensionalen Mannigfaltigkeit M mit einer allgemeinen Hyperfläche γ-ten Grades (mit lauter unbestimmten Koeffizienten) ist eine relativ zum Körper der Koeffizienten irreduzible Mannigfaltigkeit von der Dimension $a - 1$.

Wendet man diesen Satz k mal nacheinander an, so folgt:

Der Durchschnitt einer irreduziblen a-dimensionalen Mannigfaltigkeit mit k allgemeinen Hyperflächen von beliebigen Gradzahlen ist im Fall $k > a$ leer, im Fall $k = a$ ein System von endlich vielen konjugierten Punkten, im Fall $k < a$ eine irreduzible $(a - k)$-dimensionale Mannigfaltigkeit.

Insbesondere folgt, wenn man die Gradzahlen der Hyperflächen alle gleich Eins wählt:

Der Durchschnitt einer irreduziblen a-dimensionalen Mannigfaltigkeit M mit einem allgemeinen linearen Raum S_b ist im Fall $a + b < n$ leer, im Fall $a + b = n$ ein System von endlich vielen konjugierten Punkten, im Fall $a + b > n$ eine irreduzible $(a + b - n)$-dimensionale Mannigfaltigkeit.

Im Fall $a + b = n$ seien $u^{(1)}, \ldots, u^{(a)}$ die allgemeinen Hyperebenen, deren Durchschnitt S_b ist, und $\zeta^{(1)}, \ldots, \zeta^{(g)}$ die Schnittpunkte von M und S_b.

Sind dann $u_0, \ldots, u_n$ weitere Unbestimmte, so kann man nach § 4 die zugeordnete Form des Punktsystems $\xi^{(1)}, \ldots, \xi^{(g)}$ bilden:

$$F(u, u^{(1)}, \ldots, u^{(a)}) = \prod_{\nu=1}^{g} (\xi_0^{(\nu)} u_0 + \xi_1^{(\nu)} u_1 + \ldots + \xi_n^{(\nu)} u_n)^q.$$

Dabei ist $q = 1$ im Fall eines vollkommenen Grundkörpers, und $q = p^e$ im Fall eines unvollkommenen Körpers von der Charakteristik p. $F(u, u^{(1)}, \ldots, u^{(a)})$ heißt nach ZAG IX die *zugeordnete Form* der Mannigfaltigkeit M, und der Grad qg dieser Form heißt der *Grad* der Mannigfaltigkeit M. Die Anzahl g der Schnittpunkte von M mit S_b heißt der *reduzierte Grad* von M. Im Fall eines vollkommenen Grundkörpers besteht zwischen dem Grad und dem reduzierten Grad kein Unterschied.

Ein *k-facher Punkt* einer rein a-dimensionalen Mannigfaltigkeit M ist ein solcher Punkt P von M, der als Schnittpunkt von M mit einem allgemeinen, durch P gehenden linearen Raum S_{n-a} die Multiplizität k hat. Die Mannigfaltigkeit M heißt *singularitätenfrei*, wenn alle ihre Punkte einfache Punkte sind.

§ 7.

Schnitt von Mannigfaltigkeiten mit speziellen Hyperflächen.
Der Bezoutsche Satz.

Es sei C eine irreduzible Kurve, H eine allgemeine und H' eine spezielle Hyperfläche vom Grade g, wobei wir annehmen, daß H' die Kurve nicht enthält, also nur endlich viele Punkte mit ihr gemeinsam hat. $\xi^{(1)}, \ldots, \xi^{(h)}$ seien die Schnittpunkte von C und H. Bei der Spezialisierung $H \to H'$ gehen $\xi^{(1)}, \ldots, \xi^{(h)}$ relationstreu in $\eta^{(1)}, \ldots, \eta^{(h)}$ über, und jeder Schnittpunkt η von C und H' erhält bei der Spezialisierung eine eindeutig bestimmte Multiplizität, die wir die *Schnittpunktsmultiplizität* von η als Schnittpunkt von C und H' nennen.

Die Schnittpunktsmultiplizität ist stets positiv.

Beweis. Nach dem Kriterium von § 4 genügt es, zu zeigen, daß die Korrespondenz zwischen den Hyperflächen H' und ihren Schnittpunkten η' mit C irreduzibel ist. Das ist aber klar (und wurde in § 6 schon bemerkt); denn man erhält ein allgemeines Paar dieser Korrespondenz, indem man durch einen allgemeinen Punkt ξ von C die allgemeinste Hyperfläche H legt.

Bemerkung. Beim Beweis wurde die „*Methode der Problemumkehrung*" benutzt. Das ursprünglich gestellte Problem war: Gegeben eine Hyperfläche H', gesucht ihre Schnittpunkte η mit C. Um aber die Irreduzibilität der dadurch definierten Korrespondenz zu beweisen, wurde ein allgemeines Elementepaar in der Weise konstruiert, daß von einem allgemeinen Punkt ξ von C ausgegangen wurde und eine dazu passende Hyperfläche konstruiert

268

wurde. Man geht also von der Lösung des Problems aus und sucht sich die Daten des Problems dazu passend.

Genau ebenso beweist man allgemeiner, daß *beim Schnitt einer a-dimensionalen Mannigfaltigkeit M mit a Hyperflächen, die M nur in endlich vielen Punkten schneiden und durch Spezialisierung aus allgemeinen Hyperflächen entstanden gedacht werden, nur Schnittpunkte von positiver Vielfachheit auftreten.*

Aus dieser Tatsache folgt ein

Dimensionssatz. *Der Schnitt einer irreduziblen a-dimensionalen Mannigfaltigkeit M mit einer M nicht enthaltenden Hyperfläche H′ enthält nur Bestandteile von der Dimension a − 1.*

Beweis. Gesetzt, der Durchschnitt D hätte einen irreduziblen Bestandteil D_1 von einer Dimension $< a - 1$. η sei ein Punkt von D_1, der nicht einem der anderen irreduziblen Bestandteile $D_2, \ldots, D_r$ von D angehört (z. B. ein allgemeiner Punkt von D_1). Durch den Punkt η kann man $a - 1$ Hyperebenen $U_1', \ldots, U_{a-1}'$ legen, welche D außerdem nur in endlich vielen Punkten schneiden. Unter diesen Schnittpunkten nun hat der Punkt η die Vielfachheit Null. Denn wenn H eine allgemeine Hyperfläche ist und $U_1, \ldots, U_{a-1}$ allgemeine Hyperebenen sind, so kann man die relationstreue Spezialisierung $H \to H'$, $U_i \to U_i'$ in zwei Schritten vornehmen: Zuerst spezialisiert man $H \to H'$, sodann $(U_1, \ldots, U_{a-1}) \to (U_1', \ldots, U_{a-1}')$. Bei der ersten Spezialisierung gehen die Schnittpunkte $\xi^{(1)}, \ldots, \xi^{(h)}$ von $M, H, U_1, \ldots, U_{a-1}$ in Schnittpunkte $\zeta^{(1)}, \ldots, \zeta^{(h)}$ von $M, H', U_1, \ldots, U_{a-1}$, also von $D, U_1, \ldots, U_{a-1}$ über. Keiner dieser Punkte liegt auf D_1, denn D_1 hat mit den allgemeinen Hyperebenen $U_1, \ldots, U_{a-1}$ aus Dimensionsgründen keine Punkte gemeinsam. Also liegen $\zeta^{(1)}, \ldots, \zeta^{(h)}$ alle auf der Vereinigung $D_2 + \ldots + D_r$. Das bleibt aber richtig, wenn man nunmehr $U_1, \ldots, U_{a-1}$ zu $U_1', \ldots, U_{a-1}'$ spezialisiert: $\zeta^{(1)}, \ldots, \zeta^{(h)}$ gehen dabei in Punkte $\eta^{(1)}, \ldots, \eta^{(h)}$ über, die auf $D_2 + \ldots + D_r$ liegen, und unter denen daher η nicht vorkommt. — Andererseits aber hat, wie wir gesehen haben, jeder Schnittpunkt η eine positive Multiplizität. Der Widerspruch beweist, daß unsere Annahme falsch war.

Der eben bewiesene Dimensionssatz ist ein Spezialfall eines allgemeineren Satzes über den Schnitt von zwei Mannigfaltigkeiten der Dimensionen r und s mit $r + s > n$, der aber wesentlich schwieriger zu beweisen ist[11].

Wir kehren nun zum Fall einer Kurve, die mit einer Hyperfläche geschnitten wird, zurück und beweisen den auf diesen Fall bezüglichen *„Bezoutschen Satz"*:

---

[11] ZAG XII, Math. Annalen **115** (1938).

Die Anzahl der Schnittpunkte einer irreduziblen Kurve C mit einer allgemeinen Hyperfläche H ist gleich dem Produkt $g\,\gamma$ der Gradzahlen von C und H, wobei für g der reduzierte Grad zu nehmen ist.

Beweis. Da in der Dissertation von W.-L. Chow[2]) ein allgemeingültiger Beweis enthalten ist, beschränken wir uns hier auf den Fall eines vollkommenen Grundkörpers[12]). Ist dann ξ ein normierter allgemeiner Punkt, so können wir, eventuell nach Umnumerierung der Koordinaten, annehmen, daß $\xi_0 = 1$ ist und daß $\xi_2, \ldots, \xi_n$ separable algebraische Funktionen von ξ_1 sind[13]). Ist dann

$$f_k(\xi_1, \xi_k) = 0 \qquad\qquad (k = 2, 3, \ldots, n)$$

die definierende Gleichung von ξ_k und bedeutet ∂_k die partielle Ableitung nach ξ_k, so ist

$$\partial_k f_k(\xi_1, \xi_k) \neq 0.$$

Durch die Gleichungen

$$(1) \qquad (\zeta_1 - \xi_1)\,\partial_1 f_k(\xi) + (\zeta_k - \xi_k)\,\partial_k f_k(\xi) = 0$$

wird also eine Gerade durch den Punkt ξ definiert, die man die *Tangente* der Kurve C im Punkte ξ nennt. Macht man die Polynome f_k durch Einführung von ξ_0 homogen, so kann man die Gleichung (1) in der homogenen Form

$$(2) \qquad \zeta_0\,\partial_0 f_k(\xi) + \zeta_1\,\partial_1 f_k(\xi) + \zeta_k\,\partial_k f_k(\xi) = 0$$

schreiben. Die Gleichungen (2) stellen Polarhyperebenen dar (vgl. § 4, Gleichung (7)).

Nun betrachte man die irreduzible Korrespondenz, die jedem Punkte ξ' von C alle durch ξ' gehenden Hyperflächen H' zuordnet. Ein allgemeines Paar (ξ, H) der Korrespondenz erhält man entweder, indem man durch einen allgemeinen Punkt ξ von C die allgemeinste Hyperfläche legt, oder indem man von einer allgemeinen Hyperfläche H ausgeht und für ξ irgend einen der Schnittpunkte von H mit C wählt. Aus der *ersten* Bildungsweise des allgemeinen Paares (ξ, H) ersieht man, daß die Hyperebene H die Tangente der Kurve C im Punkte ξ nicht enthält, sondern mit ihr nur den Punkt ξ gemeinsam hat. Dasselbe gilt folglich auch bei der zweiten Erzeugungsweise eines allgemeinen Paares, denn die algebraischen Eigenschaften des allgemeinen Paares sind immer dieselben. Also folgt: *Eine allgemeine Hyperfläche H*

---

[12]) Übrigens läßt sich der Fall eines unvollkommenen Grundkörpers mühelos auf den eines vollkommenen zurückführen, indem man durch wiederholte Adjunktion aller p-ten Wurzeln den Körper zu einem vollkommenen erweitert.

[13]) Für einen Beweis dieses bekannten Hilfssatzes siehe die nächstfolgende Abhandlung ZAG XIV dieser Serie.

schneidet die Kurventangenten in ihren Schnittpunkten mit der Kurve C nur in diesen Punkten[14]).

Nunmehr gehen wir von der allgemeinen Hyperfläche H durch relationstreue Spezialisierung zu einer solchen Hyperfläche H' über, die in γ voneinander unabhängige allgemeine Hyperebenen $L_1, \ldots, L_\gamma$ zerfällt. Die Anzahl der Schnittpunkte η von C mit H' ist offenbar gleich $g\,\gamma$. Die Vielfachheiten dieser Schnittpunkte η sind einerseits positiv, andererseits aber nach dem Kriterium von § 4 auch nicht größer als Eins, da sonst die Polarhyperebenen (2) mit der Polarhyperebene von η in bezug auf H' eine Gerade durch den Punkt gemeinsam haben müßten. Ist η etwa ein Punkt von L_1, so ist die Polarhyperebene von η in bezug auf H' auch L_1, und L_1 hat mit der durch die Gleichung (2) definierten Tangente eben nur den einen Punkt η gemeinsam. Also sind die Vielfachheiten der Schnittpunkte η alle gleich Eins. Nach dem Prinzip der Erhaltung der Anzahl ist nun auch die Anzahl der Schnittpunkte von H und C gleich $g\,\gamma$, was zu beweisen war.

Verallgemeinerung. *Der Durchschnitt einer irreduziblen Mannigfaltigkeit M vom (reduzierten) Grade γ mit einer allgemeinen Hyperfläche vom Grade g hat den (reduzierten) Grad $g\,\gamma$.*

Beweis. M habe die Dimension a, der Durchschnitt mit H also die Dimension $a - 1$. Schneidet man M mit $a - 1$ allgemeinen Hyperebenen, so erhält man nach § 5 eine irreduzible Kurve vom Grad γ. Diese schneidet H nach dem Bezoutschen Satz in $g\,\gamma$ Punkten. Also schneidet der Durchschnitt von M und H einen allgemeinen linearen Raum S_{n-a-1} in $g\,\gamma$ Punkten, was zu beweisen war.

Wiederholte Anwendung ergibt:

Der Durchschnitt einer irreduziblen a-dimensionalen Mannigfaltigkeit vom reduzierten Grad γ mit $k \leqq a$ allgemeinen Hyperflächen von den Graden $e_1, \ldots, e_k$ hat den reduzierten Grad $\gamma e_1 e_2 \ldots e_k$. Im Fall $k = a$ besteht er also aus $\gamma e_1 e_2 \ldots e_k$ Punkten.

Geht man von den allgemeinen Hyperflächen $H_1, \ldots, H_k$ zu den speziellen Hyperflächen $H'_1, \ldots, H'_k$ über, so möge der Durchschnitt $M \cdot H'_1 \cdots H'_k$ in irreduzible Bestandteile $I_1, \ldots, I_r$ zerfallen. Keiner von diesen hat eine Dimension $< a - k$. Wir nehmen an, daß sie alle genau die Dimension $a - k$ haben. Nun nimmt man noch $a - k$ allgemeine Hyperebenen $L_1, \ldots, L_{a-k}$ hinzu, welche $I_1, \ldots, I_r$ je in einem System konjugierter Punkte schneiden. Ein solcher Punkt ξ_1, Schnittpunkt von I_1 mit $L_1, \ldots, L_{a-k}$, möge als Schnittpunkt von $M, H_1, \ldots, H_k, L_1, \ldots, L_{a-k}$ die Vielfachheit μ_1 haben. Dann haben alle konjugierten Punkte $\xi_1^{(\nu)}$ dieselbe

---

[14]) Dieser Satz ist ein Spezialfall eines Satzes von Bertini. Vgl. ZAG V (Math. Annalen **110**), sowie W.-L. Chow, ebenda **114** (1937), S. 664.

Vielfachheit μ_1. Wir nennen μ_1 die *Vielfachheit von I_1 als Bestandteil des Schnittes von M mit $H'_1, \ldots, H'_k$.* Es sei weiter g_1 der reduzierte Grad von I_1, also die Anzahl der konjugierten Schnittpunkte $\xi_1^{(\nu)}$. Entsprechende Bezeichnungen mögen für $I_2, \ldots, I_r$ gelten. Da die Summe der Vielfachheiten aller Schnittpunkte von $M, H'_1, \ldots, H'_k, L_1, \ldots, L_{a-k}$ gleich $\gamma\, e_1 \ldots e_k$ ist, so folgt

$$\mu_1 g_1 + \mu_2 g_2 + \ldots + \mu_r g_r = \gamma\, e_1, e_2, \ldots, e_k,$$

in Worten: *Die Summe der Grade der irreduziblen Bestandteile des Schnittes $M \cdot H'_1 \ldots H'_k$, multipliziert mit ihren Vielfachheiten, ist gleich dem Produkt der Gradzahlen von M und $H'_1, \ldots, H'_k$.*

Die Sätze dieses Paragraphen lassen sich, wenigstens im Falle eines Grundkörpers von der Charakteristik Null, leicht auf mehrfach projektive Räume übertragen. Da die Beweismethode und die Ergebnisse in ZAG I, § 3 ausführlich dargestellt sind, möge dieser Hinweis genügen.

(Eingegangen am 30. 10. 1937.)

272

21.

Zur algebraischen Geometrie XIV
Schnittpunktszahlen von algebraischen Mannigfaltigkeiten

Mathematische Annalen 115, 4 (1938) 619–642

1928 habe ich zum ersten Male eine Definition der Multiplizität eines Schnittpunktes von zwei Mannigfaltigkeiten M_d und M_{n-d} im projektiven Raum S_n gegeben und bewiesen, daß die Summe der Multiplizitäten der Schnittpunkte gleich dem Produkt der Gradzahlen ist[1]). Mein Beweis beruhte auf einem einfachen Grundgedanken von G. Schaake, benutzte aber daneben schwierige, meist idealtheoretische Hilfsmittel. 1933 gab Severi[2]), von derselben Multiplizitätsdefinition ausgehend, einen einfacheren, auf dem Chasles-Schubertschen Korrespondenzprinzip beruhenden Beweis. Ich werde nun im ersten Teil dieser Arbeit zeigen, daß mein ursprünglicher Beweis sich unter Beibehaltung des Grundgedankens, aber unter Vermeidung der Idealtheorie so sehr vereinfachen läßt, daß er an Kürze und Natürlichkeit nichts zu wünschen übrig läßt (§§ 1 bis 3).

Der Grundgedanke ist folgender. Durch eine singuläre projektive Transformation wird die Mannigfaltigkeit M_d in eine zerfallende M_d verwandelt, die in so viele lineare Räume zerfällt, als der Grad von M_d beträgt. Die Zahl der Schnittpunkte dieser zerfallenden Mannigfaltigkeit mit der anderen M_{n-d} ist offenbar gleich dem Produkt der Gradzahlen.

Auf Grund des Prinzips der Erhaltung der Anzahl ist also die Schnittpunktsanzahl irgend einer zu M_d projektiv äquivalenten Mannigfaltigkeit (insbesondere von M_d selbst) mit M_{n-d} ebenfalls gleich dem Produkt der Gradzahlen, wenn noch bewiesen werden kann, daß beim Übergang von einer allgemeinen zu der singulären Projektivität keine Schnittpunkte zusammenrücken. Dieses nun hatte ich früher idealtheoretisch bewiesen; es folgt aber viel einfacher aus der Tatsache, daß die zerfallende Mannigfaltigkeit M_d die andere M_{n-d} bei passender Wahl der singulären Projektivität nicht berührt.

Die Tatsache, daß jede Mannigfaltigkeit durch eine singuläre Projektivität in eine voll zerfallende übergeführt werden kann, ist auch für den zweiten Teil dieser Abhandlung grundlegend.

---

[1]) B. L. van der Waerden, Math. Annalen **99** (1928), S. 497—541.
[2]) F. Severi, Abh. Math. Sem. Hamburg **9** (1933), S. 335—364.

In § 4 wird folgender Satz bewiesen: Wenn eine algebraische Mannigfaltigkeit M_d ein algebraisches System von Mannigfaltigkeiten durchläuft, und man schneidet sie mit einer festen Mannigfaltigkeit M_e, so durchläuft die Schnittmannigfaltigkeit M_{d+e-n} ebenfalls ein algebraisches System, sofern ihre Bestandteile mit ihren Schnittmultiplizitäten gezählt werden. Der Satz ist leicht für den Fall eines linearen Raumes M_e zu beweisen, während eine beliebige M_e wieder durch eine singuläre lineare Transformation in lineare Räume zerlegt wird.

In § 5 wird dieser Satz auf Mannigfaltigkeiten M_d und M_e in einer singularitätenfreien M_n übertragen. Dadurch wird eine algebraische Begründung sowohl des Schubertschen Kalküls der abzählenden Geometrie (siehe § 6) als auch der von Severi entwickelten Theorie der Äquivalenzscharen auf algebraischen Mannigfaltigkeiten[3]) ermöglicht.

Erster Teil.

Der verallgemeinerte Bezoutsche Satz.

§ 1.

Tangentialräume von algebraischen Mannigfaltigkeiten.

ξ sei ein allgemeiner Punkt einer r-dimensionalen algebraischen Mannigfaltigkeit M. Die inhomogenen Koordinaten $\xi_1, \ldots, \xi_n$ sind also Elemente eines Körpers Ω und das System $\{\xi_1, \ldots, \xi_n\}$ hat den Transzendenzgrad r über dem Grundkörper K. K werde als vollkommener Körper (z. B. algebraisch abgeschlossen) angenommen. Dann gilt zunächst der folgende

Hilfssatz. *Man kann die ξ so umnumerieren, daß $\xi_{r+1}, \ldots, \xi_n$ separable algebraische Funktionen von $\xi_1, \ldots, \xi_r$ sind.*

Beweis. Hat K die Charakteristik 0, so ist nichts zu beweisen. Hat K die Charakteristik p und sind $\xi_1, \ldots, \xi_r$ algebraisch unabhängig, so gibt es eine über K irreduzible algebraische Gleichung, die ξ_{r+1} mit $\xi_1, \ldots, \xi_r$ verknüpft:

$$f(\xi_1, \ldots, \xi_r, \xi_{r+1}) = 0, \quad f(x_1, \ldots, x_r, x_{r+1}) \neq 0.$$

---

[3]) F. Severi, Nuovi contributi alla teoria delle serie di equivalenza etc., Mem. Reale Accad. d'Italia 4 (1933), S. 71—129. — Contributi alla teoria delle serie etc., ebenda 8 (1937), S. 387—410. — Vgl. auch die Noten desselben Verfassers in Atti Accad. naz. Lincei, Rend. (6) **17** (1933); **20** (1934) und **21** (1935).

Wenn $f(x)$ als Polynom in $x_1^p, \ldots, x_{r+1}^p$ geschrieben werden könnte, so wäre $f(x)$ die p-te Potenz eines anderen Polynoms:

$$f(x) = \Sigma\, a_{\varrho_1 \ldots \varrho_{r+1}}\, x_1^{p\,\varrho_1} \ldots x_{r+1}^{p\,\varrho_{r+1}}$$

$$= (\Sigma\, \sqrt[p]{a_{\varrho_1 \ldots \varrho_{r+1}}}\; x_1^{\varrho_1} \ldots x_{r+1}^{\varrho_{r+1}})^p,$$

entgegen der vorausgesetzten Irreduzibilität von $f(x)$. Wir können also $x_1, \ldots, x_{r+1}$ und entsprechend $\xi_1, \ldots, \xi_{r+1}$ so umnumerieren, daß f nicht als Polynom in x_{r+1}^p geschrieben werden kann. Dann ist ξ_{r+1} auf Grund der Gleichung $f(\xi) = 0$ eine separable algebraische Funktion von $\xi_1, \ldots, \xi_r$.

Nunmehr betrachten wir in derselben Weise die Gleichung, die ξ_{r+2} mit $\xi_1, \ldots, \xi_r$ verknüpft, und schließen, daß man diese ξ_ν so umnumerieren kann, daß ξ_{r+2} eine separable algebraische Funktion von $\xi_1, \ldots, \xi_r$ ist. Nach wie vor ist ξ_{r+1} separabel über dem Körper $K(\xi_1, \ldots, \xi_r, \xi_{r+2})$, dieser aber ist separabel über $K(\xi_1, \ldots, \xi_r)$; also sind ξ_{r+1} und ξ_{r+2} beide separabel über $K(\xi_1, \ldots, \xi_r)$. So fährt man fort, bis alle ξ_j erschöpft sind.

Ist $f = 0$ die Gleichung einer Hyperfläche, welche die Mannigfaltigkeit M enthält, also

$$f(\xi_1, \ldots, \xi_n) = 0,$$

so kann man die Gleichung $f = 0$ zunächst durch Einführung von ξ_0 homogen machen und dann die Polarhyperebene des Punktes ξ in bezug auf die Hyperfläche bilden:

$$(1) \qquad \eta_0 \frac{\partial f(\xi)}{\partial \xi_0} + \eta_1 \frac{\partial f(\xi)}{\partial \xi_1} + \ldots + \eta_n \frac{\partial f(\xi)}{\partial \xi_n} = 0.$$

Nach dem Eulerschen Satz gilt

$$(2) \qquad \xi_0 \frac{\partial f(\xi)}{\partial \xi_0} + \xi_1 \frac{\partial f(\xi)}{\partial \xi_1} + \ldots + \xi_n \frac{\partial f(\xi)}{\partial \xi_n} = 0.$$

Setzt man $\xi_0 = \eta_0 = 1$ und subtrahiert (2) von (1), so erhält man die Gleichung der Polarhyperebene in inhomogener Form

$$(3) \qquad (\eta_1 - \xi_1) \frac{\partial f(\xi)}{\partial \xi_1} + \ldots + (\eta_n - \xi_n) \frac{\partial f(\xi)}{\partial \xi_n} = 0.$$

Wir behaupten nun:

Satz 1. *Alle Polaren* (3) *des allgemeinen Punktes ξ in bezug auf alle M enthaltenden Hyperflächen f haben einen r-dimensionalen linearen, den Punkt ξ enthaltenden Raum S_r zum Durchschnitt, den Tangentialraum von M im*

Punkte ξ. Die Gleichungen dieses Tangentialraumes lauten, wenn die Koordinaten $\xi_1, \ldots, \xi_n$ dem Hilfssatz entsprechend numeriert werden

$$(4) \quad \begin{cases} \eta_{r+1} - \xi_{r+1} = \sum_{1}^{r} \frac{\partial \xi_{r+1}}{\partial \xi_j} (\eta_j - \xi_j) \\ \cdots \cdots \cdots \cdots \cdots \cdots \cdots \cdots \cdots \cdots \\ \eta_n - \xi_n = \sum_{1}^{r} \frac{\partial \xi_n}{\partial \xi_j} (\eta_j - \xi_j). \end{cases}$$

Beweis. Wir haben erstens zu zeigen, daß der durch (4) definierte lineare Raum in allen Hyperebenen (3) enthalten ist. Setzt man (4) in (3) ein, so ergibt sich für die linke Seite

$$\sum_{j=1}^{r} (\eta_j - \xi_j) \frac{\partial f(\xi)}{\partial \xi_j} + \sum_{k=r+1}^{n} \sum_{j=1}^{r} \frac{\partial \xi_k}{\partial \xi_j} (\eta_j - \xi_j) \frac{\partial f(\xi)}{\partial \xi_k}$$

$$= \sum_{j=1}^{r} (\eta_j - \xi_j) \left(\frac{\partial f(\xi)}{\partial \xi_j} + \sum_{k=r+1}^{n} \frac{\partial f(\xi)}{\partial \xi_k} \frac{\partial \xi_k}{\partial \xi_j} \right).$$

Durch Differentiation der Gleichung $f(\xi) = 0$ nach ξ_j sieht man aber, daß die letzte Klammer den Wert 0 hat. Also ist der durch (4) definierte Raum in (3) enthalten.

Zweitens haben wir zu zeigen, daß unter den Hyperebenen (3) $n - r$ linear unabhängige vorkommen, deren Durchschnitt genau der Raum (4) ist. Zu dem Zweck nehmen wir für $f = 0$ diejenige irreduzible Gleichung, die ξ_{r+1} mit $\xi_1, \ldots, \xi_r$ verbindet:

$$f(\xi_1, \ldots, \xi_r, \xi_{r+1}) = 0.$$

Ihre Polare (3) lautet:

$$\sum_{j=1}^{r} (\eta_j - \xi_j) \frac{\partial f}{\partial \xi_j} + (\eta_{r+1} - \xi_{r+1}) \frac{\partial f}{\partial \xi_{r+1}} = 0.$$

Dividiert man durch $\dfrac{\partial f}{\partial \xi_{r+1}}$, so ergibt sich

$$- \sum_{j=1}^{r} (\eta_j - \xi_j) \frac{\partial \xi_{r+1}}{\partial \xi_j} + (\eta_{r+1} - \xi_{r+1}) = 0;$$

das ist aber die erste Gleichung (4). Genau so erhält man alle anderen Gleichungen (4). Damit ist Satz 1 bewiesen.

Schreibt man die Gleichungen des Tangentialraums in der Gestalt (1), so sieht man, daß der Tangentialraum unabhängig von der Numerierung der Koordinaten und projektiv invariant mit der Mannigfaltigkeit M verknüpft ist. Die unendlich vielen linearen Gleichungen (1) sind einem endlichen Gleichungssystem vom Range $n - r$ äquivalent.

Als nächstes beweisen wir, daß eine Mannigfaltigkeit M_r und ein allgemeiner linearer Raum S_{n-r} in ihren Schnittpunkten keine Tangente gemeinsam haben. Genauer:

Satz 2. *Eine irreduzible r-dimensionale Mannigfaltigkeit M wird von einem allgemeinen linearen Raum S_{n-r} in endlich vielen Punkten geschnitten und die Tangentialräume von M in diesen Punkten haben mit S_{n-r} jeweils nur einen Punkt (den Berührungspunkt) gemeinsam.*

Beweis. Zunächst zeigen wir, daß es nur endlich viele Schnittpunkte geben kann. Betrachtet man den S_{n-r} als Durchschnitt von r allgemeinen Hyperebenen und schneidet M der Reihe nach mit diesen Hyperebenen, so erniedrigt sich die Dimension der irreduziblen Bestandteile der Schnittmannigfaltigkeit bei jedem Schritt um mindestens Eins; also bleibt zum Schluß nur eine nulldimensionale Schnittmannigfaltigkeit übrig. — Die Schlußweise bleibt gültig, wenn man S_{n-r} durch einen Punkt ξ' von M, sonst aber allgemein wählt.

Um nun die übrigen Behauptungen des Satzes zu beweisen, bilden wir eine Korrespondenz zwischen den Punkten von M und den Teilräumen S_{n-r} des Raumes S_n, indem wir einem Punkte ξ' von M alle durch ihn gehenden Räume S'_{n-r} zuordnen. Die Gleichungen der Korrespondenz sind unmittelbar hinzuschreiben: sie drücken aus, daß ξ' in M liegt und daß S'_{n-r} durch ξ' geht. Ein allgemeines Paar (ξ, S_{n-r}) der Korrespondenz erhält man, indem man für ξ' einen allgemeinen Punkt ξ von M wählt und diesen mit einem allgemeinen S_{n-r-1} durch einen S_{n-r} verbindet. Durch relationstreue Spezialisierung erhält man aus dem allgemeinen Paar (ξ, S_{n-r}) in der Tat alle Paare (ξ', S'_{n-r}). Die Korrespondenz ist also irreduzibel.

Das Prinzip der Konstantenzählung ergibt für die Dimension des Bildraumes den Wert

$$r + (n - r)\, r = r\, (n - r + 1);$$

denn M hat die Dimension r und durch jeden Punkt von M gehen $\infty^{(n-r)r}$ Räume S_{n-r}, während umgekehrt einem allgemeinen S_{n-r}, der mindestens einen Punkt ξ enthält, nach dem schon bewiesenen nur ∞^0 Punkte ξ' entsprechen. Die Zahl $r\,(n - r + 1)$ ist aber die Dimension der (irreduziblen) Gesamtheit aller Räume S_{n-r}. Also fällt der Bildraum der Korrespondenz mit der Gesamtheit aller S_{n-r} zusammen, d. h. jeder, und insbesondere jeder allgemeine, Raum S_{n-r} enthält in der Tat Punkte ξ von M. Ist ξ irgend ein solcher Punkt, so hängt das Paar (ξ, S_{n-r}) wirklich von $r\,(n - r + 1)$ Parametern ab, ist also ein allgemeines Paar der Korrespondenz.

Alle allgemeinen Elementepaare einer irreduziblen Korrespondenz haben bekanntlich genau dieselben algebraischen Eigenschaften. Sie können durch Körperisomorphismen ineinander übergeführt werden. Also brauchen wir die behauptete Eigenschaft über den Tangentialraum nur zu beweisen für den Fall, daß ξ ein allgemein gewählter Punkt von M und S_{n-r} der allgemeinste durch ξ gehende lineare Raum ist. Für diesen Fall ist sie aber unmittelbar klar, denn ein allgemeiner S_{n-r} durch ξ hat mit einem speziellen Raum S_r durch ξ nur den einen Punkt ξ gemeinsam.

Wir werden den durch Satz 2 ausgedrückten Sachverhalt kurz so aussprechen: Der allgemeine Raum S_{n-r} schneidet die Mannigfaltigkeit M. aber berührt sie nicht.

§ 2.

Projektive Transformation von Mannigfaltigkeiten.

$M = M_d$ sei eine irreduzible d-dimensionale algebraische Mannigfaltigkeit vom Grade g im projektiven Raum S_n; ihre Gleichungen mögen lauten

$$(1) \qquad f_\lambda(\eta_0, \eta_1, \ldots, \eta_n) = 0.$$

T sei eine projektive Transformation, die η in η^T überführt:

$$(2) \qquad \eta_j^T = \varSigma\, \tau_{jk}\, \eta_k.$$

Unter M^T verstehen wir nun die Mannigfaltigkeit mit den Gleichungen

$$(3) \qquad f_\lambda(\eta^T) = f_\lambda(\varSigma\, \tau_{jk}\, \eta_k) = 0.$$

Das heißt also: η gehört zu M^T, wenn η^T zu M gehört[4]). Die Definition soll auch für singuläre projektive Transformationen gelten (d. h. für solche mit verschwindender Determinante).

Wir konstruieren nun eine singuläre Projektivität, welche den ganzen Raum S_n auf den Unterraum S_{n-d} projiziert, folgendermaßen: $q^{(0)}, \ldots, q^{(n)}$ seien $n + 1$ linear-unabhängige Punkte von S_n. Jeder Punkt ζ kann dann linear durch $q^{(0)}, \ldots, q^{(n)}$ ausgedrückt werden:

$$(4) \qquad \eta_k = \sum_0^n q_k^{(\nu)} \lambda_\nu = \sum_0^{n-d} q_k^{(\nu)} \lambda_\nu + \sum_{n-d+1}^n q_k^{(\nu)} \lambda_\nu.$$

Die beiden angegebenen Teilsummen nennen wir $\dot\eta_k$ und $\ddot\eta_k$; dann ist also

$$\eta_k = \dot\eta_k + \ddot\eta_k.$$

Geometrisch liegt η auf der Verbindungslinie der Punkte $\dot\eta$ und $\ddot\eta$. Der Punkt $\dot\eta$ liegt in dem von $q^{(0)}, \ldots, q^{(n-d)}$ aufgespannten Raum S_{n-d}, während $\ddot\eta$ in dem von $q^{(n-d+1)}, \ldots, q^{(n)}$ aufgespannten, zu S_{n-d} windschiefen Raum S_{d-1} liegt. $\dot\eta$ ist also die Projektion von η aus S_{d-1} auf S_{n-d}.

---

[4]) Eigentlich ist M^T also die Transformierte von M mit der inversen Transformation T^{-1}. Mit Rücksicht auf die singulären Transformationen, die keine Inverse besitzen, wurde die obige Definition gewählt.

Löst man die linearen Gleichungen (4) nach den λ_ν auf und setzt diese in

$$\dot{\eta}_k = \sum_0^{n-d} q_k^{(\nu)} \lambda_\nu$$

ein, so sieht man, daß die ζ_k lineare Funktionen der η_j sind. Die Projektion, die η in $\dot{\eta}$ überführt, ist also eine singuläre lineare Transformation.

Wählen wir alle $q_k^{(\nu)}$ als unabhängige Unbestimmte, so sind S_{n-d} und S_{d-1} allgemeine lineare Räume. Der Raum S_{n-d} schneidet also die Mannigfaltigkeit M_d in endlich vielen Punkten $\sigma^{(1)}, \ldots, \sigma^{(g)}$. Ein Punkt η gehört zu $\dot{M}$, wenn die Projektion $\dot{\eta}$ zu M gehört. $\dot{\eta}$ ist aber stets ein Punkt von S_{n-d}; wenn also $\dot{\eta}$ auch noch zu M_d gehören soll, muß $\dot{\eta}$ einer der Punkte $\sigma^{(1)}, \ldots, \sigma^{(g)}$ sein. Ist etwa $\dot{\eta} = \sigma^{(1)}$, so ist $\eta = \dot{\eta} + \ddot{\eta}$ ein Punkt des Verbindungsraumes $\Sigma^{(1)}$ von S_{d-1} mit $\sigma^{(1)}$. *Demzufolge zerfällt die transformierte Mannigfaltigkeit $\dot{M}_d$ in lineare Räume $\Sigma^{(1)}, \ldots, \Sigma^{(g)}$ von der Dimension d, die alle durch S_{d-1} gehen.*

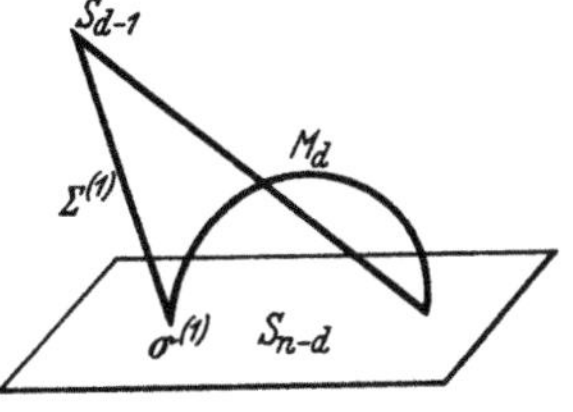

Fig. 1.

Wir wollen nun die Polaren der Punkte von $\Sigma^{(1)}$ in bezug auf die Gleichungen (3) bilden. Die Gleichung einer solchen Polaren heißt, wenn ξ ein veränderlicher Punkt der Polare ist

$$\sum \xi_k \frac{\partial f_\lambda(\eta^T)}{\partial \eta_k} = \sum \sum \xi_k \frac{\partial f_\lambda(\eta^T)}{\partial \eta_j^T} \frac{\partial \eta_j^T}{\partial \eta_k} = 0$$

oder wegen (2)

(4)
$$\sum \sum \xi_k \frac{\partial f_\lambda(\eta^T)}{\partial \eta_j^T} \tau_{jk} = 0.$$

Setzen wir nun

(5)
$$\frac{\partial f_\lambda(\eta^T)}{\partial \eta_j^T} = u_j,$$

so wird (4)

(6)
$$\sum \sum \tau_{jk} u_j \xi_k = 0.$$

u_j sind die Koeffizienten der Gleichung der Polare des Punktes η^T in bezug auf die Form $f_\lambda(\eta)$. Wir nennen diese Hyperebene kurz u. Die Gleichung (6) besagt, daß der transformierte Punkt ξ^T mit den Koordinaten

$$\xi_j^T = \sum \tau_{jk} \xi_k$$

in der Hyperebene u liegen soll, oder, was dasselbe ist, daß ξ in der transformierten Hyperebene u^T liegen soll. Daher der Satz: *Die Polare von η in bezug auf die transformierte Gleichung (3) ist die Transformierte u^T der Polare u des Punktes η^T in bezug auf die ursprüngliche Gleichung (1).*

Wenn der Punkt ξ *allen* Polaren (4) in bezug auf alle Gleichungen (3) angehört, so liegt er in dem Tangentialraum des Punktes η in bezug auf M^T.

Dazu ist notwendig und hinreichend, daß ξ^T allen Polaren des Punktes η^T in bezug auf M angehört, mit anderen Worten, daß ξ^T in dem Tangentialraum des Punktes η^T in bezug auf M liegt. Das heißt also:

Satz 3. *Der Tangentialraum von η in bezug auf M^T ist der Transformierte des Tangentialraums von η^T in bezug auf M.*

Für nichtsinguläre projektive Transformationen drückt dieser Satz nichts anderes aus als die projektive Invarianz des Tangentialraumes. Der Satz gilt aber auch für singuläre Transformationen, vorausgesetzt, daß ζ einen bestimmten Bildpunkt η^T besitzt, also daß die η_k^T nicht alle Null werden. Wenden wir den Satz an auf die Projektion S, die η in $\dot\eta$ überführt, so müssen wir für η einen Punkt nehmen, der einem der Räume $\Sigma^{(\nu)}$, etwa dem Raum $\Sigma^{(1)}$ angehört, aber nicht dem Raum S_{d-1}, da sonst der Punkt $\dot\eta$ nicht bestimmt wäre. Der Punkt $\dot\eta$ fällt dann in $\sigma^{(1)}$. Dieser Punkt besitzt, als allgemeiner Punkt von M, einen bestimmten Tangentialraum S_d. Nach § 1 hat dieser mit S_{n-d} nur den einen Punkt $\sigma^{(1)}$ gemeinsam. Der transformierte Raum $\dot S_d$ ist also der Verbindungsraum dieses Punktes $\sigma^{(1)}$ mit S_{d-1}, d. h. $\dot S_d$ fällt mit $\Sigma^{(1)}$ zusammen. Damit ist bewiesen:

Satz 4. *Die zerfallende Mannigfaltigkeit $\dot M = \Sigma^{(1)} + \ldots + \Sigma^{(g)}$ mit den Gleichungen (3) besitzt in jedem Punkt ζ von $\Sigma^{(1)}$ oder $\Sigma^{(2)}$, $\ldots$, $\Sigma^{(g)}$, der nicht zu S_{d-1} gehört, einen Tangentialraum $\dot S_d$, der mit $\Sigma^{(1)}$ (bzw. $\Sigma^{(2)}$, $\ldots$, $\Sigma^{(g)}$) zusammenfällt.*

Geometrisch ist das gefundene Ergebnis natürlich völlig einleuchtend. Der Sinn des obigen Beweises liegt in dem Nachweis, daß der geometrische Tangentialraum mit dem algebraisch definierten Durchschnitt der Polaren der Gleichungen (3) zusammenfällt, was z. B. nicht erfüllt wäre, wenn $\dot M$ mehrfach zu zählende Bestandteile enthielte. Die Vollkommenheit des Grundkörpers ist dabei eine wesentliche Voraussetzung.

Das gefundene Ergebnis bleibt natürlich gültig, wenn die Transformation $\eta \to \dot\eta$ noch mit einer nichtsingulären projektiven Transformation zusammengesetzt wird, welche bewirkt, daß jeder einzelne der Räume $\Sigma^{(1)}$, $\ldots$, $\Sigma^{(g)}$ in allgemeine Lage bezüglich einer zweiten Mannigfaltigkeit M_{n-d} gebracht wird.

Als Hauptergebnis dieses Paragraphen ist die Tatsache zu betrachten, daß *in der Familie der projektiv transformierten einer irreduziblen Mannigfaltigkeit M_d vom Grade g als Grenzgebilde auch solche Mannigfaltigkeiten vorkommen, welche in g verschiedene lineare Räume derselben Dimension d zerfallen, die einen Raum S_{d-1} gemeinsam haben.* Diese Tatsache wurde von G. Schaake[5]) zuerst bemerkt und wie oben bewiesen.

---

[5]) G. Schaake, Afbeeldingen van figuren op de punten eener lineaire ruimte, Diss. Amsterdam 1922, Kapitel 6. Vgl. auch F. Severi, Mem. Accad. Ital. **5** (1934), S. 239—283.

§ 3.

Schnittpunkte von zwei Mannigfaltigkeiten.

M_d und M_{n-d} seien zwei absolut irreduzible Mannigfaltigkeiten von den Dimensionen d und $n - d$ und den Graden g und h in S_n. T sei eine *allgemeine* projektive Transformation (mit unbestimmten Koeffizienten). Es gilt dann der folgende

S a t z A. *M_d^T und M_{n-d} haben genau $g \cdot h$ Schnittpunkte. Wenn bei einer Spezialisierung der Transformation T die Anzahl der Schnittpunkte endlich bleibt, sind die bei der Spezialisierung entstehenden Vielfachheiten dieser Schnittpunkte stets positiv.*

Haben insbesondere M_d und M_{n-d} selbst endlich viele Schnittpunkte, so erscheinen diese, wenn T zur Identität spezialisiert wird, mit positiven Vielfachheiten behaftet. Diese heißen *Schnittpunktsmultiplizitäten*. Ihre Summe ist nach dem Prinzip der Erhaltung der Anzahl gleich $g \cdot h$.

Zunächst soll gezeigt werden, daß M_d^T und M_{n-d} endlich viele Schnittpunkte haben. Wir betrachten eine Korrespondenz, welche jedem Punkt τ' des $(n^2 + 2n)$-dimensionalen Raumes mit den Koordinaten τ'_{jk} $(j, k = 0, 1, \ldots, n)$ alle Punktepaare (ξ', η') von S_n zuordnet, für welche ξ' zu M_d und η' zu M_{n-d} gehört, während die Transformation T' gerade ξ' in η' überführt:

$$(1) \qquad \lambda \xi'_j = \Sigma \, \tau'_{jk} \, \eta'_k.$$

Die Gleichungen (1) können natürlich durch Elimination von λ inhomogen gemacht werden; dazu treten dann noch die Gleichungen, welche ausdrücken, daß ξ' zu M_d und η' zu M_{n-d} gehören. Da ξ' zu M_d gehört, gehört η' zur transformierten Mannigfaltigkeit $M_d^{T'}$. η' ist also ein Schnittpunkt von $M_d^{T'}$ und M_{n-d}.

Ein allgemeines Element (τ, ξ, η) der Korrespondenz erhält man, indem man für ξ einen allgemeinen Punkt von M_d, für η einen davon unabhängigen allgemeinen Punkt von M_{n-d}, für T schließlich die allgemeinste projektive Transformation, die ξ in η überführt, wählt. Das lineare Gleichungssystem (1) in den Unbekannten λ und τ'_{jk} besitzt ja eine allgemeine Lösung, aus der jede Lösung durch Parameterspezialisierung entsteht. Die allgemeine Lösung hängt, von einem unwesentlichen Proportionalitätsfaktor abgesehen, noch von $n^2 + n$ Parametern ab. Die Existenz eines allgemeinen Elementes der Korrespondenz beweist ihre Irreduzibilität.

Nehmen wir an, daß die an der Korrespondenz beteiligten Punkte τ' eine a-dimensionale Mannigfaltigkeit bilden und daß jedem allgemeinen Punkt τ dieser Mannigfaltigkeit eine b-dimensionale Mannigfaltigkeit von Punktepaaren entspricht, so ist nach dem Prinzip der Konstantenzählung

$$a + b = d + (n - d) + (n^2 + n) = n^2 + 2n.$$

Wenn wir zeigen können, daß $b = 0$ ist, so folgt $a = n^2 + 2\,n$, d. h. die Urmannigfaltigkeit der Korrespondenz ist der ganze $(n^2 + 2\,n)$-dimensionale Raum, und einem allgemeinen Punkt τ entsprechen endlich viele Punktepaare (ξ, η). Das heißt also, M_d^T und M_{n-d} haben für allgemeine T endlich viele Schnittpunkte, wie behauptet wurde.

Gesetzt, die Behauptung wäre falsch, dann wäre $b > 0$, und jedem Element τ' der Urmannigfaltigkeit würden unendlich viele Punktepaare (ξ', η') entsprechen, d. h. $M_d^{T'}$ und M_{n-d} hätten stets unendlich viele Schnittpunkte. Das würde auch dann gelten, wenn τ' eine singuläre projektive Transformation darstellt.

Nun gibt es aber, wie wir in § 2 sahen, singuläre projektive Transformationen, die M_d in eine zerfallende Mannigfaltigkeit überführen, bestehend aus g linearen Räumen $\overset{(\nu)}{\underset{d}{\Sigma}}$, welche zu M_{n-d} allgemeine Lage haben und daher M_{n-d} je in h Punkten schneiden. Wir kommen also zu einem Widerspruch.

Es ist also gezeigt, daß M_d^T uhd M_{n-d} endlich viele Schnittpunkte haben, falls T eine allgemeine projektive Transformation ist. Bei einer Spezialisierung $T \to T'$ können Schnittpunkte zusammenrücken und jeder Schnittpunkt η' erhält eine bestimmte Multiplizität. Da aber die Korrespondenz zwischen T' und η' nach dem obigen irreduzibel ist, sind die entstehenden Multiplizitäten alle positiv.

Wählt man für T' wieder die vorhin betrachtete singuläre Transformation S, die M_d in eine in g lineare Räume zerfallende Mannigfaltigkeit M_d^S überführt, so ist die Zahl der Schnittpunkte von M_d^S und M_{n-d} gleich $g \cdot h$, da jeder der g linearen Räume $\overset{(\nu)}{\underset{d}{\Sigma}}$ genau h Schnittpunkte mit M_{n-d} hat. Wenn wir noch zeigen können, daß diese Schnittpunkte bei der Spezialisierung notwendig die Multiplizität Eins erhalten, so folgt daraus, daß auch bei allgemeinem T die Zahl der Schnittpunkte gleich $g \cdot h$ ist.

Gesetzt, die Multiplizität eines Schnittpunktes ζ von M_d^S und M_{n-d} wäre größer als Eins. Dann müßten nach ZAG V (Math. Ann. **110**) die Polaren dieses Schnittpunktes ζ in bezug auf die Gleichungen von M_d^S und M_{n-d} mindestens eine Gerade gemeinsam haben. Die Polaren von ζ bezüglich M_{n-d} haben den Tangentialraum S_{n-d} zum Durchschnitt, ebenso die bezüglich M_d^S nach § 1 den Tangentialraum $\overset{(\nu)}{\underset{d}{\Sigma}}$. Da aber $\overset{(\nu)}{\underset{d}{\Sigma}}$ allgemeine Lage zu M_{n-d} hat, haben $\overset{(\nu)}{\underset{d}{\Sigma}}$ und S_{n-d} nur einen Punkt gemeinsam (Satz 2). Damit haben wir einen Widerspruch gefunden und den Hauptsatz in allen Teilen bewiesen.

Der Hauptsatz gilt ebenso auch für zerfallende Mannigfaltigkeiten M_d und M_{n-d}, vorausgesetzt, daß deren absolut irreduzible Bestandteile alle dieselbe Dimension d bzw. $n-d$ haben. Denn die Schnittpunkte von M_d^T und M_{n-d} setzen sich zusammen aus den Schnittpunkten der absolut irreduziblen Bestandteile von M_d^T mit denen von M_{n-d}. Es gibt keine Schnittpunkte, die mehreren irreduziblen Bestandteilen von M_{n-d} gleichzeitig angehören, denn der Durchschnitt von zwei solchen Bestandteilen ist eine Mannigfaltigkeit von einer Dimension $< n-d$, welche M_d^T für allgemeine T nicht schneidet.

Zweiter Teil.

§ 4.

Schnittpunktsgruppen von Systemen von algebraischen Mannigfaltigkeiten.

In ZAG IX haben W.-L. Chow und ich[6]) den Begriff des algebraischen Systems von algebraischen Mannigfaltigkeiten definiert. Die einzelnen Elemente eines solchen Systems $\mathfrak{S}$ sind rein d-dimensionale Mannigfaltigkeiten M, deren irreduzible Bestandteile M_λ mit gewissen Vielfachheiten α_λ versehen sind. Wir schreiben dementsprechend

$$M = \varSigma\,\alpha_\lambda\,M_\lambda.$$

Der Grad von M ist

$$g = \varSigma\,\alpha_\lambda\,g_\lambda,$$

wo g_λ den Grad von M_λ bedeutet. Alle Mannigfaltigkeiten M des Systems $\mathfrak{S}$ haben denselben Grad g und dieselbe Dimension d.

Hat eine Mannigfaltigkeit M_{n-d} mit M_λ nur endlich viele Schnittpunkte, so bilden diese, mit ihren Vielfachheiten versehen, eine Schnittpunktsgruppe Q_λ. Unter der Schnittpunktsgruppe (kürzer Schnittgruppe) Q von M und M_{n-d} verstehen wir nun die Punktgruppe

$$Q = \varSigma\,\alpha_\lambda\,Q_\lambda$$

und bezeichnen sie mit $M \cdot M_{n-d}$. Wir wollen nun folgenden Satz beweisen:

Satz B_0. *Durchläuft $M = M_d$ ein irreduzibles algebraisches System von Mannigfaltigkeiten, so durchläuft auch die Schnittgruppe $Q = M_d \cdot M_{n-d}$ ein irreduzibles System von nulldimensionalen Mannigfaltigkeiten, und die Zuordnung $M \to Q$ ist eine irreduzible Korrespondenz zwischen diesen beiden Systemen.*

Beweis. 1. Es genügt, zu beweisen, daß die Paare (M, Q) eine irreduzible Korrespondenz bilden. M^* sei das allgemeine Element des irreduziblen Systems, Q^* seine Schnittgruppe mit M_{n-d}. Das Paar (M^*, Q^*) ist jedenfalls das allgemeine Element einer Mannigfaltigkeit von Paaren (M', Q'), die durch relationstreue Spezialisierung aus (M^*, Q^*) hervorgehen. Wir haben zu be-

---

[6]) Math. Annalen **113** (1937), S. 692—712.

283

weisen, daß die bei einer solchen relationstreuen Spezialisierung entstehende Punktgruppe Q' genau die Schnittgruppe von M' und M_{n-d} ist. Dabei können wir M^* und M_{n-d} als absolut irreduzibel annehmen, da sie andernfalls in ihre absolut irreduziblen Bestandteile zerlegt werden können, was eine entsprechende Zerlegung der Schnittpunktsgruppe nach sich zieht.

2. Wir beweisen unsere Behauptung zunächst für den Fall, daß M_{n-d} ein allgemeiner (von M^* und M' unabhängiger) linearer Raum S_{n-d}, Durchschnitt der allgemeinen Hyperebenen $u^{(1)}, \ldots, u^{(d)}$, ist. Ist nun M eine irreduzible d-dimensionale Mannigfaltigkeit und sind $\zeta^{(1)}, \ldots, \zeta^{(g)}$ die Schnittpunkte von M mit S_{n-d}, so ist die zugeordnete Form von M nach ZAG IX durch

$$F\,(u^{(0)},\,u^{(1)},\,\ldots,\,u^{(d)}) = \prod_{v}\,(\sum_k u_k^{(0)}\,\zeta_k^{(v)})$$

definiert; sie ist also gleichzeitig die zugeordnete Form $F\,(u^{(0)})$ der Schnittpunktsgruppe $(\zeta^{(1)}, \ldots, \zeta^{(g)})$. Durch Produktbildung folgt dieselbe Tatsache für reduzible Mannigfaltigkeiten M, insbesondere für M^* und M'. Die relationstreue Spezialisierung $M^* \to M'$ bedeutet nach Definition eine relationstreue Spezialisierung der zugeordneten Form $F\,(u^{(0)}, \ldots, u^{(d)})$, also auch eine relationstreue Spezialisierung der Schnittpunktsgruppe $(\zeta^{(1)}, \ldots, \zeta^{(g)})$, deren zugeordnete Form ja ebenfalls $F\,(u^{(0)}, \ldots, u^{(d)})$ ist. Damit ist im Falle $M_{n-d} = S_{n-d}$ die Behauptung bewiesen.

3. Nun ersetzen wir M_{n-d} durch eine projektiv transformierte M_{n-d}^T mittels einer allgemeinen Projektivität T. S sei eine singuläre Projektivität, welche (nach § 2) M_{n-d} in eine zerfallende Mannigfaltigkeit

$$M_{n-d}^S = S_{n-d}^{(1)} + \ldots + S_{n-d}^{(h)}$$

überführt. Die Schnittgruppe von M^* und M_{n-d}^T sei Q^*. Wir spezialisieren $T \to S$ und $M^* \to M'$. Machen wir zuerst die Spezialisierung $T \to S$, so rücken keine Schnittpunkte zusammen, denn M^* und M_{n-d}^S haben in ihren Schnittpunkten keine Tangente gemeinsam, und daraus folgt (wie in § 3), daß alle Spezialisierungsmultiplizitäten gleich Eins sind. Bei der nachfolgenden Spezialisierung $M^* \to M'$ gehen die Schnittpunktsgruppen der einzelnen $S_{n-d}^{(v)}$ mit M^* nach dem schon Bewiesenen in die von $S_{n-d}^{(v)}$ mit M' über. Insgesamt geht also $Q^* = (M^*, M_{n-d}^T)$ in $(M', \Sigma\, S_{n-d}^{(v)}) = (M', M_{n-d}^S)$ über.

Wird andererseits zuerst $M^* \to M'$ und dann $T \to S$ spezialisiert, so geht bei der ersten Spezialisierung Q^* in eine Punktgruppe Q'' über, von der wir beweisen wollen, daß sie mit der Schnittgruppe $M' \cdot M_{n-d}^T$ identisch ist. Q'' besteht jedenfalls aus Schnittpunkten von M' mit M_{n-d}^T mit gewissen Multiplizitäten. Es sei nun

$$(1) \qquad\qquad M' = \Sigma\, \alpha_\lambda\, M'_\lambda$$

die Zerlegung von M' in irreduzible Mannigfaltigkeiten. Da die Schnittpunkte der irreduziblen Mannigfaltigkeit M'_λ mit M^T_{n-d} alle konjugiert sind, kommen sie in Q' alle mit der gleichen Multiplizität β'_λ vor. Es ist also

$$(2) \qquad Q'' = \Sigma \beta_\lambda (M'_\lambda, M^T_{n-d}).$$

Bei der nachfolgenden Spezialisierung $T \to S$ geht (M'_λ, M^T_{n-d}) in (M'_λ, M^S_{n-d}) über (nach Definition der letzteren Punktgruppe), mithin Q'' in die Punktgruppe

$$\Sigma \beta_\lambda (M'_\lambda, M^S_{n-d}).$$

Wenn wir nun wüßten, daß die Reihenfolge der beiden Spezialisierungen gleichgültig ist, so würde folgen

$$\Sigma \beta'_\lambda (M'_\lambda, M^S_{n-d}) = (M', M^S_{n-d})$$

oder, wenn die Gruppe rechts vermöge (1) in ihre Bestandteile zerlegt wird,

$$\beta'_\lambda = \alpha'_\lambda,$$

also nach (2)

$$Q'' = \Sigma \alpha'_\lambda (M'_\lambda, M^T_{n-d}) = (M', M^T_{n-d}),$$

was zu beweisen war.

Der Nachweis, daß die Reihenfolge der beiden Spezialisierungen gleichgültig ist, kann folgendermaßen geführt werden. Das allgemeine Element M^* des irreduziblen Systems $\mathfrak{S}$ kann mit dem besonderen Element M' durch ein eindimensionales Teilsystem von $\mathfrak{S}$ (eine Kurve auf der Mannigfaltigkeit $\mathfrak{S}$) verbunden werden [7]). M' kann ein singulärer Punkt dieser Kurve sein; in diesem Falle lösen wir die Singularität auf, betrachten also M' als Bildpunkt eines einfachen Punktes P' einer Hilfskurve C, während M^* das Bild eines allgemeinen Punktes P^* der Kurve C ist. Statt M' und M^* schreiben wir also fortan $M(P')$ und $M(P^*)$. Statt nun direkt M^* zu M' zu spezialisieren, spezialisieren wir im obigen Beweis P^* zu P', was von selbst die Spezialisierung $M^* \to M'$ zur Folge hat. Mit dieser Abänderung führen wir nun den obigen Beweis durch und haben zu zeigen, daß die Reihenfolge der Spezialisierungen $T \to S$ und $P^* \to P'$ für das Ergebnis gleichgültig ist.

Wir betrachten die algebraische Korrespondenz, die zwischen den Paaren (P^*, T) einerseits und den Punkten Y der Punktgruppe Q^* andererseits besteht. Die Gleichungen der Korrespondenz drücken aus, daß P^* ein Punkt von C ist und daß Y zu $M(P^*)$ und zu M^T_{n-d} gehört. Dem speziellen Paar (P', S) entsprechen in der Korrespondenz nur endlichviele Punkte Y'. Die Urmannigfaltigkeit der Korrespondenz besteht aus allen Paaren (P'', T''), wobei P'' irgend ein Punkt der Kurve C und T'' irgend eine projektive Trans-

---

[7]) ZAG XI, Math. Annalen **114** (1937), § 1, Hilfssatz, S. 684.

formation ist. Das Paar (P', S') stellt einen einfachen Punkt der Urmannigfaltigkeit dar, denn P' ist ein einfacher Punkt der Kurve C und S ist ein, selbstverständlich einfacher, Punkt des Raumes aller projektiven Transformationen. Wenn aber einem einfachen Punkt der Urmannigfaltigkeit in der Korrespondenz endlichviele Punkte der Bildmannigfaltigkeit entsprechen, so haben alle diese Punkte nach ZAG VI [8]), § 3, Satz 1 ganz bestimmte Spezialisierungsmultiplizitäten, unabhängig davon, wie die relationstreue Spezialisierung $P^* \to P'$, $T \to S$, $Q^* \to Q'$ ausgeführt wird. Es ist also auch gleichgültig, ob man zuerst $P^* \to P'$ und dann $T \to S$ spezialisiert oder umgekehrt.

Es wäre an sich denkbar, daß die Zahlen β_λ davon abhängig wären, auf welchem Zweig der Kurve M^* gegen M' strebt, d. h. also welchen der endlich vielen Punkte P' man dem Element M' zuordnet. Der obige Beweis zeigt aber, daß die Zahlen β'_λ unabhängig von der Wahl von P' immer gleich α'_λ sind.

4. Schließlich untersuchen wir, was mit der Punktgruppe $Q^* = M^* \cdot M^T_{n-d}$ geschieht, wenn M^* zu M' und T zur Identität I spezialisiert wird. Spezialisieren wir zuerst $T \to I$, so geht Q^* in $M^* \cdot M_{n-d}$ über, denn so wurde die Punktgruppe $M^* \cdot M_{n-d}$ gerade definiert. Spezialisiert man sodann $M^* \to M'$, so geht Q^* in eine Punktgruppe Q' über, bestehend aus Schnittpunkten von M und M_{n-d} mit gewissen Multiplizitäten. Dasselbe Ergebnis muß auf Grund der obigen Schlußweise auch bei der anderen Reihenfolge der Spezialisierungen zum Vorschein kommen. Läßt man zuerst M^* in M' übergehen, so geht $Q^* = M^* \cdot M^T_{n-d}$ nach dem schon Bewiesenen in $M' \cdot M^T_{n-d}$ über. Geht nun T in die Identität I über, so geht $M' \cdot M^T_{n-d}$ in $M' \cdot M_{n-d}$ über, denn so wurden die Multiplizitäten der Punkte von $M' \cdot M_{n-d}$ gerade definiert. Vergleich mit dem Ergebnis der ersten Reihenfolge der Spezialisierungen ergibt

$$Q' = M' \cdot M_{n-d}.$$

Damit ist Satz B_0 in allen Teilen bewiesen.

Wir betrachten nun etwas allgemeiner zwei Mannigfaltigkeiten M_d und M_e mit $d + e = n + k$, von denen M_d ein irreduzibles System durchläuft. Der Durchschnitt D von M_d und M_e hat mit einem allgemeinen linearen Raum S_{n-k} jedenfalls Punkte gemeinsam, also hat D mindestens die Dimension k; wir nehmen an, daß D genau die Dimension k hat. Die irreduziblen Bestandteile von D haben dann (nach Satz 2 aus ZAG XII) genau die Dimension k. Wir zählen sie mit gewissen Vielfachheiten, die folgendermaßen definiert werden: Der allgemeine S_{n-k} schneidet M_d und M_e in Schnittmannigfaltigkeiten N_{d-k} und N_{e-k}, die endlich viele Schnittpunkte haben, alle nach dem vorangehenden mit gewissen Vielfachheiten versehen. Diese

---

[8]) Math. Annalen **110** (1934), S. 144.

Schnittpunkte sind genau die Schnittpunkte von D mit S_{n-k}, sie verteilen sich also auf die irreduziblen Bestandteile von D und jeder von ihnen gehört einem und nur einem dieser Bestandteile an. Es sei D_1 ein solcher Bestandteil. Da die Schnittpunkte von D_1 mit S_{n-k} algebraisch konjugiert sind, haben sie alle (als Schnittpunkte von N_{d-k} und N_{e-k}) dieselbe Multiplizität μ_1. Diese definieren wir als Multiplizität von D_1 als Bestandteil der Schnittmannigfaltigkeit D von M_d und M_e.

Nun kann Satz B_0 unmittelbar auf den Fall $d + e = n + k$ ausgedehnt werden:

Satz B. *Durchläuft M_d ein irreduzibles algebraisches System von Mannigfaltigkeiten, so durchläuft auch die Schnittmannigfaltigkeit $Q_k = M_d \cdot M_e$ ein irreduzibles System von k-dimensionalen Mannigfaltigkeiten, und die Zuordnung $M_d \to Q_k$ ist eine irreduzible Korrespondenz.*

Beweis. M_d^* sei wie im vorigen Beweis das allgemeine Element des gegebenen irreduziblen Systems, Q_k^* seine Schnittmannigfaltigkeit mit M_e. Das Paar (M_d^*, Q_k^*) definiert wieder eine irreduzible Mannigfaltigkeit von Paaren (M_d', Q_k'), die durch relationstreue Spezialisierung entstehen. Wir haben zu beweisen, daß Q' genau die Schnittmannigfaltigkeit $M' \cdot M_e$ ist. Dazu schneiden wir alle Mannigfaltigkeiten $M_e, M_d^*, Q_k^*, M_d', Q_k'$ mit einem allgemeinen linearen Raum S_{n-k}. Dadurch gehen sie in $M_{e-k}, M_{d-k}^*, Q_0^*, M_{d-k}', Q_0'$ über. Die einfache Überlegung im 2. Teil des Beweises von Satz B_0, die von dem Schnitt mit linearen Räumen handelt, liefert unmittelbar das Ergebnis, daß das Paar (M_d', Q_0') eine relationstreue Spezialisierung von (M_d^*, Q_0^*) ist. Da Q_0^* die Schnittgruppe von M_{d-k}^* und M_{e-k} (im Raum S_{n-k}) ist, so folgt aus Satz B_0, daß Q_0' die Schnittgruppe von M_{d-k}' und M_{e-k} ist. Daraus folgt auf Grund unserer Definitionen der Vielfachheiten der irreduziblen Bestandteile von $M_d' \cdot M_e$, daß Q_k' genau die Schnittmannigfaltigkeit $M_d' \cdot M_e$ ist, was zu beweisen war.

Bemerkung. Die Beweise der Sätze B_0 und B gelten mit einer Einschränkung auch noch dann, wenn in dem betrachteten irreduziblen System $\mathfrak{S}$ von Mannigfaltigkeiten M_d einzelne Mannigfaltigkeiten vorkommen, die einen mehr als k-dimensionalen Durchschnitt mit M_e haben ($k = d + e - n$). Es ist nur vorauszusetzen, daß das allgemeine Element M_d^* mit M_e wirklich einen nur k-dimensionalen Durchschnitt besitzt. Die Schnittmannigfaltigkeit $Q_k^* = M_d^* \cdot M_e$ definiert dann ein irreduzibles System von Mannigfaltigkeiten, welches zum System $\mathfrak{S}$ in Korrespondenz steht, derart, daß jedem allgemeinen M_d^* ein einziges Q_k^* entspricht. Der Beweis des Satzes B liefert nun folgendes Ergebnis: *Ist M_d' ein solches Element von $\mathfrak{S}$, das mit M_e einen nur k-dimensionalen Durchschnitt hat, so entspricht diesem M_d' in der Korrespondenz nur ein Bildelement Q_k', nämlich $M_d' \cdot M_e$. Hat aber M_d' einen mehr als k-dimensionalen Durchschnitt mit M_e, so können dem M_d' in der Korrespondenz* *mehrere Bildelemente Q'_k entsprechen.* Über die letzteren Bildelemente sagt der Satz nichts aus.

Man kann nach Severi[2]) die Schnittmannigfaltigkeit $M_d \cdot M_e$ auch dann, aber nicht eindeutig, definieren, wenn M_d und M_e einen mehr als k-dimensionalen Durchschnitt haben. Zu dem Zweck bettet man M_d in ein irreduzibles System ein, dessen allgemeines Element M_d^* mit M_e einen k-dimensionalen Durchschnitt Q_k^* hat, und nimmt dann wie oben eine relationstreue Spezialisierung $(M_d^*, Q_k^*) \to (M_d, Q_k)$ vor. Irgend eine der so erhaltenen rein k-dimensionalen Mannigfaltigkeiten Q_k gilt als *virtuelle Schnittmannigfaltigkeit $M_d \cdot M_e$.*

<h2 style="text-align:center">§ 5.</h2>

<h3 style="text-align:center">Schnitt von Mannigfaltigkeiten auf M_n.</h3>

Es sei M_n eine singularitätenfreie Mannigfaltigkeit im Raum S_r. Haben zwei Mannigfaltigkeiten M_d und M_e in M_n einen nur k-dimensionalen Durchschnitt D_k $(k = d + e - n)$, so definiert man nach Severi[2]) die Vielfachheiten der irreduziblen Bestandteile $D_k^{(\lambda)}$ von D_k folgendermaßen:

Verbindet man alle Punkte von M_e mit allen Punkten eines allgemeinen (also M_n nicht treffenden) linearen Raumes S_{r-n-1}, so erhält man einen projizierenden Kegel K_{e+r-n}. Enthält M_e mehrfach gezählte Bestandteile, so werden die entsprechenden Bestandteile von K_{e+r-n} ebenso oft gezählt wie jene. Jeder Bestandteil $D_k^{(\lambda)}$ von D_k ist gleichzeitig ein irreduzibler Bestandteil des Durchschnittes $M_d \cdot K_{e+r-n}$ und hat als solcher eine Multiplizität μ. Diese soll gleichzeitig als Multiplizität von $D_k^{(\lambda)}$ in $D_k = M_d \cdot M_e$ gelten.

Zur Rechtfertigung der Definition ist nachzuweisen, daß der Durchschnitt $M_d \cdot K_{e+r-n}$ tatsächlich nur Bestandteile von einer Dimension $\leqq k$ hat. Das geht so: Ist P ein nicht in D_k enthaltener Punkt von $M_d \cdot K_{e+r-n}$, so liegt P auf einer Verbindungslinie QR, wo Q auf M_e und R auf S_{r-n-1} liegt. Wir betrachten nun eine algebraische Korrespondenz zwischen P und R, bestehend aus den Punktepaaren (P, R), bei denen P zu M_d, R zu S_r gehört, während die Verbindungslinie PR einen Punkt von M_e enthält. Aus dem Prinzip der Konstantenzählung ergibt sich, daß die Dimension eines irreduziblen Bestandteils der Korrespondenz einerseits gleich $d + e + 1$, andererseits gleich $a + b$ ist, wobei a die Dimension der Bildmannigfaltigkeit (der Gesamtheit der Punkte R) ist, während einem allgemeinen Punkte R dieser Bildmannigfaltigkeit ∞^b Punkte P entsprechen. Schneidet man nun die Bildmannigfaltigkeit mit dem allgemeinen Raum S_{r-n-1}, legt also dem Punkte R noch $n + 1$ allgemeine lineare Bedingungen auf, so verringert sich die Dimension a um $n + 1$, während b ungeändert bleibt. Die Dimension der Korrespondenz wird also $(d + e + 1) - (n + 1) = k$. Die Dimension der Urmannigfaltigkeit, d. h. der Gesamtheit der Punkte P, denen Punkte R auf S_{r-n-1} entsprechen, wird nun wieder nach dem Prinzip der Konstantenzählung $\leqq k$, womit unsere Behauptung bewiesen ist. Mit genau denselben Schlüssen, bei denen M_n die Rolle von M_d übernimmt, beweist man, daß der vollständige Schnitt $M_n \cdot K_{e+r-n}$ nur Bestandteile von einer Dimension $\leqq e$ enthält. Dieser Durchschnitt besteht somit aus M_e und einem Restschnitt N_e:

$$M_n \cdot K_{e+r-n} = M_e + N_e.$$

Nun sei $e < n$, und es sei P irgend ein fester, von S_{r-n+1} unabhängiger Punkt von M_e. Wir wollen beweisen, daß P *nicht zu* N_e *gehört*.

Beim Beweis können wir M_e als irreduzibel annehmen. N_e' sei ein irreduzibler Bestandteil von N_e. Gesetzt, P würde zu N_e' gehören. P^* sei ein allgemeiner Punkt von N_e' (nicht in M_e); dann ist $P^* \to P$ eine relationstreue Spezialisierung. Verbinde P^* und P mit S_{r-n-1}; das ergibt zwei Verbindungsräume S_{r-n}^* und S_{r-n}, von denen der letzte eine relationstreue Spezialisierung des ersteren ist. S_{r-n}^* ist ein erzeugender Raum des projizierenden Kegels K_{e+r-n} von M_e; also hat S_{r-n}^* einen Punkt Q^* ($\neq P^*$) mit M_e gemeinsam. Die relationstreue Spezialisierung $P^* \to P$, $S_{r-n}^* \to S_{r-n}$ kann durch $Q^* \to Q$ erweitert werden. Q ist ein gemeinsamer Punkt von S_{r-n} und M_e (denn Q^* war ein solcher von S_{r-n}^* und M_e). Nun haben aber M_e und S_{r-n} wegen $e < n$ außer P keinen gemeinsamen Punkt[9]). Also ist $Q = P$. Das heißt, die beiden verschiedenen Schnittpunkte P^* und Q^* von S_{r-n}^* mit M_n gehen bei der relationstreuen Spezialisierung $S_{r-n}^* \to S_{r-n}$ in einen einzigen Punkt $P = Q$ über. Dieser wäre also ein doppelt zu zählender Schnittpunkt von M_n und S_{r-n}. Aber P ist ein einfacher Punkt von M_n, und der Verbindungsraum von P mit einem allgemeinen Raum S_{r-n-1} hat daher in P nur einen einfach zu zählenden Schnittpunkt mit M_n. Wir sind somit auf einen Widerspruch gestoßen.

Bei dem oben geführten Beweis ist ganz wesentlich die Voraussetzung benutzt, daß P ein *einfacher* Punkt von M_n ist. Für singuläre Punkte gilt die Behauptung im allgemeinen nicht, wie man sich durch Beispiele leicht klar machen kann.

Jetzt sind wir im Stande, den Satz B auf Mannigfaltigkeiten in M_n zu übertragen:

---

[9]) Denn der Kegel, der M_e aus P projiziert, hat nur die Dimension $e + 1 \leqq n$ und hat daher mit einem allgemeinen linearen Raum S_{r-n-1} keine Punkte gemeinsam. Er hat also auch mit dem Verbindungsraum S_{r-n} keinen erzeugenden Strahl gemeinsam. Das wäre aber doch der Fall, wenn S_{r-n} einen von P verschiedenen Punkt Q von M_e enthielte.

Satz C. *Durchläuft M_d ein irreduzibles algebraisches System von Mannigfaltigkeiten in M_n, so durchläuft die Schnittmannigfaltigkeit $D_k = M_d \cdot M_e$
mit dem festen M_e ein irreduzibles System von k-dimensionalen Mannigfaltigkeiten, und die Zuordnung $M_d \to Q_k$ ist eine irreduzible Korrespondenz. Dabei
ist vorausgesetzt, daß M_n singularitätenfrei ist und daß der Schnitt $M_d \cdot M_e$
nur Bestandteile von höchstens der Dimension $k = d + e - n$ enthält.*

Beweis. Es sei wieder (wie bei den Beweisen der Sätze B_0 und B) M_d^* das
allgemeine Element des gegebenen irreduziblen Systems, Q_k^* seine Schnittmannigfaltigkeit mit M_e, und es sei $M_d^* \to M_d'$, $Q_k^* \to Q_k'$ eine relationstreue
Spezialisierung. Wir haben zu beweisen, daß Q_k' genau die Schnittmannigfaltigkeit von M_d' und M_e ist. Wir ergänzen Q_k^* durch R_k^* zum vollständigen
Schnitt $Q_d^* + R_d^* = M_d^* \cdot K_{e+r-n}$. Die irreduziblen Bestandteile von M_d^*
kommen nach Definition mit genau derselben Multiplizität in $M_d^* \cdot K_{e+r-n}$
vor, und der vollständige Schnitt $M_d^* \cdot K_{e+r-n}$ besteht somit aus M_d^* und einigen
weiteren Bestandteilen, die nach dem obigen ebenfalls k-dimensional sind
und nicht zu M_e, also zum Restschnitt N_e gehören:

$$(1) \qquad R_k^* \subset N_e.$$

Die relationstreue Spezialisierung $M_d^* \to M_d'$, $Q_k^* \to Q_k'$ läßt sich nun durch
$R_k^* \to R_k'$ ergänzen. Die Summe $Q_d^* + R_d^* = M_d^* \cdot K_{e+r-n}$ ergibt dann bei
der relationstreuen Spezialisierung nach Satz B

$$(2) \qquad Q_k' \dotplus R_k' = M_d' \cdot K_{e+r-n},$$

wobei die Relation (1) ebenfalls erhalten bleibt:

$$R_k' \subset N_e.$$

Aus demselben Grunde ist weiter

$$Q_k' \subset M_d' \quad \text{und} \quad Q_k' \subset M_e.$$

Es sei nun I_k ein irreduzibler Bestandteil von $M_d' \cdot M_e$ und P ein fester Punkt
von I_k. Dann liegt P nach dem früher Bewiesenen nicht in N_e, also nicht
in R_k'; mithin ist I_k nicht in R_k' enthalten. Die Vielfachheit von I_k als Bestandteil von $Q_k' + R_k' = M_d' \cdot K_{e+r-n}$ ist somit gleich der Vielfachheit von I_k
als Bestandteil von Q_k'; andererseits ist sie nach Definition auch gleich der
Vielfachheit von I_k als Bestandteil von $M_d' \cdot M_e$. Da dasselbe für alle irreduziblen Bestandteile von $M_d' \cdot M_e$ gilt, so folgt die Behauptung

$$M_d' \cdot M_e = Q_k'.$$

§ 6.

Der Schubertsche Kalkül.

In § 5 haben wir nach Severi die Schnittmannigfaltigkeit $M_d \cdot M_e = M_k$ von zwei Mannigfaltigkeiten M_d und M_e in M_n, deren Durchschnitt die Dimension $k = d + e - n$ hat, definiert. M_k ist entweder leer ($M_k = 0$) oder eine Summe von irreduziblen Bestandteilen mit bestimmten positiven Multiplizitäten. Um nun das Rechnen mit dem Symbol $M_d \cdot M_e$ zu ermöglichen, haben wir zunächst das kommutative und das assoziative Gesetz der Multiplikation sowie die Distributivgesetze nachzuweisen.

1. Das kommutative Gesetz. Es genügt, den Fall $d + e = n$, $k = 0$ zu behandeln, da der allgemeine Fall sich durch Schnitt mit einem allgegemeinen S_{r-k} in S_r darauf zurückführen läßt (vgl. den Übergang von Satz B_0 zu Satz B in § 4). Weiter können wir M_d und M_e als absolut irreduzibel annehmen.

Es sei P ein Schnittpunkt von M_d und M_e. Wir haben zu beweisen, daß die Multiplizität von P als Punkt von $M_d \cdot M_e$ dieselbe ist wie die Multiplizität von P als Punkt von $M_e \cdot M_d$. Beim Beweise folgen wir Severi[2]), indem wir die in § 5 gegebene Definition der Schnittpunktsmultiplizität durch eine andere, allgemeinere ersetzen.

Wir nehmen zu M_d eine Mannigfaltigkeit N_d hinzu, welche P nicht enthält, auf M_n liegt und welche ebenso wie M_d nur endlich viele Punkte mit M_e gemeinsam hat, und wählen sie so, daß $M_d + N_d = L_d$ ein Element eines irreduziblen Systems von Mannigfaltigkeiten ist, dessen allgemeines Element L_d^* die Mannigfaltigkeit M_e nicht berührt. Daß es eine solche Mannigfaltigkeit N_d wirklich gibt, wird sich nachher zeigen. Die Multiplizitäten der Schnittpunkte von L_d^* und M_e nach der in § 4 gegebenen Definition sind dann gleich Eins; denn wenn L_d^* und M_e sich nicht berühren, so berührt L_d^* auch den projizierenden Kegel K_{e+r-n} nicht. Bei der relationstreuen Spezialisierung $L_d^* \to L_d = M_d + N_d$ gehen eine gewisse Anzahl μ von Schnittpunkten von L_d^* und N_d in den Punkt P über, und aus Satz C folgt, daß *diese Anzahl gleich der Multiplizität von P als Schnittpunkt von $M_d + N_d$ mit M_e, also auch von M_d mit M_e ist.*

Wir haben hier also eine zweite Definition der Schnittmultiplizität μ gefunden, welche den Vorteil hat, daß sie nur die Mannigfaltigkeit M_n und nicht den umgebenden Raum benutzt. Severi hat diese Definition sogar an die Spitze gestellt. Aus ihr folgt, nebenbei bemerkt, die Invarianz der Schnittpunktsmultiplizität bei birationalen Transformationen, für welche der Schnittpunkt P nicht Fundamentalpunkt ist.

Um nun zu zeigen, daß es tatsächlich Mannigfaltigkeiten N_d mit den gewünschten Eigenschaften gibt, projizieren wir M_d (wie in § 5 M_e) aus einem allgemeinen linearen Raum S_{r-n-1}. Der projizierende Kegel K_{d+r-n} schneidet M_n in M_d und einem Restschnitt N_d. Unterwerfen wir den Kegel K_{d+r-n} einer allgemeinen projektiven Transformation T, so schneidet K_{d+r-n}^T die Mannigfaltigkeit M_n nach einer Schnittmannigfaltigkeit L_d^*. Durchläuft T alle projektiven Transformationen, so durchläuft nach § 4 die Schnittmannigfaltigkeit $L_d^* = K_{d+r-n}^T \cdot M_n$ ein irreduzibles algebraisches System von Mannigfaltigkeiten. Der allgemeine Kegel K_{d+r-n}^T berührt nach § 2 die feste Mannigfaltigkeit M_e nicht, also berühren auch L_d^* und M_e sich nicht. Die Anzahl der Schnittpunkte von L_d^* und M_e oder, was dasselbe ist, von K_{d+r-n}^T und M_e, welche bei der Spezialisierung $T \to I$ in P übergehen, ist also nach dem vorhin Bewiesenen gleich der Multiplizität von P als Punkt von $M_d \cdot M_e$.

Dieselbe Zahl stellt nach der Definition des § 3 auch die Multiplizität von P als Schnittpunkt von M_e und K_{d+r-n} dar, oder nach der Definition von § 5 die Multiplizität von P als Punkt von $M_e \cdot M_d$. Damit ist das kommutative Gesetz bewiesen.

2. Das assoziative Gesetz. Offenbar stellen die Symbole $M_d \cdot (M_e \cdot M_f)$ und $(M_d \cdot M_e) \cdot M_f$ dieselbe Punktmenge dar. Wir wollen nun zeigen, daß jeder irreduzible Bestandteil von der Dimension $l = d + e + f - 2n$ in $M_d \cdot (M_e \cdot M_f)$ und in $(M_d \cdot M_e) \cdot M_f$ mit derselben Multiplizität vorkommt.

Wir betrachten zuerst den Fall dreier Mannigfaltigkeiten im projektiven Raum S_n. Wir können $d + e + f = 2n$ annehmen, da der Fall $d + e + f = 2n + l$ durch Schnitt mit einem allgemeinen linearen Raum S_{n-l} auf den ersteren zurückgeführt werden kann. Weiter dürfen M_d, M_e und M_f als absolut irreduzibel angenommen werden.

Es seien S und T zwei voneinander unabhängige, allgemeine projektive Transformationen. Dann bestehen sowohl $M_d^S \cdot (M_e \cdot M_f^T)$ als $(M_d^S \cdot M_e) \cdot M_f^T$ aus lauter einfach gezählten Punkten. Sie stimmen also überein. Bei der relationstreuen Spezialisierung $S \to I$ und der nachfolgenden Spezialisierung $T \to I$ geht (nach Satz B) $M_d^S \cdot (M_e \cdot M_f^T)$ in $M_d \cdot (M_e \cdot M_f)$ und $(M_d^S \cdot M_e) \cdot M_f^T$ in $(M_d \cdot M_e) \cdot M_f$ über. Also stimmen auch diese Punktgruppen überein.

Nun gehen wir zu dem Fall über, daß M_d, M_e, M_f in eine singularitätenfreie M_n eingebettet sind. Es seien K_{d+r-n} und K'_{f+r-n} die projizierenden Kegel von M_d und M_f aus zwei allgemeinen linearen Räumen S_{r-n-1}, S'_{r-n-1}. Nach § 5 stimmt der Schnitt $M_d \cdot M_e$ mit dem Schnitt $K_{d+r-n} \cdot M_e$ überein, wenn man von solchen Restbestandteilen von $K_{d+r-n} \cdot M_e$ absieht, welche keinen festen Punkt enthalten. Also stimmt auch $(M_d \cdot M_e) \cdot M_f$ mit $(K_{d+r-n} \cdot M_e) \cdot M_f$ und ebenso weiter mit $(K_{d+r-n} \cdot M_e) \cdot K'_{f+r-n}$ überein, abgesehen von Restbestandteilen ohne feste Punkte. Ebenso stimmt $M_d \cdot (M_e \cdot M_f)$ mit $K_{d+r-n} (M_e \cdot K'_{f+r-n})$ überein, wieder abgesehen von

Restbestandteilen ohne feste Punkte. Nach dem schon Bewiesenen gilt aber in S_r

$$(K_{d+r-n} \cdot M_e) \cdot K'_{f+r-n} = K_{d+r-n} \cdot (M_e \cdot K'_{f+r-n}),$$

also nach Weglassen der von S_{r-n-1} und S'_{r-n-1} abhängigen Restbestandteile:

$$(M_d \cdot M_e) \cdot M_f = M_d \cdot (M_e \cdot M_f).$$

3. Die Distributivgesetze:

$$M_d \cdot (M_e + M'_e) = M_d \cdot M_e + M_d \cdot M'_e,$$
$$(M_d + M'_d) \cdot M_e = M_d \cdot M_e + M'_d \cdot M_e.$$

Diese sind selbstverständlich, da der Schnitt $M_d \cdot M_e$ im Fall eines zerfallenden $M_d = \sum_\lambda M_d^{(2)}$ als Summe $\sum_\lambda M_d^{(2)} \cdot M_e$ erklärt war, und ebenso für zerfallende M_e.

Um nun von dem bisher behandelten Kalkül der Schnittmannigfaltigkeiten zum Schubertschen Kalkül der abzählenden Geometrie [10]) überzugehen, müssen wir den Begriff der *Schnittpunktszahl* $[M_d \cdot M_{n-d}]$ einführen. Haben M_d und M_{n-d} nur endlich viele Schnittpunkte, so ist $[M_d \cdot M_{n-d}]$ einfach die Summe der Multiplizitäten dieser Schnittpunkte. Haben aber M_d und M_{n-d} unendlich viele Punkte gemeinsam, so nehme man (wie unter 1.) zu M_d eine Mannigfaltigkeit N_d hinzu, welche mit M_{n-d} nur endlich viele Punkte gemeinsam hat, derart, daß $M_d + N_d = L_d$ ein Element eines irreduziblen Systems von Mannigfaltigkeiten ist, deren allgemeines Element L_d^* mit M_{n-d} nur endlich viele Schnittpunkte hat. Dann definiert man die Schnittpunktszahl $[M_d \cdot M_{n-d}]$ als die Differenz

$$[L_d^* \cdot M_{n-d}] - [N_d \cdot M_{n-d}].$$

Sie kann allerdings negativ ausfallen [11]).

Die Schnittpunktszahl $[M_d \cdot M_{n-d}]$ ist von der Wahl von N_d und L_d unabhängig und ist symmetrisch in M_d und M_{n-d}. Ist nämlich $M_d + N_d = L_d$ wie oben und ebenso $M_{n-d} + N_{n-d} = L_{n-d}$, wobei N_{n-d} und L_{n-d}^* mit L_d nur endlich viele Schnittpunkte haben, so ist

$$[M_d \cdot M_{n-d}] = [L_d^* \cdot M_{n-d}] - [N_d \cdot M_{n-d}]$$
$$= [L_d^* \cdot L_{n-d}] - [L_d^* \cdot N_{n-d}] - [N_d \cdot L_{n-d}] + [N_d \cdot N_{n-d}]$$
$$= [L_d^* \cdot L_{n-d}^*] - [L_d^* \cdot N_{n-d}] - [N_d \cdot L_{n-d}^*] + [N_d \cdot N_{n-d}]$$

---

[10]) H. Schubert, Kalkül der abzählenden Geometrie, Leipzig 1879.

[11]) Beispiel: Ist M_1 eine Gerade auf einer doppelpunktfreien kubischen Fläche M_2 des Raumes S_3 und N_1 ein Kegelschnitt, der M_1 zu einem ebenen Schnitt L_1 der Fläche ergänzt, so ist
$$[M_1 \cdot M_1] = [L_1^* \cdot M_1] - [N_1 \cdot M_1] = 1 - 2 = -1.$$
Dabei bedeutet L_1^* den Schnitt der Fläche mit einer allgemeinen Ebene.

und ebenso

$$[M_{n-d} \cdot M_d] = [L_{n-d}^* \cdot L_d^*] - [L_{n-d}^* \cdot N_d] - [N_{n-d} \cdot L_d^*] + [N_{n-d} \cdot N_d]$$
$$= [L_d^* \cdot L_{n-d}^*] - [N_d \cdot L_{n-d}^*] - [L_d^* \cdot N_{n-d}] + [N_d \cdot N_{n-d}],$$

also

$$[M_d \cdot M_{n-d}] = [M_{n-d} \cdot M_d],$$

und da die rechte Seite von der Wahl von L_d^* und N_d unabhängig ist, ist die linke es auch.

Genau analog kann man Schnittpunktszahlen von mehreren Mannigfaltigkeiten, z. B.

$$[M_d \cdot M_e \cdot M_{2n-d-e}]$$

auch in dem Falle definieren, daß diese mehr als nur endlich viele Schnittpunkte haben, und die Unabhängigkeit der Produkte von der Reihenfolge der Faktoren beweisen. Bei der nun folgenden Begründung des Schubertschen Kalküls werden wir immer nur den Fall betrachten, daß alle vorkommenden Schnittmannigfaltigkeiten $M_d \cdot M_e$ die „richtige" Dimension $d + e - n$ haben (oder leer sind). Der andere Fall läßt sich immer darauf zurückführen, indem man M_d durch eine Differenz $L_d^* - N_d$ (oder M_e durch $L_e^* - N_e$) ersetzt.

Im Schubertschen Kalkül wird jeder Mannigfaltigkeit M_d auf der festen M_n ein „Bedingungssymbol" μ (oder λ, ν, δ, ε, p, q, g, h, ...) zugeordnet: μ bedeutet die einem Punkte von M_n auferlegte Bedingung, zu M_d zu gehören. Ist die Dimension $d = n - k$, so heißt μ eine k-stufige (oder k-fache) Bedingung. Der Summe zweier Bedingungssymbole von gleicher Stufe soll die Summe der zugeordneten Mannigfaltigkeiten, dem Produkt zweier Bedingungssymbole das Produkt der Mannigfaltigkeiten (die Schnittmannigfaltigkeit) entsprechen, alles unter Berücksichtigung der Multiplizitäten der irreduziblen Bestandteile der Mannigfaltigkeiten und Schnittmannigfaltigkeiten. Das Produkt einer k-stufigen und einer l-stufigen Bedingung ist eine $(k + l)$-stufige Bedingung. Mehr als n-stufige Bedingungen werden nicht betrachtet; eine n-stufige Bedingung stellt eine endliche Zahl von Punkten dar.

Das Entscheidende ist aber die Gleichheitsdefinition. Zwei k-stufige Bedingungen werden einander gleich gesetzt, wenn die Schnittpunktszahlen der zugehörigen Mannigfaltigkeiten mit jeder beliebigen Mannigfaltigkeit Q_k auf M_n dieselben sind:

$$\mu = \nu, \text{ wenn stets } [M_{n-k} \cdot Q_k] = [N_{n-k} \cdot Q_k] \text{ gilt.}$$

Dabei genügt es, wenn die Gleichheit der Schnittpunktszahlen für diejenigen Q_k gilt, welche M_{n-k} und N_{n-k} nur in endlich vielen Punkten schneiden; sie gilt dann ganz von selbst für alle Q_k, da diese, wie angegeben, durch Differenzen ersetzt werden können.

Die Gleichheitsdefinition ist offenbar symmetrisch, reflexiv und transitiv; auch kann man symbolische Bedingungsgleichungen ohne weiteres addieren und subtrahieren. Man kann eine Gleichung zwischen Bedingungssymbolen aber auch auf beiden Seiten mit einem beliebigen Symbol multiplizieren, vorausgesetzt, daß die Stufenzahl dabei $\leqq n$ bleibt. Denn wenn

$$M_{n-k} \cdot Q_k = N_{n-k} \cdot Q_k$$

für beliebige Q_k gilt, so gilt für beliebige L_{n-h} und Q_{h+k}

$$M_{n-k} \cdot L_{n-h} \cdot Q_{h+k} = N_{n-k} \cdot L_{n-h} \cdot Q_{h+k};$$

d. h. aus $\mu = \nu$ folgt $\mu\lambda = \nu\lambda$.

Zwei n-stufige Bedingungen sind nach der Definition einander gleich, wenn sie dieselbe Anzahl von Punkten darstellen. Man kann also auch jedes n-stufige Bedingungssymbol einer Zahl gleichsetzen.

Beispiel zum Schubertschen Kalkül. M_n sei die singularitätenfreie (weil homogene) Mannigfaltigkeit aller Geraden des Raumes S_3. Es bedeute g die einer Geraden auferlegte Bedingung, eine gegebene Gerade zu schneiden oder allgemeiner einem gegebenen linearen Komplex anzugehören. Die Lage der gegebenen Geraden oder die Wahl des Komplexes ist dabei gleichgültig; da alle linearen Komplexe eine irreduzible Mannigfaltigkeit bilden und alle Bedingungen g daher im Sinne des Kalküls gleich sind. g_p bedeute die einer Geraden auferlegte Bedingung, durch einen gegebenen Punkt p zu gehen, g_e die (ebenfalls zweistufige) Bedingung, in einer gegebenen Ebene zu liegen, g_s die dreistufige Bedingung, einem gegebenen Geradenbüschel anzugehören, schließlich G die vierstufige Bedingung, mit einer gegebenen Geraden zusammenfallen. Man könnte nach der eben gemachten Bemerkung G auch durch die Zahl 1 ersetzen.

g^2 bedeutet die Bedingung, zwei gegebene Geraden zu schneiden. Wählt man die gegebenen Geraden so, daß sie sich schneiden, so gehen die Geraden, welche beide schneiden, entweder durch den Schnittpunkt oder sie liegen in der Ebene der beiden gegebenen Geraden. Da die beiden durch die Bedingungen g definierten speziellen linearen Komplexe sich auch in diesem Spezialfall nicht berühren (vgl. ZAG VIII, Math. Annalen **113**, S. 203), so sind die Bedingungen g_p und g_e, in die g^2 sich aufspaltet, beide mit der Vielfachheit Eins zu zählen. Es gilt also

$$(1) \qquad\qquad g^2 = g_p + g_e,$$

in Worten: *Die Anzahl der Geraden einer algebraischen Strahlenkongruenz, die zwei gegebene Geraden schneiden, ist gleich der Summe aus Bündelgrad und Feldgrad der Kongruenz.*

Multipliziert man (1) auf beiden Seiten mit g, so folgt, da $g \cdot g_p$ und $g \cdot g_e$ beide dasselbe bedeuten wie g_s:

$$(2) \qquad g^3 = 2\,g_s.$$

Nochmalige Multiplikation mit g ergibt:

$$(3) \qquad g^4 = 2\,G,$$

in Worten: Es gibt genau zwei Geraden, welche vier gegebene Geraden schneiden oder allgemeiner vier gegebenen linearen Komplexen angehören.

Natürlich läßt sich das Ergebnis in diesem einfachen Fall noch direkt algebraisch bestätigen. Es war mir aber nur darum zu tun, durch ein möglichst einfaches Beispiel den Schubertschen Kalkül zu erläutern.

Es sei noch bemerkt, daß der in der ersten Hälfte dieses Paragraphen behandelte Kalkül der Schnittmannigfaltigkeiten weiter reicht als der Schubertsche Kalkül, da der letztere sich nur mit Anzahlen, der erstere aber sich mit den Schnittmannigfaltigkeiten selbst befaßt. Der Kalkül der Schnittmannigfaltigkeiten kann zur Begründung der Severischen Theorie der Äquivalenzscharen auf algebraischen Mannigfaltigkeiten[3]) verwendet werden.

(Eingegangen am 30. 10. 1937.)

22.

Zur algebraischen Geometrie XV
Lösung des Charakteristikenproblems für Kegelschnitte

Mathematische Annalen 115, 5 (1938) 645–655

Chasles hat empirisch gefunden, daß die „Anzahl" der Kegelschnitte[1]) einer festen Ebene, in welchen sich ein System von ∞^1 und ein System von ∞^4 Kegelschnitten durchdringen, durch einen Ausdruck von der Form

$$(1) \qquad\qquad a\,\mu + b\,\nu$$

angegeben wird, wobei a, b Zahlen sind, die nur von dem System ∞^4, und μ, ν Zahlen, die nur von dem System ∞^1 abhängen. Genauer sind μ, ν die „Anzahlen" der Kegelschnitte des Systems ∞^1, welche irgendeinen gegebenen Punkt enthalten bzw. irgendeine gegebene Gerade berühren.

Die Formel (1) wird selbstverständlich, aber auch ziemlich nutzlos, wenn man unter einem Kegelschnitt nur ein Punktgebilde: eine Kurve 2. Ordnung versteht, denn die Kurven 2. Ordnung hängen von sechs homogenen Parametern (Gleichungskoeffizienten) ab und bilden daher einen projektiven Raum S_5; im S_5 aber gilt für die „Anzahl" der Schnittpunkte eines Systems von ∞^1 und eines Systems von ∞^4 Punkten sogar die einfachere „Bezoutsche" Formel $\gamma\mu$ (Produkt der Gradzahlen). Die Formel (1) erhält dagegen ihren vollen Sinn, wenn man nach Study[2]) den Begriff des *vollständigen Kegelschnitts* einführt. Ein solcher besteht aus einer Kurve 2. Ordnung und einer Kurve 2. Klasse, die so miteinander verknüpft sind, daß sie zusammen als Grenzgebilde einer nicht zerfallenden Kurve 2. Ordnung und deren Tangentengebilde aufgefaßt werden können.

Es gibt, wie im folgenden § 1 näher ausgeführt werden soll, drei Arten von vollständigen Kegelschnitten:

1. Eine *nicht ausgeartete* Kurve 2. Ordnung mit ihrem Tangentengebilde.

2. *δ-Ausartung.* Ein Geradenpaar und ein doppelt gezähltes Strahlenbüschel, dessen Scheitel beiden Geraden angehört.

3. *η-Ausartung.* Eine doppelt gezählte Gerade und zwei Strahlenbüschel, deren Zentren auf der Geraden liegen.

---

[1]) „Anzahl" in Anführungsstrichen bedeutet Summe der Schnittpunktsmultiplizitäten.

[2]) E. Study, Über die Geometrie der Kegelschnitte, insbesondere deren Charakteristikenproblem, Math. Annalen **26** (1886), S. 58—101.

297

Den Fällen 2 und 3 gemeinsam ist die „$\delta\eta$-Ausartung", die aus einer doppelt gezählten Geraden und einem doppelt gezählten Strahlbüschel besteht. Von diesem Fall abgesehen, genügt in den Fällen 1 und 2 die Angabe der sechs Koeffizienten der Kurve 2. Ordnung, um den vollständigen Kegelschnitt festzulegen. Im Fall 3 genügt das aber nicht, da die beiden Strahlenbüschel dann noch auf der Geraden verschoben werden können. Die vollständigen Kegelschnitte entsprechen somit nicht ausnahmslos eineindeutig den Punkten eines projektiven Raumes S_5, und davon rührt es her, daß die einfachere Bezoutsche Regel durch die zweigliedrige Formel (1) ersetzt werden muß[3]).

Study hat für die Formel (1) einen Beweis gegeben, der nur in Hinblick auf das Fehlen eines exakten Multiplizitätsbegriffs einer Ergänzung bedarf. Wir werden in § 2 die Formel (1) mit den Methoden von ZAG VI[4]) herleiten.

Unter Heranziehung der schärferen Methoden von ZAG XIV[5]) gelingt auch die Bestimmung der „Anzahl" der gemeinsamen Elemente eines Systems von ∞^2 und eines Systems von ∞^3 Kegelschnitten. Das Ergebnis ist die sogenannte Cremonasche Charakteristikenformel[6]).

§ 1.

Die Mannigfaltigkeit $\mathfrak{M}_5$ der vollständigen Kegelschnitte.

Eine allgemeine Kurve 2. Ordnung

$$(2) \qquad \Sigma a_{ik} x_i x_k = 0 \qquad\qquad (a_{ik} \text{ Unbestimmte})$$

definiert eine Kurve 2. Klasse, ihr Tangentengebilde

$$(3) \qquad \Sigma \alpha_{ik} u_i u_k = 0,$$

wobei die Matrizes (a_{ik}) und (α_{ik}) zueinander invers sind. Die $2 \cdot 6$ Koordinaten a_{ik} und α_{ik} definieren einen Punkt A in einem zweifach-projektiven Raum $S_{5,5}$. Der Punkt A hängt rational von den Unbestimmten a_{ik} ab, ist also der allgemeine Punkt einer irreduziblen rationalen Mannigfaltigkeit $\mathfrak{M}_5$ in $S_{5,5}$. Die Mannigfaltigkeit $\mathfrak{M}_5$ ist durch ihren allgemeinen Punkt A vollständig

---

[3]) Auf die scharfe Polemik zwischen Study und Zeuthen, die sich an die hier geschilderte Studysche Auffassung geknüpft hat, brauchen wir hier nicht einzugehen, da sie weniger die Sache selbst betraf als die Frage nach dem Werte anderer Untersuchungen von Halphen und die Ehre seines Andenkens. Wir stellen uns hier einfach auf den — mathematisch jedenfalls einwandfreien — Studyschen Standpunkt.

[4]) Math. Annalen **110** (1934), S. 134—160.

[5]) Math. Annalen **115** (1938), Heft 4, S. 619—642.

[6]) Sie heißt nun einmal so. In Cremonas gesammelten Werken habe ich sie nicht finden können. Eine Beweisandeutung bei Study[2]).

bestimmt. Jeder Punkt B von $\mathfrak{M}_5$ definiert eine Kurve 2. Ordnung und eine Kurve 2. Klasse:

$$(4) \qquad \Sigma b_{ik} x_i x_k = 0, \qquad \Sigma \beta_{ik} u_i u_k = 0,$$

die zueinander in der Beziehung stehen, daß sie alle (homogenen) algebraischen Relationen erfüllen, welche im allgemeinen Fall zwischen den Kurven (2), (3) bestehen. Ein solches Kurvenpaar nennen wir einen *vollständigen Kegelschnitt*. Da zu jedem Punkt von $\mathfrak{M}_5$ ein vollständiger Kegelschnitt gehört und umgekehrt, kann man $\mathfrak{M}_5$ die Mannigfaltigkeit der vollständigen Kegelschnitte nennen.

Zu den algebraischen Relationen, die im allgemeinen Fall zwischen den zueinander inversen Matrizes (a_{ik}), (α_{ik}) bestehen, gehören die folgenden:

$$\Sigma a_{ik}\alpha_{kl} = \delta_{il} \qquad (= 0 \text{ für } i \neq l \text{ und } = 1 \text{ für } i = l),$$

oder homogen geschrieben:

$$\Sigma a_{ik}\alpha_{kl} = 0 \quad \text{für} \quad k \neq l,$$
$$\Sigma a_{1k}\alpha_{k1} = \Sigma a_{2k}\alpha_{k2} = \Sigma a_{3k}\alpha_{k3}.$$

Dieselben Relationen müssen nun auch die spezialisierten b und β erfüllen:

$$\Sigma b_{ik}\beta_{kl} = 0 \quad \text{für} \quad k \neq l,$$
$$\Sigma b_{1k}\beta_{k1} = \Sigma b_{2k}\beta_{k2} = \Sigma b_{3k}\beta_{k3}.$$

Daraus folgt:

$$(5) \qquad \Sigma b_{ik}\beta_{kl} = \varepsilon \delta_{il} \qquad (\varepsilon = 0 \text{ oder} \neq 0).$$

Wir wollen nun zeigen, daß $\mathfrak{M}_5$ eine singularitätenfreie Mannigfaltigkeit ist, und daß ihre Punkte genau die drei in der Einleitung aufgezählten Arten von vollständigen Kegelschnitten definieren. Zu diesem Zwecke genügt es, *erstens* zu zeigen, daß keine anderen Kurvenpaare die Relationen (5) erfüllen als die drei aufgezählten Arten, und *zweitens* für die Kurvenpaare der angegebenen Arten in der Nachbarschaft eines vorgegebenen solchen Kurvenpaares K, K' eine eineindeutige (rationale) Parameterdarstellung anzugeben, welche für spezielle Werte der Parameter genau das jeweils vorgegebene Kurvenpaar (K, K') ergibt, für allgemeine Werte der Parameter aber das allgemeine Kurvenpaar (2), (3) darstellt.

Das erstere zeigen wir so: Die Matrix (b_{ik}) der quadratischen Form $\Sigma b_{ik} x_i x_k$ kann durch Koordinatentransformation jedenfalls auf eine der drei Gestalten

$$\begin{pmatrix} 1 & 0 & 0 \\ 0 & 1 & 0 \\ 0 & 0 & 1 \end{pmatrix}, \qquad \begin{pmatrix} 1 & 0 & 0 \\ 0 & 1 & 0 \\ 0 & 0 & 0 \end{pmatrix}, \qquad \begin{pmatrix} 1 & 0 & 0 \\ 0 & 0 & 0 \\ 0 & 0 & 0 \end{pmatrix}$$

gebracht werden. Die Relationen (5) besagen dann, daß die Matrix (β_{ik}) in diesen drei Fällen die Gestalt

$$\begin{pmatrix} \varepsilon & 0 & 0 \\ 0 & \varepsilon & 0 \\ 0 & 0 & \varepsilon \end{pmatrix}, \qquad \begin{pmatrix} 0 & 0 & 0 \\ 0 & 0 & 0 \\ 0 & 0 & \beta_{33} \end{pmatrix}, \qquad \begin{pmatrix} 0 & 0 & 0 \\ 0 & \beta_{22} & \beta_{23} \\ 0 & \beta_{32} & \beta_{33} \end{pmatrix}$$

hat. Im dritten Fall kann man durch eine weitere Koordinatentransformation auch die Matrix $\begin{pmatrix} \beta_{22} & \beta_{23} \\ \beta_{32} & \beta_{33} \end{pmatrix}$ noch auf Diagonalform bringen. Schreibt man nun in diesen drei Fällen die Kurvenpaare (4) hin, so erhält man genau die in der Einleitung unter 1., 2., 3. angegebenen geometrischen Gestalten.

Die so erhaltene gleichzeitige Diagonalform des Matrixpaares (b_{ik}), (β_{ik}) bildet auch den Ausgangspunkt bei der Herleitung der Parameterdarstellung der Kurvenpaare in der Nachbarschaft eines gegebenen. Die Matrizes (b_{ik}), (β_{ik}) des vorgegebenen Kurvenpaares K, K' seien, wenn wir aus typographischen Gründen nur ihre Hauptdiagonalen angeben:

$$\begin{array}{llll} \text{im Fall 1:} & (1, 1, 1) & \text{und} & (1, 1, 1); \\ \text{im Fall 2:} & (1, 1, 0) & \text{und} & (0, 0, 1) \quad (\delta\text{-Ausartung}); \\ \text{im Fall 3:} & (1, 0, 0) & \text{und} & (0, 1, 1) \quad (\eta\text{-Ausartung}); \\ \text{oder} & (1, 0, 0) & \text{und} & (0, 0, 1) \quad (\delta\eta\text{-Ausartung}). \end{array}$$

Wir bilden nun die Formen:

$$(6) \qquad f(x) = (x_1 + p_{12}x_2 + p_{13}x_3)^2 + q_2(x_2 + p_{23}x_3)^2 + q_2q_3x_3^2,$$

$$(7) \qquad \varphi(u) = q_2q_3u_1^2 + q_3(\pi_{21}u_1 + u_2)^2 + (\pi_{31}u_1 + \pi_{32}u_2 + u_3)^2,$$

wobei die p_{ik} und q_i Unbestimmte sind und die π_{ik} so bestimmt werden, daß die Matrizes

$$\begin{pmatrix} 1 & p_{12} & p_{13} \\ 0 & 1 & p_{23} \\ 0 & 0 & 1 \end{pmatrix} \quad \text{und} \quad \begin{pmatrix} 1 & 0 & 0 \\ \pi_{21} & 1 & 0 \\ \pi_{31} & \pi_{32} & 1 \end{pmatrix}$$

zueinander gespiegelt invers sind. Die Formen (6) und (7) sind dann zueinander kontragredient, d. h. ihre Matrizes sind zueinander invers. Außerdem läßt sich die Form (2) sehr leicht (bis auf einen Faktor a_{11}, auf den es nicht ankommt) auf die Gestalt (6) bringen; die inverse Form (3) wird dann natürlich gleich (7). (6) und (7) sind also nur andere Parameterdarstellungen des allgemeinen Formenpaares (2), (3). Gibt man aber speziell den p_{ik} und π_{ik} die Werte Null und setzt

$$\begin{array}{lll} \text{im Fall 1:} & q_2 = q_3 = 1, \\ \text{im Fall 2:} & q_2 = 1, & q_3 = 0, \\ \text{im Fall 3:} & q_2 = 0, & q_3 = 1, \\ \text{bzw.} & q_2 = q_3 = 0, \end{array}$$

so gehen (6) und (7) genau in die oben in Matrixgestalt angegebenen Normalformen der Kurvenpaare (K, K') über. Außerdem sind für jedes Formenpaar f, φ, das sich überhaupt auf die Gestalt (6), (7) bringen läßt, die Koeffizienten p_{ik} und q_i eindeutig bestimmt. Die Parameterdarstellung ist also, soweit sie überhaupt reicht, eineindeutig. Damit sind alle Behauptungen bewiesen.

Wir fassen zusammen:

Die vollständigen Kegelschnitte bilden eine singularitätenfreie rationale Mannigfaltigkeit und haben genau die in der Einleitung angegebenen Gestalten (nicht ausgearteter Kegelschnitt, δ-Ausartung und η-Ausartung).

§ 2.

Die Chaslessche Charakteristikenformel.

Die Mannigfaltigkeit $\mathfrak{M}_5$ der vollständigen Kegelschnitte ist birational und in einer Richtung eindeutig abgebildet auf den Raum S_5 der Kurven 2. Ordnung; denn jeder vollständige Kegelschnitt definiert in eindeutiger Weise eine Kurve 2. Ordnung, und jede Kurve 2. Ordnung definiert umgekehrt im allgemeinen eindeutig einen vollständigen Kegelschnitt. Die Eindeutigkeit der umgekehrten Abbildung erleidet aber eine Ausnahme, wenn die vorliegende Kurve 2. Ordnung eine doppelt gezählte Gerade ist; eine solche kann nämlich in ∞^2 Weisen durch Wahl zweier Punkte auf der Geraden zu einer η-Ausartung ergänzt werden. Die η-Ausartungen bilden eine vierdimensionale Teilmannigfaltigkeit H von $\mathfrak{M}_5$, deren Bild in S_5 eine nur zweidimensionale Mannigfaltigkeit F ist, welche man die *Veronesesche Fläche* nennt.

Ist nun $\mathfrak{C}$ eine Kurve und $\mathfrak{B}$ eine vierdimensionale Mannigfaltigkeit in $\mathfrak{M}_5$, und sind $\mathfrak{C}'$, $\mathfrak{B}'$ ihre Bildmannigfaltigkeiten in S_5, so ist nach ZAG VI, § 7, Formel (29) die Schnittpunktszahl $(\mathfrak{C} \cdot \mathfrak{B})$ gleich der Schnittpunktszahl $[\mathfrak{C}' \cdot \mathfrak{B}']$, vermindert um die b-fache Schnittpunktszahl $[\mathfrak{C} \cdot \mathsf{H}]$, wobei die ganze Zahl b nur von $\mathfrak{B}$ abhängt (und insbesondere dann gleich Null ist, wenn $\mathfrak{B}'$ die Fläche F nicht enthält):

$$[\mathfrak{C} \cdot \mathfrak{B}] = [\mathfrak{C}' \cdot \mathfrak{B}'] - b \cdot [\mathfrak{C} \cdot \mathsf{H}].$$

Nach dem Bezoutschen Satz ist $[\mathfrak{C}' \cdot \mathfrak{B}']$ gleich dem Produkt der Gradzahlen von $\mathfrak{C}'$ und $\mathfrak{B}'$. Diese seien c' und b'. Dann folgt:

$$(8) \qquad\qquad [\mathfrak{B} \cdot \mathfrak{C}] = b'c' - b\,[\mathfrak{C} \cdot \mathsf{H}].$$

c' ist offenbar gleich der Schnittpunktszahl von $\mathfrak{C}$ mit einer Mannigfaltigkeit M, die durch eine lineare Bedingung in den Koordinaten b_{ik} gegeben wird.

301

(M ist z. B. die Gesamtheit aller Kegelschnitte durch einen festen Punkt.) Also kann man statt (8) auch schreiben:

$$(9) \qquad [\mathfrak{B} \cdot \mathfrak{C}] = b'[\mathsf{M} \cdot \mathfrak{C}] - b\,[\mathsf{H} \cdot \mathfrak{C}],$$

wobei b und b' nur von $\mathfrak{B}$ abhängen.

Die Gleichung (9) hat schon die typische Gestalt einer Charakteristikenformel. Sie kann auch als symbolische Gleichung im Schubertschen Sinne geschrieben werden, indem man den Mannigfaltigkeiten $\mathfrak{B}$, M, H Symbole β, μ, η zuordnet:

$$(10) \qquad \beta = b'\mu + b\eta.$$

Die Bedeutung einer solchen symbolischen Gleichung ist nämlich die, daß sie nach Bildung der Schnittpunktszahl der zu den Symbolen β, μ, η gehörigen Mannigfaltigkeiten $\mathfrak{B}$, M, H mit einer beliebigen Kurve $\mathfrak{C}$ eine richtige Zahlengleichung (9) herauskommt.

Zweiter Beweis von (9). Wir werden zeigen, daß nicht nur die symbolische Gleichung (10), sondern darüber hinaus sogar eine Äquivalenz im Sinne der Theorie der linearen Scharen besteht:

$$(11) \qquad \mathfrak{B} + b\,\mathsf{H} \sim b' \cdot \mathsf{M}.$$

Das heißt, die Mannigfaltigkeiten $\mathfrak{B} + b\,\mathsf{H}$ und $b'\,\mathsf{M}$ gehören einer und derselben linearen Schar von vierdimensionalen Mannigfaltigkeiten auf $\mathfrak{M}_5$ an.

Die Gleichung der Hyperfläche $\mathfrak{B}'$ vom Grade b' sei $f(b_{ik}) = 0$; die lineare Gleichung von M sei $l(b_{ik}) = 0$. Wir betrachten die durch die Gleichung

$$\lambda_1 f(b_{ik}) + \lambda_2\, l(b_{ik})^{b'} = 0$$

definierte lineare Schar von vierdimensionalen Mannigfaltigkeiten auf $\mathfrak{M}_5$. Das eine Basiselement der Schar ist die b'-mal gezählte Mannigfaltigkeit M; das andere, mit der Gleichung $f(b_{ik}) = 0$, besteht aus der Mannigfaltigkeit $\mathfrak{B}$, einmal gezählt, und eventuell noch der Mannigfaltigkeit H mit einer gewissen Vielfachheit b. Dabei ist $b = 0$, wenn $\mathfrak{B}'$ die Veronesesche Fläche F nicht enthält, denn dann genügen die Punkte von H nicht der Gleichung $f(b_{ik}) = 0$.

Um die Charakteristikenformel (9) auf die Chaslessche Gestalt zu bringen, wenden wir (9) speziell auf den Fall an, daß $\mathfrak{B}$ die Mannigfaltigkeit N der Kegelschnitte ist, die eine lineare Bedingung in den Koordinaten β_{ik} erfüllen (z. B. eine gegebene Gerade berühren). Die Bildmannigfaltigkeit $\mathfrak{B}'$ ist in diesem Fall eine quadratische Hyperfläche, mithin ist $b' = 2$. Man erhält

$$(12) \qquad [\mathsf{N} \cdot \mathfrak{C}] = 2\,[\mathsf{M} \cdot \mathfrak{C}] - b_{\mathsf{N}}\,[\mathsf{H} \cdot \mathfrak{C}].$$

Um b_{N} zu bestimmen, wählen wir für $\mathfrak{C}$ speziell ein Büschel von doppeltberührenden Kegelschnitten, bestehend aus den Kegelschnitten, die in zwei

gegebenen Punkten zwei gegebene Geraden berühren. In diesem Fall wird
$M \cdot \mathfrak{C} = N \cdot \mathfrak{C} = 1$, also nach (12):

$$b_N \cdot [H \cdot \mathfrak{C}] = 1, \text{ also } b_N = 1.$$

Aus (12) wird jetzt:

$$(13) \qquad [N \cdot \mathfrak{C}] = 2\,[M \cdot \mathfrak{C}] - [H \cdot \mathfrak{C}].$$

Subtrahiert man die mit β multiplizierte Gleichung (13) von (9) und setzt
$b' - 2\,b = a$, so folgt:

$$(14) \qquad [\mathfrak{B} \cdot \mathfrak{C}] = a\,[M \cdot \mathfrak{C}] + b\,[N \cdot \mathfrak{C}].$$

Das ist die Chaslessche Charakteristikenformel (1). Sie kann auch als
symbolische Gleichung $\beta = a\,\mu + b\,\nu$ geschrieben werden.

Die Herleitung der Endformel (14) gilt zunächst nur für solche Kurven $\mathfrak{C}$,
die nicht aus lauter η-Ausartungen bestehen. Geht man aber, statt von (9)
auszugehen, von (11) aus, so erhält man an Stelle von (14) die lineare Äquivalenz

$$(15) \qquad \mathfrak{B} \sim a\,M + b\,N,$$

aus der (14) für ganz beliebige Kurven $\mathfrak{C}$ folgt.

Nebenbei folgt aus der Gleichung (13), wenn für die Mannigfaltigkeiten
M, N, H die Schubertschen Bedingungssymbole μ, ν, η eingeführt werden,
die Schubertsche Gleichung

$$\nu = 2\,\mu - \eta.$$

Dual dazu hat man

$$\mu = 2\,\nu - \delta.$$

§ 3.

Die Cremonasche Charakteristikenformel.

Wir wollen die Schnittpunktszahl eines Systems von ∞^2 und eines
Systems von ∞^3 Kegelschnitten ermitteln. Dazu soll die von Schaake[7] angegebene Methode der ausgearteten Kollineationen verwendet werden. Wir
verwenden die Begriffe und Bezeichnungen der vorangehenden Arbeit ZAG XIV.

Gegeben sei ein System $\mathfrak{S}$ von ∞^3 vollständigen Kegelschnitten. Wir
unterwerfen die Ebene der Kegelschnitte einer Punkttransformation

$$(16) \qquad \begin{cases} x_0' = x_0, \\ x_1' = x_1, \\ x_2' = \lambda\,x_2. \end{cases}$$

---

[7] G. Schaake, Diss. Amsterdam 1922. Siehe auch ZAG XIV, Math. Ann. **115**.

Die Kegelschnittkoordinaten transformieren sich dabei folgendermaßen:

$$\begin{cases} b'_{00} = b_{00} \\ b'_{01} = b_{01} \\ b'_{11} = b_{11} \\ b'_{02} = \lambda^{-1} b_{02} \\ b'_{12} = \lambda^{-1} b_{12} \\ b'_{22} = \lambda^{-2} b_{22} \end{cases} \qquad \begin{cases} \beta'_{00} = \beta_{00} \\ \beta'_{01} = \beta_{01} \\ \beta'_{11} = \beta_{11} \\ \beta'_{02} = \lambda\, \beta_{02} \\ \beta'_{12} = \lambda\, \beta_{12} \\ \beta'_{22} = \lambda^{2}\, \beta_{22} \end{cases}$$

Das System $\mathfrak{S}$ wird in ein System $\mathfrak{S}_\lambda$ transformiert, das durch einen von λ abhängigen Punkt S_λ eines Bildraumes abgebildet werden kann. Es gibt eine relationstreue Spezialisierung S_0 dieses Bildpunktes für $\lambda = 0$. S_0 ist wiederum Bildpunkt eines Systems $\mathfrak{S}_0$. Da nun $\mathfrak{S}_\lambda$ für $\lambda = 1$ in $\mathfrak{S}$ und für $\lambda = 0$ in $\mathfrak{S}_0$ übergeht, so gilt für die Schubertschen Bedingungssymbole $\sigma_\lambda, \sigma, \sigma_0$ von $\mathfrak{S}_\lambda, \mathfrak{S}, \mathfrak{S}_0$ die Gleichung

$$\sigma = \sigma_\lambda = \sigma_0.$$

Wir wollen nun $\mathfrak{S}_0$ näher untersuchen. Es sei

$$f\,(b_{00},\, b_{01},\, b_{11},\, \ldots,\, |\, \beta_{00},\, \ldots,\, \beta_{22}) = 0$$

eine der Gleichungen des Systems $\mathfrak{S}$. Dann gilt für $\mathfrak{S}_\lambda$ die Gleichung

$$(17) \quad f\,(b'_{00}, b'_{01}, b'_{11}, \lambda b'_{02}, \lambda b'_{12}, \lambda^2 b'_{22} \,|\, \lambda^2 \beta'_{00}, \lambda^2 \beta'_{01}, \lambda^2 \beta'_{11}, \lambda \beta'_{02}, \lambda \beta'_{12}, \beta'_{22}) = 0.$$

Die Tatsache, daß alle Punkte von $\mathfrak{S}_\lambda$ die Gleichung (17) erfüllen, läßt sich durch algebraische Gleichungen in den Koordinaten des Systems $\mathfrak{S}_\lambda$ ausdrücken (vgl. ZAG XI, § 2, Math. Annalen 113, S. 700, oben), welche für $\lambda = 0$ erhalten bleiben. Also gilt für $\mathfrak{S}_0$ die Gleichung

$$(18) \qquad f\,(b^{*}_{00}, b^{*}_{01}, b^{*}_{11}, 0, 0, 0 \,|\, 0, 0, 0, 0, 0, \beta^{*}_{22}) = 0.$$

Die Gleichung (18) bedeutet folgendes: Wenn der Kegelschnitt

$$K^{*}\,(b^{*}_{00},\, \ldots,\, b^{*}_{22} \,|\, \beta^{*}_{00},\, \ldots,\, \beta^{*}_{22})$$

zu $\mathfrak{S}_0$ gehören soll, so muß der Kegelschnitt

$$K\,(b^{*}_{00},\, b^{*}_{01},\, b^{*}_{11},\, 0,\, 0,\, 0 \,|\, 0,\, \ldots,\, 0,\, \beta^{*}_{22})$$

die Gleichungen von $\mathfrak{S}$ erfüllen, also zu $\mathfrak{S}$ gehören.

Wir müssen nun 2 Fälle unterscheiden:

Fall 1. $\beta^{*}_{22} \neq 0$. Der Kegelschnitt $K\,(b^{*}_{00}, b^{*}_{01}, b^{*}_{11}, 0, 0, 0 \,|\, 0, \ldots, 0, \beta^{*}_{22})$ ist eine δ-Ausartung mit Doppelpunkt in $(0, 0, 1)$, bestehend aus zwei Geraden, die diesen Punkt mit den Schnittpunkten von K^{*} und der Geraden $x_2 = 0$ verbinden. Es gibt in $\mathfrak{S}$ nur endlich viele solche Kegelschnitte, wenn $\mathfrak{S}$ nicht aus lauter δ-Ausartungen besteht; denn von den höchstens ∞^2 δ-Ausartungen des Systems $\mathfrak{S}$ haben nur endlich viele einen beliebig gegebenen Punkt zum Doppelpunkt. Diese endlich vielen δ-Ausartungen definieren je ein Schnittpunktpaar mit der Geraden $x_2 = 0$. Die Kegelschnitte K^* des Systems $\mathfrak{S}_0$ müssen je eines von diesen endlich vielen Punktpaaren enthalten.

Fall 2. $\beta_{22}^* = 0$. Die Gleichungen (18) sind dann, soweit sie in den β-Koordinaten einen nicht verschwindenden Grad haben, von selbst erfüllt. Die Gleichung $\beta_{22}^* = 0$ bedeutet, daß K^* die Gerade $x_2 = 0$ berührt.

Es folgt, daß das System $\mathfrak{S}_0$ in irreduzible Systeme zerfällt, welche *entweder* bestehen aus lauter Kegelschnitten durch zwei feste Punkte, *oder* aus lauter Kegelschnitten mit einer festen Tangente. *Jedes System $\mathfrak{S}$ von ∞^3 Kegelschnitten läßt sich somit durch eine ausgeartete Kollineation in eine Summe von Systemen überführen, deren Kegelschnitte entweder zwei feste Punkte oder eine feste Tangente haben.*

Dual dazu gilt: Jedes System $\mathfrak{S}$ von ∞^3 Kegelschnitten läßt sich durch eine ausgeartete Kollineation in eine Summe von Systemen überführen, deren Kegelschnitte entweder zwei feste Tangenten oder einen festen Punkt gemeinsam haben.

Das Schubertsche Symbol für das (irreduzible) System der Kegelschnitte mit zwei festen Tangenten ist ν^2. Das Symbol für ein dreidimensionales System $\mathfrak{P}$ mit einem festen Punkt sei π. Dann folgt also die Gleichung

$$(19) \qquad \sigma = c\nu^2 + \pi.$$

Der feste Punkt des Systems $\mathfrak{P}$ habe etwa die Koordinaten (y_0, y_1, y_2). Die Kegelschnitte des Systems genügen also der Gleichung

$$(20) \quad b_{00}\, y_0^2 + 2\, b_{01}\, y_0\, y_1 + b_{11}\, y_1^2 + 2\, b_{02}\, y_0\, y_2 + 2\, b_{12}\, y_1\, y_2 + b_{22}\, y_2^2 = 0.$$

Wir wenden nun auf das System $\mathfrak{P}$ noch einmal die Transformation (17) an und gehen wieder zu $\lambda = 0$ über. $\mathfrak{P}$ geht dabei über in ein System $\mathfrak{P}_0$, dessen Kegelschnitte entweder zwei feste Punkte oder eine feste Tangente haben. Die Gleichung (20) aber gibt, wie die obige Herleitung zeigt, zu einer Gleichung

$$(21) \qquad b_{00}^*\, y_0^2 + 2\, b_{01}^*\, y_0\, y_1 + b_{11}^*\, y_1^2 = 0$$

Anlaß, welche bedeutet, daß die Kegelschnitte des Systems $\mathfrak{P}_0$ alle den festen Punkt $(y_0, y_1, 0)$ auf der Geraden $x_2 = 0$ enthalten. $\mathfrak{P}_0$ ist also eine Summe von Systemen, welche entweder zwei feste Punkte enthalten oder einen festen Punkt $(y_0, y_1, 0)$ mit einer festen Tangente $x_2 = 0$ in diesem Punkt.

Das Symbol eines Systems mit zwei festen Punkten ist μ^2. Das Symbol für das irreduzible System der Kegelschnitte, die einen festen Punkt enthalten und in diesem Punkt eine feste Tangente besitzen, sei $\widehat{\mu\nu}$. Wie wir sehen werden, ist $\widehat{\mu\nu}$ wohl zu unterscheiden von $\mu\nu$, dem Symbol des Systems der Kegelschnitte durch einen beliebigen Punkt und mit einer beliebigen (im allgemeinen nicht durch den Punkt gehenden) Tangente.

Wir haben also auf Grund der ausgearteten Kollineation, die $\mathfrak{P}$ in $\mathfrak{P}_0$ überführt, die symbolische Gleichung

$$(22) \qquad \pi = a\mu^2 + b\,\widehat{\mu\nu}.$$

Aus (19) und (22) folgt die gesuchte Charakteristikenformel

$$(23) \qquad \sigma = a\,\mu^2 + b\,\widehat{\mu\nu} + c\,\nu^2,$$

in der die Zahlenfaktoren a, b, c vom System $\mathfrak{S}$ (von der Bedingung σ) abhängen. Multipliziert man (23) mit einem beliebigen, zu einem System $\mathfrak{T}$ gehörigen Symbol τ, so erhält man die Zahlengleichung

$$(24) \qquad \sigma\tau = a \cdot \mu^2\,\tau + b \cdot \widehat{\mu\nu} \cdot \tau + c \cdot \nu^2\tau.$$

Zwischen den Symbolen $\mu^2, \widehat{\mu\nu}, \nu^2$ besteht keine lineare Gleichung der Gestalt

$$\mu^2 + f\,\widehat{\mu\nu} + g\,\nu^2 = 0.$$

Denn andernfalls würde durch Multiplikation mit $\delta\nu^2$ folgen $e = 0$, durch Multiplikation mit $\eta\mu^2$ ebenso $g = 0$ und durch Multiplikation mit μ^3 schließlich $f = 0$. Eine weniger als dreigliedrige Charakteristikenformel ist also unmöglich.

Die Formel (23) gilt insbesondere für $\sigma = \mu\nu$. Es gilt also

$$\mu\nu = a\,\mu^2 + b\,\widehat{\mu\nu} + c\,\nu^2.$$

Multipliziert man beide Seiten mit $\delta\nu^2$ und mit $\eta\mu^2$, so folgt wieder $a = 0$ und $c = 0$. Multipliziert man nun mit μ^3, so folgt

$$\mu^4\nu = b\,\mu^3 \cdot \widehat{\mu\nu}.$$

Die linke Seite hat den Wert 2, denn das Symbol ν stellt eine quadratische Hyperfläche im Raum S_5 der Punktkegelschnitte dar. Das Produkt $\mu^3 \cdot \widehat{\mu\nu}$ rechter Hand hat aber den Wert 1, denn das Symbol $\widehat{\mu\nu}$ stellt eine lineare Schar von ∞^2 Kegelschnitten dar, die mit je drei allgemeinen Hyperebenen in S_5 nur einen Punkt mit der Multiplizität Eins gemeinsam hat. Daher ist $b = 2$ und somit

$$\mu\nu = 2\,\widehat{\mu\nu}.$$

Dieses Ergebnis ist geometrisch offenbar folgendermaßen zu interpretieren: Der Durchschnitt des Systems der Kegelschnitte durch einen festen Punkt und des Systems der Kegelschnitte mit einer festen, diesen Punkt enthaltenden Tangente ist das *doppelt gezählte* System $\widehat{\mu\nu}$. Dieses Beispiel zeigt wieder einmal, wie vorsichtig man beim Prinzip der Erhaltung der Anzahl mit den Multiplizitäten sein muß!

Der hier gegebene Beweis der Formel (23) beruhte auf der Annahme, daß das System $\mathfrak{S}$ höchstens ∞^2 η-Ausartungen und das System $\mathfrak{P}$ höchstens ∞^2 δ-Ausartungen enthält; er versagt also zunächst, wenn einer der irreduziblen Bestandteile von $\mathfrak{S}$ ganz aus η-Ausartungen oder wenn ein Bestandteil von $\mathfrak{P}$ ganz aus δ-Ausartungen besteht. Der letzte Fall ist ganz leicht: zum betreffenden irreduziblen Bestandteil von $\mathfrak{P}$ gehört dann das Symbol $\mu\delta$, und nach § 2 ist

$$\delta = 2\,\nu - \mu,$$

also

$$(25) \qquad \mu\delta = 2\,\mu\nu - \mu^2 = 4\,\widehat{\mu\nu} - \mu^2.$$

Wenn aber $\mathfrak{S}$ ganz aus η-Ausartungen besteht, so hilft man sich mit einem auch in ZAG XIV mehrmals angewandten Kunstgriff: Man stellt $\mathfrak{S}$ als Differenz von zwei Systemen dar, von denen das erste einem irreduziblen System von Mannigfaltigkeiten angehört, dessen allgemeines Element nicht aus lauter η-Ausartungen besteht, während der Subtrahend ebenfalls nicht aus lauter η-Ausartungen besteht.

In dieser Weise ergibt sich, daß die Formeln (23) und (24) ganz allgemein gelten. Allerdings können die Koeffizienten a, b, c unter Umständen negativ ausfallen, wie das Beispiel (25) zeigt.

(Eingegangen am 10. 12. 1937.)

23.

Die Bedeutung des Bewertungsbegriffs für die algebraische Geometrie

Bericht, vorgetragen auf der Tagung in Jena am 23. Okt. 1941

Jahresbericht der Deutschen Mathematiker-Vereinigung 52, 2 (1942) 161–172

Durch neuere amerikanische Untersuchungen sind die beiden miteinander zusammenhängenden Fragen der lokalen Uniformisierung der algebraischen Mannigfaltigkeiten und ihrer birationalen Transformation in singularitätenfreie um ein beträchtliches Stück ihrer vollständigen Lösung näher gebracht worden. Dabei hat sich der Bewertungsbegriff in wachsendem Maße als fundamental erwiesen. Ein zusammenhängender Bericht über diese Untersuchungen ist um so mehr erwünscht, da die amerikanischen Zeitschriften jetzt nicht allgemein zugänglich sind.

§ 1. Algebraische Funktionen einer Veränderlichen.

Die Grundbegriffe der Theorie der algebraischen Funktionen von mehreren Veränderlichen versteht man am besten, wenn man von der Theorie der algebraischen Funktionen einer Veränderlichen ausgeht.

Zu einer algebraischen Kurve im n-dimensionalen projektiven Raum gehört ein algebraischer Funktionenkörper: der Körper der rationalen Funktionen der Koordinatenverhältnisse eines veränderlichen Punktes der Kurve. Umgekehrt gehört zu jedem von endlich vielen Elementen $E_1, E_2, \ldots, E_n$ erzeugten algebraischen Funktionenkörper einer Veränderlichen eine algebraische Kurve, deren allgemeiner Punkt die homogenen Koordinaten $E_0 = 1, E_1, \ldots, E_n$ hat.[1]) Eine solche Kurve nennt man ein *projektives Modell* des Funktionenkörpers. Da die Wahl der Erzeugenden $E_1, \ldots, E_n$ in hohem Grade willkürlich ist, so gibt es zu einem vorgegebenen Funktionenkörper mehrere projektive Modelle. Da aber jedes andere Erzeugendensystem rational von $E_1, \ldots, E_n$ abhängt und umgekehrt, so entsteht jedes projektive Modell durch *birationale Transformation* aus einem einzigen solchen Modell. Die algebraische Geometrie interessiert sich besonders für die birational invarianten Eigenschaften der Kurven, das heißt also für diejenigen, die nur vom Funktionenkörper, nicht vom Modell abhängen.

---

1) Siehe etwa meine Einführung in die algebraische Geometrie, Berlin 1939, § 29.

308

Zum Funktionenkörper $K(E_1, \ldots, E_n)$ gehört, wenn der Konstantenkörper K der Körper der komplexen Zahlen ist, eine Riemannsche Fläche, auf der alle Funktionen des Körpers eindeutig und meromorph sind. Die Beziehung der Kurve zur Riemannschen Fläche ist die folgende: Zu jeder *Stelle* oder Punkt der Riemannschen Fläche gehört ein einziger Punkt der Kurve, dessen Koordinatenverhältnisse die Werte sind, die die Verhältnisse $E_0 : E_1 : \cdots : E_n$ an der betreffenden Stelle annehmen.[2]) Umgekehrt gehört zu einem einfachen Punkt der Kurve eine einzige Stelle der Riemannschen Fläche; zu einem vielfachen Punkt aber können mehrere Stellen gehören. Man sagt dann, daß durch den vielfachen Punkt mehrere *Zweige* der Kurve hindurchgehen, indem man unter einem *Zweig* ein durch reguläre analytische Funktionen vermitteltes Bild einer Umgebung von $u = 0$ in der u-Ebene versteht, wo u eine Ortsuniformisierende der betreffenden Stelle der Riemannschen Fläche ist. Zum Begriff des Zweiges gehört die Auszeichnung eines Punktes $u = 0$ in der betreffenden Umgebung; diesem entspricht ein Punkt der Kurve, der *Mittelpunkt* des Zweiges. Zweige einer algebraischen Kurve entsprechen nach dem Gesagten eineindeutig den Stellen des Funktionenkörpers oder den Punkten der Riemannschen Fläche.

Statt der Riemannschen Fläche kann man, nach einem von den italienischen Geometern vielfach befolgten Verfahren, auch ein singularitätenfreies projektives Modell des Funktionenkörpers benutzen; die Punkte eines solchen Modells entsprechen nämlich eineindeutig den Stellen des Körpers.[3])

Ein rein algebraisches Äquivalent des Stellenbegriffs ist der Dedekind-Webersche Begriff des *Ortes*[4]) eines algebraischen Funktionenkörpers. Man kommt zu ihm durch die Erwägung, daß jede Funktion η des Körpers an jeder Stelle der Riemannschen Fläche einen Wert $f(\eta)$ annimmt, der auch ∞ sein kann. Unter einem *Ort* versteht man demnach eine Zuordnung, die jeder Funktion η des Funktionenkörpers einen Wert $f(\eta)$ zuordnet, der entweder ∞ ist oder einem algebraischen

---

2) Nehmen $E_1, \ldots, E_n$ an der betreffenden Stelle endliche Werte an, so liegt der zugehörige Kurvenpunkt im endlichen. Werden einige E unendlich und wird etwa E_j unter ihnen von der höchsten Ordnung unendlich, so bleiben $E_0 : E_j, E_1 : E_j, \ldots, E_n : E_j$ alle endlich und bestimmen den Kurvenpunkt.

3) Siehe etwa § 45 des unter 1) zitierten Buches.

4) Das Wort *Ort* (zum Unterschied von *Stelle* und *Punkt*) ist von mir gewählt worden; der Begriff findet sich bei Dedekind und Weber, J. reine u. angew. Math. **92** (1882), S. 181.

Erweiterungskörper des Konstantenkörpers K angehört, mit folgenden Eigenschaften:

$$(1) \quad \begin{cases} f(\eta\zeta) \ = f(\eta) \cdot f(\zeta), & \text{wenn } f(\eta) \text{ und } f(\zeta) \text{ endlich;} \\ f(\eta + \zeta) = f(\eta) + f(\zeta), & \text{wenn } f(\eta) \text{ und } f(\zeta) \text{ endlich;} \\ f(\eta) = \infty, \text{ wenn } f(\eta^{-1}) = 0, \text{ und umgekehrt.} \end{cases}$$

Wir nehmen nun wieder an, der Konstantenkörper K sei der Körper der komplexen Zahlen. Dann gehört, wie wir gesehen haben, zu jeder Stelle des Körpers ein einziger Ort. Um auch das Umgekehrte zu beweisen, nehmen wir ein projektives Modell des Funktionenkörpers zu Hilfe und bemerken zunächst, daß zu jedem Ort ein Punkt der Kurve gehört. Wenn nämlich alle Funktionen des Körpers bestimmte Werte annehmen, so nehmen auch die Verhältnisse $E_i : E_j$ bestimmte Werte an, und diese Werte bestimmen einen Punkt. Der so definierte Punkt heißt der *Mittelpunkt* des Ortes auf der Kurve. Nimmt man das Modell als singularitätenfrei an, so ist dieser Punkt ein einfacher Punkt und seine Umgebung ein einziger Zweig; damit ist die gesuchte Zuordnung Ort $\rightarrow$ Stelle gefunden. Für diesen Beweis braucht man nicht einmal ganz die Existenz eines singularitätenfreien Modells, sondern es genügt offensichtlich bereits das folgende

Lemma: *Zu jedem vorgegebenen Ort gibt es ein solches projektives Modell des Funktionenkörpers, auf dem der Mittelpunkt des Ortes ein einfacher Punkt ist.*

Die modernen Algebraiker[5]) ziehen es vor, statt des Dedekind-Weberschen Ortsbegriffes den äquivalenten Begriff der Bewertung zugrunde zu legen, in dem die Analogie zwischen Zahl- und Funktionenkörpern so schön zum Ausdruck kommt. Man geht davon aus, daß jeder Pol und jede Nullstelle einer meromorphen Funktion auf der Riemannschen Fläche eine bestimmte ganzzahlige Vielfachheit besitzt. Geht man nun von einer bestimmten Stelle der Riemannschen Fläche aus und setzt für jede von Null verschiedene Funktion η des Körpers

$$\begin{cases} w(\eta) = + k, \text{ wenn } \eta \text{ an der Stelle eine } k\text{-fache Nullstelle,} \\ w(\eta) = - k, \text{ wenn } \eta \text{ dort einen } k\text{-fachen Pol hat,} \\ w(\eta) = 0, \text{ wenn } \eta \text{ weder } 0 \text{ noch } \infty \text{ wird,} \end{cases}$$

---

5) So F. K. Schmidt in seinem schönen Beweis des Riemann-Rochschen Satzes, Math. Z. **41** (1936), S. 415.

so hat die Zuordnung $\eta \to w(\eta)$ die folgenden Eigenschaften:

$$\begin{cases} w(\eta\zeta) = w(\eta) + w(\zeta) \\ w(\eta + \zeta) \geqq \operatorname{Min}\{w(\eta),\, w(\zeta)\} \\ w(c) = 0 \text{ für konstante } c \neq 0. \end{cases}$$

Eine solche Zuordnung nennt man eine *Bewertung* des Funktionenkörpers. Sie definiert einen *Bewertungsring* $\Re$, bestehend aus der Null und den η mit $w(\eta) \geqq 0$, und ein *Bewertungsideal* $\mathfrak{p}$ in $\Re$, bestehend aus der Null und den η mit $w(\eta) > 0$. Der Restklassenring $\Re/\mathfrak{p}$ umfaßt den Konstantenkörper und hat jedenfalls einen kleineren Transzendenzgrad als der Funktionenkörper. In unserem Fall, wo dieser den Transzendenzgrad Eins hat, ist $\Re/\mathfrak{p}$ eine algebraische Erweiterung des Konstantenkörpers K. Ordnet man nun jedem Element η von $\Re$ die zugehörige Restklasse $f(\eta)$ zu, jedem nicht zu $\Re$ gehörigen Element des Funktionenkörpers aber das Symbol ∞, so erhält man eine Zuordnung $\eta \to f(\eta)$ mit allen Eigenschaften eines *Ortes*. Also definiert jede Bewertung einen Ort.

Umgekehrt definiert jeder Ort nach dem früher bewiesenen eine Stelle auf der Riemannschen Fläche und diese wiederum eine Bewertung, wie wir ja oben sahen. Man kann aber auch unter Vermeidung dieses transzendenten Umweges direkt zu jedem Ort eine Bewertung konstruieren. Denn zu einem gegebenen Ort gehört ein Ring $\Re$, bestehend aus den η mit endlichem $f(\eta)$, und dieser Ring hat genau die von Krull[6]) geforderte Eigenschaft eines Bewertungsringes, nämlich daß der gegebene Funktionenkörper sein Quotientenkörper ist und daß in ihm von je zwei Elementen immer eines durch das andere teilbar ist. Also gibt es nach Krull eine einzige Bewertung, deren Bewertungsring $\Re$ ist. Dieser Zusammenhang zwischen Bewertung und Bewertungsring gilt nur dann allgemein, wenn man den Bewertungsbegriff so allgemein faßt, wie Krull es a. a. O. tut, d. h. wenn man als Werte $w(\eta)$ nicht nur reelle Zahlen, sondern allgemeiner Elemente irgendeiner geordneten Abelschen Wertegruppe zuläßt. Das wollen wir auch im folgenden tun. Im Fall einer Veränderlichen treten allerdings nur diskrete archimedische Bewertungen auf, so daß man als Wertegruppe die additive Gruppe der natürlichen Zahlen nehmen kann.

---

6) W. Krull, Allgemeine Bewertungstheorie, J. reine u. angew. Math. **167** (1932), S. 165, Satz 1.

Wir können das bisher Erörterte für algebraische Funktionenkörper einer Veränderlichen über dem Körper der komplexen Zahlen in den folgenden eineindeutigen Zuordnungen zusammenfassen:

$$\text{Stelle} \rightleftarrows \text{Zweig} \rightleftarrows \text{Ort} \rightleftarrows \text{Bewertung}.$$

§ 2. Algebraische Funktionen von zwei oder mehreren Veränderlichen.

Die Begriffe Zweig, Stelle, Ort, Bewertung, die im Fall einer Veränderlichen äquivalent waren, sind es bei zwei Veränderlichen nicht mehr, sondern führen zu verschiedenen Grundbegriffen: Keil, Stelle, Primdivisor, Ort.

Wir gehen aus von einer algebraischen Fläche F_0. Zu ihr gehört wieder ein Funktionenkörper, und man kann F_0 als ein projektives Modell des Funktionenkörpers auffassen.

Unter einem *Keil* auf der Fläche F_0 verstehen wir nach Walker[7] ein durch reguläre analytische Funktionen vermitteltes Bild einer offenen Menge eines (u, v)-Parameterraumes auf der Fläche. Der Begriff des Keiles entspricht dem des Kurvenzweiges im Fall einer Veränderlichen. Beispiele von Keilen erhält man, idem man auf irgendeinem Modell F des Funktionenkörpers eine genügend kleine Umgebung eines einfachen Punktes von F, der für die birationale Abbildung $F \rightarrow F_0$ nicht fundamental ist, abgrenzt; in einer solchen Umgebung sind nämlich alle Koordinaten reguläre Funktionen von zwei unter ihnen, die man dann u und v nennen kann.[8]

Zeichnet man in der (u, v)-Umgebung einen Punkt P aus, so kommt man zum Begriff einer *Stelle*. Dem Punkt P entspricht in der Abbildung auf F_0 ein Punkt P_0, der *Mittelpunkt* der Stelle. Für die meisten Untersuchungen dürfte es zweckmäßig sein, sich auf solche Stellen zu beschränken, die in der angegebenen Weise durch einfache Punkte von projektiven Modellen F bestimmt werden. Die Ortsuniformisierenden u, v können dann als Funktionen des Körpers gewählt werden. Bei

---

7) R. J. Walker, Reduction of the Singularities of an Algebraic Surface, Ann. of Math. **36** (1935), S. 336.

8) Das Wort Wedge = Keil soll wohl darauf hindeuten, daß die Bilder der (u, v)-Umgebungen auf F_0 manchmal scharfe Ecken haben. Ist z. B. F_0 die (x, y)-Ebene, und ist die Abbildung der Umgebung $|u| < \varepsilon$, $|v| < \varepsilon$ durch

$$x = u, \quad y = uv$$

gegeben, so wird der Keil durch $|x| < \varepsilon$, $|y| < \varepsilon |x|$ gegeben und hat bei $x = y = 0$ eine Ecke. Das liegt daran, daß die Gerade $u = 0$ auf den Punkt $x = y = 0$ abgebildet wird.

dieser Einschränkung ist der Begriff der Stelle ein algebraischer Begriff, im Gegensatz zum wesentlich transzendenten Begriff des Keiles.

Betrachten wir nun andererseits wie in § 1 Zuordnungen $\eta \to f(\eta)$ mit den Eigenschaften (1), so können hier (außer dem trivialen Fall, daß die Zuordnung einen Isomorphismus darstellt) zwei verschiedene Fälle eintreten:

1. Die Werte $f(\eta)$ sind algebraisch über dem Grundkörper; in diesem Fall heißt die Zuordnung ein *Ort*.

2. Die Werte $f(\eta)$ sind algebraische Funktionen einer Veränderlichen; in diesem Fall heißt die Zuordnung nach Jung[9]) *Primdivisor*.

Bei n Veränderlichen können entsprechend n verschiedene Fälle vorkommen: die Werte $f(\eta)$ können Funktionen von $0, 1, 2, \ldots, (n-1)$ Veränderlichen sein. Jedoch sind die beiden Extremfälle 0 und $n-1$ (Örter und Primdivisoren) bisher allein wichtig geworden.

Beispiele von Primdivisoren ergeben die irreduziblen Kurven auf der Fläche F_0, die vielfachen Kurven ausgenommen. Denn die Werte, die eine Funktion η des Körpers auf der Kurve annimmt, bilden jeweils eine Funktion einer Veränderlichen, die man $f(\eta)$ nennen kann, und die Eigenschaften (1) sind erfüllt. Leider erhält man in dieser Weise nicht alle Primdivisoren des Körpers; denn die einzelnen Punkte von F_0 können durch birationale Transformationen in Kurven verwandelt werden, und diese Kurven ergeben Primdivisoren des Körpers. Die Punkte dieser Kurven können ihrerseits wieder in Kurven verwandelt werden, und so läßt sich aus jedem Punkt von F_0 ein unendlicher Baum von Primdivisoren entwickeln. Man nennt diese *Punktprimdivisoren*, im Gegensatz zu den *Kurvenprimdivisoren*, die den Kurven von F_0 zugeordnet sind.

Im Fall einer Veränderlichen ergab jeder einfache Punkt P_0 von F_0 ohne weiteres einen Ort. Das ist hier nicht so; denn nicht jede Funktion η nimmt in P_0 einen bestimmten Wert $f(\eta)$ an: es gibt ja auch solche Funktionen, die in P_0 eine Unbestimmtheitsstelle haben. Beispiele von Örtern erhält man, indem man nicht von den Punkten von F_0, sondern von den Kurvenzweigen auf F_0 ausgeht. Ist $\mathfrak{z}$ ein solcher algebraischer oder transzendenter Kurvenzweig mit der Ortsuniformisierenden τ, so gehen alle Funktionen η des Funktionenkörpers auf dem Zweig $\mathfrak{z}$ in analytische Funktionen von τ über, und diese nehmen für $\tau = 0$ bestimmte Werte $f(\eta)$ an, für die (1) gilt.

---

9) H. W. E. Jung, Primteiler und ihr Verhalten bei birationalen Transformationen. Rend. Palermo **26** (1908), S. 113.

Wie man aus den angegebenen Beispielen sieht, fallen die drei Begriffe Stelle, Ort und Primdivisor im Fall der Funktionen mehrerer Veränderlichen ganz auseinander.

Ziehen wir nun die Bewertungen $w(\eta)$ des Funktionenkörpers heran, so müssen wir unterscheiden zwischen *nulldimensionalen Bewertungen*, bei denen $\mathfrak{R}/\mathfrak{p}$ algebraisch über dem Konstantenkörper K, und *eindimensionalen Bewertungen*, bei denen $\mathfrak{R}/\mathfrak{p}$ ein Funktionenkörper einer Veränderlichen ist. Ordnet man in beiden Fällen wie in § 1 jedem Element η von $\mathfrak{R}$ seine Restklasse in $\mathfrak{R}/\mathfrak{p}$, jedem nicht zu $\mathfrak{R}$ gehörigen η aber das Symbol ∞ zu, so erhält man ohne weiteres die folgenden eineindeutigen Zuordnungen:

eindimensionale Bewertung $\rightleftharpoons$ Primdivisor,

nulldimensionale Bewertung $\rightleftharpoons$ Ort.[10])

Bei den Primdivisoren kann man die Werte $w(\eta)$ wieder als ganze Zahlen annehmen.[11]) Wenn der Funktion η der ganzzahlige Wert $k = w(\eta)$ zugeordnet ist, so sagt man auch, η sei durch $\mathfrak{P}^k$ teilbar, wo $\mathfrak{P}$ der betreffende Primdivisor ist. Bei den Örtern aber muß man auch nichtarchimedische Wertegruppen mit in Betracht ziehen. Ist ein Ort nämlich durch einen algebraischen Kurvenzweig $\mathfrak{z}$ gegeben, so hat jede Funktion η, die auf $\mathfrak{z}$ nicht identisch 0 oder ∞ wird, als Funktion der Zweiguniformisierenden τ eine bestimmte ganzzahlige Ordnung $w(\eta) = \pm k$ ($+ k$ für eine k-fache Nullstelle, $- k$ für einen k-fachen Pol). Wird jedoch η auf $\mathfrak{z}$ identisch Null, so wird $w(\eta)$ unendlich groß im Vergleich zu allen ganzen Zahlen, die Wertegruppe also nichtarchimedisch. Welche Gruppen als Wertegruppen überhaupt in Betracht kommen, haben im Fall zweier Veränderlichen Zariski[11]), im Fall von n Veränderlichen MacLane und Schilling[12]) eingehend untersucht.

Eine erste Anwendung des Bewertungsbegriffs auf Funktionenkörper zweier Veränderlicher ist die von Zariski[13]) entwickelte Idealtheorie der „unendlich benachbarten Punkte".

Es sei S eine Stelle eines solchen Funktionenkörpers, etwa definiert durch einen einfachen Punkt P einer algebraischen Fläche F. Eine nulldimensionale Bewertung B heißt *zur Stelle S gehörig*, wenn alle Funktionen η des Körpers, die an der Stelle S Null werden (insbesondere also die Ortsuniformisierenden u und v) in der Bewertung positive Werte $w(\eta)$ haben. Wir betrachten nun eine feste zu S gehörige Bewertung B und bilden den Ring $\mathfrak{R}$ der an der Stelle S endlich bleibenden Funktionen des Körpers. Diejenigen

---

10) Vgl. O. Zariski, Ann. of Math. **41** (1940), S. 854.

11) O. Zariski, Ann. of Math. **40** (1939), S. 645.

12) Saunders MacLane and O. F. G. Schilling, Ann. of Math. **40** (1939), S. 507.

13) O. Zariski, Polynomial Ideals defined by infinitely near base points, Amer. J. of Math. **60** (1938), S. 151.

η aus $\Re$, deren Werte $w(\eta)$ größer sind als ein gegebener Wert w, bilden ein Primärideal $\mathfrak{q}$ in $\Re$. Solche Primärideale heißen v-*Ideale*. Sie gehören alle zum gleichen Primideal $\mathfrak{p}$ und bilden eine einzige Vielfachenkette $\mathfrak{p} = \mathfrak{q}_1 \subset \mathfrak{q}_2 \subset \mathfrak{q}_3 \ldots$ Sie lassen sich eindeutig darstellen als Produkte von unzerlegbaren v-Idealen $P_1 = \mathfrak{q}_1$, $P_2 = \mathfrak{q}_2$, $P_3 = \mathfrak{q}_j, \ldots$ Diesen entspricht eineindeutig eine Reihe von sukzessiven ,,Nachbarpunkten" $P = P_1, P_2, \ldots$ des Punktes P auf der Fläche F.[14]) Genauer: *Die Funktionen des Ideals P_k sind diejenigen an der Stelle S endlich bleibenden Funktionen, deren Nullstellenkurven auf F durch den Nachbarpunkt P_k (und daher notwendig auch durch $P_1, P_2, \ldots, P_{k-1}$) hindurchgehen.*

Verwandelt man die Nachbarpunkte $P_1, P_2, \ldots, P_k$ durch gewöhnliche quadratische Cremonatransformationen der Reihe nach in Kurven, so gehören zu diesen Kurven Primdivisoren. Auf die Fläche F bezogen, handelt es sich um Punktprimdivisoren. Man kann also die Nachbarpunkte eines Punktes P auch durch die zu diesem Punkt gehörigen Punktprimdivisoren definieren.

Für die nähere Ausführung dieser Gedankengänge muß auf die angeführte Arbeit von Zariski[13]) verwiesen werden.

§ 3. Die Auflösung der Singularitäten.

Der wichtigste der im vorigen Paragraphen eingeführten Begriffe ist offenbar der des Primdivisors. Denn wenn man, von der klassischen Theorie der algebraischen Funktionen einer Veränderlichen als Vorbild ausgehend, die Hauptprobleme einer Theorie der algebraischen Funktionen von mehreren Veränderlichen zu entwickeln sucht, so kommt man auf Fragen folgender Art: Welche Beziehung besteht zwischen den Zählerdivisoren und den Nennerdivisoren, d. h. zwischen den Nullstellen- und Poldivisoren einer algebraischen Funktion? Was kann man über lineare Divisorenscharen und ihre Dimension aussagen? Wie verhalten sich die Differentiale des Körpers besonders hinsichtlich ihrer Nullstellen- und Poldivisoren?

Die weitere Verfolgung dieser Fragen hat gezeigt, daß es erforderlich ist, den Primdivisoren eine Beschränkung aufzuerlegen. Denn wenn eine Funktion des Körpers auf einer Kurve Null wird, so wird sie auch in fast allen Punkten dieser Kurve Null, und diese Punkte können durch birationale Transformationen, wie schon erwähnt, in Kurven verwandelt werden und geben so zu schwer übersehbaren, unendlichen Mengen von Divisoren Anlaß, die alle in den Zähler der betreffenden Funktion aufgenommen werden müßten, weil die Funktion ja durch sie alle teilbar ist.

Man könnte zunächst versuchen, diese Beschränkung dadurch zu erreichen, daß man sich auf die Kurvendivisoren eines bestimmten pro-

---

14) Für den Begriff des Nachbarpunktes siehe meine Einführung in die algebraische Geometrie. Berlin 1939, § 55.

jektiven Modells F_0 beschränkt und die Punktdivisoren von der Betrachtung ausschließt. Denn auf der Fläche F_0 gibt es offenbar nur endlich viele irreduzible Kurven, auf denen eine vorgegebene Funktion η Null oder unendlich wird. Diese Beschränkung ist aber nur dann angängig, wenn die Fläche F_0 singularitätenfrei ist. Denn in einem singulären Punkt kann, wie schon Noether erkannt hat, ein Differential unendlich werden, ohne daß es auf einer durch den Punkt gehenden Kurve unendlich wird; auch kann man Schnittpunktszahlen von Kurven in singulären Punkten nicht gut definieren, und ähnliche Schwierigkeiten gibt es mehr. Aus diesen Gründen ist es von der größten Wichtigkeit, singularitätenfreie projektive Modelle F eines vorgegebenen Funktionenkörpers zu konstruieren, oder, wenn das nicht ohne weiteres gelingen sollte, wenigstens die einzelnen singulären Punkte lokal zu uniformisieren, d. h. sie und ihre Nachbarschaft durch reguläre Funktionen von Ortsuniformisierenden lückenlos darzustellen.

So kommt man auf zwei Hauptprobleme, deren Erledigung Vorbedingung für jede mehr in die Tiefe gehende Theorie ist, nämlich:

I. Problem der lokalen Uniformisierung: *Überdeckung einer vorgegebenen algebraischen Mannigfaltigkeit F_0 mit endlich vielen Keilen.*

II. Problem der Auflösung der Singularitäten: *Konstruktion eines singularitätenfreien projektiven Modells F, derart, daß die birationale Abbildung $F \to F_0$ ausnahmslos eindeutig ist.*

II ist offenbar das weitergehende Problem: seine Lösung zieht die von I unmittelbar nach sich.

Das lokale Uniformisierungsproblem I ist im Fall von zwei Veränderlichen von Black[15]) und von Jung[16]), im allgemeinen Fall von Kneser[17]) und von Zariski[18]) gelöst worden. Die Zariskische Lösung trägt weiter als die Knesersche, weil bei Zariski die Ortsuniformisierenden immer als Funktionen des Körpers gewählt werden, während Kneser den Körper durch Adjunktion von Radikalen erweitern mußte. Auf den Gedankengang des Zariskischen Beweises kommen wir noch zurück.

---

15) C. W. M. Black, The parametrie representation of the neighborhood of a singular point of an analytic surfaces, Proc. Amer. Acad. **37** (1901).

16) H. W. E. Jung, Darstellung der Funktionen eines algebraischen Körpers zweier unabhängiger Veränderlichen in der Umgebung einer Stelle. J. reine u. angew. Math. **133** (1908).

17) H. Kneser, Örtliche Uniformisierung der analytischen Funktionen mehrerer Veränderlichen, Jahresber. d. DMV. **45** (1935), S. 76.

18) O. Zariski, Local Uniformisation von Algebraic Varieties, Ann. of Math. **41** (1940), S. 852.

Das Transformationsproblem II ist bisher nur für höchstens zwei Veränderliche gelöst, und zwar von Walker[19]) und von Zariski.[20]) Der Zariskische Beweis beruht auf der abwechselnden Anwendung von zwei Schritten. Der erste Schritt, der Übergang zu einem ganz abgeschlossenen Erweiterungsring, befreit die Fläche mit einem Schlage von allen vielfachen Kurven. Der zweite Schritt, eine quadratische Transformation mit einem Fundamentalpunkt, vereinfacht die Singularität der Fläche in diesem Punkt. Daß die abwechselnde Anwendung dieser beiden Verfahren nach endlich vielen Schritten zu einer singularitätenfreien Fläche führt, wird bewiesen mit Hilfe eines Lemmas, das genau so lautet wie das in § 1 für den Fall einer Veränderlichen formulierte, nämlich so:

Lemma. *Zu jedem vorgegebenen Ort gibt es ein solches projektives Modell des Funktionenkörpers, auf dem der Mittelpunkt des Ortes ein einfacher Punkt ist.*

Dasselbe Lemma gilt nach Zariski[18]) auch im Fall von r Veränderlichen und dient ihm zum Beweis des lokalen Uniformisierungssatzes. Zum Zwecke dieses Beweises wird die Gesamtheit aller Örter des Funktionenkörpers, die „Riemannsche Mannigfaltigkeit", als topologischer Raum betrachtet. Die Örter sind ja Abbildungen des Funktionenkörpers auf eine Gaußsche Zahlenkugel. Für solche Abbildungen hat nun Tychonoff[21]) einen Umgebungsbegriff definiert: Sind $\eta_1, \ldots, \eta_m$ endlich viele Funktionen, $f(\eta_1), \ldots, f(\eta_m)$ ihre Bilder auf der Zahlenkugel, schließlich $U_1, \ldots, U_m$ Umgebungen dieser Punkte auf der Kugel, so besteht eine Umgebung der Abbildung f aus denjenigen Abbildungen, die $\eta_1, \ldots, \eta_m$ der Reihe nach in $U_1, \ldots, U_m$ hinein abbilden. Bei dieser Umgebungsdefinition bilden alle Örter, wie sich herausstellt, einen *bikompakten* topologischen Raum $\mathfrak{M}$, d. h. man kann aus jeder Überdeckung von $\mathfrak{M}$ mit offenen Mengen eine endliche Überdeckung auswählen. Nun gibt es nach dem Lemma zu jedem Ort ein solches Modell F, auf dem der Mittelpunkt P des Ortes ein einfacher Punkt ist. Ist U eine Umgebung von P auf F, so definiert diese erstens einen Keil auf F_0, zweitens aber auch eine offene Menge auf $\mathfrak{M}$, bestehend aus den Örtern, deren Mittelpunkte in U liegen. Auf Grund der Bikompaktheit müssen endlich viele solche Umgebungen den ganzen

<hr>

19) R. J. Walker, Reduction of the Singularities of an Algebraic Surface, Ann. of Math. **36** (1935), S. 336.

20) O. Zariski, The Reduction of the Singularities of an Algebraic Surface, Ann. of Math. **40** (1939), S. 639.

21) A. Tychonoff, Math. Ann. **102** (1930), S. 544.

Raum $\mathfrak{M}$ überdecken. Daraus folgt unmittelbar, daß die endlich vielen zugehörigen Keile die ganze Mannigfaltigkeit F_0 überdecken, womit I gelöst ist.

Wie in Knesers Beweis für die Ortsuniformisierung, so wird auch in Zariskis Beweis des Lemmas nicht nur die gegebene Mannigfaltigkeit F_0, sondern auch der umgebende Raum, in den sie als Hyperfläche eingebettet ist, transformiert, und zwar nach einer Cremonatransformation. Die Konstruktion dieser Cremonatransformation benutzt einen Algorithmus von Perron zur gleichzeitigen Approximation von mehreren Irrationalzahlen.

§ 4. Die Jungschen Stellendefinitionen.

Wenn es nicht gelingen sollte, die birationale Abbildbarkeit aller Mannigfaltigkeiten auf singularitätenfreie zu beweisen, so müßte man sich, wie es H. W. E. Jung immer schon getan hat, mit der lokalen Uniformisierung begnügen. Hat man die ganze Fläche F_0 mit Keilen überdeckt, so definiert jeder Punkt eines solchen Keiles eine Stelle auf F_0. Die Gesamtheit dieser Stellen wollen wir eine „Stellenüberdeckung" nennen. Was aber bei Jung zugrunde gelegt wird, ist nicht eine ganz beliebige Stellenüberdeckung, sondern eine, die in einem gewissen Sinne weder zu viel noch zu wenig Stellen enthält.[22] Um das für den Fall zweier Veränderlichen näher zu erläutern, müssen einige neue Begriffe eingeführt werden.

Für das Verhalten eines Primdivisors $\mathfrak{P}$ in bezug auf eine Stelle S gibt es drei Möglichkeiten. Betrachten wir nämlich diejenigen Funktionen des Körpers, die an der Stelle S endlich bleiben, also in Potenzreihen $\eta(u, v)$ nach den Ortsuniformisierenden u, v entwickelt werden können. Unter diesen betrachten wir besonders die durch $\mathfrak{P}$ teilbaren, für die also $w(\eta) > 0$, d. h. $f(\eta) = 0$ ist. Haben alle diese Potenzreihen $\eta(u, v)$ einen irreduziblen Faktor gemeinsam, der an der Stelle S Null wird, so nennt man den Primdivisor *von erster Art* in bezug auf die Stelle S. Haben sie keinen solchen Faktor gemeinsam, aber werden sie wohl alle Null an der Stelle S, so heißt der Primdivisor *von zweiter Art* in S. Werden schließlich die $\eta(u, v)$ nicht alle Null für $u = v = 0$, so heißt der Primdivisor *fremd* zur Stelle S.

Jung verlangt nun, wie aus seinen Beweisen hervorgeht, von einer „Stellendefinition" unter anderem folgendes: Es soll nicht vorkommen, daß ein und derselbe Primdivisor an einer Stelle von erster, an einer

---

22) Vgl. zum folgenden vor allem: H. W. E. Jung, Stellentransformation in algebraischen Körpern zweier Veränderlicher, Acta math. 68 (1937), S. 7.

anderen Stelle von zweiter Art ist, sondern er soll entweder an genau einer Stelle von zweiter Art, oder an unendlich vielen Stellen von erster Art sein.

Weiter wird man wohl verlangen müssen, daß jeder Ort des Körpers zu einer bestimmten Stelle S gehört in dem in § 2 (kleine Typen) erklärten Sinne, daß alle die Funktionen, die an der Stelle S Null werden, auch an dem betreffenden Ort Null werden.

Ob diese beiden Forderungen für eine „Stellendefinition" im Jung - schen Sinne genügen, und wie sie auf den Fall von mehr als zwei Veränderlichen zu verallgemeinern sind, wäre noch zu untersuchen.

Alle an eine Stellendefinition vernünftigerweise zu stellenden Anforderungen sind jedenfalls dann erfüllt, wenn man die Stellen durch die Punkte eines singularitätenfreien Modells definiert, sofern es ein solches Modell gibt. So legen denn auch die italienischen Geometer immer ein singularitätenfreies Modell ihren Untersuchungen zugrunde.

(Eingegangen am 4. 3. 1942.)

24.

Divisorenklassen in algebraischen Funktionenkörpern

Commentarii Mathematici Helvetici 20, 1 (1947) 68–109

H. Hasse hat eine arithmetische Theorie der Divisorenklassen eines
algebraischen Funktionenkörpers einer Veränderlichen über einem beliebigen vollkommenen Konstantenkörper Ω entwickelt[1]). Bei einigen für
die Theorie des Abelschen Funktionenkörpers grundlegenden Sätzen ist
Hasse jedoch nicht zur Durchführung der Beweise gekommen, hat aber
die Erwartung ausgesprochen, daß diese Beweise sich aus der von mir
gegebenen Begründung der algebraischen Geometrie ergeben würden. In
der Tat ist es mir gelungen, nicht nur die fraglichen Beweise mit meinen
Methoden zu erbringen, sondern darüber hinaus noch ein weiteres Problem zu lösen, das sich aus den Hasseschen Fragestellungen zwangsläufig
ergab, nämlich die Konstruktion einer *Klassenmannigfaltigkeit*, deren
Punkte eineindeutig den Divisorenklassen nullten Grades des gegebenen
algebraischen Funktionenkörpers entsprechen. Die Klassenmannigfaltigkeit ist ein ausgezeichnetes projektives Modell des Abelschen Funktionenkörpers; ihre Punkte übernehmen die Rolle der Hasseschen „X-Punkte“
(Hasse § 7, 7). Die Hasseschen Sätze § 6, 5 und § 8, 2 aber, deren Beweise
hier gegeben werden sollen, drücken im wesentlichen aus, daß *den algebraischen Operationen, die man mit Divisorenklassen vornehmen kann, auch
algebraische Operationen auf der Klassenmannigfaltigkeit entsprechen.* Diese
Operationen sind: die Addition von Klassen $X + Y = Z$ und die komplexen Multiplikationen $X = \mu Y$, die aus den algebraischen Korrespondenzen des Funktionenkörpers entstehen.

Der Darstellung dieser Ergebnisse soll eine Übersicht über die grundlegenden Begriffe und Sätze der algebraischen Geometrie vorangeschickt
werden. Ich habe mich dabei nicht auf das unbedingt notwendige Minimum an Sätzen beschränkt, sondern ich habe mich bemüht, aus diesem
einleitenden Teil einen Rechenschaftsbericht über die Begründung der

---

[1]) *H. Hasse*, Zur arithmetischen Theorie der algebraischen Funktionenkörper, Jahresber. D. M. V. 52 (1942) S. 1—48.

algebraischen Geometrie und eine Art Lehrgang der algebraischen Geometrie für Algebraiker zu machen. Die Definitionen sind vollständig angegeben, während bei den Sätzen jeweils angegeben ist, wo man ihre Beweise finden kann.

Nach dieser Einführung (§ 1 bis 10) folgt der Hauptteil (§ 11 bis 19). Der Gedankengang des Hauptteils ist folgender. Wir gehen von einem algebraischen Funktionenkörper K vom Geschlechte g aus und legen als „projektives Modell" dieses Körpers eine singularitätenfreie Kurve Γ zugrunde. Nach Erweiterung des Konstantenkörpers Ω zu einem algebraisch abgeschlossenen Körper $\overline{\Omega}$ entsprechen die Punkte von Γ eineindeutig den Stellen von K (§ 11). Die Gruppen von g Punkten auf Γ können durch Koordinaten dargestellt und so auf Punkte eines Bildraumes abgebildet werden (§ 12). Die Gesamtheit dieser Bildpunkte ist eine glatte, d. h. singularitätenfreie algebraische Mannigfaltigkeit M (§ 13). Produkte und Äquivalenzen von Punktgruppen auf Γ können durch algebraische Gleichungen zwischen ihren Koordinaten ausgedrückt werden, geben also zu algebraischen Korrespondenzen auf M Anlaß (§ 14). Setzt man dieses Ergebnis zu dem Hasseschen Begriff des $\mathfrak{X}$-Punktes in Beziehung, so erhält man den Beweis des ersten Hasseschen Satzes (§ 15). Sodann wird die erwähnte Klassenmannigfaltigkeit konstruiert, deren Punkte eineindeutig den Divisorenklassen vom Grade Null entsprechen (§ 16). Die Multiplikatoren (komplexe Multiplikationen) von K ergeben ebenfalls algebraische Korrespondenzen auf M sowie auch auf der Klassenmannigfaltigkeit (§ 17). Setzt man dieses Ergebnis wieder zum Hasseschen Begriff des $\mathfrak{X}$-Punktes in Beziehung, so ergeben sich Beweise der weiteren Sätze von Hasse (§ 18).

Erster Teil

Grundlagen der algebraischen Geometrie

§ 1. Resultantensysteme. Relationstreue Spezialisierung. Prinzip der Erhaltung der Anzahl.

1.1. Notwendig und hinreichend dafür, daß ein System homogener Gleichungen

$$f_i(y_0, \ldots, y_n) = 0 \tag{1}$$

eine von der Nullösung verschiedene Lösung in einem geeigneten (algebraischen) Erweiterungskörper des Konstantenkörpers Ω besitzt, ist das Verschwinden des Resultantensystems

$$R_j(a) = 0 \; .$$

321

Dabei sind die R_j ganzzahlige Polynome in den Koeffizienten a der Formen f_i. Beweis: Moderne Algebra II, § 80 (2. Aufl.).

1.2. Zwei Nicht-Nullösungen eines homogenen Gleichungssystems (1) werden zur gleichen *Lösungsklasse* gerechnet, wenn sie sich nur um einen Faktor λ unterscheiden. Die Lösungsklassen heißen auch *Punkte* des projektiven Raumes S_n. Die Gesamtheit aller Lösungsklassen eines Systems (1) heißt, falls nicht leer, eine *algebraische Mannigfaltigkeit* in S_n.

1.3. Ein Punkt y' heißt eine *relationstreue Spezialisierung* eines von unbestimmten Parametern $t_1, \ldots, t_r$ abhängigen Punktes y *für die Parameterspezialisierung* $t \to t'$, wenn alle in den y homogenen Gleichungen $f(t,y) = 0$ bei der Spezialisierung $t \to t'$, $y \to y'$ erhalten bleiben. Ist von Parametern t nicht die Rede, so spricht man von einer *relationstreuen Spezialisierung* $y \to y'$ schlechthin.

1.4. Zu jeder Parameterspezialisierung $t \to t'$ gehört stets mindestens eine relationstreue Spezialisierung $y \to y'$, vorausgesetzt daß t_j und y_k einem und demselben Erweiterungskörper von Ω angehören. Ebenso läßt sich jede relationstreue Spezialisierung $x \to x'$ stets zu einer ebensolchen Spezialisierung $(x,y) \to (x',y')$ fortsetzen, sofern die x und y einem und demselben Erweiterungskörper von Ω angehören. Beweis: Math. Ann. **97**, p. 761, oder Einführung alg. Geom., p. 107.

1.5. Wenn ein Gleichungssystem

$$f_i(t,x) = 0 \tag{2}$$

nur endlich viele Lösungsklassen $x^{(1)}, \ldots, x^{(q)}$ besitzt und wenn bei einer Spezialisierung $t \to t'$ der Unbestimmten t die Lösungszahl endlich bleibt, so gehört dazu eine bis auf die Reihenfolge der y eindeutige relationstreue Spezialisierung $x^{(1)} \to y^{(1)}, \ldots, x^{(q)} \to y^{(q)}$. Sie wird gefunden, indem man mit Unbestimmten $u_0, \ldots, u_n$ die Form

$$F(u) = \prod_{\nu=1}^{q} (u_0 x_0^{(\nu)} + u_1 x_1^{(\nu)} + \cdots + u_n x_n^{(\nu)})$$

bildet (oder im Fall eines Körpers von der Charakteristik p eine geeignete p^j-te Potenz dieses Produktes), sie durch Multiplikation mit einem von den u unabhängigen Faktor ganzrational in den t und primitiv in den u macht und dann in ihr $t \to t'$ spezialisiert. Die Zahl, die angibt, wie oft eine Lösungsklasse y des spezialisierten Systems $f_i(t',y) = 0$ unter den $y^{(1)}, \ldots, y^{(q)}$ vorkommt, heißt die *Multiplizität* der Lösung y für die Spezialisierung $t \to t'$.

1.6. *Prinzip der Erhaltung der Anzahl:* Die Summe der Multiplizitäten der Lösungen des spezialisierten Gleichungssystems ist gleich der Anzahl der Lösungen des ursprünglichen Problems (2).

1.7. Die Multiplizität einer Lösung y des spezialisierten Gleichungssystems ist positiv, wenn diese Lösung y durch eine relationstreue Spezialisierung für $t \to t'$ aus einer Lösung x des Systems (2) hervorgeht. Beweise zu 1.5 bis 1.7: Math. Ann. **97**, p. 762—766, oder Einf. alg. Geom. § 38.

1.8. *Kriterium für Multiplizität Eins:* Wenn die spezialisierten Gleichungen

$$f_i(t', y) = 0 \tag{3}$$

eine Lösung y besitzen, derart, daß die „Tangentialebenen" der Hyperflächen (3) im Punkt y, die durch die Gleichungen

$$x_0 \, \partial_0 f_i(t', y) + x_1 \, \partial_1 f_i(t', y) + \cdots + x_n \, \partial_n f_i(t', y) = 0 \quad \left[\partial_\kappa = \frac{\partial}{\partial y_\kappa}\right] \tag{4}$$

definiert werden, nur den Punkt y miteinander gemeinsam haben, so hat die Lösung y höchstens die Multiplizität Eins. Beweis: ZAG 5, Math. Ann. **110**, oder Einf. alg. Geom. § 39.

1.9. Die Behauptungen 1.5 bis 1.7 gelten auch in dem allgemeineren Fall, daß t nicht ein System von Unbestimmten, sondern ein allgemeiner Punkt einer irreduziblen Mannigfaltigkeit M (vgl. 2.2) und t' ein einfacher Punkt von M (vgl. 4.6) ist. Beweis: ZAG 6, Math. Ann. **110**, § 3.

§ 2. Algebraische Mannigfaltigkeiten

Beweise: Math. Ann. **96**, p. 183 oder Mod. Alg. II, § 93 oder Einf. alg. Geom. § 28.

2.1. Jede algebraische Mannigfaltigkeit ist eindeutig darstellbar als unverkürzbare Vereinigung von *irreduziblen*, d. h. nicht weiter zerlegbaren Mannigfaltigkeiten.

2.2. Jede irreduzible Mannigfaltigkeit M besitzt einen *allgemeinen* Punkt, d. h. einen solchen Punkt ξ von M, aus dem alle Punkte von M durch relationstreue Spezialisierung hervorgehen. Er ist bis auf Körperisomorphie eindeutig bestimmt. Wenn eine Mannigfaltigkeit einen allgemeinen Punkt besitzt, so ist sie irreduzibel. Jeder Punkt ξ, dessen Koordinaten einem beliebigen Erweiterungskörper des Konstantenkörpers angehören, ist allgemeiner Punkt einer irreduziblen Mannigfaltigkeit.

2.3. Die *Dimension* einer irreduziblen Mannigfaltigkeit M ist der Transzendenzgrad der Koordinatenverhältnisse eines allgemeinen Punktes von M.

2.4. Eine nulldimensionale irreduzible Mannigfaltigkeit ist ein System konjugierter Punkte.

2.5. Eine *rein r-dimensionale* Mannigfaltigkeit ist eine solche, deren irreduzible Bestandteile alle dieselbe Dimension r haben. Diese Bestandteile können mit willkürlichen positiven Vielfachheiten versehen werden.

2.6. Eine rein $(n - 1)$-dimensionale Mannigfaltigkeit in S_n ist eine *Hyperfläche*, d. h. sie wird durch eine einzige homogene Gleichung $f = 0$ gegeben. Umgekehrt ist jede Hyperfläche rein $(n - 1)$-dimensional. Ihre irreduziblen Bestandteile entsprechen den irreduziblen Faktoren der Form f und werden mit denselben Vielfachheiten versehen wie diese.

2.7. Wenn eine Mannigfaltigkeit M nicht ganz in der uneigentlichen Hyperebene $y_0 = 0$ liegt, so kann man durch die Normierung $y_0 = 1$ für die eigentlichen Punkte von M inhomogene Koordinaten $y_1, \ldots, y_n$ einführen. Das *zugehörige Ideal* von M im Polynombereich $\Omega\,[x_1, \ldots, x_n]$ ist dann die Gesamtheit aller Polynome f, die Null werden in allen Punkten von M.

2.8. Das zugehörige Ideal einer irreduziblen Mannigfaltigkeit ist prim. Jedes Primideal in $\mathfrak{o} = \Omega\,[x_1, \ldots, x_n]$ mit Ausnahme des Einheitsideals $\mathfrak{o}$ ist zugehöriges Ideal einer einzigen irreduziblen Mannigfaltigkeit.

2.9. Bei Erweiterung des Grundkörpers Ω kann eine irreduzible Mannigfaltigkeit nur in irreduzible Bestandteile von derselben Dimension zerfallen, die dann in bezug auf Ω konjugiert sind.

§ 3. Algebraische Korrespondenzen

Beweise: ZAG 6, Math. Ann. 110, S. 142, oder Einf. alg. Geom. § 33.

3.1. Eine *algebraische Korrespondenz* ist eine algebraische Mannigfaltigkeit von Punktepaaren (x, y), x in S_m, y in S_n, gegeben durch ein System von homogenen Gleichungen

$$f_\nu(x, y) = 0 \ . \tag{1}$$

3.2. Die Punkte x bilden eine algebraische Mannigfaltigkeit M, die *Urmannigfaltigkeit* der Korrespondenz, deren Gleichungen durch Elimination der y aus (1) gefunden werden. Die Punkte y bilden ebenso die *Bildmannigfaltigkeit* N der Korrespondenz. Man spricht von einer *Korrespondenz zwischen M und N*.

3.3. In einer irreduziblen Korrespondenz sind M und N beide irreduzibel. Einem allgemeinen Punkt ξ von M entspricht eine relativ zum Körper $\Omega\left(\dfrac{\xi_1}{\xi_0}, \ldots, \dfrac{\xi_m}{\xi_0}\right)$ irreduzible Teilmannigfaltigkeit N_ξ von N, ebenso einem allgemeinen Punkt η von N eine irreduzible Teilmannigfaltigkeit M_η von M.

3.4. *Prinzip der Konstantenzählung:* Ist q die Dimension einer irreduziblen Korrespondenz, a die von M, b die von N_ξ, c die von N und d die von M_η, so gilt

$$q = a + b = c + d \ .$$

3.5. Einem jeden Punkt x von M entspricht in einer irreduziblen Korrespondenz eine Teilmannigfaltigkeit N_x von N, die keine Bestandteile von kleinerer Dimension als b enthält.

3.6. Ist die Urmannigfaltigkeit M einer Korrespondenz irreduzibe und entspricht jedem Punkt von M eine irreduzible Teilmannigkeit von N, die immer dieselbe Dimension b hat, so ist die Korrespondenz irreduzibel.

§ 4. Schnitt von Mannigfaltigkeiten mit linearen Teilräumen und Hyperflächen

Beweise: ZAG 13, Math. Ann. **115**, p. 359, oder Einf. alg. Geom. § 34 und § 40—41.

4.0. Ein *linearer Teilraum* S_{n-k} des Raumes S_n wird durch k unabhängige lineare Gleichungen definiert. Sind die Koeffizienten lauter unabhängige Unbestimmte, so heißt der Teilraum *allgemein*. Ein linearer Teilraum S_m ist durch $(m+1)$ linear unabhängige Punkte bestimmt.

4.1. Der Durchschnitt einer irreduziblen a-dimensionalen Mannigfaltigkeit M mit einem allgemeinen S_{n-k} ist im Fall $k > a$ leer, im Fall $k = a$ ein System von endlich vielen konjugierten Punkten, im Fall $k < a$ eine irreduzible $(a-k)$-dimensionale Mannigfaltigkeit. Die Anzahl der Schnittpunkte im Fall $k = a$ heißt der *reduzierte Grad* von M.

4.2. Der Durchschnitt einer irreduziblen a-dimensionalen Mannigfaltigkeit vom reduzierten Grad g mit k allgemeinen Hyperflächen von den Graden $\gamma_1, \ldots, \gamma_\kappa$ ist im Fall $k > a$ leer, im Fall $k = a$ ein System von $g\,\gamma_1, \ldots, \gamma_\kappa$ konjugierten Punkten und im Fall $k < a$ eine irreduzible $(a-k)$-dimensionale Mannigfaltigkeit vom reduzierten Grad $g\,\gamma_1, \ldots, \gamma_\kappa$.

4.3. Der Durchschnitt einer irreduziblen a-dimensionalen Mannigfaltigkeit vom reduzierten Grad g mit einer Hyperfläche vom Grade γ, die sie nicht ganz enthält, ist rein $(a-1)$-dimensional und ihre irreduziblen Bestandteile können mit solchen positiven Vielfachheiten versehen werden, daß die Summe ihrer Gradzahlen $g_1, \ldots, g_r$ multipliziert mit diesen Vielfachheiten, gleich $g\gamma$ ist:

$$\mu_1 g_1 + \mu_2 g_2 + \cdots + \mu_r g_r = g\gamma \ .$$

Dabei werden die Vielfachheiten $\mu_1, \ldots, \mu_r$ im Fall einer Kurve $(a=1)$ durch relationstreue Spezialisierung definiert, indem man die Kurve zuerst mit einer allgemeinen Hyperfläche desselben Grades schneidet und dann diese Hyperfläche spezialisiert. Der allgemeine Fall wird auf den Fall $a=1$ reduziert, indem man noch $(a-1)$ allgemeine Hyperflächen hinzunimmt, die die Mannigfaltigkeit nach 4.2 in einer irreduziblen Kurve schneiden.

4.4. Der Durchschnitt einer rein a-dimensionalen Mannigfaltigkeit M mit $k \leqq a$ Hyperflächen von den Gradzahlen $\gamma_1, \ldots, \gamma_\kappa$ enthält keine Bestandteile von kleinerer Dimension als $a-k$. Falls sie auch keine Bestandteile höherer Dimension enthält, ist sie rein $(a-k)$-dimensional und ihre irreduziblen Bestandteile können mit solchen positiven Vielfachheiten $\mu_1, \ldots, \mu_r$ versehen werden, daß der gesamte Grad

$$\mu_1 g_1 + \cdots + \mu_r g_r = g\gamma_1 \cdots, \gamma_k \ .$$

ist, wo g der reduzierte Grad von M ist und $g_1, \ldots, g_r$ die reduzierten Grade der Schnittbestandteile sind.

4.5. Sind f und g Formen gleichen Grades, von denen die zweite Null wird auf M, so stimmt der Durchschnitt von M mit der Hyperfläche $f=0$ genau überein mit dem von M mit $f+\lambda g=0$, auch was die Vielfachheiten der irreduziblen Bestandteile betrifft.

4.6. Ein Punkt P einer rein a-dimensionalen Mannigfaltigkeit M heißt ein *s-facher Punkt* von M, wenn ein allgemeiner, durch P gelegter linearer Raum S_{n-a} die Mannigfaltigkeit im Punkte y mit der Multiplizität s schneidet. Ist $s=1$, so heißt P ein *einfacher Punkt* von M. Eine Mannigfaltigkeit mit lauter einfachen Punkten heißt *singularitätenfrei* oder *glatt*.

4.7. Eine a-dimensionale Mannigfaltigkeit M hat in jedem einfachen Punkt einen *Tangentialraum S_a*, Durchschnitt der Tangentialhyperebenen aller Hyperflächen durch M. Wenn umgekehrt M in P einen Tangentialraum besitzt, so ist P ein einfacher Punkt von M.

§ 5. Zugeordnete Formen und algebraische Systeme von algebraischen Mannigfaltigkeiten

Beweise: Chow und v. d. Waerden, ZAG 9, Math. Ann. 113, p. 692. Zum Teil auch Einf. alg. Geom. § 37.

5.1. Die *zugeordnete Form* einer irreduziblen nulldimensionalen Mannigfaltigkeit, bestehend aus den konjugierten Punkten $y^{(1)}, \ldots, y^{(h)}$, ist das Produkt

$$F(u) = \prod_{\nu=1}^{h} (u_0 y_0^{(\nu)} + u_1 y_1^{(\nu)} + \cdots + u_n y_n^{(\nu)})$$

oder im Fall eines unvollkommenen Körpers von der Charakteristik p eine solche p^f-te Potenz dieses Produktes, daß die Form dem Grundkörper Ω angehört.

5.2. Die *zugeordnete Form* $F(u; u^{(1)}, \ldots, u^{(r)})$ einer irreduziblen r-dimensionalen Mannigfaltigkeit M ist definiert als die zugeordnete Form der nulldimensionalen Schnittmannigfaltigkeit von M mit den allgemeinen Hyperebenen $u^{(1)}, \ldots, u^{(r)}$. Sie ist (bis auf einen Faktor ± 1) von der Reihenfolge der Hyperebenen $u, u^{(1)}, \ldots, u^{(r)}$ unabhängig. Ihr Grad heißt der *Grad* von M und ist im Fall eines vollkommenen Grundkörpers Ω gleich dem reduzierten Grad (4.1), sonst gleich diesem mal p^f.

5.3. Die zugeordnete Form einer rein r-dimensionalen Mannigfaltigkeit, deren irreduzible Bestandteile mit Vielfachheiten $e_1, \ldots, e_s$ versehen sind, ist das Produkt der zugeordneten Formen dieser Bestandteile mit Exponenten $e_1, \ldots, e_s$:

$$F = F_1^{e_1} F_2^{e_2} \ldots F_s^{e_s}.$$

5.4. Die Bedingung dafür, daß die Hyperebenen $v, v^{(1)}, \ldots, v^{(r)}$ einen Punkt mit M gemeinsam haben, lautet

$$F(v, v^{(1)}, \ldots, v^{(r)}) = 0.$$

5.5. Die zugeordnete Form bestimmt die Mannigfaltigkeit M eindeutig. Ihre Koeffizienten können als *homogene Koordinaten* von M aufgefaßt werden. Durch diese Koordinaten werden die Mannigfaltigkeiten M gegebenen Grades und gegebener Dimension eineindeutig abgebildet auf Punkte eines projektiven Bildraumes $\mathfrak{B}$.

5.6. Die Bedingung, daß ein Punkt P einer Mannigfaltigkeit M angehört, läßt sich durch algebraische Gleichungen zwischen den Koordinaten von P und denen von M ausdrücken. Ebenso läßt sich die Bedingung, daß M in einer anderen Mannigfaltigkeit N enthalten ist, durch Gleichungen zwischen den Koordinaten von M und N ausdrücken.

5.7. Ein *algebraisches System von algebraischen Mannigfaltigkeiten* ist eine solche Menge von Mannigfaltigkeiten M, deren Bildmenge in $\mathfrak{B}$ eine algebraische Mannigfaltigkeit ist. Das System heißt irreduzibel, wenn die Bildmenge es ist; die Dimension des Systems ist die Dimension der Bildmenge.

5.8. Alle Mannigfaltigkeiten gegebenen Grades und gegebener Dimension in S_n bilden ein algebraisches System.

5.9. Wenn in einer irreduziblen Korrespondenz zwischen M und N jedem Punkt x von M eine Mannigfaltigkeit N_x auf N entspricht, die 1. immer dieselbe Dimension b hat, während 2. M keine mehrfachen Punkte besitzt, so bilden diese Bildmannigfaltigkeiten N_x ein irreduzibles System $\mathfrak{A}$, und M ist auf $\mathfrak{A}$ rational abgebildet. — Läßt man die beiden Voraussetzungen 1., 2. fallen, so kann man das System $\mathfrak{A}$ und die rationale Abbildung von M auf $\mathfrak{A}$ zwar immer noch definieren, und wenn einem Punkt x von M in dieser Abbildung eine einzige Mannigfaltigkeit A_x entspricht, so ist auch $N_x = A_x$, andernfalls aber ist N_x die Vereinigungsmenge aller Bildmannigfaltigkeiten A_x, die dem Punkte x in der Abbildung entsprechen. Ist x ein einfacher Punkt von M, so ist entweder $N_x = A_x$, oder N_x hat eine höhere Dimension als b.

Bemerkung: Die Ausführungen dieses Paragraphen lassen sich ohne weiteres auf Mannigfaltigkeiten von Punktepaaren, Punkttripeln usw. übertragen. Am einfachsten geschieht das dadurch, daß die Punktepaare (x, y) auf Punkte z eines Bildraumes abgebildet werden, die durch die Koordinaten

$$z_{ik} = x_i y_k$$

definiert werden, wo i von 0 bis m und k von 0 bis n läuft, wenn x einem Raume S_m und y einem Raume S_n angehört. Diese Bildpunkte z_{ik} bilden eine algebraische Mannigfaltigkeit Z, deren Gleichungen lauten

$$z_{ik} z_{jl} - z_{il} z_{jk} = 0 \; .$$

Jeder Mannigfaltigkeit M von Punktepaaren (x, y) entspricht eineindeutig eine Teilmannigfaltigkeit M' von Z, und als zugeordnete Form von M kann man die zugeordnete Form von M' betrachten.

§ 6. Durchschnitte von algebraischen Mannigfaltigkeiten

Die Sätze dieses § werden in der vorliegenden Arbeit nicht gebraucht.
Beweise: ZAG 14, Math. Ann. 115, p 619.
Der Grundkörper Ω wird in diesem § als vollkommen vorausgesetzt.

6.1. Wenn zwei reine Mannigfaltigkeiten M_d und M_{n-d} von den
Dimensionen d und $n - d$ in S_n nur endlich viele Punkte gemeinsam
haben, so hat jeder dieser Schnittpunkte eine positive Vielfachheit und
die Summe der Vielfachheiten ist gleich dem Produkt der Gradzahlen.

6.2. Der Durchschnitt zweier Mannigfaltigkeiten M_d und M_e in S_n
hat, wenn $d + e = n + k$ ist, keine irreduziblen Bestandteile von kleinerer Dimension als k. Falls er auch keine Bestandteile höherer Dimension hat, also rein k-dimensional ist, so kann man seine irreduziblen
Bestandteile mit solchen Vielfachheiten $\mu_1, \ldots, \mu_r$ versehen, daß die
Summe ihrer Gradzahlen, multipliziert mit diesen Vielfachheiten, gleich
dem Produkt der Gradzahlen von M_d und M_e ist:

$$\mu_1 g_1 + \mu_2 g_2 + \cdots + \mu_r g_r = g\,\gamma \ .$$

6.3. Durchläuft M_d ein irreduzibles System $\mathfrak{S}$ von algebraischen
Mannigfaltigkeiten, so durchläuft auch die Schnittmannigfaltigkeit $Q_k =$
$M_d \cdot M_e$ ein irreduzibles System von k-dimensionalen Mannigfaltigkeiten
und die Zuordnung $M_d \to Q_k$ ist eine irreduzible Korrespondenz zwischen
diesen beiden Systemen. Dieses gilt unter der Voraussetzung, daß kein
M_d des Systems mit dem festen M_e einen mehr als k-dimensionalen
Durchschnitt hat. Ist diese Voraussetzung nur für das allgemeine Element
M_d^* von $\mathfrak{S}$ erfüllt, so kann man das System der Q_k und die Korrespondenz
$M_d^* \to Q_k = M_d^* \cdot M_e$ immer noch definieren, und einem solchen Element
M_d' von $\mathfrak{S}$, das mit M_e einen k-dimensionalen Durchschnitt hat, entspricht in der Korrespondenz auch nur ein Bildelement $Q_k' = M_d' \cdot M_e$;
hat aber M_d' einen mehr als k-dimensionalen Durchschnitt mit M_e, so
können diesem M_d' mehrere Bildelemente Q_k' entsprechen, von denen jede
als *virtuelle Schnittmannigfaltigkeit* $M_d' \cdot M_e$ bezeichnet werden kann.

6.4. Ist M_n eine glatte, rein n-dimensionale Mannigfaltigkeit in S_r,
so kann man jede Mannigfaltigkeit M_e auf S_r zu einem vollständigen
Schnitt

$$M_n \cdot K_{e+r-n} = M_e + N_e$$

ergänzen, wobei K_{e+r-n} so gewählt werden kann, daß der Restschnitt N_e
einen vorgegebenen Punkt P von M_n nicht enthält.

6.5. Ist D_k ein irreduzibler Bestandteil des Durchschnittes $M_d \cdot M_e$ zweier Mannigfaltigkeiten M_d und M_e auf M_n, von der Dimension $k = d + e - n$, und wird K_{e+r-n} wie in 6.4 so gewählt, daß K_{e+r-n} irgendeinen Punkt von D_k nicht enthält, so wird die Vielfachheit von D_k als Bestandteil von $M_d \cdot M_e$ auf M_n definiert als die Vielfachheit von D_k als Bestandteil von $M_d \cdot K_{e+r-n}$. Diese Vielfachheit ist von der Wahl von K_{e+r-n} unabhängig. Die Summe aller D_k mit ihren Vielfachheiten ist der Durchschnitt $M_d \cdot M_e$ auf M_n.

6.6. Für die so definierten Durchschnitte $M_d \cdot M_e = Q_k$ gilt wieder 6.3; weiter gelten das kommutative, assoziative und distributive Gesetz:

$$M_d \cdot M_e = M_e \cdot M_d$$
$$(M_d \cdot M_e) \cdot M_f = M_d \cdot (M_e \cdot M_f)$$
$$M_d \cdot (M_e + M_e') = M_d \cdot M_e + M_d \cdot M_e'$$

§ 7. Lineare Scharen

Beweise: ZAG 6, § 5—6, Math. Ann. **110**, S. 148, sowie Einf. alg. Geom. § 42—44.

7.1. Eine *virtuelle lineare Schar* von $(d-1)$-dimensionalen Teilmannigfaltigkeiten einer irreduziblen d-dimensionalen Mannigfaltigkeit M besteht aus den Schnittmannigfaltigkeiten von M mit einer linearen Schar von Hyperflächen

$$\lambda_0 F_0 + \lambda_1 F_1 + \cdots + \lambda_r F_r = 0 \;, \tag{1}$$

von denen keine M enthält, wobei zu allen diesen Schnittmannigfaltigkeiten noch eine feste $(d-1)$-dimensionale Mannigfaltigkeit, deren Bestandteile mit beliebigen positiven oder negativen Multiplizitäten versehen sind, hinzugefügt werden darf. Sind alle Mannigfaltigkeiten der Schar effektiv, d. h. haben ihre irreduziblen Bestandteile nichtnegative Vielfachheiten, so heißt die Schar *effektiv*. Die Zahl r heißt die *Dimension* der Schar. Ein Punkt, der allen Mannigfaltigkeiten einer effektiven Schar gemeinsam ist, heißt *Basispunkt* der Schar. Eine feste $(d-1)$-dimensionale Mannigfaltigkeit, die allen Elementen der Schar als Bestandteil angehört, heißt *fester Bestandteil* der Schar.

7.2. Eine lineare Formenschar (1) von der Dimension r, in der t linear unabhängige Formen vorhanden sind, die M enthalten, schneidet aus M eine lineare Schar von der Dimension $r - t$ aus.

7.3. Die Dimension r einer linearen Schar ist gleich der Anzahl der willkürlichen Punkte, durch die ein Element der Schar bestimmt ist.

7.4. Eine lineare Schar ist ein irreduzibles System von Mannigfaltigkeiten: ihre Elemente gehen durch relationstreue Spezialisierung aus einem allgemeinen Element C_λ der Schar, dessen Parameter λ Unbestimmte sind, hervor.

7.5. Das allgemeine Element einer linearen Schar ohne feste Bestandteile ist relativ zum Körper $\Omega(\lambda)$ irreduzibel. Es ist gleichgültig, ob man zuerst ein allgemeines Element C_λ der Schar und darauf einen allgemeinen Punkt ξ bestimmt oder zuerst einen allgemeinen Punkt ξ von M nimmt und durch ihn ein möglichst allgemeines C_λ der Schar legt: beide Male erhält man bis auf Isomorphie dasselbe Paar (λ, ξ). Die durch das allgemeine Paar (λ, ξ) bestimmte irreduzible Korrespondenz zwischen dem Parameterraum S_r und der Mannigfaltigkeit M ordnet jedem speziellen Punkt λ' von S_r genau die Punkte von $C_{\lambda'}$ zu.

7.6. Jede lineare Schar ohne feste Bestandteile definiert eine rationale Abbildung von M auf eine Bildmannigfaltigkeit M' in S_r, wobei den Mannigfaltigkeiten C_λ die hyperebenen Schnitte von M' in folgendem präzisen Sinne entsprechen: Liegt ein Punkt P auf C_λ, so liegt mindestens einer seiner Bildpunkte P' in der entsprechenden Hyperebene; liegt umgekehrt einer der Bildpunkte P' in dieser Hyperebene, so liegt P auf C_λ. Die *Fundamentalpunkte* der Abbildung, d. h. die Punkte P, die mehr als endlich viele Bildpunkte P' haben, sind die Basispunkte der Schar. Einem s-fachen Punkt von M, der nicht Fundamentalpunkt ist, entsprechen höchstens s Bildpunkte, insbesondere einem einfachen Punkt nur ein Bildpunkt.

7.7. Ist M rational auf M' abgebildet, so entspricht jeder linearen Schar ohne feste Bestandteile auf M' eindeutig eine ebensolche Schar auf M. Ist die Abbildung birational, so ist das Entsprechen eineindeutig.

7.8. Die Elemente einer effektiven linearen Schar von der Dimension r, die einen gegebenen einfachen Punkt P von M enthalten, bilden, sofern P nicht Basispunkt der Schar ist, eine lineare Teilschar von der Dimension $r - 1$. Hält man k einfache Punkte $P_1, \ldots, P_k$ fest $(k \leqq r)$, so erhält man eine Teilschar von der Dimension r' mit

$$r - k \leqq r' \leqq r .$$

Hält man irgendwelche $(d - 1)$-dimensionale Teilmannigfaltigkeiten fest, die nicht aus lauter einfachen Punkten von M bestehen, so bilden diejenigen Elemente der linearen Schar, die diese Teilmannigfaltigkeiten mit vorgegebener Vielfachheiten enthalten, falls es solche Elemente überhaupt gibt, eine lineare Teilschar.

7.9. Wenn zwei lineare Scharen ein Element gemeinsam haben, so sind sie in einer beide umfassenden linearen Schar enthalten. Sind beide Scharen effektiv, und ist die Mannigfaltigkeit M glatt, so ist die umfassende Schar auch effektiv.

§ 8. Divisoren und lineare Scharen auf glatten Mannigfaltigkeiten

Beweise: Einf. alg. Geom. § 46—47, sowie Chow, Math. Ann. 114, p. 655. Von jetzt an sei M eine glatte, irreduzible d-dimensionale Mannigfaltigkeit.

8.1. Werden irgendwelche $(d - 1)$-dimensionale Teilmannigfaltigkeiten von M mit beliebigen ganzzahligen Vielfachheiten versehen, so bilden sie einen *Divisor*, und wenn alle Vielfachheiten positiv sind, einen *ganzen Divisor*.

8.2. Sind $(r + 1)$ linear unabhängige Divisoren einer r-dimensionalen Schar ganz, so ist die Schar effektiv, d. h. alle ihre Divisoren sind ganz.

8.3. Jede effektive lineare Schar läßt sich zu einer eindeutig bestimmten *Vollschar* erweitern, die sich effektiv nicht mehr erweitern läßt.

8.4. Zwei Divisoren C, D heißen *äquivalent*, wenn sie in einer linearen Schar enthalten sind. Die Äquivalenz ist symmetrisch, reflexiv und transitiv. Die zu einem Divisor äquivalenten Divisoren bilden eine *Divisorenklasse*. Die ganzen Divisoren einer Klasse bilden, falls es sie gibt, eine Vollschar.

8.5. Durch Zusammenfassen der irreduziblen Bestandteile zweier Divisoren und Addition ihrer Vielfachheiten bildet man ihr *Produkt*. Die Divisoren bilden bei der Multiplikation eine abelsche Gruppe. Die zum leeren Divisor oder *Einsdivisor* äquivalenten Divisoren bilden darin eine Untergruppe: die *Einsklasse*. Die Faktorengruppe nach der Einsklasse ist die *Divisorenklassengruppe*.

8.6. Über die *Sätze von Bertini* siehe ZAG 10, Math. Ann. 113, p. 705, sowie Chow, Math. Ann. 114, p. 664.

8.7. Der zur Mannigfaltigkeit M gehörige *Funktionenkörper* besteht aus allen rationalen Funktionen

$$\varphi = \frac{f(\xi)}{g(\xi)} \, ,$$

wo ξ ein allgemeiner Punkt von M ist, während f und g Formen gleichen Grades sind. Der *Divisor* einer solchen Funktion φ ist der Quotient der von den Hyperflächen $f = 0$ und $g = 0$ auf M ausgeschnittenen Divisoren.

§ 9. Divisoren auf einer algebraischen Kurve

Beweise : W.-L. Chow, Math. Ann. 114, p. 655, oder Einf. alg. Geom. § 45. Von jetzt an sei der Konstantenkörper Ω vollkommen.

9.1. Jede irreduzible Kurve Γ kann durch birationale Transformation in eine glatte Kurve Γ' verwandelt werden. Jedem Punkt von Γ' entspricht dabei ein einziger Punkt von Γ, einem Punkt von Γ können aber mehrere Punkte von Γ' entsprechen.

9.2. Eine *Stelle* von Γ ist ein Punkt von Γ'. Ist Γ'' eine zweite glatte birationale Transformierte von Γ, so entsprechen sich die Punkte von Γ' und Γ'' eineindeutig ; der Begriff der Stelle ist also von der Wahl von Γ' unabhängig.

9.3. Ein *Divisor* von Γ ist ein Divisor von Γ'. Auch die Begriffe Äquivalenz, Divisorenklasse, Vollschar usw. werden auf Γ' definiert. Alle diese Begriffe sind von der Wahl von Γ' unabhängig.

9.4. Die Vielfachheit eines Schnittpunktes P von Γ mit einer Hyperfläche H ist eine Summe von Beiträgen der einzelnen P entsprechenden Stellen P' von Γ', die so definiert werden : Man bette H ein in die lineare Schar aller Hyperflächen gleichen Grades, deren allgemeines Element H_λ sei. Diese lineare Schar schneidet aus Γ eine lineare Schar von Punktgruppen aus, der auf Γ' wieder eine lineare Schar entspricht. Bei der Spezialisierung $H_\lambda \to H$ erhält man nicht nur auf Γ, sondern auch auf Γ' durch relationstreue Spezialisierung eine ganz bestimmte Punktgruppe, in der jeder Punkt P' mit einer gewissen Vielfachheit vorkommt. Diese heißt die *Schnittmultiplizität von H und Γ für die Stelle P'.*

9.5. Jede rationale Funktion

$$\varphi = \frac{f(\xi)}{g(\xi)}$$

im Sinne von 8.7 hat an jeder Stelle P' eine gewisse *Ordnung*, die als Differenz der Schnittvielfachheiten des Zählers und Nenners für die Stelle P' definiert wird. Ist die Ordnung positiv, so hat man eine *Nullstelle*, ist sie negativ, einen *Pol* der Funktion φ. Die Nullstellen und Pole, mit ihren Ordnungen als Vielfachheiten versehen, bilden den *Divisor* der Funktion φ im Sinne von 8.7.

9.6. Die Summe der Ordnungen einer Funktion φ ist Null.

9.7. Es gibt zu jeder Stelle eine Ortsuniformisierende τ, d. h. eine Funktion der Ordnung 1. Alle anderen Funktionen auf Γ, insbesondere die Koordinatenverhältnisse $\xi_i : \xi_0$, können in Potenzreihen nach τ entwickelt werden.

9.8. Umgekehrt gehört zu jedem System von Potenzreihenentwicklungen eines allgemeinen Punktes von Γ

$$\xi_k = a_{k0} + a_{k1}\,\tau + a_{k2}\,\tau^2 + \cdots$$

eine Stelle von Γ.

Über den Zusammenhang des Stellenbegriffs mit dem Bewertungsbegriff siehe § 11.

§ 10. Differentialklasse, adjungierte Kurven und Riemann-Rochscher Satz

Beweise : Einf. alg. Geom. § 48—51.

10.1. Sind φ und η rationale Funktionen auf einer Kurve Γ, so heißt der Ausdruck $\varphi d\eta$ ein *Differential* des Funktionenkörpers. Die Gleichheit $\varphi d\eta = \psi d\zeta$ bedeutet

$$\frac{d\eta}{d\zeta} = \frac{\psi}{\varphi}\,.$$

Ist τ Ortsuniformisierende einer Stelle P' und setzt man $\varphi d\eta = \chi d\tau$, so versteht man unter der *Ordnung* des Differentials an der Stelle P' die Ordnung von χ in P'. Ist sie positiv, so hat man eine *Nullstelle*, ist sie negativ, einen *Pol* des Differentials. Die Nullstellen und Pole, mit ihren Ordnungen als Vielfachheiten versehen, bilden den *Divisor* des Differentials. Die Divisoren aller Differentiale bilden eine *Divisorenklasse W*. Ein Differential ohne Pole heißt *Differential erster Gattung*.

10.2. Eine ebene Kurve Γ habe die Gleichung $f = 0$. Die Polare eines Punktes Q wird durch

$$\sum_0^2 q_k \partial_k f = 0 \quad \left(\partial_k = \frac{\partial}{\partial x_k}\right)$$

definiert. Sie schneide die Kurve Γ in einem Punkte P, und die Schnittmultiplizität an einer zu P gehörigen Stelle P' sei ν'. Die Verbindungsgerade PQ schneide Γ an der Stelle P' mit der Multiplizität κ'. Dann ist

$$\delta' = \nu' - (\kappa' - 1)$$

unabhängig von der Wahl von Q und nicht negativ. Der Divisor, der aus allen Stellen P' mit den Vielfachheiten δ' besteht, heißt der *Doppelpunktsdivisor* von Γ. Eine Kurve $g = 0$ heißt zu Γ *adjungiert*, wenn ihr Schnitt mit Γ durch den Doppelpunktsdivisor teilbar ist.

10.3. *Satz vom Doppelpunktsdivisor.* Wenn eine Kurve $g = 0$ aus Γ den Divisor G ausschneidet und wenn eine adjungierte Kurve $F = 0$ aus Γ mindestens den Divisor DG ausschneidet, wo D der Doppelpunktsdivisor von Γ ist, so gilt eine Identität

$$F = Af + Bg \; ,$$

wobei die Kurve $B = 0$ zu Γ adjungiert ist.

10.4. Die adjungierten Kurven irgendeines Grades schneiden aus Γ außer dem Doppelpunktsdivisor eine Vollschar aus. Insbesondere schneiden die adjungierten Kurven $(n - 3)$-Ordnung die zur Differentialschar W gehörige Vollschar aus.

10.5. Die Anzahl der linear unabhängigen Differentiale erster Gattung, oder die um 1 vermehrte Dimension der zu W gehörigen Vollschar, heißt das *Geschlecht g* von Γ. Ist $g = 0$, so ist Γ rational. Die Summe der Ordnungen eines Differentials ist $2g - 2$.

10.6. *Reduktionssatz.* Es sei C ein ganzer Divisor und P eine Stelle. Wenn es einen ganzen Divisor der Differentialklasse gibt, der C, aber nicht CP enthält, dann ist P ein fester Punkt der Vollschar $|\,C\,|$, und umgekehrt.

10.7. *Riemann-Rochscher Satz.* Ist $\dim C$ die *Dimension* einer Divisorenklasse C, d. h. die um 1 verminderte Anzahl der linear unabhängigen ganzen Divisoren der Klasse, und ist c der *Grad* der Klasse, d. h. die Summe der Vielfachheiten aller Punkte eines Divisors der Klasse, so gilt

$$\dim C = c - g + i \; ,$$

wobei der *Spezialitätsindex* $i \geqq 0$ durch

$$i = 1 + \dim \frac{W}{C}$$

definiert ist. Ist $i > 0$, so heißt die Klasse und jede ihrer Divisoren *spezial*. Ist $\dim C \geqq g$ oder $c > 2g - 2$, dann ist $i = 0$ und die Klasse ist nicht spezial.

Zweiter Teil

Anwendung der algebraischen Geometrie auf die arithmetische Theorie der algebraischen Funktionenkörper

§ 11. Die Kurve Γ

Wie Hasse gehen wir von einem algebraischen Funktionenkörper $K = \Omega(x,y)$ vom Geschlechte g mit $f(x,y) = 0$ aus. Hasse definiert nun (§ 3) mit Hilfe einer Divisorenklasse M dieses Funktionenkörpers neue homogene Erzeugende $x_0 : \ldots : x_n$, leitet ein System (G) von homogenen algebraischen Gleichungen zwischen $x_0, \ldots, x_n$ her und bemerkt dazu: In der Ausdrucksweise der algebraischen Geometrie ist der Übergang von der ursprünglich gegebenen Erzeugung $K = \Omega(x,y)$ mit $f(x,y) = 0$ zu einer Erzeugung $K = \Omega(x_0 : \ldots : x_n)$ mit (G) eine birationale Transformation der gegebenen zweidimensionalen affinen Kurve in eine n-dimensionale projektive Kurve mit lauter einfachen Punkten. In der Tat kann man in Anschluß an die Hasseschen Ausführungen beweisen, daß die Gleichungen (G) eine irreduzible algebraische Kurve Γ definieren, deren allgemeiner Punkt die Koordinaten $x_0, \ldots, x_n$ hat, daß diese Kurve zur Kurve $f(x,y) = 0$ birational äquivalent ist, daß die Schnittpunktgruppen von Γ mit den Hyperebenen des Raumes S_n den ganzen Divisoren der Klasse M entsprechen und daß die Kurve Γ lauter einfache Punkte hat.

Wesentlich einfacher werden aber die Beweise, wenn man nicht erst am Schluß der Rechnungen, sondern gleich am Anfang die Begriffe und Ausdrucksweise der algebraischen Geometrie einführt. Man gehe zunächst von der affinen zur projektiven Ebene S_2 über, indem man die Gleichung $f(x,y) = 0$ homogen macht. Die so erhaltene Kurve in S_2 kann nach 9.1 durch birationale Transformation in eine glatte Kurve Γ in einem projektiven Raum S_n verwandelt werden. Geht man den Beweis des Satzes 9.1 noch einmal durch, so sieht man leicht, daß die Dimension des Raumes S_n beliebig groß gewählt werden kann; wir können also $n \geq 2g$ annehmen. Die Hyperebenen des Raumes S_n schneiden aus der Kurve Γ eine lineare Schar von Punktgruppen aus, von der wir annehmen können, daß sie eine Vollschar ist. Der Existenzbeweis der glatten Kurve Γ kann nämlich leicht so geführt werden, daß dabei nur Vollscharen verwendet werden. Man kann auch nachträglich die die Transformation vermittelnde lineare Schar zu einer Vollschar erweitern.

Ein allgemeiner Punkt von Γ habe die Koordinaten $\xi_0, \ldots, \xi_n$, wobei wir $\xi_0 = 1$ normieren können: dann sind $\xi_0, \ldots, \xi_n$ sämtlich Elemente

336

des Funktionenkörpers K. Sie entsprechen den Hasseschen Erzeugenden $x_0, \ldots x_n$.

Ist $\bar{\mathfrak{p}}$ ein „algebraischer Punkt" von K/Ω im Sinne von Hasse, d. h. ein Primdivisor des Körpers $\bar{K} = \bar{\Omega}(x, y)$ oder ein „Punkt" im Sinne von Dedekind und Weber[1]), wo $\bar{\Omega}$ der zu Ω gehörige algebraisch abgeschlossene Körper ist, und ist ξ_j eine von den Koordinaten ξ_k, die die kleinste Ordnung an der Stelle $\bar{\mathfrak{p}}$ haben, so bleiben alle $\xi_j^{-1}\xi_k$ endlich an der Stelle $\bar{\mathfrak{p}}$, und durch

$$\bar{p}_k = \xi_j^{-1}\xi_k(\bar{\mathfrak{p}})$$

ist ein Punkt $\bar{p}$ mit den Koordinaten $\bar{p}_0, \ldots, \bar{p}_n$ definiert, die alle endlich und nicht alle Null sind. Alle homogenen Gleichungen $F(\xi) = 0$, die für die ξ_k gelten, gelten auch für $\xi_j^{-1}\xi_k$, also auch für die $\bar{p}_k$; also ist $\bar{p}$ ein Punkt der Kurve. Wir beweisen nun:

11.1. (Vgl. Hasse § 3, 2.) *Verschiedene „algebraische Punkte"* $\bar{\mathfrak{p}}, \bar{\mathfrak{q}}$ *ergeben auch verschiedene Punkte* $\bar{p}, \bar{q}$ *von Γ.*

11.2. (Vgl. Hasse § 3, 5.) *Alle Punkte $\bar{p}$ der Kurve Γ mit Koordinaten aus $\bar{\Omega}$ können in dieser Weise erhalten werden.*

Beweis von **11.1.** Gesetzt, zwei verschiedene Primdivisoren $\bar{\mathfrak{p}}, \bar{\mathfrak{q}}$ würden denselben Punkt $\bar{p}(\bar{p}_0, \ldots, \bar{p}_n)$ ergeben. Ohne Beschränkung der Allgemeinheit können wir $\bar{p}_0 = 1$ und $\xi_0 = 1$ annehmen. Es sei ξ_{n+1} eine Funktion des Körpers, die für $\bar{\mathfrak{p}}$ und $\bar{\mathfrak{q}}$ verschiedene endliche Werte $\xi_{n+1}(\bar{\mathfrak{p}})$ und $\xi_{n+1}(\bar{\mathfrak{q}})$ annimmt. Durch die homogenen Koordinaten $(\xi_0, \xi_1, \ldots, \xi_n, \xi_{n+1})$ ist ein allgemeiner Punkt ξ^* einer Kurve Γ^* in S_{n+1} definiert (2.2). Da die Koordinaten von ξ^* rationale Funktionen von denen von ξ sind und umgekehrt, so ist Γ^* birational auf Γ abgebildet, und zwar wird der einem Punkt $p^*(\bar{p}_0, \bar{p}_1, \ldots, \bar{p}_n, \bar{p}_{n+1})$ von Γ^* entsprechende Punkt $\bar{p}(\bar{p}_0, \ldots, \bar{p}_n)$ von Γ gefunden, indem die letzte Koordinate einfach weggelassen wird. Zu jedem Primdivisor $\bar{\mathfrak{p}}$ oder $\bar{\mathfrak{q}}$ des Körpers $\bar{K}$ gehört nach der oben angegebenen Vorschrift nicht nur ein Punkt von Γ, sondern auch ein Punkt von Γ^*, und zwar gehören zu $\bar{\mathfrak{p}}$ und $\bar{\mathfrak{q}}$ zwei verschiedene Punkte von Γ^*, da $\xi_{n+1}(\bar{\mathfrak{p}}) \neq \xi_{n+1}(\bar{\mathfrak{q}})$ sein sollte. Diese zwei Punkte von Γ^* stimmen aber in allen Koordinaten außer der letzten überein, daher entspricht ihnen in der birationalen Abbildung ein und derselbe Punkt $\bar{p}$ von Γ. Einem einfachen Punkt $\bar{p}$ von Γ kann aber in einer birationalen Abbildung nur ein Punkt von Γ^* entsprechen (7.6).

---

[1]) *Dedekind und Weber*, Crelle's Journal 92, (1882) p. 181.

Beweis von 11.2. Zu einem Punkt $\bar{p}$ von Γ gehört eine Bewertung des Funktionenkörpers $\overline{K}$, die folgendermaßen definiert wird: Jede Funktion des Körpers kann als Quotient von Formen gleichen Grades in $\xi_0, \ldots, \xi_n$ geschrieben werden:

$$\varphi = \frac{f(\xi_0, \ldots, \xi_n)}{g(\xi_0, \ldots, \xi_n)} \, .$$

Die Flächen $f = 0$ und $g = 0$ schneiden Γ in $\bar{p}$ mit bestimmten Multiplizitäten μ und ν. Dann wird die Bewertung durch

$$w(\varphi) = \mu - \nu$$

definiert. Die Eindeutigkeit der Definition und die Eigenschaft $w(\varphi\psi) = w(\varphi) + w(\psi)$ sind klar. Die Eigenschaft

$$w(\varphi + \psi) \geqq \mathrm{Min}\,\big(w(\varphi), w(\psi)\big)$$

kommt darauf hinaus, daß die Schnittmultiplizität von $f_1 + f_2$ oder allgemeiner von $\lambda_1 f_1 + \lambda_2 f_2$ mit Γ mindestens gleich der kleineren der Schnittmultiplizitäten von f_1 und f_2 mit Γ ist. Dies aber folgt aus 7.8.

Diese Bewertung definiert bekanntlich einen Primdivisor des Körpers $\overline{K}$ im Sinne von Dedekind und Weber. Jedes Element φ des Bewertungsringes ist nämlich modulo dem Bewertungsideal einer Konstanten ω aus $\overline{\Omega}$ kongruent, und die Zuordnung $\varphi \to \omega$ ist ein „Punkt" $\bar{\mathfrak{p}}$ im Dedekind-Weberschen Sinne. Wir schreiben $\varphi(\bar{\mathfrak{p}}) = \omega$.

Nimmt man wieder $\bar{p}_0 = 1$ und $\xi_0 = 1$ an, so haben die Funktionen $\xi_1 - \bar{p}_1, \ldots, \xi_n - \bar{p}_n$ in der eben definierten Bewertung positive Ordnungszahlen, also gehören sie dem Bewertungsideal an, mithin ist

$$\xi_0(\bar{\mathfrak{p}}) = \bar{p}_0,\, \xi_1(\bar{\mathfrak{p}}) = \bar{p}_1, \ldots, \xi_n(\bar{\mathfrak{p}}) = \bar{p}_n \, .$$

Das heißt aber: der dem „algebraischen Punkt" $\bar{\mathfrak{p}}$ entsprechende Kurvenpunkt ist genau der Punkt $\bar{p}$, von dem wir ausgegangen sind. Damit ist 11.2 bewiesen.

Nach 11.1 und 11.2 entsprechen die „algebraischen Punkte" $\bar{\mathfrak{p}}$ des Körpers K eineindeutig den Punkten $\bar{p}$ der Kurve Γ mit Koordinaten aus $\overline{\Omega}$. Wir brauchen daher von jetzt an zwischen $\bar{p}$ und $\bar{\mathfrak{p}}$ nicht mehr zu unterscheiden. Wir bezeichnen mit Hasse $\bar{p}_0, \ldots, \bar{p}_n$ als die homogenen Koordinaten des Punktes $\bar{\mathfrak{p}}$.

Als weitere Vereinfachung lassen wir von jetzt an die Querstriche, die bei Hasse zum Ausdruck bringen sollen, daß die Koordinaten $\bar{p}_k$ dem Körper $\overline{\Omega}$ angehören sollen, weg. Wir bezeichnen die Punkte von Γ also einfach mit P, p oder $\mathfrak{p}$, ihre Koordinaten mit p_k, die Punktgruppen mit $\mathfrak{A}, \mathfrak{O}$, usw.

§ 12. Die Chow-Koordinaten eines Divisors

Ein ganzer Divisor $\mathfrak{A} = \mathfrak{p}^{(1)}\mathfrak{p}^{(2)}\ldots\mathfrak{p}^{(h)}$, d. h. eine Punktgruppe auf Γ, kann nach 5.5 eindeutig charakterisiert werden durch die zugeordnete Form

$$A(u) = P^{(1)}(u)\,P^{(2)}(u)\ldots P^{(h)}(u)\;,$$

wo

$$P^{(\nu)}(u) = p_0^{(\nu)}u_0 + p_1^{(\nu)}u_1 + \cdots + p_n^{(\nu)}u_n\;.$$

Die Koeffizienten der Form $A(u)$ nennen wir die *Chow-Koordinaten* des Divisors $\mathfrak{A}$, weil Chow als erster die zugeordneten Formen systematisch als Beweismittel in die algebraische Geometrie eingeführt hat.

Hasse definiert die Koordinaten von $\mathfrak{A}$ etwas komplizierter. Für jeden Punkt $\mathfrak{p}^{(\nu)}$ werden die mit Unbestimmten $x_0,\ldots,x_n$ gebildeten Ausdrücke $p_j x_i - p_i x_j$ irgendwie als $P_0(x),\ldots,P_r(x)$ durchnumeriert und dann die Form

$$F^{(\nu)}(x,t) = P_0(x)t^r + P_1(x)t^{r-1} + \cdots + P_r(x)$$

gebildet. Multiplikation dieser Formen ergibt eine Form $\Phi(x,t)$, deren Koeffizienten die Hasse-Koordinaten von $\mathfrak{A}$ sind.

Die Beziehung zwischen den Hasse-Koordinaten und den Chow-Koordinaten ist leicht zu finden. Die Form $F^{(\nu)}(x,t)$ ist nämlich linear in $p_0,\ldots,p_n$ und entsteht folglich aus der Linearform

$$P^{(\nu)}(u) = p_0^{(\nu)}u_0 + \cdots + p_n^{(\nu)}u_n\;,$$

indem für die u_j gewisse ganzzahlige Polynome in den x und t eingesetzt werden. Folglich entsteht auch das Produkt $\Phi(x,t)$ aus dem Produkt $A(u)$ durch dieselbe Substitution. Somit sind die Koeffizienten von $\Phi(x,t)$ gewisse ganzzahlige Linearkombinationen der Koeffizienten von $A(u)$, m. a. W. die Hasse-Koordinaten sind ganzzahlige Linearkombinationen der Chow-Koordinaten.

Umgekehrt sind bei Charakteristik Null nach *Franz* (Hasse, § 3, 3) alle symmetrischen Funktionen der Koordinaten von $\mathfrak{p}^{(1)},\ldots,\mathfrak{p}^{(h)}$, also insbesondere die Chow-Koordinaten, homogene Polynome in den Hasse-Koordinaten.

Wir werden im folgenden nur mit den Chow-Koordinaten arbeiten, weil sie einfacher definiert, frei von Willkür und leichter zu handhaben sind.

§ 13. Die Mannigfaltigkeit aller Gruppen von g Punkten auf $\varGamma$

Es sei g das Geschlecht der Kurve $\varGamma$. Wenn im folgenden von *Punkt-gruppen* die Rede ist, so sind damit immer ganze Divisoren

$$\mathfrak{A} = \mathfrak{p}^{(1)}\mathfrak{p}^{(2)}\ldots,\mathfrak{p}^{(g)}$$

vom Grade g auf $\varGamma$ gemeint. Einer solchen Punktgruppe entspricht nach § 12 eine zugeordnete Form

$$A(u) = \lambda P^{(1)}(u)P^{(2)}(u)\ldots P^{(g)}(u) \ , \tag{1}$$

deren Koeffizienten die Chow-Koordinaten $a_0,\ldots,a_h$ von $\mathfrak{A}$ sind. Faßt man sie als Koordinaten eines Bildpunktes A in einem h-dimensionalen Bildraum S_h auf, so erhält man eine Abbildung der Gesamtheit aller Punktgruppen $\mathfrak{A}$ auf eine Gesamtheit von Bildpunkten in S_h. Wir wollen beweisen, daß diese Gesamtheit eine algebraische Mannigfaltigkeit M ist.

Vergleicht man in (1) links und rechts die Koeffizienten der Potenz-produkte der u, so erhält man Gleichungen der Form

$$a_j = \lambda g_j(p^{(1)},\ldots,p^{(g)}) \ , \tag{2}$$

die durch Elimination von λ homogen werden :

$$a_j g_k(p^{(1)},\ldots,p^{(g)}) - a_k g_j(p^{(1)},\ldots,p^{(g)}) = 0 \ . \tag{3}$$

Dazu kommen die Gleichungen, die ausdrücken, daß die Punkte $p^{(\nu)}$ auf $\varGamma$ liegen :

$$f_i(p^{(\nu)}) = 0 \ . \tag{4}$$

Aus den homogenen Gleichungen (3), (4) eliminieren wir $p^{(1)},\ldots,p^{(g)}$ durch Bildung des Resultantensystems

$$R_j(a_0,\ldots,a_h) = 0 \ . \tag{5}$$

Die Gleichungen (5) sind notwendig und hinreichend, damit $a_0,\ldots,a_k$ die Koordinaten einer Punktgruppe $\mathfrak{A}$ auf $\varGamma$ sind. Also bilden diese Punkt-gruppen eine algebraische Mannigfaltigkeit M.

Diese Mannigfaltigkeit M ist irreduzibel; denn alle Punktgruppen $\mathfrak{A} = \mathfrak{p}^{(1)}\ldots\mathfrak{p}^{(g)}$ entstehen durch relationstreue Spezialisierung aus einer allgemeinen Punktgruppe $\mathfrak{X} = \varPi^{(1)}\ldots\varPi^{(g)}$, wo $\varPi^{(1)},\ldots,\varPi^{(g)}$ unabhängige allgemeine Punkte von $\varGamma$ sind.

Wir wollen nun beweisen, daß die Mannigfaltigkeit M glatt ist, d. h. lauter einfache Punkte hat. Dazu dient das folgende

Kriterium. Wenn die Koordinaten $1, \xi_1, \ldots, \xi_n$ *eines allgemeinen Punktes einer g-dimensionalen Mannigfaltigkeit* M *separable Funktionen von den g algebraisch unabhängigen* $\xi_1, \ldots, \xi_g$ *sind und wenn von den verschiedenen konjugierten Punkten* $\xi^{(\nu)}$ *mit Koordinaten* $(1, \xi_1, \ldots, \xi_g, \xi^{(\nu)}_{g+1}, \ldots, \xi^{(\nu)}_n)$, *die zu den gegebenen* $\xi_1, \ldots, \xi_g$ *gehören, bei der relationstreuen Spezialisierung* $\xi^{(\nu)} \to \eta^{(\nu)}$ *nur einer in den Punkt* $\eta (1, \eta_1, \ldots, \eta_n)$ *übergeht, dann ist* η *ein einfacher Punkt von* M.

Beweis: Die Punkte $\xi^{(\nu)}$ sind die Schnittpunkte von M mit den Hyperebenen

$$x_k - x_0 \xi_k = 0 \qquad (k = 1, \ldots, g) . \tag{6}$$

Da jedes ξ_{g+i} eine separable Funktion von $\xi_1, \ldots, \xi_g$ ist, so gilt für jedes von ihnen eine Gleichung

$$f_i(\xi_1, \ldots, \xi_g, \xi_{g+i}) = 0 \tag{7}$$

mit

$$\partial_{g+i} f_i(\xi_1, \ldots, \xi_g, \xi_{g+i}) \neq 0 .$$

Die Gleichung (7) kann durch Einführung von ξ_0 homogen gemacht werden. Sie gilt für den allgemeinen Punkt, also für alle Punkte von M, d. h. die Hyperfläche $f_i = 0$ enthält M. Die Tangentialebene dieser Hyperfläche im Punkte ξ hat nach 1.8 die Gleichung

$$x_0 \partial_0 f_i(\xi) + x_1 \partial_1 f_i(\xi) + \cdots + x_g \partial_g f_i(\xi) + x_{g+i} \partial_{g+i} f_i(\xi) = 0 . \tag{8}$$

Die Gleichungen (6) und (8) bestimmen die Verhältnisse

$$x_0 : x_1 : \cdots : x_g : x_{g+1} : \cdots : x_n$$

eindeutig, denn wenn x_0 willkürlich angenommen wird, so bestimmen sich $x_1, \ldots, x_g$ aus (6), die übrigen x_{g+i} aus (8). Also zählt (nach Kriterium 1.8) der Punkt ξ als Schnittpunkt von M mit den Hyperebenen (6) einfach, und dasselbe gilt von den konjugierten Punkten $\xi^{(\nu)}$. Spezialisiert man also ein System von r allgemeinen Hyperebenen zunächst zu den Hyperebenen (6), so gehen die Schnittpunkte von M mit den allgemeinen Hyperebenen in die nur einmal gezählten Punkte $\xi^{(\nu)}$ über. Spezialisiert man nun weiter $\xi_j \to \eta_j$ (j = 1, \ldots, r), so entsteht der Punkt η bei dieser Spezialisierung nach Voraussetzung ebenfalls nur einmal. Also ist

η ein einfacher Schnittpunkt von M mit den Hyperebenen $x_j - x_0\eta_j = 0$ und somit ein einfacher Punkt von M.

Als Anwendung dieses Kriteriums beweisen wir nun, daß die Mannigfaltigkeit M der Punktgruppen $\mathfrak{A}$ lauter einfache Punkte hat.

Es sei $\mathfrak{A} = p^{(1)} \ldots p^{(g)}$ eine solche Punktgruppe. Die Punkte $p^{(1)}, \ldots, p^{(g)}$ mögen alle im endlichen liegen ($p_0^{(i)} \neq 0$) und es möge $p_0^{(i)} = 1$ gewählt werden ($i = 1, \ldots, g$). Die Tangenten von Γ in diesen Punkten mögen die uneigentliche Hyperebene $x_0 = 0$ in den Punkten $q^{(1)}, \ldots, q^{(g)}$ schneiden und die Verbindungslinien $p^{(i)}p^{(j)}$ möge dieselbe Hyperebene in $q^{(ij)}$ schneiden. Wir wählen die Koordinatenebene $x_1 = 0$ so, daß sie keinen der Punkte $q^{(i)}$, $q^{(ij)}$ und keinen der uneigentlichen Punkte von Γ enthält. Dann werden die Hyperebenen

$$x_1 = p_1^{(1)} x_0, \ldots, x_1 = p_1^{(g)} x_0 \ ,$$

die parallel zur Koordinatenebene $x_1 = 0$ sind, ebenfalls die Punkte $q^{(1)}, \ldots, q^{(g)}$ nicht enthalten, also werden sie auch die Tangenten von Γ in den Punkten $p^{(1)}, \ldots, p^{(g)}$ nicht enthalten, d. h. sie berühren Γ in diesen Punkten nicht. Auch enthält keine von ihnen zwei verschiedene Punkte $p^{(i)}$, $p^{(j)}$, denn sonst müßte deren Verbindungsgerade und somit auch der Punkt $q^{(ij)}$ in der betreffenden Ebene liegen.

Die Koordinaten von $\mathfrak{A}$ sind die Koeffizienten der Form (1), wobei wir $\lambda = 1$ annehmen dürfen. Unter ihnen heben wir besonders die Koeffizienten a_i von $u_0^{g-i} u_1^i$ hervor. Setzen wir in (1) $u_2 = \cdots = u_n = 0$, so bleiben nur diese Glieder übrig und wir erhalten

$$\sum_i a_i u_0^{g-i} u_1^i = (u_0 + p_1^{(1)} u_1)(u_0 + p_1^{(2)} u_1) \ldots (u_0 + p_1^{(g)} u_1) \ .$$

Folglich ist $a_0 = 1$ und $a_1, \ldots, a_g$ sind die elementaren symmetrischen Funktionen der ersten Koordinaten $p_1^{(1)}, \ldots, p_1^{(g)}$ der Punkte $p^{(1)}, \ldots, p^{(g)}$.

Dies gilt für jede Punktgruppe, also auch dann, wenn die Punkte $p^{(1)}, \ldots, p^{(g)}$ durch ebensoviele allgemeine Punkte $\Pi^{(1)}, \ldots, \Pi^{(g)}$ von Γ ersetzt werden, deren erste Koordinaten $\Pi_1^{(1)}, \ldots, \Pi_1^{(g)}$ unabhängige Unbestimmte sind. Von den Koordinaten $\alpha_0, \ldots, \alpha_g$ dieser allgemeinen Punktgruppe ist $\alpha_0 = 1$, und $\alpha_1, \ldots, \alpha_g$ sind die elementar-symmetrischen Funktionen der Unbestimmten $\Pi_1^{(1)}, \ldots, \Pi_1^{(g)}$.

Nun ist klar, daß die Unbestimmten $\Pi_1^{(1)}, \ldots, \Pi_1^{(g)}$ separable Funktionen sind; denn sie sind Wurzeln einer Gleichung

$$\Pi^g - \Pi^{g-1} \alpha_1 + \Pi^{g-2} \alpha_2 - \cdots = 0$$

mit lauter verschiedenen, nämlich unbestimmten Wurzeln $\Pi_1^{(1)}, \ldots, \Pi_1^{(g)}$. Die übrigen Koordinaten $\Pi_2^{(\nu)}, \ldots, \Pi_n^{(\nu)}$ eines jeden dieser Punkte sind wiederum separable Funktionen der ersten Koordinaten $\Pi_1^{(\nu)}$, wie wir nachher beweisen werden. Also sind alle $\Pi_k^{(2)}$ separable Funktionen von $\alpha_1, \ldots, \alpha_r$. Dann sind aber auch alle Koordinaten der Punktgruppe $\Pi^{(1)}, \ldots, \Pi^{(g)}$ separable Funktionen von $\alpha_1, \ldots, \alpha_g$, d. h. die erste Bedingung des Kriteriums ist erfüllt.

Um auch die zweite Bedingung zu verifizieren, setzen wir die relationstreue Spezialisierung $\alpha_1, \ldots, \alpha_g \to a_1, \ldots, a_g$ zu einer relationstreuen Spezialisierung nicht nur sämtlicher Punkte $\Pi^{(1)}, \ldots, \Pi^{(g)}$, sondern auch sämtlicher konjugierter Schnittpunkte $\Pi^{(1\nu)}, \ldots, \Pi^{(g\nu)}$ der Hyperebenen $x_1 - x_0 \Pi_1^{(1)} = 0, \ldots, x_1 - x_0 \Pi_1^{(g)} = 0$ mit der Kurve Γ fort. Kombiniert man jeden der m Punkte $\Pi^{(1\nu)}$ mit jedem der m Punkte $\Pi^{(2\nu)}$, usw. bis $\Pi^{(g\nu)}$, so erhält man m^g Punktgruppen $(\Pi^{(1\lambda)}\Pi^{(2\mu)}, \ldots, \Pi^{(g\nu)})$. Die ersten $g + 1$ Koordinaten dieser Punktgruppen, nämlich die Koeffizienten der Potenzprodukte $u_0^{g-i} u_1^i$ in ihrer zugeordneten Form, sind alle gleich den elementarsymmetrischen Funktionen $1, \alpha_1, \ldots, \alpha_g$ der Koordinaten $\Pi_1^{(1)}, \ldots, \Pi_1^{(g)}$. Diese Punktgruppen sind auch die einzigen Punktgruppen von Γ, deren erste $g + 1$ Koordinaten diese Werte haben. Wir haben nun nachzuweisen, daß bei der Spezialisierung $\alpha_1 \to a_1, \ldots, \alpha_g \to a_g$ nur eine von diesen Punktgruppen in $\mathfrak{A} = p^{(1)} \ldots p^{(g)}$ übergeht.

Bei dieser Spezialisierung mögen die Punkte $\Pi^{(1\lambda)}$ in $p^{(1\lambda)}$ übergehen. Da die ersten Koordinaten $\Pi_1^{(1\lambda)}$ alle gleich $\Pi_1^{(1)}$ sind, so sind auch nach der Spezialisierung die ersten Koordinaten $p_1^{(1\lambda)}$ alle gleich $p_1^{(1)}$, entsprechend alle $p_1^{(2\mu)} = p_1^{(2)}, \ldots$ und alle $p_1^{(g\nu)} = p_1^{(g)}$. Die Punkte $\Pi^{(1\lambda)}$ sind die Schnittpunkte der Hyperebene $x_1 - x_0 \Pi_1^{(1)} = 0$ mit der Kurve Γ; die spezialisierten Punkte $p^{(1\lambda)}$ sind also die Schnittpunkte der Hyperebene $x_1 - x_0 p_1^{(1)} = 0$ mit Γ. Gesetzt nun, es würden bei der Spezialisierung zwei von diesen Punkten $\Pi^{(1\lambda)}$ in den einen Punkt $p^{(1)}$ hineinrücken, so müßte die spezialisierte Hyperebene die Kurve im Punkte $p^{(1)}$ berühren. Dies ist aber nicht der Fall, also rückt nur einer von den Punkten $\Pi^{(1\lambda)}$ in $p^{(1)}$ hinein, ebenso nur ein $\Pi^{(2\mu)}$ in $p^{(2)}$, usw. bis $p^{(g)}$. Daraus folgt leicht, daß nur eine von den m^g Punktgruppen $\Pi^{(1\lambda)} \Pi^{(2\mu)} \cdots \Pi^{(g\nu)}$, nämlich $\Pi^{(1)}\Pi^{(2)} \cdots \Pi^{(g)}$, bei der Spezialisierung in $p^{(1)} \cdots p^{(g)}$ übergeht.

Wir haben noch den Beweis nachzuholen, daß die Koordinaten $\Pi_1, \ldots, \Pi_n$ eines Punktes von Γ, dessen erste Koordinate Π_1 eine Unbestimmte ist, separable Funktionen von Π_1 sind. Zum Beweis betrachten wir die zugeordnete Form von

$$F(u,v) = \prod_{\nu} \left(u_0 \xi_0^{(\nu)} + u_1 \xi_1^{(\nu)} + \cdots + u_n \xi_n^{(\nu)} \right) , \tag{9}$$

wo $\xi^{(\nu)}$ die Schnittpunkte der Hyperebene v mit Γ sind. Setzen wir für v speziell die Hyperebene $x_1 - \Pi_1 x_0 = 0$, so liegen diese Schnittpunkte $\xi^{(\nu)}$ alle im Endlichen; wir können also $\xi_0^{(\nu)} = 1$ annehmen. $\xi_1^{(\nu)}$ ist dann $= \Pi_1$. Alle $\xi^{(\nu)}$ sind allgemeine Punkte von Γ, also gehen sie durch Isomorphismen auseinander hervor, d. h. sie sind konjugiert über $\Omega(\Pi_1)$. Daher haben alle Linearfaktoren in der Zerlegung (9) auch gleiche Vielfachheiten. Wären diese Vielfachheiten >1, so würden sie nach der Spezialisierung $\Pi_1 \to p_1$ auch noch >1 sein. Aber die Hyperebene $x_1 - p_1 x_0 = 0$ berührt die Kurve Γ im Punkt p nicht. Also sind die Vielfachheiten in der Zerlegung (9) alle gleich Eins. Betrachtet man die Form (9) als Polynom in u_0, so hat sie lauter einfache Nullstellen $u_0 = - u_1 \xi_1^{(\nu)}$ $- \cdots - u_n \xi_n^{(\nu)}$, darunter auch die Nullstelle $-u_1 \Pi_1 - \cdots - u_n \Pi_n$. Also ist $-u_1 \Pi_1 - \cdots - u_n \Pi_n$ eine separable Funktion von $\Pi_1, u_1, \ldots, u_n$. Ersetzt man ein u_k durch eine andere Unbestimmte v_k, so ist auch $-u_1 \Pi_1 - \cdots - v_k \Pi_k - \cdots - u_n \Pi_n$ separabel. Also ist auch die Differenz $(-u_k + v_k)\Pi_k$ und somit auch Π_k selber separabel über $K(\Pi_1, u_1, \ldots, u_n, v_k)$. Da aber in der irreduziblen Gleichung für Π_k die Unbestimmten $u_1, \ldots, u_n, v_k$ gar nicht vorkommen, so folgt, daß Π_k separabel über $K(\Pi_1)$ ist, was wir beweisen wollten.

§ 14. Algebraischer Ausdruck der Aequivalenz von Divisoren auf Γ

Divisorenklassen vom Grade Null heißen nach Hasse *Nullklassen*. Jede Nullklasse kann durch einen Quotienten $\mathfrak{D}^{-1}\mathfrak{P}$ von ganzen Divisoren g-ten Grades repräsentiert werden, wobei der Nenner $\mathfrak{D}$ sogar beliebig gewählt werden kann. Die Darstellung ist eindeutig, wenn $\mathfrak{P}$ *regulär*, d. h. nicht spezial ist. Dies alles folgt leicht aus dem Riemann-Rochschen Satz (10.7). Man kann bei gegebener Nullklasse A den Nenner $\mathfrak{D}$ immer so wählen, daß der Zähler $\mathfrak{P}$ regulär, also eindeutig bestimmt ist (vgl. Hasse § 4, 3).

Der Übergang zu einem anderen Nenner $\mathfrak{D}'$ wird durch die Äquivalenz

$$\mathfrak{D}^{-1}\mathfrak{P} \sim \mathfrak{D}'^{-1}\mathfrak{P}' \quad \text{oder} \quad \mathfrak{P}\mathfrak{D}' \sim \mathfrak{P}'\mathfrak{D} \tag{1}$$

vermittelt. *Wir wollen diese Äquivalenz durch algebraische Gleichungen zwischen den Koordinaten von $\mathfrak{P}$, $\mathfrak{D}$, $\mathfrak{P}'$, $\mathfrak{D}'$ ausdrücken.*

Durch die $2g$ Punkte der Punktgruppe $\mathfrak{P}\mathfrak{D}'$ kann man, da die Dimension n des Raumes S_n größer oder gleich $2g$ sein sollte, stets eine Hyperebene v legen. Diese schneide insgesamt eine Punktgruppe $\mathfrak{R}\mathfrak{P}\mathfrak{O}'$ aus Γ aus. Ist nun $\mathfrak{P}\mathfrak{O}' \sim \mathfrak{P}'\mathfrak{O}$, also $\mathfrak{R}\mathfrak{P}\mathfrak{O}' \sim \mathfrak{R}\mathfrak{P}'\mathfrak{O}$, so muß auch die Punktgruppe $\mathfrak{R}\mathfrak{P}'\mathfrak{O}$ durch eine Hyperebene w ausgeschnitten werden, da die Hyperebenen nach der in § 11 gemachten Voraussetzung eine Vollschar ausschneiden. Wenn umgekehrt die beiden Punktgruppen $\mathfrak{R}\mathfrak{P}\mathfrak{O}'$ und $\mathfrak{R}\mathfrak{P}'\mathfrak{O}$ je durch eine Hyperebene ausgeschnitten werden, so sind sie äquivalent und es folgt (1). Die Äquivalenz (1) besagt also genau: *Es gibt zwei Hyperebenen v und w und eine Punktgruppe $\mathfrak{R}$, so daß v die Punktgruppe $\mathfrak{R}\mathfrak{P}\mathfrak{O}'$ und w die Punktgruppe $\mathfrak{R}\mathfrak{P}'\mathfrak{O}$ auf Γ ausschneidet.*

Ist $\mathfrak{A}$ die volle Schnittgruppe einer Hyperebene v mit der Kurve Γ, so ist die zugeordnete Form $G(u,v)$ von $\mathfrak{A}$ nach 5.2 ganz rational und homogen in $v_0,\ldots,v_n$. Bezeichnen nun $P(u)$, $P'(u)$, $O(u)$, $O'(u)$, $R(u)$ die zugeordneten Formen der Punktgruppen $\mathfrak{P}$, $\mathfrak{P}'$, $\mathfrak{O}$, $\mathfrak{O}'$, $\mathfrak{R}$, so kann die eben kursivierte geometrische Beziehung algebraisch durch

$$\begin{aligned} G(u,v) &= \lambda R(u) P(u) O'(u) \\ G(u,w) &= \mu R(u) P'(u) O(u) \end{aligned} \quad \Big\} \tag{2}$$

ausgedrückt werden, wobei die Proportionalitätsfaktoren λ und μ nicht von den u abhängen dürfen.

Vergleicht man in (2) links und rechts die Koeffizienten der Potenzprodukte der u, so erhält man Gleichungen der Gestalt

$$\begin{aligned} g_j(v) &= \lambda\, h_j(\mathfrak{R}, \mathfrak{P}, \mathfrak{O}') \quad, \\ g_j(w) &= \mu\, h_j(\mathfrak{R}, \mathfrak{P}', \mathfrak{O}) \quad. \end{aligned}$$

Elimination von λ und μ ergibt homogene Gleichungen

$$g_j(v)h_k(\mathfrak{R}, \mathfrak{P}, \mathfrak{O}') - g_k(v)h_j(\mathfrak{R}, \mathfrak{P}, \mathfrak{O}') = 0 \tag{3}$$
$$g_j(w)h_k(\mathfrak{R}, \mathfrak{P}', \mathfrak{O}) - g_k(w)h_j(\mathfrak{R}, \mathfrak{P}', \mathfrak{O}) = 0 \quad.$$

Dazu kommen noch die in § 13 (5) hergeleiteten homogenen Gleichungen in den Koordinaten von $\mathfrak{R}$, die ausdrücken, daß $\mathfrak{R}$ eine Punktgruppe von Γ ist

$$R_j(\mathfrak{R}) = 0 \quad.$$

Aus den homogenen Gleichungen (3), (4) eliminieren wir durch Bildung des Resultantensystems nacheinander die v, die w und die Koordinaten von $\mathfrak{R}$ und erhalten so das Eliminationsergebnis

$$S_i(\mathfrak{O}, \mathfrak{P}, \mathfrak{O}', \mathfrak{P}') = 0 \quad. \tag{5}$$

Die Gleichungen (5) sind homogen in den Chow-Koordinaten der Punktgruppen $\mathfrak{O}$, $\mathfrak{P}$, $\mathfrak{O}'$, $\mathfrak{P}'$ und drücken genau die Äquivalenz (1) aus.

§ 15. Punkte und relationstreue Spezialisierungen. Beweis des Satzes von Hasse § 7, 6

Die Ausführungen des vorigen Paragraphen reichen zum Beweis des von Hasse § 7, 6 formulierten Satzes hin. Bevor wir diesen Beweis erbringen, müssen wir aber den Hasseschen Begriff des „algebraischen Punktes" eines Funktionenkörpers K vom Transzendenzgrad g näher betrachten und mit den Grundbegriffen der algebraischen Geometrie vergleichen.

Hasse geht von einer homogenen Transzendenzbasis $X_0 : X_1 : \ldots : X_g$ des Körpers K aus, normiert $X_0 = 1$, bildet den Integritätsbereich I der in bezug auf $X_1, \ldots, X_g$ ganzen Größen von K und betrachtet die homomorphen Abbildungen von I auf algebraische Erweiterungskörper Ω^* von Ω. Diese Homomorphismen werden dann noch ausgedehnt auf diejenigen gebrochenen Elemente $\eta = \varphi\, \psi^{-1}$ von K, deren Nenner ψ beim betreffenden Homomorphismus nicht in Null übergeht. Solche Homomorphismen nennt Hasse „algebraische Punkte" von K.

Sind $\zeta_1, \ldots, \zeta_m$ Erzeugende des Integritätsbereiches I, so wird jeder Homomorphismus von I durch eine relationstreue Spezialisierung $(\zeta_1, \ldots, \zeta_m) \to (z_1, \ldots, z_m)$ gegeben. Das Element $\eta = f(\zeta)g(\zeta)^{-1}$ geht dabei in $y = f(z)g(z)^{-1}$ über.

Zu der Terminologie der algebraischen Geometrie paßt die Bezeichnung dieser Homomorphismen als „Punkte" nicht. Ein Punkt ist in der algebraischen Geometrie kein Homomorphismus, sondern eine Reihe von homogenen Koordinaten oder etwas, was durch eine solche Reihe eindeutig bestimmt wird, und an diesem Begriff „Punkt" hängen soviele andere Begriffe und Bezeichnungen, daß man dasselbe Wort unmöglich in einer anderen Bedeutung verwenden kann. Was bei Hasse „Punkt" heißt, ist in unserer Bezeichnungsweise eine *relationstreue Spezialisierung* $\zeta \to z$, der Übergang von einem allgemeinen zu einem speziellen Punkt einer algebraischen Mannigfaltigkeit.

Die Unterscheidung des Hasseschen „Punkt"-Begriffes von unserem war noch nicht nötig, solange alle Homomorphismen $\zeta \to z$ immer von *einem und demselben* allgemeinen Punkt ζ ihren Ausgang nahmen. In der Tat konnten wir (§ 11) beweisen, daß die Hasseschen „algebraischen Punkte" p des Körpers K eineindeutig den Punkten p der Kurve Γ entsprechen. Eine klare Trennung wird aber nötig, sobald Homomorphismen wie $\mathfrak{X} \to \mathfrak{P}$ und $\mathfrak{X}' \to \mathfrak{P}'$, die von verschiedenen $\mathfrak{X}$ ausgehen, gleichzeitig betrachtet werden, wie es bei Hasse (§ 7, 6) geschieht.

Wir sahen oben, wie die Spezialisierung $\zeta \to z$ die Spezialisierung

$\eta \to y$ für solche Funktionen η, deren Nenner bei der Spezialisierung nicht Null werden, induziert. Die Spezialisierung $(\zeta,\eta) \to (z,y)$ ist dann wieder relationstreu. Denn wenn $F(\zeta,\eta) = 0$ eine algebraische Relation zwischen ζ und η ist, und wenn $\eta = f(\zeta)g(\zeta)^{-1}$ in diese Relation eingesetzt und das Ergebnis durch Multiplikation mit einer Potenz von $g(\zeta)$ ganz rational gemacht wird, so bleibt das Ergebnis bei der Ersetzung $\zeta \to z$ erhalten. Dividiert man nun wieder durch dieselbe Potenz von $g(z)$ und ersetzt $f(z)g(z)^{-1}$ durch y, so erhält man $F(z,y) = 0$. Die Spezialisierung $(\zeta,\eta) \to (z,y)$ ist also eine relationstreue Fortsetzung der gegebenen Spezialisierung $\zeta \to z$, und zwar die einzig mögliche, denn die Relation

$$g(\zeta)\eta = f(\zeta)$$

muß bei jeder relationstreuen Spezialisierung erhalten bleiben.

Wir sehen also: Wenn η sich als Bruch $f(\zeta)g(\zeta)^{-1}$ so darstellen läßt, daß der Nenner für $\zeta \to z$ nicht Null wird, so läßt sich die relationstreue Spezialisierung $\zeta \to z$ eindeutig durch $\eta \to y$ fortsetzen. Von diesem Satz gilt auch die Umkehrung:

15.1. *Wenn die relationstreue Spezialisierung* $(\zeta_1,\ldots,\zeta_m) \to (z_1,\ldots,z_m)$ *sich eindeutig durch* $\eta \to y$ *fortsetzen läßt, so ist* η *als Quotient*

$$\eta = \frac{f(\zeta)}{g(\zeta)}.$$

so darstellbar, daß der Nenner für $\zeta = z$ *nicht Null wird.*

Beweis. Wir können $y \neq 0$ annehmen, da man sonst nur η durch $\eta - 1$ zu ersetzen braucht. In I gilt die Idealtheorie der ganz-abgeschlossenen Bereiche (Math. Ann. 101, p. 293 oder Mod. Alg. II, § 105). Im Sinne dieser Theorie setzen wir als gekürzten Bruch

$$\eta \sim \frac{\mathfrak{p}_1 \cdots \mathfrak{p}_r}{\mathfrak{q}_1 \cdots \mathfrak{q}_s}.$$

Die gemeinsamen Nullstellen aller Polynome $P(\zeta)$ eines Primideals $\mathfrak{p}_j$ bilden die (nach 2.9 irreduzible) Nullstellenmannigfaltigkeit M_j von $\mathfrak{p}_j$. Wir zeigen zunächst, daß M_j den Punkt z nicht enthält. Ist ξ ein allgemeiner Punkt von M_j, so kann man durch $\zeta \to \xi$ eine erste relationstreue Spezialisierung von $\Omega[\zeta_1,\ldots,\zeta_m]$ definieren. Da $\mathfrak{q}_1 \cdots \mathfrak{q}_s$ nicht durch $\mathfrak{p}_j$ teilbar ist, so gibt es im Produkt $\mathfrak{q}_1 \cdots \mathfrak{q}_s$ ein Polynom $h(\zeta)$, das nicht durch $\mathfrak{p}_j$ teilbar ist. Dann ist $\eta \cdot h(\zeta)$ quasiteilbar durch

$$\frac{p_j}{q_1 \cdots q_s} \, q_1 \cdots q_s = p_j$$

also ganz :
$$\eta \cdot h(\zeta) = p(\zeta) \tag{2}$$

und sogar teilbar durch p_j; denn in diesem Fall folgt, da p_j ein höheres Primideal ist, aus der Quasiteilbarkeit die Teilbarkeit. Da also $p(\zeta)$ durch p_j teilbar ist, aber $h(\zeta)$ nicht, so ist $p(\xi) = 0$, aber $h(\xi) \neq 0$. Bei der Spezialisierung $\zeta \to \xi$ muß (2) erhalten bleiben, also muß η bei dieser Spezialisierung in Null übergehen. Gesetzt nun, z wäre eine Nullstelle von p_j, dann könnte man an diese erste Spezialisierung $\zeta \to \xi$ eine zweite ebenfalls relationstreue $\xi \to z$ reihen. Dabei bleibt η Null; die erhaltene relationstreue Spezialisierung ist also verschieden von der angenommenen Spezialisierung $\eta \to y \neq 0$, entgegen der vorausgesetzten Eindeutigkeit. Also können die Nullstellenmannigfaltigkeiten von $p_1, \ldots, p_r$ den Punkt z nicht enthalten.

Es gibt also in $p_1 \cdots p_r$ Polynome $f_1(\zeta), \ldots, f_r(\zeta)$, die für $\zeta = z$ nicht Null werden. Ihr Produkt $f(\zeta)$ ist durch $p_1 \cdots p_r$ teilbar und es ist $f(z) \neq 0$. Da nun $f(\zeta)$ durch η quasiteilbar ist, so ist $f(\zeta)$ auch durch η teilbar :
$$f(\zeta) = \eta \cdot g(\zeta) \ .$$

In dieser Gleichung führe man die relationstreue Spezialisierung $\zeta \to z$, $\eta \to y$ durch. Da die linke Seite nicht Null wird, wird die rechte Seite auch nicht Null, also ist $g(z) \neq 0$. Damit ist die gewünschte Darstellung

$$\eta = \frac{f(\zeta)}{g(\zeta)} \qquad \text{mit} \quad g(z) \neq 0$$

gefunden.

Nun seien $\mathfrak{O}$ und $\mathfrak{O}'$ zwei feste Punktgruppen auf $\varGamma$ im Sinne von § 13. Mit jeder allgemeinen oder in der Bezeichnung von Hasse höchsttranszendenten Punktgruppe $\mathfrak{X}$ ist eine zweite $\mathfrak{X}'$ verbunden durch die Äquivalenz

$$\frac{\mathfrak{X}}{\mathfrak{O}} \sim \frac{\mathfrak{X}'}{\mathfrak{O}'} \ .$$

Der Hassesche Satz § 7, 6 kann jetzt so formuliert werden :

Ist

$$\frac{\mathfrak{P}}{\mathfrak{O}} \sim \frac{\mathfrak{P}'}{\mathfrak{O}'} \tag{4}$$

und ist $\mathfrak{P}'$ regulär, d. h. nicht spezial (10.7), so läßt sich die relationstreue Spezialisierung $\mathfrak{X} \to \mathfrak{P}$ eindeutig durch $\mathfrak{X}' \to \mathfrak{P}'$ fortsetzen. Ist umgekehrt

$(\mathfrak{X}, \mathfrak{X}') \to (\mathfrak{P}, \mathfrak{P}')$ *eine Fortsetzung der relationstreuen Spezialisierung* $\mathfrak{X} \to \mathfrak{P}$, *so gilt* (4).

Beweis. Nach § 14 wird die Äquivalenz (3) durch homogene algebraische Gleichungen

$$S_i(\mathfrak{O}, \mathfrak{X}, \mathfrak{O}', \mathfrak{X}') = 0 \tag{5}$$

ausgedrückt. Bei jeder relationstreuen Spezialisierung müssen diese Gleichungen erhalten bleiben, also folgt für jede solche Spezialisierung $(\mathfrak{X}, \mathfrak{X}') \to (\mathfrak{P}, \mathfrak{P}')$ die Äquivalenz (4). Die Fortsetzung einer gegebenen relationstreuen Spezialisierung $\mathfrak{X} \to \mathfrak{P}$ ist nach 1.4 immer möglich und wenn dabei $\mathfrak{X}'$ etwa in $\mathfrak{P}''$ übergeht, so gilt (4) auch für $\mathfrak{P}''$; da aber die reguläre Punktgruppe $\mathfrak{P}'$ durch (4) eindeutig bestimmt ist, so kann $\mathfrak{P}''$ nur mit $\mathfrak{P}'$ zusammenfallen. Damit ist alles bewiesen.

§ 16. Die Mannigfaltigkeit der Nullklassen

Bei gegebenem $\mathfrak{O}$ kann man aus jeder Nullklasse A einen Repräsentanten $\dfrac{\mathfrak{P}}{\mathfrak{O}}$ auswählen, aber diese Darstellung ist nicht invariant und auch nicht immer eindeutig. Macht man aber (wie Hasse § 6, 3) $\mathfrak{O}$ variabel und betrachtet die Gesamtheit aller Repräsentanten $\dfrac{\mathfrak{S}}{\mathfrak{R}}$ der Nullklasse A, so ist diese Gesamtheit eindeutig bestimmt. Nach § 14 ist diese Gesamtheit von Paaren $(\mathfrak{R}, \mathfrak{S})$ durch ein System algebraischer Gleichungen

$$S_i(\mathfrak{O}, \mathfrak{P}, \mathfrak{R}, \mathfrak{S}) = 0$$

gegeben, sie ist also eine algebraische Mannigfaltigkeit N_A. Durchläuft nun A alle Nullklassen, so durchläuft N_A, wie wir zeigen werden, ein algebraisches System von algebraischen Mannigfaltigkeiten. Das heißt: Bildet man die Mannigfaltigkeiten N_A auf Punkte ab, so bilden diese Punkte eine irreduzible algebraische Mannigfaltigkeit $\mathfrak{K}$: die *Klassenmannigfaltigkeit*. Die Punkte von $\mathfrak{K}$ entsprechen eineindeutig den Nullklassen A.

Die Gesamtheit aller Paare $(\mathfrak{R}, \mathfrak{S})$, derart daß

$$\frac{\mathfrak{P}}{\mathfrak{O}} \sim \frac{\mathfrak{S}}{\mathfrak{R}} \tag{1}$$

gilt, ist durch das Gleichungssystem

$$R_i(\mathfrak{O}, \mathfrak{P}, \mathfrak{R}, \mathfrak{S}) = 0 \tag{2}$$

gegeben. Halten wir $\mathfrak{D}$ fest und lassen $\mathfrak{P}$ die ganze in § 13 untersuchte glatte Mannigfaltigkeit M durchlaufen, so definiert (2) eine algebraische Korrespondenz zwischen M und der Mannigfaltigkeit N aller Paare $(\mathfrak{R}, \mathfrak{S})$. Wir wollen nun beweisen:

16.1. *Die Korrespondenz* (2) *ist irreduzibel.*

16.2. *Jedem $\mathfrak{P}$ entspricht in der Korrespondenz eine Mannigfaltigkeit $N_{\mathfrak{P}}$ von Paaren $(\mathfrak{R}, \mathfrak{S})$, die immer dieselbe Dimension g besitzt.*

Aus 16.1 und 16.2 folgt nach 5.9:

16.3. *Die Mannigfaltigkeiten $N_{\mathfrak{P}}$ bilden ein irreduzibles System von algebraischen Mannigfaltigkeiten.*

Das ist aber genau das, was wir beweisen wollten. Da nämlich N offensichtlich nur von der Divisorenklasse $A = \left\{\dfrac{\mathfrak{P}}{\mathfrak{D}}\right\}$ abhängt, so kann man statt $N_{\mathfrak{P}}$ auch N_A schreiben. *Die N_A sind den Nullklassen A eineindeutig zugeordnet und bilden ein irreduzibles System $\mathfrak{R}$ von algebraischen Mannigfaltigkeiten.*

Beweis von 16.1. Wir schreiben statt (1)

$$\mathfrak{P}\,\mathfrak{R} \sim \mathfrak{D}\,\mathfrak{S} . \tag{3}$$

Die Äquivalenz (3) definiert zunächst eine Korrespondenz zwischen den $\mathfrak{S}$ und den Produkten $\mathfrak{T} = \mathfrak{P}\,\mathfrak{R}$. Diese ist nach dem Kriterium 3.6 irreduzibel, denn erstens bilden die $\mathfrak{S}$ nach § 13 eine irreduzible Mannigfaltigkeit, und zweitens entspricht jedem $\mathfrak{S}$ in der Korrespondenz (3) eine lineare Schar von Punktgruppen $\mathfrak{P}\,\mathfrak{R} = \mathfrak{T}$, die als lineare Schar selbstverständlich irreduzibel ist und nach dem Riemann-Rochschen Satz immer dieselbe Dimension g hat.

Ein allgemeines Paar dieser irreduziblen Korrespondenz erhält man, indem man $\mathfrak{S}$ durch eine allgemeine Punktgruppe $\mathfrak{S}^*$ ersetzt und $\mathfrak{P}\,\mathfrak{R}$ durch die allgemeine Punktgruppe $\mathfrak{T}^*$ der zugehörigen linearen Schar

$$\mathfrak{T}^* \sim \mathfrak{D}\,\mathfrak{S}^* . \tag{4}$$

Alle Tripel $(\mathfrak{P}, \mathfrak{R}, \mathfrak{S})$ der Korrespondenz (3) erhält man nun, indem man zunächst das Paar $(\mathfrak{T}^*, \mathfrak{S}^*)$ relationstreu zu $(\mathfrak{T}, \mathfrak{S})$ spezialisiert und $\mathfrak{T}$ in allen möglichen Weisen in $\mathfrak{T} = \mathfrak{P}\,\mathfrak{R}$ zerlegt. Diese relationstreue Spezialisierung kann man aber fortsetzen zu einer relationstreuen Spezialisierung aller einzelnen Punkte der Punktgruppe $\mathfrak{T}^*$. Also geht die Zerlegung $\mathfrak{T} = \mathfrak{P}\,\mathfrak{R}$ aus irgendeiner der möglichen Zerlegungen $\mathfrak{T}^* = \mathfrak{P}^*\,\mathfrak{R}^*$ durch relationstreue Spezialisierung hervor. Demnach kann man die Tripel $(\mathfrak{P},\ \mathfrak{R},\ \mathfrak{S})$ auch so erhalten: Zunächst wird das allgemeine Paar $(\mathfrak{T}^*,\ \mathfrak{S}^*)$ der Korrespondenz (4) gebildet, dann wird $\mathfrak{T}^*$ in allen möglichen Weisen in zwei Faktoren $\mathfrak{P}^*\,\mathfrak{R}^*$ zerlegt, schließlich wird relationstreu spezialisiert: $\mathfrak{P}^* \to \mathfrak{P},\ \mathfrak{R}^* \to \mathfrak{R},\ \mathfrak{S}^* \to \mathfrak{S}$.

Wir wollen nun zeigen, daß die Zerlegungen $\mathfrak{T}^* = \mathfrak{P}^*\,\mathfrak{R}^*$ alle untereinander konjugiert sind, d. h. daß sie durch Körperisomorphismen — auch eine Art von relationstreuen Spezialisierungen — aus einer solchen Zerlegung hervorgehen. Wenn das gezeigt ist, so folgt, daß alle Tripel $(\mathfrak{P},\ \mathfrak{R},\ \mathfrak{S})$ aus einem einzigen Tripel $(\mathfrak{P}^*,\ \mathfrak{R}^*,\ \mathfrak{S}^*)$ durch relationstreue Spezialisierung hervorgehen. Die Mannigfaltigkeit aller Tripel $(\mathfrak{P},\ \mathfrak{R},\ \mathfrak{S})$ besitzt also ein allgemeines Tripel $(\mathfrak{P}^*,\ \mathfrak{R}^*,\ \mathfrak{S}^*)$, woraus ihre Irreduzibilität folgt.

Wir betrachten die Vollschar $|\,\mathfrak{T}^*\,| = |\,\mathfrak{D}\,\mathfrak{S}^*\,|$. Die Punktgruppe $\mathfrak{S}^*$ ist allgemein, also nicht spezial. Es sei $\mathfrak{D} = q_1 q_2 \cdots q_g$. Die Restschar von $q_1 q_2 \cdots q_k\ (1 \leq k \leq g)$ in bezug auf $|\,\mathfrak{T}^*\,|$ ist $|\,q_{k+1} \cdots q_g\,\mathfrak{S}^*\,|$; diese Schar ist mithin auch nicht spezial. Für $k < g$ hat sie auch keine festen Punkte. Denn wenn etwa q_{k+1} ein fester Punkt wäre, so wäre $|\,q_{k+2} \cdots q_g\,\mathfrak{S}^*\,|$ nach dem Reduktionssatz 10.6 eine Spezialschar, was nicht der Fall ist; und wenn einer der Punkte $s_1, s_2, \ldots, s_g$, aus denen $\mathfrak{S}^*$ besteht, etwa s_g ein fester Punkt wäre, so wäre $|\,q_{k+1} \cdots q_g s_1 \cdots s_{g-1}\,|$ ebenfalls nach 10.6 spezial, also wäre auch $|\,q_g s_1 \cdots s_{g-1}\,|$ spezial. Das ist aber auch nicht der Fall; denn der Punkt q_g stellt den Punktgruppen der Differentialschar, die ihn enthalten sollen, eine lineare Bedingung und die hinzukommenden allgemeinen Punkte $s_1, \ldots, s_{g-1}$ stellen ihnen noch $g - 1$ davon unabhängige lineare Bedingungen; das macht insgesamt g lineare Bedingungen, während doch die Differentialschar nur die Dimension $g - 1$ hat. Damit ist gezeigt, daß die Restschar von $q_1 \cdots q_k$ in bezug auf $|\,\mathfrak{T}^*\,|$ für $k \leq g$ nicht spezial ist und für $k < g$ auch keine festen Punkte hat.

Diese Eigenschaften gelten um so mehr, wenn $q_1, \ldots, q_k$ durch ebensoviele allgemeine, von $\mathfrak{S}^*$ unabhängige Punkte ersetzt werden. Nun zerlegen wir $\mathfrak{T}^*$ irgendwie in $\mathfrak{T}^* = \mathfrak{P}^*\,\mathfrak{R}^*$ und setzen $\mathfrak{P}^* = p_1 p_2 \cdots p_g$ und $\mathfrak{R}^* = r_1 r_2 \cdots r_g$. Unter den Punkten $p_1, \ldots, p_g$ seien etwa $p_1, \ldots, p_k$ algebraisch unabhängig untereinander und von $\mathfrak{S}^*$, während die übrigen $p_{k+1}, \ldots, p_g$ algebraisch von ihnen und von $\mathfrak{S}^*$ abhängen. Dann folgt, daß der Rest von $p_1 \cdots p_k$ in bezug auf $|\,\mathfrak{T}^*\,|$ für $k \leq g$ nicht spezial ist und für $k < g$ auch keine festen Punkte hat.

Nun benutzen wir, daß $\mathfrak{T}^* = \mathfrak{P}^*\,\mathfrak{R}^*$ eine allgemeine Punktgruppe der

Vollschar $|\mathfrak{T}^*|$ ist. Hält man nun die algebraisch unabhängigen Punkte $p_1, \ldots, p_k$ fest, so muß der Rest $p_{k+1} \cdots p_g \, \mathfrak{R}^*$ eine allgemeine Punktgruppe der Restschar $|\mathfrak{T}^*| : p_1 \cdots p_k$ sein. Wäre nun $k < g$, so wäre diese Restschar nach dem eben Bewiesenen eine Schar ohne feste Punkte. Andererseits wäre p_{k+1} ein solcher fester Punkt, denn p_{k+1} ist algebraisch abhängig von $\mathfrak{S}^*$ und $p_1, \ldots, p_k$. Also muß $k = g$ sein, d. h. die Punkte $p_1, \ldots, p_g$ sind algebraisch unabhängig. Weiter ist die Restschar $|\mathfrak{R}^*|$ von $p_1 \cdots p_g$ in bezug auf $|\mathfrak{T}^*|$ nicht spezial, also hat sie die Dimension Null, d. h. die Punktgruppe $\mathfrak{R}^*$ ist durch $\mathfrak{S}^*$ und $\mathfrak{P}^*$ eindeutig bestimmt.

Die Punktgruppen $\mathfrak{S}^*$, $\mathfrak{P}^*$ und $\mathfrak{R}^*$ haben also folgende Strukturen: $\mathfrak{S}^*$ und $\mathfrak{P}^*$ bestehen aus lauter unabhängigen allgemeinen Punkten, und $\mathfrak{R}^*$ ist durch die Äquivalenz

$$\mathfrak{D} \, \mathfrak{S}^* \sim \mathfrak{P}^* \, \mathfrak{R}^*$$

eindeutig bestimmt.

Nun ist klar, daß jedes System von $2g$ unabhängigen allgemeinen Punkten von Γ aus jedem anderen ebensolchen System durch einen Körperisomorphismus hervorgeht. Also gibt es bis auf Körperisomorphie nur ein Tripel $(\mathfrak{P}^*, \mathfrak{R}^*, \mathfrak{S}^*)$, aus dem alle Tripel $(\mathfrak{P}, \mathfrak{R}, \mathfrak{S})$ relationstreu hervorgehen. Damit ist 16.1 bewiesen.

Beweis von 16.2. Es sei $\mathfrak{P}$ gegeben und $(\mathfrak{R}, \mathfrak{S})$ irgendeine Lösung der Äquivalenz (3). Wir haben zu beweisen, daß das Paar $(\mathfrak{R}, \mathfrak{S})$ höchstens den Transzendenzgrad g hat.

Es seien $r_1, \ldots, r_k$ die algebraisch unabhängigen unter den Punkten $r_1, \ldots, r_g$ von $\mathfrak{R}$. Weiter sei d die Dimension der durch (3) definierten linearen Schar $|\mathfrak{S}|$ bei gegebenen $r_1, \ldots, r_g$. Dann ist das Paar $(\mathfrak{R}, \mathfrak{S})$ von höchstens $k + d$ unabhängigen Parametern abhängig. Wir haben zu beweisen: $k + d \leqq g$.

Nach dem Riemann-Rochschen Satz ist d gleich dem Spezialitätsindex von $|\mathfrak{S}|$, d. h. gleich der Anzahl der linear unabhängigen Divisoren der Restschar

$$\left| \frac{\mathfrak{D}}{\mathfrak{S}} \right| = \left| \frac{\mathfrak{D}\mathfrak{D}}{\mathfrak{D}\mathfrak{S}} \right| = \left| \frac{\mathfrak{D}\mathfrak{D}}{\mathfrak{P}\mathfrak{R}} \right| ,$$

wo $\mathfrak{D}$ irgendein Divisor der Differentialschar ist. Setzt man

$$\left| \frac{\mathfrak{D}\mathfrak{D}}{\mathfrak{P}} \right| = |\mathfrak{D}_1| ,$$

so ist d die Anzahl der linear unabhängigen Divisoren der Restschar $|\mathfrak{D}_1| : \mathfrak{R}$.

Die Anzahl der linear unabhängigen Divisoren der Schar $|\,\mathfrak{D}_1\,|$ vom
Grade $2g - 2$ ist nach dem Riemann-Rochschen Satz höchstens g. Die
algebraisch unabhängigen Punkte $r_1, \ldots, r_k$ von $\mathfrak{R}$ legen den Punkt-
gruppen von $|\,\mathfrak{D}_1\,|$, die sie enthalten sollen, ebenso viele linear unab-
hängige Bedingungen auf. Also ist die Anzahl der linear unabhängigen
Divisoren der Restschar höchstens $g - k$, mithin

$$d \leqq g - k$$

oder

$$d + k \leqq g \; ,$$

was wir beweisen wollten.

Als *Koordinaten der Klasse A* bezeichnet man naturgemäß die Chow-
Koordinaten der Mannigfaltigkeit N_A. Es kann keine Kollision entstehen, wenn der Bildpunkt von A auf der Klassenmannigfaltigkeit $\mathfrak{R}$,
dessen Koordinaten eben die Koordinaten von A sind, ebenfalls mit dem
Buchstaben A bezeichnet wird.

Die Mannigfaltigkeit N_A besteht aus allen Paaren $(\mathfrak{R}, \mathfrak{S})$, so daß
$\mathfrak{R}^{-1}\mathfrak{S}$ zur Klasse A gehört. Drückt man die Tatsache, daß $(\mathfrak{R}, \mathfrak{S})$ zu
N_A gehört, nach 5.6 durch algebraische Gleichungen zwischen den Koor-
dinaten von $\mathfrak{R}$, $\mathfrak{S}$ und A aus, so erhält man ein Gleichungssystem

$$G_i(\mathfrak{R}, \mathfrak{S}, A) = 0 \; , \tag{5}$$

welches ausdrückt, daß der Divisor $\mathfrak{R}^{-1}\mathfrak{S}$ zur Klasse A gehört.

Wenn ein zweiter Divisor $\mathfrak{S}^{-1}\mathfrak{T}$ zur Klasse B gehört:

$$G_i(\mathfrak{S}, \mathfrak{T}, B) = 0 \; , \tag{6}$$

so gehört der Produktdivisor $\mathfrak{R}^{-1}\mathfrak{T}$ zur Produktklasse $AB = C$:

$$G_i(\mathfrak{R}, \mathfrak{T}, C) = 0 \; . \tag{7}$$

Durch Elimination von $\mathfrak{R}$, $\mathfrak{S}$, $\mathfrak{T}$ aus (5), (6), (7) erhält man ein
Gleichungssystem

$$H_k(A, B, C) = 0 \; . \tag{8}$$

Umgekehrt, wenn (8) erfüllt ist, so gibt es drei Divisoren $\mathfrak{R}$, $\mathfrak{S}$, $\mathfrak{T}$, die
(5), (6), (7) erfüllen, d. h. A ist die Klasse von $\mathfrak{R}^{-1}\mathfrak{S}$, B die von $\mathfrak{S}^{-1}\mathfrak{T}$
und C die von $\mathfrak{R}^{-1}\mathfrak{T}$, mithin ist dann $C = AB$. Das Gleichungssystem
(8) ist also der Produktrelation $C = AB$ äquivalent.

§ 17. Korrespondenzen und Multiplikatoren

Es sei μ eine algebraische Korrespondenz zwischen Γ und Γ'. Die Korrespondenz darf reduzibel sein, aber wir setzen voraus, daß jede ihrer irreduziblen Bestandteile einem allgemeinen Punkt ξ von Γ endlich viele Punkte $\eta_1, \ldots, \eta_h$ von Γ' zuordnet. Geht man durch relationstreue Spezialisierung $\xi \to p$ zu einem speziellen Punkt p über, so gehen $\eta_1, \ldots, \eta_h$ in (abgesehen von der Reihenfolge) eindeutig bestimmte Punkte $q_1, \ldots, q_h$ über. Die Korrespondenz μ ordnet also jedem Punkt p eine eindeutig bestimmte Punktgruppe $\mu p = \mathfrak{q} = q_1 \ldots q_h$ zu. Einer Punktgruppe $\mathfrak{a} = p_1 \ldots p_m$ wird demnach die Punktgruppe $\mu \mathfrak{a} = (\mu p_1)(\mu p_2) \ldots (\mu p_m)$ zugeordnet. Man beweist leicht: Wenn $\mathfrak{a}$ eine rationale Punktgruppe ist, so ist auch $\mu \mathfrak{a}$ rational, und wenn $\mathfrak{a}$ eine lineare Schar durchläuft, so durchläuft auch $\mu \mathfrak{a}$ eine lineare Schar. Schließlich kann man die Zuordnung auch auf nicht ganze Divisoren $\mathfrak{a} \mathfrak{b}^{-1}$ ausdehnen, indem man $\mu(\mathfrak{a} \mathfrak{b}^{-1}) = (\mu \mathfrak{a})(\mu \mathfrak{b})^{-1}$ setzt. Auch jetzt entspricht einer linearen Schar wieder eine lineare Schar, folglich entsprechen äquivalenten Divisoren auch äquivalente Divisoren, und einer Divisorenklasse B entspricht eine Divisorenklasse μB.

Die Korrespondenzen μ heißen nach Hasse auch *Multiplikatoren*, und zwar reguläre Multiplikatoren, wenn sie einer allgemeinen Nullklasse $Y = \{\mathfrak{Y}\, \mathfrak{U}^{-1}\}$ (wo $\mathfrak{U}$ eine feste, $\mathfrak{Y}$ eine allgemeine Punktgruppe ist) wieder eine allgemeine Nullklasse $X = \mathfrak{X}\, \mathfrak{O}^{-1}$ zuordnen.

Wir wollen nun die Beziehung

$$\mu(\mathfrak{P}\, \mathfrak{U}^{-1}) \sim \mathfrak{S}\, \mathfrak{O}^{-1} \tag{1}$$

durch algebraische Gleichungen zwischen den Koordinaten von $\mathfrak{P}$, $\mathfrak{U}$, $\mathfrak{S}$ und $\mathfrak{O}$ ausdrücken. Wir zerlegen sie in

$$\mu\, \mathfrak{P} = \mathfrak{Q} \; , \tag{2}$$

$$\mu\, \mathfrak{U} = \mathfrak{B} \; , \tag{3}$$

$$\mathfrak{Q}\, \mathfrak{B}^{-1} \sim \mathfrak{S}\, \mathfrak{O}^{-1} \; . \tag{4}$$

Die Systeme $(p, q_1, \ldots, q_h)$ gehen durch relationstreue Spezialisierung aus $(\xi, \eta_1, \ldots, \eta_h)$ hervor; sie bilden also eine irreduzible Mannigfaltigkeit, deren Gleichungen

$$f_j(p, q_1, \ldots, q_h) = 0 \tag{5}$$

heißen mögen. Schreibt man die Gleichungen (5) für alle Punkte $p_1, \ldots, p_\varrho$ der Punktgruppe $\mathfrak{P}$ auf, fügt man die homogenen Gleichungen

$$\pi_j g_k(p_1, \ldots, p_g) - \pi_k g_j(p_1, \ldots, p_g) = 0 \tag{6}$$

hinzu, die nach § 13 ausdrücken, daß die Punktgruppe $\mathfrak{P}$ aus den Punkten $p_1, \ldots, p_g$ besteht, dann noch die entsprechenden Gleichungen für die Punktgruppe $\mathfrak{Q} = (q_{11} \ldots q_{1h})(q_{21} \ldots q_{2h}) \ldots (q_{g1} \ldots q_{gh})$ und die Gleichungen, die ausdrücken, daß alle diese Punkte p_i und q_{ij} auf Γ liegen, und eliminiert man nunmehr die Koordinaten der Punkte p_i und q_{ij} aus allen diesen Gleichungen, so erhält man ein System von Gleichungen

$$R_i(\mathfrak{P}, \mathfrak{Q}) = 0 \ , \tag{7}$$

das die Beziehung (2) ausdrückt.

Ebenso drückt das System

$$R_i(\mathfrak{U}, \mathfrak{B}) = 0 \tag{8}$$

die Beziehung (3) aus.

Schließlich haben wir noch die Äquivalenz (4) durch algebraische Gleichungen auszudrücken. Zu diesem Zweck zerlegen wir $\mathfrak{Q}$ und $\mathfrak{B}$ in lauter Gruppen von g Punkten:

$$\mathfrak{Q} = \mathfrak{Q}_1 \ldots \mathfrak{Q}_h \ , \tag{9}$$

$$\mathfrak{B} = \mathfrak{B}_1 \ldots \mathfrak{B}_h \ . \tag{10}$$

Dann ist (4) gleichwertig mit

$$\mathfrak{D} \, \mathfrak{Q}_1 \ldots \mathfrak{Q}_h \sim \mathfrak{B}_1 \ldots \mathfrak{B}_h \, \mathfrak{S} \ .$$

Diese Bedingung zerlegen wir in

$$\mathfrak{D} \, \mathfrak{Q}_1 \sim \mathfrak{B}_1 \, \mathfrak{D}_2 \, , \tag{11}$$

$$\mathfrak{D}_2 \, \mathfrak{Q}_2 \sim \mathfrak{B}_2 \, \mathfrak{D}_3 \ , \tag{12}$$

$$\ldots$$

$$\mathfrak{D}_{h-1} \, \mathfrak{Q}_{h-1} \sim \mathfrak{B}_{h-1} \, \mathfrak{D}_h \ , \tag{13}$$

$$\mathfrak{D}_h \, \mathfrak{Q}_h \sim \mathfrak{B}_h \, \mathfrak{S} \ , \tag{14}$$

wo $\mathfrak{D}_2, \mathfrak{D}_3, \ldots, \mathfrak{D}_h$ lauter Gruppen von g Punkten sind, die durch die Äquivalenzen (11) bis (13) der Reihe nach bestimmbar sind. Nach § 10 können alle Äquivalenzen (11) bis (14) durch algebraische Gleichungen ausgedrückt werden. Ebenso kann man (9) und (10) nach der Methode des § 14 durch algebraische Gleichungen ausdrücken [vgl. § 14, (2) und (3)]. Eliminiert man aus allen diesen Gleichungen die Koordinaten der Hilfsgruppen $\mathfrak{Q}_1, \ldots, \mathfrak{Q}_h, \mathfrak{B}_1, \ldots, \mathfrak{B}_h, \mathfrak{D}_2, \ldots, \mathfrak{D}_h,$ so erhält man schließlich ein System von Gleichungen

$$S_i(\mathfrak{D}, \mathfrak{Q}, \mathfrak{S}, \mathfrak{V}) = 0 \,, \tag{15}$$

das die Äquivalenz (4) ausdrückt.

Eliminiert man endlich aus (7), (8), (15) die Koordinaten der Punktgruppen $\mathfrak{Q}$ und $\mathfrak{V}$, so erhält man die gesuchten Gleichungen

$$T_i(\mathfrak{D}, \mathfrak{P}, \mathfrak{U}, \mathfrak{S}) = 0 \,, \tag{16}$$

welche die Äquivalenz (1) ausdrücken.

Zum Schluß vollziehen wir den Übergang auf die Klassenmannigfaltigkeit. Der zu Klasse A gehörige Punkt der Klassenmannigfaltigkeit möge ebenfalls mit A bezeichnet werden. Dann drückt das Gleichungssystem

$$G_i(\mathfrak{D}, \mathfrak{S}, A) = 0 \tag{17}$$

nach § 16, Schluß aus, daß der Divisor $\mathfrak{D}\,\mathfrak{S}^{-1}$ zur Klasse A gehört; ebenso drückt das System

$$G_i(\mathfrak{U}, \mathfrak{P}, B) = 0 \tag{18}$$

aus, daß der Divisor $\mathfrak{P}\,\mathfrak{U}^{-1}$ zur Klasse B gehört. Eliminiert man aus (16), (17), (18) und den Gleichungen, die ausdrücken, daß $\mathfrak{D}$, $\mathfrak{P}$, $\mathfrak{S}$, $\mathfrak{U}$ zu M gehören, die Koordinaten von $\mathfrak{D}$, $\mathfrak{P}$, $\mathfrak{S}$, $\mathfrak{U}$, so erhält man ein Gleichungssystem

$$H(A, B) = 0 \,, \tag{19}$$

welches die Beziehung $A = \mu B$ algebraisch zum Ausdruck bringt.

§ 18. Beweis der Hasseschen Sätze § 8, 2

Die Ausführungen des vorigen Paragraphen reichen zum Beweis der von Hasse in § 8, 4 formulierten Sätze hin. Bevor wir diese Beweise aber geben, müssen wir die in diesen Sätzen steckenden Begriffe, insbesondere die Zerlegung der Hasseschen „Punkte" bei Erweiterung des Funktionenkörpers, näher betrachten und mit den Begriffen der algebraischen Geometrie vergleichen.

Was geschieht mit den „algebraischen Punkten" $\mathfrak{X} \to \mathfrak{P}$ eines Funktionenkörpers K bei Erweiterung dieses Funktionenkörpers? Hasse beruft sich (§ 7, 3) für diese Frage auf eine von Lorenzen ausgearbeitete, aber nicht publizierte Erweiterung der Krullschen Zerlegungstheorie. Da wir in dieser Arbeit alles Behauptete auch beweisen wollen, können wir uns nicht auf diese unpublizierte Theorie stützen. Außerdem ist die Formu-

356

lierung, die sich aus der algebraischen Geometrie ergibt, einfacher und in gewisser Hinsicht auch allgemeiner als die von Hasse verwendete.

Hasse benutzt für den Körper K und den separablen normalen Erweiterungskörper K^* eine und dieselbe homogene Transzendenzbasis $X_0 : X_1 : \ldots : X_g$. Wenn es sich um solche Homomorphismen handelt, bei denen X_0 nicht in Null übergeht, kann man wieder $X_0 = 1$ normieren. Ein Erzeugendensystem der ganzen Größen von K sei $\zeta_1, \ldots, \zeta_m$, von K^* ebenso $\zeta_1^*, \ldots, \zeta_q^*$. Die ζ^* sind ganze algebraische Funktionen von den ζ, die ζ ganze rationale Funktionen von den ζ^*. Das Punktepaar (ζ, ζ^*) ist allgemeines Paar einer algebraischen Korrespondenz zwischen M und M^*, wobei ζ allgemeiner Punkt von M und ζ^* allgemeiner Punkt von M^* ist. Durch relationstreue Spezialisierung $(\zeta, \zeta^*) \to (z, z^*)$ erhält man aus dem allgemeinen Punktepaar alle Punktepaare der Korrespondenz. Die Frage der „Zerlegungstheorie" ist nun: In wievielen verschiedenen Weisen kann eine gegebene relationstreue Spezialisierung $\zeta \to z$ zu einer ebensolchen $(\zeta, \zeta^*) \to (z, z^*)$ fortgesetzt werden?

Die Methode der algebraischen Geometrie gestattet es, dieselbe Frage gleich viel allgemeiner zu stellen. Die Voraussetzung, daß die ζ^* ganze algebraische Funktionen von den ζ und die ζ ganze rationale Funktionen von den ζ^* sind, die bei den Anwendungen nicht von vornherein erfüllt ist, kann fallengelassen werden.

Wir brauchen bloß vorauszusetzen:

1) M sei eine irreduzible, glatte g-dimensionale Mannigfaltigkeit.

2) $\Re$ sei eine irreduzible Korrespondenz zwischen M und N, in der einem allgemeinen Punkt ξ von M nur endlich viele konjugierte Punkte $\eta_1, \ldots, \eta_h$ von N entsprechen. (Die Zahl h ist natürlich gleich dem reduzierten Körpergrad von $\Omega(\xi, \eta)$ über $\Omega(\xi)$, wo η einer von den konjugierten Punkten η_ν ist.)

3) Einem speziellen Punkt p von M entsprechen in der Korrespondenz auch nur endlich viele Punkte q von N.

Unter diesen Voraussetzungen kann man die relationstreue Spezialisierung $\xi \to p$ zu einer relationstreuen Spezialisierung $(\xi, \eta_1, \ldots, \eta_h) \to (p, q_1, \ldots, q_h)$ fortsetzen, und die Punkte $q_1, \ldots, q_h$ sind nach § 1.9 bis auf die Reihenfolge eindeutig durch die Spezialisierung $\xi \to p$ allein bestimmt. Die Multiplizität e_i eines Punktes q_i ist die Zahl, die angibt, wie oft q_i unter den $q_1, \ldots, q_h$ vorkommt. Konjugierte Punkte in bezug auf $\Omega(p)$ haben natürlich dieselbe Multiplizität. Ist f_i die Anzahl der verschiedenen Konjugierten eines Punktes q_i, oder, was dasselbe ist, der reduzierte Körpergrad von $\Omega(p, q_i)$ über $\Omega(p)$, so ist

$$h = \sum e_i f_i \tag{1}$$

die Gesamtzahl der Punkte q_i, jeder mit seiner Vielfachheit gezählt.

Nun machen wir die weiteren Voraussetzungen, die in den Anwendungen nachher auch erfüllt sind:

4) N sei glatt;

5) Den Punkten $q_1, \ldots, q_h$ entsprechen in der Korrespondenz nur endlich viele Punkte von M;

6) $\Omega(\eta_1)$ sei normal über $\Omega(\xi)$, d. h. ξ und $\eta_1, \ldots, \eta_h$ seien rationale Funktionen von η_1.

Dann können wir beweisen, daß in (1) alle e_i und f_i gleich sind, so daß wir statt (1) schreiben können

$$h = e f r \ , \tag{2}$$

wo r die Anzahl der Systeme konjugierter Punkte q_i ist.

Zum Beweis bemerken wir zunächst, daß nach 6) die umgekehrte Korrespondenz $\mathfrak{K}$ eine rationale Abbildung von N auf M ist. Wegen 4) und 5) folgt daraus, daß jedem der Punkte q_i nur ein Punkt von M, nämlich der Punkt p entspricht.

Nunmehr betrachten wir die Korrespondenz zwischen den Punkten η_1 einerseits und den Systemen $(\xi, \eta_1, \ldots, \eta_h)$ andererseits. Die Korrespondenz ist eine birationale Abbildung, deren Urmannigfaltigkeit N doppelpunktfrei ist, und dem Punkt q_1 entsprechen nur endlich viele Systeme $(p, q_1, \ldots, q_h)$, nämlich der einzige Punkt p und dazu endlich viele Möglichkeiten für $q_1, \ldots, q_h$. Also entspricht dem Punkt q_1 nur ein einziges System $(p, q_1, \ldots, q_h)$, d. h. die relationstreue Spezialisierung $\eta_1 \to q_1$ läßt sich nur in einer Weise zu $\xi \to p$, $\eta_1 \to q_1, \ldots, \eta_h \to q_h$ fortsetzen.

Wenn nun q_1 die Vielfachheit e hat und wenn etwa $\eta_1, \ldots, \eta_e$ bei der Spezialisierung in q_1 hineinrücken, so wollen wir zunächst beweisen, daß alle anderen q_i auch die Vielfachheit e haben. Es seien $S_1, \ldots, S_e$ die Substitutionen der Galoisschen Gruppe, die η_1 in $\eta_1, \ldots, \eta_e$ überführen. Setzt man sie zusammen mit der relationstreuen Spezialisierung, die $\eta_1, \ldots, \eta_e$ in q_1 überführt, so erhält man lauter relationstreue Spezialisierungen, die η_1 in q_1 überführen, also mit der gegebenen Spezialisierung $\eta_1, \ldots, \eta_h \to q_1, \ldots, q_h$ übereinstimmen müssen.

Wir üben nun diese zusammengesetzte relationstreue Spezialisierung auf η_{e+1} aus. Die Substitutionen $S_1, \ldots, S_e$ mögen η_{e+1} in $\eta_{e+1}, \ldots, \eta_{2e}$ überführen. Die darauffolgende relationstreue Spezialisierung $\eta_i \to q_i$

führt $\eta_{e+1}, \ldots, \eta_{2e}$ in $q_{e+1}, \ldots, q_{2e}$ über. Die resultierende Spezialisierung ist aber nach dem vorigen Absatz immer dieselbe; also müssen $q_{e+1}, \ldots, q_{2e}$ zusammenfallen. In dieser Weise fallen also je e von den Punkten $q_1, \ldots, q_n$ zusammen, d. h. jeder Punkt q_i hat mindestens dieselbe Vielfachheit wie q_1. Umgekehrt hat q_1 auch mindestens dieselbe Vielfachheit wie q_i; mithin haben die Punkte q_i alle dieselbe Vielfachheit e.

Wie leicht ersichtlich, bilden die Substitutionen $S_j (j = 1, \ldots, e)$ mit der Eigenschaft, daß $S_j \eta_1$ bei der relationstreuen Spezialisierung in q_1 übergeht, eine Gruppe: die *Trägheitsgruppe* der Spezialisierung $\eta_1 \to q_1$. Dies nebenbei.

Nun mögen $q_1 = \cdots = q_e$, $q_{e+1} = \cdots = q_{2e}$, usw. bis q_{fe} ein System von f konjugierten Punkten bilden. Es gibt dann zu jedem dieser Punkte q_k einen Körperisomorphismus, der q_k in q_1 überführt, und dieser läßt sich zu einer isomorphen relationstreuen Spezialisierung des ganzen Punktsystems $q_1, \ldots, q_h$ fortsetzen, die die q_i irgendwie untereinander permutiert.

Übt man nun nacheinander aus: zuerst die Substitution S_k der Galoisschen Gruppe, die η_1 in η_k überführt, dann die relationstreue Spezialisierung $(\eta_1, \ldots, \eta_h) \to (q_1, \ldots, q_h)$, die η_k in q_k überführt, schließlich die eben konstruierte Permutation P_k, die q_k in q_1 überführt, so erhält man eine relationstreue Spezialisierung, die η_1 in q_1 überführt, also mit der gegebenen $(\eta_1, \ldots, \eta_h) \to (q_1, \ldots, q_h)$ identisch sein muß.

Die Substitutionen S_k $(k = 1, \ldots, e f)$ mögen η_{ef+1} in η_{ef+k} überführen. Übt man danach die Spezialisierung $\eta_{ef+k} \to q_{ef+k}$ und die Permutation P_k aus, so muß schließlich q_{ef+1} herauskommen:

$$P_k \, q_{ef+k} = q_{ef+1} \, .$$

Die Punkte $q_{ef+1}, \ldots, q_{ef+ef}$ gehen also alle durch Isomorphismen P_{k}^{-1} aus dem einen Punkt q_{ef+1} hervor, d. h. sie sind zu q_{ef+1} konjugiert. Es gibt somit mindestens ebenso viele zu q_{ef+1} konjugierte Punkte als zu q_1 konjugierte. Selbstverständlich gilt auch das Umgekehrte, also gibt es zu jedem Punkt q_i gleich viele konjugierte Punkte wie zu q_1, d. h. alle f_i sind untereinander gleich.

Nebenbei bemerkt, bilden die S_k wieder eine Gruppe: die *Zerlegungsgruppe* der Spezialisierung $\eta_1 \to q_1$. Erweitert man den Grundkörper Ω durch Adjunktion von q_1, so wird $f = 1$, d. h. alle Punkte q_i sind rational durch q_1 ausdrückbar. Der Körper $\Omega(p, q_1)$ ist also normal über $\Omega(p)$ und mit den Körpern $\Omega(p, q_i)$ identisch. Seine Galois'sche Gruppe ist die Faktorgruppe der Zerlegungs- nach der Trägheitsgruppe.

Wie man sieht, ist die Voraussetzung, daß die η ganz algebraisch von
den ξ abhängen, bei allen diesen Beweisen nicht nötig, was die Anwendung, wie wir sehen werden, sehr vereinfacht. An die Stelle der von
Hasse, Krull und Lorenzen gemachten Voraussetzung der ganzen Abgeschlossenheit der Ringe $\Omega[\xi_1, \ldots, \xi_n]$ und $\Omega[\eta_1, \ldots, \eta_n]$ tritt allerdings die etwas schärfere, daß die algebraischen Mannigfaltigkeiten M
und N glatt sein sollen.

Auf Grund der hier entwickelten Begriffe können wir nun die Hasseschen Sätze I, II, III so formulieren:

18.1. *Sind $\mathfrak{D}$, $\mathfrak{A}$ zwei feste Gruppen von g Punkten auf Γ, ist μ ein
Multiplikator, $\mathfrak{Y}$ eine allgemeine Punktgruppe und $\mathfrak{X}$ durch*

$$\frac{\mathfrak{X}}{\mathfrak{D}} \sim \mu \frac{\mathfrak{Y}}{\mathfrak{A}} \tag{3}$$

*definiert, so bleibt diese Äquivalenz bei jeder relationstreuen Spezialisierung
$(\mathfrak{X}, \mathfrak{Y}) \to (\mathfrak{P}, \mathfrak{Q})$ erhalten:*

$$\frac{\mathfrak{P}}{\mathfrak{D}} \sim \mu \frac{\mathfrak{Q}}{\mathfrak{A}} \tag{4}$$

18.2. *Jede relationstreue Spezialisierung $\mathfrak{X} \to \mathfrak{P}$ läßt sich zu $(\mathfrak{X}, \mathfrak{Y}) \to$
$(\mathfrak{P}, \mathfrak{Q})$ fortsetzen, wobei nach 18.1 wieder (4) gilt.*

18.3. *Jede relationstreue Spezialisierung $\mathfrak{Y} \to \mathfrak{Q}$ läßt sich zu $(\mathfrak{X}, \mathfrak{Y}) \to$
$(\mathfrak{P}, \mathfrak{Q})$ fortsetzen. Ist dabei $\mathfrak{P}$ regulär, so ist $\mathfrak{P}$ durch (4) eindeutig bestimmt.*

Beweis. 18.1 folgt unmittelbar aus § 17. Sodann sind 18.2 und 18.3
nach 1.4 selbstverständlich.

Daß die Hassesche Formulierung dieser Sätze erheblich komplizierter
ist, liegt daran, daß Hasse die Paare $(\mathfrak{P}, \mathfrak{Q})$ nicht direkt durch relationstreue Spezialisierung definiert, sondern von $\mathfrak{X} \to \mathfrak{P}$ zuerst auf den „darin steckenden Punkt" $\mathfrak{X}^* \to \mathfrak{P}^*$ und von da aus erst auf den davon
„induzierten Punkt" $\mathfrak{Y} \to \mathfrak{Q}$ übergeht. Sein Ansatz zwingt ihn dazu, da
$\mathfrak{X}$ und $\mathfrak{Y}$ nicht von einer gemeinsamen Transzendenzbasis abhängig sind
und somit die Krull-Lorenzensche Zerlegungstheorie der „Punkte" nicht
ohne weiteres auf $\mathfrak{X}$ und $\mathfrak{Y}$ anwendbar ist. Aus diesem Grunde wird ein
$\mathfrak{X}^*$ zwischengeschoben, das von derselben Transzendenzbasis ganz abhängig ist wie $\mathfrak{X}$, und von welchem $\mathfrak{Y}$ seinerseits rational abhängt. In
unserem Aufbau ist das nicht nötig.

Wir wollen zum Schluß noch kurz zeigen, daß unsere Sätze 18.1 bis 3 tatsächlich dasselbe leisten wie die entsprechenden Sätze von Hasse, indem wir dieselben Folgerungen daraus ziehen. Der Multiplikator μ sei *regulär*, d. h. die durch (3) definierte Punktgruppe $\mathfrak{X}$ habe ebenso wie die allgemeine Punktgruppe $\mathfrak{Y}$ den Transzendenzgrad g. Dann ist $\mathfrak{X}$ auch regulär, also durch (3) eindeutig bestimmt. Da $\mathfrak{X}$ rational von $\mathfrak{Y}$ abhängt und beide denselben Transzendenzgrad haben, so hängt $\mathfrak{Y}$ umgekehrt algebraisch von $\mathfrak{X}$ ab, und bei gegebenem allgemeinem $\mathfrak{X}$ gibt es nur endlich viele Lösungen $\mathfrak{Y}$ von (3). Das kann man auch so formulieren: Die Gleichung

$$X = \mu\, Y \tag{5}$$

hat, wenn X eine allgemeine Nullklasse ist, nur endlich viele Lösungen. Die Anzahl dieser Lösungen ist der reduzierte Körpergrad $n = n(\mu)$ von $\Omega(\mathfrak{Y})$ über $\Omega(\mathfrak{X})$. Ist $\Omega(\mathfrak{Y})$ separabel über $\Omega(\mathfrak{X})$, so ist der reduzierte Körpergrad $n(\mu)$ gleich dem Körpergrad $N(\mu)$.

Die Lösungen von (5) gehen aus einer einzigen Lösung Y durch Multiplikation mit den Lösungen der Gleichung

$$1 = \mu\, D \tag{6}$$

hervor. Also hat diese Gleichung ebenfalls genau $n(\mu)$ Lösungen.

Nach 18.2 ist die Gleichung

$$A = \mu\, B \tag{7}$$

bei gegebener Nullklasse $A = \left|\dfrac{\mathfrak{P}}{\mathfrak{O}}\right|$ stets lösbar. Da ihre Lösungen aus einer einzigen durch Multiplikation mit den Lösungen von (6) hervorgehen, so hat die Gleichung (7) stets genau $n(\mu)$ Lösungen $B_1, \ldots, B_n$.

Wählt man nun bei gegebenem A und $\mathfrak{O}$ den Divisor $\mathfrak{A}$ so, daß die Klassen $B_1\mathfrak{A}, \ldots, B_n\mathfrak{A}$ alle regulär sind, so sind die Repräsentanten $\mathfrak{Q}_1, \ldots, \mathfrak{Q}_n$ dieser Klassen eindeutig bestimmt, und die Gleichung (4) hat nur endlich viele Lösungen $\mathfrak{Q}_1, \ldots, \mathfrak{Q}_n$. Die Voraussetzungen unserer „Zerlegungstheorie" sind dann anwendbar: die relationstreue Spezialisierung $\mathfrak{X} \to \mathfrak{P}$ ist eindeutig zu $(\mathfrak{X}, \mathfrak{Y}_1, \ldots, \mathfrak{Y}_n) \to (\mathfrak{P}, \mathfrak{Q}_1, \ldots, \mathfrak{Q}_n)$ fortsetzbar, die $\mathfrak{Q}$ zerfallen in r Klassen zu je f konjugierter $\mathfrak{Q}_i$ mit der Vielfachheit e. Nach 18.2 kommt aber jede von den n verschiedenen Lösungen der Gleichung (4) unter diesen $\mathfrak{Q}_i$ vor. Also muß die Vielfachheit $e = 1$ sein.

Damit sind alle Ergebnisse von Hasse § 7, 3—4 neu hergeleitet, und zwar sehr viel einfacher als bei Hasse.

(Eingegangen den 22. Oktober 1946.)

25.

Über einfache Punkte von algebraischen Mannigfaltigkeiten

Mathematische Zeitschrift 51, 4 (1948) 497–501

Neuerdings hat ABELLANAS[1]) für einfache Punkte einer algebraischen Mannigfaltigkeit Reihenentwicklungen der Koordinaten nach Ortsuniformisierenden hergeleitet und mit ihrer Hilfe bewiesen, daß meine Definition des Begriffes einfacher Punkt[2]) einer von ZARISKI[3]) gegebenen Definition äquivalent ist. Wir wollen nun zeigen, daß die Reihenentwicklung etwas einfacher hergeleitet werden kann, wenn man nach LIOUVILLE statt der einzelnen Koordinaten $\xi_1, \ldots, \xi_n$ eine Linearkombination $u_1\xi_1 + \cdots + u_n\xi_n$ einführt und diese in eine Potenzreihe entwickelt. Mit derselben Methode zeigen wir, daß die Koordinaten eines allgemeinen Punktes von M differenzierbare Funktionen der Ortsuniformisierenden eines einfachen Punktes η sind, deren Ableitungen an der Stelle η endlich bleiben. Schließlich geben wir einen direkten Beweis für die Äquivalenz unserer Einfachheitsdefinition mit der von ZARISKI. Der Grundkörper wird immer als perfekt vorausgesetzt.

§ 1.

Herleitung der Potenzreihenentwicklung.

Wir gehen von der zugeordneten Form $F(u^{(0)}, u^{(1)}, \cdots, u^{(d)})$ einer irreduziblen d-dimensionalen Mannigfaltigkeit M in S_n aus.[4]) Sie ist ganzrational und symmetrisch in den Hyperebenen $u^{(0)}, \ldots, u^{(d)}$ und zerfällt als Funktion der $u = u^{(0)}$ in Primfaktoren, die den Schnittpunkten der Hyperebenen $u^{(1)}, \ldots, u^{(d)}$ mit M entsprechen:

$$(1) \qquad F(u; u^{(1)}, \ldots, u^{(d)}) = \prod_\nu (u_0\xi_0^{(\nu)} + u_1\xi_1^{(\nu)} + \cdots + u_n\xi_n^{(\nu)}).$$

Ersetzen wir $u^{(1)}, \ldots, u^{(d)}$ durch spezielle Hyperebenen, die die Mannigfaltigkeit M in endlich vielen Punkten $\eta^{(\nu)}$ schneiden, so gilt die Zerlegung (1) immer noch:

$$(2) \qquad F(u; v^{(1)}, \ldots, v^{(d)}) = \prod_\nu (u_0\eta_0^{(\nu)} + u_1\eta_1^{(\nu)} + \cdots + u_n\eta_n^{(\nu)})$$

---

[1]) P. ABELLANAS, Rev. Acad. Ci. exact. Madrid **36**, 482—499 (1942).

[2]) Siehe etwa B. L. VAN DER WAERDEN, Einführung in die algebraische Geometrie, § 38.

[3]) O. ZARISKI, Ann. of Math. (2) **40**, 639—689 (1939).

[4]) CHOW und VAN DER WAERDEN, ZAG 9, Math. Ann. **113**, (1937) S. 692.

362

und die $\eta^{(\nu)}$ erscheinen rechts mit ihren richtigen Vielfachheiten. Ist nun η ein einfacher Punkt von M und wählt man η als Koordinatenanfangspunkt $(1, 0, \ldots, 0)$, so kann man die Koordinaten so numerieren, daß die Hyperebenen $x_1 = 0, \ldots, x_d = 0$ mit dem Tangentialraum von M nur den Punkt η gemeinsam haben. Setzt man dann in (2) für $v^{(1)}, \ldots, v^{(d)}$ diese Hyperebenen $x_1 = 0, \ldots, x_d = 0$, so spaltet $F(u; v^{(1)}, \ldots, v^{(d)})$ einen einfachen Faktor u_0, dem einfachen Schnittpunkt $\eta = (1, 0, \ldots, 0)$ entsprechend, ab. Setzt man in (2) allgemeiner für $v^{(1)}, \ldots, v^{(d)}$ die Hyperebenen

$$(3) \qquad x_1 - t_1 x_0 = 0, \quad x_2 - t_2 x_0 = 0, \quad \ldots, \quad x_d - t_d x_0 = 0,$$

so erhält man aus (2) eine Form

$$(4) \qquad F(u_0, \ldots, u_n; t_1, \ldots, t_d) = \prod_\nu (u_0 \eta_0^{(\nu)} + u_1 \eta_1^{(\nu)} + \cdots + u_n \eta_n^{(\nu)}),$$

die für $t_1 = \cdots = t_d = 0$ einen einfachen Faktor u_0 abspaltet:

$$F(u_0, \ldots, u_n; 0, \ldots, 0) = u_0 \cdot h(u_0, \ldots, u_n).$$

Nun wenden wir eine Verallgemeinerung des HENSELschen Lemmas an:

Lemma: *Wenn ein Polynom $F(u_0; t_1, \ldots, t_d)$ für $t_1 = \cdots = t_d = 0$ in zwei teilerfremde Faktoren $g(u_0) \cdot h(u_0)$ zerfällt, so zerfällt es für unbestimmte $t_1, \ldots, t_d$ in zwei Faktoren $G(u_0) \cdot H(u_0)$ deren Koeffizienten Potenzreihen in $t_1, \ldots, t_d$ sind und die für $t_1 = \cdots = t_d = 0$ in $g(u_0)$ und $h(u_0)$ übergehen, wobei $G(u_0)$ denselben Grad hat wie $g(u_0)$.*

Den Beweis dieses Lemmas findet man bei ABELLANAS; er verläuft genau so wie im Fall einer Veränderlichen t_1. Der Beweis ist übrigens besonders einfach in dem hier vorliegenden besonderen Fall $g(u_0) = u_0$, da man dann nur anzusetzen braucht

$$G(u_0) = u_0 - (P_1 + P_2 + \cdots)$$

wo $P_1, P_2, \ldots$ Polynome vom Grade $1, 2, \ldots$ in $t_1, \ldots, t_d$ sind und so bestimmt werden, daß die Potenzreihe $P_1 + P_2 + \cdots$ eine Nullstelle von $F(u_0; t_1, \ldots, t_d)$ wird:

$$(5) \qquad F(P_1 + P_2 + \cdots; t_1, \ldots, t_d) = 0.$$

Entwickelt man in (5) beide Seiten nach Gliedern steigenden Grades in $t_1, \ldots, t_d$, so erhält man eine Reihe von Gleichungen der Gestalt

$$\alpha P_1 + Q_1 = 0$$
$$\alpha P_2 + \beta P_1^2 + \cdots = 0$$
$$\ldots\ldots$$

aus denen man sukzessiv $P_1, P_2, \ldots$ rational ausrechnen kann. Als Nenner erscheint dabei nur der Koeffizient α des Gliedes u_0 in $F(u_0; 0, \ldots, 0)$, der nach Voraussetzung nicht verschwindet.

Unser Beweis ergibt noch folgenden Zusatz: Enthält F außer u_0, noch $u_1, \ldots, u_n$, so sind auch $P_1, P_2, \ldots$ rationale Funktionen von $u_1, \ldots, u_n$.

In unserem Fall sind, wie sich aus der Zerlegung (4) ergibt, alle Nullstellen von $F(u_0; t_1, \ldots, t_d)$ Linearformen in $u_1, \ldots, u_n$; also muß auch $P_1 + P_2 + \cdots$ eine Linearform in $u_1, \ldots, u_n$ sein:

$$P_1 + P_2 + \cdots = \sum_1^n - \xi_k u_k$$

wo $\xi_1, \ldots, \xi_n$ natürlich wieder Potenzreihen in $t_1, \ldots, t_d$ sind. Setzt man noch $\xi_0 = 1$, so sieht man, daß das Polynom (4) einen Linearfaktor

$$u_0 - (P_1 + P_2 + \cdots) = \xi_0 u_0 + \xi_1 u_1 + \cdots + \xi_n u_n$$

hat, wobei $\xi_1, \ldots, \xi_n$ Potenzreihen in $t_1, \ldots, t_d$ sind, die für $t_1 = \cdots = t_d = 0$ in Null übergehen. Der Punkt ξ ist dabei einer von den Schnittpunkten $\eta^{(\nu)}$ der Hyperebene (3) mit M; daraus folgt wegen $\xi_0 = 1$ unmittelbar:

$$\xi_1 = t_1, \quad \xi_2 = t, \quad \ldots, \quad \xi_d = t_d.$$

Damit ist bewiesen:

Satz 1. *Die Mannigfaltigkeit M besitzt einen allgemeinen Punkt dessen Koordinaten Potenzreihen in den „Ortsuniformisierenden" $t_1, \ldots, t_d$ sind, welche für $t_1 = 0, \ldots, t_d = 0$ in die Koordinaten des gegebenen einfachen Punktes η übergehen, und zwar können bei geeigneter Numerierung der Koordinaten mit der Normierung $\xi_0 = 1$ die ersten d Koordinaten $\xi_1, \ldots, \xi_d$ selbst als Ortsuniformisierende verwendet werden.*

§ 2.

Differentiation nach den Ortsuniformisierenden.

Wie wir sahen, hat $F(u_0, u_1, \ldots, u_n; t_1, \ldots, t_d)$ als Polynom in u_0 eine einfache Nullstelle $\vartheta = -\xi_1 u_1 - \cdots - \xi_n u_n$. Da die Nullstelle einfach ist, ist die Ableitung $\dfrac{\partial F}{\partial u_0}(\xi) \neq 0$, und das bleibt richtig für $t_1 = 0, \ldots$, $t_d = 0$. Demnach ist ϑ eine separable Funktion von $u_1, \ldots, u_n; t_1, \ldots, t_d$ und man kann die partiellen Ableitungen

$$\frac{\partial \vartheta}{\partial t_k} = -\frac{\dfrac{\partial F}{\partial t_k}(\vartheta)}{\dfrac{\partial F}{\partial u_0}(\vartheta)}$$

definieren. Ersetzt man eine Unbestimmte u_i durch $u_i - 1$, bildet also

$$\vartheta + \xi_i = -\xi_1 u_1 - \cdots - \xi_i (u_i - 1) - \cdots - \xi_n u_n,$$

364

so ist $\vartheta + \xi_i$ ebenfalls eine separable, differenzierbare Funktion, und dasselbe gilt folglich von der Differenz

$$(\vartheta + \xi_i) - \vartheta = \xi_i.$$

Die partiellen Ableitungen der ξ_i nach den t_k existieren also sämtlich, und man kann sie als rationale Funktionen von $u_1, \ldots, u_n, \xi_1, \ldots, \xi_n$ darstellen, deren Nenner für $\xi = \eta$ nicht verschwinden:

$$\frac{\partial \xi_i}{\partial t_k} = \frac{g(u_1, \ldots, u_n; \xi_1, \ldots, \xi_n)}{h(u_1, \ldots, u_n; \xi_1, \ldots, \xi_n)}.$$

Die linke Seite hängt von $u_1, \ldots, u_n$ gar nicht ab. Ordnet man also rechts Zähler und Nenner nach Potenzprodukte der $u_1, \ldots, u_n$, so sind die Koeffizienten dieser Potenzprodukte proportional und man kann die linke Seite als Quotient irgend zweier solcher Koeffizienten darstellen.

$$(6) \qquad \frac{\partial \xi_i}{\partial t_k} = \frac{g_1(\xi)}{h_1(\xi)}.$$

Man kann diese Darstellung sogar so erwählen, daß der Nenner nicht Null wird für $\xi = \eta$:

$$h_1(\eta) \neq 0.$$

Damit ist bewiesen:

Satz 2. *Die Koordinaten ξ_i sind separable algebraische Funktionen von den Ortsuniformisierenden t_k. Ihre partiellen Ableitungen lassen sich als Quotienten (6) so darstellen, daß der Nenner an der Stelle η nicht Null wird.*

§ 3.

Die Zariskische Definition eines einfachen Punktes.

Die ZARISKISCHE Definition heißt so: η ist ein einfacher Punkt der d-dimensionalen Mannigfaltigkeit M mit dem zugehörigen Primideal $\mathfrak{m}$, wenn in dem Restklassenring $\mathfrak{o}/\mathfrak{m}$ das zum Punkt η gehörige Primideal $\bar{\mathfrak{p}} = \mathfrak{p}/\mathfrak{m}$ sich als isolierte Primärkomponente eines Primideals mit einer d-gliedrigen Basis darstellen läßt:

$$(7) \qquad (\bar{f}_1, \ldots, \bar{f}_d) = [\bar{\mathfrak{p}}, \bar{\mathfrak{q}}_2, \ldots, \bar{\mathfrak{q}}_r] \qquad \text{mit } \bar{\mathfrak{q}}_k \not\equiv 0\,(\bar{\mathfrak{p}}).$$

Löst man die Restklassen nach $\mathfrak{m}$ in ihre Elemente auf, so kann man statt (7) auch schreiben

$$(\mathfrak{m}, f_1, \ldots, f_d) = [\mathfrak{p}, \mathfrak{q}_1, \ldots, \mathfrak{q}_r] \qquad \text{mit } \mathfrak{q}_k \not\equiv 0\,(\mathfrak{p}).$$

wobei $\mathfrak{p}$ das zum Punkt η gehörige Primideal ist:

$$\mathfrak{p} = (x_1 - \eta_1, \ldots, x_n - \eta_n).$$

Dazu ist nach Mod. Alg. II, § 96 notwendig und hinreichend, daß in dem Ideal $(\mathfrak{m}, f_1, \ldots, f_d)$ solche Formen vorkommen, deren Tangentialhyperebenen im Punkt η nur den Punkt η gemeinsam haben. Wenn

365

das der Fall ist, so hat der Punkt η als Schnittpunkt von M mit den Hyperflächen $f_1 = 0, \ldots, f_d = 0$ nach einem bekannten Kriterium (ZAG 5, Math. Ann. 110, oder Einf. alg. Geom. § 39) die Multiplizität Eins, also ist η dann auch ein einfacher Punkt in unserem Sinne.

Nun sei umgekehrt η ein einfacher Punkt in unserem Sinne. Wir wählen für η wieder den Koordinatenanfangspunkt und numerieren die Koordinaten so, daß die Hyperebenen $x_1 = 0, \ldots, x_d = 0$ mit dem Tangentialraum von M nur den Punkt η gemeinsam haben. Der Tangentialraum S_d war (Einf. alg. Geom. § 40) definiert als Durchschnitt aller Tangentialhyperebenen von Hyperflächen durch M. Unter diesen Tangentialhyperebenen gibt es $n-d$ linear Unabhängige, deren Durchschnitt genau S_d ist. Diese seien etwa die Tangentialhyperebenen der Hyperflächen $f_{d+1} = 0, \ldots, f_n = 0$.

Weiter setzen wir $f_1 = x_1, \ldots, f_d = x_d$. Dann haben die Tangentialhyperebenen der Hyperflächen $f = 0, \ldots, f_d = 0, f_{d+1} = 0, \ldots, f_n = 0$ im Punkt η nur den Punkt η miteinander gemeinsam. Daraus folgt nach dem zitierten Satz (Mod. Alg. II, § 96), daß $\mathfrak{p} = (x_1, \ldots, x_n)$ eine isolierte Primärkomponente des Ideals $(\mathfrak{m}, f_1, \ldots, f_d)$ ist, d. h. η ist einfach im Sinne der Definition von ZARISKI.

(Eingegangen am 5. März 1945.)

26.

Birationale Transformation von linearen Scharen auf algebraischen Mannigfaltigkeiten

Mathematische Zeitschrift 51, 4 (1948) 502–523

Einleitung.

Wenn man die grundlegenden Abhandlungen der italienischen Schule über die Geometrie auf einer algebraischen Fläche liest, wo von linearen Scharen und Vollscharen, von virtuellen und effektiven Multiplizitäten von Basispunkten die Rede ist, so kommt man zwangsläufig auf die Frage, wie man die darin vorkommenden Begriffe einfach, exakt und verallgemeinerungsfähig fassen könnte. Die Lösung dieses Problems kann ich heute vorlegen. Sie beruht auf der Bewertungstheorie.

Es ist leicht, eine lineare Schar von Kurven ohne feste Bestandteile auf einer algebraischen Fläche bei einer birationalen Transformation so zu transformieren, daß die transformierte Schar wieder eine lineare Schar ohne feste Bestandteile ist. Aber nun stößt man auf die paradoxe Erscheinung, daß eine Vollschar nicht ohne weiteres in eine Vollschar übergeht. Bei einer quadratischen Transformation der Ebene z. B.:

$$x_0 : x_1 : x_2 = x_1' x_2' : x_2' x_0' : x_0' x_1'$$

geht die lineare Schar aller Geraden der Ebene

$$\lambda_0 x_0 + \lambda_1 x_1 + \lambda_2 x_2 = 0$$

in eine Kegelschnittschar mit 3 Basispunkten

$$\lambda_0 x_1' x_2' + \lambda_1 x_2' x_0' + \lambda\, x_0' x_1' = 0$$

über, die offensichtlich keine Vollschar ist, da sie sich zur fünfgliedrigen Schar aller Kegelschnitte der Ebene erweitern läßt.

Um die Situation zu retten, muß man den Begriff einer *linearen Schar mit vorgegebenen Basispunkten* einführen. Schreibt man nämlich in unserem Beispiel den Kegelschnitten der Schar ihre drei festen Punkte vor, so ist die Schar vollständig. In anderen Fällen muß man vorschreiben, daß die Kurven der Schar in einem Basispunkt eine gegebene Tangente haben oder durch gegebene Punkte doppelt oder dreifach hindurchgehen sollen.

Nun kann es vorkommen, daß die Kurven einer linearen Schar in einem Basispunkt effektiv eine höhere Multiplizität haben als die vorgeschriebene; daher unterscheidet man zwischen der *effektiven* und

367

der *virtuellen* (vorgeschriebenen) Multiplizität. Ist die effektive Multiplizität eines Basispunktes größer als die virtuelle, so muß man, wenn dieser Basispunkt durch eine birationale Transformation in eine Kurve übergeführt wird, nach ENRIQUES[1]) diese Kurve als festen Bestandteil zu sämtlichen Kurven der transformierten Schar hinzunehmen, mit einer Vielfachheit gleich der Übermultiplizität, d. h. gleich der Differenz von effektiver und virtueller Multiplizität vor der Transformation.

Will man diese Begriffe erschöpfend erklären, so muß man nicht nur gewöhnliche Basispunkte, sondern auch unendlich benachbarte in Betracht ziehen. Die Angelegenheit wird dann alsbald so kompliziert, daß eine Verallgemeinerung auf mehr als 2 Dimensionen fast aussichtslos erscheint. Jedoch gelingt diese Verallgemeinerung durch den *Bewertungsbegriff*. Dieser bildet einen vollständigen Ersatz für den Begriff des unendlich benachbarten Punktes und macht jedes Eingehen auf die Struktur dieser Nachbarpunkte überflüssig. Er ermöglicht eine birational invariante Theorie der linearen Scharen und Vollscharen auf algebraischen Mannigfaltigkeiten, wie in § 4—8 auseinandergesetzt werden soll.

Vorher möge in enger Anlehnung an KÄHLER[2]) eine einfachere Art, birationale Invarianten zu gewinnen, behandelt werden, die den Bewertungsbegriff noch nicht voraussetzt. Beide Theorien beruhen auf einem Satz über Funktionen auf einer Mannigfaltigkeit, den wir jetzt herleiten wollen.

§ 1.

Ein Satz über rationale Funktionen auf einer Mannigfaltigkeit.

Das folgende Lemma gehört dem Gedankenkreis des NOETHERschen Fundamentalsatzes an:

Lemma. *Ist M eine glatte (d. h. singularitätenfreie) d-dimensionale Mannigfaltigkeit und sind alle Schnittbestandteile von M mit einer Hyperfläche $f = 0$ mit mindestens derselben Vielfachheit auch in dem Schnitt von M mit einer zweiten Hyperfläche $F = 0$ enthalten, so gilt eine Identität*

$$(1) \qquad\qquad x_0^\varrho F = Qf + R$$

wobei die Hyperfläche $R = 0$ die Mannigfaltigkeit M enthält.

Beweis. Die Gesamtheit aller Formen G, deren Produkt mit F sich in der gewünschten Weise darstellen läßt:

$$GF = Qf + R$$

ist ein H-Ideal. Wenn wir beweisen können, daß dieses H-Ideal keine

---

[1]) Siehe Enzyklopaedie d. math. Wiss. III C 6 b und die dort angeführte Literatur.

[2]) E. KÄHLER, Forme differenziali e funzioni algebriche, Mem. Accad. Ital. Mat. **3**, 3, 1—19.

Nullstelle außer der trivialen $(0, \ldots, 0)$ besitzt, so sind wir fertig. Daraus folgt nämlich nach dem HILBERTschen Nullstellensatz, daß eine geeignete Potenz von x_0 in dem H-Ideal enthalten ist, also (1) gilt.

Es sei also P ein Punkt des Raumes. Wir haben dann zu beweisen, daß es in dem H-Ideal eine Form G gibt. die in P nicht Null wird. Das bedeutet nämlich, daß P keine Nullstelle des H-Ideals ist.

Wenn P nicht auf M liegt, so gibt es eine Form G, die auf M überall Null wird, aber nicht in P. Dann ist auch GF Null auf M; wir können also

$$GF = R = 0 \cdot f + R$$

setzen und schließen, daß die Form G zum H-Ideal gehört. Damit ist dieser Fall erledigt: von jetzt an liege P auf M.

Es sei S_{n-d-1} ein linearer Raum, der weder M noch den Tangentialraum von M in P trifft. Beide Forderungen sind erfüllt, wenn S_{n-d-1} *allgemein* gewählt wird, und das wollen wir auch mit Rücksicht auf den folgenden Beweis tun.

Dann wird der Verbindungsraum S_{n-d} von S_{n-d-1} mit P die Mannigfaltigkeit M nur in endlich vielen Punkten treffen, unter denen P mit der Multiplizität 1 vorkommt. Wir wählen das Koordinatensystem so, daß S_{n-d-1} die Gleichungen $x_0 = 0$, $x_1 = 0$, $\ldots$, $x_d = 0$ hat.

Für unbestimmte $\xi_0, \xi_1, \ldots, \xi_d$ hat der durch S_{n-d-1} gehende lineare Raum

$$x_0 \xi_1 - x_1 \xi_0 = 0, \; x_0 \xi_2 - x_2 \xi_0 = 0, \; \ldots, \; x_0 \xi_d - x_d \xi_0 = 0$$

ebenfalls nur endlich viele Schnittpunkte mit M. Diese Punkte $\xi^{(\nu)} = (\xi_0, \ldots, \xi_d, \xi_{d+1}^{(\nu)}, \ldots, \xi_n^{(\nu)})$ sind dann allgemeine Punkte von M, also in Bezug auf den Körper k der rationalen Funktionen von $\xi_0, \ldots, \xi_d$ konjugiert. Einer von ihnen heiße ξ. Alle seine Koordinaten $\xi_0, \ldots, \xi_n$ sind dann ganze algebraische Funktionen der Unbestimmten $\xi_0, \ldots, \xi_d$, und der Grad des Körpers $k(\xi)$ über k ist gleich dem Grad γ der Mannigfaltigkeit M.

In diesem Körper bilden wir nun die Normen $n(\xi)$ und $N(\xi)$ der Körperelemente $f(\xi)$ und $F(\xi)$. Aus der Definition

$$N(\xi) = \prod_{\nu=1}^{\gamma} F(\xi^{(\nu)})$$

folgt leicht, daß $N(\xi)$ ein homogenes Polynom in $\xi_0, \ldots, \xi_d$ vom Grade $\beta\gamma$ ist, wo β der Grad der Form F ist. Ebenso ist $n(\xi)$ homogen in $\xi_0, \ldots, \xi_d$.

Wir wollen nun *erstens* zeigen, daß $n(\xi)$ durch $f(\xi)$ und $N(\xi)$ durch $F(\xi)$ teilbar ist:

$$(2) \qquad\qquad n(\xi) = g(\xi)\, f(\xi)$$

$$(3) \qquad\qquad N(\xi) = G(\xi)\, F(\xi),$$

zweitens, daß die Form N durch n teilbar ist:

$$(4) \qquad\qquad N(\xi) = A(\xi)\, n(\xi)$$

drittens, daß die in (3) vorkommende Form G im Punkte P nicht Null wird:

$$(5) \qquad\qquad G(P) \neq 0.$$

Aus (2), (3), (4) folgt:

$$G(\xi)\, F(\xi) = A(\xi)\, g(\xi)\, f(\xi)$$

oder, wenn $A \cdot g = Q$ gesetzt wird,

$$G(\xi)\, F(\xi) - Q(\xi)\, f(\xi) = 0.$$

Folglich ist $R = GF - Qf$ Null auf M und wir haben

$$GF = Qf + R.$$

Somit gehört G zum H-Ideal, womit alles bewiesen ist. Also brauchen wir nur noch (2), (3), (4), (5) zu beweisen.

$F(\xi)$ genügt einer algebraischen Gleichung vom Grade γ, deren letztes Glied gerade $N(\xi)$ ist:

$$F(\xi)^\gamma + a_1 F(\xi)^{\gamma-1} + \ldots + a_{\gamma-1} F(\xi) \pm N(\xi) = 0$$

Löst man $N(\xi)$ aus dieser Gleichung auf, so erhält man gerade die Gleichung (3). Ebenso beweist man (2).

Um (4) zu beweisen, müssen wir die Hyperflächen $N = 0$ und $n = 0$ näher charakterisieren. Wir behaupten:

A. Die Hyperfläche $N = 0$ ist der projizierende Kegel des Durchschnittes von $F = 0$ mit M aus dem linearen Raum S_{n-d-1}, und ihre irreduziblen Bestandteile haben dieselben Vielfachheiten wie die entsprechenden Bestandteile dieses Durchschnittes. Entsprechend verhält sich $n = 0$ zu $f = 0$.

Nehmen wir diese Behauptung für einen Augenblick als bewiesen an, so folgt daraus, da die irreduziblen Bestandteile des Durchschnittes von $f = 0$ mit M mit mindestens derselben Vielfachheit auch im Durchschnitt von $F = 0$ mit M enthalten sind, daß der projizierende Kegel $n = 0$ in dem projizierenden Kegel $N = 0$ enthalten ist. Also ist die Form $N(x)$ durch $n(x)$ teilbar, womit (4) bewiesen ist.

Aber auch (5) folgt jetzt leicht. Nach ZAG 14, § 5 (Math. Anm. **115**, S. 635) setzt der Durchschnitt des projizierenden Kegels $N = 0$ mit M sich zusammen aus dem Durchschnitt von $F = 0$ mit M und einem Restschnitt S, der den Punkt P nicht enthält. Nun folgt aus (3), daß $N - GF$ Null ist auf M, also ist der Durchschnitt von $N = 0$ mit M derselbe wie der von $GF = 0$ mit M, also zusammengesetzt aus dem Durchschnitt von $F = 0$ mit M und dem von $G = 0$ mit M. Also ist der Restschnitt S genau der Durchschnitt der Hyperfläche $G = 0$ mit M. Da P nicht zu diesem Durchschnitt gehört, ist $G(P) \neq 0$.

Es kommt also nur noch darauf an, die Behauptung A für beliebige Hyperflächen $F = 0$ zu beweisen. Es sei also I ein irreduzibler Bestandteil des Durchschnittes von $F = 0$ mit M mit der Vielfachheit μ, und es sei K der projizierende Kegel von I, d. h. die Gesamtheit aller Punkte, deren Verbindungsraum mit S_{n-d-1} einen Punkt von I enthält. Wir haben zu beweisen, daß K mit derselben Vielfachheit μ in $N = 0$ enthalten ist.

Die Form $N(x)$ hängt nur von $x_0, \ldots, x_d$ ab. Zerlegen wir sie in irreduzible Faktoren

$$N(x) = N_1 N_2 \ldots N_h,$$

so hängen auch diese Faktoren nur von $x_0, \ldots, x_d$ ab. Wenn also ein Punkt Q von M einer Hyperfläche $N_i(x) = 0$ angehört, so gehört auch der ganze Verbindungsraum von Q mit S_{n-d-1} der Hyperfläche $N_i(x) = 0$ an, und wenn die irreduzible Mannigfaltigkeit I einer Hyperfläche $N_i(x) = 0$ angehört, so gehört der ganze irreduzible Verbindungskegel K dieser Hyperfläche an, also ist dann K mit dieser Hyperfläche $N_i = 0$ identisch.

Aus (3) folgt, wie schon bemerkt, daß $N = 0$ dieselbe Schnittmannigfaltigkeit mit M hat wie $GF = 0$. In dieser Schnittmannigfaltigkeit ist I als Bestandteil enthalten, und zwar mindestens mit der Multiplizität μ. Also muß mindestens einer von den irreduziblen Bestandteilen $N_i = 0$ die Mannigfaltigheit I enthalten, und zwar müssen so viele von ihnen I enthalten, daß insgesamt mindestens die Schnittmultiplizität μ herauskommt. Wir haben aber gesehen, daß die Hyperflächen $N_i = 0$, die I enthalten, mit dem Kegel K identisch sind. Also muß der Kegel K so oft als Bestandteil in $N = 0$ enthalten sein, daß insgesamt die Schnittmultiplizität μ für I herauskommt.

Nun folgt aber aus dem angeführten Satz ZAG 14, § 5, daß I nur einmal gezählt als Bestandteil in dem Durchschnitt von M und K enthalten ist. Also muß der Kegel K mindestens μ mal in der Hyperfläche $N = 0$ enthalten sein, damit die erforderliche Schnittmultiplizität μ herauskommt.

Jetzt müssen wir noch zeigen, daß K nicht mehr als μ mal als Bestandteil in der Hyperfläche $N = 0$ enthalten ist.

Der Durchschnitt von $F = 0$ und M hat, wenn seine irreduziblen Bestandteile I mit ihren Vielfachheiten μ gezählt werden, den Grad $\beta\gamma$. Also hat auch der projizierende Kegel dieses Durchschnittes, wenn seine irreduziblen Bestandteile K mit ebendenselben Vielfachheiten μ gezählt werden, den Grad $\beta\gamma$. Die Hyperfläche $N = 0$ hat aber auch nur den Grad $\beta\gamma$. Also kann sie die Kegel K nicht öfter als je μ mal als Bestandteile enthalten. Damit ist die Behauptung A bewiesen und der Beweis des Lemmas beendet.

Wir betrachten nun rationale Funktionen auf M

$$\varphi(\xi_0, \ldots, \xi_n) = \frac{f(\xi_0, \ldots, \xi_n)}{g(\xi_0, \ldots, \xi_n)}$$

wo ξ ein allgemeiner Punkt von M ist und wo f und g Formen gleichen Grades sind. Die λ-Stellen dieser Funktion werden durch die Gleichung

$$f - \lambda g = 0$$

gegeben. Statt dessen kann man, wenn $\lambda = -\lambda_2 : \lambda_1$ gesetzt wird auch schreiben

$$(8) \qquad \lambda_1 f + \lambda_2 g = 0.$$

Diese Gleichung definiert eine lineare Schar von Mannigfaltigkeiten M_{d-1} auf M. Die festen Bestandteile dieser linearen Schar können außer Betracht bleiben. Die übrig bleibenden Teile definieren die *Niveaumannigfaltigkeiten* der Funktion φ, insbesondere für $\lambda_2 = 0$ die *Nullmannigfaltigkeit* und für $\lambda_1 = 0$ die *Polmannigfaltigkeit* von φ Alle diese sind rein $(d-1)$-dimensial, und jeder ihrer irreduziblen Bestandteile hat eine wohldefinierte Vielfachheit. Die Vielfachheit eines irreduziblen Bestandteils $\mathfrak{P}$ in der Polmannigfaltigkeit z. B. ist gleich der Vielfachheit von $\mathfrak{P}$ als Bestandteil der Schnittmannigfaltigkeit von $g = 0$ mit M, vermindert um das Minimum der Vielfachheiten von $\mathfrak{P}$ als Bestandteil der Schnittmannigfaltigkeiten von $f = 0$ und $g = 0$ mit M.

S a t z 1. *Wenn ein Punkt P von M nicht auf der Polmannigfaltigkeit der Funktion* $\varphi = \dfrac{f}{g}$ *liegt, so kann man den Nenner g so wählen, daß er nicht Null wird im Punkte P.* Man sagt dann, daß die Funktion φ *endlich bleibt* in P.

B e w e i s. Die Hyperflächen $f = 0$ und $g = 0$ mögen aus M Schnittmannigfaltigkeiten AC und BC ausschneiden, wobei B die Polmannigfaltigkeit von φ ist, zu der P nach Voraussetzung nicht gehört. Wir denken uns die Numerierung der Koordinaten so, daß $x_0 \neq 0$ in P. Nun legen wir durch die Mannigfaltigkeit B eine Hyperfläche $G = 0$, die nicht durch P geht. Sie möge M in BD schneiden. Die Hyperfläche $fG = 0$ schneidet dann M in $ABCD$. Da diese Schnittmannigfaltigkeit BC als Bestandteil enthält, so folgt nach dem Lemma

$$x_0^\varrho f G = Q g + R$$

also

$$\xi_0^\varrho f(\xi) G(\xi) = Q(\xi) g(\xi)$$

$$\frac{f(\xi)}{g(\xi)} = \frac{Q(\xi)}{\xi_0^\varrho G(\xi)}$$

wobei $x_0^\varrho G(x)$ im Punkte P nicht Null wird.

Ich habe an anderer Stelle[3]) ausgeführt, daß es in jedem einfachen Punkt P von M ein System von Ortsuniformisierenden $u_1, \ldots, u_d$ gibt, die dem Körper der rationalen Funktionen auf M angehören und für die man sogar geeignete Verhältnisse wie $\xi_0^{-1}\xi_k$ wählen kann, derart,

---

[3]) In der vorangehenden Arbeit, Über einfache Punkte von algebraischen Mannigfaltigkeiten.

daß, wenn etwa $\xi_0 = 1$ normiert wird, alle Ableitungen $\dfrac{d\,\xi_i}{d\,u_j}$ an der Stelle P endlich bleiben. Bei geeigneter Numerierung der Koordinaten sind $\xi_0^{-1}\xi_1, \xi_0^{-1}\xi_2, \ldots, \xi_0^{-1}\xi_d$ Ortsuniformisierende für alle Punkte von M, mit Ausnahme einer höchstens $(d-1)$-dimensionalen Teilmannigfaltigkeit.

S a t z 2. *Wenn* $u_1, \ldots, u_d$ *Ortsuniformisierende für* P *sind und wenn eine rationale Funktion* φ *an der Stelle* P *endlich bleibt, so bleiben auch alle partiellen Ableitungen* $\dfrac{\partial\,\varphi}{\partial\,u_j}$ *an der Stelle* P *endlich.*

B e w e i s. Nach Satz 1 ist φ als Quotient

$$\varphi = \frac{f(\xi)}{g(\xi)} = \frac{f(1,\xi_1,\ldots,\xi_n)}{g(1,\xi_1,\ldots,\xi_n)}$$

darstellbar, wobei $g(P) \neq 0$ ist. Nach der Regel für totale Differentiation ist nun

$$\frac{\partial\,\varphi}{\partial\,u_j} = \sum_k \left(\frac{\partial}{\partial\,\xi_k}\frac{f(\xi)}{g(\xi)}\right)\frac{\partial\,\xi}{\partial\,u_j} = \sum_k \frac{g(\xi)\,\partial_k f(\xi) - f(\xi)\,\partial_k g(\xi)}{g(\xi)^2}\frac{\partial\,\xi_k}{\partial\,u_i}$$

Rechts bleibt alles endlich an der Stelle P, also bleibt auch die linke Seite endlich für P.

§ 2.
Die Kählerschen Tensoren.

Die inhomogenen Koordinaten $\xi_1, \ldots, \xi_n$ eines allgemeinen Punktes einer glatten d-dimensionalen Mannigfaltigkeit M seien separable algebraische Funktionen von $\xi_1, \ldots, \xi_d$. Es sei weiter $\varphi_{\lambda\mu\ldots\pi}$ ein System von rationalen Funktionen auf M, wobei jede der Indices $\lambda\mu\ldots\pi$ von 1 bis d läuft. Ein solches System heißt ein *Tensor* φ auf M, und die $\varphi_{\lambda\mu\ldots\pi}$ heißen die *Komponenten* des Tensors in Bezug auf $\xi_1, \ldots, \xi_d$.

Die Tensorkomponenten können Symmetriebedingungen von der Art

$$\sum_S \alpha_S\, S\, \varphi_{\lambda\mu\ldots\pi} = 0$$

unterworfen werden, wo die α_S Konstante und die S Permutationen der Indices sind; z. B. kann verlangt werden, daß die Tensoren symmetrisch oder antisymmetrisch sind. Ein antisymmetrischer oder alternierender Tensor mit s Indices definiert ein s-faches Differential auf M:

$$(1) \qquad d\,w = \sum \varphi_{\lambda\mu\ldots\pi}\, d\,\xi_\lambda\, d\,\xi_\mu \ldots d\,\xi_\pi$$

wobei über jeden zweimal vorkommenden Index summiert wird.

Sind $u_1, \ldots, u_d$ Ortsuniformisierende eines Punktes, sodaß alle ξ_k differenzierbare Funktionen von $u_1, \ldots, u_d$ sind, so werden die Komponenten des Tensores φ in Bezug auf $u_1, \ldots, u_d$ durch

$$(2) \qquad \varphi_{\alpha\beta\ldots\varepsilon} = \sum \varphi_{\lambda\mu\ldots\pi} \frac{\partial \xi_\lambda}{\partial u_\alpha} \frac{\partial \xi_\mu}{\partial u_\beta} \cdots \frac{\partial \xi_\pi}{\partial u_\varepsilon}$$

definiert, wobei wieder über jeden zweimal vorkommenden Index zu summieren ist.

Die Transformationsformel (2) ist transitiv: transformiert man den Tensor φ zuerst auf $u_1, \ldots, u_d$, dann weiter auf $v_1, \ldots, v_d$, so erhält man dasselbe Ergebnis wie wenn man direkt auf $v_1, \ldots, v_d$ transformiert:

$$(3) \qquad \varphi_{\alpha'\beta'\ldots\varepsilon'} = \sum \varphi_{\alpha\beta\ldots\varepsilon} \frac{\partial u_\alpha}{\partial v_{\alpha'}} \cdots \frac{\partial u_\varepsilon}{\partial v_{\varepsilon'}}$$

$$= \sum \varphi_{\lambda\mu\ldots\pi} \left(\frac{\partial \xi_\lambda}{\partial u_\alpha} \frac{\partial u_\alpha}{\partial v_{\alpha'}} \right) \left(\frac{\partial \xi_\mu}{\partial u_\beta} \frac{\partial u_\beta}{\partial v_{\beta'}} \right) \cdots \left(\frac{\partial \xi_\pi}{\partial u_\varepsilon} \frac{\partial u_\varepsilon}{\partial v_{\varepsilon'}} \right)$$

$$= \sum \varphi_{\lambda\mu\ldots\pi} \frac{\partial \xi_\lambda}{\partial v_{\alpha'}} \frac{\partial \xi_\mu}{\partial v_{\beta'}} \cdots \frac{\partial \xi_\pi}{\partial v_{\varepsilon'}}.$$

Ein Vektor φ heißt *endlich in* P, wenn nach Transformation auf die Ortsuniformisierenden des Punktes P die Vektorkomponenten alle in P endlich bleiben. Diese Eigenschaft ist von der Wahl der Ortsuniformisierenden unabhängig. Denn wenn in (3) die $\varphi_{\alpha\beta\ldots\varepsilon}$ alle endlich bleiben in P und $u_1, \ldots, u_d$ ebenso, so bleiben nach Satz 2 auch die Ableitungen $\frac{\partial u_\alpha}{\partial v_{\alpha'}}$ usw. endlich in P, also die ganze rechte Seite von (3) und somit auch die linke. Ebenso beweist man durch Vertauschung der u_α und $v_{\alpha'}$, daß aus der Endlichkeit der $\varphi_{\alpha'\beta'\ldots\varepsilon'}$ die der $\varphi_{\alpha\beta\ldots}$ folgt.

Ist ein Tensor φ nicht endlich in P, so geht mindesten eine irreduzible Polmannigfaltigkeit $\mathfrak{P}$ von einer Funktion $\varphi_{\varkappa\lambda\ldots\nu}$ durch P. Ist k die höchste Vielfachheit von $\mathfrak{P}$ als Bestandteil einer Polmannigfaltigkeit einer Tensorkomponente $\varphi_{\varkappa\lambda\ldots\nu}$, so zählt $\mathfrak{P}$ als *Polmannigfaltigkeit von der Vielfachheit k des Tensors* φ. Man beweist wie oben aus (3), daß diese Vielfachheit von der Wahl der Ortsuniformisierenden des Punktes P unabhängig ist.

Die Vielfachheit k ändert sich aber auch nicht, wenn P durch einen allgemeinen Punkt der Polmannigfaltigkeit $\mathfrak{P}$ ersetzt wird; denn die Ortsuniformisierenden des Punktes P sind gleichzeitig Ortsuniformisierende des allgemeinen Punktes. Daraus folgt, daß die eben difinierte Vielfachheit nicht von der Wahl des Punktes P, sondern nur von $\mathfrak{P}$ abhängt.

Jeder Tensor hat nur endlich viele Polmannigfaltigkeiten $\mathfrak{P}$. Denn wenn man zunächst wieder ein bestimmtes System von unabhängigen veränderlichen $\xi_1, \ldots, \xi_d$ dem Funktionenkörper zugrunde legt, so können alle Vektorkomponenten $\varphi_{\lambda\mu\ldots\pi}$ als Brüche $f_{\lambda\mu\ldots\pi}(\xi) : g(\xi)$ mit einem und demselben Hauptnenner $g(\xi)$ geschrieben werden. Als Punkte der Polmannigfaltigkeiten $\mathfrak{P}$ kommen nun erstens diejenigen in Betracht, in denen $g = 0$ wird, zweitens diejenigen, für die $\xi_1, \ldots, \xi_d$ nicht Ortsuniformisierende sind. Diese zwei Arten von Punkten verteilen sich aber auf je endlich viele $(d{-}1)$-dimensionale irreduzible Mannigfaltigkeiten.

Wir fassen zusammen: *Jeder Tensor besitzt eine rein $(d{-}1)$-dimensionale Polmannigfaltigkeit, bestehend aus endlich vielen irreduziblen Bestandteilen $\mathfrak{P}$ mit bestimmten Vielfachheiten. Außerhalb von dieser Polmannigfaltigkeit bleibt der Tensor in jedem Punkt endlich.*

Fehlt die Polmannigfaltigkeit, so heißt der Tensor *überall endlich.* Die überall endlichen alternierenden Tensoren s-ter Stufe definieren nach (1) die *überall endlichen s-fachen Differentiale* der Mannigfaltigkeit M.

§ 3.

Transformation der Tensoren.

Ist eine Mannigfaltigkeit M' rational auf M abgebildet, so entspricht jedem allgemeinen Punkt ξ' von M' ein allgemeiner Punkt ξ von M:

$$\xi_0 : \xi_1 : \ldots : \xi_n = F_0(\xi') : F_1(\xi') : \ldots : F_n(\xi').$$

Bekanntlich wird die rationale Abbildung von M' auf M durch eine lineare Schar $\{C_\lambda\}$ vermittelt, die man erhält, wenn man die Formenschar

$$F_\lambda = \lambda_0 F_0 + \lambda_1 F_1 + \ldots + \lambda_n F_n$$

mit M' schneidet und den von λ unabhängigen Teil D der Schnittmannigfaltigkeit wegläßt. Die Fundamentalpunkte der Abbildung sind die Basispunkte der Schar; sie bilden eine höchstens $(d{-}2)$-dimensionale Teilmannigfaltigkeit B' von M'. Allen anderen Punkten von M' entspricht, wenn M' glatt ist, nur ein Bildpunkt auf M [4].

Einem Tensor φ auf M entspricht ein Tensor φ' auf M' vermöge der Transformationsformel

$$\varphi'_{\lambda'\mu'\ldots\pi'} = \sum \varphi_{\lambda\mu\ldots\pi} \frac{\partial\xi_\lambda}{\partial\xi'_{\lambda'}} \frac{\partial\xi_\mu}{\partial\xi'_{\mu'}} \cdots \frac{\partial\xi_\pi}{\partial\xi'_{\pi'}}.$$

Dabei soll wieder $\xi_0 = \xi'_0 = 1$ sein, alle ξ_i separable Funktionen von $\xi_1, \ldots, \xi_d$ und alle ξ'_i separable Funktionen von $\xi'_1, \ldots, \xi'_d$. Sind $u'_{1'}, \ldots, u'_{d'}$ Ortsuniformisierende auf M', so werden die Komponenten von φ in Bezug auf sie durch

---

[4]) Siehe etwa B. L. v. d. WAERDEN, ZAG 6, Math. Anm. **110**, § 6.

375

$$\varphi'_{\alpha'\beta'\ldots\varepsilon'} = \sum \varphi'_{\lambda'\mu'\ldots\pi'}\,\frac{\partial\,\xi'_{\lambda'}}{\partial\,u'_{\alpha'}}\cdots\frac{\partial\,\xi'_{\pi'}}{\partial\,u'_{\varepsilon'}}$$

$$= \sum \varphi_{\lambda\mu\ldots\pi}\left(\frac{\partial\,\xi'_{\lambda'}}{\partial\,u'_{\alpha'}}\,\frac{\partial\,\xi_{\lambda}}{\partial\,\xi'_{\lambda'}}\right)\cdots\left(\frac{\partial\,\xi'_{\pi'}}{\partial\,u'_{\varepsilon'}}\,\frac{\partial\,\xi_{\pi}}{\partial\,\xi'_{\pi'}}\right)$$

$$= \sum \varphi_{\lambda\mu\ldots\pi}\,\frac{\partial\,\xi_{\lambda}}{\partial\,u'_{\alpha'}}\cdots\frac{\partial\,\xi_{\pi}}{\partial\,u'_{\varepsilon'}}$$

gegeben.

Wir beweisen nun nach Kähler:

Satz 3. *Wenn der Tensor* φ *auf* M *überall endlich ist, so ist auch* φ' *auf* M' *überall endlich.*

Beweis. Gesetzt, φ' hätte auf M' eine Polmannigfaltigkeit $\mathfrak{P}'$. Wir wählen auf $\mathfrak{P}'$ einen Punkt P', der nicht Fundamentalpunkt der Abbildung ist. Das ist möglich, weil $\mathfrak{P}'$ die Dimension $d-1$ hat. Für den Bildpunkt P auf M können wir $p_0 \neq 0$ annehmen. Dann bleiben die Funktionen

$$\xi_i = \frac{\xi_i}{\xi_0} = \frac{F_i(\xi')}{F_0(\xi')}$$

im Punkt P' alle endlich. Wir können die Veränderliche so numerieren daß $\xi_1,\ldots,\xi_d$ Ortsuniformisierende für P sind. $u'_1,\ldots,u'_{d'}$ seien Ortsuniformisierende für P'. Dann bleiben in der Formel

$$\varphi'_{\alpha'\beta'\ldots\varepsilon'} = \sum \varphi_{\lambda\mu\ldots\pi}\,\frac{\partial\,\xi_{\lambda}}{\partial\,u'_{\alpha'}}\,\frac{\partial\,\xi_{\mu}}{\partial\,u'_{\beta'}}\cdots\frac{\partial\,\xi_{\pi}}{\partial\,u'_{\varepsilon'}}$$

alle Größen rechter Hand endlich an der Stelle P'. Also bleibt auch die linke Seite endlich, q. e. d.

Ist die Abbildung birational, so folgt auch umgekehrt aus der Endlichkeit von φ' die von φ. Die überall endlichen Tensoren mit gegebenen Symmetriebedingungen bilden also jeweils eine birational invariante lineare Schar. Insbesondere gilt das von den ein- bis d-fachen Differentialen 1. Gattung. Auch die Dimensionen dieser linearen Scharen sind birational invariant. Daraus folgt speziell die Invarianz des geometrischen Geschlechtes p_g und der Irregularität $p_g - p_a$ einer algebraischen Fläche, der Mehrgeschlechter usw.

§ 4.

Bewertungen und Primdivisoren.

Es sei wieder M eine glatte, irreduzible Mannigfaltigkeit von der Dimension d. Die Teilmannigfaltigkeiten von M bezeichnen wir fortan als *Teile* von M, insbesondere die rein $(d-1)$-dimensionalen als *Höchstteile*, die höchstens $(d-2)$-dimensionalen als *niedere Teile*.

Jedes irreduzible Höchstteil (*Primhöchstteil*) $\mathfrak{P}$ von M definiert eine Bewertung des zu M gehörigen Funktionenkörpers. Ist nämlich ξ ein allgemeiner Punkt von M und

$$\varphi = \frac{f(\xi_0, \ldots, \xi_n)}{g(\xi_0, \ldots, \xi_n)}$$

eine rationale Funktion auf M, so schneiden die Hyperflächen $f = 0$ und $g = 0$ aus M gewisse Höchstteile aus. Wenn nun $\mathfrak{P}$ darin mit den Vielfachheiten a und b vorkommt, so wird der Wert von φ in der zu $\mathfrak{P}$ gehörigen Bewertung durch

$$w(\varphi) = a - b$$

definiert. Wird dieselbe Funktion noch in anderer Weise dargestellt

$$\varphi = \frac{f(\xi)}{g(\xi)} = \frac{f_1(\xi)}{g_1(\xi)},$$

so ist $fg_1 - f_1g$ Null auf M, also schneidet fg_1 denselben Höchstteil aus M aus wie f_1g, also ist $a + b_1 = a_1 + b$, also $a - b = a_1 - b_1$.

Die Bewertungseigenschaften

$$w(\varphi\psi) = w(\varphi) + w(\psi),$$
$$w(\varphi + \psi) \geqq \mathrm{Min}\,\{w(\varphi),\ w(\psi)\}$$

sind erfüllt. Die erste ist selbstverständlich, die letzte folgt daraus, daß jede Hyperfläche der linearen Schar $\lambda_1 f_1 g_2 + \lambda_2 f_2 g_1 = 0$, insbesondere die Hyperfläche $f_1 g_2 + f_2 g_1 = 0$, die Mannigfaltigkeit M in $\mathfrak{P}$ mit mindestens dieselbe Vielfachheit schneidet, wie die kleinste der beiden Schnittvielfachheiten der Basishyperflächen $f_1 g_2 = 0$ und $f_2 g_1 = 0$ (vgl. etwa meine Einführung in die algebr. Geometrie, § 44, Satz 4).

Die so definierten Bewertungen heißen im zweidimensionalen Fall *Kurvenprimdivisoren*; im allgemeinen Fall nennen wir sie *höhere Primdivisoren*. Wir bezeichnen sie mit großen deutschen Typen $\mathfrak{P}, \mathfrak{Q}, \ldots$. Wir unterscheiden nicht zwischen dem Höchstteil $\mathfrak{P}$ (speziell auf einer Fläche: der Kurve $\mathfrak{P}$) und dem durch ihn bestimmten Primdivisor $\mathfrak{P}$.

Unter einem (höheren) *Divisor* $\mathfrak{D}$ verstehen wir nunmehr eine endliche Gesamtheit von höheren Primdivisoren $\mathfrak{P}_1, \mathfrak{P}_2, \ldots, \mathfrak{P}_r$ mit beliebigen ganzzahligen Exponenten $e_1, \ldots, e_r$, die auch negativ sein dürfen, und schreiben

$$\mathfrak{D} = \mathfrak{P}_1^{e_1} \mathfrak{P}_2^{e_2} \ldots \mathfrak{P}_r^{e_r}$$

Sind alle Exponenten $e_1, \ldots, e_r \geqq 0$, so heißt der Divisor $\mathfrak{D}$ *ganz*. Wir unterscheiden nicht zwischen dem ganzen Divisor $\mathfrak{D}$ und dem entsprechenden Höchstteil von M, bestehend aus den Primhöchstteilen $\mathfrak{P}_1, \ldots, \mathfrak{P}_r$ mit den Vielfachheiten $e_1, \ldots, e_r$.

Das *Produkt* von zwei Divisoren wird, wie man sich denken kann, durch Addition der Exponenten gebildet. Bei dieser Produktdefinition bilden die Divisoren $\mathfrak{D}$ offensichtlich eine ABELsche Gruppe. Das Einselement dieser Gruppe ist der leere Divisor $\mathfrak{E} = \mathfrak{P}^0$.

Zu jeder Funktion φ gehört ein Divisor $(\varphi) = \mathfrak{B}\mathfrak{C}^{-1}$, wobei $\mathfrak{B}$ die Nullmannigfaltigkeit und $\mathfrak{C}$ die Polmannigfaltigkeit von φ ist.

Außer den höheren Primdivisoren brauchen wir noch andere Bewertungen, die durch birationale Transformation in höhere Primdivisoren übergehen können. Zunächst ein Beispiel!

Ein einfacher Punkt P einer algebraischen Fläche M definiert eine Bewertung $\mathfrak{p}$. Ist nämlich wie oben

$$\varphi = \frac{f(\xi)}{g(\xi)}$$

und hat die Schnittkurve von M mit $f = 0$ einen α-fachen Punkt in P und die von M mit $g = 0$ einen β-fachen, so kann man definieren

$$w_P(\varphi) = \alpha - \beta.$$

Die Eigenschaften der Bewertung gelten wie oben. Solche Bewertungen nennen wir *Punktbewertungen* oder *Punktprimdivisoren*.

Man kann eine Punktbewertung in eine Kurvenbewertung verwandeln, indem man durch eine birationale Transformation den Punkt P in eine Kurve überführt. Man braucht nur die lineare Kurvenschar, die die rationale Abbildung definiert, so zu wählen, daß sie in P einen einfachen Basispunkt hat.

In ähnlicher Weise definiert auch jeder zu P unmittelbar benachbarte Punkt P_1 (d. h. jede Tangentenrichtung von M in P) eine Bewertung. Wenn nämlich die Schnittkurve von $f = 0$ mit M im Punkt P einen α-fachen und im Nachbarpunkt P_1 einen α_1-fachen Punkt hat[5]), so kann man definieren

$$w_1(f) = \alpha + \alpha_1,$$
$$w_1(\varphi) = w_1(f) - w_1(g).$$

Durch die birationale Transformation, die P in eine Kurve $\mathfrak{P}$ verwandelt, wird die Bewertung w_1 in eine Punktbewertung w_Q transformiert, wo Q ein Punkt von $\mathfrak{P}$ ist. Transformiert man sodann Q weiter in eine Kurve $\mathfrak{Q}$, so kann man die Bewertung w_1 sogar in eine Kurvenbewertung verwandeln.

Analog gehören auch zu den höheren Nachbarpunkten von P Bewertungen, die durch geeignete birationale Transformationen in höhere Primdivisoren verwandelt werden können.

Im mehrdimensionalen Fall gibt es keine Theorie der unendlich benachbarten Punkte, Kurven usw., aber der Bewertungsbegriff gibt uns einen vollwertigen Ersatz für diese fehlende Theorie. Wir definieren einfach:

Ein Primdivisor ist eine solche ganzzahlige Bewertung des Funktionenkörpers, in welcher die Konstanten den Wert Null haben und deren Restklassenring den Transzendenzgrad $n-1$ hat.

---

[5]) Für die Definition dieser Begriffe siehe meine Einführung in die algebr. Geom., § 55.

Man kann leicht beweisen, daß alle solche Bewertungen sich birational in höhere Primdivisoren transformieren lassen, aber wir brauchen nicht einmal das. — Alle Primdivisoren, die nicht höhere Primdivisoren sind, heißen *niedere Primdivisoren*.

Die Werte $w(\varphi)$ sind zunächst nur für Funktionen φ des Körpers definiert. Wir wollen nun $w(f)$ auch für Formen $f(x_0, \ldots, x_n)$ definieren, und schließlich auch $w(\mathfrak{D})$ für beliebige (höhere) Divisoren $\mathfrak{D}$.

Wir definieren zunächst eine teilweise Ordnung der Veränderlichen $x_0, x_1, \ldots, x_n$ durch die Festsetzung: x_i ist früher als x_k, wenn $w(\xi_i^{-1}\xi_k) > 0$ ist. Wenn x_i früher ist als x_k und x_k früher als x_l, so ist auch x_i früher als x_l. Wir können nun die Numerierung des $x_0, \ldots, x_n$ so abändern, daß frühere x_i auch eine kleinere Nummer haben. Dann ist insbesondere $w(\xi_0^{-1}\xi_k) \geqq 0$ für alle x_k.

Nun definieren wir für jede Form $f(x_0, \ldots, x_n)$ vom Grade g:

$$w(f) = w\left\{\xi_0^{-g} f(\xi_0, \ldots, \xi_n)\right\} = w\left\{f\left(1, \frac{\xi_1}{\xi_0}, \ldots, \frac{\xi_n}{\xi_0}\right)\right\}$$

Die Definition ist unabhängig von der Wahl von x_0, solange nur die Bedingung erfüllt bleibt, daß kein x_k früher ist als x_0. Denn wenn man x_0 etwa durch x_1 ersetzt und weder x_1 früher ist als x_0 noch x_0 früher als x_1, so ist $w(\xi_0^{-1}\xi_1) = 0$, also $w\{\xi_0^{-g}f(\xi)\} = w\{\xi_1^{-g}f(\xi)\}$. Man hätte ξ_0 auch durch eine Linearkombination $\eta_0 = a_0\xi_0 + \cdots + a_n\xi_n$ ersetzen können, vorausgesetzt daß $w(\eta_0^{-1}\xi_k) \geqq 0$ für alle k und daher auch $w(\eta_0^{-1}\eta) \geqq 0$ für jede Linearkombination η von $\xi_0, \ldots, \xi_n$. Die Definition von $w(f)$ ist also projektiv invariant.

Noch einfacher wird die Definition von $w(f)$, wenn man durch die Normierung $\xi_0 = 1$ inhomogene Koordinaten einführt, was wir von jetzt an tun wollen. Dann ist nämlich jede Form $f(\xi)$ selber eine Funktion des Körpers und es ist

$$w(f) = w\{f(\xi)\}.$$

Vor der Normierung muß man sich aber davon überzeugen, daß die Bedingung $w(\xi_0^{-1}\xi_k) \geqq 0$ für alle k erfüllt ist.

Der so definierte Wert $w(f)$ ist offenbar niemals negativ. Diejenigen f, für die $w(f) > 0$ ist, bilden im Polynombereich von $x_0, \ldots, x_n$ ein Prim-H-Ideal, und zwar einen echten Teiler des die Mannigfaltigkeit M definierenden Primideals $\mathfrak{M}$. Die zu diesem Prim-H-Ideal gehörige irreduzible Mannigfaltigkeit N ist also ein echter Teil von M. Wir nennen N die *Nullmannigfaltigkeit der Bewertung* w. Sie hat die Eigenschaft, daß $w(f)$ größer oder gleich Null ist, je nachdem ob die Hyperfläche $f = 0$ sie enthält oder nicht.

Satz 4. *Wenn N die Dimension $d-1$ hat, so ist w der zu N gehörige höhere Divisor.*

Beweis: Es sei W der zu N gehörige höhere Divisor, also $W(fg^{-1}) = a - b$, wenn die Hyperflächen $f = 0$ und $g = 0$ aus M den

Höchstteil N mit den Vielfachheiten a und b ausschneiden. Wir zeigen zunächst: wenn $W(\varphi) \geqq 0$, so ist auch $w(\varphi) \geqq 0$, und wenn $W(\varphi) > 0$, dann auch $w(\varphi) > 0$.

Wenn $W(\varphi) \geqq 0$, so kann man nach Satz 1 die Funktion φ so als Quotient $f g^{-1}$ darstellen, daß die Hyperfläche $g = 0$ die Nullmannigfaltigkeit N nicht enthält. Dann ist $w(g) = 0$, also $w(f g^{-1}) = w(f) - w(g) \geqq 0$. Ist insbesondere $W(\varphi) > 0$, so enthält $f = 0$ die Nullmannigfaltigkeit N, also ist $w(f) > 0$ und daher auch $w(f g^{-1}) > 0$. Ersetzt man nun φ durch φ^{-1}, so folgt ebenso: ist $W(\varphi) \leqq 0$, so ist auch $w(\varphi) \leqq 0$, und wenn $W(\varphi) < 0$, dann auch $w(\varphi) < 0$. Ist nun $W(\varphi) = 0$, so ist $w(\varphi)$ sowohl $\geqq 0$ als $\leqq 0$, also $= 0$. Also sind w und W beide gleichzeitig positiv, Null oder negativ. Daraus folgt aber leicht, daß die Bewertungen w und W äquivalent sind.

Die Umkehrung von Satz 4 ist trivial. Wir können also die höheren Primdivisoren auch charakterisieren als solche, deren Nullmannigfaltigkeiten Höchstteile von M, die niederen als solche, deren Nullmannigfaltigkeiten niedere Teile von M sind.

Nun zur Definition von $w(\mathfrak{D})$! Wir wählen einen beliebigen Punkt P_0 auf der Nullmannigfaltigkeit N des Primdivisors w. Zu jedem ganzen Divisor $\mathfrak{A} = \mathfrak{P}_1^{e_1} \mathfrak{P}_2^{e_2} \ldots \mathfrak{P}_\mu^{e_\mu}$ auf M gibt es nach SEVERI[6]) eine Hyperfläche $f = 0$, die aus M den Höchstteil $\mathfrak{A}\mathfrak{B}$ ausschneidet, wobei der Rest $\mathfrak{B}$ den Punkt P_0 und folglich die Mannigfaltigkeit N nicht enthält. Nun definieren wir

$$w(\mathfrak{A}) = w(f).$$

Die Definition ist von der Wahl der Hyperfläche $f = 0$ unabhängig. Denn wenn $g = 0$ eine zweite ebensolche Hyperfläche ist, die den Höchstteil $\mathfrak{A}\mathfrak{C}$ ausschneidet, wobei $\mathfrak{C}$ wieder N nicht enthält, so können wir zunächst, indem wir eventuell unwesentliche Faktoren x_0^h hinzufügen, f und g als Formen gleichen Grades annehmen. Dann ist $\varphi = f g^{-1}$ eine Funktion des Körpers und

$$w(f) = w(\varphi g) = w(\varphi) + w(g).$$

Zur Funktion $\varphi = f g^{-1}$ gehört der Divisor

$$\mathfrak{A}\mathfrak{B}(\mathfrak{A}\mathfrak{C})^{-1} = \mathfrak{B}\mathfrak{C}^{-1}.$$

Ein allgemeiner Punkt P von N ist weder in $\mathfrak{B}$ noch in $\mathfrak{C}$ enthalten. Nach Satz 1 können wir also φ auch so als Quotient $F G^{-1}$ darstellen, daß G und daher auch F nicht Null werden in P. Dann ist

$$w(\varphi) = w(F) - w(G) = 0,$$

mithin

$$w(f) = w(g).$$

<hr>

[6]) F. SEVERI, Abh. Math. Sem. Hamburg 9 (1933) S. 335 ff.

Die so definierte Funktion $w(\mathfrak{A})$ hat wieder die Eigenschaft, daß $w(\mathfrak{A}) > 0$ oder $= 0$, je nachdem ob der Höchstteil $\mathfrak{A}$ die Nullmannigfaltigkeit N enthält oder nicht. Weiter gilt offenbar

$$w(\mathfrak{A}\mathfrak{B}) = w(\mathfrak{A}) + w(\mathfrak{B}).$$

Wenn die Hyperfläche $f = 0$ genau den Hauptteil $\mathfrak{A}$ ausschneidet, so ist nach Definition $w(\mathfrak{A}) = w(f)$.

Für nichtganze Divisoren $\mathfrak{A}\mathfrak{B}^{-1}$ definieren wir schließlich

$$w(\mathfrak{A}\mathfrak{B}^{-1}) = w(\mathfrak{A}) - w(\mathfrak{B}).$$

Mit der ursprünglichen Bewertung $w(\varphi)$ ist diese erweiterte Bewertung $w(\mathfrak{D})$ in der Weise verknüpft, daß $w(\mathfrak{D}) = w(\varphi)$, wenn $\mathfrak{D}$ der zugehörige Divisor der Funktion φ ist, d. h. $\mathfrak{D} = \mathfrak{A}\mathfrak{B}^{-1}$, $\mathfrak{A}$ der Nulldivisor, $\mathfrak{B}$ der Poldivisor von φ.

Die Funktion $w(\mathfrak{D})$ bietet einen Ersatz für den Begriff der Vielfachheit einer Kurve in einem Punkt oder in einem unendlich benachbarten Punkt.

§ 5.
Virtuelle und effektive Multiplizitäten.

Es seien eine Anzahl *niedere* Primdivisoren $\mathfrak{p}_1, \ldots, \mathfrak{p}_r$ und dazu gewisse *Mindestwerte* oder *vorgegebene Multiplizitäten* $a_1, \ldots, a_r$ gegeben. Wir betrachten die Gesamtheit aller ganzen Divisoren, die in den Bewertungen $\mathfrak{p}_1, \ldots, \mathfrak{p}_r$ mindestens die Werte $a_1, \ldots, a_r$ haben.

Diese Mindestwerte bedingen im allgemeinen andere als *Folgewerte*. Alle ebenen Kurven z. B., die in einem vorgegebenen Punkt P eine Vielfachheit $a \geq 3$ haben, haben in allen den Bewertungen, die zu den Nachbarpunkten P_1 von P gehören, von selbst auch den Mindestwert 3; denn der Wert in der zu P_1 gehörigen Bewertung ist die Summe $a + b$ der Vielfachheiten in P und P_1, und aus $a \geq 3$ folgt $a + b \geq 3$[7]).

Wir definieren also: wenn alle ganzen Divisoren, die in $\mathfrak{p}_1, \ldots, \mathfrak{p}_r$ die Mindestwerte $a_1, \ldots, a_r$ haben, in $\mathfrak{p}$ von selbst den Mindestwert a, so heißt dieser Mindestwert ein *Folgewert* der vorgegebenen Mindestwerte. Alle diese (im allgemeinen unendlich viele) Folgewerte bilden zusammen ein *System von virtuellen Werten*.

Haben alle Divisoren einer linearen Schar in der Bewertung $\mathfrak{p}$ Werte $\geq e$ und ist e das genaue Minimum dieser Werte, so heißt $e = e(\mathfrak{p})$ der *effektive* Wert der linearen Schar in der Bewertung $\mathfrak{p}$. Der effektive Wert kann auch definiert werden als das Minimum der

---

[7]) Ein anderes Beispiel: aus $a + b \geq 3$ folgt $a \geq 2$; denn die Kurven, die nicht durch P gehen, gehen auch nicht durch P_1, und die nur einfach durch P gehen, können auch nur einfach durch P_1 gehen.

Werte der Basisdivisoren der Schar in der Bewertung $\mathfrak{p}$. Ist insbesondere $\mathfrak{p}$ ein höherer Primdivisor $\mathfrak{P}$, so enthalten alle Divisoren der Schar den gemeinsamen Faktor $\mathfrak{P}^e$.

Ist zu einer linearen Schar ein System von virtuellen Werten $v(\mathfrak{p})$ gegeben und erfüllen alle Divisoren der Schar auch wirklich die Bedingung, daß ihre Werte in der Bewertung $\mathfrak{p}$ mindestens $v(\mathfrak{p})$ betragen, so ist $e(\mathfrak{p}) \geqq v(\mathfrak{p})$. Die Differenz $e(\mathfrak{p}) - v(\mathfrak{p})$ heißt die *Übermultiplizität* oder der *Überwert* der Schar in der Bewertung $\mathfrak{p}$.

Beispiel. Die ebenen Kurven 3. Ordnung durch 8 gegebene Punkte $P_1, \ldots, P_8$ gehen alle noch durch einen neunten Punkt P_9 (der auch zu einem der gegebenen unendlich benachbart sein kann). Betrachtet man nun die 8 zu $P_1, \ldots, P_8$ gehörigen Mindestwerte 1 und ihre Folgewerte als virtuelle Werte für das Kurvenbüschel, so hat das Büschel in $P_1, \ldots, P_8$ keine Überwerte, aber in P_9 den Überwert 1.

Liegen aber von den 8 Punkten vier auf einer Geraden $\mathfrak{G}$, so enthalten alle Kurven des Büschels diese Gerade als Bestandteil. Zu der Bewertung $\mathfrak{G}$ gehört also der Überwert 1 und ebenso zu jedem Punkt von $\mathfrak{G}$, mit Ausnahme der 4 Punkte P_i auf $\mathfrak{G}$.

Im folgenden betrachten wir niemals eine lineare Schar allein, sondern immer eine lineare Schar mit einem System von virtuellen Werten, wobei wir annehmen, daß die Divisoren der Schar ganz sind und mindestens diese Werte haben. Ein Divisor $\mathfrak{D}_\lambda = \mathfrak{D}\mathfrak{C}_\lambda$ der Schar besteht aus einem festen Bestandteil $\mathfrak{D}$, der allen Divisoren der Schar gemeinsam ist, und einem von λ abhängigen Bestandteil $\mathfrak{C}_\lambda$. Die irreduziblen Bestandteile $\mathfrak{P}$ von $\mathfrak{D}$ haben, falls $\mathfrak{D}$ nicht 1 ist, positive effektive Multiplizitäten $e(\mathfrak{P})$; diese sind sämtlich als Übermultiplizitäten zu betrachten, denn virtuelle Werte $v(\mathfrak{p})$ gibt es nach der anfangs gegebenen Definition nur für niedere, nicht für höhere Divisoren.

§ 6.

Transformation der linearen Scharen.

Eine lineare Schar $\{\mathfrak{C}_\lambda\}$ ohne feste Bestandteile ist definiert durch eine lineare Schar von Funktionen

$$(1) \qquad \varphi_\lambda = \lambda_0 \varphi_0 + \cdots + \lambda_r \varphi_r$$

in folgender Weise: die zu φ_λ gehörigen Divisoren (φ_λ) werden zunächst mit einem ganzen Divisor $\mathfrak{A}$ multipliziert, damit sie alle ganz werden, und dann mit $\mathfrak{B}^{-1}$, um sie von gemeinsamen festen Bestandteilen zu befreien:

$$\mathfrak{C}_\lambda = (\varphi_\lambda)\,\mathfrak{A}\mathfrak{B}^{-1}.$$

Fügt man nun noch einen beliebigen festen Divisor $\mathfrak{D}$ als Faktor hinzu, so erhält man die allgemeinste lineare Kurvenschar

$$\mathfrak{D}_\lambda = \mathfrak{C}_\lambda\mathfrak{D} = (\varphi_\lambda)\,\mathfrak{A}\mathfrak{B}^{-1}\mathfrak{D} = (\varphi_\lambda)\,\mathfrak{F}.$$

382

Zu dieser Kurvenschar sei wie oben ein System virtueller Werte gegeben. Wir wollen das Verhalten der Schar bei einer birationalen Transformation, die M wieder in eine glatte Mannigfaltigkeit M' überführt, untersuchen.

Ist ξ ein allgemeiner Punkt von M und ξ' ein allgemeiner Punkt von M', so sind die ξ' rationale Funktionen von den ξ und umgekehrt, also kann man jede rationale Funktion von den ξ' auch durch die ξ ausdrücken und umgekehrt, also entspricht jeder rationalen Funktion $\varphi(\xi)$ eindeutig und isomorph eine rationale Funktion $\varphi'(\xi')$. Auf Grund dieses Körperisomorphismus entspricht auch jeder Bewertung von M eine Bewertung von M' und umgekehrt. Dabei kommt es nur endlich oft vor, daß einem höheren Primdivisor von M ein niederer Primdivisor auf M' entspricht oder umgekehrt; denn nur endlich viele Primhöchstteile von M werden in niedere Teile von M' transformiert (vgl. wieder ZAG **6**, Math. Ann. 110, § 6), und zu jedem dieser Primhöchstteile gehört nach Satz 4 nur ein höherer Primdivisor.

Der linearen Funktionenschar (1) entspricht auf M' wieder eine lineare Funktionenschar

$$\varphi'_\lambda = \lambda_0\,\varphi'_0 + \cdots + \lambda_r\,\varphi'_r$$

und dieser wieder eine lineare Divisorenschar ohne feste Bestandteile

$$\mathfrak{C}'_\lambda = (\varphi'_\lambda)\,\mathfrak{A}'\mathfrak{B}'^{-1}.$$

Am einfachsten ist die Transformation der linearen Schar $\{\mathfrak{D}_\lambda\}$ zu erklären, wenn diese *keine Überwerte*, also insbesondere auch keine festen Bestandteile $\mathfrak{D}$ hat. Dann ist $\mathfrak{D}_\lambda = \mathfrak{C}_\lambda$, und wir erklären die transformierte Schar ebenso als $\mathfrak{D}'_\lambda = \mathfrak{C}'_\lambda$, *ohne feste Bestandteile und ohne Überwerte*. Das heißt also, wir setzen die virtuellen Werte für die transformierte Schar $\mathfrak{D}'_\lambda = \mathfrak{C}'_\lambda$ gleich den effektiven Werten.

Etwas komplizierter wird die Transformationsregel, wenn Überwerte vorhanden sind. Das Prinzip ist, daß die Überwerte erhalten bleiben sollen:

$$(2) \qquad\qquad e(\mathfrak{p}) - v(\mathfrak{p}) = e(\mathfrak{p}') - v(\mathfrak{p}'),$$

wenn dem Primdivisor $\mathfrak{p}$ der Primdivisor $\mathfrak{p}'$ entspricht, gleichgültig ob es sich dabei um höhere oder um niedere Primdivisoren handelt; nur daß für die höheren $v(\mathfrak{P}) = 0$ zu setzen ist. Man verfährt demnach so: zunächst wird $\mathfrak{C}'_\lambda$ wie oben bestimmt. Sodann wird $\mathfrak{D}'$ zusammengesetzt aus denjenigen höheren Primdivisoren $\mathfrak{P}'$, denen vor der Transformation solche höhere oder niedere Primdivisoren $\mathfrak{P}$ oder $\mathfrak{p}$ entsprechen, die in $\{\mathfrak{D}_\lambda\}$ einen Überwert haben, mit Exponenten gleich diesen Überwerten, damit (2) für diese $\mathfrak{P}'$ erfüllt ist. Solche Divisoren $\mathfrak{P}'$ gibt es nur endlich viele; denn es gibt nur endlich viele in $\mathfrak{D}$ aufgehende höhere $\mathfrak{P}$ und auch nur endlich viel niedere $\mathfrak{p}$, die bei der Transformation in höhere übergehen. Jetzt sind $\mathfrak{D}'$ und $\mathfrak{C}'_\lambda$, also auch

$\mathfrak{D}'_\lambda = \mathfrak{C}'_\lambda \mathfrak{D}'$ bekannt und somit auch die effektiven Werte $e(\mathfrak{p}')$ für alle niederen Primdivisoren $\mathfrak{p}'$. Schließlich bestimmt man die virtuellen Multiplizitäten $v(\mathfrak{p}')$ so, daß (2) stets erfüllt ist.

Es kann vorkommen, daß die so definierte virtuelle Multiplizität $v(\mathfrak{p}')$ negativ ausfällt. Besteht z. B. die Schar $\{\mathfrak{D}_\lambda\}$ aus der einen Geraden $x_0 = 0$ in der Ebene mit den Punkten $(0, 1, 0)$ und $(0, 0, 1)$ als vorgeschriebene Basispunkte und wendet man die in der Einleitung angegebene CREMONA-Transformation an, so besteht die transformierte Schar nur aus dem Einheitsdivisor $\mathfrak{C}$, und der Punkt $(1, 0, 0)$, der der Geraden $x_0 = 0$ in der Transformation entspricht, hat die effektive Multiplizität $e(\mathfrak{p}') = 0$ und den Überwert $e(\mathfrak{p}') - v(\mathfrak{p}') = e(\mathfrak{p}) - 0 = 1$, also die virtuelle Multiplizität $v(\mathfrak{p}') = -1$.

Die negativen virtuellen Multiplizitäten werden durch den Begriff der Folgewerte, den wir am Anfang des § 5 zur Definition der virtuellen Werte verwendeten, nicht erfaßt: die Folgewerte sind ihrer Definition nach stets positiv oder Null. Wir müssen den Begriff des virtuellen Wertsystems also erweitern, indem wir außer den Systemen von Folgewerten $v(\mathfrak{p})$ auch noch solche Systeme von virtuellen Werten $v(\mathfrak{p}')$ zulassen, die durch birationale Transformation nach der obigen Vorschrift aus Systemen von Folgewerten $v(\mathfrak{p})$ hervorgehen. In § 7 werden wir die zulässigen Systeme von virtuellen Werten noch anders charakterisieren lernen.

Wir können die Definition der transformierenden Schar $\{\mathfrak{D}'_\lambda\}$ noch etwas anders formulieren. Setzt man nämlich wie oben

$$\mathfrak{D}_\lambda = (\varphi_\lambda) \cdot \mathfrak{F}$$

und entsprechend

$$\mathfrak{D}'_\lambda = (\varphi'_\lambda) \cdot \mathfrak{F}',$$

so wird der Divisor $\mathfrak{F}'$ direkt durch die Forderung bestimmt, daß für jeden höheren Divisor $\mathfrak{P}'$ die Forderung (2) der Erhaltung der Überwerte erfüllt sein soll, wobei natürlich $v(\mathfrak{P}') = 0$ zu setzen ist. Man kennt also für jedes $\mathfrak{P}'$ die effektive Vielfachheit, mit der $\mathfrak{P}'$ im Produkt $(\varphi'_\lambda) \cdot \mathfrak{F}'$ aufgehen muß, und da man auch (φ'_λ) kennt, so ist damit $\mathfrak{F}'$ bekannt.

Wenn eine einzelne Funktion φ_λ in einer (höheren oder niederen) Bewertung $\mathfrak{p}$ einen um k höheren Wert hat als den Mindestwert aller Funktionen φ_λ, so hat auch die transformierte Funktion φ'_λ für $\mathfrak{p}'$ einen um k höheren Wert als den Mindestwert aller Funktionen φ'_λ; denn der Wert von φ_λ in der Bewertung $\mathfrak{p}$ ist gleich dem von φ'_λ in der Bewertung $\mathfrak{p}'$.

Ebenso sieht man: wenn eine lineare Teilschar der Funktionenschar $\{\varphi_\lambda\}$ einen um k höheren Mindestwert hat als die ganze Schar, so gilt dasselbe für die transformierte Schar. Die Hinzufügung der von λ unabhängigen Faktoren $\mathfrak{F}$ und $\mathfrak{F}'$ ändert daran nichts. Beim Übergang von den Scharen $\{\mathfrak{D}_\lambda\}$ und $\{\mathfrak{D}'_\lambda\}$ zu zwei sich entsprechenden

Teilscharen erhöhen sich also in (2) beide Seiten um denselben Betrag:
die v bleiben nämlich ungeändert und die e erhöhen sich beide um k.
Daher gilt die Relation (2) auch für die Teilschar, sobald sie für die
ganze Schar gilt. Damit ist bewiesen:

Satz 5. *Eine lineare Schar und eine Teilschar gehen wieder in
eine lineare Schar und ihre entsprechende Teilschar über, wenn beide
nach (2) transformiert werden.*

Durch diesen Satz wird erst klar, wie vernünftig die Transformationsvorschrift (2) ist.

Nach bekannten Schlüssen (vgl. etwa meine Einf. alg. Geom. § 46)
ist jede lineare Schar in einer eindeutig bestimmten *Vollschar* enthalten, d. h. in einer solchen, die sich unter Beibehaltung der virtuellen Werte nicht mehr erweitern läßt. Nunmehr folgt aus Satz 5 unmittelbar die *Invarianz* des Begriffes der Vollschar:

*Eine Vollschar geht bei der Transformation nach (2) wieder in
eine Vollschar über.*

§ 7.

Produkte und Quotienten von linearen Scharen.

Sind $\{\mathfrak{A}_\lambda\}$ und $\{\mathfrak{B}_\mu\}$ lineare Scharen:

$$\mathfrak{A}_\lambda = (\varphi_\lambda) \cdot \mathfrak{F}; \quad \varphi_\lambda = \lambda_1 \varphi_1 + \ldots + \lambda_r \varphi_r$$

$$\mathfrak{B}_\mu = \{\psi_\mu\} \cdot \mathfrak{G}; \quad \psi_\mu = \mu_1 \psi_1 + \ldots + \mu_s \psi_s$$

so ist die kleinste lineare Schar $\{\mathfrak{C}_\lambda\}$, die alle Produkte $\mathfrak{A}_\lambda \mathfrak{B}_\mu$ enthält,
durch

$$\mathfrak{C}_\lambda = (\chi_\lambda) \mathfrak{F} \mathfrak{G}$$

$$\chi_\lambda = \lambda_1 \varphi_1 \psi_1 + \lambda_2 \varphi_1 \psi_2 + \ldots + \lambda_{rs} \varphi_r \psi_s$$

gegeben. Diese lineare Schar $\{\mathfrak{C}_\lambda\}$ heißt die *Produktschar* $\{\mathfrak{A}_\lambda\} \cdot \{\mathfrak{B}_\mu\}$.
Als virtuelle Mindestwerte für die Produktschar gelten die Summen
der virtuellen Werte der beiden Faktoren für jeden niederen Primdivisor $\mathfrak{p}$.

Jetzt wollen wir die *Restschar* einer Schar $\{\mathfrak{A}_\lambda\}$ in Bezug auf eine
Vollschar $\{\mathfrak{C}_\lambda\}$ definieren. Wir wählen eine Kurve $\mathfrak{A}_0$ aus $\{\mathfrak{A}_\lambda\}$ willkürlich aus und setzen vorraus, daß es in der Vollschar $\{\mathfrak{C}_\lambda\}$ solche
Divisoren $\mathfrak{C}$ gibt, die durch $\mathfrak{A}_0$ teilbar sind: $\mathfrak{C} = \mathfrak{A}_0 \mathfrak{B}$. Wenn das der
Fall ist, so bilden die Divisoren $\mathfrak{B}$, die mit $\mathfrak{A}_0$ multipliziert Divisoren
der Vollschar $\{\mathfrak{C}_\lambda\}$ ergeben, eine lineare Schar $\{\mathfrak{C}_\lambda\} : \mathfrak{A}_0$. Diese Schar
ist aber von der Wahl von $\mathfrak{A}_0$ unabhängig, denn wenn $\mathfrak{B}$ zu der Schar
gehört, so ist nicht nur $\mathfrak{A}_0 \mathfrak{B}$, sondern die ganze lineare Schar $\{\mathfrak{A}_\lambda \mathfrak{B}\}$
in der Vollschar $\{\mathfrak{C}_\lambda\}$ enthalten[8]). Die Schar der $\mathfrak{B}$ hängt somit nur

---

[8]) Wenn eine lineare Schar $\{\mathfrak{A}_\lambda\} \mathfrak{B}$ mit einer Vollschar $\{\mathfrak{C}_\lambda\}$ ein Exemplar
$\mathfrak{A}_0 \mathfrak{B}$ gemeinsam hat, so ist sie ganz in der Vollschar enthalten.

von den Scharen $\{\mathfrak{C}_\lambda\}$ und $\{\mathfrak{A}_\lambda\}$ ab und wird als Quotient $\{\mathfrak{C}_\lambda\}:\{\mathfrak{A}_\lambda\}$ oder als *Rest* von $\{\mathfrak{A}_\lambda\}$ in Bezug auf $\{\mathfrak{C}_\lambda\}$ bezeichnet. Als virtuelle Werte der Quotientenschar werden die Differenzen der virtuellen Werte der Scharen $\{\mathfrak{C}_\lambda\}$ und $\{\mathfrak{A}_\lambda\}$ genommen. Die Quotientenschar kann auch definiert werden als die umfassendste Schar $\{\mathfrak{B}_\mu\}$, deren Produkt mit $\{\mathfrak{A}_\lambda\}$ eine Teilschar von $\{\mathfrak{C}_\lambda\}$ ergibt. Sie ist offensichtlich wieder eine Vollschar.

Satz 6. *Jede Vollschar läßt sich als Quotient von zwei Vollscharen ohne Überwerte darstellen.*

Beweis. Zunächst seien die virtuellen Werte $v(\mathfrak{p})$ der Vollschar $\{\mathfrak{B}_\mu\}$ nach § 5 als Folgewerte von irgendwelchen vorgegebenen Mindestwerten definiert. Die Formen F, die in allen Bewertungen $\mathfrak{p}$ mindestens die Werte $v(\mathfrak{p})$ haben, bilden ein H-Ideal, das nach dem HILBERTschen Basissatz eine endliche Basis $(F_1, \ldots, F_r)$ besitzt. Für jedes $\mathfrak{p}$ sind alle Werte $w(F_1), \ldots, w(F_r)$ mindestens gleich $v(\mathfrak{p})$. Wir wollen zeigen, daß das Minimum $v(\mathfrak{p})$ wirklich erreicht wird:

$$(1) \qquad \text{Min } \{w(F_1), \ldots, w(F_r)\} = v(\mathfrak{p}).$$

Nach der Definition des Begriffes Folgewert ist $v(\mathfrak{p})$ das Minimum der Werte $w(\mathfrak{A})$ für alle Divisoren $\mathfrak{A}$, die die vorgegebenen Mindestwerte haben. Also ist $v(\mathfrak{p}) = w(\mathfrak{A})$ für ein geeignetes solches $\mathfrak{A}$. Nach der Definition von $w(\mathfrak{A})$ aber (§ 4) ist $w(\mathfrak{A})$ gleich einem geeigneten $w(F)$, wo F ebenso wie $\mathfrak{A}$ mindestens die vorgeschriebenen Mindestwerte hat, also dem H-Ideal angehört. Wären nun alle $w(F_1), \ldots, w(F_\mu)$ größer als $v(\mathfrak{p})$, so wäre auch $w(\mathfrak{A})$ größer als $v(\mathfrak{p})$, was nicht geht. Also ist (1) richtig.

(1) gilt übrigens nicht nur für niedere, sondern auch für höhere Primdivisoren, nur daß dann natürlich $v(\mathfrak{p}) = 0$ ist.

Die Formen $G(x)$, deren Durchschnitte mit M den Divisor $\mathfrak{B}_0$ als Faktor enthalten, bilden wieder ein H-Ideal mit einer endlichen Basis $(G_1, \ldots, G_s)$. Man beweist wie oben, daß

$$(2) \qquad \text{Min } \{w(G_1), \ldots, w(G_s)\} = w(\mathfrak{B}_0)$$

für jeden Primdivisor $\mathfrak{p}$ gilt.

Nun sei m das Maximum der Gradzahlen von $F_1, \ldots, F_\mu, G_1, \ldots, G_s$. Dann gibt es zu jedem $\mathfrak{p}$ in jedem der beiden vorhin erwähnten H-Ideale Formen F oder G vom Grade m, für welche gilt

$$(3) \qquad\qquad w(F) = v(\mathfrak{p})$$

$$(4) \qquad\qquad w(G) = w(\mathfrak{B}_0)$$

Die Formen F vom Grade m, die in allen Bewertungen $\mathfrak{p}$ mindestens die Werte $v(\mathfrak{p})$ haben, schneiden aus M eine lineare Schar $\{\mathfrak{C}_\lambda\}$ aus, deren effektiver Mindestwert für jedes $\mathfrak{p}$ genau gleich dem virtuellen $v(\mathfrak{p})$ ist, also eine lineare Schar ohne Überwerte. Ebenso schneiden die Formen G vom Grade m, die durch $\mathfrak{B}_0$ teilbar sind, aus M eine lineare Schar $\{\mathfrak{A}_\lambda \mathfrak{B}_0\}$ aus, deren effektiver Mindestwert für jedes $\mathfrak{p}$ gleich $w(\mathfrak{B}_0)$ ist. Läßt man den festen Bestandteil $\mathfrak{B}_0$ aus dieser Schar weg, so bleibt die Schar $\{\mathfrak{A}_\lambda\}$ übrig, deren effektive Mindestwerte demnach Null sind. Setzt man die virtuellen Mindestwerte dieser Schar auch Null, so hat man wieder eine Schar ohne Überwerte.

Nun ist aber klar, daß die Schar $\{\mathfrak{A}_\lambda \mathfrak{B}_0\}$ als Teilschar in $\{\mathfrak{C}_\lambda\}$ enthalten ist. Die Quotientenschar $\{\mathfrak{C}_\lambda\} : \{\mathfrak{A}_\lambda\}$ enthält also $\mathfrak{B}_0$, hat dieselben virtuellen Werte wie $\{\mathfrak{B}_\mu\}$ und ist daher mit der Vollschar $\{\mathfrak{B}_\mu\}$ identisch. Damit ist Satz 6 für diesen Fall bewiesen.

Sind nun zweitens die virtuellen Werte $v(\mathfrak{p})$ nicht als Folgewerte, sondern nach § 6 durch Transformation definiert, so kann man vor der Transformation den eben dargestellten Beweis geben. Das Ergebnis ist birational invariant, gilt also auch nach der Transformation.

Auf Grund von Satz 6 läßt sich jedes zulässige System von virtuellen Werten darstellen als Differenz zweier Systeme von effektiven Werten von wirklichen linearen Scharen ohne feste Bestandteile. Diese Charakterisierung der virtuellen Wertsysteme ist birational invariant.

<h2 style="text-align:center">§ 8.</h2>

<h3 style="text-align:center">Virtuelle Vollscharen und Divisorenklassen.</h3>

Eine Vollschar $|\mathfrak{A}|$ ist bestimmt durch einen ganzen Divisor $\mathfrak{A}$ und ein System von virtuellen Werten $u(\mathfrak{p})$, wobei $\mathfrak{A}$ mindestens diese Werte $u(\mathfrak{p})$ haben muß:

$$w(\mathfrak{A}) \geqq u(\mathfrak{p}) \qquad\qquad \text{für jedes } \mathfrak{p}.$$

Zwei Vollscharen $|\mathfrak{A}|$, $|\mathfrak{B}|$ werden multipliziert, indem man die Repräsentanten $\mathfrak{A}$, $\mathfrak{B}$ multipliziert und die virtuellen Werte $u(\mathfrak{p})$, $v(\mathfrak{p})$ addiert. Die Division ist nur dann möglich, wenn der Quotient $|\mathfrak{A}| : |\mathfrak{B}|$ existiert; dann ist sie aber eindeutig. Die Vollscharen bilden also eine ABELsche Halbgruppe.

Man kann diese Halbgruppe leicht zu einer Gruppe erweitern, indem man formal Brüche einführt, mit den üblichen Festsetzungen

$$(1) \qquad \frac{|\mathfrak{A}|}{|\mathfrak{B}|} = \frac{|\mathfrak{C}|}{|\mathfrak{D}|}, \text{ wenn } |\mathfrak{A}| \cdot |\mathfrak{D}| = |\mathfrak{B}| \cdot |\mathfrak{C}|$$

$$(2) \qquad \frac{|\mathfrak{A}|}{|\mathfrak{B}|} \cdot \frac{|\mathfrak{C}|}{|\mathfrak{D}|} = \frac{|\mathfrak{A}| \cdot |\mathfrak{C}|}{|\mathfrak{B}| \cdot |\mathfrak{D}|}$$

$$(3) \qquad \frac{|\mathfrak{A}|}{|\mathfrak{B}|} = |\mathfrak{C}|, \text{ wenn } |\mathfrak{A}| = |\mathfrak{B}| \cdot |\mathfrak{C}|$$

Diese Brüche nennt man *virtuelle Vollscharen*. Eine virtuelle Vollschar ist demnach bestimmt durch einen gebrochenen Divisor $\mathfrak{A}\mathfrak{B}^{-1}$ und ein System von virtuellen Werten, die Differenzen der virtuellen Werte der Zähler- und Nennerschar. Dabei darf der gebrochene Divisor

$\mathfrak{A}\mathfrak{B}^{-1}$ durch jeden äquivalenten Divisor $\mathfrak{C}\mathfrak{D}^{-1}$ ersetzt werden: zwei Divisoren $\mathfrak{F}$, $\mathfrak{G}$ heißen bekanntlich äquivalent, wenn ihr Quotient $\mathfrak{F}\mathfrak{G}^{-1}$ der Divisor einer Funktion φ des Funktionenkörpers ist. Die virtuellen Werte der virtuellen Vollschar aber dürfen nicht durch andere ersetzt werden: ändert man sie, so erhält man eine andere virtuelle Vollschar. Das alles folgt leicht aus der Gleichheitsdefinition (1).

Rechnet man alle zu einem (gebrochenen) Divisor $\mathfrak{F}$ äquivalenten Divisoren zu einer *Divisorenklasse*, so kann man das Gesagte auch so ausdrücken: *Eine virtuelle Vollschar ist bestimmt durch eine Divisorenklasse und ein System virtueller Werte.* Das System virtueller Werte ist dabei nicht willkürlich, sondern es muß sich als Differenz von zwei Systemen effektiver Mindestwerte von wirklichen Vollscharen ohne Überwerte darstellen lassen. Aus Satz 6 folgt nämlich, daß jede virtuelle Vollschar sich als Quotient von effektiven Vollscharen ohne Überwerte darstellen läßt:

$$\frac{|\mathfrak{A}|}{|\mathfrak{B}|} = \frac{|\mathfrak{C}| : |\mathfrak{D}|}{|\mathfrak{G}| : |\mathfrak{H}|} = \frac{|\mathfrak{C}| \cdot |\mathfrak{H}|}{|\mathfrak{D}| \cdot |\mathfrak{G}|},$$

wo $|\mathfrak{C}|$, $|\mathfrak{D}|$, $|\mathfrak{G}|$, $|\mathfrak{H}|$ effektive Vollscharen ohne Überwerte sind.

Der Begriff der Divisorenklasse ist nicht birational invariant, wohl aber der Begriff „virtuelle Vollschar".

(Eingegangen am 5. März 1945.)

27.

Zur algebraischen Geometrie 16
Vielfältigkeiten von abstrakten Ketten

Mathematische Annalen 125 (1953) 314–324

Vorbemerkung zur Terminologie.

ANDRÉ WEIL hat in seinem Buch „Foundations of algebraic Geometry" die algebraische Geometrie auf eine neue Basis gestellt und dabei auch die alte Terminologie durch eine radikal neue ersetzt. Im Zentrum stehen bei ihm — mit Recht — die absolut irreduziblen Mannigfaltigkeiten, die er kurzerhand Varieties nennt. Er unterscheidet scharf zwischen einer Vereinigungsmenge solcher Varieties (bunch of varieties) und einer formalen Summe von Varieties derselben Dimension (cycle). Die Unterscheidung dieser beiden ganz verschiedenen Begriffe, die bisher durch dasselbe Wort Mannigfaltigkeit (= variety) gedeckt wurden, scheint mir unbedingt geboten; die Wahl der Worte kommt mir aber nicht glücklich vor. Ich möchte daher den folgenden Vorschlag machen, der durch Gespräche und Vorträge in Paris und Lüttich schon vorbereitet wurde.

Vorschlag	Engl. Äquivalent	ANDRÉ WEIL	bisher üblich
Vielfältigkeit	variety	bunch of varieties	algebr. Mannigfaltigk.
Teil	part		Teilmannigfaltigkeit
unteilbare V.	indivisible var.	variety	absolut irreduzible M.
V. über einem Körper K	var. over a field K	regular algebr. bunch over K	
irreduzible V.	irreducible var.		irreduzible M.
K gut für V	K good for V	K field of definition for V	
Kette	chain	cycle	algebr. Mannigfaltigk.
Kurve	curve	bunch of curves	algebr. Kurve
Fläche	surface	bunch of surfaces	algebr. Fläche
Vielfältigkeit von Ketten	variety of chains		algebr. System von Mannigfaltigkeiten
Spezialisierung	specialization	specialization	relationstreue Spezialisierung

Der Grund, warum das Wort "cycle" durch "chain" ersetzt wurde, ist dieser. WEIL versteht unter "cycles" nur solche Ketten, deren Bestandteile auf der betrachteten Vielfältigkeit nicht singulär sind. Diese Einschränkung ist aber in einer allgemeinen Theorie der Ketten unzweckmäßig.

Einleitung.

ANDRÉ WEIL hat neben den gewöhnlichen unteilbaren Vielfältigkeiten in affinen oder projektiven Räumen noch "abstract Varieties" eingeführt, die dachziegelartig von gewöhnlichen Vielfältigkeiten V_α überdeckt werden,

389

so daß jeder Punkt P der abstrakten Vielfältigkeit V von mindestens einem Punkt P_α eines Bausteines V_α bedeckt wird. Er stellte mir 1947 die Frage, ob der von Chow und mir definierte Begriff der Vielfältigkeit von Ketten sich auch für Ketten auf einer abstrakten Vielfältigkeit definieren ließe. Nach einem mißglückten Versuch gelang es mir, eine Antwort zu finden, die ich im Oktober 1947 Weil brieflich mitgeteilt habe. Etwas vereinfacht und mit ausführlichen Beweisen versehen, soll diese Lösung hier dargestellt werden.

Der Grundgedanke dieser Lösung ist folgender. Statt eine Vielfältigkeit durch Gleichungen zu definieren, kann man sie auch als Vereinigung von irreduziblen Vielfältigkeiten definieren, wobei die Punkte einer irreduziblen Vielfältigkeit durch Spezialisierung aus einem allgemeinen Punkt hergeleitet werden. Genau so kann man auch eine Vielfältigkeit von Ketten als Vereinigung von irreduziblen Vielfältigkeiten definieren, wobei die Ketten einer irreduziblen Vielfältigkeit durch Spezialisierung einer allgemeinen Kette hervorgehen. Es handelt sich also nur noch darum, den Begriff Spezialisierung für Ketten auf einer abstrakten Vielfältigkeit zu definieren. Das soll in § 4 geschehen.

In § 1—3 sollen die Grundbegriffe der bisherigen Theorie der Ketten unter Zugrundelegung der neuen Terminologie erklärt werden. Dabei habe ich mich dem Standpunkte von Weil weitgehend genähert: nicht die irreduziblen, sondern die unteilbaren Vielfältigkeiten stehen im Zentrum der Aufmerksamkeit. Der Grundkörper wird jeweils so erweitert, daß alle vorkommenden Vielfältigkeiten in unteilbare zerfallen und daß keine störenden inseparablen Körpererweiterungen auftreten. Dazu genügt es, den Körper so zu erweitern, daß die zugeordneten Formen dieser Vielfältigkeiten in absolut irreduzible Faktoren zerfallen.

§ 1. Grundbegriffe.

Als *Grundkörper* nehmen wir einen beliebigen Körper k. Man kann mit Hodge und Pedoe k als Primkörper annehmen, aber es ist nicht unbedingt notwendig. Der *universelle Erweiterungskörper* Ω wird durch Adjunktion von abzählbar vielen Unbestimmten und algebraischem Abschluß aus k erhalten. Alle Koordinaten von Punkten und alle Koeffizienten von Gleichungen werden immer aus Ω entnommen.

Statt des festen Universalkörpers Ω könnte man auch einen „wachsenden Erweiterungskörper" betrachten, der nach Bedarf durch Adjunktion von Unbestimmten und algebraischen Funktionen erweitert werden kann. Demgegenüber hat der feste Universalkörper den Vorteil, eine *Menge* zu sein. Die Lösungen eines Gleichungssystems in Ω bilden also auch eine Menge. Ein weiterer Vorteil ist, daß man zu je zwei Teilkörpern von Ω immer einen Vereinigungskörper bilden kann. Mit K sollen Zwischenkörper bezeichnet werden, die k umfassen und in Ω enthalten sind. Diese Zwischenkörper sollen immer durch Adjunktion von endlich vielen Elementen an k erzeugt werden können.

Ein Zwischenkörper L heißt *separabel erzeugbar* über K, wenn L aus K erzeugt werden kann durch Adjunktion von algebraisch unabhängigen Elementen und separablen algebraischen Funktionen dieser Elemente.

Ein *Punkt des affinen Raumes R_n* ist eine Reihe von n Koordinaten $p_1, \ldots, p_n$ aus Ω. Ein *Punkt des projektiven Raumes S_n* ist ein *Strahl* des affinen Raumes R_{n+1}, bestehend aus allen Punkten $(\omega\, p_0, \omega\, p_1, \ldots, \omega\, p_n)$, wobei $(p_0, \ldots, p_n)$ ein fester von $(0, \ldots, 0)$ verschiedener Punkt von R_{n+1} ist und ω alle Elemente von Ω durchläuft.

Die Menge aller Punkte von R_n oder S_n, die einem endlichen System von algebraischen Gleichungen

$$f_k\,(p_1,\,\ldots,\,p_n) = 0 \quad \text{oder} \quad f_k\,(p_0,\,\ldots,\,p_n) = 0$$

genügen, wobei die f_k im ersten Fall Polynome, im zweiten Fall Formen mit Koeffizienten aus Ω sein sollen, heißt eine *Vielfältigkeit*. Dabei wird vorausgesetzt, daß die Menge nicht leer ist. Die übliche Bezeichnung „algebraische Mannigfaltigkeit" möchte ich vermeiden, erstens weil sie zu lang ist, zweitens weil in der Topologie der Begriff Mannigfaltigkeit (manifold) eine andere Bedeutung hat. Man könnte mit SCHLÄFLI statt Vielfältigkeit auch Totalität sagen.

Eine Vielfältigkeit, die sich als Vereinigung von zwei echten *Teilen* (d. h. Teilvielfältigkeiten) darstellen läßt, heißt *zerlegbar*. Ist eine solche Zerlegung unmöglich, so heißt die Vielfältigkeit *unteilbar*.

Der Begriff unteilbar ist nicht mit *irreduzibel* zu verwechseln. Eine Vielfältigkeit, die durch Gleichungen mit Koeffizienten aus einem bestimmten Körper K definiert ist, möge eine *Vielfältigkeit über K* heißen. Läßt sie sich nicht zerlegen in echte Teile, die wieder Vielfältigkeiten über K sind, so heißt sie *irreduzibel über K*. Eine unteilbare Mannigfaltigkeit aber bleibt irreduzibel bei jeder Körpererweiterung: sie ist *absolut irreduzibel*.

Ein Punkt P heißt eine *Spezialisierung* eines Punktes X in bezug auf einen Körper K, wenn alle Gleichungen $f\,(x_1,\,\ldots,\,x_n) = 0$, mit Koeffizienten aus K, oder im projektiven Falle alle homogenen Gleichungen $f\,(x_0,\,x_1,\,\ldots,\,x_n) = 0$, die für den Punkt X gelten, richtig bleiben bei Ersetzung von X durch P. Eine über K irreduzible Vielfältigkeit V besitzt einen *allgemeinen Punkt X*, aus dem alle Punkte von V durch Spezialisierung in bezug auf K hervorgehen. Der allgemeine Punkt ist bis auf Isomorphismen über K eindeutig durch V bestimmt.

Im affinen Fall heißt das, daß die Koordinaten $x_1,\,\ldots,\,x_n$ eindeutig bestimmt sind bis auf einen auf alle x_k anzuwendenden Körperisomorphismus, der die Elemente von K unverändert läßt. Im projektiven Fall sind natürlich die x_k nur bis auf einen gemeinsamen Faktor ω eindeutig bestimmt. Numeriert man aber die Koordinaten so, daß $x_0 \neq 0$ ist, und normiert ω dann so, daß $x_0 = 1$ wird, so sind die inhomogenen Koordinaten $x_1,\,\ldots x_n$ des Punktes X bis auf einen Isomorphismus eindeutig bestimmt. Die Anzahl der algebraisch unabhängigen unter den so normierten x_k heißt die *Dimension* von V.

Eine irreduzible, also insbesondere eine unteilbare Vielfältigkeit kann auch definiert werden als die Menge aller Punkte P, die aus einem gegebenen Punkt X durch Spezialisierung entstehen. So geht A. WEIL vor. Eine beliebige Vielfältigkeit kann dann als Vereinigung von endlich vielen unteilbaren Vielfältigkeiten definiert werden.

WEIL führt außerdem den Begriff der *abstrakten Vielfältigkeit* ein. Ein solches *Abstraktum*, wie ich es kurz nennen will, ist definiert durch ein endliches System von unteilbaren Vielfältigkeiten V_α, zwischen denen birationale Abbildungen $T_{\beta\alpha}$ gegeben sind. WEIL nimmt die V_α als affine Vielfältigkeiten an, aber es macht für die Definition keinen Unterschied, wenn wir von vornherein die V_α als Vielfältigkeiten in projektiven Räumen annehmen. Auf jeder V_α sei ein echter Teil (d. h. Teilvielfältigkeit) F_α gegeben, die *Grenze* von V_α. Die Grenzen werden bei der Konstruktion des Abstraktums nicht mit benutzt, sondern nur die Reste $V_\alpha - F_\alpha$. Weiter seien alle V_α birational äquivalent;

es gebe also birationale Abbildungen T_β von V_1 auf V_β. Definiert man nun

$$T_{\beta\alpha} = T_\beta\, T_\alpha^{-1},$$

so ist $T_{\beta\alpha}$ eine birationale Abbildung von V_α auf V_β. Die $T_{\beta\alpha}$ genügen den Bedingungen

$$T_{\gamma\beta}\, T_{\beta\alpha} = T_{\gamma\alpha}$$
$$T_{\alpha\alpha} = 1$$
$$T_{\beta\alpha}^{-1} = T_{\alpha\beta}.$$

Eine rationale Abbildung heißt *regulär* in einem Punkt P, wenn die rationalen Funktionen, die die Koordinaten des Bildpunktes eines allgemeinen Punktes X definieren, so gewählt werden können, daß ihre Nenner im Punkt P nicht Null werden. Ist nun (P_α, P_β) ein solches Paar der Transformation $T_{\beta\alpha}$, daß P_α nicht zu F_α und P_β nicht zu F_β gehört, so wird verlangt, daß dann $T_{\beta\alpha}$ regulär ist im Punkt P_α und $T_{\alpha\beta}$ regulär im Punkt P_β.

Unter dieser Bedingung gibt es zu jedem Punkt P_α von· $V_\alpha - F_\alpha$ ein maximales System von Punkten $(P_\alpha, P_\beta, P_\gamma, \ldots)$, zu denen mindestens P_α selbst gehört, derart, daß P_β nicht zu F_β gehört, P_γ nicht zu F_γ, usw., und daß je zwei von diesen Punkten, etwa P_β und P_γ, immer ein Paar in der Transformation $T_{\gamma\beta}$ bilden.

Jedes solche maximale System $(P_\alpha, P_\beta, \ldots)$ heißt nun ein *Punkt* des Abstraktums

$$V = \{V_\alpha, T_{\alpha\beta}; F_\alpha\}.$$

V selbst ist die Menge aller ihrer Punkte.

Eine Spezialisierung eines „Punktes" $(X_\alpha, X_\alpha, \ldots)$ in bezug auf einen Körper K wird so definiert: man spezialisiere irgendeines der X_λ, etwa X_α zu P_α, wobei P_α nicht auf der Grenze F_α liegen möge. Dann ergänze man P_α zu einem maximalen System $(P_\alpha, P_\beta, \ldots)$ von zugeordneten Punkten. Dieses maximale System P ist dann eine Spezialisierung des vorgelegten Systems X.

Die Menge aller Spezialisierungen eines „Punktes" X von V ist ein irreduzibler Teil von V in bezug auf K. Jedem irreduziblen Teil W_α eines V_α, der nicht auf der Grenze F_α liegt, entspricht ein einziger irreduzibler Teil W von V, von derselben Dimension. Ist W_α unteilbar, so ist W es auch. In dieser Weise übertragen sich alle wichtigen Begriffe von den einzelnen V_α auf das Abstraktum V.

Eine Vereinigung von endlich vielen unteilbaren Teilen von V ist ein *Teil* oder eine *Teilvielfältigkeit* von V.

§ 2. Die Cayleyform einer irreduziblen Vielfältigkeit.

Es sei V eine über K irreduzible Vielfältigkeit von der Dimension d im projektiven S_n. Es seien $u^{(1)}, \ldots, u^{(d)}$ Hyperebenen mit unbestimmten Koordinaten $u_k^{(\nu)}$. Die Unbestimmten $u_k^{(\nu)}$ sollen algebraisch unabhängig über K sein. Die d Hyperebenen schneiden V in endlich vielen über K konjugierten Punkten $X^{(1)}, \ldots, X^{(g)}$. Nimmt man nun eine weitere Reihe von Unbestimmten u_k $(k = 0, \ldots, n)$ hinzu, so ist das Produkt

$$(1) \qquad P = \prod_1^g (u_0\, X_0^{(\nu)} + u_1\, X_1^{(\nu)} + \cdots + u_n\, X_n^{(\nu)})$$

eine symmetrische Funktion der Punkte $X^{(1)}, \ldots, X^{(g)}$. Im Fall der Charakteristik Null ist also das Produkt rational in $K(u, u^{(1)}, \ldots, u^{(d)})$; in diesem Fall setzen wir

$$P = Q(u, u^{(1)}, \ldots, u^{(d)}).$$

Im Fall der Charakteristik p ist eine p^e-te Potenz des Produktes P rational, und wir setzen

$$(2) \qquad P^q = Q(u, u^{(1)}, \ldots, u^{(d)}) \qquad (q = p^e).$$

In jedem der beiden Fälle ist Q ganz in u und rational in $u^{(1)}, \ldots, u^{(d)}$. Man kann also setzen

$$(3) \qquad Q = \frac{A}{B} F(u, u^{(1)}, \ldots, u^{(d)}),$$

wobei A und B nur von $u^{(1)}, \ldots, u^{(d)}$ abhängen, während F ganz in $u, u^{(1)}, \ldots, u^{(d)}$ ist und keinen nur von $u^{(1)}, \ldots, u^{(d)}$ abhängigen Faktor mehr enthält.

Die so definierte Form F haben Chow und ich die *zugeordnete Form* von V genannt[1]. Bei einer Vertauschung der Variablenreihen $u, u^{(1)}, \ldots, u^{(d}$ bleibt F bis auf Faktoren ± 1 ungeändert. Für den Beweis siehe Ch-W. § 1.

Aus der Formel (2) folgt, daß Q die Unbestimmten $u_0, \ldots, u_n$ nur in der q-ten Potenz enthält. Dasselbe gilt nach (3) für F. Wegen der Vertauschungsmöglichkeit folgt daraus, daß F auch die übrigen $u_k^{(v)}$ nur in der q-ten Potenz enthält. Also ist F eine q-te Potenz einer Form in den u_k und $u_k^{(v)}$ mit Koeffizienten aus einem Körper K_0, der aus K durch Adjunktion der q-ten Wurzeln aller Koeffizienten von F entsteht:

$$(4) \qquad F = F_0^q.$$

Die Form F_0 hat gegenüber F den Vorteil, daß sie von der Wahl des Körpers K unabhängig ist. Sie möge die *Cayleyform* der Vielfältigkeit V heißen.

Die Cayleyform ist, von einem unwesentlichen Faktor C abgesehen, mit dem Produkt (1) identisch. Um das einzusehen, denken wir uns die Schnittpunkte $X^{v)}$ durch $X_0^{v)} = 1$ normiert. Das Produkt (1) enthält dann ein Glied u_0^g, die Form (2) ein Glied u_0^{qg}. Die Glieder von Q haben also keinen gemeinsamen von $u^{(1)}, \ldots, u^{(d)}$ abhängigen Faktor: der Zähler A in (3) ist Eins. Statt (3) kann man nun schreiben

$$BQ = F$$

oder

$$(5) \qquad BP^q = F_0^q.$$

Setzt man in (5) links und rechts $u_0 = 1, u_1 = 0, \ldots, u_n = 0$, so sieht man, daß B ebenfalls eine q-te Potenz ist

$$(6) \qquad B = C^q,$$

wobei C eine Form in $u^{(1)}, \ldots, u^{(d)}$ mit Koeffizienten aus K_0 ist. Setzt man (6) in (5) ein, so folgt schließlich

$$C^q P^q = F_0^q$$

oder, da die p^e-te Wurzel eindeutig ist,

$$(7) \qquad CP = F_0.$$

---

[1] Chow, W.-L., u. B. L. v. d. Waerden: Math. Ann. **113**, 692 (1937). Zu zitieren als Ch-W.

Wir sehen also, daß man durch eine geeignete Erweiterung des Konstanten-körpers K, nämlich durch Adjunktion von p^e-ten Wurzeln, immer erreichen kann, daß nicht nur eine p^e-te Potenz des Produktes P, sondern das Produkt P selbst rational in $K\,(u,\,u^{(1)},\,\ldots,\,u^{(d)}\,)$ ist. Die zugeordnete Form F ist dann frei von vielfachen Faktoren, und der Unterschied zwischen der Cayleyform und der zugeordneten Form verschwindet.

Im folgenden wird immer vorausgesetzt, daß diese Körpererweiterung vollzogen ist, daß also der Körper K die Koeffizienten der Cayleyform enthält. Wie CHOW[2]) bemerkt hat, ist diese Voraussetzung gleichbedeutend mit der von WEIL gemachten Voraussetzung, daß der Körper $K\,(X)$, der aus K durch Adjunktion der Koordinatenverhältnisse eines allgemeinen Punktes X von V entsteht, separabel erzeugbar ist.

Die Cayleyform $F = F_0$ ist durch V eindeutig bestimmt. Umgekehrt bestimmt sie V eindeutig, denn durch die Faktorenzerlegung (1) erhält man g allgemeine Punkte von V, aus denen durch Spezialisierung alle Punkte von V erhalten werden können. Wie man die Gleichungen von V erhält, ist in Ch.-W angegeben.

Wenn bei einer Erweiterung des Grundkörpers die Vielfältigkeit V zerfällt, so zerfällt auch die Cayleyform in Faktoren, und umgekehrt. Eine endliche Körpererweiterung genügt immer, um die zugeordnete Form in absolut irre-duzible Faktoren zu zerlegen (ZAG 10, Math. Ann. 113, 706). Daher zerfällt auch V nach einer endlichen Körpererweiterung in unteilbare Vielfältigkeiten.

Wird V von vornherein als unteilbar und $K\,(X)$ als separabel erzeugbar vorausgesetzt, so ist die zugeordnete Form absolut irreduzibel, und umgekehrt. Unter diesen Voraussetzungen nennt WEIL den Körper K "a field of definition" für V. Mir scheint dieser Ausdruck nicht glücklich gewählt, denn man denkt dabei unwillkürlich an einen Körper, in dem die Vielfältigkeit V (etwa durch Gleichungen) definiert ist. Ich möchte daher lieber vorschlagen, einen solchen Körper K *gut für* V zu nennen, und umgekehrt V *gut für* K.

Die Cayleyform $F = F_0$ zerfällt nach (7) in lauter verschiedene Faktoren

$$u_0\,X_0^{(\nu)} + u_1\,X_1^{(\nu)} + \cdots + u_n\,X_n^{(\nu)}.$$

Normiert man $X_0^{(\nu)} = 1$, so erhält man als Faktoren

$$u_0 + u_1\,X_1^{(\nu)} + \cdots + u_n\,X_n^{(\nu)};$$

folglich ist

$$\vartheta = u_1\,X_1^{(\nu)} + \cdots + u_n\,X_n^{(\nu)}$$

separabel in bezug auf den Körper $K\,(u_1,\,\ldots,\,u_n,\,u^{(1)},\,\ldots,\,u^{(d)})$. Aus dem Beweis des Satzes vom primitiven Element ist bekannt, daß $X_1^{(\nu)},\,\ldots,\,X_n^{(\nu)}$ rational durch das primitive Element ϑ ausgedrückt werden können. Also sind die Koordinaten der Schnittpunkte $X^{(\nu)}$ separable algebraische Funktionen von $u^{(1)},\,\ldots,\,u^{(d)}$. Solche Funktionen sind aber differenzierbar. Durch eine bekannte Überlegung[3]) kann man nun in jedem dieser Punkte einen *Tangential-raum* an V konstruieren. Von diesem Tangentialraum werden wir in § 4 Gebrauch machen.

---

[2]) CHOW, W. L.: Amer. J. Math. 72, 252 (1950).
[3]) V. D. WAERDEN, B. L.: ZAG 14, Math. Ann. 115, 621 (1938), Satz 1.

§ 3. Ketten im projektiven Raum und ihre Koordinaten.

Eine *Kette* C besteht aus endlich vielen unteilbaren Vielfältigkeiten $V_1, \ldots, V_s$ von derselben Dimension d, mit beliebigen ganzzahligen Vielfachheiten $e_1, \ldots, e_s$ versehen. In dieser Untersuchung sollen $e_1, \ldots, e_s$ als positive ganze Zahlen angenommen werden. Ketten werden als Summen geschrieben:

$$C = e_1 V_1 + \cdots + e_s V_s.$$

Sind $F_1, \ldots, F_s$ die Cayleyformen von $V_1, \ldots, V_s$, so heißt

$$F = F_1^{e_1} \ldots F_s^{e_s}$$

die Cayleyform der Kette C. Der Grad der Form F in u ist $g = e_1 g_1 + \cdots + e_s g_s$, wobei g_i der Grad von F_i ist. g heißt der *Grad* der Kette C.

In Ch-W wurde bewiesen, daß die Koeffizienten der Form F gewisse homogene Gleichungen erfüllen müssen, die auch hinreichend sind, damit F die Cayleyform einer Kette von der Dimension d und vom Grade g ist.

Die Koeffizienten der Cayleyform F gehören einem Körper K an, dem Vereinigungskörper der Koeffizientenkörper der Formen $F_1, \ldots, F_s$. Manchmal genügt auch ein kleinerer Körper, aber darauf kommt es hier nicht an.

Die Koeffizienten von F heißen die *Koordinaten* der Kette C. Nur die Koordinatenverhältnisse sind eindeutig durch C bestimmt; es handelt sich also um homogene Koordinaten $c_0, \ldots c_N$.

Faßt man $c_0, \ldots, c_N$ als Punktkoordinaten in einem Bildraum S_N auf, so erhält man den *Bildpunkt* der Kette C. Der Bildpunkt wird zweckmäßig mit demselben Buchstaben C bezeichnet wie die Kette. Das führt nicht einmal dann zu Verwechslungen, wenn die Kette C selber aus einem einzigen Punkt des Raumes S_n besteht; denn in diesem Fall sind die Koeffizienten der Cayleyform gerade die homogenen Koordinaten $c_0, \ldots, c_n$ dieses Punktes.

Eine *Vielfältigkeit von Ketten* derselben Dimension d und desselben Grades g ist die Menge aller Ketten, deren Koordinaten einem endlichen System von homogenen Gleichungen genügen. Die Vielfältigkeiten von Ketten entsprechen also eineindeutig den Vielfältigkeiten von Punkten im Bildraum S_N. Die Begriffe Dimension, irreduzibel, unteilbar usw. können ohne weiteres von den Vielfältigkeiten im Bildraum auf die Vielfältigkeiten von Ketten übertragen werden. Die Ketten einer irreduziblen Vielfältigkeit können alle durch Spezialisierung einer allgemeinen Kette derselben Vielfältigkeit erhalten werden.

Insbesondere bilden alle Ketten der Dimension d und des Grades g im S_n eine Vielfältigkeit. Aber auch die Ketten, die auf einer gegebenen Vielfältigkeit W des S_n liegen, bilden eine Vielfältigkeit. Die wichtigsten Relationen zwischen Ketten, z. B., daß eine Kette der Trägervielfältigkeit einer anderen Kette angehört, können durch algebraische Gleichungen zwischen den Kettenkoordinaten ausgedrückt werden. Das alles wurde in Ch-W bewiesen.

§ 4. Spezialisierung von abstrakten Ketten.

Es sei V ein Abstraktum, definiert durch die unteilbaren Vielfältigkeiten V_α mit den Grenzen F_α und den Abbildungen $T_{\beta\alpha}$ von V_α auf V_β. Es sei C eine Kette auf V, d. h. eine Summe von unteilbaren d-dimensionalen Teilen von V mit Koeffizienten e_k. Wir wollen den Begriff der Spezialisierung der Kette C definieren. Dabei beschränken wir uns der Einfachheit halber zunächst auf den Fall, daß C unteilbar ist. Der Konstantenkörper, in bezug auf den die

Spezialisierung stattfindet, möge K heißen. Der Körper K möge die Koeffizienten der rationalen Abbildungen $T_{\beta\alpha}$ sowie der Cayleyformen aller V_α enthalten.

Die unteilbare Vielfältigkeit C auf V ist definiert als ein maximales System von unteilbaren Vielfältigkeiten C_α auf V_α, deren allgemeine Punkte P_α nicht zu F_α gehören und sich in den Transformationen $T_{\beta\alpha}$ entsprechen:

$$T_{\beta\alpha} P_\alpha = P_\beta .$$

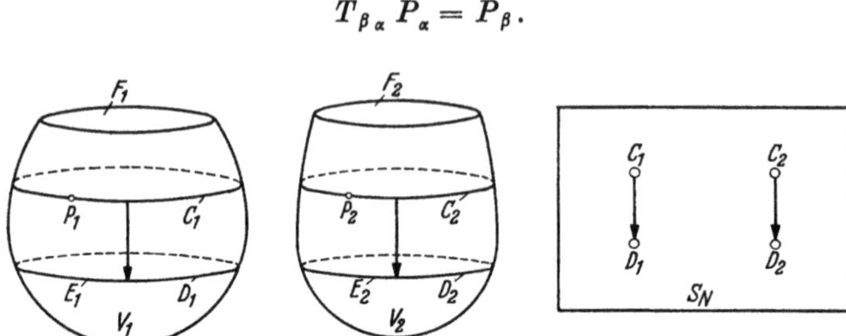

Die Teile C_α haben je einen Bildpunkt C_α in einem projektiven Raum S_N. Es seien etwa $C_1, \ldots, C_h$ diese Bildpunkte.

Nun sei $C_1 \to D_1, \ldots, C_h \to D_h$ eine gleichzeitige Spezialisierung der Punkte $C_1, \ldots, C_h$. Das bedeutet, daß alle homogenen Gleichungen zwischen den Koordinaten von $C_1, \ldots, C_h$ mit Koeffizienten aus K auch für $D_1, \ldots, D_h$ gelten. Zu diesen homogenen Gleichungen gehören auch diejenigen, die ausdrücken, daß der Punkt C_α Bildpunkt einer Kette C_α ist und daß diese auf V_α liegt. Also stellen $D_1, \ldots, D_h$ wieder d-dimensionale Ketten auf $V_1, \ldots, V_h$ dar.

Es kann sein, daß einige unteilbare Komponenten von D zur Grenze F_α gehören. Diese Komponenten mögen aus D_α weggelassen werden. Was übrig bleibt, ist eine Kette E_α (möglicherweise die Nullkette, wenn alle Komponenten weggelassen werden müssen). Wir beweisen nun den entscheidenden

Hilfssatz. Sind A_α und A_β zwei unteilbare d-dimensionale Teile von V_α und V_β, die sich in der Transformation $T_{\beta\alpha}$ entsprechen, wobei A_α nicht zur Grenze F_α und ebenso A_α nicht zu F_β gehören möge, so hat A_α in D_α denselben Koeffizienten wie A_β in D_β.

Beweis. Die Transformation $T_{\beta\alpha}$ induziert eine Korrespondenz $C_{\beta\alpha}$ zwischen C_α und C_β. Diese Korrespondenz, betrachtet als eine Vielfältigkeit von Punktepaaren, kann wieder durch einen Bildpunkt $C_{\beta\alpha}$ in einem projektiven Raum S_N dargestellt werden (wobei N größer sein kann als das frühere N). Die simultane Spezialisierung von C_α und C_β zu D_α und D_β kann nun zu einer Spezialisierung der drei Punkte C_α, C_β, $C_{\beta\alpha}$ zu D_α, D_β, $D_{\beta\alpha}$ erweitert werden.

Alle Relationen zwischen C_α, C_β und $C_{\beta\alpha}$, die durch algebraische Gleichungen zwischen den Koordinaten dieser drei Ketten ausgedrückt werden können, bleiben bei der Spezialisierung erhalten. Zu diesen Relationen gehören die folgenden:

1. Die Beziehung, die darin besteht, daß von den Punktepaaren von $C_{\beta\alpha}$ immer der erste Punkt zu C_α und der zweite zu C_β gehört. Die zugehörigen Gleichungen können genau so hergeleitet werden, wie die, welche ausdrücken, daß eine Kette auf der Trägervielfältigkeit einer anderen liegt.

2. Die Beziehung, die besagt, daß jedem Punkt von C_α oder C_β mindestens ein Punktepaar von $C_{\beta\alpha}$ entspricht. Diese Gleichungen können so erhalten werden: aus den Gleichungen $f(x, y) = 0$ von $C_{\beta\alpha}$ eliminiert man die Koordinaten y des zweiten Punktes; so erhält man ein System von Hyperflächen; nun verlangt man, daß C_α auf allen diesen Hyperflächen liegt.

3. Die Beziehung, die besagt, daß alle Punktepaare von $C_{\beta\alpha}$ auch zu $T_{\beta\alpha}$ gehören.

Alle diese Beziehungen bleiben bei der Spezialisierung erhalten. Also gehört von den Punktepaaren von $D_{\beta\alpha}$ der erste Punkt immer zu D_α und der zweite zu D_β, jeder Punkt von D_α oder D_β gehört mindestens einem Punktepaar von $D_{\beta\alpha}$ an, und alle Punktepaare von $D_{\beta\alpha}$ gehören zu $T_{\beta\alpha}$.

Nun seien, wie vorhin, A_α und A_β zwei unteilbare d-dimensionale Teile von V_α und V_β, die nicht zur Grenze F_α oder F_β gehören und die sich in der Transformation $T_{\beta\alpha}$ entsprechen. Die Transformation $T_{\beta\alpha}$ induziert eine Korrespondenz $A_{\beta\alpha}$ zwischen A_α und A_β. Wenn A_α eine Komponente von D_α ist, so ist auf Grund der obigen Beziehungen $A_{\beta\alpha}$ eine Komponente von $D_{\beta\alpha}$ und A_β eine Komponente von D_β. Wir haben zu beweisen, daß A_α in D_α mit derselben Vielfachheit auftritt wie A_β in D_β. Dazu genügt es zu beweisen, daß A_α in D_α mit derselben Vielfachheit auftritt wie $A_{\beta\alpha}$ in $D_{\beta\alpha}$.

Wir schneiden C_α und D_α mit einer allgemeinen Hyperebene H_1 des Raumes S_n, in dem V_α liegt. Die Hyperebene H_1 enthält keine Komponente von C_α oder D_α; die Schnitte $H_1 \cdot C_\alpha$ und $H_1 \cdot D_\alpha$ haben also die richtige Dimension $d - 1$. Nun ist D_α eine Spezialisierung von C_α, also ist $H_1 \cdot D$ eine Spezialisierung von $H_1 \cdot C$, und zwar handelt es sich um eine simultane Spezialisierung[4]). Man kann dieselbe Hyperebene H_1 auch als Hyperfläche im zweifach-projektiven Raum der Punktepaare von V_α und V_β auffassen: die Gleichung dieser Hyperfläche enthält eben nur die Koordinaten des ersten Punktes. In einem beliebig fest gewählten Punkt einer Komponente von $C_{\beta\alpha}$ oder $D_{\beta\alpha}$ ist diese Gleichung natürlich nicht erfüllt, weil H_1 ja allgemein gewählt war. Also haben die Durchschnitte $H_1 \cdot C_{\beta\alpha}$ und $H_1 \cdot D_{\beta\alpha}$ wieder die „richtige" Dimension $d - 1$ und $H_1 \cdot D_{\beta\alpha}$ ist eine Spezialisierung von $H_1 \cdot C_{\beta\alpha}$.

Nimmt man in derselben Weise eine zweite Hyperebene H_2 hinzu, usw. bis H_d, und setzt zur Abkürzung $H_1 H_2 \ldots H_d = L$ (linearer Raum), so findet man, daß

$$D_\alpha, \; LD_\alpha, \; D_{\beta\alpha}, \; LD_{\beta\alpha}$$

simultane Spezialisierungen von

$$C_\alpha, \; LC_\alpha, \; C_{\beta\alpha}, \; LC_{\beta\alpha}$$

sind.

LC_α ist eine Summe von endlich vielen Punkten $P_1, \ldots, P_g$, wo g der Grad von C_α ist. Der allgemeine lineare Raum L hat in diesen Punkten keine Tangente mit C_α gemeinsam; jeder Schnittpunkt hat also Vielfachheit Eins. Die Schnittpunkte liegen nicht auf F_α, die Transformation $T_{\beta\alpha}$ ist also regulär in diesen Punkten. Jedem P_ν entspricht somit ein einziges Punktepaar (P_ν, P'_ν) von $C_{\beta\alpha}$. Diese Punktepaare haben in der Kette $LC_{\beta\alpha}$ wieder die Vielfachheit Eins, weil L und $C_{\beta\alpha}$ keine Tangente gemeinsam haben; die Transformation $T_{\beta\alpha}$ ist ja differenzierbar und führt Tangentenrichtungen in Tangentenrichtungen über.

---

[4]) Siehe ZAG 14, Math. Ann. **115**, 633 (1938), Satz B.

Genau dieselbe Überlegung gilt für LA_α und $LA_{\beta\alpha}$. Der Durchschnitt LA_α besteht aus Punkten $Q_1, \ldots, Q_\nu$ mit der Vielfachheit Eins, und jedem Q_λ entspricht ein Punktepaar (Q_λ, Q'_λ), das in $LA_{\beta\alpha}$ wieder Vielfachheit Eins hat.

Bei der simultanen Spezialisierung gehen die Punkte $P_1, \ldots, P_q$, aus denen LC_α besteht, in gewisse Punkte $R_1, \ldots, R_g$ über, die zusammen die Kette LD_α bilden. Ebenso gehen die Punktepaare (P_ν, P'_ν) in ebensoviele Punktepaare (R_ν, R'_ν) über, die zusammen die Kette $LD_{\beta\alpha}$ bilden. Die Zahl f, die angibt, wie oft Q_1 unter den R_ν vorkommt, ist gleich der Zahl f', die angibt, wie oft das Paar (Q_1, Q'_1) unter den (R_ν, R'_ν) vorkommt. Die Zahl f ist aber nach Definition die Vielfachheit von A_α als Bestandteil von LD_α; ebenso ist f' die Vielfachheit von $A_{\beta\alpha}$ als Bestandteil von $LD_{\beta\alpha}$. Also sind diese Vielfachheiten gleich, was zu beweisen war.

Auf Grund des nunmehr bewiesenen Hilfssatzes ist die Vielfachheit f, mit der A_α in D_α oder (was dasselbe ist) in E_α vorkommt, vom Index α unabhängig. Faßt man nun ein maximales System von zugeordneten Teilen $(A_\alpha, A_\beta, \ldots)$ zu einem d-dimensionalen unteilbaren Teil A von V zusammen, so kann man diesem Teil die Vielfachheit f zuordnen. Macht man dasselbe für alle Teile A_α oder $A_\beta, \ldots$, die überhaupt in einem der endlich vielen $D_\alpha, D_\beta, \ldots$ als Komponenten vorkommen, so erhält man endlich viele Teile A mit ganz bestimmten Vielfachheiten f. Diese kann man nun zu einer Kette

$$D = \Sigma f A$$

zusammenfassen. Diese Kette D heißt eine *Spezialisierung der Kette C in bezug auf den Körper K.*

Da die Spezialisierung $C \to D$ durch simultane Spezialisierungen $C_1 \to D_1$, $C_2 \to D_2, \ldots$ auf den projektiven Vielfältigkeiten $V_1, V_2, \ldots$ definiert war, wobei jede nicht zur Grenze F_α gehörige Komponente A_α in D_α genau denselben Koeffizienten f hat, wie A in D, so lassen sich die Eigenschaften der abstrakten Spezialisierungen unmittelbar aus den Eigenschaften der projektiven Spezialisierungen ablesen.

Die Beschränkung auf unteilbare Ketten C ist ganz unwesentlich. Ist

$$C = \Sigma e_\nu\, C^{(\nu)}$$

eine beliebige Kette, so nehme man simultane Spezialisierungen

$$C^{(1)} \to D^{(1)}, \;\; C^{(2)} \to D^{(2)}, \ldots$$

vor und bilde

$$D = \Sigma e_\nu\, D^{(\nu)}.$$

Eine Spezialisierung einer Spezialisierung ist wieder eine Spezialisierung.

§ 5. Vielfältigkeiten von Ketten.

Wir definieren nun: Alle Spezialisierungen einer Kette C in bezug auf einen Grundkörper K bilden *eine über K irreduzible Vielfältigkeit von Ketten.*

Die Spur einer Spezialisierung $C \to D$ auf V_α ist eine Spezialisierung $C_\alpha \to D_\alpha$. Umgekehrt läßt jede Spezialisierung $C_\alpha \to D_\alpha$ sich zu einer simultanen Spezialisierung $C_1 \to D_1, \ldots, C_h \to D_h$ erweitern, die ihrerseits eine Spezialisierung $C \to D$ definiert. Die Gesamtheit der Ketten D_α, die durch Spezialisierung aus C_α entstehen, ist eine irreduzible Vielfältigkeit von Ketten auf V_α. So entspricht jeder irreduziblen Vielfältigkeit von Ketten auf V eine ebensolche von Ketten auf jedem V_α.

Eine *Vielfältigkeit von Ketten* ist eine Vereinigung von endlich vielen irreduziblen Vielfältigkeiten.

Eine irreduzible Vielfältigkeit von Ketten kann bei Erweiterung des Grundkörpers zerlegbar werden. Bleibt die Vielfältigkeit aber bei jeder Erweiterung von K irreduzibel, so heißt sie *unteilbar*. Jede Vielfältigkeit von Ketten kann als Vereinigung von endlich vielen unteilbaren Vielfältigkeiten dargestellt werden. Läßt man die überflüssigen weg, so ist die Darstellung sogar eindeutig. Alle diese Sätze sind für projektive Ketten bekannt und übertragen sich ohne weiteres auf abstrakte Ketten.

Die *Dimension* einer irreduziblen Vielfältigkeit von Ketten ist der Transzendenzgrad der allgemeinen Kette C, aus der alle anderen durch Spezialisierung hervorgehen. Die allgemeine Kette C möge durch

$$C = \Sigma\, e_\nu\, C^{(\nu)}$$

gegeben sein, wobei jede der unteilbaren Ketten $C^{(\nu)}$ wieder als maximale Systeme zugeordneter $C_\alpha^{(\nu)}$ definiert sind. Die Koordinatenverhältnisse der Cayleyform von $C_\alpha^{(\nu)}$ erzeugen einen Körper $K_\alpha^{(\nu)}$. Der Vereinigungskörper aller $K_\alpha^{(\nu)}$ sei H. Der Transzendenzgrad dieses Körpers H über K ist die Dimension der Vielfältigkeit von Ketten. Sie ist genau die Anzahl der unabhängigen Unbestimmten, von denen das allgemeine Element C der Vielfältigkeit abhängt.

Im Falle einer einzigen projektiven Vielfältigkeit $V = V_1$ reduzieren alle diese Begriffe sich auf die bekannten projektiven Begriffe. Auch wenn die projektive Vielfältigkeit willkürlich als abstrakte Vielfältigkeit aufgefaßt wird, indem man h Exemplare $V_1, \ldots, V_h$ von ihr bildet und auf diesen Exemplaren beliebige Grenzen $F_1, \ldots, F_h$ annimmt, auch dann kommt der neue Begriff der Vielfältigkeit von Ketten genau auf den alten hinaus.

(Eingegangen am 6. Juni 1952.)

28.

Zur algebraischen Geometrie 17
Lokale Dimension und Satz von Eckmann

Mathematische Annalen 128 (1954) 128–134

Unter der *lokalen Dimension* eines Polynomideals H (oder der Nullstellenvielfältigkeit V des Ideals) in einem Punkte O versteht man die höchste Dimension einer irreduziblen Vielfältigkeit, die O enthält und in V enthalten ist. Ist die lokale Dimension 0 oder -1, so bedeutet das, daß O ein isolierter Punkt oder gar kein Punkt von V ist. Dafür soll in § 1 ein algebraisches Kriterium hergeleitet werden. Das Kriterium heißt

$$(H, P^r) = (H, P^{r+1}),$$

wobei P das zu O gehörige Primideal und r ein genügend hoher Exponent ist.

Mit Hilfe dieses Kriteriums wird in § 2 bewiesen, daß die lokale Dimension eines Ideals in O bei einer Spezialisierung der Koeffizienten der Basispolynome des Ideals niemals abnehmen kann.

Ein zweiter, ganz kurzer Beweis dieses Satzes verwendet keine Idealtheorie, dafür aber den Begriff Cayleyform einer Kette.

Auf Grund dieses Spezialisierungssatzes soll in § 3 ein Satz von ECKMANN[1]), spezialisiert für Polynome, aber verallgemeinert für beliebige algebraisch abgeschlossene Körper, bewiesen werden. Es seien $h_1, \ldots, h_n$ Polynome in $x_1, \ldots, x_n$, welche die Bedingungen

$$x_1 h_1 + x_2 h_2 + \cdots + x_n h_n = 0$$

identisch erfüllen. Ist der Grundkörper K der Körper der komplexen Zahlen und ist n ungerade, so besagt der Satz von ECKMANN, daß es auf jeder Sphäre

$$\sum \bar{x}_j x_j = R^2$$

eine gemeinsame Nullstelle der h_j gibt. Der Satz gilt sogar für stetige Funktionen h_j der x_k, die nur auf der Sphäre definiert sind. Sind die h_j aber Polynome, so folgt aus dem Satz von ECKMANN, daß der Koordinatenanfangspunkt O keine isolierte Nullstelle des Ideals $H = (h_1, \ldots, h_n)$ sein kann, d. h. daß die lokale Dimension von H in O mindestens Eins beträgt. Umgekehrt, wenn die lokale Dimension mindestens Eins beträgt, so gibt es einen irreduziblen Teil der Nullstellenvielfältigkeit V, der sich zusammenhängend von O bis ins Unendliche erstreckt und daher mit jeder Sphäre um O einen Punkt gemeinsam hat. Der Satz von ECKMANN für Polynome ist also gleichbedeutend mit der Aussage, daß die lokale Dimension mindestens Eins beträgt.

Diese Aussage hat aber nicht nur im komplexen Körper, sondern in jedem algebraisch abgeschlossenen Körper K einen Sinn. Für den Fall homogener

---

[1]) B. ECKMANN: Systeme von Richtungsfeldern in Sphären und stetige Lösungen komplexer linearer Gleichungen. Comment. Math. Helv. 15 (1942), S. 1, Satz IV.

Formen $h_1, \ldots, h_n$ hat W. HABICHT sie bewiesen[2]). Sie soll nun für beliebige Polynome als richtig nachgewiesen werden.

In § 1 wird eine bemerkenswerte Beweismethode von DEDEKIND verwendet. Sie dient bei DEDEKIND und ebenso später bei KRULL[3]) dazu, unter gewissen Voraussetzungen aus $P^r = P^{r+1}$ auf $P^r = 0$, oder allgemeiner aus $A = A P$ auf $A = 0$ zu schließen, wobei A ein beliebiges Ideal und P ein vom Einheitsideal E verschiedenes Primideal ist. Wird der DEDEKINDsche Beweis auf den Restklassenring E/Q' angewandt, wo Q' ein durch P teilbares Primärideal ist, so folgt, daß man aus $P^r \equiv 0 \ (P^{r+1}, Q')$ und $P \neq E$ auf $P^r \equiv 0 \ (Q')$ schließen kann.

Der Beweis in § 3 ist dem Beweise von HABICHT sehr ähnlich. Auch hier wird der Satz zuerst für den „allgemeinen Fall", sodann durch Spezialisierung der Koeffizienten für jeden speziellen Fall bewiesen. Für die Spezialisierung verwendet HABICHT Resultantensysteme, während hier der Spezialisierungssatz von § 2 zur Anwendung kommt.

§ 1. Ein Kriterium für positive lokale Dimension.

Es sei $E = L\,[x_1, \ldots, x_n]$ ein Polynombereich über einem beliebigen Körper L. Es seien $h_1, \ldots, h_m$ Polynome aus E, und es sei H das Ideal $(h_1, \ldots, h_m)$. Wird H als Durchschnitt von Primäridealen dargestellt:

$$(1) \qquad\qquad H = [Q_1, Q_2, \ldots, Q_t],$$

so kann man die *lokale Dimension* von H in O definieren als die höchste Dimension eines Ideals Q_k in (1), das O als Nullstelle hat. Dabei können wir O als Nullpunkt der Koordinaten annehmen; das zugehörige Primideal ist dann

$$P = (x_1, \ldots, x_n)\,.$$

Es gibt nun zwei Fälle:

Fall 1. Diejenigen Q_k, die O als Nullstelle haben, sind nulldimensional und haben daher O als einzige Nullstelle. In diesem Fall ist H *höchstens nulldimensional* in O. Auch der Fall, daß kein Q_k die Nullstelle O hat, fällt darunter.

Fall 2. Mindestens ein Q_k ist mehr als nulldimensional und hat O als Nullstelle. In diesem Fall ist H *mindestens eindimensional in O.*

Wir wollen nun ein algebraisches Kriterium aufstellen, das es gestattet, zwischen den Fällen 1 und 2 zu entscheiden. Das Kriterium ist im Grunde längst bekannt, aber, wie ich glaube, in der Literatur nicht leicht zu finden. Das Kriterium ist in den folgenden beiden Sätzen enthalten:

Satz 1. Ist H in O höchstens nulldimensional, so ist für genügend großes r

$$(2) \qquad\qquad (H, P^r) = (H, P^{r+1}).$$

Satz 2. Gilt umgekehrt (2) für ein r, so ist H in O höchstens nulldimensional.

---

[2]) W. HABICHT: Über die Lösbarkeit gewisser algebraischer Gleichungssysteme. Comment. Math. Helv. 18 (1946), S. 154.

[3]) W. KRULL: Idealtheorie. Ergebn. d. Math. IV 3, S. 36.

Beweis 1. In (1) mögen etwa $Q_1, \ldots, Q_s$ die Nullstelle O haben, die übrigen Q_k nicht. Dann können wir statt (1) schreiben

$$(3) \qquad H = [Q, R]$$

mit

$$Q = [Q_1, \ldots, Q_s], \quad R = [Q_{s+1}, \ldots, Q_t].$$

Eine genügend hohe Potenz P^r ist durch $Q_1, \ldots, Q_s$, also durch Q teilbar:

$$(4) \qquad P^r \equiv 0 \ (Q).$$

P^r und R haben keine gemeinsame Nullstelle; ihr größter gemeinsamer Teiler ist also das Einheitsideal:

$$(5) \qquad E = (P^r, R).$$

Multipliziert man (5) mit Q, so folgt

$$Q = (P^r Q, R Q)$$

und daher

$$Q \equiv 0 \ (P^r, H).$$

Umgekehrt ist (P^r, H) teilbar durch Q, also hat man [wie in Mod. Alg. II, 2. Aufl., § 86, Gleichung (2)]

$$(6) \qquad Q = (P^r, H)$$

und ebenso

$$(7) \qquad Q = (P^{r+1}, H).$$

Aus (6) und (7) folgt unmittelbar die Behauptung (2).

Beweis 2. Aus (2) folgt

$$(8) \qquad P^r \equiv 0 \ (H, P^{r+1}).$$

Nun sei Q' irgendeines von den Primäridealen Q_i aus (1), das O als Nullstelle hat. Wir wollen zeigen, daß Q' ein Teiler von P^r ist. Daraus folgt dann, daß Q' keine Nullstellen außer O hat.

Aus (8) folgt

$$(9) \qquad P^r \equiv 0 \ (Q', P^{r+1}).$$

Nun sei $P^r = (u_1, u_2, \ldots, u_h)$. Dann kann man statt (9) schreiben

$$u_i \equiv \Sigma \ p_{ik} u_k \ (\mathrm{mod}\, Q')$$

oder

$$(10) \qquad \Sigma \ (\delta_{ik} - p_{ik}) \, u_k \equiv 0 \quad (\mathrm{mod}\, Q'),$$

wobei die p_{ik} zu P gehören. Ist D die Determinante der $\delta_{ik} - p_{ik}$, so folgt

$$(11) \qquad D u_k \equiv 0 \quad (\mathrm{mod}\, Q').$$

Entwicklung der Determinante D ergibt

$$D = 1 + \cdots,$$

wobei die Glieder ... alle durch P teilbar sind. Also folgt, daß D nicht Null wird im Punkte O, und daraus weiter, daß D nicht teilbar ist durch das zu Q' gehörige Primideal P'. Somit folgt aus (11)

$$u_k \equiv 0 \ (Q'),$$

d. h.

$$P^r \equiv 0 \ (Q').$$

Damit ist alles bewiesen.

Das durch die Sätze 1 und 2 gegebene Kriterium soll nun in ein Rangkriterium umgeformt werden. Die Anzahl der modulo (H, P^r) linear unabhängigen Polynome ist immer endlich, da es nur endlich viele Monome $X_1, \ldots, X_\omega$ vom Grade $< r$ gibt und alle höheren Monome $\equiv 0 \pmod{P^r}$ sind. Diese Anzahl sei δ_r. Wenn (H, P^r) ein echter Teiler von (H, P^{r+1}) ist, so ist natürlich $\delta_r < \delta_{r+1}$. Das Kriterium (2) kann somit auch als

$$\delta_r = \delta_{r+1}$$

geschrieben werden, und die Sätze 1 und 2 können so ausgedrückt werden:

Satz 3. Ist H in O höchstens nulldimensional, so ist δ_r von einem gewissen Index r an konstant:

$$\delta_r = \delta_{r+1} = \delta_{r+2} = \cdots.$$

Satz 4. Ist dagegen H in O mindestens eindimensional, so wächst δ_r immerfort an:

$$\delta_1 < \delta_2 < \cdots.$$

Wird mit Polynomen *modulo P^r* gerechnet, so bedeutet das, daß alle Glieder vom Grade r und höher vernachlässigt werden. Die Monome $X_1, \ldots, X_\omega$ sind linear unabhängig $(\bmod P^r)$ und bilden eine Basis für alle Polynome $(\bmod P^r)$. Der lineare Rang der Algebra E/P^r ist also gleich der Anzahl ω, genauer $\omega\,(r)$.

Alle Polynome im Ideal $H \pmod{P^r}$ werden erhalten, indem man $h_1, \ldots, h_m$ mit allen Monomen $X_1, \ldots, X_\omega$ multipliziert, die entstehenden $m\omega$-Produkte durch Weglassung der Glieder vom Grade r und höher mod P^r reduziert und aus den so reduzierten Polynomen $Y_1, \ldots, Y_{m\omega}$ Linearkombinationen $\Sigma\, c_j Y_j$ bildet. Die Anzahl der linear-unabhängigen unter den Y_j sei $\varrho\,(r)$. Man kann $\varrho\,(r)$ als Rang einer Matrix A erhalten, indem man für $j = 1, 2, \ldots, m\omega$

$$(12) \qquad\qquad Y_j = \Sigma\, a_{jk} X_k$$

schreibt.

Die a_{jk} sind die Elemente der „dialytischen Matrix" A. Sylvester und Hurwitz haben in ihrer Resultantentheorie die dialytische Matrix schon benutzt. Macaulay war geradezu ein Meister auf diesem Instrument[4]).

Die Anzahl δ_r der modulo (H, P^r) linear unabhängigen Polynome ist nun offenbar die Differenz

$$(13) \qquad\qquad \delta_r = \omega\,(r) - \varrho\,(r).$$

Dabei ist $\varrho\,(r)$ der Rang der Matrix A und $\omega\,(r)$ die Anzahl ihrer Spalten.

§ 2. Das Verhalten der lokalen Dimension bei Spezialisierung.

Die Koeffizienten der Polynome $h_1, \ldots, h_m$ mögen einem Erweiterungskörper L des Grundkörpers K angehören. Wir nennen die Koeffizienten t und nehmen an, es sei eine Spezialisierung $t \to t'$ gegeben, bei der alle algebraischen Relationen $f\,(t) = 0$ mit Koeffizienten aus K erhalten bleiben. Man

---

[4]) F. S. Macaulay: Algebraic Theory of Modular Systems. Cambridge Tracts in Math. **19** (1916).

kann sich z. B. vorstellen, daß die Koeffizienten t Polynome in Unbestimmten u mit Koeffizienten aus K sind und daß die Spezialisierung $t \to t'$ dadurch erhalten wird, daß die u durch irgendwelche Elemente u' eines Erweiterungskörpers L' von K ersetzt werden.

Satz 5. Bei einer Spezialisierung $t \to t'$ wird die lokale Dimension von V in irgendeinem Punkte B niemals kleiner.

Der Satz gilt auch dann, wenn bei einer Spezialisierung $t \to t'$. der Punkt B mitspezialisiert wird, so daß alle algebraischen Gleichungen $f(t, B) = 0$ erhalten bleiben. Wir wollen aber der Bequemlichkeit halber für B den Koordinatenanfangspunkt O wählen. Der allgemeine Fall kann darauf zurückgeführt werden, indem der Anfangspunkt nach B verschoben wird.

Beweis. Die lokale Dimension von V in O sei vor der Spezialisierung d, nach der Spezialisierung d'. Wir nehmen $d' < d$ an und wollen einen Widerspruch herleiten.

Ist $d' = 0$, so folgt der Widerspruch sofort aus der Formel (13) des § 1. Der Rang einer Matrix kann nämlich bei einer Spezialisierung niemals größer werden, also kann δ_r niemals kleiner werden. Ist nun $d' = 0$ und $d > 0$, so strebt nach Satz 4 δ_r gegen ∞, während nach Satz 3 δ'_r beschränkt bleiben würde. Das ist ein Widerspruch.

Ist $d' > 0$, so kann man d und d' je um eine Einheit vermindern, indem man zu den Polynomen $h_1, \ldots, h_m$ noch eine allgemeine Linearform hinzufügt. Setzt man das Verfahren fort, so wird schließlich $d' = 0$ und man erhält wie oben einen Widerspruch.

Wir geben nun noch einen zweiten *Beweis des Satzes 5*, der die Idealtheorie vermeidet, dafür aber mit der Cayleyform arbeitet.

Wir machen zunächst die Polynome $h_1, \ldots, h_m$ homogen in $x_0, x_1, \ldots, x_n$. Ist nun die lokale Dimension gleich d, so gibt es eine d-dimensionale Kette z im projektiven Raum, die den Punkt O enthält und in den Hyperflächen $h_i = 0$ enthalten ist. Sind $z_0, \ldots, z_N$ die Koordinaten dieser Kette (d. h. die Koeffizienten ihrer zugeordneten Form), so kann man die Bedingungen, daß z eine Kette darstellt und daß diese den Punkt O enthält und in allen $h_i = 0$ enthalten ist, durch homogene Gleichungen in den z und t ausdrücken[5]):

$$(14) \qquad\qquad F_j(z, t) = 0.$$

Die Spezialisierung $t \to t'$ kann zu einer gleichzeitigen Spezialisierung $(z, t) \to (z', t')$ fortgesetzt werden, bei der (14) erhalten bleibt:

$$(15) \qquad\qquad F_j(z', t') = 0.$$

Es gibt also eine Kette z' von der Dimension d, die den Punkt O enthält und in den spezialisierten Hyperflächen $h'_j = 0$ enthalten ist. Das heißt aber: die lokale Dimension von V' in O ist mindestens d.

---

[5]) Siehe CHOW und VAN DER WAERDEN: ZAG 9 (Math. Ann. **113**) oder ZAG 18 in diesem Annalenheft.

§ 3. Der Satz von ECKMANN für Polynome.

Als Vorbereitung dient

Satz 6. Polynome $h_1, \ldots, h_n$ in $x_1, \ldots, x_n$, welche die Bedingung

$$(16) \qquad x_1 h_1 + x_2 h_2 + \cdots + x_n h_n = 0$$

erfüllen, lassen sich immer so darstellen:

$$(17) \qquad h_i = \Sigma f_{ik} x_k \ \text{mit} \ f_{ik} = -f_{ki} \ \text{und} \ f_{ii} = 0.$$

Beweis. Der von W. HABICHT für Formen gegebene Beweis (l. c.[2], S. 167) gilt ohne weiteres für Polynome.

Werden die Polynome f_{ik} als allgemeine Polynome irgendeines Grades, also mit lauter unbestimmten Koeffizienten angesetzt, so sprechen wir vom *allgemeinen Fall*, sonst von einem *speziellen Fall*.

Der Satz von ECKMANN bezieht sich auf ungerade Dimensionen n. Wir betrachten aber zunächst den Fall, daß n gerade ist.

Satz 7. Für gerade n im allgemeinen Fall ist die lokale Dimension des Ideals $(h_1, \ldots, h_n)$ im Koordinatenanfangspunkt O Null.

Beweis: Wäre die lokale Dimension im allgemeinen Fall positiv, so müßte sie nach Satz 5 bei jeder Spezialisierung der Koeffizienten der Polynome f_{ik} positiv bleiben. Wir spezialisieren nun die Polynome $f_{ik} = -f_{ki}$ zu Konstanten $c_{ik} = -c_{ki}$, deren Determinante nicht Null ist. Da die Dimension gerade ist, ist das immer möglich. Die Gleichungen

$$(18) \qquad \Sigma c_{ik} x_k = 0$$

haben dann nur die Nullösung, also ist die lokale Dimension Null. Also muß sie auch vor der Spezialisierung Null gewesen sein.

Nach diesen Vorbereitungen kann der Satz von ECKMANN für Polynome bewiesen werden:

Satz 8. Für ungerade n ist die lokale Dimension des Ideals $(h_1, \ldots, h_n)$ im Koordinatenanfangspunkt stets positiv.

Wir brauchen nur den allgemeinen Fall zu behandeln; durch Spezialisierung folgt der Satz dann für jeden speziellen Fall nach Satz 5.

Die f_{ik} seien also allgemeine Polynome irgendeines Grades und die h_i seien durch (17) definiert.

Die Gleichungen

$$(19) \qquad h_2 = 0, \ h_3 = 0, \ \ldots, h_n = 0$$

haben sicher die Lösung $(0, \ldots, 0)$, die wir O genannt haben. Die Gleichung $h_2 = 0$ hat eine $(n-1)$-dimensionale Lösungsvielfältigkeit. Nimmt man noch $h_3 = 0$ hinzu, so sind alle Bestandteile des Schnittes mindestens $(n-2)$-dimensional. So weiterschließend, findet man schließlich, daß alle Bestandteile der durch (19) definierten Vielfältigkeit. mindestens eindimensional sind. Der Punkt O ist also in einem mindestens eindimensionalen irreduziblen Teil J der Vielfältigkeit (19) enthalten.

Dieser Teil J liegt nicht in der Hyperebene $x_1 = 0$. Setzt man nämlich in (17) $x_1 = 0$ und beschränkt man sich auf $h_2, \ldots, h_n$, so hat man genau den „allgemeinen Fall" in $(n-1)$-Dimensionen vor sich, und nach Satz 7 ist in diesem Fall die lokale Dimension des Ideals $(h_2, \ldots, h_n)$ im Punkte O Null.

Auf J sind $h_2, h_3, \ldots, h_n$ alle Null. Nach (16) ist dann auch das Produkt $x_1 h_1$ Null. Aber x_1 ist nicht Null auf J, wie wir eben gesehen haben. Also gilt $h_1 = 0$ auf J. Die lokale Dimension des Ideals $(h_1, h_2, \ldots, h_n)$ im Punkte O ist also positiv, q.e.d.

(Eingegangen am 11. März 1954.)

29.

Zur algebraischen Geometrie 18
Ketten in mehrfach-projektiven Räumen

Mathematische Annalen 128 (1954) 135–137

Ein *zweifach projektiver Raum* $S_{m,n}$ ist die Gesamtheit aller Punktepaare (x, y), wobei x aus einem projektiven S_m und y aus einem projektiven S_n entnommen wird. Eine *Vielfältigkeit* in $S_{m,n}$ wird durch homogene Gleichungen definiert, die in den x und y einzeln homogen sind. Diese Begriffe wurden in ZAG 1 (Math. Ann. **108**, S. 121) schon eingeführt. Die Vielfältigkeiten in $S_{m,n}$ heißen auch *Korrespondenzen*. Irreduzibilität und Dimension werden für sie genau so definiert wie für Vielfältigkeiten im S_n. Siehe auch ZAG 6 (Math. Ann. **110**, S. 138).

Aus irreduziblen Vielfältigkeiten von derselben Dimension kann man *Ketten*, d. h. formale Linearkombinationen mit ganzzahligen Koeffizienten bilden. Solche Ketten in $S_{m,n}$ wurden in ZAG 16 (Math. Ann. **125**, S. 321) als Hilfsmittel benutzt. Es wurde gesagt, daß auch diese Ketten ihre Koordinaten haben, für welche ähnliche Sätze gelten wie für die Koordinaten von Ketten in S_n. Insbesondere wurden in ZAG 16, § 4 unter 1., 2., 3. die folgenden drei Sätze benutzt:

Satz 1. Ist K eine Kette in $S_{m,n}$ und C eine Kette in S_m, so kann die Relation zwischen K und C, die darin besteht, daß von den Punktepaaren (x, y) von K immer der erste Punkt zu C gehört, durch homogene Gleichungen zwischen den Koordinaten von K und C ausgedrückt werden.

Satz 2. Die Bedingung, daß ein Punktepaar (x, y) auf einer Kette K liegt, kann durch homogene Gleichungen zwischen den x und y und den Koordinaten von K ausgedrückt werden.

Satz 3. Sind K und L Ketten in $S_{m,n}$, so kann die Relation, die besagt, daß alle Punkte von K zu L gehören, durch homogene Gleichungen zwischen den Koordinaten von K und L ausgedrückt werden.

In allen diesen Fällen sind unter „homogene Gleichungen" universell gültige homogene Gleichungen mit ganzzahligen Koeffizienten, die nur von den Gradzahlen der Ketten abhängen, zu verstehen. Die in diesen Gleichungen vorkommenden Ketten und Punkte dürfen also beliebig spezialisiert werden, ohne daß die Gleichungen ihre Gültigkeit verlieren.

Eine formale Definition der Kettenkoordinaten wurde in ZAG 16 nicht gegeben und die Beweise der Sätze 1—3 wurden nur angedeutet. Einem von CHEVALLEY (Math. Reviews **14**, S. 1012) geäußerten Wunsch entsprechend, sollen diese Definition und diese Beweise nun gegeben werden.

Es ist bekannt, daß durch die Abbildung

$$(1) \qquad z_{ik} = x_i y_k$$

der Raum $S_{m,n}$ eineindeutig auf eine Bildvielfältigkeit U' in S_{mn+m+n} abgebildet werden kann. Die Gleichungen von U' lauten

$$(2) \qquad z_{ik}\, z_{jl} = z_{il}\, z_{jk}.$$

Einer Vielfältigkeit V in $S_{m,n}$ entspricht dabei wieder eine Vielfältigkeit V' auf U'. Ist nämlich

$$(3) \qquad f_k(x,\, y) = 0$$

ein System von Gleichungen für V, so kann man zunächst dafür sorgen, daß die Gleichungen (3) den gleichen Grad in den x wie in den y haben. Hat z. B. eine der Gleichungen (3) den Grad 3 in den x und 5 in den y, so multipliziert man die Gleichung der Reihe nach mit x_0^2, mit x_1^2, ..., mit x_m^2 und erhält so $m + 1$ neue Gleichungen, die in den x und in den y den gleichen Grad 5 haben und zusammen der Ausgangsgleichung äquivalent sind. Ersetzt man nun in diesen Gleichungen die Produkte $x_i y_k$ durch z_{ik} und fügt noch die Gleichungen (2) hinzu, so erhält man die Gleichungen von V'.

Umgekehrt entspricht jeder V' auf U' eine V in $S_{m,n}$. Ein System von Gleichungen für V erhält man nämlich durch Einsetzen von (1) in die Gleichungen von V'.

Einer Kette K in $S_{m,n}$ entspricht demnach eine Kette K' auf U' und umgekehrt. Wir definieren nun: *Die Koordinaten von K sind die von K'.*

Die Sätze 2 und 3 folgen nun unmittelbar aus den entsprechenden Sätzen für den projektiven Raum:

Satz 2'. Die Bedingung, daß ein Punkt z auf einer Kette K' liegt, kann durch homogene Gleichungen zwischen den Koordinaten des Punktes und der Kette ausgedrückt werden.

Satz 3'. Die Bedingung, daß eine Kette K' auf einer Kette L' liegt, kann durch homogene Gleichungen zwischen den Koordinaten von K' und L' ausgedrückt werden.

Satz 2' ist genau Satz 3 der gemeinsamen Arbeit ZAG 9 von Chow und mir (Math. Ann. **113**, S. 699). Die Gleichungen mögen heißen

$$(4) \qquad g_j(k',\, z) = 0.$$

Um Satz 2 zu erhalten, braucht man nur (1) in (4) einzusetzen.

Der Beweis von Satz 3' ist im zweiten Absatz von § 2 der erwähnten Arbeit ZAG 9 enthalten. Da dieselbe Beweismethode auch Satz 1 ergeben wird, will ich den Beweis hier noch einmal geben.

Es seien $p^{(1)}, \ldots, p^{(g)}$ die Schnittpunkte von K' mit d allgemeinen Hyperebenen $u^{(1)}, \ldots, u^{(d)}$, wobei g der Grad und d die Dimension der Kette K' ist. Die p und u haben die drei Bedingungen von ZAG 9, Satz 1 zu erfüllen, die durch die dortigen Gleichungen (2), (3), (5) ausgedrückt werden. Nun füge man noch die Bedingungen hinzu, die nach Satz 2' ausdrücken, daß die Punkte $p^{(1)}, \ldots, p^{(g)}$ alle auf L' liegen. Diese Bedingungen garantieren, daß alle Punkte von K' auf L' liegen. Sodann bilde man aus dem gesamten homogenen Gleichungssystem nach Mod. Alg. II, § 80 das Resultantensystem zunächst nach $p^{(1)}$, dann von den Resultanten wieder die Resultanten nach $p^{(2)}$, usw. bis alle $p^{(i)}$ eliminiert sind. Schließlich ordnet man die erhaltenen

Resultanten nach Potenzprodukten der Unbestimmten $u^{(1)}, \ldots, u^{(d)}$ und setzt alle Koeffizienten Null. So erhält man die gesuchten Gleichungen zwischen den Koordinaten der Ketten K' und L'.

Aus Satz 3' ergibt sich unmittelbar Satz 3.

Der Beweis des Satzes 1 wird genau so geführt, wie der von Satz 3, nur muß man vorher die Gleichungen von C nach Satz 2' aufstellen. Sie mögen lauten

$$(5) \qquad h_i(c,\, x) = 0,$$

wobei die c die Koordinaten von C sind. Diese Gleichungen kann man zunächst wieder gleichgradig in den x und y machen und dann die Produkte $x_i y_k$ durch z_{ik} ersetzen. So erhält man Gleichungen

$$(6) \qquad H_j(c,\, z_{ik}) = 0,$$

die, mit (2) kombiniert, die Gleichungen des Bildes C' der Gesamtheit aller der Punktepaare $(x,\, y)$ ergeben, deren Punkte x auf C liegen. In diese Gleichungen setzt man die Punkte $p^{(1)}, \ldots, p^{(g)}$ von K' ein und verfährt weiter wie im Beweis des Satzes 3'.

(Eingegangen am 11. März 1954.)

30.

Zur algebraischen Geometrie 19
Grundpolynom und zugeordnete Form

Mathematische Annalen 136 (1958) 139–155

Heinrich Behnke zum 60. Geburtstag

Es sei k ein beliebiger Konstantenkörper und V eine über k irreduzible Varietät im projektiven Raum P_n. Die *zugeordnete Form* von V wurde von Chow und mir[1]) folgendermaßen definiert. Man schneide V mit r allgemeinen Hyperebenen $U_1, \ldots, U_r$. Die Koordinaten der Hyperebenen U_i sind Unbestimmte $U_{ij}(1 \leqq i \leqq r, 0 \leqq j \leqq n)$. Wenn V nicht in der Hyperebene $y_0 = 0$ liegt, so liegen die Schnittpunkte $y^{(\nu)}$ auch nicht in dieser Hyperebene, und man kann ihre Koordinaten durch $y_0 = 1$ eindeutig normieren. Nun bildet man mit neuen Unbestimmten $U_{0j}(j = 0, \ldots, n)$ die Linearformen

$$(1) \qquad (U_0 y^{(\nu)}) = \sum_0^n U_{0j} y_j^{(\nu)}.$$

Ihr Produkt (wenn nötig in eine p^e-te Potenz erhoben, falls k die Charakteristik p hat) ist eine Form F in den U_{0j}, die rational von den übrigen $U_{1j}, \ldots, U_{rj}$ abhängt. Macht man sie durch Multiplikation mit einem nur von den Variablenreihen $U_1, \ldots, U_r$ abhängigen Faktor ganz rational und primitiv in diesen Variablenreihen, so erhält man die zugeordnete Form

$$(2) \qquad F(U) = F(U_0, U_1, \ldots, U_r).$$

Diese Definition scheint für die geometrischen Anwendungen die zweckmäßigste zu sein. Für die algebraische Untersuchung eignet sich aber eine andere Definition besser, auf die man so geführt wird:

Es sei x ein allgemeiner Punkt von V. Mit der Normierung $x_0 = 1$ seien $x_1, \ldots, x_n$ die inhomogenen Koordinaten von x. Nun bildet man, mit unbestimmten Koeffizienten u_{ij}, die $r + 1$ Linearkombinationen

$$z_i = - \sum_1^n u_{ij} x_j \qquad (i = 0, \ldots, r).$$

Man zeigt dann leicht, daß $z_1, \ldots, z_r$ in bezug auf den Koeffizientenkörper $k(u)$ algebraisch unabhängig sind, dagegen $z_0, \ldots, z_r$ abhängig. Es gibt also ein einziges irreduzibles Polynom $F(Z_0, \ldots, Z_r)$ mit Koeffizienten aus $k(u)$ mit der Eigenschaft

$$F(z_0, \ldots, z_r) = 0.$$

Das Polynom F kann als ganz rational in den u_{ij} angenommen werden.

---

1) W.-L. Chow und B. L. van der Waerden, ZAG 9. Math. Annalen 113 (1937), S. 692.

Ferner kann man annehmen, daß es keinen nur von den u_{ij} abhängigen Faktor enthält. Das nunmehr bis auf einen konstanten Faktor eindeutig bestimmte Polynom $F(Z_i; u_{ij})$ heißt nach W. Krull[2]) das *Grundpolynom* der Varietät V oder des zugehörigen Primideals p. Das Grundpolynom stimmt mit der von K. Hentzelt[3]) eingeführten und von E. Noether[4]) eingehend studierten „Elementarteilerform" des Primideals p überein.

Aus dem Grundpolynom erhält man die zugeordnete Form einfach, indem man die Z_i in U_{i0} und die u_{ij} in $U_{ij}(j > 0)$ umbenennt. Das soll in § 1 gezeigt werden.

In § 2 wird bewiesen, daß $F(U)$ bei einer Permutation der Variablenreihen $U_0, \ldots, U_r$ bis auf das Vorzeichen in sich übergeht und daß $F(U)$ sich im Falle der Charakteristik p als p^e-te Potenz eines Polynoms $F_0(U_0, \ldots, U_r)$ schreiben läßt, wobei F_0, als Polynom in U_0 betrachtet, ein Produkt von lauter verschiedenen Linearfaktoren (1) ist. Der Exponent $q = p^e$ heißt der *Inseparabilitätsgrad* der Form F.

Im Fall eines vollkommenen Grundkörpers k ist $q = 1$ und $F = F_0$. Allgemein gilt $q = 1$ dann und nur dann, wenn $F(U)$ als Polynom in $Z_0 = U_{00}$ separabel ist (siehe dazu Krull[2]), Satz 6 u. 7). In § 3 wird gezeigt, daß das genau dann der Fall ist, wenn $k(x)$ im Sinne von A. Weil[5]) separabel erzeugbar über k ist. Dieses Ergebnis hat S. V. K. Hegde[6]) bereits erhalten. Der hier zu gebende Beweis ist etwas einfacher als der von Hegde, beruht aber auf demselben Grundgedanken.

In § 4 wird untersucht, unter welcher Bedingung $F(U)$ absolut prim ist. Das ist genau dann der Fall, wenn $k(x)$ im Sinne von A. Weil[5]) regulär über k ist, oder, wie wir auch sagen werden, wenn k *gut für* V ist. Dieser Satz ist äquivalent dem Satze 9 von Krull[2]).

In § 5 wird zunächst gezeigt, daß alle x_j *ganze* algebraische Funktionen von $z_1, \ldots, z_r$ über $k(u_{11}, \ldots, u_{rn})$ als Konstantenkörper sind. Darauf gestützt, wird eine Methode entwickelt, die es gestattet, alle Punkte von V zu finden, indem man für $z_1, \ldots, z_r$ speziell Werte $t_1, \ldots, t_r$ einsetzt und t_0 als Lösung der Gleichung

$$F(t_0, t_1, \ldots, t_r) = 0$$

bestimmt[7]). Ferner wird gezeigt, daß ein Punkt y dann und nur dann auf V liegt, wenn die Linearformen

$$t_i = -\sum_1^n u_{ij} y_j$$

[2]) W. Krull: Parameterspezialisierung in Polynomringen. II. Archiv der Math. 1 (1948), S. 129.

[3]) K. Hentzelt: Zur Theorie der Polynomideale und Resultanten. Math. Annalen 88 (1923) S. 53.

[4]) E. Noether: Eliminationstheorie und allgemeine Idealtheorie. Math. Annalen 90 (1923) S. 229.

[5]) A. Weil: Foundations of Algebraic Geometry (1946), zu zitieren als "Foundations".

[6]) S. V. Keshava Hegde: The associated form of a variety over a field of prime characteristic. Commentarii math. Helvetici 30, (1956) S. 132.

[7]) Die Methode wurde bereits in § 1 meiner Arbeit ZAG 3, Math. Annalen 108 (1933) entwickelt.

identisch in den u_{ij} die Gleichung $F(t_0, \ldots, t_r) = 0$ erfüllen. Man kann also, wenn das Grundpolynom $F(Z_i; u_{ij})$ bekannt ist, ohne weiteres die Gleichungen von V im affinen Raum A_n aufstellen. Für den projektiven Raum ist eine kleine Modifikation erforderlich, die in der Arbeit[1]) angegeben ist.

In § 6 wird untersucht, wie eine Primvarietät V bei Erweiterung des Grundkörpers k zerfallen kann. Es wird gezeigt, daß die irreduziblen Teile alle dieselbe Dimension haben und eineindeutig den Primfaktoren der Form $F(U)$ entsprechen. Dieses Ergebnis steht zwar schon in § 1 meiner Arbeit ZAG 10 (Math. Annalen 113, S. 706) und bei HEDGE[6]), aber gewisse Einzelheiten des Beweises sind weder in ZAG 10 noch bei HEGDE ganz ausgeführt.

Anschließend ergibt sich, ebenfalls in § 6, eine einfache Charakterisierung der Körper k, die gut für V sind, sowie eine Konstruktion des kleinsten guten Körpers für eine gegebene unteilbare Varietät V.

Zur Bequemlichkeit des Lesers wird aus früheren Arbeiten nur das folgende als bekannt vorausgesetzt:

1) Der Begriff „allgemeiner Punkt einer über k irreduziblen Varietät" und dessen einfachste Eigenschaften,

2) die Sätze aus Chapter I von WEILs "Foundations", die in meiner „Neubegründung"[8]) neu hergeleitet wurden,

3) der Begriff „Spezialisierung" und die Tatsache, daß eine Spezialisierung $z \to t$ über einem Körper L sich immer zu einer Spezialisierung des Paares (z, x) fortsetzen läßt. Im allgemeinen muß man dabei auch unendliche Spezialisierungen zulassen oder den projektiven Raum heranziehen.

§ 1. Algebraische und geometrische Definition der Form $F(U)$

A. Die algebraische Definition

Das System $(x) = (x_1, \ldots, x_n)$ habe den Transzendenzgrad r über dem Grundkörper k. Im Fall $r = n$ gibt es keine zugeordnete Form; wir nehmen also $r < n$ an. Wir adjungieren dem Grundkörper n^2 von den x_j unabhängige Unbestimmte u_{ij} und bilden

$$(3) \qquad z_i = -\sum_1^n u_{ij} x_j \qquad (i = 0, 1, \ldots, n-1) .$$

Es sei

$(\dot{z})$ bzw. $(\dot{u})$ das System der z_i bzw. u_{ij} mit $1 \leq i \leq r$,

$(\ddot{z})$ bzw. $(\ddot{u})$ das System der z_i bzw. u_{ij} mit $0 \leq i \leq r$,

(z) bzw. (u) das System aller z_i bzw. v_{ij}, wobei der Index j immer von 1 bis n geht. Dann gilt

Satz 1. *$z_1, \ldots, z_r$ sind algebraisch unabhängig über $k(u)$, also auch über $k(\dot{u})$ und $k(\ddot{u})$. Dagegen sind $z_0, \ldots, z_r$ algebraisch abhängig über $k(\ddot{u})$.*

Beweis. Wären $z_1, \ldots, z_r$ abhängig über $k(u)$, so wären, da man die z_i beliebig permutieren kann, je r aus (z) ausgewählte z_i abhängig über $k(u)$.

---

[8]) B. L. VAN DER WAERDEN: Über A. WEILs Neubegründung der algebr. Geom. Abhandlungen math. Sem. Univ. Hamburg (1958).

Das System (z) hätte also über $k(u)$ einen Transzendenzgrad kleiner als r. Aus (3) kann man die x_j auflösen, also hätte das System (x) auch einen Transzendenzgrad kleiner als r über $k(u)$, was unseren Voraussetzungen über (x) und (u) widerspricht. Also sind $z_1, \ldots, z_r$ unabhängig über $k(u)$.

Wegen (3) hat das System $\ddot{z}$ höchstens den Transzendenzgrad r über $k(\ddot{u})$. Also sind $z_0, \ldots, z_r$ abhängig über $k(\ddot{u})$. Damit ist Satz 1 bewiesen.

Aus Satz 1 folgt, daß das System $(\ddot{z})$ genau den Transzendenzgrad r über $k(\ddot{u})$ hat. Es gibt also ein einziges Primpolynom $F(Z_0, \ldots, Z_r)$ mit Koeffizienten aus $k(\ddot{u})$ mit der Eigenschaft

$$(4) \qquad F(z_0, \ldots, z_r) = 0 \,.$$

Durch Multiplikation mit einem nur von den u_{ij} abhängigen Faktor kann man $F(Z)$ ganzrational in allen u_{ij} machen. Falls das so erhaltene Polynom in den Z_i und u_{ij} einen nur von den u_{ij} abhängigen Faktor enthält, kann man diesen weglassen. So erhält man ein Primpolynom

$$(5) \qquad F(Z; \ddot{u}) = F(Z_0, \ldots, Z_r; u_{01}, \ldots, u_{rn}) \,,$$

das bis auf einen konstanten Faktor eindeutig bestimmt ist. Wir nennen es nach Krull das *Grundpolynom* der Varietät V über dem Körper k.

B. Faktorzerlegung des Grundpolynoms

Ersetzt man in $F(Z_0, \ldots, Z_r)$ die Unbestimmten $Z_1, \ldots, Z_r$ durch die Körperelemente $z_1, \ldots, z_r$, die nach Satz 1 über $k(\ddot{u})$ unabhängig sind, so erhält man ein Polynom in einer Unbestimmten $Z_0 = U_{00}$ mit Koeffizienten aus $k(\ddot{u}, \dot{z})$:

$$(6) \qquad F(Z_0) = F(Z_0, \dot{z}; \ddot{u}) \,.$$

Das Polynom ist irreduzibel in $k(\ddot{u}, \dot{z})[Z_0]$ und hat die Nullstelle

$$(7) \qquad z_0 = -\sum_1^n u_{0j} x_j \,.$$

In einem geeigneten algebraischen Erweiterungskörper von $k(\ddot{u}, \dot{z})$ zerfällt das Polynom (6) ganz in Linearfaktoren $(Z_0 - z_0^{(\nu)})$, wobei die $z_0^{(\nu)}$ konjugiert zu z_0 in bezug auf $k(\ddot{u}, \dot{z})$ sind. Übt man auf (7) die Isomorphismen aus, die z_0 in seine Konjugierten $z_0^{(\nu)}$ überführen, so erhält man, da die u_{0j} bei diesen Isomorphismen fest bleiben,

$$z_0^{(\nu)} = -\sum_1^n u_{0j} x_j^{(\nu)} \,.$$

Also zerfällt $F(Z_0)$ in einem algebraischen Erweiterungskörper von $k(\ddot{u}, \dot{z})$ ganz in Linearfaktoren

$$(8) \qquad Z_0 - z_0^{(\nu)} = Z_0 + \sum_1^n u_{0j} x_j^{(\nu)} \,.$$

Ist z_0 separabel über $k(\ddot{u}, \dot{z})$, so ist $F(Z_0)$ einfach das Produkt der Linearfaktoren (8), multipliziert mit einem von Z_0 unabhängigen Faktor. Ist z_0

inseparabel über $k(\ddot{u}, \dot{z})$ und ist p die Charakteristik, so kommt jeder Linearfaktor $q = p^e$ mal vor und man hat

$$(9) \qquad F(Z_0) = f(\dot{z}, \ddot{u}) \prod_{\nu} (Z_0 + \sum_{1}^{n} u_{0j} x_j^{(\nu)})^q.$$

Setzt man im separablen Fall $q = 1$, so gilt die Formel (9) allgemein. Der Exponent q heißt der *Separabilitätsgrad* des Grundpolynoms. Er ist gleich Eins im Fall eines separablen und gleich p^e im Fall eines inseparablen Grundpolynoms.

Nach Satz 1 hat der Körper $k(\dot{u})(\dot{z})$ den Transzendenzgrad r über $k(\dot{u})$, also denselben Transzendenzgrad wie der umfassendere Körper $k(\dot{u})(x)$. Also ist $k(\dot{u})(x)$ algebraisch über $k(\dot{u})(\dot{z})$. Die x_j und ebenso die konjugierten Größen $x_j^{(\nu)}$ sind also algebraisch über $k(\dot{u}, \dot{z})$.

Wir schreiben (9) als

$$F(Z_0, \dot{z}; \ddot{u}) = f(\dot{z}, \ddot{u}) \, P(Z_0, \dot{z}; \ddot{u})$$

oder, nach Potenzen von Z_0 geordnet, als

$$f_0 Z_0^g + \cdots + f_{g-1} Z_0 + f_g = f(\dot{z}, \ddot{u}) \cdot (Z_0^g + \cdots + p_{g-1} Z_0 + p_g).$$

Der Vergleich der Koeffizienten der Potenzen von Z_0 links und rechts zeigt erstens, daß $f(\dot{z}, \ddot{u}) = f_0$ ein Polynom in $(\dot{z})$ und $(\ddot{u})$ ist, und zweitens, daß die Koeffizienten p_i des Polynoms P gleich Quotienten f_i/f_0, also rationale Funktionen von $(\dot{z})$ und $(\ddot{u})$ sind.

Andererseits ist P ein Produkt von Linearfaktoren in den Z_0 und den u_{0j} mit Koeffizienten, die algebraische Funktionen von $(\dot{z})$ und $(\dot{u})$ sind. Das Produkt ist somit ein Polynom in Z_0 und den u_{0j} mit Koeffizienten, die rationale Funktionen von $(\dot{z})$ und $(\dot{u})$ sind. Man kann also schreiben

$$P(Z_0, \dot{z}; \ddot{u}) = \frac{Q(Z_0, \dot{z}, \ddot{u})}{R(\dot{z}, \dot{u})},$$

wobei der Nenner R nur noch $(\dot{z})$ und $(\dot{u})$, nicht mehr die u_{0j} enthält. Der Faktor $f(\dot{z}, \ddot{u})$, mit dem man P zu multiplizieren hat, um eine ganz rationale Funktion in allen Variablen zu erhalten, enthält also nur $(\dot{z})$ und $(\dot{u})$ und kann $f(\dot{z}, \dot{u})$ genannt werden.

Somit vereinfacht sich (9) zu

$$(10) \qquad F(Z_0, \dot{z}; \ddot{u}) = f(\dot{z}, \dot{u}) \prod_{\nu} (Z_0 + \Sigma \, u_{0j} x_j^{(\nu)})^q.$$

C. Übergang zur geometrischen Definition

Wir betrachten eine Korrespondenz K zwischen einem affinen Raum $A_{r(n+1)}$, in dem ein Punkt v durch Koordinaten

$$v_{ij} (i = 1, \ldots, r, \; j = 0, \ldots, n)$$

gegeben ist, und einem affinen Raum A_n, in dem ein Punkt y durch Koordinaten

$$y_1, \ldots, y_n$$

gegeben wird.

Die Gleichungen der Korrespondenz K seien:

erstens die Gleichungen, die ausdrücken, daß y auf der Primvarietät V liegt,

zweitens diejenigen, die ausdrücken, daß y in den Hyperebenen $v_1, \ldots, v_r$ liegt:

$$(11) \qquad v_{i0} + \Sigma\, v_{ij} y_j = 0 .$$

Nun wird behauptet:

Die Korrespondenz K ist irreduzibel über k.

Beweis: Ein allgemeines Paar $(w; x)$ der Korrespondenz wird erhalten, indem man in (11) statt (y) einen allgemeinen Punkt (x) von V und statt der $v_{ij}(j > 0)$ Unbestimmte w_{ij} einsetzt, die w_{i0} aber so bestimmt, daß (11) erfüllt bleibt:

$$(12) \qquad w_{i0} = - \Sigma\, w_{ij} x_j .$$

Daß jede Gleichung $g(w; x) = 0$ erfüllt bleibt, wenn man das allgemeine Paar $(w; x)$ durch irgend ein Paar $(v; y)$ der Korrespondenz ersetzt, ist trivial; denn man kann in $g(w; x) = 0$ zunächst für die w_{i0} ihre Ausdrücke (12) einsetzen:

$$g(- \Sigma\, w_{ij} x_j \, , \; w_{ij}; x) = 0 ,$$

sodann die x durch y ersetzen:

$$g(- \Sigma\, w_{ij} y_j \, , \; w_{ij}; y) = 0$$

und schließlich die Unbestimmten w_{ij} durch v_{ij} ersetzen, was wegen (11) ohne weiteres

$$g(v_{i0} \, , \; v_{ij}; y) = 0$$

ergibt.

Wenn man die Unbestimmten w_{ij} $(j > 0)$ in u_{ij} umbenennt, so gehen die w_{i0} in die früher durch (3) eingeführten z_i über. Also definieren auch die z_i, u_{ij} und x_i ein allgemeines Paar $(\dot z, \dot u; x)$ der Korrespondenz.

Die Dimension der Korrespondenz K ist

$$r(n + 1) = r + r\,n ,$$

denn (x) hat die Dimension r und die Zahl der Unbestimmten w_{ij} oder u_{ij} ist $r\,n$.

Nach Satz 1 sind die z_i und u_{ij} $(1 \leq i \leq r)$ algebraisch unabhängig. Ihre Anzahl ist wieder $r(n + 1)$. Also kann man ein allgemeines Paar (U', y') der Korrespondenz auch so erhalten: Für die $r(n + 1)$ Koordinaten von U' nimmt man lauter unabhängige Unbestimmte

$$U_{ij}(i = 1, \ldots, r;\ j = 0, \ldots, n) ,$$

und für y' nimmt man irgend eine dazu gehörige Lösung der Gleichungen der Korrespondenz, d. h. irgend einen Schnittpunkt von V mit den Hyperebenen $U_1, \ldots, U_r$. Alle so erhaltenen Paare (U', y') haben den Transzendenzgrad $r(n + 1)$, sind also allgemeine Paare der Korrespondenz. Da (y') algebraisch über $k(U')$ ist, so gibt es in einem normalen algebraischen Erweiterungskörper von $k(U')$ nur endlich viele Lösungen y', die wir auch mit $y^{(\nu)}$ bezeichnen können. Weil alle allgemeinen Paare einer irreduziblen Korrespondenz äquivalent sind, so folgt:

Alle Schnittpunkte $y^{(\nu)}$ von V mit den allgemeinen Hyperebenen $U_1, \ldots, U_r$ sind konjugiert in bezug auf den Körper

$$k(U') = k(U_1, \ldots, U_r) \, .$$

Ist y' irgend einer von diesen Schnittpunkten, so gibt es einen Isomorphismus, der das früher konstruierte allgemeine Paar $(\dot{z}, \dot{u}; x)$ in (U', y') überführt. Wir erweitern nun das System $\dot{u}$ zu $\ddot{u}$, indem wir noch die Unbestimmten u_{0j} $(j = 1, \ldots, n)$ hinzunehmen und verabreden, daß sie beim Isomorphismus in neue Unbestimmte U_{0j} übergehen. So erhalten wir

$$k(\dot{z}, \ddot{u}; x) \cong k(U', U_{0j}; y') \, .$$

Schließlich nehmen wir links noch die Unbestimmte Z_0, rechts U_{00} hinzu und erhalten isomorphe Polynombereiche

$$k(\dot{z}, \ddot{u}; x)\,[Z_0] \cong k(U', U_{0j}; y')\,[U_{00}] \, .$$

Unter B hatten wir das Primpolynom

$$F(Z_0) = F(Z_0, \dot{z}; \ddot{u})$$

mit der Nullstelle

(13)
$$z_0 = -\, \Sigma\, u_{0j}\, x_j$$

betrachtet. Ihm entspricht im Isomorphismus ein Primpolynom

$$F(U_{00}) = F(U_{00}, U_{0j}; U') = F(U)$$

mit der Nullstelle

(14)
$$z_0' = -\, \Sigma\, U_{0j}\, y_j' \, .$$

In einem geeigneten algebraischen Erweiterungskörper von $k(U_{0j}, U')$ zerfällt das Polynom $F(U)$ natürlich wieder in Linearfaktoren

$$U_{00} - z_0^{(\nu)}\, ,$$

wobei die $z_0^{(\nu)}$ in bezug auf $k(U_{0j}, U')$ zu z_0' konjugiert sind. Daraus folgen wie unter B die Gleichungen

(15)
$$z_0^{(\nu)} = -\, \Sigma\, U_{0j}\, y_j^{(\nu)}$$

und die zu (10) analoge Faktorzerlegung

(16)
$$F(U) = f(U')\, \prod_\nu (U_0\, y^{(\nu)})^q$$

mit

$$(U_0\, y^{(\nu)}) = U_{00} + \sum_1^n U_{0j}\, y_j^{(\nu)} \, .$$

In (16) sind die $y^{(\nu)}$ die durch $y_0 = 1$ normierten Schnittpunkte der Hyperebenen $U_1, \ldots, U_r$ mit der Varietät V. Ferner bedeutet:

U das System aller $U_{ij}\,(0 \leq i \leq r\,,\ 0 \leq j \leq n)\,,$
U_0 die Reihe der $U_{0j}\,(0 \leq j \leq n)\,,$
U' das System der übrigen $U_{ij}\,(i \neq 0)\,.$

Damit ist der Übergang zur geometrischen Definition der zugeordneten Form vollzogen.

Im folgenden werden wir abwechselnd die geometrische und die algebraische Ausdrucksweise gebrauchen, wie es jeweils zweckmäßig erscheint. Die Variabelnreihe

$$(U_{i0}, U_{i1}, \ldots, U_{in}) \text{ oder } (Z_i, u_{i1}, \ldots, u_{in})$$

wird immer mit U_i bezeichnet. Die Ausdrücke „Grundpolynom", „zugeordnete Form" und Form $F(U)$ sollen als Synonyme gelten.

Die Bezeichnung „Form" rechtfertigt sich dadurch, daß $F(U)$ in der Variabelnreihe U_0 homogen vom Grade

$$g = g_0 q$$

ist, wie man aus (16) ohne weiteres entnimmt. Dabei bedeutet g_0 die Anzahl der verschiedenen Linearfaktoren rechts in (16), oder den (reduzierten) Grad der Varietät V.

§ 2. Die Vertauschungseigenschaft und der Separabilitätsgrad

Wir gehen von der algebraischen Definition der Form

$$F(U) = (Z_0, \ldots, Z_r; \ddot{u})$$

aus, die so lautete: F ist ein Primpolynom in $Z_0, \ldots, Z_r$ und den $u_{ij} = U_{ij}$ $(j > 0)$ mit der Eigenschaft (4).

Vertauscht man irgend zwei von den Variablenreihen $u_0, \ldots, u_r$, etwa u_i und u_l, so werden nach (3) auch z_i und z_l vertauscht. Alle algebraischen Eigenschaften der z_i und u_{ij} bleiben bei dieser Vertauschung erhalten, also bleibt auch (4) erhalten. Da aber F bis auf einen konstanten Faktor das einzige Primpolynom mit der Eigenschaft (4) ist, so geht F bei der Vertauschung der Indizes i und l in cF über. Vertauscht man noch einmal, so erhält man

$$c^2 F = F \, ,$$

also muß $c = \pm 1$ sein. Da jede Permutation ein Produkt von Transpositionen ist, so folgt:

Bei beliebigen Permutationen der Variablenreihen $U_0, \ldots, U_r$ geht $F(U)$ bis auf das Vorzeichen in sich über.

Aus dieser Vertauschungseigenschaft folgt, daß $F(U)$ in jeder einzelnen Variablenreihe U_i homogen ist und in jeder den Grad $g = g_0 q$ hat.

Der Inseparabilitätsgrad q ist 1 im separablen, p^e im inseparablen Fall. In beiden Fällen ist die q-te Potenz einer Summe gleich der Summe der q-ten Potenzen der Glieder. Also folgt aus (16), daß $F(U)$ die Variabeln U_{0j} der Reihe U_0 nur als q-te Potenzen enthält. Wegen der Vertauschungseigenschaft kann $F(U)$ auch die übrigen Variabeln U_{ij} nur als q-te Potenzen enthalten. Also hat man

$$(17) \qquad\qquad F(U) = F_0(U)^q,$$

wobei F_0 eine Form in den Variabelnreihen $U_0, \ldots, U_r$ mit Koeffizienten aus dem q-ten Wurzelkörper von k ist.

Nach (10) ist $f(\dot{z}, \dot{u})$ gleich dem Koeffizienten der höchsten Potenz von Z_0 in $F(Z_0, \dot{z}; \ddot{u})$. Also enthält auch $f(\dot{z}, \dot{u})$ die Variabeln u_{ij} und z_i nur als q-te Potenzen und man kann schreiben

$$(18) \qquad f(\dot{z}, \dot{u}) = f_0(\dot{z}, \dot{u})^q \, .$$

Setzt man (17) und (18) in (10) ein und zieht auf beiden Seiten die q-te Wurzel, so folgt

$$(19) \qquad F_0(Z_0, \dot{z}; \ddot{u}) = f_0(\dot{z}, \dot{u}) \prod_{\nu} \left(Z_0 + \sum_1^n u_{0j} x_j^{(\nu)} \right) \, .$$

§ 3. Die Separabilität von F (Z_0)

Der folgende Satz 2 umfaßt die Sätze 8 und 9 von HEGDE[6]).

Satz 2. *Ist $k(x)$ separabel erzeugbar über k, so ist z_0 separabel über $k(\ddot{u}, \dot{z})$ und umgekehrt. Ferner ist dann*

$$(20) \qquad k(\ddot{u}, x) = k(\ddot{u}, \ddot{z}) \, .$$

Beweis. Wir brauchen nur den Fall der Charakteristik p zu behandeln. Der p-te Wurzelkörper von k werde allgemein mit k' bezeichnet.

1. Es sei $k(x)$ separabel erzeugbar über k. Dann ist $k(u, x)$ auch separabel erzeugbar über $k(u)$. Nun ist aber, da die lineare Transformation (3) umkehrbar ist,

$$k(u, x) = k(u, z) \, ,$$

also ist $k(u, z)$ separabel erzeugbar über $k(u)$.

Nach WEILs "Foundations" oder nach § 3 meiner „Neubegründung" folgt daraus, daß der Wurzelkörper $k(u)'$ und $k(u, z)$ linear disjunkt sind. Nach Satz 5 meiner „Neubegründung" kann man einige von den z_i als separierende Variabeln wählen. Da die Indices i in u_{ij} und z_i beliebig permutiert werden können, kann man auch $z_1, \ldots, z_r$ als separierende Variabeln wählen. Also ist z_0 separabel über

$$k(u, z_1, \ldots, z_r) = k(u, \dot{z}) \, .$$

Nun ist z_0 aber algebraisch über $k(\ddot{u}, \dot{z})$, d. h. in der irreduziblen Gleichung für z_0 über $k(u, \dot{z})$ kommen nur die $\ddot{u}$ und $\dot{z}$ vor. Also ist z_0 separabel über $k(\ddot{u}, \dot{z})$.

2. Nun sei z_0 separabel über $k(\ddot{u}, \dot{z})$, also auch über $k(u, \dot{z})$. Was für z_0 gilt, gilt auch für $z_{r+1}, \ldots, z_n$; also sind alle z_i separabel über $k(u, \dot{z})$. Somit ist $k(u, z)$ separabel erzeugbar über k. Daraus folgt, daß k' und $k(u, z)$ linear disjunkt sind. Um so mehr sind k' und $k(x)$ linear disjunkt, also ist $k(x)$ separabel erzeugbar über k.

3. Ist das der Fall, so folgt wie unter 1, daß $z_0, \ldots, z_{n-1}$ und daher auch $x_1, \ldots, x_n$ separabel über $k(u, \dot{z})$ sind. Nach § 1 sind alle x_i algebraisch über $k(\ddot{u}, \dot{z})$, d. h. in den Minimalgleichungen für $x_1, \ldots, x_n$ über $k(u, \dot{z})$ kommen nur die $\ddot{u}$ und $\dot{z}$ wirklich vor. Also sind $x_1, \ldots, x_n$ separabel über $k(\ddot{u}, \dot{z})$.

Die Isomorphismen des Körpers $k(\ddot{u}, x)$, die die Elemente von $k(\ddot{u}, \dot{z})$ fest lassen, führen (x) in konjugierte Systeme $(x^{(\nu)})$ und z_0 in

$$z_0^{(\nu)} = - \sum_1^n u_{0j} x_j^{(\nu)}$$

über. Die $(x^{(\nu)})$ sind algebraisch über $k(\dot{u}, \dot{z})$ und die u_{0j} sind algebraisch unabhängig über diesem Körper. Daraus folgt: wenn ein $(x^{(\nu)})$ von (x) verschieden ist, so ist auch $z^{(\nu)}$ von z_0 verschieden. Wenn also ein Isomorphismus der eben betrachteten Art z_0 fest läßt, so muß er auch (x) fest lassen. Somit bleiben die x_i fest bei allen den Isomorphismen von $k(\ddot{u}, x)$, die die Elemente von $k(\ddot{u}, \ddot{z})$ fest lassen. Da die x_i außerdem separabel über $k(\ddot{u}, \ddot{z})$ sind, müssen sie diesem Körper angehören. Daraus folgt (20), und Satz 2 ist vollständig bewiesen.

Wir denken uns den Grundkörper k durch Adjunktion der Unbestimmten $\ddot{u}$ zu $K = k(\ddot{u})$ erweitert. Die Gleichung (20) besagt dann, daß $K(x)$ aus K erzeugt wird durch Adjunktion der $r + 1$ Größen $z_0, \ldots, z_r$. Dabei sind $z_1, \ldots, z_r$ unabhängig über K, und z_0 ist eine separable Funktion von $z_1, \ldots, z_r$, die mit $z_1, \ldots, z_r$ durch die Gleichung (4) verbunden ist.

In meiner „Neubegründung" wurde der separabel erzeugbare Körper $k(x)$ durch $r + 1$ Elemente $y_1, \ldots, y_{r+1}$ erzeugt, wobei y_{r+1} eine separable Funktion der unabhängigen $y_1, \ldots, y_r$ war. Wir sehen jetzt, daß über dem neuen Grundkörper $K = k(\ddot{u})$ die Linearkombinationen $z_0, \ldots, z_r$ die Rolle der $y_1, \ldots, y_{r+1}$ übernehmen können. Die Rolle des früher benutzten Primpolynoms $f(Y)$ mit der Eigenschaft $f(y) = 0$ spielt jetzt das Primpolynom $F(Z_0, \ldots, Z_r)$, d. h. die zugeordnete Form. Auf diesem Grundgedanken beruhen die folgenden Beweise.

Es sei k_v der kleinste vollkommene Körper über k. Ist k selbst vollkommen, so ist $k_v = k$; andernfalls entsteht k_v aus k durch Adjunktion aller p^m-ten Wurzeln $(m = 1, 2, \ldots)$. Die Koeffizienten der in § 2 definierten Form $F_0(U)$ liegen in k_v. Wir beweisen nun:

$F_0(U)$ *ist über* k_v *irreduzibel.*

Beweis. Im Falle $k_v = k$ ist die Behauptung klar. Andernfalls nehmen wir an, $F_0(U)$ wäre zerlegbar:

$$(21) \qquad\qquad F_0(U) = G(U) H(U) .$$

Da F_0, als Form in U_0 betrachtet, nach (19) ein Produkt von lauter verschiedenen Linearfaktoren ist, sind die Faktoren $G(U)$ und $H(U)$ sicher teilerfremd. Wir erheben nun beide Seiten von (21) in eine solche q'-te Potenz $(q' = p^m)$, daß alle Koeffizienten links und rechts in k liegen. Man erhält links eine Potenz von $F(U)$, aber $F(U)$ ist prim in $k(U)$, also müssen rechts beide Faktoren Potenzen von F sein. Das widerspricht aber der Teilerfremdheit von G und H. Damit ist die Behauptung bewiesen.

Mit derselben Methode beweist man, daß die Varietät V über k_v irreduzibel bleibt. Würde nämlich V über k_v zerfallen, so gäbe es in $k_v[X]$ ein Produkt

$$G(X) H(X) ,$$

das Null wäre auf V, während $G(X)$ und $H(X)$ es nicht sind. Bestimmt man

$q' = p^m$ so, daß die q'-ten Potenzen von $G(X)$ und $H(X)$ beide in $k[X]$ liegen, so erhält man wie vorhin einen Widerspruch.

Die zugeordnete Form von V über k_v ist offensichtlich $F_0(U)$.

§ 4. Die absolute Irreduzibilität der Form F (U)

Satz 3. *Die Form $F(U)$ ist dann und nur dann absolut prim, wenn $k(x)$ regulär über k ist, d. h. wenn k und $k(x)$ linear disjunkt sind.*

Beweis. 1. $F(U)$ sei absolut prim. Aus (17) folgt dann, daß $q = 1$ sein muß. Nach Satz 2 ist also $k(x)$ separabel erzeugbar über k und es gilt

$$k(\ddot{u}, x) = k(\ddot{u}, \ddot{z}) \, .$$

Das System $(\ddot{u}, \ddot{z})$ besteht aus $(r + 1)\, n + r + 1$ Elementen und hat nur den Transzendenzgrad $(r + 1)\, n + r$, also hat das Ideal $\mathfrak{P}_{(u, z)/k}$ eine Basis aus nur einem Primpolynom (für die Bezeichnungen siehe WEILs "Foundations" oder meine „Neubegründung"). Dieses Primpolynom ist eben die zugeordnete Form $F(U)$. Diese ist nach Voraussetzung absolut prim, bleibt also prim nach der Erweiterung von k zu $\bar{k}$. Daraus folgt, daß die Basis von $\mathfrak{P}_{(\ddot{u}, \ddot{z})/\bar{k}}$ aus eben demselben Primpolynom $F(U)$ besteht. Nach Theorem 3 aus WEILs "Foundations" (Satz 3 meiner „Neubegründung") sind also $\bar{k}$ und $k(\ddot{u}, \ddot{z})$ linear disjunkt. Um so mehr sind $\bar{k}$ und $k(x)$ linear disjunkt, d. h. $k(x)$ ist regulär über k.

Der hier angewandte Schluß ist genau derselbe, der in meiner „Neubegründung" zum Beweis des Satzes 6 führte. Die Form $F(U)$ tritt eben an die Stelle des damals benutzten Polynoms $f(Y)$.

2. Nun sei $k(x)$ regulär über k. Der Körper $K = k(\ddot{u})$ ist frei in bezug auf (x), also ist $K(x)$ regulär über K ("Foundations", Theorem 5, oder „Neubegründung", Satz 6). Nach Satz 2 ist aber

$$K(x) = k(\ddot{u}, x) = k(\ddot{u}, \ddot{z}) = K(\ddot{z}) \, ,$$

also ist $K(\ddot{z})$ regulär über K.

Wir wollen zeigen, daß $F(U) = F(Z_0, \ldots, Z_r, \ddot{u})$ als Polynom in $Z_0, \ldots, Z_r$ absolut prim ist. Wäre das nicht der Fall, so würde $F(U)$ in $\bar{K}[Z_0, \ldots, Z_r]$ zerfallen:

$$F(U) = G(Z) \cdot H(Z) \, .$$

Ersetzt man $Z_0, \ldots, Z_r$ durch $z_0, \ldots, z_r$, so wird F Null, also muß G oder H dann Null werden, etwa

$$(22) \qquad\qquad G(z_0, \ldots, z_r) = 0 \, .$$

Die in $G(z_0, \ldots, z_r)$ vorkommenden Potenzprodukte der z_i wären dann linear abhängig über $\bar{K}$, aber nicht über K. Das geht nicht, weil $\bar{K}$ und $K(\ddot{z})$ linear disjunkt sind. Also ist $F(U)$ als Polynom in $Z_0, \ldots, Z_r$ absolut prim.

Wäre nun $F(U)$ als Funktion sämtlicher U_{ij} nicht absolut prim, so müßte es in der Zerlegung

$$(23) \qquad\qquad F(U) = G(U)\, H(U)$$

einen Faktor, etwa $H(U)$, geben, der die $Z_i = U_{i0}$ nicht enthält, sondern nur die U_{ij} mit $j > 0$.

Wir betrachten nun beide Seiten von (23) als Funktionen von $Z_0 = U_{00}$ allein und wenden links die Zerlegung (16) an. Der erste Faktor $f(U')$ dieser Zerlegung enthält $U_{01}, \ldots, U_{0n}$ nicht. Die übrigen Faktoren sind Linearfaktoren $U_{00} + \cdots$, die keinen von U_{00} unabhängigen Faktor enthalten. Also muß $H(U)$ als Faktor in $f(U')$ enthalten sein, also enthält $H(U)$ die U_{0j} nicht.

Vertauscht man nun die Variabelnreihen $U_0, \ldots, U_r$, so folgt, daß $H(U)$ keinen der U_{ij} enthalten kann, d. h. $H(U)$ ist eine Konstante und $F(U)$ absolut irreduzibel.

Ist $k(x)$ regulär über k, so heißt der Körper k *gut für die Varietät* V. A. Weil nennt k in diesem Fall a field of definition of V. Aus Satz 3 folgt unmittelbar:

Satz 4. *Der Körper k ist dann und nur dann gut für V, wenn die zugeordnete Form $F(U)$ absolut prim ist.*

§ 5. Ganzheitseigenschaften. Die Gleichungen von V

A. Die x_j als ganze Funktion von $z_1, \ldots, z_r$

Wir beweisen zunächst:

Satz 5. *Wenn das Grundpolynom $F(Z)$ als Polynom in $Z_0, \ldots, Z_r$ den Grad h hat, so enthält es ein Glied $c\, Z_0^h$ mit $c \neq 0$.*

Beweis. Wir schreiben die ersten $r + 1$ Gleichungen (3) in Matrixform

$$\begin{pmatrix} z_0 \\ \vdots \\ z_r \end{pmatrix} = - \begin{pmatrix} u_{01} \ldots u_{0n} \\ \vdots \quad\ \vdots \\ u_{r1} \ldots u_{rn} \end{pmatrix} \begin{pmatrix} x_1 \\ \vdots \\ x_n \end{pmatrix}$$

oder kurz

$$(24) \qquad\qquad z = -u\,x\,.$$

Mit $(r+1)^2$ Unbestimmten t_{ij} bilden wir eine quadratische Matrix T und setzen

$$(25) \qquad\qquad z' = T^{-1}z \text{ oder } z = T z'\,.$$

Durch die Transformation $Z = T Z'$ möge das Grundpolynom $F(Z)$ in $F'(Z')$ übergehen:

$$(26) \qquad\qquad F(TZ') = F'(Z')\,.$$

In F' kommt das Glied $Z_0'^h$ mit einem von Null verschiedenen Koeffizienten vor, denn die Elemente der ersten Spalte der Matrix T sind Unbestimmte und F ist nicht identisch Null.

Ersetzt man in (26) die Z' durch z', so wird die linke Seite und daher auch die rechte Seite Null. Also ist $F'(Z')$ das Primpolynom in $Z_0', \ldots, Z_r'$ mit der Eigenschaft $F'(z') = 0$.

Setzt man (24) in (25) ein und setzt $T^{-1}u = u'$, so erhält man

$$(27) \qquad\qquad z' = -u'x\,.$$

Die Gleichung $F(z) = 0$ gilt, wenn man $z = -ux$ in ihr einsetzt, identisch in den u. Sie bleibt also gelten, wenn u überall durch u' und z durch z' ersetzt wird. Also kann man das Polynom $F'(Z')$ mit der Eigenschaft $F'(z') = 0$ einfach dadurch erhalten, daß man in $F(Z)$ die u durch u' und die Z durch Z' ersetzt.

Würde nun F kein Glied mit Z_0^h enthalten, so würde das durch die Substitution $u \to u'$ erhaltene Polynom F' kein Glied mit $Z_0'^h$ enthalten. Wir haben aber gesehen, daß ein solches Glied vorhanden ist. Also enthält F ein Glied mit Z_0^h.

Aus Satz 5 folgt, daß das Polynom $f(\dot z, \dot u)$ in (10) die Variabeln $z_1, \ldots, z_r$ nicht enthalten kann. Sonst hätte nämlich die rechte Seite von (10) in allen Variabeln $Z_0, z_1, \ldots, z_r$ zusammen einen Grad größer als $g = g_0 q$, enthielte aber Z_0 nur in der g-ten Potenz, was dem Satz 5 widerspricht.

Nebenbei folgt jetzt $g = h$ und $c = f(\dot u)$.

Aus Satz 5 folgt weiter, daß z_0 eine *ganze* algebraische Funktion von $z_1, \ldots, z_r$ über $k(\ddot u)$ als Konstantenkörper ist. Diese Funktion erfüllt die Gleichung

$$(28) \qquad\qquad F(z_0) = 0$$

oder ausführlicher

$$(29) \qquad\qquad F(z_0, z_1, \ldots, z_r; \ddot u) = 0 \, .$$

Setzt man $z_0 = -\Sigma u_{0j} x_j$ in (29) ein und hebt unter den $\ddot u$ die u_{0j} besonders hervor, so erhält man

$$(30) \qquad\qquad F(-\Sigma u_{0j} x_j, z_1, \ldots, z_r; u_{0j}, \dot u) = 0 \, .$$

Die u_{0j} sind Unbestimmte, die von den $\dot u$, x_j und z_i unabhängig sind. Die Gleichung (30) gilt identisch in den u_{0j}. Man kann also $u_{01} = 1$ und die übrigen $u_{0j} = 0$ setzen. So erhält man eine Gleichung für x_1:

$$(31) \qquad\qquad f(\dot u) \, x_1^h + f_1(\dot z; \dot u) \, x_1^{h-1} + \cdots + f_h(\dot z; \dot u) = 0$$

und analoge Gleichungen für $x_2, \ldots, x_n$. Daher:

Die Koordinaten $x_1, \ldots, x_n$ des allgemeinen Punktes x von V sind ganze algebraische Funktionen von $z_1, \ldots, z_r$ über $k(\dot u)$ als Konstantenkörper.

B. Die Gleichungen von V

In ZAG 3 (Math. Ann. **108**) wurde eine Methode entwickelt, durch Faktorzerlegung des Grundpolynoms $F(Z_0, \ldots, Z_r; \ddot u)$ für spezielle Werte von $Z_1, \ldots, Z_r$ alle Punkte von V zu erhalten. Auf Grund dieser Methode kann man auch sehr leicht die Bedingungen herleiten, die ein Punkt y zu erfüllen hat, damit er auf V liegt. Diese Methode und diese Bedingungen wollen wir jetzt, leicht modifiziert, neu herleiten.

Es sei K ein beliebiger Erweiterungskörper von k. Wir nehmen an, daß die u_{ij} $(0 \leq i \leq r, 1 \leq j \leq n)$ Unbestimmte über K sind. Wir bezeichnen mit L die algebraisch abgeschlossene Hülle von $K(\dot u)$, wobei $(\dot u)$ wie immer das System der u_{ij} mit $1 \leq i \leq r$ bezeichnet. Wir setzen uns zum Ziel, alle Punkte y von V mit Koordinaten aus L zu bestimmen.

Setzt man in (29) für alle z_i ihre Ausdrücke (3) ein, so erhält man

$$(32) \qquad F(-\Sigma u_{ij} x_j;\ \ddot{u}) = 0 .$$

Diese Gleichungen gelten (identisch in den u_{ij}) für den allgemeinen Punkt x von V und daher für jeden Punkt y von V:

$$(33) \qquad F(-\Sigma u_{ij} y_j;\ \ddot{u}) = 0 .$$

Setzt man also

$$(34) \qquad t_i = -\Sigma u_{ij} y_j \qquad\qquad (i = 0, \ldots, r) ,$$

so erhält man

$$(35) \qquad F(t_0, t_1, \ldots, t_r;\ \ddot{u}) = 0 .$$

Wenn die y_j in L liegen, liegen $t_1, \ldots, t_r$ ebenfalls in L.

Jetzt kehren wir den Zusammenhang um und nehmen $t_1, \ldots, t_r$ als gegebene Elemente von L an. Wir bestimmen t_0 als irgend eine Lösung der Gleichung (35) im algebraisch abgeschlossenen Universalkörper Ω. Solche Lösungen gibt es immer, denn (35) enthält nach Satz 5 ein Glied $c\,t_0^h$ mit $c \neq 0$.

Das Polynom $F(Z_0, Z_1, \ldots, Z_r;\ \ddot{u})$ ist über $k(\ddot{u})$ irreduzibel. Die Gleichung (35) definiert also eine über $k(\ddot{u})$ irreduzible Varietät im affinen $(r+1)$-dimensionalen Raum der Punkte $t = (t_0, \ldots, t_r)$. Ein allgemeiner Punkt dieser Varietät ist der durch (3) definierte Punkt $z = (z_0, \ldots, z_r)$, denn $z_1, \ldots, z_r$ sind nach Satz 1 algebraisch unabhängig über $k(\ddot{u})$. Also ist t eine Spezialisierung von z über $k(\ddot{u})$.

Eine solche Spezialisierung läßt sich immer zu einer Spezialisierung

$$(36) \qquad (z, x) \to (t, y) \text{ über } k(\ddot{u})$$

fortsetzen. Unendliche Spezialisierungen sind dabei von vornherein ausgeschlossen, da $x_1, \ldots, x_n$ nach (31) ganze algebraische Funktionen von $z_1, \ldots, z_r$ über $k(\ddot{u})$ sind.

Bei der Spezialisierung (36) bleibt die Gleichung (31) erhalten. Also ist jedes y_j algebraisch über L und daher in L enthalten.

Ferner ist y eine Spezialisierung von x über k und daher ein Punkt von V.

Auch die Gleichungen (3) bleiben bei der Spezialisierung erhalten; daher gelten wieder die Gleichungen (34). Das heißt: Wenn wir, von den t_i ausgehend, den Punkt y bestimmen und aus y nach (34) wieder die t_i berechnen, so kommen wir genau auf die t_i zurück, von denen wir ausgegangen sind. Insbesondere folgt aus der ersten Gleichung (34)

$$t_0 = -\Sigma u_{0j} y_j ,$$

daß die Lösungen t_0 der Gleichung (35) samt und sonders dem Körper $L(u_{0j})$ angehören und Linearformen in den Unbestimmten $u_{01}, \ldots, u_{0n}$ sind. Die Koeffizienten dieser Linearformen sind die y_j; diese sind also durch t_0 jeweils *eindeutig bestimmt*. Zu jeder Lösung t_0 der Gleichung (35) gehört ein einziger Punkt y.

Will man Punkte y mit Koordinaten aus der algebraisch abgeschlossenen Hülle $\bar{K}$ des vorgegebenen Körpers K erhalten, so muß man, wie das in ZAG 3 getan wurde, für $(\dot{u})$ ein System von Elementen aus $\bar{K}$ so wählen, daß $f(\dot{u}) \neq 0$ wird. Dann wird $L = \bar{K}$ und die eben erklärte Methode ergibt alle Punkte y von V mit Koordinaten aus $\bar{K}$. Die u_{0j} bleiben nach wie vor Unbestimmte.

In dieser Arbeit bleiben wir aber dabei, daß sämtliche u_{ij} Unbestimmte sind.

Wir haben gesehen: Wenn y auf V liegt und $t_0, \ldots, t_r$ durch (34) definiert werden, so erfüllen diese t_i die Gleichung (35).

Es gilt aber auch umgekehrt: *Wenn die durch (34) definierten t_i die Gleichung (35) erfüllen, so liegt y auf V.* Man kann dann nämlich nach der eben geschilderten Methode durch Fortsetzung der Spezialisierung $z \to t$ zu $(z, x) \to$ $\to (t, y')$ einen Punkt y' von V erhalten, für den

$$t_0 = - \sum u_{0j} y_j'$$

gilt. Vergleicht man das mit

$$t_0 = - \sum u_{0j} y_j \, ,$$

so folgt $y_j = y_j'$, weil die u_{0j} über L unabhängig sind. Also ist y ein Punkt von V.

Daraus folgt:

Satz 6. *Ein Punkt y mit Koordinaten aus L liegt dann und nur dann auf V, wenn die Linearformen (34) die Gleichung (35) erfüllen:*

$$(37) \qquad F(- \sum u_{ij} y_j; \ddot{u}) = 0 \, .$$

Dieser Satz gilt insbesondere für Punkte y mit Koordinaten aus $\bar{K}$. Über $\bar{K}$ sind die u_{ij} unabhängige Unbestimmte; also folgt aus (36), daß die Koeffizienten aller Potenzprodukte der u_{ij} gleich Null sein müssen. So erhält man ein explizites Gleichungssystem für V:

$$F_h(y_1, \ldots, y_n) = 0 \, .$$

Die Koeffizienten dieser Gleichungen sind ganzzahlige Linearkombinationen der Koeffizienten der zugeordneten Form F.

Die geometrische Bedeutung der Formel (37) ist ganz einfach. Legt man durch einen Punkt y möglichst allgemein $r + 1$ Hyperebenen $w_0, \ldots, w_r$, so haben die Koeffizienten w_{ij} dieser Hyperebenen w_i die Bedingungen

$$w_{i0} + \sum_1^n w_{ij} y_j = 0$$

zu erfüllen. Daraus ergibt sich

$$(38) \qquad w_{i0} = - \sum w_{ij} y_j \, .$$

Damit nun y auf V liegt, müssen die Hyperebenen $w_0, \ldots, w_r$ die Bedingung

$$(39) \qquad F(w_0, \ldots, w_r) = 0$$

erfüllen, wo F die zugeordnete Form ist. Dabei kann man die w_{ij} mit $i \neq 0$ als Unbestimmte u_{ij} wählen und die w_{i0} durch (38) definieren. So ergibt sich (37).

Im projektiven Fall ist das Prinzip gleich, aber die Rechnung etwas komplizierter. Die allgemeinste Hyperebene w, die durch einen Punkt $(y_0, \ldots, y_n)$ im projektiven Raum geht, ist durch

$$w_j = \Sigma\, s_{jk} y_k \qquad (s_{jk} = -\, s_{kj},\, s_{jj} = 0)$$

gegeben. Nimmt man wieder $r+1$ solche Hyperebenen $u_0, \ldots, u_r$ und setzt die Bedingungsgleichung

$$(40) \qquad\qquad F(w_0, \ldots, w_r) = 0$$

identisch in den s_{jk} an, so erhält man nach[1]) die Gleichungen der Varietät V im projektiven Raum.

§ 6. Zerfall einer Varietät bei Erweiterung des Grundkörpers

Es gibt zwei Methoden, das Verhalten einer Primvarietät bei Erweiterung des Grundkörpers k zu untersuchen. Man kann, wie das in den Arbeiten [2]) und [8]) geschehen ist, das Verhalten des zu V gehörigen Primideals p bei einer Erweiterung von k untersuchen. Einfacher ist es jedoch, die Idealtheorie ganz beiseite zu lassen und die Zerlegung von V aus der Faktorzerlegung der Form $F(U)$ zu erschließen.

Der Grundkörper k kann für die Zwecke dieser Untersuchung immer als vollkommen angenommen werden. Ist er es nämlich nicht, so kann man von k zum vollkommenen Wurzelkörper k_v übergehen. Auf die Reduzibilität von V hat diese Körpererweiterung keinen Einfluß (siehe § 3, Schluß).

Ist k vollkommen, so ist der Separabilitätsgrad $q = 1$ und die Gleichung (16) vereinfacht sich zu

$$(41) \qquad F(U) = f(U')\prod_{\nu} (U_0 y^{(\nu)}) = f(U')\, G(U)\,.$$

Dabei sind $y^{(\nu)}$ die Schnittpunkte von V mit den allgemeinen Hyperebenen $U_1, \ldots, U_r$; $f(U')$ ist eine Funktion von $U_1, \ldots, U_r$ allein und $G(U)$ ist das Produkt der Linearfaktoren $(U_0 y^{(\nu)})$.

Nun sei K ein beliebiger Erweiterungskörper von k. Wir nehmen an, daß die U_{ij} auch über K noch Unbestimmte sind. Die $y^{(\nu)}$ brauchen über $K(U_1, \ldots, U_r)$ nicht mehr alle konjugiert zu sein, sondern sie können in Systeme konjugierter Punkte zerfallen. Entsprechend zerfällt das Produkt $G(U)$ in Teilprodukte

$$(42) \qquad\qquad G(U) = G_1 G_2 \ldots G_s\,,$$

wobei jeder Faktor G_i ein Produkt von über $K(U)$ konjugierten Linearfaktoren $(U_0 y^{(\nu)})$ ist.

Jeder der über $K(U)$ konjugierten, zu G_i gehörigen Punkten $y^{(\nu)}$ hat über K den Transzendenzgrad r und ist daher allgemeiner Punkt einer über K irreduziblen, r-dimensionalen Varietät V_i.

Macht man die G_i durch Multiplikation mit geeigneten Faktoren $f_i(U')$ ganzrational auch in $U_1, \ldots, U_r$, so erhält man die zugeordneten Formen $F_i(U)$ der Varietäten V_i. Aus (42) folgt, wenn beide Seiten ganz rational in allen Variabelnreihen gemacht und alle nur von $U_1, \ldots, U_r$ abhängigen Faktoren links und rechts weggekürzt werden,

$$(43) \qquad F(U) = F_1 F_2 \ldots F_s .$$

Nun wird behauptet:

V ist die Vereinigung von $V_1, \ldots, V_s$.

Beweis. Nach § 5 B gehört ein Punkt y dann und nur dann zu V, wenn die $r + 1$ Hyperebenen $w_0, \ldots, w_r$, die durch

$$w_{i0} = - \Sigma w_{ij} y_j$$

mit unbestimmten w_{ij} $(i \neq 0)$ definiert sind, identisch in diesen Unbestimmten die Gleichung

$$F(w_0, \ldots, w_r) = 0$$

erfüllen. Ist das für F der Fall und gilt (43), so ist es auch für ein F_i der Fall und umgekehrt. Also liegt ein Punkt y dann und nur dann auf V, wenn er auf einer der V_i liegt. Damit ist die Behauptung bewiesen.

Nach dem eben Bewiesenen entsprechen die Primfaktoren F_i von $F(U)$ eineindeutig den irreduziblen Teilen V_i von V. Ist also F absolut prim, so ist auch V absolut irreduzibel oder „unteilbar".

Bei vollkommenen Grundkörpern k gilt auch das Umgekehrte: Ist V unteilbar, so ist F absolut prim. Ist k unvollkommen, so können wir nur schließen, daß F eine q-te Potenz einer absoluten Primform F_0 ist.

Damit haben wir

Satz 7. *Die Varietät V ist dann und nur dann unteilbar, wenn ihre zugeordnete Form $F(U)$ eine q-te Potenz einer absoluten Primform $F_0(U)$ ist, wobei $q = 1$ oder $q = p^e$ ist.*

Aus Satz 4 und Satz 7 folgt:

Satz 8. *Der Körper k ist dann und nur dann gut für V, wenn V unteilbar und $q = 1$, d. h. $F(U) = F_0(U)$ ist.*

Dieser Satz gibt uns, wenn eine unteilbare Varietät gegeben ist, ein einfaches Mittel an die Hand, den kleinsten Körper k_0 zu konstruieren, der gut für V ist. Man normiert die zugeordnete Form $F(U)$ so, daß ein Koeffizient gleich Eins wird. Im Fall der Charakteristik p sieht man zu, ob F sich als Polynom in den U_{ij}^q schreiben läßt, wobei $q = p^e$ als möglichst hohe Potenz von p zu bestimmen ist. Sodann bildet man $F_0(U)$, indem man aus allen Koeffizienten von $F(U)$ die q-te Wurzel auszieht. Für Charakteristik Null setze man $F_0 = F$. Die Koeffizienten von F_0 erzeugen den gewünschten Körper k_0.

(Eingegangen am 30. April 1958)

31.

Invariants Birationnels

Atti del Convegno Internazionale di Geometria Algebrica Torino 1961 (1962) 35–47

Dans cette conférence, le mot *variété* signifiera: Variété indivisible dans un espace projectif S_n. Pour chaque variété, on aura donc: 1) un corps de définition K supposé algébriquement clos; 2) un point générique x qui génère la variété; 3) un corps universel Ω, c'est à dire un corps algébriquement clos d'un degré de transcendance infinie sur K. Les points de la variété sont toutes les spécialisations de x sur K dans $S_n(\Omega)$.

Partons d'une variété V_0 de dimension d. Un *modèle* de V sera: une variété V dans un espace projectif et une correspondance birationnelle T qui transforme V_0 en V. Entre deux modèles V_1 et V_2 on aura donc une correspondance birationnelle bien définie

$$T_{12} = T_1 T_2^{-1}$$

qui transforme V_2 en V_1.

Notre but est de trouver des invariants birationnels de V_0.

I. - LES CHAMPS DE TENSEURS.

Une première méthode de définir des invariants est due, dans des cas spéciaux, à MAX NOETHER et ENRIQUES, et dans toute généralité à KAEHLER. C'est la *méthode des formes différentielles*.

Choisissons, dans le corps de fonctions $K(x)$, des variables indépendantes u^1, ..., u^d, de sorte que $K(x)$ soit séparable sur $K(u)$. Un *champ de tenseurs* sur V_0 sera définie, par rapport aux coordonnées u^a, par un système de composantes

$$t_{\alpha\beta}\ldots$$

qui sont des fonctions rationnelles sur V_0, c'est à dire des éléments de $K(x)$. Les composantes du même tenseur par rapport à un autre système de variables indépendantes v^μ seront définies par des formules comme

$$(1) \qquad t'_{\mu\nu} = \Sigma\, t_{\alpha\beta}(\partial u^\alpha\, \partial v^\mu)(\partial u^\beta\, \partial v^\nu)\ .$$

Au lieu de « champ de tenseurs » nous dirons aussi « tenseur » tout court. Aux composantes $t_{\alpha\beta}\ldots$ on peut imposer des conditions de symétrie tout à fait arbitraires

$$\underset{\pi}{\Sigma}\, c_\pi \pi t_{\alpha\beta}\ldots = 0$$

où les π sont des permutations des indices et les c_π des nombres entiers.

Une question importante est: Comment définir la notion « tenseur partout fini »?

Soit V un modèle sans points multiples, s'il y a un tel modèle. Pour chaque point p de V il y a un système d'uniformisantes locales v^1, ..., v^d. Or, si les $t'_{\mu\nu}\ldots$ définies par (1) sont finies en p et si cela est vrai pour tous les points p de V, on dit que le tenseur t est *partout fini* ou simplement *fini*.

Les tenseurs antisymmétriques définissent les différentielles simples, doubles, etc.:

$$\Sigma\, t_\alpha du^\alpha \qquad .\qquad \Sigma\, t_{\alpha\beta} du^\alpha du^\beta\ ,\qquad \ldots$$

et si les tenseurs sont partout finis, les différentielles seront de première espèce.

Le nombre maximal de tenseurs finis linéairement indépendants à un nombre donné d'indices est toujours fini. Si on se restreint aux tenseurs antisymmétriques, on obtient les genres géométriques g_1, g_2, ..., g_d. D'autres conditions de symétrie mènent aux plurigenres, etc.

Or, la notion « tenseur fini » est *invariante*. Si t est fini sur un modèle sans points multiples V_1, t sera fini en tous les points simples de tout autre modèle V_2. La raison est la suivante.

Si t avait des pôles sur V_2, les variétés-pôles seraient des *diviseurs*, c'est à dire des sous-variétés de dimension $(d-1)$. Or, un point générique p_2 d'une telle sous-variété ne peut être fondamental pour T_{12}. Donc, la transformation T_{12} sera régulière en p_2. Il s'ensuit que les $\partial u^\alpha/\partial v^\mu$ sont finis en p_2, et puisque les $t_{\alpha\beta}$ sont partout finis, les $t'_{\mu\nu}$, définis par (1), seront finis en p_2. Il n'y a donc pas de pôles.

Ce raisonnement est dû à KAEHLER.

La question se pose: Comment définir la notion « tenseur partout fini », si on n'a pas un modèle sans singularités?

La question a été envisagée par MAX NOETHER dans un mémoire classique de 1870 (Vol. 2 des Math. Annalen). Soit $F = 0$ une surface dans l'espace projectif S_3, F étant une forme de degré m en x, y, z, w. Pour les points à distance finie, on peut poser $w = 1$. Une différentielle double, finie sur F, peut être écrite

$$\omega = \frac{G(x, y, z, 1)}{F_z(x, y, z, 1)} \, dx \, dy$$

où $G(x, y, z, w)$ est une *forme adjointe* à F du degré $(m-4)$ et F_z la dérivée de F par rapport à z.

La notion « forme adjointe » est définie, dans les cas les plus simples, par les conditions: 1) une courbe multiple de multiplicité s sur F doit avoir la multiplicité $s-1$ au moins sur G; 2) un point multiple isolé de multiplicité s sur F doit avoir la multiplicité $s-2$ au moins sur G.

Pour voir la raison d'être de la dernière condition, supposons que O n'a pas de points infiniment voisins multiples sur F. Dans ce cas on peut transformer O en une courbe simple C' sur un modèle V' par une dilatation, comme l'appelle B. SEGRE. Or, si G a multiplicité $s-2$ au moins dans O, la différentielle ω restera finie sur le diviseur C', c.à.d. C' ne sera pas un pôle de ω.

Essayons de généraliser cette méthode. Même si nous n'avons pas un modèle sans points multiples, nous pouvons toujours considérer *tous* les points simples sur *tous* les modèles V de V_0. J'ai proposé d'appeler ces points *Stellen*. En Français, on pourrait dire *places*. La notion de place est invariante par rapport aux transformations birationnelles. Pour chaque place, il existe un système d'uniformisants v^μ, auquel on peut référer les vecteurs t par les formules (1).

Maintenant, on peut définir: Un tenseur t est partout fini si les composantes $t'_{\mu\nu}$ sont finies à chaque place. S'il existe un modèle V sans points multiples, cette définition équivaut à l'ancienne. On démontre cela par la méthode de KAEHLER.

La question se pose: Peut-on construire un nombre fini de modèles V_i tel que chaque tenseur qui est fini dans tous les points simples des V_i est partout fini?

NOTE, AJOUTÉE APRÈS LA CONFÉRENCE.

La réponse à cette question est affirmative si le corps K a charactéristique zéro. Avant de démontrer cela, il faut que j'explique d'abord quelques notions. Soit w une valuation (exponentielle) du corps $K(x)$ tel que $w(a) = 0$ pour tous les a dans K. L'*anneau* d'une valuation w est l'ensemble des fonctions η tel que $w(\eta) \geqq 0$. L'*idéal* de la valuation est l'ensemble des η tel que $w(\eta) > 0$. Le *corps des résidus* de la valuation est l'anneau des résidus

$$A' = A/I \;,$$

où A est l'anneau et I l'idéal de la valuation. La *dimension* de la valuation est le degré de transcendance du corps A' par rapport au corps des constantes K.

Soit U un modèle projectif de V_0 et $x_0, \ldots, x_n$ les coordonnées d'un point générique de U. On peut supposer que les x_i sont des éléments de $K(x)$. On peut diviser les x_i par un x_j à valeur minimale $w(x_j)$. Après la division, les x_i sont toutes dans l'anneau A, mais pas toutes dans l'idéal I. Soient $y_0, \ldots, y_n$ les images des x_i dans l'homomorphisme naturel $A \to A'$. Les y_i définissent un point y dans l'espace projectif, qui est point générique d'une sous-variété indivisible W de U. Cette variété W est le *centre* de la valuation w sur U. La dimension de W ne peut pas surpasser la dimension d' de la valuation.

Dans un mémoire très important sur la résolution locale des singularités (Annals of Math. 41, p. 852), ZARISKI a démontré, pour le cas de charactéristique zéro, le théorème suivant:

(A) Il existe un nombre fini de modèles V_1, ..., V_m tel que chaque valuation *zérodimensionnelle* a un centre simple sur un de ces modèles.

Ceci donné, retournons à la question de l'existence d'un système de modèles V_1, ..., V_m tel que chaque tenseur fini dans tous les points simples des V_i soit partout fini. Après la conférence, j'ai demandé à M. ZARISKI s'il pouvait démontrer cette existence au moyen du théorème de la résolution locale, c.à.d. au moyen du théorème (A). Dans sa réponse, il énonça le théorème de la résolution locale de la manière suivante:

(B) Il existe un nombre fini de modèles V_1, ..., V_m tel que chaque valuation w a un centre simple sur un de ces modèles.

Or, on voit aisément que le théorème (B) est une conséquence de (A). En effet, une valuation w à dimension d' définit une application du corps $K(x)$ sur un corps de fonctions $K(y)$ de d' variables augmenté du symbole ∞. Pour les éléments de l'anneau de la valuation cette application est l'homomorphisme naturel $A \to A'$, et l'image de tous les η qui ne sont pas dans A est ∞. De même, on peut appliquer le corps $K(y)$ sur le corps K augmenté du symbole ∞. Combinant les deux applications, on obtient une application du corps $K(x)$ sur le corps K augmenté du symbole ∞. Cette application correspond à une valuation w_0 zéro-dimensionnelle. Le centre de w_0 est contenu dans le centre de w. Donc, si le centre de w_0 est simple sur un modèle V_i, le centre de w le sera aussi.

Après cette remarque de ZARISKI, la solution du problème posé était très simple. En effet, les modèles V_1, ..., V_m du théorème (B) ont déjà la propriété désirée: *Chaque tenseur t qui est fini dans tous les points simples de V_1, ..., V_m est partout fini.*

En effet, si le tenseur t avait une variété-pôle simple sur un autre modèle U, cette variété-pôle W aurait la dimension $d-1$, et elle définirait une valuation w, $(d-1)$-dimensionnelle, dont le centre sur U est exactement la variété W. Sur une des variétés V_i, le centre de cette valuation w serait une variété simple W_i, image de W dans la correspondance birationnelle entre U et V_i. Dans un point générique de W_i, le tenseur t est fini par hypothèse. La démonstration de KAEHLER montre que t est aussi fini dans un point générique de W, ce qui contredit l'hypothèse que W est une variété-pôle.

II. - Les systèmes linéaires.

Un autre moyen d'obtenir des invariants birationnels est fourni par la théorie des systèmes linéaires. Pour les surfaces, cette théorie fut créée par Enriques et Castelnuovo. Examinons les notions fondamentales de cette théorie.

On considère, sur une surface U sans points multiples, un système linéaire de diviseurs, c.à.d. de courbes sur la surface. On suppose que les courbes C du système ont, en les points-base P du système, des multiplicités v qui ne sont pas inférieures à certaines multiplicités prescrites, qu'on appelle *multiplicités virtuelles* v_0. La *multiplicité effective* v peut surpasser v_0. Le minimum de la différence $v - v_0$ pour toutes les courbes du système est appelé l'*excès* du système en P. Le système $|C|$ est dit *complet* s'il est impossible de construire un système plus ample $|D|$ sans violer les conditions $v \geqq v_0$.

Or, il ne suffit pas de prescrire des multiplicités pour des points ordinaires. On doit parfois prescrire des multiplicités pour des points infiniment voisins. Considérons, par example, dans le plan le système des coniques passant par deux points P et Q et ayant une tangente donnée t au point P. On peut transformer les coniques en droites par une transformation de Crémona. Le système transformé est complet, mais le système des coniques est complet seulement si l'on prescrit une multiplicité 1 en P et en Q et une multiplicité 1 en un point P_1 infiniment voisin à P dans la direction de la tangente t. Plus précisément, il faut imposer aux coniques la condition que la somme des multiplicités en P et P_1 soit 2 au moins et que la multiplicité en Q soit 1 au moins.

On peut généraliser cette théorie aux variétés de dimension d arbitraire. J'ai fait cela dans les Acta Salamanticensia II (1947).

Pour obtenir cette généralisation, j'ai introduit, au lieu de la notion « point infiniment voisin », la notion de *valuation*. Il suffit de considérer des valuations $(d-1)$-dimensionnelles de $K(x)$. La signification géométrique d'une telle valuation est la suivante. Sur un modèle U de V_0 considérons un diviseur premier, c.à.d. une sous-variété E de dimension $d-1$ définie sur K, indivisible et simple sur U. Chaque fonction $\eta \neq 0$ est un quotient de formes :

$$\eta = F(x)/G(x) \ .$$

Dans l'intersection de l'hypersurface F avec U, la composante E aura une multiplicité a, et dans l'intersection de G avec U, elle aura une multiplicité b. Alors

$$v(\eta) = a - b$$

est une valuation $(d-1)$-dimensionnelle, et toute valuation $(d-1)$-dimensionnelle peut être obtenu de cette manière. Bien entendu, on suppose toujours que les valeurs $v(\eta)$ des constantes a soient zéro, et on ne distingue pas entre deux valuations équivalentes.

Un point simple P de U définit aussi une valuation, car par une dilatation on peut transformer P en un diviseur simple sur un autre modèle V.

De même, un point P_i infiniment voisin d'ordre i de P peut être transformé, par $i + 1$ dilatations successives, en un diviseur simple sur un modèle V.

Ceci nous permet de disposer entièrement de la notion « point infiniment voisin ». La notion « valuation » suffit. Dans la suite, le mot valuation signifiera toujours : valuation $(d-1)$-dimensionnelle.

Cela posé, la théorie générale des systèmes linéaires s'établit de la manière suivante. Soit U un modèle de V_0 sans points multiples. Soient x_0, ..., x_n les coordonnées homogènes d'un point générique de U. Soit v une valuation de $K(x)$. D'abord, $v(\eta)$ n'est définie que pour les fonctions η, mais on peut aussi définir $v(F)$ pour chaque forme $F(x_0, ..., x_n)$ de degré g. La définition est :

$$v(F) = \operatorname*{Max}_{i} v(F/x_i^g) \ .$$

Puis, soit D un diviseur de U, c.à.d. un cycle de dimension $(d-1)$ sur U. Si D est un diviseur effectif, on définit : $v(D)$ est le minimum de $v(F)$ pour toutes les formes F qui sont divisibles par D, c.à.d. dont l'intersection avec U est la somme de D et d'un diviseur effectif. Pour les diviseurs $C - D$ non effectifs on définit

$$(2) \qquad v(C-D) = v(C) - v(D) \ .$$

Pour motiver ces définitions, considérons les cas les plus simples.

1°. La valuation v soit définie par un diviseur premier E sur U.

Alors, si

$$D = cE + \text{autres termes}$$

on aura $v(D) = c$.

2°. La valuation v soit définie par un point simple P de U. Alors, si la multiplicité du diviseur effectif D au point P est r, on aura $v(D) = r$. Cet example montre que les valeurs $v(D)$ nous fournissent la généralisation désirée des « multiplicités effectives » d'ENRIQUES.

3°. La valuation v soit définie par un point infiniment voisin P_1 d'un point simple P d'une surface U. Alors, si les multiplicités d'une courbe D en P et en P_1 sont r et s, on aura $v(D) = r + s$. Dans le cas des coniques, nous avions imposé aux coniques la condition $r + s \geqq 2$. Cette condition est donc équivalente à

$$v(D) \geqq 2 \quad .$$

Si on a un système linéaire $|D|$ sur U on peut définir, pour chaque valuation v, la *valeur effective* v_0 de D comme le minimum des valeurs $v(D)$ pour tous les diviseurs D du système. C'est la généralisation naturelle de la notion multiplicité effective d'une courbe base ou d'un point base d'un système de courbes sur une surface.

D'autre part, les valeurs virtuelles $v_0 \leqq v_e$ peuvent être assignées arbitrairement, suivant certaines conventions. Les conventions sont:

1°. Pour les valuations v correspondant aux diviseurs de U on pose toujours $v_0 = 0$.

2°. Pour les autres valuations on prend les v_0 égales aux différences

$$v_e(B) - v_e(C)$$

des valeurs effectives de deux systèmes B et C sans diviseurs fixes sur U.

Ces deux conventions sont aussi adoptées par ENRIQUES.

Pour transformer un système linéaire $|D|$ par une transformation birationnelle $U \rightarrow U'$, on se sert de la *règle de l'invariance de l'excès* ou règle d'ENRIQUES généralisée:

$$(3) \qquad v'_e - v'_0 = v_e - v_0 \qquad \text{pour toute valuation } v.$$

Cette règle sert à déterminer 1° les multiplicités des diviseurs fixes du système transformé, et 2° les valeurs virtuelles v'_0 pour toutes les valuations qui ne sont pas définies par des diviseurs de U'.

Si toutes les différences $v_e - v_0$ sont zéro, on a un *système sans excès*. Un tel système n'a pas de diviseurs fixes et les systèmes transformés sur tous les modèles U' sont de nouveau sans excès.

Si un système $|C|$ est (totalement) contenu dans un système $|D|$ ayant les mêmes valeurs virtuelles v_0, le système transformé $|C'|$ sera contenu dans $|D'|$. Donc, si $|C'|$ est complet, $|C|$ le sera aussi, et inversement.

Les systèmes complets forment un sémi-groupe abélien qui est invariant par rapport aux transformations birationnelles. Chaque système complet est la différence $|C| - |D|$ de deux systèmes complets sans excès.

Voici les théorèmes principaux dont on trouve la démonstration dans Acta Salamanticensia II et aussi dans Math. Zeitschrift 51, p. 502.

De nouveau, la question se pose: Comment définir la notion « Système complet » si on n'a pas de modèle sans points multiples?

J'ai résolu cette question dans une petite note « The invariant theory of linear sets on an algebraic variety », Proc. Congrès int. Amsterdam III 542.

L'idée de cette solution est la suivante.

D'abord, puisque chaque système complet est une différence de deux systèmes sans excès, il suffit de considérer les systèmes sans excès.

Prenons un modèle U sans sous-variété multiple de dimension $(d-1)$. Un tel modèle existe toujours. Un système $|D|$ sans excès sur U est l'intersection de U et d'un système de formes

$$F(\lambda) = \lambda_1 F_1 + \ldots + \lambda_r F_r$$

moins les diviseurs fixes de cette intersection. Or, la partie variable de cette intersection reste la même si toutes les formes F_i sont multipliées ou divisées par un seul facteur F_0. Nous pouvons donc diviser $F(\lambda)$ par une forme F_0 du même degré. Ainsi, nous obtenons un système linéaire de fonctions

$$f(\lambda) = \lambda_1 f_1 + \ldots + \lambda_r f_r \ ,$$

c'est à dire un K-module de fonctions à base finie dans le corps $K(x)$.

Au lieu de « K-module à base finie » nous dirons simplement *module*. Donc: Chaque module M dans $K(x)$ détermine un système sans excès $|D|$ sur U. Deux modules M et fM (où le facteur f est une fonction) définissent le même systeme $|D|$, et si deux modules M et M' définissent le même système, on a $M' = fM$.

Or, la notion « module » est invariante par rapport aux transformations birationnelles. Donc, si au lieu de considérer les systèmes $|D|$, on considère les modules de fonctions, on aura une théorie invariante.

Soit v une valuation. La notion de valuation, elle aussi, est invariante. On définit la *valeur effective* v_e d'un module M dans la valuation v comme le minimum des $v(f)$ pour toutes les fonctions f de M:

$$(4) \qquad v_e = v_e(M) = \operatorname*{Min}_{f \in M} v(f)$$

Un module M est *complet*, s'il est *impossible* de trouver un module plus ample sans violer les conditions

$$(5) \qquad v(f) \geqq v_e \ .$$

Le nombre entier v_e est une fonction de la valuation v. Appelons cette fonction φ et écrivons

$$v_e = \varphi v \ .$$

Une telle fonction φ à valeurs entières φv sera appelée une *distribution* de valeurs entières. On voit donc que chaque module M définit une distribution φ_M. Inversement, la distribution φ_M détermine un module $\overline{M}$, le module complet contenant M. Ce module complet $\overline{M}$ est simplement l'ensemble de toutes les f satisfaisant aux inégalités (5).

Au produit MN de deux modules correspond la somme des distributions:

$$\varphi_{MN}v = \varphi_M v + \varphi_N v \ .$$

Les distributions φ_M forment donc un sémigroupe abélien. Ce sémigroupe est contenu dans un groupe qu'on obtient en formant des différences $\varphi = \varphi_M - \varphi_N$ définies par

$$(6) \qquad\qquad \varphi v = \varphi_M v - \varphi_N v \ .$$

Une telle différence définit aussi un module complet, formé par toutes les fonctions f satisfaisant aux inégalités

$$(7) \qquad\qquad v(f) \geqq \varphi v \ .$$

La dimension de ce module est un invariant numérique de la distribution φ. Il peut se passer que les valeurs effectives v_e de ce module surpassent les valeurs virtuelles φv. Alors la différence $v_e - \varphi v$ s'appelle l'*excès* du module pour la valuation v par rapport à la distribution φ.

Voilà donc un système de définitions invariantes, indépendant de l'existence d'un modèle sans points multiples, et qui peut servir de base à une théorie invariante des modules et des distributions. Mais la théorie est encore à faire.

Comme vous voyez, un module complet est déterminé par un ensemble infini d'inégalités (5) ou plus généralement (7): une inégalité pour chaque valuation. Or, cet ensemble peu maniable peut être remplacé par un sous-ensemble équivalent plus commode. On prend un modèle U sans sous-variété multiple de dimension $d - 1$. On considère d'abord toutes les valuation v^+ qui correspondent à des diviseurs premiers sur U. Les valeurs φv^+ pour les valuations v^+ sont toutes zéro, à l'exception d'un nombre fini parmi elles. Puis, on ajoute un nombre fini de valuation $v_1^-, ..., v_m^-$, correspondant à des diverseurs premiers sur d'autres modèles $U_1, ..., U_m$. On écrit les inégalités (7) pour toutes les valuations v^+ et pour les valuations $v_1^-, ..., v_m^-$. Cela suffit: les autres inégalités en sont des conséquences.

La démonstration est assez simple. Les inégalités (7) pour les valuations v^+ définissent un module de fonctions à dimension finie. Chaque inégalité correspondant à une valuation v^-, qui n'est pas une conséquence des autres inégalités, diminue la dimension du module d'une unité au moins. Donc, un nombre fini d'inégalités suffit.

Si on impose aux fonctions f seulement les inégalités (7) pour les valuations v^+, on obtient *le module des fonctions multiplès d'un diviseur D*. La dimension de ce module est déterminé, dans le cas classique, par le « théorème de Riemann-Roch » de Hirzebruch. Il serait désirable d'étendre ce théorème au cas général d'un module complet et d'obtenir ainsi un théorème invariant de Riemann-Roch.

Seconde note, ajoutée après la conférence.

Une méthode encore plus générale d'obtenir des invariants birationnels se base sur la considération des *modules de tenseurs*. Si les tenseurs t_1, ..., t_r ont le même nombre d'indices et la même symétrie, on peut former les tenseurs

$$t(\lambda) = \lambda_1 t_1 + ... + \lambda_r t_r$$

qui forment un module de tenseurs à base finie.

La *valeur* d'un tenseur dans une valuation $(n-1)$-dimensionnelle peut être définie de la manière suivante. Supposons que la valuation v soit définie par une sous-variété simple E sur un modèle U. Pour un point générique de E soient u^1, ..., u^d les uniformisantes locales. Soient $t_{\alpha\beta\,...}$ les composantes du tenseur t par rapport aux variables u^α. Alors la valeur $\mathrm{v}(t)$ sera, par définition, le minimum des valeurs $v(t_{\alpha\beta\,...})$ des composantes $t_{\alpha\beta\,...}$.

Ceci posé, on peut définir la valeur effective v_e d'un module de tenseurs dans une valuation v comme le minimum des valeurs $v(t)$ pour tous les tenseurs du module. On peut définir la notion *module complet* comme auparavant. Pour chaque module de tenseurs M, on peut considérer la distribution φ des valeurs effectives $\varphi v = v_e$. Pour chaque distribution φ on peut définir un module complet par les inégalités (7). Si on suppose $\varphi v^+ = 0$ pour toutes

438

les valuations v^+ à l'exception d'un nombre fini entre eux (où les v^+ sont définies par les diviseurs premiers d'un modèle U comme auparavant), le module complet défini par les inégalités (7) aura une dimension finie. Cette dimension est un invariant de la distribution φ.

Dans le cas spécial où toutes les valeurs φv sont zéro, on revient aux invariants de Kaehler. D'autre part, si les tenseurs sont des scalaires sans indices, on revient à la théorie des modules de fonctions.

Zürich, Juillet 1961.

32.

The Theory of Equivalence Systems of Cycles on a Variety(*)

Istituto Nazionale di Alta Matematica, Symposia Matematica V (1970) 255-262

In 1932 the first paper of Severi on equivalence systems of groups of points on algebraic surfaces appeared [1]. Soon afterwards, Severi extended his theory to equivalence systems of virtual varieties C^k, i.e. k-dimensional cycles on a variety V^r without singularities.

In 1941 Severi's lectures on this subject were published in a book « Serie, sistemi d'equivalenza e corrispondenze algebriche sulle varietà algebriche », based upon lecture notes taken by Conforto and Martinelli. This book contains a systematic exposition of the theory of equivalence systems.

In September 1954, in a Symposium on Algebraic Geometry at the Amsterdam Congres, Severi gave a new exposition of the foundations of his theory. In the discussion which followed this lecture, André Weil severely critized Severi's definitions. His main criticism was directed against Severi's definition of the virtual intersection of a set of cycles $A_1, ..., A_s$ of dimensions $h_1, ..., h_s$ on V^r. If the point-set intersection of the carriers of the cycles $A_1, ..., A_s$ has the normal dimension

$$(1) \qquad k = h_1 + ... + h_s - (s-1)r$$

then it is known that the intersection cycle or product

$$C^k = A_1 ... A_s$$

is uniquely defined. A rigorous definition of the intersection cycle was given by Severi [2] in 1933 and worked out by van der Waerden [3] in 1938. Equivalent definitions were given by Chevalley (1945) and by André Weil in his Foundations (1946). So this point was settled to the satisfaction of all.

(*) Conferenza tenuta il 13 aprile 1970.

440

However, if the dimension of the point-set intersection is larger than k, the product symbol $A_1 \dots A_s$ is defined, according to Severi himself, only as a member of a variety of algebraic equivalent cycles. The definition is as follows. We can write A_1 as a difference $B_1 - C_1$, where B_1 and C_1 are variable in irreducible algebraic systems of cycles $|\bar{B}_1|$ and $|\bar{C}_1|$. Let $\bar{B}_1$ and $\bar{C}_1$ be generic cycles of these systems. Just so, $\bar{B}_2$ and $\bar{C}_2$, etc. can be formed. Now the intersection cycle

$$(2) \qquad \bar{P}j = (\bar{B}_1 - \bar{C}_1)(\bar{B}_2 - \bar{C}_2) \dots (\bar{B}_s - \bar{C}_s)$$

is well-defined, because the point-set intersections of the carriers all have the right dimension k. If we specialize $\bar{B}_1 \to B_1$, $\bar{C}_1 \to C_1$, etc., the cycle (2) has a specialization P^k, defined as a member of a connected algebraic system. This was Severi's definition. Severi did not use the expression « specializations »; he used expressions like « $\bar{B}_1$ tends to B_1 ». If the ground field is the field of complex numbers, the limit of a cycle $\bar{B}_1$ is at the same time a specialization, and conversely, so the use of expressions like « $\bar{B}$ tends to B » is unobjectionable.

André Weil's objection against Severi's definitions may be formulated as follows. The cycle $A_1 \dots A_s$ is defined only as a member of a connected variety of cycles, i.e. only up to algebraic equivalence. Now, rational equivalence is stricter than algebraic equivalence. Hence, in setting up a definition of rational equivalence of cycles, the product $A_1 \dots A_s$ is well-defined only if the point-set intersection of the carriers has the normal dimension k.

This was Weil's criticism, and it was perfectly justified, but Severi was very angry. In a dramatic session of the Amsterdam colloquium, Samuel tried to save the situation by giving a rigorous definition of the notion « rational equivalence of cycles ». After the congress, Samuel and I corresponded about this definition, but Samuel himself dropped it, because it turned out that if X and Y are cycles and X is equivalent to zero according to Samuel's definition, it does not follow that $X \cdot Y$ is equivalent to zero.

In the Proceedings of the Amsterdam congress [4], I reported about my correspondence with Severi and Samuel, and I made the proposal to return to the definition given by Severi on page 70 of his book, but to formulate this definition more precisely than Severi did. I guess very few people ever read this Summary. Therefore I shall explain my proposal anew, starting from first principles. It seems to me that Severi's theory is a highly important generalization of the classical theory of linear systems of cycles C^{r-1} on V^r. I very much hope that some young man, with fresh energy, will take up the subject, and work out the theory anew, using the notions and tools of modern algebraic geometry. Severi's ideas were very fertile, and in some of his papers he expressed himself with complete clarity and precision, but in other cases his exposition was not rigorous.

Let us start with an algebraic surface $V = V^2$ without singularities, and consider two linear systems of one-dimensional cycles $|A|$ and $|B|$. To be more explicit, let A be the intersection cycle of the form

$$F = F_0 + \lambda_1 F_1 + \ldots + \lambda_q F_q$$

with V, minus the fixed part of this intersection. The λ_i are indeterminates, hence A is a generic cycle of the linear system $|A|$. Just so, B is the intersection cycle of the form

$$G_\mu = G_0 + \mu_1 G_1 + \ldots + \mu_t G_t$$

with V, minus the fixed part of this intersection. It is known [5] that the carrier of A is an irreducible curve over $k(\lambda)$, and the carrier of B irreducible over $k(\mu)$.

The λ and μ are supposed to be independent indeterminates. Hence the carriers of the one-dimensional cycles A and B have no curve in common, and the intersection $A \cdot B$ is a well-defined group of points, depending rationally on the parameters λ and μ. Among these points, there may be fixed points, independent of the λ and μ. There also may be semi-fixed points, which are variable on a curve C, independent of the λ and μ. Let us agree to leave out all fixed and semi-fixed points and to retain only those points of $A \cdot B$ which are variable over the whole surface V^2. Algebraically, this means that any point of the remaining group of points has degree of transcendency 2 with respect to the ground field k.

The remaining variable part

$$D = (AB)_{\text{var}} = (VF_\lambda G_\mu)_{\text{var}}$$

is uniquely defined as a cycle on V. The Chow-coordinates of the cycle D are rational functions of the λ and μ. Hence D is the generic cycle of a rational variety $|D|$ of cycles. This variety will be called an *intersection system of full variability on* V. The cycles of this system are the specializations of D over k.

We can also consider a linear system of groups of points on a curve C on V. Leaving out all fixed points of the system, we obtain an *intersection system of restricted variability on* V.

Finally, we may consider a fixed group of points on V.

These three kinds of varieties of zero-dimensional cycles are called, according to Severi, *elementary systems* (or series) *of groups of points on* V.

Now let $|D_1|, |D_2|, ..., |D_m|$ be elementary systems of groups of points on V. Let D_1 be a generic cycle of $|D_1|$, and D_2 a generic cycle of $|D_2|$, totally independent of D_1, etc. The sum or difference

$$E = \pm D_1 \pm D_2 \pm ... \pm D_m$$

is a generic element of a rational variety of cycles. Any subvariety of this rational variety, whether reducible or not, is called a *system* (or series) *of equivalence of groups of points on* V.

Equivalence systems of cycles C^k on $V = V^r$ may be defined in the same way, but the definition is slightly more complicated, because intersections of higher dimension than k may occur. Let $|A_1|, ..., |A_{r-k}|$ be linear systems of divisors on $V = V^r$, and let $A_1, A_2, ...$ be independent generic divisors of these systems. As before, A_1 may be defined as the intersection cycle of the form

$$F = F_0 + \lambda_1 F_1 + ... + \lambda_q F_q$$

with V minus the fixed part of this intersection cycle. Just so, A_2 is defined by a form G_μ, etc. The $\lambda, \mu, ...$ are independent indeterminates.

The point-set intersection of $A_1, A_2, ..., A_{r-k}$ is an algebraic variety over the field

$$K = k(\lambda, \mu, ...) ,$$

hence it is a union of irreducible varieties U_i over K. Consider any one of these parts U_i and its generic point X over K. The degree of transcendency of X over k may be r or less than r. If it is less, U_i lies on a subvariety V' of V of dimension less than r, which is defined over k and hence does not depend on the $\lambda, \mu, ...$ In this case we shall say that U_i has *restricted variability*. On the other hand, if the degree of transcendency of X over k is r, we shall say that U_i has *full variability*.

Now it can be shown that the parts U_i of full variability always have the « right » dimension k, the lowest possible dimension an intersection of $r-k$ divisors can have. In fact, only one such part U_1 of full variability can occur. It is irreducible over K, but it may be reducible over the algebraically closed extension $\bar{K}$. Every indivisible, i.e. absolutely irreducible part of U_1 has a well-defined intersection multiplicity according to André Weil. Counting the indivisible parts with these multiplicities, we obtain an intersection cycle

$$B^k = (A_1 A_2 ... A_{r-k})_{\text{var}} = (V F_\lambda G_\mu ...)_{\text{var}} .$$

Once more, the coordinates of the cycle B^k are rational functions of the indeterminates $\lambda, \mu, \ldots$. Hence B^k is the generic cycle of a rational variety $|B^k|$ of cycles. This variety of cycles will be called an *intersection system of full variability on* V.

The proof that there is only one part U_1 of full variability and that it has just the right dimension k can be given as follows. Instead of forming the intersection $VF_\lambda G_\mu \ldots$ at once, we can take one form after the other and consider the sequence of intersections

$$V, \quad VF_\lambda, \quad VF_\lambda G_\mu, \quad \ldots \, ,$$

retaining at every step only those parts of the intersection that have full variability.

The intersection cycle VF_λ minus its fixed part is just A_1, the generic cycle of a linear system of divisors on V. It is known that the carrier of A_1 is irreducible over $k(\lambda)$. Hence the process can be continued: we can form the intersection cycle $A_1 \cdot A_2$ of dimension $r-2$. It may contain a fixed part, not depending on the μ. The points of this fixed part satisfy the equation

$$G_\mu = 0 \quad \text{identically in } \mu \, .$$

This equation implies $G_0 = 0$, hence the fixed part is contained in a proper subvariety of V defined over k. This means: the fixed part of $A_1 \cdot A_2$ does not contain points of full variability over k. Subtracting the fixed part, we are left with a cycle

$$(A_1 \cdot A_2)_{\text{var}} = (V \cdot F_\lambda \cdot G_\mu)_{\text{var}} \, .$$

If this cycle is not zero, its carrier is an irreducible variety of dimension $r-2$ over $k(\lambda, \mu)$. A generic point X of this variety has full variability in A_1 if the λ are kept fixed, and hence full variability in V if the λ and μ are both made variable. This means: X has degree of transcendency $r-2$ over $k(\lambda, \mu)$, hence degree of transcendency $r-1$ over $k(\lambda)$, hence degree of transcendency r over k.

In the same way we may proceed. At the end we are left with a variable part $B^k = (VF_\lambda G_\mu \ldots)_{\text{var}} = (A_1 A_2 \ldots)_{\text{var}}$ whose carrier is irreducible over $k(\lambda, \mu, \ldots)$ and which varies, with varying $\lambda, \mu, \ldots$, over the whole variety V.

The specializations of the cycle B^k form a rational variety of cycles over k, which is called, as I have said already, an intersection system of full variability on V.

In the same way, if α is an irreducible subvariety of dimension

$h \geqq k$ on V, we may define intersection systems of full variability on W. The generic cycle B^k of such a system is the variable part of the intersection of W with $h - k$ forms F_λ, G_μ, ... formed with independent indeterminates $\lambda, \mu, ...$:

$$(WA_1 ... A_{k-h})_{\mathrm{var}} = (WF_\lambda G_\mu ...)_{\mathrm{var}} \; .$$

Such a system B^k is called an *intersection system of restricted variability on V*. In the case $h = k$, the system consists of the fixed cycle W only.

Both kinds of intersection systems, of full and of restricted variability, are called, according to Severi, *elementary systems of species k on V*.

Let $|B_1|, |B_2|, ..., |B_m|$ be elementary systems of species k on V. Let B_1 be a generic cycle of $|B_1|$, and B_2 a generic cycle of $|B_2|$, etc., each B_i being formed by means of a new set of indeterminates λ, μ, etc. The sum or difference

$$(3) \qquad C = \pm B_1 \pm B_2 \pm ... \pm B_m$$

is the generic element of a rational variety of k-dimensional cycles. Any subvariety of this rational variety is called a *system of equivalence of cycles C^k on V*.

Two cycles C_1 and C_2 of dimension k are called *equivalent*, if they belong to one system of equivalence. Obviously this relation is reflexive and symmetric. I shall now prove that it is transitive.

Let C_1 and C_2 be specializations of a cycle C of the form (3), and let C_2 and C_3 be specializations of a cycle C' of the same form (3). We may suppose that in the expressions

$$C = \pm B_1 \pm B_2 \pm ... \pm B_m$$
$$C' = \pm B_1' \pm B_2' \pm ... \pm B_n'$$

the terms B_i and B_j' are all formed with new indeterminates. Now among the specializations of the cycle

$$C + C' - C_2$$

we have

$$C_1 + C_2 - C_2 = C_1$$

and also

$$C_2 + C_3 - C_2 = C_3 \; .$$

Hence C_1 is rationally equivalent to C_3.

It follows that an *equivalence class* can be defined as the class of all cycles rationally equivalent to a given cycle C^k.

Severi has given outlines of proofs of the following theorems:

I) *Every cycle C^h in projective space S^d is a complete intersection of $d-h$ divisors i.e. cycles of dimension $d-1$ in S^d:*

$$C^h = H_1 \cdot H_2 \ldots H_{d-h}$$

(Severi, p. 64).

The divisors H_i are differences of positive divisors

$$H_i = P_i - Q_i .$$

If P_i and Q_i each vary in a linear system of divisors, the difference H_i is said to vary in a *virtual linear system.*

Now consider the intersection

$$V \cdot H_1 \ldots H_{r-k}$$

where $V = V^r$ is a variety without singularities in S^d, whereas $H_1, \ldots, H_{r-k}$ are the generic elements of virtual linear systems of divisors in S^d. Subtracting from the intersection all parts of restricted variability, we are left with a cycle C^k of full variability in V^r. This cycle

$$C^k = (V \cdot H_1 \ldots H_{r-k})_{\text{var}}$$

is the generic cycle of a variety of cycles $|C^k|$. Such a variety $|C^k|$ is called a *virtual elementary system of species k.* Severi's second theorem now states (Severi, p. 84):

II) *The necessary and sufficient condition for a variety of cycles C^k on S^r to be an equivalence system is that it be contained in a virtual elementary system of species k.*

The third theorem is Severi's « Teorema fondamentale » (p. 112):

III) *Every unirational variety of cycles C^k on V^r is a system of equivalence.*

I sincerely hope that someone will elaborate Severi's theory and give rigorous proofs of his main theorems. I am quite sure that at least Theorems I and II are valid for an arbitrary perfect field k.

Testo pervenuto il 2 aprile 1970.
Bozze licenziate il 29 giugno 1970.

REFERENCES

[1] F. SEVERI, *La serie canonica e la teoria delle serie principali di gruppi di punti sopra una superficie algebrica*, Commentarii Mathematici Helvetici, 4 (1932), 268.

[2] F. SEVERI, *Über die Grundlagen der algebraischen Geometrie*, Abh. Math. Sem. Hamburg, 9 (1933), 335.

[3] B. L. VAN DER WAERDEN, *Zur algebraischen Geometrie 14*. Math. Annalen, 115 (1938), 635.

[4] B. L. VAN DER WAERDEN, *On the definition of rational equivalence of cycles on a variety*, Proceedings Internat. Congress of Mathematicians, 1954, Amsterdam, vol. III, p. 545.

[5] B. L. VAN DER WAERDEN, *Einführung in die algebraische Geometrie*, § 42, Satz 5.

33.

Zur algebraischen Geometrie 20
Der Zusammenhangssatz und der Multiplizitätsbegriff

Mathematische Annalen 193 (1971) 89–108

Als Grundlage einer Theorie der Schnittmultiplizitäten habe ich 1926 den Begriff Spezialisierungsmultiplizität entwickelt und unter gewissen Voraussetzungen die Eindeutigkeit dieser Multiplizität bewiesen [1]. Der Gedankengang war so:

Ein „Normalproblem" sei durch Gleichungen

$$G_j(\xi, \eta) = 0 \tag{A}$$

definiert, die homogen in den Unbekannten $\eta_0, \ldots, \eta_n$ sein sollen, während die ξ Unbestimmte sind. Das Problem (A) habe für unbestimmte ξ endlich viele Lösungen $\eta^{(1)}, \ldots, \eta^{(h)}$. Jede Spezialisierung $\xi \to x$ läßt sich zu einer Spezialisierung

$$(\xi, \eta^{(1)}, \ldots, \eta^{(h)}) \to (x, y^{(1)}, \ldots, y^{(h)})$$

fortsetzen. Die $y^{(v)}$ sind Lösungen des spezialisierten Problems

$$G_j(x, y) = 0 . \tag{B}$$

Die Eindeutigkeit der spezialisierten Lösungen $y^{(1)}, \ldots, y^{(h)}$ wurde in [1] unter der Voraussetzung bewiesen, daß das spezialisierte Problem (B) bei gegebenem x nur endlich viele Lösungen hat. Wenn eine Lösung y des Problems (B) genau μ-mal unter den $y^{(v)}$ vorkommt, so sagt man, sie habe bei der Spezialisierung $\xi \to x$ die *Multiplizität* μ.

Unter allgemeineren Voraussetzungen hat Weil [2] die Eindeutigkeit der Multiplizität μ einer Lösung y des spezialisierten Problems bewiesen. Weil verlangt nämlich nicht, daß es nur endlich viele Lösungen des Problems (B) gibt, sondern nur, daß eine Lösung y isoliert ist. Northcott [3] und Leung [4] haben die Voraussetzungen von Weil noch weiter abgeschwächt.

Der Beweis für die Eindeutigkeit der Multiplizität einer isolierten Lösung y ist bei Weil recht kompliziert: er beruht auf der analytischen Theorie der lokalen Ringe. Auch Northcott und Leung benutzen diese analytische Theorie. Im folgenden soll ein Beweis mit klassischen Hilfsmitteln gegeben werden, der auf einem Spezialfall des Zusammenhangssatzes beruht.

Um den *Zusammenhangssatz* zu erklären, müssen wir einige Definitionen vorausschicken.

Eine Varietät V im projektiven Raum P^n heißt *zusammenhängend*, wenn es auch nach Erweiterung des Körpers k, über dem V definiert ist, unmöglich ist, V in zwei disjunkte, nicht leere Teilvarietäten zu zerlegen. Eine Varietät V heißt *linear zusammenhängend*, wenn je zwei Punkte innerhalb V durch eine Kette von rationalen Kurven $C_1, \ldots, C_h$ miteinander verbunden werden können. Offensichtlich folgt daraus, daß V zusammenhängend ist.

Ein *Zykel* Z^d ist eine formale Summe von absolut irreduziblen Varietäten V^d, die alle dieselbe Dimension d haben, mit ganzzahligen Koeffizienten e_i. Wir betrachten nur positive Zykel, bei denen alle e_i positiv oder Null sind. Die Vereinigung aller V^d mit $e_i \neq 0$ heißt der *Träger* des Zykels.

Nun sei ξ ein allgemeiner und x ein spezieller Punkt einer über k irreduziblen Varietät U. Dem Punkte ξ sei ein Zykel Z zugeordnet, der über $k(\xi)$ definiert und zusammenhängend ist. Die Varietät U sei im Punkte x analytisch irreduzibel, d. h. die komplette Erweiterung des lokalen Ringes sei nullteilerfrei. Zur Spezialisierung $\xi \to x$ gehören Spezialisierungen Z_x des Zykels Z_ξ. Dann besagt der Zusammenhangssatz, daß die Vereinigung der Träger aller Z_x zusammenhängend ist.

Der Zusammenhangssatz wurde von Zariski [5] zuerst aufgestellt und im Rahmen seiner Theorie der holomorphen Funktionen auf algebraischen Varietäten bewiesen. Eine Verallgemeinerung und einen einfacheren Beweis hat Chow [6] gegeben.

Für die Anwendung, die wir hier im Auge haben, ist ein Spezialfall des Zusammenhangssatzes von besonderer Bedeutung. Der Zykel Z_ξ sei ein einzelner Punkt η, der rational von ξ abhängt. Ferner sei x ein einfacher Punkt von U. In diesem Fall kann man, wie Chow gezeigt hat, den Zusammenhangssatz verschärfen und zeigen, daß alle zu $\xi \to x$ gehörigen Spezialisierungen y von η eine *linear zusammenhängende* Varietät bilden.

In der vorliegenden Arbeit wird zunächst angenommen, daß U ein projektiver Raum P^m ist. Der Satz, der hier bewiesen werden soll, lautet also (§ 5, Satz 2):

Bei einer rationalen Abbildung von P^m in P^n ist die Bildvarietät V_A eines jeden Punktes A linear zusammenhängend.

In § 1 werden nur Grundbegriffe erklärt. In § 2 und § 3 werden bekannte Ergebnisse der klassischen algebraischen Geometrie mit klassischen Methoden bewiesen. Für den Kenner enthalten § 1 bis § 3 nichts Neues.

In § 4 wird der obige Satz in den Fällen $U = P^1$ und $U = P^2$ bewiesen. In § 5 wird der Fall $U = P^m$ auf den Fall $U = P^2$ zurückgeführt.

In § 6 wird gezeigt, daß die Eindeutigkeit der Spezialisierungsmultiplizität einer isolierten Lösung y im Falle $U = P^m$ unmittelbar aus Satz 2 folgt.

Hat man einmal die Spezialisierungsmultiplizität, so kann man nach A. Weil auch Schnittmultiplizitäten von Zykeln definieren. Das soll in § 7 näher ausgeführt werden.

In § 8 wird der lineare Zusammenhangssatz für den Fall bewiesen, daß x ein einfacher Punkt einer beliebigen Varietät U ist. Durch eine Parallelprojektion wird der allgemeine Fall auf den Fall $U = P^m$ zurückgeführt.

§ 1. Grundbegriffe

Es sei k ein fester Grundkörper und Ω ein *Universalkörper* im Sinne von Weil [2], d. h. ein algebraisch abgeschlossener Erweiterungskörper von unendlichem Transzendenzgrad über k. Eine über k definierte *Korrespondenz K* zwischen P^m und P^n ist die Gesamtheit der Punktepaare (x, y) mit Koordinaten aus Ω, die einem System von homogenen Gleichungen

$$G_j(x, y) = 0 \tag{1}$$

mit Koeffizienten aus k genügen. Im x-Raum werden wir häufig $x_0 = 1$ setzen und inhomogene Koordinaten $x_1, \ldots, x_n$ einführen, aber der y-Raum soll immer der volle projektive Raum P^n sein.

Wenn die Korrespondenz K über k irreduzibel ist, so besteht sie aus allen Paaren (x, y), die Spezialisierungen eines allgemeinen Paares (ξ, η) sind. Man nennt (x, y) eine *Spezialisierung* von (ξ, η) über k, wenn alle homogenen Gleichungen $H(\xi, \eta) = 0$ mit Koeffizienten aus k, die für das Paar (ξ, η) gelten, auch für das Paar (x, y) gelten. Die Spezialisierungen von ξ bilden die *Urvarietät U*, die von η die *Bildvarietät V* der Korrespondenz, und man spricht von einer *Korrespondenz K zwischen U und V*.

Die Korrespondenz K heißt eine *rationale Abbildung*, wenn die Koordinatenverhältnisse von η rationale Funktionen von denen von ξ sind. Man hat dann

$$\beta \eta_k = F_k(\xi) . \tag{2}$$

Dabei sind die F_k Formen gleichen Grades in $\xi_0, \ldots, \xi_m$, die nicht alle Null sind.

Zu den homogenen Gleichungen $H(\xi, \eta) = 0$, die für das allgemeine Paar (ξ, η) der rationalen Abbildung (2) gelten, gehören die folgenden:

$$\eta_j F_k(\xi) - \eta_k F_j(\xi) = 0 . \tag{3}$$

Also muß für jedes Paar (x, y) gelten

$$y_j F_k(x) - y_k F_j(x) = 0 \tag{4}$$

und daher

$$\gamma y_k = F_k(x) . \tag{5}$$

Sind die $F_k(x)$ nicht alle Null, so nennt man den Punkt x für die Abbildung *regulär*; er hat dann einen einzigen, durch (5) bestimmten Bildpunkt y. Sind alle $F_k(x)$ Null, so ist der Punkt x für die Abbildung *singulär*; er kann dann mehrere Bildpunkte haben. In jedem Fall bilden die Bildpunkte y eines gegebenen Punktes x eine Varietät V_x; man nennt sie die *Bildvarietät* des Punktes x in der durch (2) definierten Abbildung.

In § 2 wird gezeigt, wie man die einzelnen Punkte y der Bildvarietät V_x durch Reihenentwicklungen erhalten kann. Diese Reihenentwicklungen bilden das wichtigste Hilfsmittel beim Beweis des linearen Zusammenhangs der Bildvarietät V_x in § 4 und § 5.

Wir brauchen einige Begriffe aus der Theorie der algebraischen Kurven. Eine Kurve C in P^m kann durch eine birationale Transformation auf eine Kurve C' ohne mehrfache Punkte in P^n abgebildet werden (siehe etwa meine Einführung in die algebraische Geometrie, § 45). Einem Punkt A von C können mehrere Punkte A' von C' entsprechen; jeder von diesen definiert einen *Zweig* $\mathfrak{z}$ der Kurve C im Punkte A. Man nennt A den *Anfangspunkt* des Zweiges $\mathfrak{z}$. Ist C'' ein anderes birationales Bild von C, so sind die Punkte von C'' eineindeutig denen von C' zugeordnet, also ist der Zweigbegriff von der Wahl der Bildkurve C' unabhängig.

Zu jedem Zweig $\mathfrak{z}$ kann man eine Ortsuniformisierende τ wählen und die Koordinaten eines allgemeinen Punktes von C in Potenzreihen nach τ entwickeln:

$$X_i(\tau) = a_i + b_i \tau + \cdots. \tag{6}$$

Dabei sind die a_i die Koordinaten des Anfangspunktes A des Zweiges $\mathfrak{z}$.

Wir schneiden nun die Kurve C mit einer Hyperfläche $F = 0$ des Raumes P^m und fragen, wie man die Schnittmultiplizität $(C \cdot F)_A$ im Punkt A erhalten kann. Die Hyperfläche $F = 0$ sei durch eine Form

$$F(X) = F(X_0, X_1, \ldots, X_m)$$

definiert. Will man sich genau ausdrücken, so muß man zwischen der *Hyperfläche* $F = 0$ und dem $(m - 1)$-dimensionalen *Zykel* F unterscheiden, der so erhalten wird: Die Form F wird über dem algebraisch abgeschlossenen Körper $\bar{k}$ in Primfaktoren P_i mit Exponenten e_i zerlegt. Jeder dieser Primfaktoren P_i definiert eine unteilbare Hyperfläche $P_i = 0$, die kurz als Hyperfläche P_i bezeichnet werden möge. Werden diese Hyperflächen e_i mal gezählt und addiert, so erhält man den $(m - 1)$-dimensionalen Zykel

$$F = \Sigma \, e_i P_i.$$

Es ist zweckmäßig, denselben Buchstaben F für die Form $F(X)$ und den Zykel F zu verwenden.

Die *Schnittmultiplizität* $(\mathfrak{z} \cdot F)_A$ eines Zweiges $\mathfrak{z}$ mit dem Zykel F im Punkte A kann bestimmt werden, indem man die Potenzreihen $X_i(\tau)$ des Zweiges in die Form F einsetzt. Ist

$$F(X_i(\tau)) = c \tau^\mu + \cdots \qquad (c \neq 0),$$

so ist μ die Schnittmultiplizität:

$$(\mathfrak{z} \cdot F)_A = \mu.$$

Die Schnittmultiplizität $(C \cdot F)_A$ der Kurve C mit dem Zykel F im Punkte A ist die Summe der Schnittmultiplizitäten der einzelnen Zweige $\mathfrak{z}$ der Kurve mit F in A:

$$(C \cdot F)_A = \Sigma \, (\mathfrak{z} \cdot F)_A. \tag{7}$$

Für den Beweis der Formel (7) siehe meine Einführung in die algebraische Geometrie, § 20. Für die Zwecke der vorliegenden Arbeit würde es übrigens genügen, die Formel (7) als Definition der Schnittmultiplizität einer Kurve C mit einem $(m-1)$-dimensionalen Zykel F zu betrachten.

§ 2. Wie erhält man alle Bildpunkte eines Punktes x^0 in einer rationalen Abbildung?

Eine irreduzible Korrespondenz K sei durch Gl. (1) definiert. Durch die Substitution

$$z_{ik} = x_i y_k \tag{8}$$

kann man K auf eine Bildvarietät in einem projektiven Raum P_N abbilden. Wir bezeichnen diese Bildvarietät wieder mit K; ihre Dimension sei d.

Ist speziell K eine rationale Abbildung (2), so bilden die *singulären Paare* (x, y), für die alle $F_k(x)$ Null sind, eine Teilvarietät K' von höchstens der Dimension $d-1$. Es sei (x^0, y^0) mit Bildpunkt z^0 ein singuläres Paar. Durch z^0 legen wir eine Hyperebene u, die keinen $(d-1)$-dimensionalen Teil von K' enthält. Durch den Schnitt mit u werden die Dimensionen von K und K' um je eine Einheit vermindert. Dieses Verfahren wird wiederholt; nach $d-1$ Schritten erhält man eine Kurve C^* auf K, die durch z^0 geht und mit K' nur endlich viele Punkte gemeinsam hat. Es sei C ein durch z^0 gehender absolut irreduzibler Bestandteil einer solchen Kurve C^*. Ein allgemeiner Punkt von C sei Z. Das entsprechende Paar (X, Y) ist regulär, weil Z nicht in K' liegt.

Mindestens ein Zweig $\mathfrak{z}$ der Kurve C hat seinen Ursprung in z^0. Ist τ eine Ortsuniformisierende für diesen Zweig $\mathfrak{z}$, so sind die Koordinaten der Punkte X und Y als Potenzreihen in τ darstellbar.

In einer Umgebung des Punktes x^0 kann man inhomogene Koordinaten $x_1, \ldots, x_n$ einführen, mit x^0 als Koordinatenanfangspunkt. Die Reihenentwicklungen für die Koordinaten der Punkte X und Y sehen dann so aus:

$$X_i(\tau) = a_i \tau + b_i \tau^2 + \cdots \quad (i = 1, \ldots, m), \tag{9}$$

$$Y_k(\tau) = c_k + d_k \tau + e_k \tau^2 + \cdots \quad (k = 0, 1, \ldots, n). \tag{10}$$

Da das Paar $X(\tau)$, $Y(\tau)$ regulär ist, müssen die Gleichungen

$$\gamma Y_k(\tau) = F_k(X(\tau)) \quad \text{mit} \quad \gamma \neq 0 \tag{11}$$

gelten. Sind die Potenzreihen $X_i(\tau)$ gegeben, so kann man die rechte Seite von (11) ausrechnen. Wenn die so erhaltenen Potenzreihen mit τ^r anfangen, kann man schreiben

$$F_k(X(\tau)) = p_k \tau^r + q_k \tau^{r+1} + \cdots, \tag{12}$$

wobei die p_k nicht alle Null sind. Auf der linken Seite von (11) kann man $\gamma = \tau^r$ wählen und erhält

$$\tau^r Y_k(\tau) = p_k \tau^r + q_k \tau^{r+1} + \cdots$$

oder

$$Y_k(\tau) = p_k + q_k \tau + \cdots. \tag{13}$$

Den Anfangspunkt y^0 dieses Kurvenzweiges erhält man, indem man $\tau = 0$ setzt:

$$y_k^0 = p_k. \tag{14}$$

Daraus folgt das gewünschte Ergebnis:

Satz 1. *Jeder zu x^0 gehörige Bildpunkt y^0 kann folgendermaßen erhalten werden. Man wählt auf U einen Kurvenzweig mit Anfangspunkt x^0 so, daß ein allgemeiner Punkt $X(\tau)$ dieses Zweiges regulär für die Abbildung ist. Man entwickelt die Koordinaten $X_i(\tau)$ in Potenzreihen nach τ. Man setzt diese Potenzreihen in die Polynome F_k ein und entwickelt nach (12). Dann sind die Anfangskoeffizienten p_k die Koordinaten eines Bildpunktes y^0.*

Wir werden das Verfahren im folgenden kurz so beschreiben: „Man nähert sich auf dem Zweige $\mathfrak{z}$ dem Punkte x^0 und erhält als Grenzpunkt den Bildpunkt y^0."

Selbstverständlich kann das gleiche Verfahren auch für reguläre x^0 angewandt werden; nur ist das Ergebnis in diesem Fall von der Wahl des Kurvenzweiges unabhängig.

Ist insbesondere U der ganze Raum P^m, so kann man die Koeffizienten $a_i, b_i, \ldots$ in den Potenzreihen (9) ganz beliebig wählen, sofern nur die $F_k(X(\tau))$ nicht alle identisch in τ Null werden. Diese Freiheit in der Wahl der Koeffizienten kann man dazu benutzen, die Potenzreihen (9) nach dem Glied mit τ^r abzubrechen: die weiteren Glieder haben auf y^0 sowieso keinen Einfluß. Man kann dann also die $X_i(\tau)$ als Polynome in τ wählen.

§ 3. Reduktion der Schnittmultiplizitäten ebener Kurven durch Cremonatransformationen

Max Noether hat in seinen grundlegenden Arbeiten über die Auflösung der Singularitäten der algebraischen Kurven gezeigt, daß man durch eine geeignete Cremonatransformation die Schnittmultiplizität $(F, G)_A$ von zwei ebenen Kurven F und G im Punkte A immer kleiner machen kann. Noether verwendete die Cremonatransformation

$$x_0 = z_1 z_2, \quad x_1 = z_2 z_0, \quad x_2 = z_1 z_2. \tag{15}$$

Wenn F in A einen r-fachen Punkt und G einen s-fachen Punkt hat, so wird die Schnittmultiplizität $(F, G)_A$ nach Noether durch die Transformation um rs verkleinert.

Statt der Transformation (15) kann man auch die etwas einfachere Transformation

$$x_1 = z_1, \quad x_2 = z_1 z_2 \tag{16}$$

benutzen. Die inhomogenen Koordinaten x_1, x_2 werden so gewählt, daß der Punkt A die Koordinaten $(0, 0)$ hat und die Achse $x_2 = 0$ die Kurven F und G in A nicht berührt. Die Kurve F mit der Gleichung $F(x_1, x_2) = 0$ wird so transformiert:

$$F(z_1, z_1 z_2) = \Phi(z_1, z_2) = z_1^r \, \varphi(z_1, z_2) \, . \tag{17}$$

Die Kurve φ in der z-Ebene heißt die *reduzierte Transformierte* der Kurve F. Ebenso erhält man für die Kurve G die reduzierte Transformierte ψ. Die Kurven φ und ψ können einen oder mehrere Schnittpunkte α auf der Geraden $z_1 = 0$ haben. Bei der Transformation $z \to x$ gehen diese Schnittpunkte α alle in den einen Punkt A über und es gilt die Formel

$$(F, G)_A = rs + \sum_\alpha (\varphi, \psi)_\alpha \, . \tag{18}$$

Für den Beweis dieser Formel macht es nichts aus, ob man die Cremonatransformation (15) oder die einfachere Transformation (16) benutzt. Der Beweis wird wohl am einfachsten so geführt, daß man die Kurve F zunächst mit den einzelnen Zweigen $\mathfrak{z}$ der Kurve G im Punkte A schneidet. Die Potenzreihen eines solchen Zweiges $\mathfrak{z}$ können so geschrieben werden:

$$X_1(\tau) = a_1 \tau^v + b_1 \tau^{v+1} + \cdots$$
$$X_2(\tau) = a_2 \tau^v + b_2 \tau^{v+1} + \cdots$$

mit $a_1 \neq 0$. Der Exponent v heißt die *Vielfachheit* des Zweiges $\mathfrak{z}$ im Punkte A. Die Transformation $x \to z$ führt $\mathfrak{z}$ in einen Zweig $\mathfrak{z}'$ in der z-Ebene über, der seinen Anfangspunkt α auf der Achse $z_1 = 0$ hat. Nun soll gezeigt werden

$$(F, \mathfrak{z})_A = rv + (\varphi, \mathfrak{z}')_\alpha \, . \tag{19}$$

Durch Summation über alle Zweige $\mathfrak{z}$ der Kurve G im Punkte A erhält man aus (19) ohne weiteres (18).

Die Formel (19) ist ein Spezialfall einer allgemeineren Formel für die Schnittmultiplizität einer Hyperfläche F mit einem Kurvenzweig $\mathfrak{z}$ im Raume P^m, die in meiner Arbeit "On infinitely near points" [7] bewiesen wurde. In unserem ebenen Spezialfall verläuft der Beweis so:

Die Reihenentwicklung des transformierten Zweiges $\mathfrak{z}'$ sei

$$\begin{aligned} Z_1(\tau) &= a_1 \tau^v + b_1 \tau^{v+1} + \cdots \\ Z_2(\tau) &= p + q\tau + \cdots \, . \end{aligned} \tag{20}$$

Durch die Transformation (16) erhält man daraus die Reihenentwicklung des Zweiges $\mathfrak{z}$:

$$X_1 = X_1(\tau) = Z_1(\tau)$$
$$X_2 = X_2(\tau) = Z_1(\tau) \, Z_2(\tau) \, .$$

Setzt man diese in F ein und beachtet (17), so erhält man

$$F(X_1, X_2) = F(Z_1, Z_1 Z_2)$$
$$= Z_1^r \, \varphi(Z_1, Z_2) \, . \tag{21}$$

Wenn die Entwicklung von $\varphi(Z_1, Z_2)$ nach Potenzen von τ mit τ^w anfängt, so fängt die Entwicklung von $F(X_1, X_2)$, wie man aus (21) sieht, mit

$$\tau^{rv+w}$$

an. Daraus ergibt sich (19) und daraus weiter (18).

Aus (18) folgt, daß jeder einzelne Term rechts kleiner ist als die linke Seite:

$$(\varphi, \psi)_\alpha < (F, G)_A \, . \tag{22}$$

Das ist alles, was wir für das folgende brauchen.

§ 4. Beweis des linearen Zusammenhangssatzes für $U = P^1$ und $U = P^2$

A. Der Fall $U = P^1$

Eine rationale Abbildung der projektiven Geraden P^1 auf eine Bildvarietät R sei durch

$$\beta \eta_k = F_k(\tau_0, \tau_1) \tag{23}$$

gegeben, wobei die F_k Formen gleichen Grades sind. Ist $D(\tau)$ der größte gemeinsame Teiler der Formen F_k, so kann man $\beta = D(\tau)$ setzen und den Faktor $D(\tau)$ links und rechts wegkürzen; so erhält man

$$\eta_k = f_k(\tau_0, \tau_1) \, . \tag{24}$$

Die Formen f_k haben jetzt keinen gemeinsamen Teiler, also keine gemeinsame Nullstelle. Daher ist jeder Punkt t der projektiven Geraden für die Abbildung regulär und der Bildpunkt y durch

$$y_k = f_k(t_1, t_2) \tag{25}$$

eindeutig bestimmt. Der lineare Zusammenhangssatz ist in diesem Fall trivial.

Die Gleichungen der Korrespondenz K lauten

$$y_i f_k(t) - y_k f_j(t) = 0 \, . \tag{26}$$

Die Bildvarietät R hat die Dimension 1 oder 0. Ist die Dimension 1, so ist R eine rationale Kurve mit der Parameterdarstellung (25). Ist sie Null, so ist R ein einzelner Punkt.

B. Der Fall $U = P^2$

Eine rationale Abbildung von P^2 in P^n sei durch

$$\beta \eta_k = F_k(\xi_0, \xi_1, \xi_2) \tag{27}$$

gegeben. Man kann wieder annehmen, daß die Formen F_k keinen gemeinsamen Teiler haben. Zu beweisen ist, daß die Bild-Varietät V_A eines Punktes A von P^2 linear zusammenhängend ist.

Der Punkt A habe die Koordinaten $(1, 0, 0)$. Man kann wieder $x_0 = 1$ setzen und inhomogene Koordinaten x_1, x_2 einführen. Die Formen $F_k(X)$ gehen durch die Substitution $X_0 = 1$ in inhomogene Polynome $F_k(X_1, X_2)$ über. Wir schreiben sie als Summen von homogenen Polynomen mit steigenden Gradzahlen $r, r+1, \ldots$:

$$F_k = L_k + M_k + \cdots . \tag{28}$$

Um alle Bildpunkte zu erhalten, hat man die Reihenentwicklungen aller Zweige $\mathfrak{z}$, die in A ihren Ursprung haben, in die Polynome F_k einzusetzen. Die Reihenentwicklung eines solchen Zweiges $\mathfrak{z}$ sei durch

$$X_1(\tau) = a_1 \tau^v + b_1 \tau^{v+1} + \cdots$$
$$X_2(\tau) = a_2 \tau^v + b_2 \tau^{v+1} + \cdots \tag{29}$$

gegeben, wobei a_1 und a_2 nicht beide Null sind. Das Verhältnis $a_2 : a_1$ bestimmt die *Tangentenrichtung* des Zweiges $\mathfrak{z}$.

Setzt man nun (29) in (28) ein, so erhält man

$$F_k(X(\tau)) = \tau^{rv} L_k(a_1, a_2) + \cdots . \tag{30}$$

Es gibt nur endlich viele Tangentenrichtungen, für die alle $L_k(a_1, a_2)$ Null werden. Wir nennen sie *schwierige Tangentenrichtungen*. Für alle anderen Richtungen ist der Bildpunkt einfach durch

$$y_k = L_k(a_1, a_2) \tag{31}$$

gegeben. Die Gl. (31) bestimmen im Bildraum eine rationale Kurve R oder einen einzelnen Punkt R. Diese Kurve oder dieser Punkt gehört jedenfalls zur Bildvarietät V_A. Alle anderen Teile von V_A rühren von den schwierigen Tangentenrichtungen her.

Mit $2(n+1)$ unabhängigen Unbestimmten λ_k und μ_k kann man zwei Linearkombinationen der F_k bilden:

$$F_\lambda = \Sigma \, \lambda_k F_k , \tag{32}$$

$$F_\mu = \Sigma \, \mu_k F_k . \tag{33}$$

Da die F_k keinen Faktor gemeinsam haben, haben F_λ und F_μ auch keinen Faktor gemeinsam. Die Zykel F_λ und F_μ haben also im Punkt A eine bestimmte Schnittmultiplizität m. Ist $m = 0$, so sind die F_k nicht alle Null im Punkt A. Der Punkt A ist dann regulär, V_A ist ein einzelner Punkt und der lineare Zusammenhang von V_A ist trivial. Wir nehmen also $m > 0$ an und setzen eine vollständige Induktion nach m an. Für einen bestimmten Wert von m soll die Behauptung, daß V_A linear zusammenhängend ist, bewiesen werden unter der Induktionsvoraussetzung, daß sie für alle kleineren Werte von m richtig ist.

Das Koordinatensystem in der x-Ebene sei so gewählt, daß die Achse $x_1 = 0$ nicht mit einer schwierigen Tangentenrichtung zusammenfällt. Die Gleichungen

$$x_1 = s_1 , \quad x_2 = s_1 s_2 \tag{34}$$

definieren wie in § 3 eine rationale Abbildung $s \to x$. Die Zusammensetzung der Abbildung $s \to x$ mit der vorgegebenen rationalen Abbildung $x \to y$ ergibt eine rationale Abbildung $s \to y$. Einem allgemeinen Punkt (S_1, S_2) der s-Ebene mit unbestimmten Koordinaten S_1, S_2 entspricht in dieser Abbildung der Punkt η mit Koordinaten

$$\beta \eta_k = F_k(S_1, S_1 S_2) = \Phi_k(S_1, S_2) . \tag{35}$$

Wie in § 3 kann man

$$\Phi_k(S_1, S_2) = S_1^r \varphi_k(S_1, S_2) \tag{36}$$

setzen, wobei die φ_k nicht alle durch S_1^r teilbar sind. Aus (35) und (36) folgt

$$\beta \eta_k = F_k(S_1, S_1 S_2) = S_1^r \varphi_k(S_1, S_2)$$

oder, wenn $\beta = \gamma S_1^r$ gesetzt wird,

$$\gamma \eta_k = \varphi_k(S_1, S_2) . \tag{37}$$

Die Abbildung $s \to y$ wird also durch die Polynome φ_k vermittelt.

Multipliziert man (35) und (36) mit λ_k und summiert über k, so erhält man

$$F_\lambda(S_1, S_1 S_2) = \Sigma \lambda_k \Phi_k(S_1, S_2) = S_1^r \Sigma \lambda_k \varphi_k(S_1, S_2)$$

oder, wenn die Summe rechts φ_λ genannt wird,

$$F_\lambda(S_1, S_1 S_2) = S_1^r \varphi_\lambda(S_1, S_2)$$

und ebenso

$$F_\mu(S_1, S_1 S_2) = S_1^r \varphi_\mu(S_1, S_2) .$$

Die Zykel φ_λ und φ_μ sind also die reduzierten Transformierten der Zykel F_λ und F_μ im Sinne von § 3. Also ist die Schnittmultiplizität m' von φ_λ und φ_μ in irgend einem Punkt α der Achse $s_1 = 0$ kleiner als die Schnittmultiplizität m von F_λ und F_μ im Punkte A. Nach der Induktionsvoraussetzung ist also die Bildvarietät W_α des Punktes α in der Abbildung $s \to y$ linear zusammenhängend.

Nun ist es leicht, den linearen Zusammenhang von V_A zu beweisen. V_A ist die Gesamtheit aller Bildpunkte y, die von den einzelnen Zweigen $\mathfrak{z}$ in der x-Ebene mit Anfangspunkt A geliefert werden. Die Zweige mit nicht schwierigen Tangentenrichtungen ergeben Punkte der rationalen Kurve R. Die Zweige $\mathfrak{z}$, die eine bestimmte schwierige Tangentenrichtung haben, gehen bei der Um-

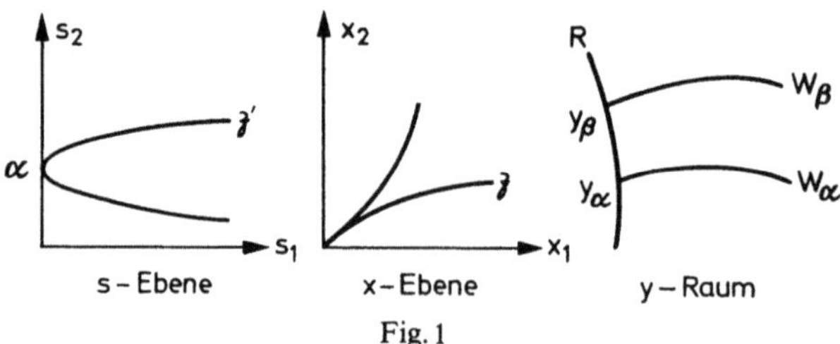

Fig. 1

kehrtransformation $x \to s$ in Zweige $\mathfrak{z}'$ über, die ihren Anfangspunkt je in einem bestimmten Punkt α auf der Achse $s_1 = 0$ haben. Statt auf die Zweige $\mathfrak{z}$ die Transformation $x \to y$ anzuwenden, kann man ebenso gut auf die $\mathfrak{z}'$ die Transformation $s \to y$ anwenden: Das Ergebnis y ist dasselbe. Also bilden die Punkte y, die man aus jenen Zweigen $\mathfrak{z}$ erhält, genau die Bildvarietät W_α. Daher ist V_A die Vereinigung der rationalen Kurve R mit den endlich vielen linear zusammenhängenden Varietäten W_α, die zu den schwierigen Tangentenrichtungen gehören.

Wir zeigen nun, daß R mit jeder Varietät W_α einen Punkt gemeinsam hat. Daraus folgt dann ohne weiteres, daß V_A linear zusammenhängend ist.

Die Achse $s_1 = 0$, ergänzt durch einen unendlich fernen Punkt, heiße g. In der Abbildung $s \to x$ entsprechen den einzelnen Punkten von g Tangentenrichtungen $a_1 : a_2$ im Punkte A. Diesen entsprechen in der Abbildung $x \to y$ wiederum Punkte von R. Also bildet die Abbildung $s \to y$ die Gerade g auf die rationale Kurve R ab.

Nähert man sich nun auf der Geraden g einem Punkte α, so erhält man in der Abbildung $s \to y$ einen Grenzpunkt y_α, der sowohl in R als auch in W_α liegt. Also hat R mit jedem W_α einen Punkt y_α gemeinsam. Da R und alle W_α linear zusammenhängend sind, ist ihre Vereinigung V_A linear zusammenhängend.

§ 5. Beweis des linearen Zusammenhangssatzes für $U = P^m$

Eine rationale Abbildung von P^m in P^n sei durch

$$\beta \eta_k = F_k(\xi_0, \ldots, \xi_n) \tag{38}$$

gegeben. Der Punkt A in P^m habe die Koordinaten $(1, 0, \ldots, 0)$. Wir wollen beweisen, daß die Bildvarietät V_A linear zusammenhängend ist, d. h., daß je zwei Punkte y' und y'' von V_A sich innerhalb V_A durch eine Kette von rationalen Kurven verbinden lassen.

Es gibt nach § 2 in P^m Zweige $\mathfrak{z}'$ und $\mathfrak{z}''$ mit Anfangspunkt A, denen in der Abbildung Zweige in P^n entsprechen, deren Anfangspunkte gerade y' und y'' sind. In der Reihenentwicklung des Zweiges $\mathfrak{z}'$ kann man alle $X_i(\tau)$ durch $X_0(\tau)$ dividieren, so daß nachher $X_0(\tau) = 1$ wird. Schreibt man σ statt τ und x_i'

statt X_i, so erhält man für den Zweig $\mathfrak{z}'$ die Reihen

$$x_i'(\sigma) = a_i\sigma + b_i\sigma^2 + \cdots \quad (i = 1, \ldots, m)\,.$$

Nach § 2 kann man die Reihenentwicklung abbrechen, ohne daß der Bildpunkt y' sich ändert. Man kann also

$$x_i'(\sigma) = a_i\sigma + b_i\sigma^2 + \cdots + e_i\sigma^q \tag{39}$$

annehmen. Ebenso für $\mathfrak{z}''$:

$$x_i''(\tau) = g_i\tau + h_i\tau^2 + \cdots + k_i\tau^r\,. \tag{40}$$

Dabei sind σ und τ Unbestimmte.

Bildet man jetzt die Summe

$$x_i(\sigma, \tau) = x_i'(\sigma) + x_i''(\tau)\,, \tag{41}$$

so behält diese ihren Sinn, wenn für σ und τ irgendwelche Körperelemente s und t eingesetzt werden, und man erhält eine rationale Abbildung

$$(s, t) \to x$$

der affinen (s, t)-Ebene in den Raum P^m. Alle Punkte der affinen Ebene sind für diese Abbildung regulär. Der s-Achse ($t = 0$) entspricht in der Abbildung der Zweig $\mathfrak{z}'$, der t-Achse der Zweig $\mathfrak{z}''$.

Setzt man nun die Abbildung $(s, t) \to x$ mit der Abbildung $x \to y$ zusammen, so erhält man eine rationale Abbildung der affinen (s, t)-Ebene in den Raum P^n. Das vollständige Bild W_Q des Punktes $Q(0, 0)$ in dieser Abbildung erhält man, indem man alle Zweige $\mathfrak{z}$, die in Q ihren Anfangspunkt haben, zunächst durch $(s, t) \to x$ in P^m und dann durch $x \to y$ in P^n abbildet; die Anfangspunkte der so erhaltenen Zweige bilden W_Q. Offenbar ist W_Q in V_A enthalten. Nähert man sich dem Punkte Q auf der s-Achse, so erhält man als Bildpunkt im y-Raum den Punkt y'. Nähert man sich auf der t-Achse, so erhält man den Bildpunkt y''. Nach § 4 ist W_Q linear zusammenhängend, also kann man y' mit y'' innerhalb W_Q durch eine Kette von rationalen Kurven miteinander verbinden.

Damit ist bewiesen:

Satz 2. *Bei einer rationalen Abbildung von $U = P^m$ in P^n ist die Bildvarietät V_A eines jeden Punktes A linear zusammenhängend.*

Es ist klar, daß der Beweis auch dann gilt, wenn U der affine Raum A^m ist.

§ 6. Die Eindeutigkeit der Multiplizität im Fall $U = P^m$ oder $U = A^m$

Ein Normalproblem möge durch die Gleichungen

$$G_j(x, y) = 0 \tag{42}$$

definiert werden, wobei der Punkt x den projektiven Raum $U = P^m$ oder den affinen $U = A^m$ durchläuft. Für einen allgemeinen Punkt ξ dieses Raumes habe das Problem endlich viele Lösungen $\eta^{(1)}, \ldots, \eta^{(h)}$. Die Spezialisierung

$\xi \rightarrow x$ läßt sich zu einer Spezialisierung

$$(\xi, \eta^{(1)}, \ldots, \eta^{(h)}) \rightarrow (x, y^{(1)}, \ldots, y^{(h)}) \tag{43}$$

fortsetzen. Wir wollen beweisen:

Satz 3. *Wenn die Varietät V_x der zum Punkt x gehörigen Lösungen y in zwei fremde Teile M und N zerfällt, so ist die Anzahl der spezialisierten Lösungen $y^{(v)}$, die in M hineinfallen, eindeutig bestimmt.* Man kann diese Anzahl die *Multiplizität* von M nennen.

Zum Beweis bilde man mit Unbestimmten U_i das Produkt

$$P(U) = \gamma \prod_v (U_0 \eta_0^{(v)} + \cdots + U_n \eta_n^{(v)}), \tag{44}$$

wobei γ eine beliebige von Null verschiedene Konstante ist. Wenn die Koordinaten η normiert werden (etwa durch $\eta_s = 1$, wenn η_s die erste von Null verschiedene Koordinate ist), so ist das Produkt rechts in (44) eindeutig bestimmt und symmetrisch in den $\eta^{(1)}, \ldots, \eta^{(h)}$. Sind die $\eta^{(v)}$ separabel über $k(\xi)$, so ist das Produkt rational in den ξ. Sind die $\eta^{(v)}$ nicht separabel, so ist eine q-te Potenz des Produktes rational in den ξ, wobei $q = p^e$ eine Potenz der Charakteristik des Körpers k ist.

Im folgenden werden die Beweise nur für den separablen Fall durchgeführt werden. Die Modifikationen im inseparablen Fall sind trivial: Man braucht nur alle Formeln in eine geeignete p^e-te Potenz zu erheben. Wir nehmen also an, daß das Produkt $P(U)$ rational in den ξ ist. Durch geeignete Wahl von γ kann man sogar erreichen, daß $P(U)$ ganz rational in den ξ wird.

Die Koeffizienten der Form (44) seien $\zeta_0, \ldots, \zeta_N$. Sie sind die Koordinaten des Zykels ζ, der aus den Punkten $\eta^{(1)}, \ldots, \eta^{(h)}$, je einmal gezählt, besteht. Die ζ_i können auch als Koordinaten eines Bildpunktes ζ in einem projektiven Raum P^N aufgefaßt werden.

Vergleicht man die Koeffizienten der Potenzprodukte der U_j links und rechts in (44), so erhält man

$$\zeta_k = \gamma g_k(\eta^{(1)}, \ldots, \eta^{(h)}) . \tag{45}$$

Aus (45) folgt

$$\zeta_j g_k(\eta) - \zeta_k g_j(\eta) = 0 . \tag{46}$$

Aus (46) folgen umgekehrt (45) und (44). Die Gl. (46) drücken also aus, daß der Zykel ζ genau aus den Punkten $\eta^{(1)}, \ldots, \eta^{(h)}$ besteht.

Jede Spezialisierung (43) läßt sich zu einer Spezialisierung

$$(\xi, \eta^{(1)}, \ldots, \eta^{(h)}, \zeta) \rightarrow (x, y^{(1)}, \ldots, y^{(h)}, z) \tag{47}$$

fortsetzen. Dabei bleiben die Gl. (46) erhalten, also besteht der Zykel z genau aus den Punkten $y^{(1)}, \ldots, y^{(h)}$. Insbesondere folgt, daß

$$(\xi, \zeta) \rightarrow (x, z) \tag{48}$$

eine Spezialisierung ist. Umgekehrt läßt sich jede Spezialisierung (48) zu einer Spezialisierung (47) fortsetzen, also stellt jeder Punkt z, der durch die Spezialisierung (48) erhalten wird, einen Zykel dar, der aus Punkten $y^{(1)}, \ldots, y^{(h)}$ besteht. Mit anderen Worten:

Das Problem, alle Spezialisierungen (43) zu erhalten, ist genau dasselbe wie das, alle Spezialisierungen (48) zu erhalten.

Wir haben gesehen, daß die Koeffizienten des Produktes (44), also die Koordinaten ζ_k des Zykels ζ, rationale Funktionen der ξ_i sind. Es existiert also eine rationale Abbildung

$$\beta\zeta_k = F_k(\xi). \tag{49}$$

Die Paare (x, z) dieser Abbildung sind genau die Spezialisierungen des allgemeinen Paares (ξ, η). So wird also das Problem, bei gegebenem x alle spezialisierten Lösungssysteme $y^{(1)}, \ldots, y^{(h)}$ zu erhalten, zurückgeführt auf das Problem, zu einem gegebenen Punkt x alle Bildpunkte z in der rationalen Abbildung (49) zu erhalten. Nach § 5 ist die Gesamtheit W_x dieser Bildpunkte z eine linear zusammenhängende Varietät.

Die Voraussetzung des Satzes 3 besagt, daß die Varietät V_x der zum gegebenen Punkt x gehörigen Lösungen y in zwei disjunkte Teile M und N zerfällt. Gesetzt nun, die Anzahl der Punkte $y^{(\nu)}$, die in M hineinfällt, wäre nicht eindeutig bestimmt. Dann würde es eine Spezialisierung $y^{(1)}, \ldots, y^{(h)}$ geben, bei der nur r Punkte $y^{(\nu)}$ in M hineinfallen und die übrigen $h - r$ in N, während bei einer anderen Spezialisierung mindestens $r + 1$ Punkte $y^{(\nu)}$ in M hineinfallen.

Ich definiere eine Korrespondenz K zwischen $y^{(1)}, \ldots, y^{(h)}$ und z durch die folgenden Gleichungen:

a) die Gleichungen $H_j(x, y^{(1)}, \ldots, y^{(h)}) = 0$, die ausdrücken, daß

$$(\xi, \eta^{(1)}, \ldots, \eta^{(h)}) \to (x, y^{(1)}, \ldots, y^{(h)})$$

eine Spezialisierung ist. In diesen Gleichungen ist x fest; es sind also Gleichungen in $y^{(1)}, \ldots, y^{(h)}$ allein. Unter diesen Gleichungen kommen auch die Gl. (42) mit $y = y^{(1)}$ oder ... oder $y = y^{(h)}$ vor;

b) die analog zu (46) gebildeten Gleichungen

$$z_j g_k(y^{(1)}, \ldots, y^{(h)}) - z_k g_j(y^{(1)}, \ldots, y^{(h)}) = 0, \tag{50}$$

die ausdrücken, daß der Zykel z genau aus den Punkten $y^{(1)}, \ldots, y^{(h)}$ besteht.

Die Korrespondenz K ist die Vereinigung von zwei fremden Teilen K_1 und K_2. Dabei besteht K_1 aus den Systemen $(y^{(1)}, \ldots, y^{(h)}, z)$, bei denen mindestens $r + 1$ Punkte $y^{(\nu)}$ in M liegen, und K_2 aus den Systemen, bei denen mindestens $n - r$ Punkte $y^{(\nu)}$ in N liegen. Daß K_1 und K_2 fremd sind, ist klar. Wir haben noch zu zeigen, daß K_1 und K_2 Korrespondenzen sind, d. h. daß sie durch algebraische Gleichungen definiert werden können.

Wir wählen eine Kombination κ von $r + 1$ Punkten $y^{(\nu)}$ und nehmen zu den Gleichungen a) und b) noch die Gleichungen hinzu, die besagen, daß diese $r + 1$ Punkte zu M gehören. So erhält man zu jeder Kombination κ eine Korrespondenz $K_1^{(\kappa)}$. Die Vereinigung dieser $K_1^{(\kappa)}$ ist K_1. Analog schließt man für K_2. Also hat man in der Tat eine Zerlegung von K in zwei fremde Teilkorrespondenzen K_1 und K_2.

K, K_1 und K_2 sind Korrespondenzen zwischen dem h-fach projektiven Raum $P^{n,n,\cdots,n}$ der Punktsysteme $(y^{(1)}, \ldots, y^{(h)})$ und dem projektiven Raum P^N der Punkte z. Jede solche Korrespondenz hat eine Urvarietät in $P^{n,\cdots,n}$ und eine Bildvarietät in P^N. Die zu K bzw. K_1 oder K_2 gehörige Bildvarietät W bzw. W_1 oder W_2 besteht aus allen Punkten z von P^N, zu denen Systeme $(y^{(1)}, \ldots, y^{(h)}, z)$ von K bzw. K_1 oder K_2 gehören. Da K die Vereinigung von K_1 und K_2 ist, ist W die Vereinigung von W_1 und W_2. Diese sind zueinander fremd, denn ein Zykel z aus nur h Punkten kann unmöglich $r+1$ Punkte in M und $h-r$ Punkte in N enthalten. Also zerfällt W in fremde Teile W_1 und W_2.

Andererseits besteht W genau aus den Punkten z, die man durch die Spezialisierungen (48) erhält. Jede solche Spezialisierung läßt sich nämlich, wie oben bemerkt wurde, zu einer Spezialisierung (47) fortsetzen. Also ist W genau die vorhin mit W_x bezeichnete Bildvarietät des Punktes x in der rationalen Abbildung (49). Diese ist aber zusammenhängend, also kann sie nicht in W_1 und W_2 zerfallen.

Damit ist Satz 3 bewiesen.

Wenn V_x in maximal zusammenhängende Teile $M_1, \ldots, M_t$ zerfällt, so hat jeder dieser Teile M_i nach Satz 3 eine bestimmte Multiplizität μ_i. Die Summe dieser Multiplizitäten ist gleich der Anzahl der Lösungen im allgemeinen Fall:

$$h = \mu_1 + \cdots + \mu_t.$$

Diese Aussage nennt man „das Prinzip der Erhaltung der Anzahl".

Als Spezialfall erhält man aus Satz 3:

Satz 4. *Die Multiplizität einer isolierten Lösung y ist eindeutig bestimmt.*

Wenn das Normalproblem insbesondere darin besteht, daß eine d-dimensionale irreduzible Varietät V mit d Hyperebenen geschnitten wird, die als Spezialisierungen von allgemeinen Hyperebenen aufgefaßt werden, so erhält jeder isolierte Schnittpunkt y nach Satz 4 eine eindeutig bestimmte Multiplizität. Ist ein Punkt A gegeben, so kann man d möglichst allgemeine Hyperebenen durch A legen und sie mit U schneiden. Ist dann die Multiplizität des Schnittpunktes A gleich Eins, so heißt A ein *einfacher Punkt* von V.

§ 7. Schnittmultiplizitäten

Wenn zwei irreduzible Varietäten V^r und V^{n-r} auf der irreduziblen Varietät W^n einen isolierten Schnittpunkt A haben, der ein einfacher Punkt von W^n ist, so kann man die Schnittmultiplizität von A als Schnittpunkt von V^r und V^{n-r} nach Severi und van der Waerden [8] oder nach Weil [2] definieren. Die erstere Definition wurde ursprünglich nur für den Fall gegeben, daß V^r und V^{n-r} nur endlich viele Schnittpunkte haben, aber auf Grund des allgemeinen Begriffes der Spezialisierungsmultiplizität, der hier in § 6 erklärt wurde, kann die Definition ohne weiteres auf den Fall eines isolierten Schnittpunktes ausgedehnt werden. Die Definition von Weil war von Anfang an auf den allgemeinen Fall eines isolierten Schnittpunktes zugeschnitten.

Severi und van der Waerden gehen in zwei Schritten vor. Zunächst wird angenommen, daß V^r und V^{n-r} im projektiven Raum P^n liegen. Dann wird V^r durch eine projektive Transformation mit unbestimmten Koeffizienten τ_{ik} in eine allgemeine Lage zu V^{n-r} gebracht. Sind

$$F_v(x_0, \ldots, x_n) = 0$$

die Gleichungen von V^r, so sind

$$F_v(\Sigma \, \tau_{ik} x_k) = 0$$

die Gleichungen der transformierten Varietät $(V^r)^T$. Diese schneidet V^{n-r} in endlich vielen Punkten. Wird die Matrix $T = (\tau_{ik})$ zur Einheitsmatrix spezialisiert, so erhält der isolierte Schnittpunkt P eine bestimmte Multiplizität. Diese wird als Schnittmultiplizität des Schnittpunktes A definiert.

Zweitens seien V^r und V^{n-r} Varietäten auf W^n, wobei W^n in einen projektiven Raum P^N eingebettet ist. Verbindet man die Punkte von V^r durch Geraden mit den Punkten eines allgemeinen linearen Teilraumes L^{N-n-1} von P_N, so erhält man einen projizierenden Kegel K^{N+r-n}. Die Multiplizität von P als Schnittpunkt von K^{N+r-n} mit V^{n-r} wird nun als Schnittmultiplizität von A als Schnittpunkt von V^r und V^{n-r} definiert.

Weil geht auch in zwei Schritten vor. Zunächst definiert er die Schnittmultiplizität eines isolierten Schnittpunktes von V^r mit einem linearen Teilraum L^{n-r} im affinen Raum A^n, indem er L^{n-r} als Spezialisierung eines allgemeinen linearen Teilraumes auffaßt. Ist nun zweitens ein isolierter Schnittpunkt A von V^r und V^{n-r} in W^n gegeben, wobei W_n im affinen Raum A^N liegt und A ein einfacher Punkt von W^n ist, so bildet Weil das Produkt $V^r \times V^{n-r}$ in $A^N \times A^N$. Der Punkt $A \times A$ ist ein isolierter Schnittpunkt von $V^r \times V^{n-r}$ mit der Diagonale Δ von $A^N \times A^N$. Diese Diagonale ist ein linearer Raum in $A^N \times A^N$, gegeben durch die Gleichungen

$$X_k - X_k' = 0 \qquad (k = 1, \ldots, N) \,.$$

Aus diesen N Gleichungen bildet Weil n Linearkombinationen

$$F_i(X - X') = 0 \qquad (i = 1, \ldots, n) \,, \tag{51}$$

wobei die Linearformen F_i so gewählt werden, daß der lineare Teilraum L^{N-n} von A^N, der durch die Gleichungen $F_i(X - P) = 0$ definiert wird, mit dem Tangentialraum von W^n im Punkte A nur A gemeinsam hat. Ein solches System von n Linearformen F_i nennt Weil "a uniformizing set of linear forms for U^n at A". Sind die Linearformen F_i so gewählt, so definieren die Gl. (51) einen linearen Teilraum in $A^N \times A^N$, der mit $V^r \times V^{n-r}$ den isolierten Schnittpunkt $A \times A$ hat. Die Multiplizität dieses Schnittpunktes wird als Schnittmultiplizität von A als Schnittpunkt von V^r mit V^{n-r} definiert.

Leung [4] hat bewiesen, daß die Definition von Severi und van der Waerden mit der von Weil äquivalent ist.

Eine andere Definition der Schnittmultiplizität wurde von Chevalley [9] gegeben. Samuel [10] hat bewiesen, daß die Definition von Chevalley mit der von Weil äquivalent ist.

§ 8. Der lineare Zusammenhangssatz für einfache Punkte auf beliebigen Varietäten U

Durch eine Parallelprojektion kann der Satz vom linearen Zusammenhang für einfache Punkte x auf U auf den Spezialfall $U = P^m$ zurückgeführt werden. Die Methode ist dieselbe, die in einer früheren Arbeit [11] zum Beweis der Eindeutigkeit der Spezialisierungsmultiplizitäten angewandt wurde.

Der Punkt x habe die homogenen Koordinaten

$$x_0 = 1, x_1 = 0, \ldots, x_N = 0 \, .$$

Für einen allgemeinen Punkt ξ von U gilt $\xi_0 \neq 0$; man kann also $\xi_0 = 1$ setzen und inhomogene Koordinaten $\xi_1, \ldots, \xi_N$ einführen. Eine rationale Abbildung $U \to V$ sei durch

$$\beta \eta_k = F_k(\xi_1, \ldots, \xi_N) \tag{52}$$

gegeben. Wir haben zu beweisen, daß die Bildvarietät V_x des einfachen Punktes x linear zusammenhängend ist.

Der Grundkörper k sei vollkommen. Adjungiert man N^2 Unbestimmte u_{ik}, so erhält man einen neuen Grundkörper

$$K = k(u_{ik}) \, .$$

Durch die lineare Koordinatentransformation

$$\xi_i^* = \Sigma \, u_{ik} \xi_k$$

erreicht man, daß $\xi_1^*, \ldots, \xi_m^*$ algebraisch unabhängig über K sind und alle ξ_i^* separable Funktionen von $\xi_1^*, \ldots, \xi_m^*$ werden. Für den Beweis siehe etwa [12], § 6. Jetzt kann man den ganzen affinen Raum A^N und insbesondere die Varietät U auf den Teilraum A^m, der von den ersten m Koordinatenachsen aufgespannt wird, projizieren, indem man die ersten m Koordinaten

$$\tau_1 = \xi_1^*, \ldots, \tau_m = \xi_m^*$$

beibehält und die übrigen ξ_i^* durch Null ersetzt oder besser ganz wegläßt. Die Projektion ist also durch die Formeln

$$\tau_i = \sum_1^N u_{ik} \xi_k \qquad (i = 1, \ldots, m) \tag{53}$$

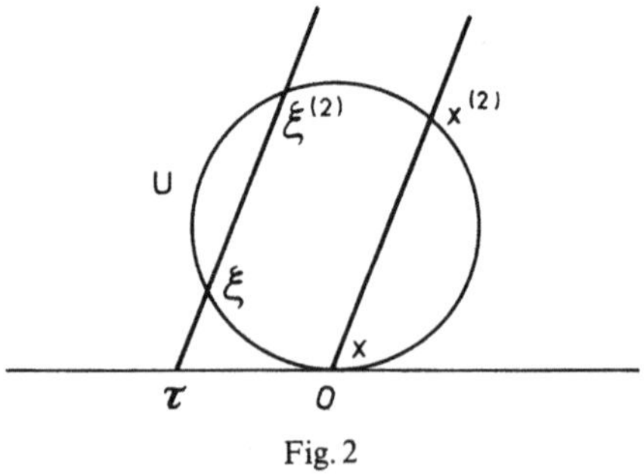

Fig. 2

464

definiert. Die τ_i sind algebraisch unabhängig über K, können also auch als unabhängige Unbestimmte aufgefaßt werden. Die ξ_k sind lineare Funktionen von den ξ_i^*, also separable algebraische Funktionen von $\tau_1, \ldots, \tau_m$.

Jetzt suchen wir bei gegebenen τ diejenigen Punkte X von U, die bei der Projektion den gleichen Punkt τ ergeben. Sie müssen die Gleichungen von U erfüllen und außerdem die linearen Gleichungen

$$\Sigma\, u_{ik} X_k = \tau_i \tag{54}$$

oder homogen geschrieben

$$\Sigma\, u_{ik} X_k - \tau_i X_0 = 0 \,. \tag{55}$$

Unendlich ferne Punkte, die (55) erfüllen und auf U liegen, gibt es nicht, denn der $(m-1)$-dimensionale Durchschnitt von U mit der unendlich fernen Hyperebene $X_0 = 0$ hat mit den m allgemeinen Hyperebenen $\Sigma\, u_{ik} X_k = 0$ keine Punkte gemeinsam. Wir können also in (55) wieder $X_0 = 1$ setzen, und zur inhomogenen Gleichungsform (54) zurückkehren.

Jede Lösung X von (54), die in U liegt, hat den maximalen Transzendenzgrad m über K, denn im Körper $K(X)$ liegen nach (54) die m unabhängigen Körperelemente $\tau_1, \ldots, \tau_m$. Also sind die X allgemeine Punkte von U über K. Es gibt also für jeden Punkt X einen Isomorphismus über K, der ξ in X überführt. Wegen (53) und (54) läßt dieser Isomorphismus die Elemente τ_i fest, also sind ξ und X konjugiert über $K(\tau)$. Es gibt also nur endlich viele Punkte X, nämlich die zu ξ konjugierten Punkte

$$\xi = \xi^{(1)}, \xi^{(2)}, \ldots, \xi^{(g)} \,.$$

Die Anzahl g ist der Grad von U: die Anzahl der Schnittpunkte von U mit m allgemeinen Hyperebenen (55). Alle symmetrischen Funktionen von $\xi^{(1)}, \ldots, \xi^{(g)}$ sind rational in den τ_i.

Bei der Abbildung (52) haben die Punkte $\xi = \xi^{(1)}, \ldots, \xi^{(g)}$ bestimmte Bildpunkte $\eta = \eta^{(1)}, \ldots, \eta^{(g)}$. Alle symmetrischen Funktionen von $\eta^{(1)}, \ldots, \eta^{(g)}$, insbesondere die Koordinaten ζ_k des aus diesen Punkten gebildeten Zykels ζ, sind rationale Funktionen von $\tau_1, \ldots, \tau_m$:

$$\gamma \zeta_k = G_k(\tau_1, \ldots, \tau_m) \,. \tag{56}$$

Um alle Bildpunkte y des einfachen Punktes $x = (0, \ldots, 0)$ in der Abbildung (52) zu erhalten, muß man die Spezialisierung $\xi \to x$ in allen möglichen Weisen zu Spezialisierungen

$$(\xi, \eta) \to (x, y) \tag{57}$$

fortsetzen. Bei der Spezialisierung $\xi \to x$ gehen die durch (53) definierten τ_i selbstverständlich in Null über. Man kann also nur so fortsetzen:

$$(\tau, \xi, \eta) \to (O, x, y) \,. \tag{58}$$

Dabei ist O der Punkt mit Koordinaten Null im τ-Raum.

Sodann kann man (58) zu

$$(\tau, \xi, \xi^{(2)}, \ldots, \xi^{(g)}; \eta, \eta^{(2)}, \ldots, \eta^{(g)}, \zeta)$$
$$\to (O, x, x^{(2)}, \ldots, x^{(g)}; y, y^{(2)}, \ldots, y^{(g)}, z) \tag{59}$$

fortsetzen. Die Gleichungen, die ausdrücken, daß der Zykel ζ aus den Punkten $\eta = \eta^{(1)}, \ldots, \eta^{(g)}$ besteht, bleiben bei der Spezialisierung erhalten, also besteht der Zykel z aus den Punkten $y = y^{(1)}, \ldots, y^{(g)}$. Der Punkt $\eta^{(\nu)}$ ist Bildpunkt von $\xi^{(\nu)}$ in der Abbildung (52), also ist $y^{(\nu)}$ Bildpunkt von $x^{(\nu)}$ in derselben Abbildung. Ferner ist ζ Bildpunkt von τ in der Abbildung (56), also ist z Bildpunkt von O in eben dieser Abbildung. Die Punkte $x^{(1)}, \ldots, x^{(g)}$ entstehen durch Spezialisierung aus $\xi^{(1)}, \ldots, \xi^{(g)}$, also sind sie die Schnittpunkte von U mit dem linearen Raum

$$\Sigma u_{ik} X_k = 0 \tag{60}$$

mit den richtigen Spezialisierungsmultiplizitäten. Da x ein einfacher Punkt von U ist, kommt x unter den Schnittpunkten $x^{(1)}, \ldots, x^{(g)}$ nur einmal vor, d. h. $x^{(2)}, \ldots, x^{(g)}$ sind von $x^{(1)} = x$ verschieden. Nun wird behauptet:

Die Punkte $x^{(2)}, \ldots, x^{(g)}$ sind für die Abbildung (52) regulär.

Beweis. Die nicht regulären Punkte bilden auf U eine Teilvarietät von höchstens der Dimension $(m-1)$. Schneidet man diese nacheinander mit den m durch x gehenden Hyperebenen (60), deren Koeffizienten u_{ik} unabhängige Unbestimmte sind, so erniedrigt sich bei den ersten $m-1$ Schritten die Dimension jedesmal um 1. Nach $m-1$ Schritten erhält man endlich viele Schnittpunkte $x, x', x'', \ldots$. Nimmt man noch die m-te Hyperebene hinzu, so bleibt nur noch der Punkt x übrig. Die von x verschiedenen Punkte $x^{(2)}, \ldots, x^{(g)}$ sind also regulär und ihre Bildpunkte $y^{(2)}, \ldots, y^{(g)}$ eindeutig bestimmt. Nur der eine Bildpunkt y kann variieren, und zwar durchläuft er genau die Bildvarietät V_x des Punktes x. Daraus folgt unmittelbar:

Zu jedem Punkt y von V_x gehört ein bestimmter Zykel

$$z = y + y^{(2)} + \cdots + y^{(g)}, \tag{61}$$

so daß der zugehörige Punkt z in der Bildvarietät W_O des Punktes O in der rationalen Abbildung (56) liegt.

Das Umgekehrte gilt aber auch:

Zu jedem Punkt z von W_O gehört ein Zykel z, der genau die Form (61) hat, wobei y in V_x liegt.

Beweis. Jede Spezialisierung $(\tau, \zeta) \to (O, z)$ läßt sich zu einer Spezialisierung (59) fortsetzen. Dabei ist es möglich, daß nicht $\xi = \xi^{(1)}$, sondern ein anderes $\xi^{(\nu)}$ zu x spezialisiert wird. Da aber die $\xi^{(\nu)}$ alle zu ξ konjugiert sind, kann man durch einen Automorphismus des Körpers

$$K(\xi^{(1)}, \ldots, \xi^{(n)})$$

$\xi^{(\nu)}$ in ξ überführen und so erreichen, daß ξ bei der Spezialisierung in x und die übrigen $\xi^{(\nu)}$ in irgendeiner Reihenfolge in $x^{(2)}, \ldots, x^{(g)}$ übergehen. Dann gehen auch $\eta^{(2)}, \ldots, \eta^{(g)}$ in die eindeutig bestimmten Punkte $y^{(2)}, \ldots, y^{(g)}$ über, und der Zykel z hat die Form (61).

Durch (61) ist also eine eineindeutige Abbildung zwischen V_x und W_O definiert: Jedem y in V_x entspricht ein z in W_O und umgekehrt.

Diese Abbildung ist eine Projektivität.

Beweis. Die Koordinaten z_i des Zykels z sind nach § 6 die Koeffizienten des Produktes

$$P(U) = \gamma(\Sigma \, U_k y_k) \cdot \prod_2^g (\Sigma \, U_k y_k^{(\nu)}),$$

also sind die z_i lineare homogene Funktionen von den y_k. Die Abbildung $y \to z$ ist also eine Projektivität.

Nun ist W_O linear zusammenhängend, also ist auch V_x linear zusammenhängend, was zu beweisen war.

Literatur

1. Waerden, B. L. van der: Der Multiplizitätsbegriff der algebraischen Geometrie. Math. Ann. **97**, 756 (1927).
2. Weil, A.: Foundations of algebraic geometry. Amer. Math. Soc. Colloquium publications, Vol. 29. First edition, 1946. Second edition, 1962.
3. Northcott, D.: Specializations over a local domain. Proc. London Math. Soc. **1** (3) (1951).
4. Leung, K.-T.: Die Multiplizitäten in der algebraischen Geometrie. Math. Ann. **135**, 170 (1958).
5. Zariski, O.: Theory and applications of holomorphic functions on algebraic varieties. Memoirs Amer. Math. Soc. **5** (1951).
6. Chow, W.-L.: On the connectedness theorem in algebraic geometry. Amer. J. of Math. **81**, 1033 (1959).
7. Waerden, B. L. van der: Infinitely near points. Proceedings Akad. Amsterdam **53**, 401 (1950).
8. — Zur algebraischen Geometrie 14. Math. Ann. **115**, 621 (1938).
9. Chevalley, C.: Intersection of algebraic and algebroid varieties. Trans. Amer. Math. Soc. **57**, 1 (1945).
10. Samuel, P.: La notion de multiplicité. Journal de Math. **30**, 159 (1951).
11. Waerden, B. L. van der: Zur algebraischen Geometrie 6, § 3. Math. Ann. **110**, 144 (1934).
12. — Verallgemeinerung des Bézoutschen Theorems. Math. Ann. **99**, 497 (1928).

Professor B. L. van der Waerden
CH-8006 Zürich/Schweiz, Bionstr. 18

(Eingegangen am 20. August 1970)

Berichtigung

zu der Arbeit von Bartel L. van der Waerden, „Eine Verallgemeinerung des Bézoutschen Theorems", Math. Annalen 99 (1928), S. 497–541.

Fräulein Emmy Noether machte mich in liebenswürdiger Weise auf folgenden Fehler im Schlußparagraphen (§ 12) der oben genannten Arbeit aufmerksam. Nachdem der Hauptsatz bewiesen war, sollte im Fall $r = 1$ die in § 10 bewiesene Ungleichung $\chi \gtrless m$ (χ = charakteristische Funktion des Ideals $(\mathfrak{p}, \mathfrak{p}')_Y$; m = Multiplizität des Schnittpunktes Y) zu $\chi = m$ verschärft werden. Die Ungleichung wird zu dem Zweck fälschlich als $\chi \leqq m$ zitiert und dann wird richtig $\Sigma \chi \geqq \Sigma m$ bewiesen, was natürlich ungenügend ist. Um den Beweis zu vervollständigen, hat man $\Sigma \chi = \Sigma m$ zu beweisen. Verfolgt man den Beweisgang in § 12, so sieht man, daß an zwei Stellen nur gezeigt wird, daß gewisse Summanden positiv sind, während für den vollständigen Beweis erforderlich ist, daß diese Summanden $= 1$ sind, und zwar handelt es sich beide Male um Summanden der Gestalt $\chi(\mathfrak{p}, \mathfrak{p}')_Y$, wo das eine Mal $\mathfrak{p}$ das zu einem allgemeinen linearen $(n-1)$-Raum gehörige und $\mathfrak{p}'$ das zu einer eindimensionalen Mannigfaltigkeit M' gehörige Primideal ist, während das andere Mal umgekehrt $\mathfrak{p}'$ zu einem allgemeinen linearen eindimensionalen Raum und $\mathfrak{p}$ zu einer $(n-1)$-dimensionalen Mannigfaltigkeit M gehört. Also ist alles zurückzuführen auf die folgende Behauptung: Wenn $\mathfrak{p}$ das zu einer r-dimensionalen Mannigfaltigkeit gehörige homogene Primideal und $\mathfrak{p}' = (l_1, \ldots, l_r)$ das zu einem allgemeinen linearen $(n-r)$-Raum gehörige homogene Primideal ist ($l_1, \ldots, l_r$ allgemeine Linearformen), so hat das Ideal $(\mathfrak{p}, \mathfrak{p}')$ als Primärkomponenten lauter Primideale, die zu je einem Schnittpunkt Y gehören. Diese Behauptung nun folgt unmittelbar aus Satz 27, 3 (§ 6).

Publikationen von B. L. von der Waerden
bis Ende 1982

Zeitschriftenbeiträge

1. Determinanten aus Formenkoeffizienten, Proceedings Royal Academy Amsterdam, *25* (1922), p. 354–358.
2. Konkomitanten von ternären quadratischen Formen, Proceedings Royal Academy Amsterdam, *26* (1923).
3. Über die Fundamentalen Identitäten der Invariantentheorie, Math. Ann. *95* (1926), p. 706–735.
*4. Zur Nullstellentheorie der Polynomideale, Math. Ann. *96* (1926), p. 183–208.
5. Ein algebraisches Kriterium für die Lösbarkeit von homogenen Gleichungen, Proceedings Royal Academy Amsterdam, *28* (1926).
*6. Der Multiplizitätsbegriff der algebraischen Geometrie, Math. Ann. *97* (1927), p. 756–774.
7. Ein Satz über Klasseneinteilungen von endlichen Mengen, Abh. Math. Sem. Univ. Hamb., *5* (1927), p. 185–188.
8. Neue Begründung der Eliminations- und Resultantentheorie, Nieuw Arch. Wiskunde, *15* (1927), p. 2–20.
9. Beweis einer Vermutung von Baudet, Nieuw Arch. Wiskunde, *15* (1927), p. 212–216.
10. Ein logarithmenfreier Beweis des Dirichletschen Einheitensatzes, Abh. Math. Sem. Univ. Hamb., *6* (1928), p. 259–262.
11. (mit D. Van Dantzig) Über metrisch homogene Räume, Abh. Math. Sem. Univ. Hamb., *6* (1928), p. 367–376.
12. Die Alternative bei nichtlinearen Gleichungen, Nachr. Ges. Wiss. Göttingen (1928), p. 1–11.
13. On Hilbert's Function, series of composition for ideals and the Theorem of Bezout, Proceedings Royal Academy Amsterdam, *31* (1928).
*14. Eine Verallgemeinerung des Bézoutschen Theorems, Math. Ann. *99* (1928), p. 497–501.
15. Die Automorphismen der projektiven Gruppen (mit O. Schreier), Abh. Math. Sem. Univ. Hamb., *6* (1928), p. 303–322.
16. De strijd om de abstraktie, Antrittsrede, P. Noordhoff, Groningen 1928.
17. Spinoranalyse, Nachr. Ges. Wiss. Göttingen 1929, p. 100–109.
18. Zur Produktzerlegung der Ideale in ganz-abgeschlossenen Ringen, Math. Ann. *101* (1929), p. 293–308.
19. Zur Idealtheorie der ganz-abgeschlossenen Ringe, Math. Ann. *101* (1929), p. 309–311.

Mit * markierte Beiträge sind in diesem Buch abgedruckt.

*20. Topologische Begründung des Kalküls der abzählenden Geometrie, Math. Ann. *102* (1929), p. 337–362.

21. Ein einfaches Beispiel einer nicht-differenzierbaren stetigen Funktion. Math. Z., *32* (1930), p. 474–475.

22. Der Zusammenhang zwischen den Darstellungen der symmetrischen und der linearen Gruppe, Math. Ann. *104* (1930), p. 92–95.

23. Kombinatorische Topologie, Jahresbericht Deutsche Math.-Ver. *39* (1930), p. 121–139.

24. Generalization of a theorem of Kronecker, Bull. Am. Math. Soc. *37* (1931), p. 427–428.

*25. Zur Begründung des Restsatzes mit dem Noetherschen Fundamentalsatz, Math. Ann. *104* (1931), p. 472–475.

26. (mit F. Levi). Über eine besondere Klasse von Gruppen, Abh. Math. Sem. Univ. Hamb., *9* (1932), p. 154–158.

27. Stetigkeitssätze für halbeinfache Liesche Gruppen, Math. Z. *36* (1933), p. 780–786.

28. Die Klassifikation der einfachen Lieschen Gruppen, Math. Z. *37* (1933), p. 446–462.

29. Die Seltenheit der Gleichungen mit Affekt, Math. Ann. *109* (1933), p. 13–16.

30. (mit L. Infeld). Die Wellengleichung des Elektrons in der allgemeinen Relativitätstheorie, Preuss. Akad. Wiss. Berlin 1933, p. 3–25.

*31. Zur algebraischen Geometrie, I. Gradbestimmung von Schnittmannigfaltigkeiten einer beliebigen Mannigfaltigkeit mit Hyperflächen, Math. Ann. *108* (1933), p. 113–125.

*32. Zur algebraischen Geometrie, II. Die geraden Linien auf den Hyperflächen des P_n, Math. Ann. *108* (1933), p. 253–259.

*33. Zur algebraischen Geometrie, III. Über irreduzible algebraische Mannigfaltigkeiten, Math. Ann. *108* (1933), p. 694–698.

*34. Zur algebraischen Geometrie, IV. Die Homologiezahlen der Quadriken und die Formeln von Halphen der Liniengeometrie, Math. Ann. *109* (1933), p. 7–12.

35. Noch eine Bemerkung zu der Arbeit „Zur Arithmetik der Polynome" von U. Wegner, in Math. Ann. *105*, p. 628–631, Math. Ann. *109* (1934), p. 679–680.

*36. Zur algebraischen Geometrie, V. Ein Kriterium für die Einfachheit von Schnittpunkten. Math. Ann. *110* (1934), p. 128–133.

*37. Zur algebraischen Geometrie, VI. Algebraische Korrespondenzen und rationale Abbildungen, Math. Ann. *110* (1934), p. 134–160.

38. Elementarer Beweis eines zahlentheoretischen Existenztheorems. J. Reine Angew. Math. *171* (1934), p. 1–3.

39. (mit L. J. Smid). Eine Axiomatik der Kreisgeometrie und der Laguerregeometrie, Math. Ann. *110* (1935), p. 753–776.

40. (mit H. Casimir). Algebraischer Beweis der vollständigen Reduzibilität der Darstellungen halbeinfacher Liescher Gruppen, Math. Ann. *111* (1935), p. 1–12.

41. Die Zerlegungs- und Trägheitsgruppe als Permutationsgruppen, Math. Ann. *111* (1935), p. 731–733.

42. Eine einfache Herleitung der Fourierschen Reihe, Prace Mat.-Fiz. *44* (1935), p. 1–5.

43. Nachruf auf Emmy Noether, Math. Ann. *111* (1935), p. 469–476.

*44. Zur algebraischen Geometrie, VII. Ein neuer Beweis des Restsatzes, Math. Ann. *111* (1935), p. 432–437.

45. Empirische Bestimmungen von Wahrscheinlichkeiten und physiologische Konzentrationsauswertung. Berichte Verh. sächs. Akad. Leipzig, *87* (1935), p. 353–364.

46. Reihenentwicklungen und Überschiebungen in der Invariantentheorie, insbesondere im quaternären Gebiet, Math. Ann. *113* (1936), p. 14–35.

*47. Zur algebraischen Geometrie, Berichtigung und Ergänzungen, Math. Ann. *113* (1936), p. 36–39.

*48. Zur algebraischen Geometrie, VIII. Der Grad der Grassmannschen Mannigfaltigkeit der linearen Räume S_m in S_n, Math. Ann. *113* (1936), p. 199–205.

49. Messung von Wahrscheinlichkeiten, insbesondere Mortalität von Krankheiten, Operationen usw. Berichte Verh. sächs. Akad. Leipzig, *88* (1936), p. 21–30.

50. Die Seltenheit der reduziblen Gleichungen und der Gleichungen mit Affekt, Monatshefte Math. Phys. *43* (1936), p. 133–147.

51. Richtige Auswertung von Erfolgsstatistiken, Klinische Wochenschrift, *15* (1936).

52. Eine einfache Herleitung der Stirlingschen Formel $n! \sim n^n e^{-n} \sqrt{2\,\pi\,n}$, Nieuw . Arch. Wiskunde, *18* (1936), p. 40–45.

*53. (mit Wei-Liang Chow). Zur algebraischen Geometrie, IX. Über zugeordnete Formen und algebraische Systeme von algebraischen Mannigfaltigkeiten, Math. Ann. *113* (1937), p. 692–704.

*54. Zur algebraischen Geometrie, X. Über lineare Scharen von reduziblen Mannigfaltigkeiten, Math. Ann. *113* (1937), p. 705–712.

*55. Zur algebraischen Geometrie, XI. Projektive und birationale Äquivalenz und Moduln von ebenen Kurven, Math. Ann. *114* (1937), p. 683–699.

56. Arithmetik und Rechentechnik der Ägypter, Berichte Verh. sächs. Akad. Leipzig, *89* (1937), p. 171–172.

*57. Zur algebraischen Geometrie, XII. Ein Satz über Korrespondenzen und die Dimension einer Schnittmannigfaltigkeit, Math. Ann. *115* (1938), p. 330–332.

*58. Zur algebraischen Geometrie, XIII. Vereinfachte Grundlagen der algebraischen Geometrie, Math. Ann. *115* (1938), p. 359–378.

*59. Zur algebraischen Geometrie, XIV. Schnittpunktzahlen von algebraischen Mannigfaltigkeiten, Math. Ann. *115* (1938), p. 619–642.

60. Die Entstehungsgeschichte der ägyptischen Bruchrechnung. Quellen und Studien Gesch. Math. *B 4* (1938), p. 359–382.

*61. Zur algebraischen Geometrie, XV. Lösung des Charakteristikenproblems für Kegelschnitte, Math. Ann. *115* (1938), p. 645–655.

62. Über die Bestimmung eines Dreiecks aus seinen Winkelhalbierenden, J. Reine Angew. Math. *179* (1938), p. 65–68.

63. Nachruf auf Otto Hölder, Math. Ann. *116* (1939), p. 157–165.

64. Vertrauensgrenzen für unbekannte Wahrscheinlichkeiten, Berichte Verh. sächs. Akad. Wiss. Leipzig, *91* (1939), p. 213–228.

65. Zenon und die Grundlagenkrise der griechischen Mathematik, Math. Ann. *117* (1940), p. 141–161.

66. Die Voraussage von Finsternissen bei den Babyloniern, Berichte Verh. sächs. Akad. Wiss. Leipzig, *92* (1940), p. 107–114.

67. Bericht über die Arbeit von H. Fitting, Beiträge zur Theorie der Gruppen endlicher Ordnung, J. Reine Angew. Math. *182* (1940), p. 215.

471

68. Biologische Konzentrationsauswertung, Berichte Verh. sächs. Akad. Wiss. Leipzig, *92* (1940), p. 41–44.
69. Die Astronomie der Pythagoreer und die Entstehung des geozentrischen Weltbildes, Himmelswelt, *14* (1940).
70. Zur babylonischen Planetenrechnung, Eudemus, *1* (1941), p. 23–48.
71. Zur pythagoreischen Algebra: Quadratwurzel und Kubikwurzel, Math. Ann. *118* (1941), p. 286–288.
72. Topologie und Uniformisierung der Riemannschen Flächen, Berichte sächs. Akad. Leipzig, *93* (1941).
*73. Die Bedeutung des Bewertungsbegriffs für die algebraische Geometrie. Jahresbericht Deutsche Math.-Ver. *52* (1942), p. 161–172.
74. Das χ^2-Kriterium in der math. Statistik, Berichte sächs. Akad. Leipzig, *95* (1942).
75. Die Berechnung der ersten und letzten Sichtbarkeit von Mond und Planeten, Berichte sächs. Akad. Leipzig, *94* (1943), p. 23–56.
76. Die Harmonielehre der Pythagoreer, Hermes, *78* (1943), p. 163–199.
77. (mit M. Gildemeister). Die Zulässigkeit des X^2-Kriteriums für kleine Versuchszahlen, Berichte sächs. Ak. Wiss. *95* (1943), p. 145.
78. Plaudereien zur babylonischen Astronomie, I. Die Himmelswelt 1943, Heft 10–12.
79. Plaudereien zur babylonischen Astronomie, II. Die Himmelswelt 1944, Heft 7–9.
80. Die Astronomie des Heraklides von Pontos. Berichte sächs. Akad. Wiss. *96* (1944), p. 47–56.
81. The foundation of the invariant theory of linear systems of curves on an algebraic surface. Proceedings Royal Academy Amsterdam, *49* (1946), p. 223–226.
*82. Divisorenklassen in algebraischen Funktionenkörpern, Comment. Math. Helv. *20* (1947), p. 68–109.
83. Birational Invariants of Algebraic Manifolds. Acta Salamantic. *2* (1947), p. 1–56.
84. Egyptian „Eternal Tables", Proceedings Royal Academy Amsterdam, *5* (1947), I p. 536–547, II p. 782–788.
85. Babylonian Astronomy, I. The Venus tablets of Ammisuduqa. Jaarbericht Ex Oriente Lux, *10* (1948), p. 414.
86. Over een bewering van Euclides (mit H. Freudenthal) Simon Stevin, *1* (1948), p. 115–121.
87. Rezension André Weil, Foundations of Algebraic Geometry. Nieuw Arch. Wiskunde 1948.
88. Free products of Groups, Am. J. Math. *70* (1948), p. 527–528.
89. The foundation of Algebraic Geometry. Courant Anniv. Vol. (1948), p. 437–449.
90. Die Arithmetik der Pythagoreer, Math. Ann. *120* (1948), I p. 127–153, II p. 676–700.
*91. Über einfache Punkte von algebraischen Mannigfaltigkeiten. Math. Z. *51* (1948), p. 497–501.
*92. Birationale Transformation von linearen Scharen auf algebraischen Mannigfaltigkeiten. Math. Z. *51* (1948), p. 502–523.
93. Babylonian Astronomy, II. The 36 Stars. J. Near Eastern Studies, *8* (1949), p. 6–26.

94. Pythagoras. De Gids, *112* (1949), p. 106–126 und 184–204.
95. Les variétés de chaînes sur une variété abstraite, Colloque de géom. algebr. Liège 1949, p. 79–85.
96. Le théorème de Bézout pour les hypersurfaces. Annali di Matematica (4) *30* (1949), p. 73–74.
97. Rez. Parker-Dubberstein, Bibliotheca Orientalis, *6* (1949), p. 17–18.
98. Über Landau's Beweis des Primzahlsatzes. Math. Z. *52* (1950), p. 649–653.
99. Dauer der Nacht und Leuchtzeit des Mondes in den Tafeln des Nabu-zuqup-GI·NA, Zeitschr. f. Assyriologie 1950, p. 291–312.
100. The Optimum Adjustment of Regulators (with P. Hazebroek) Trans. American Society of Mechanical Engineers, April 1950, p. 309–322.
101. Buchrezensionen: Reidemeister und Mugler, Gnomon *22* (1950), p. 61–65.
102. De stelling van Jordan-Brouwer, Centrumreeks *1* (1950), p. 80–84.
103. Over de Ruimte, Intreerede, P. Noordhoff, Groningen 1950.
104. Hoe moet wiskunde gegeven worden en wat moet er gegeven worden met het oog op de maatschappy? Faraday 1950, p. 10–18.
105. Infinitely near Points, Proceedings Royal Academy Amsterdam, *53* (1950), p. 401–410.
106. Les valuations en géométrie algébrique, Algèbre et Théorie des nombres, Paris 1950, p. 117–122.
107. Polygone mit maximalem Flächeninhalt, Elem. Math. *5* (1950), p. 121–125.
108. Babylonian Astronomy, III. The earliest astronomical computations, J. Near Eastern Studies, *10* (1951), p. 20–34.
109. Der Begriff Wahrscheinlichkeit, Studium Generale, *4* (1951), p. 65–68.
110. Hoeveel punten hebben op een bol plaats? Simon Stevin, *1* (1951), p. 193–195.
111. On the method of Saddle Points. Appl. sci. Research, *B 2* (1951), p. 33–45.
112. (mit K. Schütte). Auf welcher Kugel haben 5, 6, 7, 8 oder 9 Punkte mit Mindestabstand Eins Platz? Math. Ann. *123* (1951), p. 96–124.
113. (mit W. Habicht). Lagerung von Punkten auf der Kugel, Math. Ann. *123* (1951), p. 223–234.
114. Example d'un groupe avec deux générateurs, contenant un sous-groupe commutatif sans système fini de générateurs, Nieuw Arch. Wiskunde, *23* (1951), p. 190.
115. Die Astronomie der Pythagoreer, Verhandelingen Royal Acad. Amsterdam, Afd. Natuurkunde, *20* (1951), Nr. 1.
116. Das Große Jahr und die ewige Wiederkehr. Hermes, *80* (1952), p. 129–155.
117. Punkte auf der Kugel. Drei Zusätze. Math. Ann. 125 (1952), p. 213–222.
*118. Zur algebraischen Geometrie, 16. Vielfältigkeiten von abstrakten Ketten. Math. Ann. 125 (1952), p. 314–324.
119. Order tests for the two-sample problem and their power, Proceedings Royal Academy Amsterdam, A *55* (1952), p. 453–458.
120. Die Bewegung der Sonne nach griechischen und indischen Tafeln, Sitzungsber. Bayer. Akademie der Wiss. (Math.-naturw. Kl.) 1952, p. 219–232.
121. (mit K. Schütte). Das Problem der dreizehn Kugeln, Math. Ann. 125 (1953), p. 325–334.
122. Testing a Distribution Function, Proceedings Royal Academy Amsterdam, A *55* (1953), p. 453–456.
123. Ein neuer Test für das Problem der zwei Stichproben. Math. Ann. 126 (1953), p. 93–107.

124. Order Tests for the two-sample problem II und III, Proc. Ned. Akad. *56* (1953), p. 303–316.
125. Einfall und Überlegung in der Mathematik I + II + III. Elem. Math. *8* (1953), p. 121–129; *9* (1954), p. 1–9 und 49–56.
126. History of the Zodiac, Archiv f. Orientforschung, *16* (1953), p. 216–230.
127. Das Große Jahr des Orpheus, Hermes, *81* (1953), p. 481–484.
128. Bemerkungen zu den Handlichen Tafeln des Ptolemaios, Sitzungsber. Bayer. Akad. München 1953, p. 261–272.
129. Die Sichtbarkeit der Sterne in der Nähe des Horizontes, Vierteljschr. naturf. Ges. Zürich, *99* (1954), p. 20–39.
130. Zur Konstruktion des Resultantensystems für homogene Gleichungen, Arch. Math. *5* (1954), p. 371–375.
*131. Zur algebraischen Geometrie, 17. Lokale Dimension und Satz von Eckmann, Math. Ann. *128* (1954), p. 128–134.
*132. Zur algebraischen Geometrie, 18. Ketten in mehrfach-projektiven Räumen, Math. Ann. *128* (1954), p. 135–137.
133. Denken ohne Sprache. Symposium Thinking and Speaking, edited by G. Révész, Amsterdam 1954, p. 165–174.
134. Eine byzantinische Sonnentafel. Sitzungsber. Bayer. Akad. München 1954, p. 159–168.
135. Levensbericht van Hendrik de Vries. Jaarboek Kon. Ned. Akad. Amsterdam 1953–54.
136. Diophantische Gleichungen und Planetenperioden in der indischen Astronomie, Vierteljschr. naturf. Ges. Zürich, *100* (1955), p. 153–170.
137. Die Cohomologietheorie der Polyeder, Math. Ann. *130* (1955), p. 87–101.
138. Les mathématiques appliquées dans l'antiquité. L'enseignement math. *1* (1955), p. 44–55.
139. Berichtigung und Ergänzung zu 137, Math. Ann. *132* (1956), p. 130–133.
140. Tamil Astronomy, Centaurus, *4* (1956), p. 221–234.
141. Babylonische Planetenrechnung in Ägypten und Indien, Bibliotheca Orientalia, *13* (1956), p. 108–110.
142. Rez. O. Neugebauer, Astron. cun. texts, Z. D. Morgenländ. Ges. *106* (1956), p. 371.
143. The invariant theory of linear sets on an algebraic variety, Proceedings International Congress Amsterdam 1954, III, (1956), p. 542–544.
144. On the definition of rational equivalence of cycles on a variety, Proceedings International Congress Amsterdam 1954, III, (1956), p. 545–549.
145. Die Reduktionstheorie der positiven quadratischen Formen, Acta math. *96* (1956), p. 265–309.
146. Tables for the Egyptian and Alexandrian Calendar, Isis, *47* (1957), p. 387–390.
147. Babylonische Planetenrechnung, Vierteljschr. naturf. Ges. Zürich, *102* (1957), p. 39–60.
148. Über die Einführung des Logarithmus im Schulunterricht, Elem. Math. *12* (1957), p. 1–8.
149. Rez. Kennedy, Survey of Islamic Tables. Bibliotheca Orientalis, *14* (1957), p. 109.
150. The Computation of the χ-Distribution, Proceedings 3rd Berkeley symposium on statistics and probability, *1* (1956), p. 207–208.
151. Rez. Taton, Science antique et médiévale: Gnomon 1958, p. 87.

152. Rez. Sachs-Pinches, Late Babylonian Astronomical Texts, Z. f. Assyr. *57*
(1958), p. 339–342.
153. Drei umstrittene Mondfinsternisse bei Ptolemaios, Mus. Helv., *15* (1958), p.
106–109.
154. Rez. Neugebauer (Almanac for 348), Bibliotheca Orientalis, *151* (1958).
155. Rez. Parker-Dubberstein (Babylonian Chronology), Bibliotheca Orientalis, *151*
(1958).
156. Über André Weils Neubegründung der algebraischen Geometrie, Abh. Math.
Sem. Univ. Hamb., *22* (1958), p. 158–170.
157. Zur Theorie der Trennschaukel. Zeitschrift für Naturforschung, *12a* (1957),
p. 583–598.
*158. Zur algebraischen Geometrie, 19. Grundpolynom und zugeordnete Form.
Math. Ann. *136* (1958), p. 139–155.
159. The astronomical Papyrus Ryland 27. Centaurus, *5* (1958), p. 177–191.
160. La démonstration dans les sciences exactes de l'antiquité. Bull. Soc. math.
Belgique, *9* (1958), p. 8–20.
161. Art. Klaudios Ptolemaios II–III in Real-Encyclopaedie, Neue Bearbeitung
XXIII, Kol. 1793–1831 und 1839–1853.
162. Die Handlichen Tafeln des Ptolemaios. Osiris, *13* (1959), p. 54–78.
163. Ein diophantisches Problem von O. Perron. Arch. Math. *9* (1958), p. 54–58.
164. Babylonische Methoden in ägyptischen Planetentafeln. Vierteljschr. naturf.
Ges. Zürich, *105* (1960), p. 97–144.
165. Grosse Terz, Oktave und Harmonie. Mus. Helv. *17* (1960), p. 111–114.
166. Sampling inspection as a minimum loss problem. Ann. Stat. *31* (1960), p. 369–
384.
167. Exclusion Principle and Spin. Pauli Memorial Volume (1960), p. 199–244.
168. Greek Astronomical Calendars and the Athenian Civil Calendar. Journal of
Hellenic Studies, *80* (1960), p. 168–180.
169. Ausgleichspunkt, „Methode der Perser" und indische Planetenrechnung. Arch.
Hist. Exact Sc. *1* (1961), p. 107–121.
170. Pollenkörner, Punktverteilungen auf der Kugel und Informationstheorie.
Naturwissenschaften, *48* (1961), p. 189–192.
171. Secular terms and fluctuations in the motions of the sun and the moon.
Astronomical Journal, *66* (1961), p. 138–147.
*172. Invariants birationnels. Atti del Convegno Internazionale di Geometria Alge-
brica Torino 1961 (1962), p. 35–47.
173. Over logische en verzamelingstheoretische symbolen. Euclides, *37* (1962), p.
183–186.
174. Das Erbe der Antike: Art. Naturwissenschaften. Artemis-Verlag 1963.
175. Das Alter der babylonischen Mondrechnung. Archiv für Orientforschung, *20*
(1963), p. 97–102.
176. Basic Ideas and Methods of Babylonian and Greek Astronomy. In: Scientific
Change, ed. by A. C. Crombie (Heinemann, London 1963), p. 42–60.
177. Berichtigung zu der Arbeit „Punkte auf der Kugel, drei Zusätze" (Math. Ann.
125, p. 213). Math. Ann. *152* (1963), p. 94.
178. Zur Quantentheorie der Wellenfelder. Helv. phys. acta, *36* (1963), p. 945–962.
179. Pythagoreische Wissenschaft. Real-Encyclopaedie XXIV$_1$ (1964), p. 277–300.
180. Rez. M. Destombes, Un astrolabe carolingien et l'origine de nos chiffres arabes.
Byzantinische Zeitschrift 1964, p. 385–386.

181. (with E. S. Kennedy). The World-Year of the Persians. J. Amer. Oriental Soc. *83* (1963), p. 315–327.

182. Wie der Beweis der Vermutung von Baudet gefunden wurde. Abh. Math. Sem. Univ. Hamb., *28* (1965), p. 6–15.

183. Synthetische Urteile a priori. Acta Philos. Fennica, fasc. *18* (1965), p. 277–291.

184. Vergleich der mittleren Bewegungen in der babylonischen, griechischen und indischen Astronomie. Centaurus, *11* (1965), p. 1–18.

185. Sequentielle Qualitätskontrolle als Minimumproblem. Z. Wahrscheinlichkeitsth. Verw. Geb. *4* (1965), p. 187–202.

186. (mit K. Egg und H. Rüst). Die Irrtumswahrscheinlichkeit des χ^2-Testes im Grenzfall der Poissonverteilung. Z. Wahrscheinlichkeitsth. Verw. Geb. *4* (1965), p. 204–260.

187. (mit W. Meretz). Statistische Theorie der aequalen Zellteilung. Die Naturwissenschaften, *53* (1966), p. 8–11.

188. On Measurements in Quantum Mechanics. Zeitschrift für Physik, *190* (1966), p. 99–109.

189. Thesen zur Reform des Mathematikunterrichtes an den Mittelschulen. Bull. Verein Schweiz. Math.- u. Physiklehrer, *1* (1966), p. 7–10.

190. On Clifford Algebras. Proceedings Royal Academy Amsterdam, *A69* (1966), p. 78–83.

191. Die Algebra seit Galois. Jahresbericht Deutsche Math.-Ver. *68* (1966), p. 155–165.

192. The invariant theory of linear systems on varieties. Rendiconti di Mat. *25* (1966), p. 87–93.

193. Klassische und moderne Axiomatik. Elem. Math. *22* (1967), p. 1–4.

194. Art. Astronomie in Der Kleine Pauly I, Col. 664–667.

195. Mendel's Experiments. Centaurus, *12* (1968), p. 275–288.

196. The Date of Invention of Babylonian Planetary Theory. Arch. Hist. Exact Sc. *5* (1968), p. 70–78.

197. Schriften des Pythagoras. Real-Encyclopaedie, Ergänzungsband zum vierundzwanzigsten Band (1968), Col. 843–864.

198. Eupalinos and his tunnel. Isis, *59* (1968), p. 82–83.

199. Mathematische Modelle in der Biologie. Nova Acta Leopoldina, *33* (1968), p. 65–72.

200. (mit J. J. Burckhardt). Das astronomische System der Persischen Tafeln I. Centaurus, *13* (1968), p. 1–28.

201. Platon et les sciences exactes des Pythagoriciens. Bull. Soc. Belge de Logique, *21* (1969), p. 115–123.

202. Das Minimum von $D/f_{11}\ldots f_{55}$ für reduzierte positive quinäre quadratische Formen. Aequationes Math. *2* (1969), p. 233–247.

203. The foundation of Algebraic Geometry. Rendiconti del Seminario Mat. Milano, *39* (1969), p. 3–11.

204. Das heliozentrische System in der griechischen, persischen und indischen Astronomie. Neujahrsbl. Naturforschende Ges. Zürich 1970 (55 p.).

205. Berichtigung zu „Vergleich der mittleren Bewegungen" (Centaurus, *11*). Centaurus, *15* (1970), p. 21–25.

206. Ein Satz über räumliche Fünfecke. Elem. Math. *25* (1970), p. 73–78.

*207. The Theory of Equivalence Systems of Cycles on a Variety. Symposia math. Istituto Nazionale di Alta Matem. V (1970), p. 255–262.

208. Rezension A. J. E. M. Smeur: De verhandeling over de cirkelkwadratuur van Franco van Luik. Centaurus, *15* (1970), p. 107–8.

*209. The Foundation of Algebraic Geometry from Severi to André Weil. Arch. Hist. Exact Sc. *7* (1971), p. 171–180.

*210. Zur algebraischen Geometrie 20. Der Zusammenhangssatz und der Multiplizitätsbegriff. Math. Ann. *193* (1971), p. 89–108.

211. Synthetische Urteile a priori. In: Quanten und Felder, physikalische und philosophische Betrachtungen zum 70. Geburtstag von W. Heisenberg, p. 51–65 (1971).

212. Ägyptische Planetenrechnung, Centaurus, *16* (1971), p. 65–91.

213. How the Proof of Baudet's Conjecture was found. Studies in Pure Mathematics. Presented to Richard Rado (London 1971), p. 251–260.

214. Zum Gedenken an Andreas Speiser, Elem. Math. *26* (1971), p. 98–102.

215. Art. Klaudios Ptolemaios in „Der kleine Pauli" IV, Col. 1224–1229 (1972).

216. Die „Ägypter" und die „Chaldäer". Sitzungsber. Heidelberger Akad. (Math.) 1972, 5. Abhandlung, p. 201–227.

217. Die Erkenntnistheorie der Pythagoreer. Rete *1* (1973), p. 209–244.

218. Die Galois-Theorie von Heinrich Weber bis Emil Artin. Arch. Hist. Exact Sc. *9* (1973), p. 240–248.

219. Ein Briefwechsel zwischen P. Finsler und B. L. van der Waerden über die Grundlegung der algebraischen Geometrie. Arch. Hist. Exact Sc. *9* (1973), p. 249–256.

220. Über die Wechselwirkung von Mathematik und Physik. Elem. Math. *28,* p. 33–41 (1973).

221. From Matrix Mechanics and Wave Mechanics to Unified Quantum Mechanics. In: The Physicist's Conception of Nature, ed. J. Mehra (Reidel, Dordrecht 1973), p. 276–293.

222. Récension Kudlek and Mickler: Solar and Lunar Eclipses 3000 B.C. to 0. Bibliotheca Orientalis, *30* (1973).

223. Hamiltons Entdeckung der Quaternionen. Veröffentlichungen der Joachim Jungius – Ges. der Wiss., Hamburg 1973, p. 1–14.

224. Die Vorgänger des Copernicus im Altertum. Philosophia naturalis, *14* (1973), p. 407–415.

225. The Game of Rectangles. Annali di Matematica (4), *98* (1974), p. 63–75.

226. Recension R. A. Parker: Demotic Mathematical Papyri. Isis, *65* (1974), p. 110–111.

227. The Earliest Form of Epicycle Theory. Journal for History of Astronomy, *5* (1974), p. 175–185.

228. On the Sources of my Book *Moderne Algebra*. Hist. Math., *2* (1975), p. 31–40.

229. Rezension Krafft: Geschichte der Naturwissenschaft I. Gnomon 1975, p. 202–203.

230. Robert Böker neunzigjährig. Académie internationale d'histoire des sciences, *25* (1975), p. 97.

231. Rezension M. Kline: Mathematical Thought from Ancient to Modern Times. Bibliotheca Orientalis, *32* (1975), p. 1.

232. Topologische Algebra. Nieuw Arch. Wiskunde, *23* (1975), p. 212–227.

233. D. Pingree's Review of my Book „Science Awakening II". Hist. Math. *3* (1976), p. 209–211.

234. On the Motion of Venus in the Pancasiddhāntikā. Centaurus, *20* (1976), p. 35–43.
235. Gedenkrede Werner Heisenberg, Berichte Max Planck – Gesellschaft, Sonderheft (1976), p. 17–22.
236. Defense of a „Shocking" Point of View. Arch. Hist. Exact Sc. *15* (1976), p. 199–210.
237. Review of W. R. Knorr: The Evolution of the Euclidean Elements. Hist. Math. *3* (1976), p. 497–499.
238. Die vier Wissenschaften der Pythagoreer. Vorträge Rheinisch-Westfälische Akad. der Wiss. N *268* (1977), p. 7–24.
239. The „Babylonians" and the „Persians". Prismata, Festschrift für Willy Hartner (1977), p. 431–439.
240. Pell's Equation in Greek and Hindu Mathematics. Russ. Math. Surv., *31* : 5 (1976), p. 210–225.
241. Über die Methode der Kleinsten Quadrate. Nachr. Akad. Wiss. Göttingen 1977, p. 75–87.
242. Die Postulate und Konstruktionen in der frühgriechischen Geometrie. Arch. Hist. Exact Sc. *18* (1978), p. 343–347.
243. The Great Year in Greek, Persian and Hindu Astronomy, Arch. Hist. Exact Sc. *18* (1978), p. 359–383.
244. On Varieties in Multiple-Projective Spaces. Proceedings Royal Academy Amsterdam, *81* (1978), p. 303–312.
245. (with W. K. Pritchett). Thucydidean Time-Reckoning and Euctemon's Seasonal Calendar. Bulletin de Correspondance hellénique, *85* (1961), p. 17–52.
246. Rezension M. Pasch: Vorlesungen über Neuere Geometrie. Hist. Math. *5* (1978), p. 107.
247. Gerrit Mannoury and his „Surfaces-Images". Nieuw Arch. Wiskunde, *26* (1978), p. 108–115.
248. On the Motion of the Planets According to Heraclides of Pontus. Arch. internat. Histoire Sci. *28* (1978), p. 167–182.
249. Ewige Tafeln. Arithmos-Arrythmos, Festschrift für Joachim Otto Fleckenstein, München 1979, p. 285–293.
250. Die genetische Methode und der Mittelwertsatz der Differentialrechnung. Praxis Math. *22* (1980), p. 52–53.
251. The (2:n) Table in the Rhind Papyrus. Centaurus, *24* (1980), p. 259–274.
252. Two Treatises on Indian Astronomy. Journal History of Astronomy, *11* (1980), p. 50–58.
253. The Conjunction of 3102 B.C. Centaurus, *24* (1980), p. 117–131.
254. On Pre-Babylonian Mathematics I and II. Arch. Hist. Exact Sc. *23* (1980), p. 1–46.
255. Rezension Thekla Horovitz: Von Logos zur Analogie. Hist. Math. *7* (1981), p. 452.
256. Die gemeinsame Quelle der erkenntnistheoretischen Abhandlungen von Iamblichos und Proklos. Sitzungsberichte Heidelberger Akademie 1980, 12. Abhandlung.
257. Geminos als Erkenntnistheoretiker. In: Physik, Philosophie und Politik, Festschrift für C. F. von Weizsäcker (1982), p. 245–252.
258. The Motion of Venus, Mercury and the Sun in Early Greek Astronomy. Arch. Hist. Exact Sc. *26* (1982), p. 99–113.

Artikel in der Neuen Zürcher Zeitung

1. Die Geometrie der Pythagoreer, 4. Sept. 1966, Nr. 3701.
2. Die Harmonielehre der Pythagoreer, 3. Januar 1967, Nr. 77.
3. Die Arithmetik der Pythagoreer, 29. Januar 1967, Nr. 384.
4. Die Astronomie der Pythagoreer, 18. Februar 1968.
5. Die Erkenntnistheorie der Pythagoreer, 28. Juli 1968, Nr. 458.
6. Platon und die Pythagoreer, 4. August 1968, Nr. 473.
7. Vorstufen der Ideenlehre Platons, 16. Februar 1968, Nr. 101.
8. Pythagoras und die orientalischen Religionen, 1. Juni 1969, Nr. 325.
9. Die Reden des Pythagoras, 18. Januar 1970, Nr. 27.
10. Die Religionspolitik der Perserkönige, 28. November 1971, Nr. 326.
11. Darius und die Magier, 20. Februar 1972, Nr. 85.

Bücher

1. De algebraiese grondslagen der meetkunde van het aantal. Dissertation Amsterdam 1926.
2. Moderne Algebra I. Springer, Berlin 1930.
3. Moderne Algebra II. Springer, Berlin 1931.
   (2. und 3. unter dem Titel „Algebra" später in „Heidelberger Taschenbücher")
   Übersetzungen: portugiesisch 1948, englisch 1949, russisch 1948/1949, chinesisch 1964.
4. Die Gruppentheoretische Methode in der Quantenmechanik. Springer, Berlin 1932.
5. Gruppen von linearen Transformationen. Springer (Ergebnisse d. Math.), Berlin 1935.
6. De logische grondslagen van de Euklidische Meetkunde. Groningen, Noordhoff 1937.
7. Einführung in die algebraische Geometrie. Springer, Berlin 1939.
8. Ontwakende wetenschap, Noordhoff. Groningen 1950.
   Englische Übersetzung: Science wakening I. Noordhoff, Groningen 1954.
   Deutsche Übersetzung: Erwachende Wissenschaft I. Birkhäuser, Basel 1956.
9. Differentiaalrekening 1951. In: Servire's Encyclopaedie, Den Haag.
10. Integraalrekening 1958. In: Servire's Encyclopaedie, Den Haag.
11. Mathematische Statistik. Springer, Berlin 1957.
    Übersetzungen: chinesisch, russisch 1960, französisch 1967, englisch 1969.
12. Erwachende Wissenschaft II: Die Anfänge der Astronomie. Noordhoff, Groningen 1955 und Birkhäuser, Basel 1968.
    Englische Übersetzung: Science Awakening II. Noordhoff, Leiden und Oxford University Press, New York 1974.
13. Sources of Quantum Mechanics. North-Holland Publ. Co., Amsterdam 1967 und Dover Publ., New York 1968.
14. (mit H. Gross). Studien zur Theorie der quadratischen Formen. Birkhäuser, Basel 1968.
15. Group Theory and Quantum Mechanics. Springer, Berlin Heidelberg New York 1974.
16. Mathematik für Naturwissenschaftler. Bibliographisches Institut, Mannheim 1975.
17. Die Pythagoreer. Artemis, Zürich 1978.

479

Emmy Noether

Gesammelte Abhandlungen – Collected Papers

Herausgeber: **N. Jacobson**
1982. Etwa 775 Seiten (26 Seiten in Englisch)
ISBN 3-540-11504-8

This volume contains what are believed to be all of Emmy Noether's published mathematical papers, beginning with the published version of her dissertation and ending with a posthumous paper published in Crelle in 1950, fifteen years after her death. The editor also had access to a number of sets of notes of Emmy Noether's lectures at Göttingen. With the exception of one set, the new results contained in these subsequently appeared in her papers and therefore are not included in this volume. The exception is the set of notes prepared by Professor Max Deuring of Noether's lectures in the Summer Semester of 1929 which contained the first account of her theory of crossed products.
As is quite well known, it was Emmy Noether who persuaded Paul Alexandroff and Heinz Hopf to introduce group theory into combinatorial topology and to formulate the then existing simplicial homology theory in group theoretic terms in place of the more concrete setting of incidence matrices. Undoubtedly, this and the consequent transformation of the subject into the broader one of algebraic topology were inevitable. Nevertheless, how long might they have been delayed had Alexandroff and Hopf not spent the summers of 1926 and 1927 in Göttingen and thereby benefitted from Noether's vision and from the persuasiveness of her personality? On the other side of the coin, what a rich return algebra has received in the development of homological algebra that was a dividend of Emmy Noether's "investment" in combinatorial topology!
Also included are a memorial address by Professor Paul Alexandroff, and an instructive introduction by Professor Nathan Jacobson to her most important papers containing an account by Professor Feza Gaersey of Noether's contributions to physics. Both of these are both in English.

Diese gesammelten Abhandlungen enthalten alle bekannten Arbeiten von Emmy Noether, von ihrer unter Paul Gordan verfaßten Doktorarbeit bis hin zu einem Artikel, der erst 15 Jahre nach ihrem Tode puibliziert wurde. Zwei in englischer Sprache verfaßte Beiträge: ein Nachruf auf Emmy Noether von Professor Paul Alexandroff und eine wegweisende Einführung in ihre wichtigsten Arbeiten von Professor Nathan Jacobson geben einen Überblick über ihr Leben und Werk.

Springer-Verlag
Berlin
Heidelberg
New York
Tokyo